Design of Wood Structures— ASD/LRFD

Donald E. Breyer, P.E.
Professor Emeritus
Department of Engineering Technology
California State Polytechnic University
Pomona, California

Kenneth J. Fridley, Ph.D.
Professor and Head
Department of Civil, Construction, and Environmental Engineering
University of Alabama
Tuscaloosa, Alabama

Kelly E. Cobeen, S.E.
Principal
Cobeen & Associates Structural Engineering
Lafayette, California

David G. Pollock, Ph.D., P.E.
Associate Professor
Department of Civil and Environmental Engineering
Washington State University
Pullman, Washington

Sixth Edition

McGraw-Hill

New York   Chicago   San Francisco   Lisbon   London   Madrid
Mexico City   Milan   New Delhi   San Juan   Seoul
Singapore   Sydney   Toronto

Copyright © 2007, 2003, 1998, 1993, 1988, 1980 by The McGraw-Hill
Companies, Inc. All rights reserved. Printed in the United States of America.
Except as permitted under the United States Copyright Act of 1976, no part
of this publication may be reproduced or distributed in any form or by any
means, or stored in a data base or retrieval system, without the prior written permission of the publisher.

1 2 3 4 5 6 7 8 9 0   DOC/DOC   0 1 2 1 0 9 8 7 6

ISBN-13: 978-0-07-145539-8
ISBN-10: 0-07-145539-6

*The sponsoring editor for this book was Larry S. Hager and the production
supervisor was Pamela A. Pelton. It was set in Century Schoolbook by
International Typesetting and Composition. The art director for the cover
was Anthony Landi.*

Printed and bound by RR Donnelley.

McGraw-Hill books are available at special quantity discounts to use as premiums and sales promotions, or for use in corporate training programs. For
more information, please write to the Director of Special Sales, McGraw-Hill
Professional, Two Penn Plaza, New York, NY 10121-2298. Or contact your local
bookstore.

This book is printed on acid-free paper.

Contents

Preface

The purpose of this book is to introduce engineers, technologists, and architects to the design of wood structures. It is designed to serve either as a text for a course in timber design or as a reference for systematic self-study of the subject.

The book will lead the reader through the complete design of a wood structure (except for the foundation). The sequence of the material follows the same general order that it would in actual design:

1. Vertical design loads and lateral forces

2. Design for vertical loads (beams and columns)

3. Design for lateral forces (horizontal diaphragms and shearwalls)

4. Connection design (including the overall tying together of the vertical- and lateral-force-resisting systems)

The need for such an overall approach to the subject became clear from experience gained in teaching timber design at the undergraduate and graduate levels.

This text pulls together the design of the various elements into a single reference. A large number of practical design examples are provided throughout the text. Because of their widespread usage, buildings naturally form the basis of the majority of these examples. However, the principles of member design and diaphragm design have application to other structures (such as concrete formwork and falsework).

This book relies on practical, current industry literature as the basis for structural design. This includes publications of the American Forest and Paper Association (AF&PA), the International Codes Council (ICC), the American Society of Civil Engineers (ASCE), APA—The Engineered Wood Association, and the American Institute of Timber Construction (AITC).

In the writing of this text, an effort has been made to conform to the spirit and intent of the reference documents. The interpretations are those of the authors and are intended to reflect current structural design practice. The material presented is suggested as a guide only, and final design responsibility, lies with the structural engineer.

The sixth edition of this book was promoted by five major developments:

1. Publication of new dual-format (ASD/LRFD) wood design criteria in the 2005 *National Design Specification for Wood Construction* (*NDS*).

2. Publication of the new *Special Design Provisions for Wind and Seismic* (SDPWS) Supplement to the NDS.

3. Publication of the comprehensive ASD/LRFD *Manual for Engineered Wood Construction.*

4. Publication and increased adoption nationally of the 2006 *International Building Code.*

5. Publication of updated load standards in the 2005 edition of *Minimum Design Loads for Buildings and Other Structures* (ASCE 7-05).

The *National Design Specification* (NDS) is published by the American Forest & Paper Association (AF&PA) and represents the latest structural design recommendations by the wood industry. The 2005 NDS presents both traditional allowable stress design (ASD) provisions as well as new load and resistance factor design (LRFD) provisions. The inclusion of the LRFD provisions is new to the NDS for the 2005 edition. As such, the 2005 NDS is considered a dual-format design specification. While ASD has been and may continue to be the method of choice for many designers of wood buildings, the acceptance and use of LRFD for wood design is increasing.

The 2006 *ASD/LRFD Manual for Engineered Wood Construction* includes design supplements, guidelines, and manuals helpful for wood engineering design. It includes design information for sawn lumber, structural glued laminated timber, structural-use panels, shearwalls and diaphragms, poles and piles, I-joists, structural composite lumber, structural connections (nails, bolts, screws), and pre-engineered metal connectors. The Manual was first introduced in 1999 for the 1997 NDS, and has evolved into a comprehensive design support document.

The *International Building Code* (IBC) is a product of the International Codes Council (ICC). The ICC brought together the three regional model building code organizations to develop and administer a single national building code. The first edition of the IBC was published in 2000, and now nearly all regions of the U.S. have adopted all or part of the IBC at either the state or local level.

Traditionally, the NDS has been based on the principles of what is termed *allowable stress design* (ASD). In ASD *allowable stresses* of a material are compared to calculated *working stresses* resulting from *service loads*. Recently, the wood industry and design community completed the development of a *load and resistance factor design* (LRFD) specification for wood construction. In LRFD, *adjusted nominal capacities* (resistance) are compared to the effect of *factored loads*. The factors are developed for both resistance and loads such that uncertainty and consequence of failure are explicitly recognized. The LRFD approach to wood design is now included in the 2005 edition of the NDS. This sixth edition of *Design of Wood Structures* presents both ASD and LRFD guidelines as provided in the NDS. In many examples, both ASD and LRFD approaches are presented to allow the reader a direct, side-by-side comparison of the two methods.

If this book is used as a text for a formal course, an Instructor's Manual is available. Requests on school letterhead should be sent to: Civil Engineering Editor, McGraw-Hill Professional, 2 Penn Plaza, New York, NY 10121-2298.

Questions or comments about the text or examples may be addressed to any of the authors. Direct any correspondence to:

Prof. Emeritus Donald E. Breyer
Department of Engineering Technology
California State Polytechnic University
3801 West Temple Avenue
Pomona, CA 91768

Prof. David G. Pollock
Department of Civil and Environmental
  Engineering
Washington State University
P.O. Box 642910
Pullman, WA 99164-2910

Prof. Kenneth J. Fridley
Department of Civil, Construction, and
  Environmental Engineering
University of Alabama
Box 870205
Tuscaloosa, AL 35487-0205

Ms. Kelly E. Cobeen
Cobeen & Associates Structural
  Engineering
251 Lafayette Circle, Suite 230
Lafayette, CA 94549

Acknowledgment and appreciation for help in writing this text are given to Philip Line and Bradford Douglas of the American Forest and Paper Association; Jeff Linville of the American Institute of Timber Construction; John Rose, Thomas Skaggs, and Thomas Williamson of APA—The Engineered Wood Association; and Kevin Cheung of the Western Wood Products Association. Numerous other individuals also deserve recognition for their contributions to various editions of the text, including Rosdinah Baharin, Russell W. Krivchuk, William A. Baker, Michael Caldwell, Thomas P. Cunningham, Jr., Mike Drorbaugh, John R. Tissell, Ken Walters, B. J. Yeh, Thomas E. Brassell, Frank Stewart, Lisa Johnson, Edwin G. Zacher, Edward F. Diekmann, Lawrence A. Soltis, Robert Falk, Don Wood, William R. Bloom, Frederick C. Pneuman, Robert M. Powell, Sherm Nelson, Bill McAlpine, Karen Colonias, and Ronald L. Carlyle. Suggestions and information were obtained from many other engineers and suppliers, and their help is gratefully recognized.

Dedication

To our families:

Matthew, Kerry, Daniel, and Sarah

Paula, Justin, Connor, and Alison

Chris and Matthew

Lynn, Sarah, and Will

Donald E. Breyer, P.E.
Kenneth J. Fridley, Ph.D.
Kelly E. Cobeen, S.E.
David G. Pollock, Ph.D., P.E.

Nomenclature

Organizations

AF&PA
American Forest and Paper Association
American Wood Council (AWC)
1111 19th Street, NW, Suite 800
Washington, DC 20036
www.afandpa.org
www.awc.org

AITC
American Institute of Timber
  Construction
7012 South Revere Parkway, Suite 140
Centennial, CO 80112
www.aitc-glulam.org

ALSC
American Lumber Standard Committee,
  Inc.
P.O. Box 210
Germantown, MD 20875-0210
www.alsc.org

APA
APA—The Engineered Wood Association
P.O. Box 11700
Tacoma, WA 98411-0700
www.apawood.org

ASCE
American Society of Civil Engineers
1801 Alexander Bell Drive
Reston, VA 20191
www.asce.org

ATC
Applied Technology Council
201 Redwood Shores Parkway,
  Suite 240
Redwood City, CA 94065
www.atcouncil.org

AWPA
American Wood-Preservers' Association
P.O. Box 388
Selma, AL 36702-0388
www.awpa.com

BSSC
Building Seismic Safety Council
National Institute of Building
  Sciences
1090 Vermont Avenue, N.W.,
  Suite 700
Washington, DC 20005
http://www.bssconline.org/

CANPLY
Canadian Plywood Association
735 West 15 Street
North Vancouver, British Columbia,
  Canada V7M 1T2
www.canply.org

CWC
Canadian Wood Council
99 Bank Street, Suite 400
Ottawa, Ontario, Canada K1P 6B9
www.cwc.ca

CPA–CWC
Composite Panel Association
Composite Wood Council
18922 Premiere Court
Gaithersburg, MD 20879-1574
301-670-0604
www.pbmdf.com

FPL
U.S. Forest Products Laboratory
USDA Forest Service
One Gifford Pinchot Drive
Madison, WI 53726-2398
www.fpl.fs.fed.us

ICC
International Codes Council
5203 Leesburg Pike
Suite 600
Falls Church, VA 22041
www.iccsafe.org

ISANTA
International Staple, Nail and Tool
  Association
512 West Burlington Avenue, Suite 203
La Grange, IL 60525-2245
www.isanta.org

MSRLPC
MSR Lumber Producers Council
P.O. Box 6402
Helena, MT 59604
www.msrlumber.org

NFBA
National Frame Builders Association
4840 West 15th Street, Suite 1000
Lawrence, KS 66049-3876
www.postframe.org

NHLA
National Hardwood Lumber Association
P.O. Box 34518
Memphis, TN 38184-0518
www.natlhardwood.org

NLGA
National Lumber Grades Authority
#406 First Capital Place
960 Quayside Drive
New Westminster, British Columbia,
  Canada V3M 6G2
www.nlga.org

NELMA
Northeastern Lumber Manufacturers
  Association
272 Tuttle Road
P.O. Box 87A
Cumberland Center, ME 04021
www.nelma.org

NSLB
Northern Softwood Lumber Bureau
272 Tuttle Road
P.O. Box 87A
Cumberland Center, ME 04021
www.nelma.org

NWPA
Northwest Wood Products Association
149 SE 9th, #2
Bend, OR 97702
www.nwpa.org

PLIB
Pacific Lumber Inspection Bureau
33442 First Way South, #300
Federal Way, WA 98003-6214
www.plib.org

SEAOC
Structural Engineers Association of
  California
1414 K Street, Suite 260
Sacramento, CA 95814
www.seaoc.org

SLMA
Southeastern Lumber Manufacturers
  Association
P.O. Box 1788
Forest Park, GA 30298-1788
www.slma.org

SFPA
Southern Forest Products Association
P.O. Box 641700
Kenner, LA 70064-1700
Street address: 2900 Indiana Avenue
Kenner, LA 70065
www.sfpa.org
www.southernpine.com

SPIB
Southern Pine Inspection Bureau, Inc.
4709 Scenic Highway
Pensacola, FL 32504-9094
www.spib.org

SBA
Structural Board Association
25 Valleywood Drive, Unit 27
Markham, Ontario, Canada L3R 5L9
www.osbguide.com

TPI
Truss Plate Institute
218 N. Lee Street, Suite 312
Alexandria, VA 22314
www.tpinst.org

WCLIB
West Coast Lumber Inspection Bureau
P.O. Box 23145
Portland, OR 97281-3145
www.wclib.org

WRCLA
Western Red Cedar Lumber Association
1501-700 West Pender Street
Vancouver, British Columbia, Canada
 V6C 1G8
www.wrcla.org

WWPA
Western Wood Products Association
522 Southwest Fifth Avenue, Suite 500
Portland, OR 97204-2122
www.wwpa.org

WIJMA
Wood I-Joist Manufacturing Association
200 East Mallard Drive
Boise, ID 83706
www.i-joist.org

WTCA
Wood Truss Council of America
One WTCA Center
6300 Enterprise Lane
Madison, WI 53719
www.woodtruss.com

Publications

ASCE 7: American Society of Civil Engineers (ASCE). 2006. *Minimum Design Loads for Buildings and Other Structures* (ASCE 7-05), ASCE, Reston, VA.

ASD/LRFD Manual: American Forest and Paper Association (AF&PA). 2006. *ASD/LRFD Manual for Engineered Wood Construction*, 2005 ed., AF&PA, Washington, DC.

IBC: International Codes Council (ICC). 2006. *International Building Code* (*IBC*), 2006 ed., ICC, Falls Church, VA.

NDS: American Forest and Paper Association (AF&PA). 2005. *National Design Specification* (*NDS*) *for Wood Construction*, ANSI/AF&PA NDS-2005, AF&PA, Washington, DC.

SDPWS: American Forest & Paper Association (AF&PA). 2005. *Special Design Provisions for Wind and Seismic* (SDPWS) Supplement to the NDS, AF&PA, Washington, DC.

TCM: American Institute of Timber Construction (AITC). 2005. *Timber Construction Manual*, 5th ed., John Wiley & Sons Inc., Hoboken, NJ.

Additional publications given at the end of each chapter.

Units

ft	foot, feet	mph	miles per hour
ft^2	square foot, square feet	pcf	pounds per cubic foot (lb/ft^3)
in.	inch, inches	plf	pounds per lineal foot (lb/ft)
in.^{2}	square inch, square inches	psf	pounds per square foot (lb/ft^2)
k	1000 lb (kip, kilopound)	psi	pounds per square inch (lb/in.^{2})
ksi	kips per square inch (k/in.^{2})	sec	second

Abbreviations

Adj.	adjusted	MDO	medium density overlay (plywood)
Allow.	allowable	MEL	machine evaluated lumber
ASD	allowable stress design	min.	minimum
B&S	Beams and Stringers	MSR	machine stress rated lumber
c.-to-c.	center to center	NA	neutral axis
cg	center of gravity	o.c.	on center
DF-L	Douglas Fir-Larch	OM	overturning moment
Ecc.	eccentric	OSB	oriented strand board
EMC	equilibrium moisture content	PL	plate
FBD	free-body diagram	P&T	Posts and Timbers
FS	factor of safety	PSL	parallel strand lumber
FSP	fiber saturation point	Q/A	quality assurance
glulam	structural glued laminated timber	Ref	reference
ht	height	Req'd	required
IP	inflection point (point of reverse curvature and point of zero moment	RM	resisting moment
		S4S	dressed lumber (surfaced four sides)
J&P	Joists and Planks	Sel. Str.	Select Structural
lam	lamination	SCL	structural composite lumber
LF	Light Framing	SJ&P	Structural Joists and Planks
LRFD	load and resistance factor design	SLF	Structural Light Framing
		Tab.	tabulated
LFRS	lateral-force-resisting system	T&G	tongue and groove
LVL	laminated veneer lumber	TL	total load (lb, k, lb/ft, k/ft, psf)
max.	maximum	trib.	tributary
MC	moisture content based on oven-dry weight of wood	TS	top of sheathing
		WSD	working stress design

Symbols

A	area (in.^{2}, ft^2)
a	acceleration (ft/s^2)
a	length of end zone for wind pressure calculations
A_g	gross cross-sectional area of a tension or compression member (in.^{2})
$A_{\text{group-net}}$	net cross-sectional area between outer rows of fasteners for a wood tension member (in.^{2})
A_h	projected area of hole caused by drilling or routing to accommodate bolts or other fasteners at net section (in.^{2})
A_m	gross cross sectional area of main member (in.^{2})
A_n	cross-sectional area of member at a notch (in.^{2})
A_n	net cross-sectional area of a tension or compression member at a connection (in.^{2})
a_p	in-structure component amplification factor
A_s	area of reinforcing steel (in.^{2})
A_s	sum of gross cross-sectional areas of side member(s) (in.^{2})
A_T	tributary area for a structural member or connection (ft^2)
A_{web}	cross-sectional area of the web of a steel W-shaped beam or wood I joist (in.^{2})
A_x	diaphragm area immediately above the story being considered (ft^2)
b	length of shearwall parallel to lateral force; distance between chords of shearwall (ft)
b	width of horizontal diaphragm; distance between chords of horizontal diaphragm (ft)
b	width of rectangular beam cross section (in.)
C	compression force (lb, k)
c	buckling and crushing interaction factor for columns
c	distance between neutral axis and extreme fiber (in., ft)
C_b	bearing area factor
C_D	load duration factor (ASD only)
C_d	seismic deflection amplification factor
C_{di}	diaphragm factor for nail connections
C_e	exposure factor for snow load
C_{eg}	end grain factor for connections
C_F	size factor
C_{fu}	flat use factor for bending design values
C_G	grade and construction factor for wood structural panels
C_g	group action factor for multiple-fastener connections with D $\geq \frac{1}{4}$ in.
C_i	incising factor for Dimension lumber
C_L	beam stability factor

C_M	wet service factor
C_P	column stability factor
C_r	repetitive-member factor for bending design values
C_s	roof slope factor for snow load
C_s	seismic response coefficient
C_s	panel size factor for wood structural panels
C_{st}	metal side plate factor for 4-in. shear plate connections
C_T	buckling stiffness factor for 2×4 and smaller Dimension lumber in trusses
C_t	seismic coefficient depending on type of LFRS used to calculate period of vibration T
C_t	temperature factor
C_t	thermal factor for snow load
C_{tn}	toenail factor for nail connections
C_V	volume factor
C_{vx}	seismic vertical distribution factor
C_Δ	geometry factor for connections with fastener D $\geq \frac{1}{4}$ in.
D	dead load (lb, k, lb/ft, k/ft, psf)
D	diameter (in.)
d	cross-sectional dimension of rectangular column associated with axis of column buckling (in.)
d	depth of rectangular beam cross section (in.)
d	dimension of wood member for shrinkage calculation (in.)
d	pennyweight of nail or spike
d_1	shank diameter of lag bolt (in.)
d_2	pilot hole diameter for the threader portion of lag bolt (in.)
d_e	effective depth of member at a connection (in.)
d_n	effective depth of member remaining at a notch (in.)
d_x	width of rectangular column parallel to y axis, used to calculate column slenderness ratio about x axis
d_y	width of rectangular column parallel to x axis, used to calculate column slenderness ratio about y axis
E	earthquake force (lb, k)
E	length of tapered tip of lag bolt (in.)
E, E'	reference and adjusted modulus of elasticity (psi, ksi)
e	eccentricity (in., ft)
E_{axial}	modulus of elasticity of glulam for axial deformation calculation (psi)
E_m	modulus of elasticity of main member (psi)
$E_{\text{min}}, E'_{\text{min}}$	reference and adjusted modulus of elasticity for ASD stability calculations (psi)

$E_{\text{min-}n}$, $E'_{\text{min-}n}$	nominal and adjusted modulus of elasticity for LRFD stability calculations (ksi)
E_s	modulus of elasticity of side member (psi)
E_x	modulus of elasticity about x axis (psi, ksi)
E_y	modulus of elasticity about y axis (psi, ksi)
EA, EA'	reference and adjusted axial stiffness design value for structural panels (lb/ft width, k/ft width)
EI, EI'	reference and adjusted bending stiffness design value for structural panels (lb-in^2/ft width, k-in^2/ft width)
F	fluid load (lb, k, plf, klf, psf)
F	force or load (lb, k)
F	roof slope in inches of rise per foot of horizontal span
f_1	live load coefficient for special seismic load combinations
f_1	live load coefficient in LRFD load combinations when other transient loads are dominant
f_2	snow load coefficient in LRFD load combinations for various roof configurations
F_a	flood load (lb, k, plf, klf, psf)
F_a	acceleration-based seismic site coefficient at 0.3 second period
f_b	actual (computed) bending stress (psi)
F_b, F'_b	reference and adjusted ASD bending design value (psi)
F_b^*	reference ASD bending design value multiplied by all applicable adjustment factors except C_L, C_V, and C_{fu} (psi)
F_b^{**}	reference ASD bending design value multiplied by all applicable adjustment factors except C_V (psi)
F_{bE}	critical ASD buckling (Euler) value for bending member (psi)
F_{bn}, F'_{bn}	nominal and adjusted LRFD bending design value (ksi)
F_{bn}^*	nominal LRFD bending design value multiplied by all applicable adjustment factors except C_L, C_V, and C_{fu} (ksi)
F_{bn}^{**}	nominal LRFD bending design value multiplied by all applicable adjustment factors except C_V (ksi)
F_{bEn}	nominal LRFD buckling (Euler) value for bending member (ksi)
f_{bx}	actual (computed) bending stress about the x-axis (psi)
F_{bx}, F'_{bx}	reference and adjusted ASD bending design value about the x-axis (psi)
F_{bxn}, F'_{bxn}	nominal and adjusted LRFD bending design value about the x-axis (ksi)
f_{by}	actual (computed) bending stress about the y-axis (psi)
F_{by}, F'_{by}	reference and adjusted ASD bending design value about the y-axis (psi)
F_{byn}, F'_{byn}	nominal and adjusted LRFD bending design value about the y-axis (ksi)

F_c	out-of-plane seismic forces for concrete and masonry walls (lb, k, plf, klf)
f_c	actual (computed) compression stress parallel to grain (psi)
F_c, F'_c	reference and adjusted ASD compression design value parallel to grain (psi)
F_c^*	reference ASD compression design value parallel to grain multiplied by all applicable adjustment factors except C_P (psi)
F_{cE}	critical ASD buckling (Euler) value for compression member (psi)
F_{cn}, F'_{cn}	nominal and adjusted LRFD compression design value parallel to grain (ksi)
F_{cn}^*	nominal LRFD compression design value parallel to grain multiplied by all applicable adjustment factors except C_P (ksi)
F_{cEn}	nominal LRFD buckling (Euler) value for compression member (ksi)
$f_{c\perp}$	actual (computed) compression stress perpendicular to grain (psi)
$F_{c\perp}, F'_{c\perp}$	reference and adjusted ASD compression design value perpendicular to grain (psi)
$F_{c\perp 0.02}$	reduced ASD compression design value perpendicular to grain at a deformation limit of 0.02 in. (psi)
$F_{c\perp n}, F'_{c\perp n}$	nominal and adjusted LRFD compression design value perpendicular to grain (ksi)
F_e	dowel bearing strength (psi)
$F_{e\parallel}$	dowel bearing strength parallel to grain for fasteners with $D \geq 1/4$ in. (psi)
$F_{e\perp}$	dowel bearing strength perpendicular to grain for fasteners with $D \geq 1/4$ in. (psi)
$F_{e\theta}$	dowel bearing strength at angle to grain θ for fasteners with $D \geq 1/4$ in. (psi)
F_{em}	dowel bearing strength for main member (psi)
F_{es}	dowel bearing strength for side member (psi)
F_p	seismic force on a component of a structure (lb, k, plf, klf)
F_{px}	seismic story force at level x for designing the horizontal diaphragm (lb, k)
F'_θ	adjusted ASD bearing design value at angle to grain θ (psi)
$F'_{\theta n}$	adjusted LRFD bearing design value at angle to grain θ (ksi)
f_s	stress in reinforcing steel (psi, ksi)
f_t	actual (computed) tension stress in a member parallel to grain (psi)
F_t, F'_t	reference and adjusted ASD tension design value parallel to grain (psi)
F_{tn}, F'_{tn}	nominal and adjusted LRFD tension design value parallel to grain (ksi)
F_u	ultimate tensile strength for steel (psi, ksi)
F_v	velocity-based seismic site coefficient at 1.0 second period

f_v	actual (computed) shear stress parallel to grain (horizontal shear) in a beam using full design loads (psi)
F_v, F_v'	reference and adjusted ASD shear design value parallel to grain (horizontal shear) in a beam (psi)
F_{vn}, F_{vn}'	nominal and adjusted LRFD shear design value parallel to grain (horizontal shear) in a beam (ksi)
f_v^*	actual (computed) shear stress parallel to grain (horizontal shear) in a beam obtained by neglecting the loads within a distance d of face of support (psi)
F_x	seismic story force at level x for designing vertical elements (shearwalls) in LFRS (lb, k)
F_y	yield strength (psi, ksi)
F_{yb}	bending yield strength of fastener (psi, ksi)
F_bS, F_bS'	reference and adjusted ASD bending strength design value for structural panels (lb-in./ft width)
$(F_bS)_n, (F_bS)_n'$	nominal and adjusted LRFD bending strength design value for structural panels (k-in./ft width)
F_cA, F_cA'	reference and adjusted ASD axial compression strength design value for structural panels (lb/ft width)
$(F_cA)_n, (F_cA)_n'$	nominal and adjusted LRFD axial compression strength design value for structural panels (k/ft width)
F_tA, F_tA'	reference and adjusted ASD axial tension strength design value for structural panels (lb/ft width)
$(F_tA)_n, (F_tA)_n'$	nominal and adjusted LRFD axial tension strength design value for structural panels (k/ft width)
$F_s(Ib/Q), F_s(Ib/Q)'$	reference and adjusted ASD in-plane shear design value for structural panels (lb/ft width)
$(F_sIb/Q)_n, (F_sIb/Q)_n'$	nominal and adjusted LRFD in-plane shear design value for structural panels (k/ft width)
F_vt_v, F_vt_v'	reference and adjusted ASD shear through the thickness design value for structural panels (lb/in of shear-resisting panel length)
$(F_vt_v)_n, (F_vt_v)_n'$	nominal and adjusted LRFD shear through the thickness design value for structural panels (k/in of shear-resisting panel length)
G	specific gravity of wood or a wood-based member
g	acceleration of gravity (ft/s^2)
G_m	specific gravity of main member
G_s	specific gravity of side member
G_vt_v, G_vt_v'	reference and adjusted shear rigidity through the thickness design value for structural panels (lb/in of panel depth, k/in of panel depth)
H	soil, hydrostatic pressure, or bulk material load (lb, k, plf, klf, psf)
h	building height or height of wind pressure zone (ft)
h	height of shearwall (ft)
h_i, h_x	height above base to level i or level x (ft)
h_{mean}	mean roof height above ground (ft)

h_n	height above base to nth or uppermost level in building (ft)
h_x	the elevation at which a component is attached to a structure relative to grade, for design of portions of structures (ft)
I	moment of inertia (in.⁴)
I	importance factor for seismic, snow or wind load
I_p	seismic component importance factor
K	Code multiplier for dead load for use in beam deflection calculations to account for creep effects.
k	exponent for vertical distribution of seismic forces related to the building period
K_D	diameter coefficient for connections with fastener D $< \frac{1}{4}$ in.
K_e	effective length factor for column end conditions (buckling length coefficient for columns)
K_F	format conversion factor (LRFD only)
K_f	column stability coefficient for bolted and nailed built-up columns
K_{LL}	live load element factor for influence area
K_θ	angle to grain coefficient for connections with fastener D $\geq \frac{1}{4}$ in.
K_{zt}	topographical factor for wind pressure calculations
L	live load (lb, k, lb/ft, k/ft, psf)
L	beam span length (ft)
L	length (ft)
l	length (in.)
l	length of bolt in main or side members (in.)
l	length of fastener (in.)
l	unbraced length of column (in.)
L_0	unreduced floor or roof live load (lb, k, plf, klf, psf)
l/D	bolt slenderness ratio
l_b	bearing length (in.)
L_c	cantilever length in cantilever beam system (ft)
l_e	effective unbraced length of column (in.)
l_e/d	slenderness ratio of column
$(l_e/d)_x$	slenderness ratio of column for buckling about strong (x) axis
$(l_e/d)_y$	slenderness ratio of column for buckling about weak (y) axis
l_e	effective unbraced length of compression edge of beam (in.)
l_m	dowel bearing length of fastener in main member (in.)
L_r	roof live load (lb, k, lb/ft, k/ft, psf)
l_s	dowel bearing length of fastener in side member(s) (in.)
l_u	laterally unbraced length of compression edge of beam (in.)
l_x	unbraced length of column considering buckling about strong (x) axis (in.)

l_y	unbraced length of column considering buckling about weak (y) axis (in.)
M	bending moment (in.-lb, in.-k, ft-lb, ft-k)
M	mass
M'_n	adjusted LRFD moment resistance (in.-k, ft-k)
M_p	plastic moment capacity (in.-lb, in.-k)
M_u	ultimate (factored) bending moment (in.-lb, in.-k, ft-lb, ft-k)
M_y	yield moment (in.-lb, in.-k)
N	normal reaction (lb, k)
N	number of fasteners in connection
N, N'	reference and adjusted ASD lateral design value at angle to grain θ for a single split ring or shear plate connector (lb)
n	number of fasteners in row
n	number of stories (seismic forces)
N_n, N'_n	nominal and adjusted LRFD lateral design value at angle to grain θ for a single split ring or shear plate connector (k)
n_{row}	number of rows of fasteners in a fastener group
P	concentrated load or force (lb, k)
P, P'	reference and adjusted ASD lateral design value parallel to grain for a single split ring or shear plate connector (lb)
p	parallel-to-grain component of lateral force z on one fastener
p	penetration depth of fastener in wood member (in.)
p_f	flat roof snow load (psf)
p_g	ground snow load (psf)
P_n, P'_n	nominal and adjusted LRFD lateral design value parallel to grain for a single split ring or shear plate connector (k)
P'_n	adjusted LRFD axial compression resistance (k)
p_{net}	net design wind pressure for components and cladding (psf)
p_{net30}	net design wind pressure for components and cladding at a height of 30 ft in Exposure B conditions (psf)
p_s	sloping roof snow load (psf)
p_s	simplified design wind pressure for main wind force-resisting systems (psf)
p_{s30}	simplified design wind pressure for main wind force-resisting systems at a height of 30 ft in Exposure B conditions (psf)
P_u	collapse load (ultimate load capacity)
P_u	ultimate (factored) concentrated load or force (lb, k)
Q	static moment of an area about the neutral axis (in.[3])
Q, Q'	reference and adjusted ASD lateral design value perpendicular to grain for a single split ring or shear plate connector (lb)
q	perpendicular-to-grain component of lateral force z on one fastener

Q_E	effect of horizontal seismic forces (lb, k)
Q_n, Q'_n	nominal and adjusted LRFD lateral design value perpendicular to grain for a single split ring or shear plate connector (k)
R	rain load (lb, k, plf, klf, psf)
R	reaction force (lb, k)
R	seismic response modification factor
r	radius of gyration (in.)
R_1	roof live load reduction factor for large tributary roof areas
R_1	seismic force generated by mass of wall that is parallel to earthquake force being considered
R_2	roof live load reduction factor for sloped roofs
R_B	slenderness ratio of laterally unbraced beam
r_i	portion of story force resisted by a shearwall element.
R_p	seismic response modification factor for a portion of a structure
R_u	ultimate (factored) reaction force (lb, k)
R_{u1}	ultimate (factored) seismic force generated by mass of wall that is parallel to earthquake force being considered (lb, k)
S	snow load (lb, k, plf, klf, psf)
S	section modulus (in.^{3})
S	shrinkage of wood member (in.)
s	center-to-center spacing between adjacent fasteners in a row (in.)
S	length of unthreaded shank of lag bolt (in.)
S_1	mapped maximum considered earthquake spectral acceleration at 1 one-second period (g)
s_{crit}	critical spacing between fasteners in a row (in.)
S_{D1}	design spectral response acceleration at a one-second period (g)
S_{DS}	design spectral response acceleration at short periods (g)
S_{MS}	maximum considered earthquake spectral response acceleration at short periods (g)
S_{M1}	maximum considered earthquake spectral response acceleration at a one-second period (g)
S_S	mapped maximum considered earthquake spectral acceleration at short periods (g)
SV	shrinkage value for wood due to 1 percent change in moisture content (in./in.)
T	self-straining load (lb, k)
T	fundamental period of vibration of structure in direction of seismic force under consideration (sec)
T	tension force (lb, k)
t	thickness (in.)
T_a	approximate fundamental building period (sec)

t_m	thickness of main member (in.)
T'_n	adjusted LRFD tension resistance (k)
T_0	response spectrum period at which the S_{DS} plateau is reached (sec)
T_S	response spectrum period at which the S_{DS} and S_{D1}/T curves meet (sec)
t_s	thickness of side member (in.)
T_u	ultimate (factored) tension force (lb, k)
t_{washer}	thickness of washer (in.)
U	wind uplift resultant force (lb, k)
V	basic wind speed (mph)
V	seismic base shear (lb, k)
V	shear force in a beam, diaphragm, or shearwall (lb, k)
v	unit shear in diaphragm or shearwall (lb/ft)
V^*	reduced shear in beam determined by neglecting load within d from face of supports (lb, k)
v_2	unit shear in second-floor diaphragm (lb/ft)
v_{12}	unit shear in shearwall between first- and second-floor levels (lb/ft)
v_{2r}	unit shear in shearwall between second-floor and roof levels (lb/ft)
V'_n	adjusted LRFD shear resistance parallel to grain in a beam (k)
V_{px}	diaphragm forces created by the redistribution of forces between vertical elements
v_r	unit shear in roof diaphragm (lb/ft)
V_{story}	story shear force (lb, k)
V_u	ultimate (factored) shear force in a beam, diaphragm, or shearwall (lb, k)
v_u	ultimate (factored) unit shear in horizontal diaphragm or shearwall (lb/ft)
V_u^*	reduced ultimate (factored) shear in beam determined by neglecting load within d from face of supports (lb, k)
v_{u2}	ultimate (factored) unit shear in second-floor diaphragm (lb/ft)
v_{u12}	ultimate (factored) unit shear in shearwall between first- and second-floor levels (lb/ft)
v_{u2r}	ultimate (factored) unit shear in shearwall between second-floor and roof levels (lb/ft)
v_{ur}	ultimate (factored) unit shear in roof diaphragm (lb/ft)
V_{wall}	wall shear force in the wall with the highest unit shear (lb, k)
W	lateral force due to wind (lb, k, lb/ft, psf)
W	weight of structure or total seismic dead load (lb, k)
w	reference withdrawal design value for single fastener (lb/in. of penetration)
w	uniformly distributed load or force (lb/ft, k/ft, psf, ksf)

W, W'	reference and adjusted ASD withdrawal design value for single fastener (lb)
W_1	dead load of 1-ft-wide strip tributary to story level in direction of seismic force (lb/ft, k/ft)
W_2	total dead load tributary to second-floor level (lb, k)
W'_2	that portion of W_2 which generates seismic forces in second-floor diaphragm (lb, k)
w_2	uniform load to second-floor diaphragm (lb/ft, k/ft)
W_D	dead load of structure (lb, k)
W_{foot}	dead load of footing or foundation (lb, k)
w_i, w_x	tributary weight assigned to story level i or level x (lb, k)
W_n, W'_n	nominal and adjusted LRFD withdrawal design value for single fastener (lb, k)
W_p	weight of portion of structure (element or component) (lb, k, lb/ft, k/ft, psf)
w_{px}	uniform load to diaphragm at level x (lb/ft, k/ft)
W_r	total dead load tributary to roof level (lb, k)
w_r	uniform load to roof diaphragm (lb/ft, k/ft)
W'_r	that portion of W_r which generates seismic forces in roof diaphragm (lb, k)
w_u	ultimate (factored) uniformly distributed load or force (lb/ft, k/ft, psf, ksf)
w_{u2}	ultimate (factored) uniform load to second-floor diaphragm (lb/ft, k/ft)
w_{upx}	ultimate (factored) uniform load to diaphragm at level x (lb/ft, k/ft)
w_{ur}	ultimate (factored) uniform load to roof diaphragm (lb/ft, k/ft)
x	exponent dependent on structure type used in calculation of the approximate fundamental period
x	width of triangular soil bearing pressure diagram (ft)
Z	plastic section modulus (in.^{3})
z	lateral force on one fastener in wood connection (lb)
Z, Z'	reference and adjusted ASD lateral design value for single fastener in a connection (lb)
Z'_α	adjusted resultant design value for lag bolt subjected to combined lateral and withdrawal loading (lb)
Z_n, Z_n'	nominal and adjusted LRFD lateral design value for single fastener in a connection (lb, k)
Z'_{GT}	adjusted ASD connection capacity due to group tear-out failure in a wood member (lb)
Z'_{NT}	adjusted ASD connection capacity due to net tension failure in a wood member (lb)

Z'_{RT}	adjusted ASD connection capacity due to row tear-out failure in a wood member (lb)
Z'_{nGT}	adjusted LRFD connection capacity due to group tear-out failure in a wood member (lb, k)
Z'_{nNT}	adjusted LRFD connection capacity due to net tension failure in a wood member (lb, k)
Z'_{nRT}	adjusted LRFD connection capacity due to row tear-out failure in a wood member (lb, k)
$Z'_{RT\text{-}1}$	adjusted ASD row tear-out capacity for first row of fasteners in a fastener group (lb)
$Z'_{RT\text{-}2}$	adjusted ASD row tear-out capacity for second row of fasteners in a fastener group (lb)
$Z'_{RT\text{-}n}$	adjusted ASD row tear-out capacity for nth row of fasteners in a fastener group (lb)
$Z'_{nRT\text{-}1}$	adjusted LRFD row tear-out capacity for first row of fasteners in a fastener group (lb, k)
$Z'_{nRT\text{-}2}$	adjusted LRFD row tear-out capacity for second row of fasteners in a fastener group (lb, k)
$Z'_{nRT\text{-}n}$	adjusted LRFD row tear-out capacity for n^{th} row of fasteners in a fastener group (lb, k)
$Z_{\parallel}$	reference lateral design value for single bolt or lag bolt in connection with all wood members loaded parallel to grain (lb)
$Z_{\perp}$	reference lateral design value for single bolt or lag bolt in wood-to-metal connection with wood member(s) loaded perpendicular to grain (lb)
$Z_{m\perp}$	reference lateral design value for single bolt or lag bolt in wood-to-wood connection with main member loaded perpendicular to grain and side member loaded parallel to grain (lb)
$Z_{s\perp}$	reference lateral design value for single bolt or lag bolt in wood-to-wood connection with main member loaded parallel to grain and side member loaded perpendicular to grain (lb)
Δ	deflection (in.)
Δ	design story drift (amplified) at center of mass, $\delta_x - \delta_{x-1}$ (in.)
Δ_a	anchor slip contribution to shearwall deflection (in.)
Δ_b	bending contribution to shearwall deflection (in.)
Δ_D	deflection of diaphragm
Δ_{MC}	change in moisture contend of wood member (percent)
Δ_n	nail slip contribution to shearwall deflection (in.)
Δ_s	deflection of shearwall (in.)
Δ_v	sheathing shear deformation contribution to shearwall deflection (in.)
δ_x	amplified deflection at level x, determined at the center of mass at and above level x (in.)

δ_{xe}	deflection at level x, determined at the center of mass at and above level x using elastic analysis (in.)
ϕ	resistance factor (LRFD only)
ϕ_b	resistance factor for bending (LRFD only)
ϕ_c	resistance factor for compression (LRFD only)
ϕ_s	resistance factor for stability (LRFD only)
ϕ_t	resistance factor for tension (LRFD only)
ϕ_v	resistance factor for shear (LRFD only)
ϕ_z	resistance factor for connections (LRFD only)
γ	load/slip modulus for a connection (lb/in.)
λ	height and exposure factor for wind pressure calculations
λ	time effect factor (LRFD only)
μ	coefficient of static friction
ρ	redundancy/reliability factor for seismic design
θ	angle between direction of load and direction of grain (longitudinal axis of member) (degrees)
θ_m	angle of load to grain θ for main member (degrees)
θ_s	angle of load to grain θ for side member (degrees)
Ω_o	overstrength factor for seismic design

Wood Buildings and Design Criteria

1.1 Introduction

There are probably more buildings constructed with wood than any other structural material. Many of these buildings are single-family residences, but many larger apartment buildings as well as commercial and industrial buildings also use wood framing.

The widespread use of wood in the construction of buildings has both an economic and an aesthetic basis. The ability to construct wood buildings with a minimal amount of specialized equipment has kept the cost of wood-frame buildings competitive with other types of construction. On the other hand, where architectural considerations are important, the beauty and the warmth of exposed wood are difficult to match with other materials.

Wood-frame construction has evolved from a method used in primitive shelters into a major field of structural design. However, in comparison with the time devoted to steel and reinforced concrete design, timber design is not given sufficient attention in most colleges and universities.

This book is designed to introduce the subject of timber design as applied to wood-frame building construction. Although the discussion centers on building design, the concepts also apply to the design of other types of wood-frame structures. Final responsibility for the design of a building rests with the structural engineer. However, this book is written to introduce the subject to a broad audience. This includes engineers, engineering technologists, architects, and others concerned with building design. A background in statics and strength of materials is required to adequately follow the text. Most wood-frame buildings are highly redundant structures, but for design simplicity are assumed to be made up of statically determinate members. The ability to analyze simple trusses, beams, and frames is also necessary.

1.2 Types of Buildings

There are various types of framing systems that can be used in wood buildings. The most common type of wood-frame construction uses a system of horizontal diaphragms and vertical shearwalls to resist lateral forces, and this book deals specifically with the design of this basic type of building. At one time building codes classified a shearwall building as a *box system*, which was a good physical description of the way in which the structure resists lateral forces. However, building codes have dropped this terminology, and most wood-frame shearwall buildings are now classified as *bearing wall systems*. The distinction between the shearwall and diaphragm system and other systems is explained in Chap. 3.

Other types of wood building systems, such as glulam arches and post-frame (or pole) buildings, are beyond the specific scope of this book. It is felt that the designer should first have a firm understanding of the behavior of basic shearwall buildings and the design procedures that are applied to them. With a background of this nature, the designer can acquire from currently available sources (e.g., Refs. 1.4 and 1.10) the design techniques for other systems.

The basic bearing wall system can be constructed entirely from wood components. See Fig. 1.1. Here the *roof, floors, and walls* use wood framing. The calculations necessary to design these structural elements are illustrated throughout the text in comprehensive examples.

In addition to buildings that use only wood components, other common types of construction make use of wood components in combination with some other type or types of structural material. Perhaps the most common mix of structural materials is in buildings that use *wood roof and floor systems* and *concrete tilt-up* or *masonry (concrete block or brick) shearwalls*. See Fig. 1.2. This type of construction is common, especially in one-story commercial and industrial buildings. This construction is economical for small buildings, but its economy improves as the size of the building increases. Trained crews can erect large areas of *panelized* roof systems in short periods of time. See Fig. 1.3.

Design procedures for the wood components used in buildings with concrete or masonry walls are also illustrated throughout this book. The connections between wood and concrete or masonry elements are particularly important and are treated in considerable detail.

This book covers the *complete* design of wood-frame *box*-type buildings from the roof level down to the foundation. In a complete building design, *vertical loads and lateral forces* must be considered, and the design procedures for both are covered in detail.

Wind and seismic (earthquake) are the two lateral forces that are normally taken into account in the design of a building. In recent years, design for lateral forces has become a significant portion of the design effort. The reason for this is an increased awareness of the effects of lateral forces. In addition, the building codes have substantially revised the design requirements for both wind and seismic forces. These changes are the result of extensive research in wind engineering and earthquake-resistant design.

Figure 1.1 Multi-story wood-frame buildings. (*Photo Courtesy of Southern Pine Council*)

Figure 1.2a *Foreground:* Office portion of wood-frame construction. *Background:* Warehouse with concrete tilt-up walls and wood roof system. (*Photo Courtesy of Mike Hausmann.*)

Figure 1.2b Building with reinforced-masonry block walls and a wood roof system with plywood sheathing. (*Photo Courtesy of Mark Williams.*)

1.3   Required and Recommended References

The sixth edition of this book was prompted primarily by the publication of the 2005 edition of the *National Design Specification for Wood Construction* (Ref. 1.1), which for the first time is in a *dual format*, including both *allowable stress design* (ASD) and the new *load and resistance factor design* (LRFD) provisions. All previous editions of both the *National Design Specification* (NDS) and this book have been based on the ASD format. Accordingly, the sixth edition of this book has been updated to include both the ASD and LRFD provisions of the 2005 NDS.

The NDS is published by the American Wood Council (AWC) of the American Forest and Paper Association (AF&PA) and represents the latest structural design recommendations by the wood industry. In addition to basic design provisions for both ASD and LRFD, the 2005 NDS contains chapters specific to sawn lumber, glued-laminated timber, poles and piles, wood I-joists, structural

Figure 1.3 Interior of a building with a panelized roof system. (*Photo courtesy of Southern Pine Council*)

composite lumber, wood structural panels, mechanical connections, dowel-type fasteners, split ring and shear plate connectors, timber rivets, shearwalls and diaphragms, special loading conditions, and fire design. While some of the basic provisions have been changed and some new provisions added, the major advancement for the 2005 NDS is the integration of the new LRFD provisions with the traditional ASD provisions.

The NDS is actually the formal design section of what is a series of interrelated design documents. There are two primary companion documents that support and complete the dual-format *National Design Specification for Wood Construction*. The first companion document is the *NDS Supplement: Design Values for Wood Construction*, which is often referred to simply as the *Supplement* or the *NDS Supplement* as this was the original and for many years the only supplement to the NDS. The NDS Supplement contains all the reference design values for various species groupings of structural lumber and glued-laminated timber.

The second companion design document to the NDS is the *NDS Supplement: Special Design Provisions for Wind and Seismic*, also called the *Wind and Seismic Supplement* or *SDPWS*. The Wind and Seismic Supplement is the newest supplement and is maintained as a separate document due to the unique requirements related to wind- and seismic-resistant design. Included in the SDPWS are reference design values for shearwalls and diaphragms, which comprise the primary lateral-force-resisting system (LFRS) in most wood structures.

The NDS along with both the Design Values Supplement and the Wind and Seismic Supplement comprise the core of what is needed to design engineered wood structures. Because of the subject matter, the reader must have a copy of the NDS to properly follow this book. Additionally, the numerous tables of member properties, design values, fastener capacities, and unit shears for shearwalls and diaphragms are lengthy. Rather than reproduce these tables in this book, the reader is better served to have a copy of both the supplements as well. Having a copy of the *NDS*, the *NDS Supplement: Design Values for Wood*

Construction, and the *NDS Supplement: Special Design Provisions for Wind and Seismic* is analogous to having a copy of the AISC Steel Manual (Ref. 1.3) in order to be familiar with structural steel design.

In addition to the NDS and its two supplements, two other associated documents are available from the AF&PA American Wood Council. The first of these is the *NDS Commentary*, which provides additional guidance and other supporting information for the design provisions included in the NDS. The second associated document is the *ASD/LRFD Manual for Engineered Wood Construction*.

The ASD/LRFD Manual for Engineered Wood Construction was first introduced for ASD in 1999 for the 1997 NDS, and for the first time brought together all necessary information required for the design of wood structures. Prior to this, the designer referred to the NDS for the design of solid sawn lumber and glulam members, as well as the design of many connection details. For the design of other wood components and systems, the designer was required to look elsewhere. For example, the design of shearwalls and diaphragms was not covered in the NDS, but through various other sources including publications by APA—The Engineered Wood Association. The 2005 ASD/LRFD Manual for Engineered Wood Construction is significantly rewritten from the previous 2001 Manual for Engineered Wood Construction for ASD. The 2001 Manual included several additional supplements and guidelines, and duplicated much of the information presented elsewhere. The 2005 Manual contains supporting information for both LRFD and ASD, including nonmandatory design information such as span tables, load tables, and fire assemblies. The Manual is organized to parallel the NDS.

All or part of the design recommendations in the NDS will eventually be incorporated into the wood design portions of most building codes. However, the code change process can take considerable time. This book deals specifically with the design provisions of the 2005 NDS, and the designer should verify local building code acceptance before basing the design of a particular wood structure on these criteria.

This book also concentrates heavily on understanding the loads and forces required in the design of a structure. Emphasis is placed on both gravity loads and lateral forces. Toward this goal, the design loads and forces in this book are taken from the 2006 *International Building Code* (IBC) (Ref. 1.7). The IBC is published by the International Codes Council (ICC), and it is highly desirable for the reader to have a copy of the IBC to follow the discussion in this book. However, the IBC is not used in all areas of the country, and a number of the IBC tables that are important to the understanding of this book are reproduced in Appendix C. If a copy of the IBC is not available, the tables in Appendix C will allow the reader to follow the text.

Frequent references are made in this book to the NDS, the NDS Supplements, the ASD/LRFD Manual for Engineered Wood Construction and the IBC. In addition, a number of cross references are made to discussions or examples in this book that may be directly related to a particular subject. The reader should clearly understand the meaning of the following references:

Examples reference	Refers to	Where to look
NDS Sec. 15.1	Section 15.1 in 2005 NDS	2005 NDS (required reference)
NDS Supplement Table 4A	Table 4A in 2005 NDS Supplement	2005 NDS Supplement (comes with NDS)
SDPWS Supplement Table 4.2A	Table 4.2A in the 2005 Wind and Seismic Supplement	NDS Supplement: Special Design Provisions for Wind and Seismic
IBC Chap. 16	Chapter 16 in 2006 IBC	2006 IBC (recommended reference)
IBC Table 1607.1	Table 1607.1 in 2006 IBC	2006 IBC (recommended reference) or Appendix C of this book
Section 4.15	Section 4.15 of this book	Chapter 4 in this book
Example 9.3	Example 9.3 in this book	Chapter 9 in this book
Figure 5.2	Figure 5.2 in this book	Chapter 5 in this book

Another reference that is often cited in this book is the *Timber Construction Manual* (Ref. 1.4), abbreviated TCM. This handbook can be considered the basic reference on structural glued-laminated timber. Although it is a useful reference, it is not necessary to have a copy of the TCM to follow this book.

1.4 Building Codes and Design Criteria

Cities and counties across the United States typically adopt a building code to ensure public welfare and safety. Until recently, most local governments used one of the three *regional model codes* as the basic framework for their local building code. The three regional model codes were the

1. *Uniform Building Code* (Ref. 1.8)
2. The BOCA *National Building Code* (Ref. 1.6)
3. *Standard Building Code* (Ref. 1.9)

Generally speaking, the *Uniform Building Code* was used in the western portion of the United States, *the BOCA National Building Code* in the north, and the *Standard Building Code* in the south. The model codes were revised and updated periodically, usually on a 3-year cycle.

While regional code development had been effective, engineering design now transcends local and regional boundaries. The International Codes Council (ICC) was created in 1994 to develop a single set of comprehensive and coordinated national model construction codes without regional limitations. The *International Building Code* (IBC) is one of the products of the ICC. The ICC includes representation from the three regional model building code organizations: the Building Officials and Code Administrators International, Inc. (BOCA), which previously maintained the National Building Code; the International Conference of Building Officials (ICBO), which oversaw the Uniform Building Code; and the Southern Building Code Congress International, Inc. (SBCCI), which administered the Standard Building Code. The first edition of the IBC was published in 2000, with a second edition in 2003 and a third edition in 2006. Most regions of the U.S. have adopted all or part of the IBC at either the state or local level.

The standard *Minimum Design Loads for Buildings and Other Structures* (Ref. 1.5) is commonly referred to as ASCE 7-05 or simply ASCE 7. It serves as the basis for some of the loading criteria in the IBC and the regional model codes. The IBC directly references ASCE 7, as does this book.

The IBC is used throughout the text to establish the loading criteria for design. The IBC was selected because it is widely used throughout the United States, and because it represents the latest national consensus with respect to load and force criteria for structural design.

Throughout the text reference is made to the *Code* and the *IBC*. As noted in the previous section, when references of this nature are used, the design criteria are taken from the 2006 edition of the *International Building Code*. Users of other codes will be able to verify this by referring to IBC tables reproduced in Appendix C. By comparing the design criteria of another code with the information in Appendix C, the designer will be able to determine quickly whether the two are in agreement. Appendix C will also be a helpful cross-reference in checking future editions of the IBC against the values used in the text.

Although the NDS is used in this book as the basis for determining the allowable (for ASD) or ultimate (for LRFD) loads for wood members and their connections, note that the IBC also has a chapter that deals with these subjects. However, the latest design criteria are typically found in consensus-approved design specifications such as the NDS.

The designer should be aware that the local building code is the legal authority, and the user should verify acceptance by the local code authority before applying new principles. This is consistent with general practice in structural design, which is to follow an approach that is both rational and conservative. The objective is to produce structures which are economical and safe.

1.5   ASD and LRFD

The *allowable stress design* (ASD) format compares *allowable stresses* of a material to calculated *working stresses* resulting from *service loads*. A single *factor of safety* is applied to the *nominal design value* to arrive at the allowable design value. In the *load and resistance factor design* (LRFD) method, *adjusted capacities* (resistance) are compared to the effects of *factored loads*. The factors are developed for both resistance and loads such that uncertainty and consequences of failure are explicitly recognized.

As mentioned in Sec. 1.3, the sixth edition of this book was prompted primarily by the publication of the 2005 edition of the *National Design Specification for Wood Construction*, which for the first time includes both ASD and the new LRFD provisions. Basic behavioral equations form the basis for both ASD and LRFD provisions. Therefore, the basic behavior of wood is presented in this text first, followed by ASD and LRFD provisions. The reader should be careful when referencing any equations or examples and confirm that the correct format, whether ASD or LRFD, is being reviewed.

To assist the reader in distinguishing between ASD and LRFD, the text uses boxed and coded shading. Examples that relate specifically to ASD only are boxed with a bold line. LRFD examples are presented in light grey shaded boxes. When examples are not boxed or shaded, they are more fundamental in nature and may be considered to apply to both ASD and LRFD.

1.6   Organization of the Text

The text has been organized to present the complete design of a wood-frame building in an orderly manner. The subjects covered are presented roughly in the order that they would be encountered in the design of a building.

In a building design, the first items that need to be determined are the design loads. The Code requirements for vertical loads and lateral forces are reviewed in Chap. 2, and the distribution of these in a building with wood framing is described in Chap. 3.

Following the distribution of loads and forces, attention is turned to the design of wood elements. As noted previously, there are basically two systems that must be designed, one for *vertical loads* and one for *lateral forces*.

The vertical-load-carrying system is considered first. In a wood-frame building this system is basically composed of beams and columns. Chapters 4 and 5 cover the characteristics and design properties of these wood members. Chapter 6 then outlines the design procedures for beams, and Chap. 7 treats the design methods for columns and members subjected to combined axial and bending loads.

As one might expect, some parts of the vertical-load-carrying system are also part of the lateral-force-resisting system (LFRS). The sheathing for wood roof and floor systems is one such element. The sheathing distributes the vertical loads to the supporting members, and it also serves as the *skin* or *web* of the diaphragm for resisting lateral forces. Chapter 8 introduces the grades and properties of wood-structural panels and essentially serves as a transition from the vertical-load- to the lateral-force-resisting system. Chapters 9 and 10 deal specifically with the LFRS. In the typical bearing wall type of buildings covered in this text, the LFRS is made up of a diaphragm that spans horizontally between vertical shear-resisting elements known as shearwalls.

After the design of the main elements in the vertical-load- and lateral-force-resisting systems, attention is turned to the design of the connections. The importance of proper connection design cannot be overstated, and design procedures for various types of wood connections are outlined in Chaps. 11 through 14.

Chapter 15 describes the anchorage requirements between horizontal and vertical diaphragms. Basically, anchorage ensures that the horizontal and vertical elements in the building are adequately tied together.

The text concludes with a review of building code requirements for seismicly irregular structures. Chapter 16 also expands the coverage of overturning for shearwalls.

1.7   Structural Calculations

Structural design is at least as much an *art* as it is a *science*. This book introduces a number of basic structural design principles. These are demonstrated through a large number of practical numerical examples and sample calculations. These should help the reader understand the technical side of the problem, but the application of these tools in the design of wood structures is an art that is developed with experience.

Equation-solving software or *spreadsheet* application programs on a personal computer can be used to create a *template* that can easily generate the solutions of many wood design equations. Using the concept of a template, the design equations need to be entered only once. Then they can be used, time after time, to solve similar problems by changing certain variables.

Equation-solving software and spreadsheet applications relieve the user of many of the tedious programming tasks associated with writing dedicated software. Dedicated computer programs certainly have their place in wood design, just as they do in other areas of structural design. However, equation-solving software and spreadsheets have leveled the playing field considerably. Templates can be simple, or they can be extremely sophisticated. Regardless of programming experience, it should be understood that a simple template can make the solution of a set of bolt equations easier than looking up a design value in a table.

It is highly recommended that the reader become familiar with one of the popular equation-solving or spreadsheet application programs. It is further recommended that a number of the sample problems be solved using such applications. With a little practice, it is possible to create templates which will solve problems that are repetitive and tedious on a hand-held calculator.

Even with the assurance given about the relatively painless way to implement the design equations for wood, some people will remain unconvinced. For those who simply refuse to accept or deal with the computer, or for those who have only an occasional need to design a wood structure, the NDS contains tables that cover a number of common applications. The advantage of the equations is that a wider variety of design problems can be handled, but the NDS tables can accommodate a number of frequently encountered problems.

Although the power and convenience of equation-solving and spreadsheet applications should not be overlooked, all the numerical problems and design examples in this book are shown as complete hand solutions. Laptop computers may eventually replace the hand-held calculator, but the problems in this book are set up for evaluation by calculator.

With this in mind, an expression for a calculation is first given in general terms (i.e., a formula is first stated), then the numerical values are substituted in the expression, and finally the result of the calculation is given. In this way, the designer should be able to readily follow the sample calculation.

Note that the conversion from pounds (lb) to kips (k) is often made without a formal notation. This is common practice and should be of no particular concern to the reader. For example, the calculations below illustrate the adjusted axial load capacity of a tension member:

$$T = F'_t A$$

$$= (1200 \text{ lb/in.}^2)(20 \text{ in.}^2)$$

$$= 24 \text{ k}$$

where T = tensile force
  F'_t = adjusted tensile design value
  A = cross-sectional area

The following illustrates the conversion for the above calculations, which is normally done mentally:

$$T = F'_t A$$

$$= (1200 \text{ lb/in.}^2)(20 \text{ in.}^2)$$

$$= (24{,}000 \text{ lb}) \left(\frac{1 \text{ k}}{1000 \text{ lb}} \right)$$

$$= 24 \text{ k}$$

The appropriate number of significant figures used in calculations should be considered by the designer. When structural calculations are done on a calculator or computer, there is a tendency to present the result with too many significant figures. Variations in loading and material properties make the use of a large number of significant figures inappropriate. A false degree of *accuracy* is implied when the stress in a wood member is recorded in design calculations with an excessive number of significant figures.

As an example, consider the bending stress in a wood beam. If the calculated stress as shown on the calculator is 1278.356 · · · psi, it is reasonable to report 1280 psi in the design calculations. Rather than representing sloppy work, the latter figure is more realistic in presenting the degree of accuracy of the problem.

Although the calculations for problems in this text were performed on a computer or calculator, intermediate and final results are generally presented with three or four significant figures.

An attempt has been made to use a consistent set of symbols and abbreviations throughout the text. Comprehensive lists of symbols and abbreviations, and their definitions, follow the Contents. A number of the symbols and abbreviations are unique to this book, but where possible they are in agreement with those accepted in the industry. The NDS uses a comprehensive notation system for many of the factors used in the design calculations for wood structures. This notation system is commonly known as the *equation format* for wood design and is introduced in Chap. 4.

The units of measure used in the text are the U.S. Customary System units. The abbreviations for these units are also summarized after the Contents. Factors for converting to SI units are included in Appendix D.

1.8 Detailing Conventions

With the large number of examples included in this text, the sketches are necessarily limited in detail. For example, a number of the building plans are shown without doors or windows. However, each sketch is designed to illustrate certain structural design points, and the lack of full details should not detract from the example.

One common practice in drawing wood structural members is to place an × in the cross section of a *continuous* wood member. A *noncontinuous* wood member is shown with a single diagonal line in the cross section. See Fig. 1.4.

1.9 Fire-Resistive Requirements

Building codes place restrictions on the materials of construction based on the occupancy (i.e., what the building will house), area, height, number of occupants, and a number of other factors. The choice of materials affects not only the initial cost of a building, but the recurring cost of fire insurance premiums as well.

The fire-resistive requirements are important to the building designer. This topic can be a complete subject in itself and is beyond the scope of this book. However, several points that affect the design of wood buildings are mentioned here to alert the designer.

Wood (unlike steel and concrete) is a combustible material, and certain *types of construction* (defined by the Code) do not permit the use of combustible materials.

CONTINUOUS FRAMING

NON - CONTINUOUS WOOD MEMBER SUCH AS BLOCKING

Figure 1.4 Typical timber drafting conventions.

There are arguments for and against this type of restriction, but these limitations do exist.

Generally speaking, the *unrestricted use* of wood is allowed in buildings of limited floor area. In addition, the height of these buildings without automatic fire sprinklers is limited to one, two or three stories, depending upon the occupancy.

Wood is also used in another type of construction known as *heavy timber*. Experience and fire endurance tests have shown that the tendency of a wood member to ignite in a fire is affected by its cross-sectional dimensions. In a fire, large-size wood members form a protective coating of char which insulates the inner portion of the member. Thus, large wood members may continue to support a load in a fire long after an unprotected steel member has collapsed because of the elevated temperature. This is one of the arguments used against the restrictions placed on "combustible" building materials. (Note that properly protected steel members can perform adequately in a fire.)

The minimum cross-sectional dimensions required to qualify for the heavy timber fire rating are set forth in building codes. As an example, the IBC states that the minimum nominal cross-sectional dimension for a wood column is 8 in. Different minimum nominal dimensions apply to different types of wood members, and the Code should be consulted for these values. Limits on maximum allowable floor areas are much larger for wood buildings with heavy timber members, compared with buildings without wood members of sufficient size to qualify as heavy timber.

For additional information on fire design of wood members, the reader is referred to NDS Chap. 16.

1.10 Industry Organizations

A number of organizations are actively involved in promoting the proper design and use of wood and related products. These include the model building code groups as well as a number of industry-related organizations. The names and addresses of some of these organizations are listed after the Contents. Others are included in the list of references at the end of each chapter.

1.11 References

[1.1] American Forest and Paper Association (AF&PA). 2005. *National Design Specification for Wood Construction*, ANSI/AF&PA NDS-2005, AF&PA, Washington DC.

[1.2] American Forest and Paper Association (AF&PA). 2005 ASD/LRFD *Manual for Engineered Wood Construction*. 2005 ed., AF&PA, Washington DC.

[1.3] American Institute of Steel Construction (AISC). 2005. *Steel Construction Manual*—13th ed., AISC, Chicago, IL.

[1.4] American Institute of Timber Construction (AITC). 2005. *Timber Construction Manual*, 5th ed., John Wiley & Sons Inc., Hoboken, NJ.

[1.5] American Society of Civil Engineers (ASCE). 2005. *Minimum Design Loads for Buildings and Other Structures, ASCE 7-05*, ASCE, New York, NY.

[1.6] Building Officials and Code Administrators International, Inc. (BOCA). 1999. *The BOCA National Building Code/1999*, 13th ed., BOCA, Country Club Hills, IL.

[1.7]   International Codes Council (ICC). 2006. *International Building Code*, 2006 ed., ICC, Falls Church, VA.

[1.8]   International Conference of Building Officials (ICBO). 1997. *Uniform Building Code*, 1997 ed., ICBO, Whittier, CA.

[1.9]   Southern Building Code Congress International, Inc. (SBCCI). 1997. *Standard Building Code*, 1997 ed., SBCCI, Birmingham, AL.

[1.10] Walker, J.N., and Woeste, F.E. (eds.) 1992. *Post-Frame Building Design Manual*, ASAE—The Society for Engineering in Agricultural, Food, and Biological Systems, St. Joseph, MI.

Chapter 2

Design Loads

2.1 Introduction

Calculation of design loads for buildings is covered in this chapter and Chap. 3. This chapter introduces design loads, load combinations, and serviceability criteria for design of wood structures. Chapter 3 is concerned with distribution of these loads throughout the structure.

The types of loads that the designer is asked to consider are defined in American Society of Civil Engineers (ASCE) Standard 7-05, *Minimum Design Loads for Buildings and Other Structures* (Ref. 2.5), and the 2006 edition of the *International Building Code* (IBC, Ref. 2.9). Some of these load types act vertically, some act laterally (horizontally), and some include both vertical and horizontal components. The important differentiation for most wood structures is that the vertical portion of the load (acting either up or down) is generally carried by systems of joists, beams, and posts or walls, while loads acting laterally are carried primarily by sheathed shearwalls. For this reason, this text discusses the two primary groups of loads: (1) vertical loads—loads acting on the vertical-load-carrying system and (2) lateral loads—loads acting on the lateral-load-resisting system.

Certain members may function only as vertical-load-carrying members or only as lateral-load-carrying members, while a few members may be subject to a combination of vertical and lateral loads. For example, a member may function as a beam when subjected to vertical loads and as an axial tension or compression member under lateral loads (or vice versa).

Gravity loads offer a natural starting point, as little introduction is required. "Weight" is something with which most people are familiar, and the design for vertical loads is often accomplished first. The reason for starting here is twofold. First, gravity loading is an ever-present load, and quite naturally it has been the basic, traditional design concern. Second, in the case of lateral seismic loads, it is necessary to know the magnitude of the vertical loads before the design seismic loads can be estimated.

The terms load and force are often used interchangeably. Both are used to refer to a vector quantity with U.S. Customary System units of pounds (lb) or kips (k). In keeping with use in ASCE 7, this text primarily uses the term *loads* to indicate vertical, lateral, and other external environmental effects being applied to the structure. The term *forces* will primarily refer to the demand on a member as a result of applied loads. Some variation in use may occur within the text; such variation in use is not intended to affect the meaning.

Also note that the design of structural framing members usually follows the reverse order in which they are constructed in the field. That is, design starts with the lightest framing member on the top level and proceeds downward, and construction starts at the bottom with the largest members and proceeds upward.

Before starting discussion of each load type, one last item bears mentioning. An engineered design approach involves calculating a demand due to loads, and comparing the demand with the capacity of the member or element under consideration. An alternative approach to construction of wood light-frame buildings and some other systems is available in the *International Building Code* (IBC, Ref. 2.9) and the *International Residential Code* (IRC, Ref. 2.10). Other standards may be applicable in high-wind regions. This approach is referred to as *prescriptive construction* because the construction requirements are directly prescribed in code provisions or tables, rather than using calculations of demand and capacity. These provisions permit construction of a limited number of small building types, with the primary intent being one and two family dwellings. Gravity design uses tabulated joist and rafter tables that have been precalculated. Requirements for wind and seismic loads have their basis in historic building practices; minimum lengths of bracing walls and load path connections are prescriptively specified. The wind and seismic bracing provisions will generally result in different construction than would be required for an engineered design. Detailed discussion of prescriptive construction is beyond the scope of this text. The reader is referred to the IBC and IRC and their commentaries for more information on this alternate design approach.

Design loads are the subject of both the IBC and ASCE 7. It is suggested that the reader accompany the remaining portions of this chapter with a review of ASCE 7 and Chap. 16 of the IBC. The load information provided in the IBC is intended to allow the determination of general loading parameters (such as basic wind speed and Seismic Design Category) while not necessarily including full details of design loads. This information is placed in the building code to provide easy access to both designers and building officials. It is intended that the designer start with information provided in the IBC, and proceed to more specific information in ASCE 7 for design. There are a few instances, however, when the IBC description of load is more specific than or different from ASCE 7. For convenience, a number of figures and tables from the IBC are reproduced in Appendix C.

The loadings specified in ASCE 7 and the IBC represent minimum criteria; if the designer has knowledge that the actual loads will exceed the specified minimum loads, the higher values must be used for design. In addition, it is required that the structure be designed for loading that can be reasonably anticipated for a given occupancy and structure configuration. ASCE 7 and its commentary provide information on some loads and load combinations that are not addressed in the IBC.

As noted in Chap. 1, this text addresses the national model building code (IBC) and the national consensus standard (ASCE 7). When a design is submitted to a building department, the design must conform to the building code as adopted by that jurisdiction. This will often be either the IBC, or the IBC with further amendments adopted at the state or local level. It is necessary to find out from the building department what code and amendments are adopted.

2.2   Dead Loads

The notation D is used to denote dead loads. Dead loads are addressed in ASCE 7 Sec. 3.1 and IBC Sec. 1606. Included in dead loads are the weights of all materials permanently attached to the structure. In the case of a wood roof or wood floor system, this would include the weight of the roof or floor covering, sheathing, framing, insulation, ceiling (if any), and any other permanent materials such as piping, automatic fire sprinkler, and ducts.

Another dead load that must be included is the weight of fixed equipment. Mechanical or air-conditioning equipment on a roof is one example that is easily overlooked. Often this type of load is supported by a built-up member (two or three joists side by side) as shown in Fig. 2.1. The alternative is to design special larger and deeper beams to carry these isolated equipment loads.

The magnitudes of dead loads for various construction materials can be found in a number of references. A fairly complete list of weights is given in Appendix B, and additional tables are given in Refs. 2.3 and 2.5.

Because most building dead loads are estimated as uniform loads in terms of pounds per square foot (psf), it is often convenient to convert the weights of framing members to these units. For example, if the weight per lineal foot of a wood framing member is known, and if the center-to-center spacing of parallel members is also known, the dead lead in psf can easily be determined by dividing the weight per lineal foot by the center-to-center spacing. For example, if 2 × 12 beams weighing 4.3 lb/ft are spaced at 16 in. on center (o.c.), the equivalent uniform load is 4.3 lb/ft ÷ 1.33 ft = 3.2 psf. A table showing these equivalent uniform loads for typical framing sizes and spacings is given in Appendix A.

It should be pointed out that in a wood structure, the dead load of the *framing members* usually represents a fairly minor portion of the total design load. For this reason a small error in estimating the weights of framing members (either lighter or heavier) typically has a negligible effect on the final member choice. *Slightly* conservative (larger) estimates are preferred for design.

The estimation of the dead load of a structure requires some knowledge of the methods and materials of construction. A "feel" for what the unit dead loads of a wood-frame structure should total is readily developed after exposure to several

ROOF FRAMING PLAN

SECTION A

Figure 2.1 Support of equipment loads by additional framing.

buildings of this type. The dead load of a typical wood floor or roof system usually ranges between 7 and 20 psf, depending on the materials of construction, span lengths, and whether a ceiling is suspended below the floor or roof. For wood wall systems, values might range between 4 and 20 psf, depending on stud size and spacing and the type of wall sheathing used (for example, $\frac{3}{8}$-in. plywood weighs approximately 1 psf whereas $\frac{7}{8}$-in. stucco weighs 10 psf of wall surface area). Typical load calculations provide a summary of the makeup of the structure. See Example 2.1.

The dead load of a wood structure that differs substantially from the typical ranges mentioned above should be examined carefully to ensure that the various individual dead load (D) components are in fact correct. It pays in the long run to stand back several times during the design process and ask, "Does this figure seem reasonable compared with typical values for other similar structures?"

EXAMPLE 2.1   Sample Dead Load D Calculation Summary

Roof Dead Loads

Roofing (5-ply with gravel)	= 6.5 psf
Reroofing	= 2.5
$\frac{1}{2}$-in. plywood (3 psf × $\frac{1}{2}$ in.)	= 1.5
Framing (estimate 2 × 12 at 16 in. o.c.)	= 2.9
Insulation	= 0.5
Suspended ceiling (acoustical tile)	= 1.0
Roof dead load D	= 14.9 psf
Say Roof D	= 15.0 psf

Floor Dead Loads

Floor covering (lightweight concrete $1\frac{1}{2}$ in. at 100 lb/ft^3)	= 12.5
$1\frac{1}{8}$-in. plywood (3 psf × $1\frac{1}{8}$ in.)	= 3.4
Framing (estimate 4 × 12 at 4 ft-0 in. o.c.)	= 2.5
Ceiling supports (2 × 4 at 24 in. o.c.)	= 0.6
Ceiling ($\frac{1}{2}$-in. drywall, 5 psf × $\frac{1}{2}$ in.)	= 2.5
Floor dead load D	= 21.5 psf
Say Floor D	= 22.0 psf

In the summary of roof dead loads in Example 2.1, the load titled *reroofing* is sometimes included to account for the weight of roofing that may be added at some future time. Subject to the approval of the local building official, new roofing materials may sometimes be applied without the removal of the old roof covering. Depending on the materials (e.g., built-up, asphalt shingle, wood shingle), one or two overlays may be permitted.

Before moving on to another type of loading, the concept of the *tributary area* of a member should be explained. The area that is assumed to load a given member is known as the tributary area A_T. For a beam or girder, this area can be calculated by multiplying the *tributary width* times the span of the member. See Example 2.2. The tributary width is generally measured from midway between members on one side of the member under consideration to midway between members on the other side. For members spaced a uniform distance apart, the tributary width is equivalent to the spacing between members. Since tributary areas for adjacent members do not overlap, all distributed loads are assumed to be supported by the nearest structural member. When the load to a member is uniformly distributed, the load per foot can readily be determined by taking the unit load in psf times the tributary width (lb/ft^2 × ft = lb/ft). The concept of tributary area will play an important role in the calculation of many types of loads.

Note, however, that a tributary area approach should only be used when the loading is uniform. Where loading varies, a more detailed calculation is required.

EXAMPLE 2.2   Tributary Areas

In many cases, a uniform spacing of members is used throughout the framing plan. This example is designed to illustrate the *concepts* of tributary area rather than typical framing layouts. See Fig. 2.2.

Tributary Area Calculations

	A_T = tributary width $\times$ span	$A_T \times K_{LL}$
Joist J1	$A_T = 2 \times 12 = 24$ ft^2	$24 \times 2 = 48$ ft^2
Joist J2	$A_T = 2 \times 14 = 28$ ft^2	$28 \times 2 = 56$ ft^2
Girder G1	$A_T = (^{12}/_2 + {}^{14}/_2)20 = 260$ ft^2	$260 \times 2 = 520$ ft^2
Girder G2	$A_T = (^{12}/_2 + {}^{14}/_2)24 = 312$ ft^2	$312 \times 2 = 624$ ft^2
Interior column C1	$A_T = (^{12}/_2 + {}^{14}/_2)(^{20}/_2 + {}^{24}/_2) = 286$ ft^2	$286 \times 4 = 1144$ ft^2
Exterior column C2	$A_T = (^{12}/_2)(^{20}/_2 + (^{24}/_2) = 132$ ft^2	$132 \times 4 = 528$ ft^2
Corner column C3	$A_T = (^{14}/_2)(^{20}/_2) = 70$ ft^2	$70 \times 2 = 140$ ft^2

ROOF OR FLOOR FRAMING PLAN

Figure 2.2   Concepts of tributary area.

For the purpose of determining required live load, an area larger than the tributary area A_T is assumed to influence the structural performance of a member. This area is known as the *influence area* and is calculated in ASCE 7 and the IBC as the *live load element factor* K_{LL} times A_T. K_{LL}, found in ASCE 7 Table 4-2 and IBC Table

1607.9.1, varies between 4 and 1. The influence area $K_{LL} \times A_T$ typically includes the full area of all members that are supported by the member under consideration. As discussed previously, the tributary area approach assumes that each load on a structure is supported entirely by the nearest structural member. In contrast, the influence area recognizes that the total load experienced by a structural member may be influenced by loads applied outside the tributary area of the member. For example, any load applied between two beams is recognized to influence both beams, even though the load is located within the tributary area of just one of the beams. For most columns, the influence area is $4 \times A_T$ and for most beams the influence area is $2 \times A_T$. Thus, the area represented by influence area will overlap for adjacent beam members or column members. Further discussion of the influence area can be found in the ASCE 7 Commentary.

2.3   Live Loads

The term L_r is used to denote *roof live loads*. The symbols L and L_0 are used for live loads other than roof. Included in *live loads* are loads associated with *use* or *occupancy* of a structure. Roof live loads are generally associated with maintenance of the roof. While dead loads are applied permanently, live loads tend to fluctuate with time. Typically included are people, furniture, contents, and so on. ASCE 7 and IBC specify the minimum roof live loads L_r and minimum floor live loads L that must be used in the design of a structure. For example ASCE 7 Table 4-1 and IBC Table 1607.1 specify unit floor live loads L in psf for use in the design of floor systems, and unit roof live loads L_r. Note that this book uses the italicized terms L, L_1, and L_2 to denote span. The variable L denoting a span will always be shown in italics, while L and L_r denoting live loads will be shown in standard text.

The minimum live loads in ASCE 7 and the IBC are, with some exceptions, intended to address only the use of the structure in its final and occupied configuration. Construction means and methods, including loading and bracing during construction, are generally not taken into account in the design of the building. This is because these loads can typically only be controlled by the contractor, not the building designer. In wood frame structures, the construction loading can include stockpiling of construction materials on the partially completed structure. It is incumbent on the contractor to ensure that such loading does not exceed the capacity of the structural members.

For reduction of both roof live loads L_r and floor live loads L, the area contributing load to the member under design consideration is taken into account. For roof live load L_r, the tributary area is used. For floor live load L, the influence area is used. The concept that the area should be considered in determining the magnitude of the unit uniform live load, not just the total load is as follows:

> If a member has a small area contributing live load, it is likely that a fairly high unit live load will be imposed over that relatively small area. On the other hand, as the area contributing live load becomes large, it is less likely that this large area will be uniformly loaded by the same high unit load considered in the design of a member with small area.

Therefore, the consideration of the area contribution load (tributary area A_T or influence area $K_{LL} \times A_T$) in determining the unit live load has to do with the probability that high unit loads are likely to occur over small areas, but that high unit loads will probably not occur over large areas.

It should be pointed out that no reduction is permitted where live loads exceed 100 psf, in areas of public assembly, or per IBC, in parking garages. Reduction is not allowed because an added measure of safety is desired in these critical structures. In warehouses with high storage loads and in areas of public assembly (especially in emergency situations), it is possible for high unit loads to be distributed over large plan areas. However, for the majority of wood frame structures, reductions in live loads will be allowed.

Floor live loads

As noted earlier, minimum floor unit live loads are specified in ASCE 7 Table 4-1 and IBC Table 1607.1. These loads are based on the occupancy or use of the building. Typical occupancy or use floor live loads range from a minimum of 30 psf for residential structures to values as high as 250 psf for heavy storage facilities. These unit live loads, denoted L, are for members supporting small influence areas. A small influence area is defined as 400 ft^2 or less. From previous discussion of influence areas, it will be remembered that the magnitude of the unit live load can be reduced as the size of the influence area increases. For members with an influence area of $K_{LL} \times A_T \geq 400$ ft^2, the reduced live load L is determined as follows:

$$L = L_0\left(0.25 + \frac{15}{\sqrt{K_{LL} \times A_T}}\right)$$

where L_0 = unreduced floor live load specified in IBC Table 1607.1 or ASCE 7 Table 4-1. The reduced live load L is not permitted to be less than $0.5L_0$ for members supporting loads from only one floor of a structure, nor less than $0.4L_0$ for members supporting loads from two or more floors.

The calculation of reduced floor live loads is illustrated in Example 2.3.

EXAMPLE 2.3   Reduction of Floor Live Loads

Determine the total axial force required for the design of the interior column in the floor framing plan shown in Fig. 2.3. The structure is an apartment building with a floor dead load D of 10 psf and, from ASCE 7 Table 4-1, a tabulated floor live load of 40 psf. Assume that roof loads are not part of this problem and the load is received from one level.

Floor Live Load

$$\text{Tributary } A = A_T = 20 \times 20 = 400 \text{ ft}^2$$

$$K_{LL}A_T = 4(400) = 1600 \text{ ft}^2 > 400 \text{ ft}^2$$

FLOOR FRAMING PLAN

Figure 2.3 Interior column tributary area.

∴ floor live load can be reduced

$$L_0 = 40 \text{ psf}$$

$$L = L_0\left(0.25 + \frac{15}{\sqrt{K_{LL}\cdot A_T}}\right) = 40\left(0.25 + \frac{15}{\sqrt{1600}}\right) = (40)(0.625) = 25\,\text{psf}$$

Check:

$$0.5L_0 = 0.5(40) = 20 \text{ psf} < 25 \text{ psf} \quad OK$$

Use L = 25 psf.

Total Load

$$TL = D + L = 10 + 25 = 35 \text{ psf}$$
$$P = 35 \times 400 = 14{,}000 \text{ lb} = 14.0 \text{ k}$$

In addition to basic floor uniform live loads in psf, ASCE 7 and the IBC provide special alternate concentrated floor loads. The type of live load, uniform or concentrated, which produces the more critical condition in the required load combinations (Sec. 2.17) is to be used in sizing the structure.

Concentrated floor live loads other than vehicle wheel loads can be distributed over an area $2\frac{1}{2}$ ft square ($2\frac{1}{2}$ ft by $2\frac{1}{2}$ ft). Their purpose is to account for miscellaneous nonstationary equipment loads which may occur. Vehicle loads are required to be distributed over an area of 20.25 in.^{2} (4.5 in. by 4.5 in.), which is approximately the contact area between a typical car jack and the supporting floor.

It will be found that the majority of designs will be governed by the uniform live loads. However, both the concentrated loads and the uniform loads should be checked. For certain wood framing systems, NDS Sec. 15.1 (Ref. 2.1) provides a method of distributing concentrated loads to adjacent parallel framing members.

IBC Sec. 1607.9.2 also provides an alternate method of calculating floor live load reductions based on the tributary area of a member. The formula for calculating the live load reduction is different from the formula for the influence area approach.

In buildings where partitions will likely be erected, use of a uniform partition live load is required in addition to the floor live load per ASCE 7 Sec. 4.2.2 and IBC Sec. 1607.5. The partition live load applies whether or not partitions are shown on the plans. This creates an allowance for addition or reconfiguration of partitions, which occurs frequently as part of tenant improvements of office, retail, and similar spaces. ASCE 7 and IBC require a minimum live load of 15 psf in addition to floor live load. Where the weight of planned partitions is greater than 15 psf, the actual weight should be used in design. ASCE 7 and IBC do not require the partition live load where the design floor live load is 80 psf or greater. In past codes this allowance for partitions has been defined as a dead load. Live load reduction does not apply to the partition live load, as both ASCE 7 and IBC are specific that live load reduction applies only to loads specified in ASCE 7 Table 4-1 or IBC Table 1607.1.

Roof live loads

ASCE 7 and the IBC specify minimum unit live loads that are to be used in the design of a roof system. The live load on a roof is usually applied for a relatively short period of time during the life of a structure. This fact is normally of no concern in the design of structures other than wood. However, as will be shown in subsequent chapters, the length of time for which a load is applied to a *wood structure* does have an effect on the capacity (resistance).

Roof live loads are specified to account for the miscellaneous loads that may occur on a roof. These include loads that are imposed during the roofing process. Roof live loads that may occur after construction include reroofing operations, air-conditioning and mechanical equipment installation and servicing, and, perhaps, loads caused by fire-fighting equipment. Wind forces and snow loads are not normally classified as live loads, and they are covered separately.

Unit roof live loads are calculated based on the provisions of ASCE 7 Secs. 4.2 and 4.9 and IBC Sec. 1607.11. The standard roof live load for small tributary areas on flat roofs is 20 psf. A reduced roof live load may be determined based on the slope or pitch of the roof and the tributary area of the member being designed. The larger the tributary area, the lower the unit roof live load. As discussed for floor live load reductions, the consideration of tributary area has to do with the reduced probability of high unit loads occurring over large areas. Consideration of roof slope also relates to the probability of loading. On a roof that is relatively flat, fairly high unit live loads are likely to occur. However, on a steeply pitched roof much smaller unit live loads will be probable. Reduced roof live loads may be calculated for tributary areas A_T greater than 200 ft^2 and for roof slopes F steeper than 4 in./ft, as illustrated in the following formula:

$$L_r = L_0 R_1 R_2 \quad \text{and} \quad 12 \le L_r \le 20 \text{ psf}$$

$$\text{where } R_1 = \begin{cases} 1 & \text{for } A_T \le 200 \text{ ft}^2 \\ 1.2 - 0.001 A_T & \text{for } 200 \text{ ft}^2 < A_T < 600 \text{ ft}^2 \\ 0.6 & \text{for } A_T \ge 600 \text{ ft}^2 \end{cases}$$

$$R_2 = \begin{cases} 1 & \text{for } F \le 4 \\ 1.2 - 0.05F & \text{for } 4 < F < 12 \\ 0.6 & \text{for } F \ge 12 \end{cases}$$

A_T = tributary area supported by structural member, ft^2
F = the number of inches of rise per foot for a sloped roof
L_0 = minimum uniform live load per ASCE 7 Table 4-1 or IBC Table 1607.1

The above calculations and limits are for what are noted as "ordinary" roofs. Where roofs serve special functions, uniform live loads as high as 100 psf are required.

Example 2.4 illustrates the determination of roof live loads for various structural members based on the tributary area of each member.

EXAMPLE 2.4   Calculation of Roof Live Loads

Determine the uniformly distributed roof loads (including dead load and roof live load) for the purlins and girders in the building shown in Fig. 2.4. Also determine the total load on column C1. Assume that the roof is flat (except for a minimum slope of $\frac{1}{4}$ in./ft for drainage). Roof dead load D = 8 psf.

Tributary Areas

Purlin P1: $A_T = 4 \times 16 = 64 \text{ ft}^2$

Girder G1: $A_T = 16 \times 20 = 320 \text{ ft}^2$

Column C1: $A_T = 16 \times 20 = 320 \text{ ft}^2$

Figure 2.4   Example building plan.

ROOF FRAMING PLAN

Roof Loads

Flat roof:

$$\therefore R_2 = 1$$

a. Purlin

$$A_T = 64 \text{ ft}^2 < 200 \text{ ft}^2$$

$$\therefore R_1 = 1$$

$$\therefore L_r = 20 \text{ psf}$$

$$w = (D + L_r) \text{ (tributary width)}$$

$$= [(8 + 20) \text{ psf}] \, (4 \text{ ft}) = 112 \text{ lb/ft}$$

b. Girder

$$200 \text{ ft}^2 < (A_T = 320 \text{ ft}^2) < 600 \text{ ft}^2$$

$$R_1 = 1.2 - 0.001A_T = 1.2 - 0.001(320) = 0.88$$

$$L_r = 20R_1R_2 = (20)(0.88)(1) = 17.6 \text{ psf}$$

$$w = [(8 + 17.6) \text{ psf}] \, (16 \text{ ft}) = 409.6 \text{ lb/ft}$$

c. Column

$$A_T = 320 \text{ ft}^2 \qquad \text{same as girder}$$

$$\therefore L_r = 17.6 \text{ psf}$$

$$P = [(8 + 17.6) \text{ psf}](320 \text{ ft}^2) = 8192 \text{ lb}$$

It should be pointed out that the unit live loads specified in ASCE 7 and the IBC are applied on a horizontal plane. Therefore, roof live loads on a flat roof can be added directly to the roof dead load. In the case of a sloping roof, the dead load would probably be estimated along the sloping roof; the roof live load, however, would be on a horizontal plane. In order to be added together, the roof dead load or live load must be converted to a load along a length consistent with the load to which it is added. Note that both the dead load and the live load are gravity loads, and they both, therefore, are *vertical* (not inclined) vector resultant forces. See Example 2.5.

In certain framing arrangements, unbalanced live loads (or snow loads) can produce a more critical design situation than loads over the entire span. Should this occur, ASCE 7 and the IBC require that unbalanced loads be considered.

EXAMPLE 2.5   Combined D + L$_r$ on Sloping Roof (Fig. 2.5)

END ELEVATION

Figure 2.5   Combined D + L$_r$ on sloping roof.

The total roof load (D + L$_r$) can be obtained either as a distributed load along the roof slope or as a load on a horizontal plane. The lengths L_1 and L_2 on which the loads are applied must be considered.

Equivalent total roof loads (D + L_r):

Load on horizontal plane:

$$w_{TL} = w_D\left(\frac{L_1}{L_2}\right) + w_{L_r}$$

Load along roof slope:

$$w_{TL} = w_D + w_{L_r}\left(\frac{L_2}{L_1}\right)$$

Special live loads

IBC Sec. 1607 also requires design for special loads. Because these loads have to do with the occupancy and use of a structure and tend to fluctuate with time, they are identified as live loads. It should be noted that the direction of these live loads is horizontal in some cases. Examples of special live loads include ceiling vertical live loads and live loads to handrails (which are applied both horizontally and vertically). The notation L is generally used for all live loads other than roof live loads L_r.

2.4   Snow Loads

The notation S is used for snow loads. Snow loads are addressed in ASCE 7 Chap. 7 and IBC Sec. 1608. The IBC Sec. 1608 merely directs the user to ASCE 7 Chap. 7 for determination of design snow loads, and reprints ASCE 7 ground snow load maps for the convenience of the IBC user. The discussion in this section will be based on the ASCE 7 provisions.

Snow load is another type of gravity load that primarily affects roof structures. In addition, certain types of floor systems, including balconies and decks may be subjected to snow loads.

The magnitude of snow loads can vary greatly over a relatively small geographic area. As an example of how snow loads can vary, the design snow load in a certain mountainous area of southern California is 100 psf, but approximately 5 mi away at the same elevation, the snow load is only 50 psf. This emphasizes the need to be aware of local conditions. The ASCE 7 Fig. 7-1 provides ground snow loads for many areas of the United States; however, there are significant areas on the map with the notation CS, indicating that a site specific study of ground snow loads is required. Often the local building department will have established guidance for minimum design snow loads. It is generally a good idea to verify design snow loads with the building department, even beyond CS regions.

Snow loads can be extremely large. For example, a ground snow load of 240 psf is required in an area near Lake Tahoe, California. It should be noted that the specified design snow loads are on a horizontal plane (similar to roof live loads).

Unit snow loads (psf), however, are not subject to the tributary area reductions that can be used for roof live loads.

This section will provide an introduction to calculation of snow loads for flat and sloped roofs. For those involved in building design for snow load, there are a number of nuances for snow load calculations; the reader is referred to the full text of ASCE 7 Chap. 7.

Calculation of snow load for flat and sloped roofs uses completely different formulas. Calculation of snow load for flat roofs uses the formula:

$$S = p_f = 0.7C_eC_tIp_g \quad \text{and} \quad p_f \geq p_gI \text{ where } p_g \leq 20 \text{ psf}$$

$$p_f \geq 20I \text{ where } p_g > 20 \text{ psf}$$

where C_e = exposure factor
C_t = thermal factor
I = snow importance factor
p_g = ground snow load (psf)

Each of the expressions in this equation are explained in further detail as follows:

P_g = ground snow load.   Ground snow loads are determined in accordance with ASCE 7 Fig. 7-1, IBC Fig. 1608.2, by the local building department, or by a site-specific study per requirements of ASCE 7. The snow loads on the maps are associated with an annual probability of exceedence of 0.02 (mean recurrence interval of 50 years). Ground snow loads shown on the map for many locations in the western and northwestern regions of the United States are applicable only below specified elevations. Snow loads for higher regions should be determined based on site-specific data and historical records. In the formula given above for design snow load, ground snow load is reduced by a factor of 0.7 to account for the fact that snow accumulation on the ground is greater than at the roof level for most structures.

I = Snow importance factor.   Snow importance factors are a fairly recent development in the determination of design loads. An importance coefficient was first included in seismic force calculations, and has more recently been incorporated into the calculation of snow and wind design loads. The concept behind the snow importance factor is that certain structures should be designed for larger loads than ordinary structures. ASCE 7 lists snow importance factors in Table 7-4 as a function of the building's *occupancy category*. Understanding importance factors requires an explanation of occupancy categories. Explanations of occupancy category can be found in two locations: ASCE 7 Sec. 1.5 and Table 1-1, and IBC Sec. 1604.5 and Table 1604.5. Unlike other instances where the same information appears in both ASCE 7 and IBC, there are some differences between occupancy category descriptions in ASCE 7 and IBC. The differences do not often impact standard occupancy buildings, so this book uses the ASCE 7 categories. The reader is reminded that the IBC assignments of occupancy category must be used where conformance to the IBC is required.

Occupancy categories vary between I and IV, and give an indication of the relative hazard to life that may be posed, based on the occupants and materials contained in the building as well as the anticipated building use. Occupancy Category I is applicable to buildings that are not continuously occupied such as agricultural buildings and nonhazardous storage. As the occupancy category increases, the potential hazard increases. Occupancy Category III includes buildings in which groups of people congregate, those where occupants may not be able to freely exit in case of emergency, and buildings housing hazardous substances. This can often include wood frame school buildings and community centers. Occupancy category IV includes primarily essential facilities, relied on to be operational following extreme load events. This can include wood frame fire stations. Buildings not assigned to Occupancy Category I, III, or IV are assigned to II. Occupancy Category II includes the majority of wood frame buildings.

Going back to ASCE 7 Table 7-4, importance factors for snow increase with increasing occupancy category. This results in higher occupancy category facilities being designed to support heavier snow loads, therefore providing a higher factor of safety against failure. ASCE 7 importance factors are

Occupancy Category	Importance Factor
I	0.8
II	1.0
III	1.1
IV	1.2

A snow importance factor of 1.0 will be used for the majority of wood frame buildings, corresponding to an occupancy category of II. Note that ASCE 7 uses the notation I for snow importance factor, wind importance factor, and seismic importance factor. The numerical values of I can vary for different load types for a given building. The reader is cautioned to keep careful track of varying I values.

C_e = snow exposure factor.   ASCE 7 lists snow exposure factors in Table 7-2. The snow exposure factor varies as a function of the *terrain category* and exposure of the roof. This is because an exposed roof on an exposed building is likely to have snow blow off, whereas a sheltered roof on a sheltered building is likely to have snow accumulate.

There are five terrain categories listed. The first three terrain categories (B, C and D) are the same as *exposure categories* used in determining the building's wind exposure, and are described in ASCE 7 Sec. 6.5.6.3. In order to use the exposure category definitions, *surface roughness categories* must first be determined. These are also designated as B, C, and D, and are defined in ASCE 7 Sec. 6.5.6.2. Surface Roughness B has numerous closely spaced obstructions (i.e., urban and suburban areas, or wooded areas). Surface Roughness C has open terrain with scattered obstructions with heights generally less than 30 ft (i.e., flat open country, grasslands), and also applies to all water surfaces in hurricane prone regions. Surface Roughness D has flat unobstructed areas (i.e., mudflats, salt flats, and unbroken ice), and also applies to water surfaces outside of hurricane prone regions.

Exposures and terrain categories are determined as follows:

B applies where the ground surface roughness condition, as defined by Surface Roughness B, prevails in the upwind direction for a distance of at least 2600 ft or 20 times the height of the building, whichever is greater.

D applies where the ground surface roughness, as defined by Surface Roughness D, prevails in the upwind direction for a distance greater than 5000 ft or 20 times the building height, whichever is greater.

C applies where B or D is not applicable.

For practical purposes, this means that unless there are significant obstructions for at least half a mile or wide open spaces for at least a mile, an exposure or terrain category of C will likely be assigned. The fourth terrain category is "above the tree line in windswept mountain areas," and the fifth applies "in Alaska, in areas where trees do not exist within a 2-mi radius of the site." Further explanation of these is not required.

In addition, the exposure of the roof must be chosen from one of the three possible categories: fully exposed, partially exposed, and sheltered. The ASCE 7 commentary indicates that the category is intended to be separately selected for each roof on a building that has more than one roof. Footnote a of Table 7-2 provides explanation of these exposures. Fully exposed occurs when roofs are exposed on all sides without shelter from higher terrain, mechanical equipment, or conifer trees with leaves in the wintertime. A sheltered roof is defined as roof located tight in among conifers that qualify as obstructions. Partially exposed roofs are those that do not qualify as fully exposed or sheltered.

As a function of the terrain category and exposure of the roof, the snow exposure factor can range between 0.7 and 1.2.

C_t = thermal factor.   As the name implies, the thermal factor varies based on the thermal condition of the roof of a structure. Thermal factors are provided in ASCE 7 Table 7-3. For unheated structures, or for buildings with well-ventilated roofs that have high thermal resistance (R-values) and will remain relatively cold during winter months, thermal factors greater than unity are specified since heat transfer from inside the structure will not melt much of the snow on the roof. For continuously heated greenhouses with roofs that have low thermal resistance (R-values), a thermal factor of $C_t = 0.85$ is specified since heat transfer from within the structure will tend to melt substantial amounts of snow on the roof. All other structures are assigned a thermal factor of $C_t = 1.0$.

Calculation of snow load for sloped roofs uses the formula:

$$S = p_s = C_s p_f$$

where C_s = slope factor
$\quad\quad p_f$ = flat roof snow load (psf)

C_s = roof slope factor.   The roof slope factor is specified in ASCE 7 Sec. 7.4 and provides reduced snow loads based on roof slope, type of roof surface, and thermal condition of the roof. The roof slope factor is intended to address the likelihood of snow sliding to the ground from a sloped roof. Roof surfaces are classified as either "unobstructed slippery surfaces" (e.g., metal, slate, glass, or membranes with smooth surfaces) that facilitate snow sliding from the roof, or as "all other surfaces" (including asphalt shingles, wood shakes or shingles, and membranes with rough surfaces). The thermal condition of a roof is categorized as either warm (roofs with $C_t \leq 1.0$) or "cold" (roofs with $C_t > 1.0$) per ASCE 7 Table 7.3. As provided in ASCE 7 Fig. 7-2, for each category of roof the roof slope factor varies linearly between unity and zero for a specific range of roof slopes. For example, warm roofs that are not slippery or unobstructed (a typical condition for many wood-frame structures) are assigned the following C_s values:

$C_s = 1$ for roof slopes less than 30 degrees (slopes of approximately 7 in./ft or less)

$C_s = 0$ for roof slopes greater than 70 degrees

C_s varies linearly between 1 and 0 for roof slopes between 30 degrees and 70 degrees.

Calculation of the snow load for a typical structure based on IBC and ASCE 7 provisions is illustrated in Example 2.6. This example also illustrates the effects of using a load on a horizontal plane in design calculations.

EXAMPLE 2.6   Snow Loads

Assuming that the basic design snow load is not specified by the local building official, determine the total design dead load plus snow load for the rafters in the building illustrated in Fig. 2.6a. The building is a standard residential occupancy located near Houghton, Michigan in Exposure C terrain, with trees providing shelter on all sides of the structure. The building is heated, the rafters are sloped at 6 in./ft, and the roof covering consists of cement asbestos shingles. Determine the shear and moment for the rafters under dead plus snow loads if they are spaced 4 ft-0 in. o.c. Roof dead load D has been estimated as 14 psf along the roof surface.

Snow Load

Ground snow load:	$p_g = 80$ psf	from ASCE 7 Fig. 7-1
Importance factor:	$I = 1.0$	from ASCE 7 Table 7-4 Category II
Snow exposure factor:	$C_e = 1.1$	from ASCE 7 Table 7-2 for "sheltered" roof
Thermal factor:	$C_t = 1.0$	from ASCE 7 Table 7-3
Roof slope factor:	$C_s = 1.0$	from ASCE 7 Fig. 7-2a at 6:12 roof slope

Design snow load:

$$S = p_s = C_s p_f$$

$$= C_s 0.7 C_e C_t I p_g$$

$$= (1.0)(0.7)(1.1)(1.0)(1.0)(80) = 61.6 \text{ psf}$$

TYPICAL SECTION

Figure 2.6a   Example building section for roof loads.

Combining Loads

In computing the combined load to the rafters in the roof, the different lengths of the dead and snow loads must be taken into account. In addition, the shear and moment in the rafters may be combined using the sloping beam method or the horizontal plane method. In the *sloping beam method*, the gravity load is resolved into components that are parallel and perpendicular to the member. The values of shear and moment are based on the normal (perpendicular) component of load and a span length equal to the full length of the rafter. In the *horizontal plane method*, the gravity load is applied to a beam with a span that is taken as the horizontal projection of the rafter. Both methods are illustrated, and the maximum values of shear and moment are compared (see Fig. 2.6b).

$$TL = D + S$$

$$= 14 + 61.6 \left(\frac{18}{20.12} \right)$$

$$= 69 \text{ psf}$$

$$TL = D + S$$

$$= 14 \left(\frac{20.12}{18} \right) + 61.6$$

$$= 77.2 \text{ psf}$$

$$w = 69 \text{ psf} \times 4 \text{ ft} \qquad\qquad w = 77.2 \text{ psf} \times 4 \text{ ft}$$

$$= 276 \text{ lb/ft} \qquad\qquad\qquad = 309 \text{ lb/ft}$$

Use load normal to roof and rafter span parallel to roof.

Use total vertical load and projected horizontal span.

$$V = \frac{wL}{2} = \frac{0.247(20.12)}{2} \qquad\qquad V = \frac{wL}{2} = \frac{0.309(18)}{2}$$

$$= 2.48 \text{ k} \qquad\qquad\qquad\qquad = 2.78 \text{ k} \quad \text{(conservative)}$$

$$M = \frac{wL^2}{8} = \frac{0.247(20.12)^2}{8} \qquad\qquad M = \frac{wL^2}{8} = \frac{0.309(18)^2}{8}$$

$$= 12.5 \text{ ft-k} \qquad\qquad\qquad\qquad = 12.5 \text{ ft-k} \quad \text{(same)}$$

NOTE: The horizontal plane method is commonly used in practice to calculate design values for inclined beams such as rafters. This approach is convenient and gives equivalent design moments and conservative values for shear compared with the sloping beam analysis. (By definition *shear* is an internal force *perpendicular* to the longitudinal axis of a beam. Therefore, the calculation of shear using the sloping beam method in this example is theoretically correct.)

Figure 2.6b Comparison of *sloping beam method* (left) and *horizontal plane method* (right) for determining shears and moments in an inclined beam.

In addition to these basic guidelines for snow loads on flat or sloped roofs, ASCE 7 provides more comprehensive procedures for evaluating snow loads under special conditions. For example, ASCE 7 provisions include consideration of drifting snow and unbalanced snow loads, sliding snow from higher roof surfaces, rain-on-snow surcharge loads for flat roofs, and minimum design snow loads for low-slope roofs (slope ≤ 5 degrees).

2.5   Soil Loads and Hydrostatic Pressure

The notation H is used for lateral soil loads, loads due to hydrostatic pressure, and the pressure of bulk materials. Soil lateral loads and hydrostatic pressure are introduced in ASCE 7 Sec. 3.2 and IBC Sec. 1610.

Soil lateral loading most commonly occurs at retaining walls. It is relatively unusual for wood to be directly loaded by retained soils. One notable exception to this is permanent wood foundations, used in some regions of the United States. While soil lateral loading will most often come from a geotechnical investigation report, ASCE 7 Table 3-1 provides design lateral soil pressures (in psf, per foot of soil depth) for a range of soil classifications. Where retaining walls are provided, it is possible to develop hydrostatic pressure in addition. Hydrostatic pressure is most often avoided by providing drains behind retaining walls. In cases where it is not possible to provide drains, design for combined soil and hydrostatic lateral pressures is required.

In conditions where hydrostatic lateral pressures can develop, it is possible to also have upward hydrostatic pressures on adjacent floor slabs. These upward pressures would also use the notation H in load combinations.

The notation H is also defined in ASCE 7 and the IBC to include pressure of bulk materials. Although no discussion of this use is provided, it is thought to include pressure due to storage of grain, aggregates, or other bulk solids that can exert lateral pressures.

2.6   Loads due to Fluids

The notation F is used for loads due to fluids with well-defined pressures and maximum heights. There is no specific discussion of this use in either ASCE 7 or IBC. It is clearly not intended to address flood loads or hydrostatic pressure, as these are covered in other load types. This leaves other fluids that might occur in building structures. Where fluids are contained in non-building structures or nonstructural components, the reader is cautioned that a wide range of other standards might be applicable. ASCE 7 Chaps. 13 and 15 may provide some guidance for these types of structures.

2.7   Rain Loads

The notation R is used for rain loads. Rain loads are discussed in ASCE 7 Chap. 8 and IBC Sec. 1611. The concept of rain load is primarily applicable to low slope roofs that are surrounded by parapets. Where this is the case, internal drainage systems are generally provided to collect rainwater falling on the roof. It is also generally required that this type of roof have secondary drains or scuppers to provide drainage in case the primary drainage system is not operable. The rain load R is calculated as

$$R = 5.2 \, (d_s + d_h)$$

where R = rain load in psf
     5.2 = weight of water per inch
     d_s = the height of water at the inlet to the secondary drainage system, based on an undeflected roof
     d_h = the hydraulic head that develops at the inlet at its design flow

Where roof deflection might result in additional water weight due to ponding, this must also be considered.

While the possible weight of water on the roof may seem small, every year roof collapses occur during heavy rain storms, attesting to the importance of design for rain water. This load type need not be considered for sloped roofs that cannot develop water buildup.

2.8 Flood Loads

The notation F_a is used for flood loads. Flood loads are addressed in ASCE 7 Chap. 5 and IBC Sec. 1612. The provisions in IBC Sec. 1612 introduce terminology used in design for flood loads, provide direction to the jurisdiction regarding establishment of flood hazard areas, identify needed flood hazard documentation, and refer the user to ASCE 24 (Ref. 2.6) for design and construction requirements.

Because explanation of design for flood loads requires some amount of detail, and because wood structures themselves will generally be elevated above design flood elevations rather than being designed to resist flood loading, this book will not describe details of flood loads. The reader is referred to ASCE 7 Chap. 5 and ASCE 24 for information.

2.9 Self-straining Loads

The notation T is used for self-straining loads. The only available discussion of self-straining loads is found in the lengthy IBC description of the notation. It is: "Self-straining force arising from contraction or expansion resulting from temperature change, moisture change, creep in component materials, movement due to differential settlement, or combinations thereof."

Self-straining loads primarily occur due to dimensional change of the structural element itself, or movement of the support for the member. Where members are free to move with dimensional or support change, forces do not develop. Where elements are restrained against movement, dimensional or support change will induce internal stresses. One example of this behavior is the shrinkage and shortening of post-tensioned concrete slab and beam systems; where the structural system permits unrestrained shrinkage, minimal force will result. Where the structural system restrains free shrinkage, significant forces can result, which if not accounted for can damage structures. Another example is the temperature-driven expansion and contraction of aluminum mullions for curtain wall systems; where not properly allowed for, mullion connections to the structure have been failed by self-straining forces T. It is unusual for self-straining forces to need to be considered in design of wood structures.

2.10   Wind Loads—Introduction

The notation W is used for wind loads. Wind loads are addressed in ASCE 7 Chap. 6 and IBC Sec. 1609. The IBC refers the user to ASCE 7 provisions for determination of wind loads; however, ASCE 7 provisions for determining basic wind speed and exposure are repeated in the IBC for the convenience of the user. The IBC also contains requirements for protection of openings in wind-borne debris regions. The discussion in this section will be focused on ASCE 7 provisions.

ASCE 7 provisions are based on the results of extensive research regarding wind loads on structures and components of various sizes and configurations in a wide variety of simulated exposure conditions. ASCE 7 presents three methods for determination of wind loads on structures: Method 1 – simplified procedure, Method 2 – analytical procedure, and Method 3 – wind tunnel procedure. The wind tunnel procedure is used for complex buildings that might be anticipated to have unusual dynamic behavior; use is limited to a small group of buildings for which the time and expense of a detailed study can be justified. Wind load on wood buildings will be analyzed using either the simplified procedure or the analytical procedure.

This book illustrates the use of Method 1, simplified procedure, because this method may be used for the large majority of wood frame buildings. Method 2, the analytical procedure, can be used for determining wind loads on structures of all sizes, configurations, and exposure conditions. Use of Method 2 requires the defining of more variables than Method 1, but is not significantly more complicated to use. After following the discussion of Method 1 in this book, the reader is encouraged to look at the use of Method 2. A commentary and handbook on wind provisions have been developed by the Structural Engineers Association of Washington (SEAW) (Ref. 2.15 and 2.16). The 2004 editions reflect the wind provisions of the 2003 IBC and ASCE 7-02. An update to reflect the wind design provisions of the 2006 IBC and ASCE 7-05 will be available in 2007.

ASCE 7 places a series of limitations on the use of Method 1, which can be found in its Sec. 6.4.1. These limitations can be better understood once a few terms are introduced. The *main wind-force-resisting system* (MWFRS) is a system providing wind resistance for the overall structure. In wood frame buildings, the MWFRS most commonly consists of shearwalls (sheathed walls that resist in-plane loads) and roof and floor diaphragms (sheathed floor and roof assemblies that transmit in-plane loads to the shearwalls). We will see later in this chapter that the MWFRS almost always also serves as the seismic-force-resisting system. This book uses the term MWFRS when discussing wind loads only, and the term lateral-force-resisting system (LFRS) when discussing resistance to both wind and seismic forces. We will see later in this chapter that this is generally also the seismic-force resisting system. *Components and cladding* are the members making up the exterior envelope of the building, including wall and roof framing, sheathing, and finish materials. ASCE 7 Sec. 6.4.1 addresses limitations for the MWFRS and components and cladding separately.

In order for the MWFRS to be designed using method 1, the following conditions must be met:

1. It is a simple diaphragm building—a building in which both windward and leeward loads are transmitted through the floor and roof diaphragms to the same MWFRS.

2. The building is low rise—it has a mean roof height less than 60 ft, and has a least horizontal dimension not less than the mean roof height.

3. The building is enclosed—and meets requirements for wind-borne debris protection, if applicable.

4. The building is regular—has no unusual geometrical irregularity in spatial form.

5. The building is not classified as a flexible building—has a fundamental frequency greater than 1 hertz (fundamental period less than 1 second).

6. The building does not have response characteristics that create unusual loading (such as galloping or vortex shedding) and is not sited in a location where unusual wind load effects might occur.

7. The building has an approximately symmetrical cross section in each direction, and has a flat roof or a gable or hip roof with slope less than or equal to 45 degrees (12 in 12 pitch).

8. The building is exempted from the torsional load cases of ASCE 7 Figure 6-10 (per footnote 5, one- and two-story wood-frame construction is exempted), or the torsional load cases do not control design.

While this is a rather daunting list of limitations, Method 1 can still be applied to most one and two-story wood-frame buildings, in most locations. Buildings with large variations in the amount of opening in different exterior walls should be carefully evaluated to see if they qualify as enclosed buildings per the ASCE definitions. Also, checking of torsional load cases will be necessary for buildings with three or more stories. In general, the components and cladding can also be designed per Method 1 in buildings where the MWFRS cannot.

For design of the MWFRS, the basic formula for calculating design wind pressure p_s, found in ASCE 7 Sec. 6.4.2.1 is

$$p_s = \lambda K_{zt} I p_{s30}$$

For design of the components and cladding, the basic formula for calculating design wind pressure p_{net}, found in ASCE 7 Sec. 6.4.2.2 is

$$p_{net} = \lambda K_{zt} I p_{net30}$$

Each of the terms in these equations is defined as follows:

P_{s30} = **simplified design wind pressure for MWFRS.** The simplified design wind pressure is defined as the wind pressure applied over the horizontal or vertical projection of the building surface at a height of 30 ft in Exposure B conditions (described in Sec. 2.4). Simplified design wind pressures for the MWFRS are provided in ASCE 7 Fig. 6-2 as a function of the *basic wind speed*, the roof slope, and the region or zone on the building surface. Horizontal simplified design wind pressures combine the effects of windward and leeward pressures that occur on opposite sides of a structure exposed to wind loads. For roof slopes steeper than 25 degrees, two load cases must be considered for vertical loads applied to the horizontal projection of the roof surface.

The simplified design wind pressures for MWFRS essentially apply when one is considering the structure as a whole in resisting wind forces. According to ASCE 7, the MWFRS typically supports forces that are caused by loads on multiple surfaces of a structure. The LFRS used in typical wood frame buildings is described in Chap. 3.

The basic wind speed V (in miles per hour) can be read from a map of the United States given in ASCE 7 Fig. 6-1. Previous versions of wind speed maps were based on the "fastest mile" wind speed, which was defined as the highest recorded wind velocity averaged over the time it takes a mile of air to pass a given point. However, since short-term velocities due to gusts may be much higher, ASCE 7 provides maps based on 3-sec gust wind speeds. The wind speed maps in ASCE 7 show "special wind regions", which indicate that there may be a need to account for locally higher wind speeds in certain areas.

The basic wind speed is measured at a standard height of 33 ft above ground level with Exposure C (defined in Sec. 2.4) conditions and is associated with an annual probability of exceedence of 0.02 (mean recurrence interval of 50 years). The minimum velocity to be considered in designing for wind is 85 mph, and a linear interpolation between the wind speed contours in ASCE 7 Fig. 6-1 may be used. Once the designer has determined the basic wind speed from ASCE 7 Fig. 6-1 (or from the local building official in special wind regions), the simplified design wind pressure p_{s30} (in psf) can be read from ASCE 7 Fig. 6-2.

p_{net30} = **net design wind pressure for components and cladding.** The net design wind pressure is defined as the wind pressure applied normal to a building surface at a height of 30 ft in Exposure B conditions (described in Sec. 2.4). The net design wind pressure is used to determine wind forces on individual structural elements *(components and cladding)* that directly support a tributary area *(effective wind area)* of the building surface. Greater wind effects due to gusts tend to be concentrated on smaller tributary areas. Consequently, the magnitudes of p_{net30} for components and cladding are typically larger than the magnitudes of p_{s30} for MWFRS. However, according to the ASCE 7 Sec. 6.2 definition, the *effective wind area* for individual structural members *(components and cladding)* need not be taken smaller than the square of the span length divided by 3.

Net design wind pressures for components and cladding are provided in ASCE 7 Fig. 6-3 tables as a function of the basic wind speed V, the effective wind area supported by the structural element, the region or zone on the building surface (described in Example 2.10), and the roof slope. The net design wind pressures for components and cladding combine the effects of internal and external pressures that occur on individual structural elements. Two design wind pressures must be considered separately for each structural element: a positive wind pressure acting inward and a negative wind pressure acting outward. According to ASCE 7 Sec. 6.4.2.2.1, the minimum positive net design wind pressure and the maximum negative net design wind pressure for components and cladding are $p_{net30} = \pm 10$ psf.

I = **importance factor.**   As discussed in Sec. 2.4, the concept behind the importance factor is that certain structures should be designed for higher force levels than ordinary structures. Except for the default value of 1.0, note that the importance factors for snow, wind, and seismic forces are not equal.

The importance factor I for wind load is provided in ASCE 7 Table 6-1, as a function of the occupancy category (introduced in Sec. 2.4). The occupancy categories are found in ASCE 7 Table 1-1 or IBC Table 1604.5. This book uses the ASCE 7 tables; however, the reader is reminded that the IBC table must be checked when conformance with the IBC is required.

Standard occupancy structures will fall under Occupancy Category II and have a wind importance factor I of 1.0. Structures with high numbers of occupants or housing essential facilities or hazardous materials will fall into Occupancy Categories III or IV and have a wind importance factor I of 1.15. This value of I was selected because it represents an approximate conversion of the 50-year wind-speed recurrence interval in ASCE 7 Fig. 6-1 to a 100-year recurrence interval.

K_{zt} = **topographical factor.**   This factor, found in ASCE 7 Sec. 6.5.7, accounts for significantly higher wind speeds at sites located on the upper half of an exposed hill, ridge, or escarpment. The topographical factor is taken as 1.0 except for very exposed sites. ASCE 7 provides a series of five criteria, all of which must be met in order to require evaluation for a topographical factor of other than 1.0. These criteria are

1. The hill, ridge, or escarpment is isolated and unobstructed upwind by features of comparable height for a distance of 100 times the height of the feature, or 2 mi, whichever is less.

2. The feature protrudes above the height of other features in the upwind direction by a factor of 2 or more.

3. The structure is located in the upper half of a hill or ridge or near the crest of an escarpment.

4. The slope in the top half of the feature is greater than or equal to a 1:10 rise to run.

5. The height of the feature is greater than or equal to 60 ft for Exposure B and 15 ft for Exposure C.

The equation for calculating the topographic factor is

$$K_{zt} = (1 + K_1 K_2 K_3)^2$$

where K_1, K_2 and K_3 are defined in ASCE 7 Table 6-4. Because there are a number of locations where the topographic factor must be used, calculation of the factor is illustrated in the following example.

EXAMPLE 2.7 Topographic Factor for Hillside Site

A building is to be located at the peak of a 500-ft-high ridge. The ridge is 10 times as high as any obstructions in the upwind direction for 5 mi. The slope of the ridge in the upper half is one vertical in four horizontal.

$$K_{zt} = (1 + K_1 K_2 K_3)^2$$

From ASCE 7 Table 6-4

$H = 500$ ft (height of ridge)
$L_h = 1000$ ft (run of ridge over top half of height)
$x = 0$ ft (distance from top of ridge to building site)
$z = 500$ ft (height of building site above local ground level)
$K_1 = 0.72$ (for $H/L_h = 0.5$)
$K_2 = 1.00$ (for $x/L_h = 0$)
$K_3 = 0.22$ (for $z/L_h = 0.5$)
$K_{zt} = (1 + 0.72 \times 1.00 \times 0.22)^2 = 1.34$

For other examples in this book, the factor K_{zt} will be taken as 1.0.

λ = height and exposure factor. As the name implies, the effects of building height and exposure to wind have been combined into one coefficient. Values of λ are obtained from a table at the end of either ASCE 7 Fig 6-2 or Fig. 6-3 (note that these tables are identical), given the mean roof height above ground h_{mean} and the exposure condition of the site. The wind pressure increases with height above ground.

Turbulence caused by built-up or rough terrain can cause a substantial reduction in wind speed. ASCE 7 references three types of exposure, which are intended to account for the effects of different types of terrain on wind and snow loads (see Sec. 2.4).

2.11 Wind Forces—Main Wind Force Resisting System

Two methods are given in ASCE 7 for determining the design wind forces for the MWFRS. The two methods are a comprehensive *analytical method* and a *simplified method*.

The analytical method provides a more accurate description of the wind forces, but the simplified method produces satisfactory designs for many structures. A problem with the simplified method is that it gives incorrect joint moments in gable rigid frames. Consequently, the simplified method is not applied to these types of structures. Note that many wood-frame structures have a gable profile, but the primary lateral-force-resisting system (LFRS) is usually made up of a system of horizontal diaphragms and shearwalls. Therefore, most wood-frame structures do not use gable *rigid frames*, and the simplified method can be applied. (A gable glulam arch is an example of a wood rigid frame structure which would require the analytical method to determine wind forces for MWFRS.)

In the analytical method, inward pressures are applied to the windward wall, and outward pressures (suction forces) are applied to the leeward wall. The forces on a sloping roof are directed outward on the leeward side, and the force to the windward side will act either inward or outward, depending on the slope of the roof. In the simplified method, horizontal wind forces are applied to the vertical projected area of the building, and vertical forces are applied to the horizontal projected area of the building. See Example 2.8.

It should be noted that the simplified method for determination of design wind pressures applies only for fully enclosed structures. Partially enclosed and unenclosed (open) structures tend to have more complex wind pressure distributions, and therefore must be designed to resist wind forces based on the comprehensive analytical method in ASCE 7. ASCE 7 Sec 6.2 provides extensive descriptions of partially enclosed buildings and open buildings.

Door and window openings are not required to be considered an opening for purposes of categorizing the building as open or partially open as long as:

1. The door or window can reasonably be expected to be closed during a design wind event, and

2. The door or window is designed for component and cladding wind loads per ASCE 7, and

3. If located in a windborne debris region, the door or window glazing is impact resistant or protected in accordance with the requirements of IBC Sec. 1609.1.2 or ASCE 7 Sec. 6.5.9.3.

Wind forces on MWFRS are computed using the wind pressure formula introduced in Sec. 2.10 ($p_s = \lambda K_{zt} I p_{s30}$). The importance factor I specified in ASCE 7 Table 6-1 applies for all wind forces on a given building. The height and exposure factor λ is provided in ASCE 7 Fig. 6-2 based on the mean height of the roof h_{mean}.

The simplified method of analysis for wind forces requires consideration of wind loads acting normal to the longitudinal walls of a building, as well as wind loads acting normal to the transverse walls (end walls) of a building. Since wind pressures vary with roof slope and may be larger near one end of a structure due to wind directionality and aerodynamic effects, ASCE 7 Fig. 6-2 separates the vertical projected surface of a building into four zones for evaluating lateral wind forces on the main wind-force-resisting system. Similarly, the horizontal projection of the roof surface is divided into four zones for evaluating vertical wind forces on the overall structure. Each zone may have a different magnitude wind pressure p_{s30} specified in ASCE 7 Fig. 6-2, and the combined effects of wind pressures acting simultaneously in all eight zones must be considered in the design of the main wind-force-resisting system for each direction of wind loading. Tables in ASCE 7 Fig. 6-2 also specify larger vertical wind pressures at overhanging eaves or rakes located on the windward side of structures.

Since simplified design wind pressures on the vertical projection of a sloped roof may sometimes be negative (outward pressure), footnote 7 to ASCE 7 Fig. 6-2 indicates that the overall wind force due to lateral loads must also be checked with no horizontal pressure applied to the vertical projection of the roof. In addition, ASCE 7 Sec. 6.4.2.1.1 states that the overall lateral wind force on main wind-force-resisting systems must be no smaller than the force associated with a uniform horizontal design wind pressure of $p_s = 10$ psf applied over the entire vertical projected building surface.

The vertical projected area of a building is divided into Zones A, B, C, and D for the application of horizontal design wind pressures. Zones A and B are designated the wall end zone and the roof end zone, respectively, since they are located near one end of the structure. Zones C and D encompass the remainder of the vertical projected area of the structure and are designated the wall interior zone and the roof interior zone, respectively. Larger horizontal wind pressures are given in ASCE 7 Fig. 6-2 for end zones A and B, versus interior Zones C and D. End Zones A and B are assumed to include the portion of the vertical projected area located within a distance $2a$ from one end of the structure. The dimension a is defined as 0.1 times the least width of the structure or 0.4 times the mean roof height h_{mean}, whichever is smaller. However, the dimension a may not be taken less than 3 ft, or less than 0.04 times the least width of the structure.

The horizontal projected area of a building is divided into Zones E, F, G, and H for the application of vertical design wind pressures. Zones E and F are located near one end of the structure and are designated the roof end zones on the windward and leeward sides of the building, respectively. Zones G and H include the remainder of the horizontal projected area of the structure and are designated the roof interior zones on the windward and leeward sides of the building, respectively. Larger vertical wind pressures are given in ASCE 7 Fig. 6-2 for end Zones E and F versus interior Zones G and H. As with wall end zones, roof end Zones E and F are assumed to include the portion of the horizontal projected area located within a distance $2a$ from one end of the structure. The dividing line between windward and leeward zones is located at the mid-length of the structure in the direction the wind is assumed to be blowing.

The comprehensive analytical method for evaluating wind forces in ASCE 7 explicitly addresses the fact that lateral wind pressure is lower for portions of the structure near ground level and increases with height above ground. However, in the simplified method of analysis, the design wind pressure p_s is assumed constant over the entire height of each zone on the projected building surface.

EXAMPLE 2.8 Wind Forces for Main Windforce-Resisting Systems (MWFRS)

Determine the design wind pressures based on the simplified method for the primary LFRS for the building in Fig. 2.7. This is a gable structure that uses a system of diaphragms and shearwalls for resisting lateral forces. The building is a standard occupancy enclosed structure located near Fort Worth, Texas. Exposure C is to be used. K_{zt} is 1.0.

Wind forces for designing MWFRS are obtained based on p_{s30} from ASCE 7 Fig. 6-2. End zone and interior zone locations to be considered for horizontal pressures on the vertical projection of the building surface include (Figs. 2.7a and b)

    Zone A (wall end zone)

    Zone B (roof end zone)

    Zone C (wall interior zone)

    Zone D (roof interior zone)

End zone and interior zone locations to be considered for vertical pressures on the horizontal projection of the building surface include (Figs. 2.7a and b)

    Zone E (windward roof end zone)

    Zone F (leeward roof end zone)

    Zone G (windward roof interior zone)

    Zone H (leeward roof interior zone)

Figure 2.7a Wind pressure zones on vertical and horizontal projections of building surfaces for main wind-force-resisting systems; wind direction parallel to transverse walls (end walls).

Figure 2.7b Wind pressure zones on vertical and horizontal projections of building surfaces for main windforce-resisting systems; wind direction perpendicular to transverse walls (end walls).

The building in this example does not have roof overhangs. An illustration of wind pressure zones with overhangs is provided at the end of this example.

The design wind pressures in each zone for design of the MWFRS are determined as follows.

Wind speed:

$$V = 90 \text{ mph} \qquad \text{ASCE 7 Fig. 6-1}$$

Importance factor:

$$I = 1.0 \qquad \text{ASCE 7 Table 6-1}$$

Height and exposure factor:

The total height of the building is 19 ft.

The eave height is 12 ft.

Therefore, the mean roof height is

$$h_{\text{mean}} = \frac{12 + 19}{2} = 15.5 \text{ ft} > 15 \text{ ft}$$

Using linear interpolation between $\lambda = 1.21$ and $\lambda = 1.29$ for mean roof heights of 15 and 20 ft ASCE 7 Fig. 6-2:

$$\lambda = 1.22 \qquad \text{for } h_{\text{mean}} = 15.5 \text{ ft}$$

Convert a rise of 7 ft in a run of 21 ft to a roof angle θ of approx. 20 degrees.

Horizontal Wind Pressures on Vertical Projection of Building

Wall forces—End zone A for roof slope of 20 degrees

$$p_{s30} = 17.8 \text{ psf}$$

Design wind pressure:

$$p_s = \lambda K_{zt} I p_{s30} = 1.22(1.0)(1.0)(17.8) = 21.7 \text{ psf} \qquad \text{(inward pressure)}$$

Wall forces—Interior Zone C for roof slope of 20 degrees

$$p_{s30} = 11.9 \text{ psf}$$

Design wind pressure:

$$p_s = \lambda K_{zt} I p_{s30} = 1.22(1.0)(1.0)(11.9) = 14.5 \text{ psf} \qquad \text{(inward pressure)}$$

Roof forces—End Zone B for roof slope of 20 degrees

$$p_{s30} = -4.7 \text{ psf}$$

Design wind pressure:

$$p_s = \lambda K_{zt} I p_{s30} = 1.22(1.0)(1.0)(-4.7) = -5.7 \text{ psf} \qquad \text{(outward pressure)}$$

Roof forces—Interior Zone D for roof slope of 20 degrees

$$p_{s30} = -2.6 \text{ psf}$$

Design wind pressure:

$$p_s = \lambda K_{zt} I p_{s30} = 1.22(1.0)(1.0)(-2.6) = -3.2 \text{ psf} \qquad \text{(outward pressure)}$$

Vertical Wind Pressures on Horizontal Projection of Building

Roof forces—Windward End Zone E for roof slope of 20 degrees

$$p_{s30} = -15.4 \text{ psf}$$

Design wind pressure:

$$p_s = \lambda K_{zt} I p_{s30} = 1.22(1.0)(1.0)(-15.4) = -18.8 \text{ psf} \qquad \text{(upward pressure)}$$

Roof forces—Leeward End Zone F for roof slope of 20 degrees

$$p_{s30} = -10.7 \text{ psf}$$

Design wind pressure:

$$p_s = \lambda K_{zt} I p_{s30} = 1.22(1.0)(1.0)(-10.7) = -13.1 \text{ psf} \qquad \text{(upward pressure)}$$

Roof forces—Windward Interior Zone G for roof slope of 20 degrees

$$p_{s30} = -10.7 \text{ psf}$$

Design wind pressure:

$$p_s = \lambda K_{zt} I p_{s30} = 1.22(1.0)(1.0)(-10.7) = -13.1 \text{ psf} \qquad \text{(upward pressure)}$$

Roof forces—Leeward Interior Zone H for roof slope of 20 degrees

$$p_{s30} = -8.1 \text{ psf}$$

Design wind pressure:

$$p_s = \lambda K_{zt}Ip_{s30} = 1.22(1.0)(1.0)(-8.1) = -9.9 \text{ psf} \qquad \text{(upward pressure)}$$

The distance from one end of the building for which higher end zone pressures are applicable is as follows:

$$0.4h_{\text{mean}} = 0.4(15.5) = 6.2 \text{ ft}$$

$$0.1 \text{ (least width of structure)} = 0.1b = 0.1(42) = 4.2 \text{ ft}$$

$$a = \text{lesser of } (0.4h_{\text{mean}} \text{ or } 0.1b) = 4.2 \text{ ft}$$

$$2a = 2(4.2) = 8.4 \text{ ft}$$

The wind pressures for design of main wind-force-resisting systems in this structure are shown in Figs. 2.8a and 2.8b. Note that footnote 7 of ASCE 7 Fig. 6-2 would require that in Fig. 2.8a, the total horizontal load would need to be considered with the load in Zones B and D set to zero, since the load in these Zones is negative (counteracts load in zones A and C).

The upward forces shown in Fig. 2.8 are referred to as *uplift* forces. In addition to other considerations, both horizontal and uplift forces must be used in the moment stability analysis (known as a check on *overturning*) of the structure.

END ELEVATION

Figure 2.8a   Wind pressures on vertical and horizontal projections of building surfaces; wind direction parallel to transverse walls (end walls).

SIDE ELEVATION

Figure 2.8b   Wind pressures on vertical and horizontal projections of building surfaces; wind direction perpendicular to transverse walls (end walls).

Figures 2.9a, 2.9b and 2.9c illustrate wind zones on a building with a roof overhang. Descriptions of the wind zones for the overhang can be found in the footnotes to ASCE 7 Fig. 6-2. Note that the 2a dimension used to locate the end zone loads is always measured from the corner of the building wall, whether or not a roof overhang occurs.

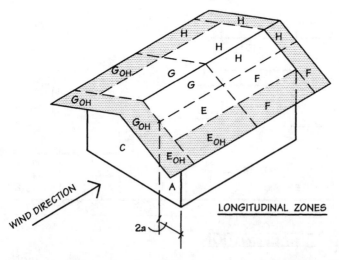

Figure 2.9a   Horizontal and vertical pressure zones for longitudinal wind load.

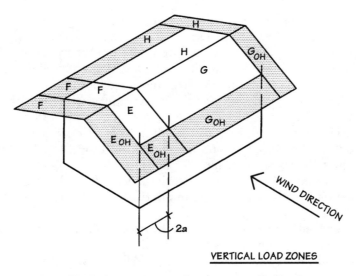

Figure 2.9b   Vertical pressure zones for transverse wind load.

Figure 2.9c   Horizontal pressure zones for transverse wind load.

Design for wind uplift involves two major areas of consideration. These areas will be described in a general fashion here and revisited later in this book once load combinations have been introduced (Sec. 2.17). The first is the direct transfer of the uplift forces from the roof down through the structure. Wind uplift loads occurring at the roof are transferred by the roof covering and sheathing to roof-framing members. Roof sheathing, its fastening, and roof-framing members require design for uplift where a net uplift occurs. Where net uplift occurs on the roof framing, connections of the roof-framing members to the supporting walls or beams and posts will have to be designed for uplift loads. Likewise the wall framing, beams, and posts and their connection to the structure below must be designed for direct uplift. This continues until the dead load of the structure is sufficient to result in zero net uplift, or until reaching the foundation. If net uplift still occurs, the foundation will need to be sized to provide dead load adequate to resist uplift forces. Each of the members and connections in this load path must be adequate to resist net uplift loads, or that member or connection could become the weak link in the load path and permit premature failure of the structure.

Net uplift loads can have a significant effect on the design of roof trusses. Members that are normally in tension due to dead and live loads will often have the load reverse to become a compression load, and vise versa. This is important not only in design of the truss members and connections, but also in providing adequate buckling restraint for slender compression members.

The second area of consideration relates to the overturning design of the structure. Depending on how a building is framed, the added requirement for simultaneous application of the horizontal and vertical wind loads could substantially affect the overturning design of the shearwalls. Horizontal wind loads create in-plane loads acting at the top of the shearwalls. The overturning moment (OM) is the load on the wall times the height of the wall. The OM tends to make the wall overturn. Resistance to the overturning (resisting moment, RM) often comes from the dead load of the structure. Where no net uplift occurs on the roof, the dead load is available to resist overturning of shearwalls. Where net uplift occurs on the roof, little or no dead load may be available to resist wall overturning. This can result in significant forces in the shearwall anchorage to the foundation and in the foundation. Figure 2.10 provides an illustration of these forces acting simultaneously.

When load combinations are discussed, it will be noted that only a portion of the dead load is permitted to be used to counter wind uplift and overturning forces.

EXAMPLE 2.9   Overall Moment Stability

Horizontal and vertical wind forces are shown acting on the shearwall in Fig. 2.10. The vertical component may or may not occur, depending on how the roof is framed. Roof framing can transmit the uplift force to the wall or some other element in the structure.

For the Fig. 2.10 loading, the gross overturning moment OM and resisting moment RM taken about the noted pivot point can be calculated as:

$$OM = P(h) + U(L)$$

$$RM = D(L)$$

SHEAR WALL

Figure 2.10 Shearwall load diagram.

The combining of the wind load and resisting dead load, however, must be in accordance with IBC or ASCE 7 load combinations. This topic is introduced in Sec. 2.17.

In addition to the vertically upward wind pressure, the term *uplift* is sometimes used to refer to the anchorage tie-down force T.

2.12 Wind Forces—Components and Cladding

The forces to be used in designing the MWFRS are described in Sec. 2.11. These are to be applied to the structure acting as a unit (i.e., to the horizontal diaphragms and shearwalls) in resisting lateral forces. The ASCE 7 wind force provisions require that special higher wind pressures be considered in the design of various *structural elements* (components and cladding) when considered *individually* (i.e., not part of the primary LFRS). In other words, when a roof beam or wall stud functions as part of the primary LFRS, the design forces will be determined in accordance with Sec. 2.11. However, when the design of these same members is considered independently, the higher wind pressures for components and cladding are to be used.

The forces on components and cladding are computed using the wind pressure formula introduced in Sec. 2.10 ($p_{net} = \lambda K_{zt} I p_{net30}$). The importance factor I will be the same for all wind forces on a given building. The height and exposure factor λ is again taken from ASCE 7 Fig. 6.3 based on the mean height of the roof h_{mean}.

Proximity to *discontinuities* on the surface of a structure affects the magnitude of wind pressure for components and cladding. Surface discontinuities include changes in geometry of a structure such as wall corners, eaves, rakes (at the ends of gable roof systems), and ridges (for roofs sloped steeper than 7 degrees) that cause locally high wind pressures to develop. Wind tunnel tests and experience have shown that significantly larger wind pressures occur at these discontinuities versus at interior regions of wall and roof surfaces. The net design wind pressures p_{net30} range from lower pressures in interior zones (Zone 1 for roofs and Zone 4 for walls) to higher pressures in end zones (Zone 2 for roofs and Zone 5 for walls) and corner zones where end zones overlap (Zone 3 for roofs only). See Example 2.10. Net design wind pressures on roof elements also vary with roof slope. Particularly high wind pressures are specified in ASCE 7 Fig. 6-3 for structural elements at overhanging eaves and rakes (at the ends of gable roof systems).

The areas to which the higher local wind pressures are applied for end zones and corner zones may or may not cover the entire tributary area of a member. Per Fig. 6-3 footnote 5, the higher pressure is to be applied over a distance from the discontinuity of 0.1 times the least width of the structure or 0.4 times the mean roof height h_{mean}, whichever is smaller. However, the distance from a discontinuity may not be taken less than 3 ft, or less than 0.04 times the least width of the structure.

ASCE 7 wind load provisions recognize that wind pressures are larger on small surface areas due to localized gust effects. Net design wind pressures p_{net30} are provided in ASCE 7 Fig. 6-3 for structural elements that support effective wind areas (tributary areas) of 10, 20, 50, or 100 ft^2. Wind pressures are also listed for wall elements that support effective wind areas of 500 ft^2. Linear interpolation is permitted for intermediate areas. Structural roof elements supporting effective wind areas in excess of 100 ft^2 should use the net design wind pressure for a 100 ft^2 area. Similarly, structural wall elements supporting effective wind areas in excess of 500 ft^2 should use the net design wind pressure for a 500 ft^2 area. The net design wind pressure for an effective wind area of 10 ft^2 applies to structural elements supporting smaller tributary areas. For more information regarding ASCE 7 wind forces, see Ref. 2.12.

EXAMPLE 2.10   Wind Forces—Components and Cladding

The basic wind pressure formula ($p_{net} = \lambda K_{zt} I p_{net30}$) is used to define forces for designing roof and wall elements and their connections. These pressures are larger than the pressures used to design the MWFRS.

Wind forces for designing individual elements and components are based on net design wind pressures p_{net30} from ASCE 7 Fig. 6-3. Typical locations to be considered are (Fig. 2.11*a*)

 a. Roof area.
 b. Wall area.
 c. Wall area.

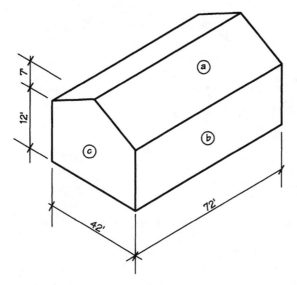

Figure 2.11a General wind load areas for members in interior zones away from discontinuities.

At a rise to run of 7 in 21, the roof slope is greater than seven degrees. Wind forces for designing individual elements near discontinuities are obtained based on p_{net30} from ASCE 7 Fig. 6-3. End zone and corner zone locations to be considered include (Fig. 2.11*b*)

 d. Wall corners (end zone 5)
 e. Eaves without an overhang (end zone 2)
 f. Rakes without an overhang (end zone 2)
 g. Roof ridge when roof slope exceeds 7 degrees (end zone 2)
 h. Overlap of roof ridge zone and roof rake zone when roof slope exceeds 7 degrees (corner zone 3)
 i. Overlap of roof eave zone and roof rake zone (corner zone 3)

The building in this example does not have a roof overhang. Overhangs at the eaves or rakes represent additional areas that require larger design wind pressures in accordance with ASCE 7 Fig. 6-3.

Example 2.8 illustrated the computation of wind pressures for designing the MWFRS (see the summary in Fig. 2.8). Wind pressures required for the design of components and cladding in the same building are evaluated in the remaining portion of this example. The areas considered are those shown in Fig. 2.11*a* and *b*. Design conditions (location and exposure condition) are the same as in Example 2.8.

Figure 2.11b Wind force areas for members in end zones or corner zones at or near discontinuities.

Information from previous example:

Exposure C

$$I = 1.0$$

$$K_{zt} = 1.0$$

$$\lambda = 1.22$$

Roof slope = 4:12 = 18.43 degrees

$$h_{\mathrm{mean}} = 15.5 \text{ ft}$$

The structure is an enclosed structure.

Components and Cladding—Away from Discontinuities

Roof forces—region a (interior zone 1) for effective wind area of 10 ft^2 and roof slope between 7 degrees and 27 degrees:

$$p_{\mathrm{net30}} = \begin{cases} 8.4 \text{ psf} & \text{(inward pressure)} \\ -13.3 \text{ psf} & \text{(outward pressure)} \end{cases}$$

Design wind pressure:

$$p_{\mathrm{net}} = \lambda K_{zt}Ip_{\mathrm{net30}} = \begin{cases} 1.22(1.0)(1.0)(8.4) = 10.2 \text{ psf} & \text{(inward pressure)} \\ 1.22(1.0)(1.0)(-13.3) = -16.2 \text{ psf} & \text{(outward pressure)} \end{cases}$$

Wall forces—regions b and c (interior zone 4) for effective wind area of 10 ft^2:

$$p_{\mathrm{net30}} = \begin{cases} 14.6 \text{ psf} & \text{(inward pressure)} \\ -15.8 \text{ psf} & \text{(outward pressure)} \end{cases}$$

Design wind pressure:

$$p_{\text{net}} = \lambda K_{zt} I p_{\text{net30}} = \begin{cases} 1.22(1.0)(1.0)(14.6) = 17.8 \text{ psf} & \text{(inward pressure)} \\ 1.22(1.0)(1.0)(-15.8) = -19.3 \text{ psf} & \text{(outward pressure)} \end{cases}$$

Components and Cladding—Near Discontinuities

Distance from discontinuity for which higher pressures are applicable:

$$0.4 h_{\text{mean}} = 0.4(15.5) = 6.2 \text{ ft}$$

$$0.1 \text{ (least width of structure)} = 0.1b = 0.1(42) = 4.2 \text{ ft}$$

$$a = \text{lesser of } (0.4 h_{\text{mean}} \text{ or } 0.1b) = 4.2 \text{ ft}$$

Roof forces—regions e, f, and g (end zone 2) for effective wind area of 10 ft^2 and roof slope between 7 degrees and 27 degrees:

$$p_{\text{net30}} = \begin{cases} 8.4 \text{ psf} & \text{(inward pressure)} \\ -23.2 \text{ psf} & \text{(outward pressure)} \end{cases}$$

Design wind pressure:

$$p_{\text{net}} = \lambda K_{zt} I p_{\text{net30}} = \begin{cases} 1.22(1.0)(1.0)(8.4) = 10.2 \text{ psf} & \text{(inward pressure)} \\ 1.22(1.0)(1.0)(-23.2) = -28.3 \text{ psf} & \text{(outward pressure)} \end{cases}$$

Roof forces—regions h and i (corner zone 3) for effective wind area of 10 ft^2 and roof slope between 7 degrees and 27 degrees:

$$p_{\text{net30}} = \begin{cases} 8.4 \text{ psf} & \text{(inward pressure)} \\ -34.3 \text{ psf} & \text{(outward pressure)} \end{cases}$$

Design wind pressure:

$$p_{\text{net}} = \lambda K_{zt} I p_{\text{net30}} = \begin{cases} 1.22(1.0)(1.0)(8.4) = 10.2 \text{ psf} & \text{(inward pressure)} \\ 1.22(1.0)(1.0)(-34.3) = -41.8 \text{ psf} & \text{(outward pressure)} \end{cases}$$

Wall forces—region d (end zone 5) for effective wind area of 10 ft^2:

$$p_{\text{net30}} = \begin{cases} 14.6 \text{ psf} & \text{(inward pressure)} \\ -19.5 \text{ psf} & \text{(outward pressure)} \end{cases}$$

Design wind pressure:

$$p_{\text{net}} = \lambda K_{zt} I p_{\text{net30}} = \begin{cases} 1.22(1.0)(1.0)(14.6) = 17.8 \text{ psf} & \text{(inward pressure)} \\ 1.22(1.0)(1.0)(-19.5) = -23.8 \text{ psf} & \text{(outward pressure)} \end{cases}$$

All of the wind pressures determined in this example apply to structural elements supporting effective wind areas of 10 ft^2 or less. For larger effective wind areas, the pressures would be smaller in accordance with ASCE 7 Fig. 6-3.

2.13 Seismic Forces—Introduction

Many designers have a good understanding of the type of loads covered thus far. However the loads that develop during an earthquake may not be as widely understood, and for this reason a fairly complete introduction to seismic (earthquake) loads E and seismic design requirements is given.

The ASCE provisions for seismic design are much broader than determining loads for which the structure must be designed. There are a number of requirements relating to the systems that may be used to resist seismic loads, to the detailing used in those systems, and to structural configuration issues. For this reason, the discussion will refer to the broad topic of seismic design requirements rather than just earthquake loads.

An explanation of terminology is also appropriate. In the ASCE 7 treatment of load combinations, the term E is used to designate earthquake loads. In the chapters addressing earthquake loads and other related design requirements, the term seismic force is almost exclusively used. These are intended to mean the same thing. Past practice has been to use the term force, rather than load, because earthquakes cause forces internal to the structure rather than loads applied to the surfaces of the structure. This book mirrors the terminology used by ASCE 7.

The Structural Engineers Association of California (SEAOC) pioneered the work in the area of seismic (earthquake) design. Various editions of the SEAOC publication *Recommended Lateral Force Requirements and Commentary* (Ref. 2.14) (commonly referred to as the *Blue Book*), have served as the basis for earthquake design requirements for many editions of the *Uniform Building Code.* Starting in the 1980s, a second code resource document, the National Earthquake Hazard Reduction Program's (NEHRP) *Recommended Provisions for Seismic Regulations for New Buildings* (Ref. 2.8) was developed to address seismic hazards and design requirements on a nationwide basis. In addition the IBC merged the code provisions of the three national model building codes, the *Uniform Building Code* (UBC) (Ref. 2.11), the *Standard Building Code* (SBC) (Ref. 2.13), and the Building Officials and Code Administrator's (BOCA) *National Building Code* (Ref. 2.7).

In the 2006 IBC, there is a significant change is the presentation of seismic design provisions. As was done for snow and wind, the IBC refers the user to ASCE 7 for determination of design loads. The IBC includes enough material so that the Seismic Design Category (SDC) can be determined. This material is duplicated from ASCE 7 for the convenience of the IBC user.

The seismic design provisions of ASCE 7-05 substantially incorporate the 2003 NEHRP Provisions (Ref. 2.8). The 2003 NEHRP Provisions Commentary (Ref. 2.8) is an excellent resource for the ASCE 7 seismic design provisions.

Much of the remaining portion of Chap. 2 deals with the basic concepts of earthquake engineering, and it is primarily limited to a review of the new seismic code as it applies to structurally *regular* wood-frame buildings. Many of the new requirements in the seismic code deal with added requirements for *irregular* structures. These more advanced topics are covered in later chapters (Chaps. 15 and 16) after the fundamentals of horizontal diaphragms and shearwalls are thoroughly understood.

Courses in structural dynamics and earthquake engineering deal at length with the subject of seismic forces. From structural dynamics it is known that a number of different forces act on a structure during an earthquake. These forces include inertia forces, damping forces, elastic forces, and an equivalent forcing function (mass times ground acceleration).

A range of approaches has been developed to analyze the loads and displacements that occur in buildings due to earthquakes. Wood-frame buildings are complex combinations of structural and nonstructural materials, all of which influence the response of a structure to earthquake (seismic) ground motion. The purpose of analysis for earthquake (seismic) loads is to make sure that the structure being designed meets the minimum strength and deflection (drift) requirements of the IBC and ASCE 7. This can be done using simple, intermediate, or complex analysis approaches.

Some of the analysis approaches use sophisticated modeling of the nonlinear behavior of the structural system, while others try to predict nonlinear behavior using simplified linear models. Some look at the complex dynamic behavior of a real building system, while others predict behavior using simplified static load procedures. The most sophisticated approach, nonlinear time-history analysis, combines the more sophisticated modeling of the structural system with the more sophisticated modeling of the dynamic loading. This is an involved and time-consuming process that takes significant expertise, and is most often reserved for the most complicated buildings. Use of this approach for wood frame buildings is rare. Intermediate levels of sophistication are available in linear dynamic and nonlinear static analysis. Use of these approaches for wood-frame buildings is also rare, but expanding. The majority of wood-frame buildings are designed using a linear static analysis approach. This is called the *equivalent lateral force procedure,* and can be found in ASCE 7 Sec. 12.8. This book primarily focuses on this type of seismic analysis.

There is one additional seismic analysis approach provided in ASCE 7 that bears discussion. ASCE 7 Sec. 12.14 is titled *Simplified Alternative Structural Design Criteria for Simple Bearing Wall or Building Frame Systems.* This section provides a simplified version of the equivalent lateral force procedure, for use on a limited scope of buildings. Use of the simplified procedure should result in roughly equivalent results. Because the scope of the buildings is limited, there are a number of checks in the general procedure, which do not need to be considered in the simplified procedure. ASCE 7 Sec. 12.14.1.1 provides a list of 12 scoping criteria that the building must meet in order to be designed using the simplified procedure. Among the 12 there is one that will be problematic for a number of wood-frame buildings. Item 11 says the simplified approach is not permitted in structures with system irregularities due to in-plane or out-of-plane offsets of lateral force resisting elements. An exception permits offsets in shearwalls in two-story buildings of light-frame construction, provided detailing requirements are met. Because offsets in shearwalls from story to story are so common in wood-frame buildings, there are likely to be a good number that do not qualify for this method. For this reason this book addresses the equivalent lateral force procedure instead.

There are structure configurations for which ASCE 7 will require the use of an analysis method more sophisticated than the equivalent lateral force procedure. This information can be found in ASCE 7 Sec. 12.6 and Table 12.6-1. The equivalent lateral force procedure may, however, always be used for light-frame construction.

Experience has proven that *regular structures* (i.e., symmetric structures and structures without discontinuities) perform much better in an earthquake than irregular structures. Therefore, even if the equivalent lateral force procedure is used in design, ASCE 7 penalizes *irregular structures* in areas of high seismic risk with *additional design requirements*. As previously noted, the definition of an irregular structure and a summary of some of the penalties that may be required in the design of an irregular wood building are covered in ASCE 7 Chap. 12.

Rather than attempting to define all of the forces acting during an earthquake, the equivalent lateral force procedure takes a simplified approach. This empirical method is one that is particularly easy to visualize. The earthquake force is treated as an inertial problem only. Before the start of an earthquake, a building is in static equilibrium (i.e., it is at rest). Suddenly, the ground moves, and the structure attempts to remain stationary. The key to the problem is, of course, the length of time during which the movement takes place. If the ground displacement were to take place slowly, the structure would simply ride along quite peacefully. However, because the ground movement occurs quickly, the structure lags behind and "seismic" forces are generated. See Example 2.11.

EXAMPLE 2.11   Building Subjected to Earthquake

1. Original static position of the building before earthquake
2. Position of building if ground displacement occurs very slowly (i.e., in a static manner)
3. Deflected shape of building because of "dynamic" effects caused by rapid ground displacement

Figure 2.12 Building deflected shape when subjected to earthquake.

The force P in Fig. 2.12 is an "equivalent static" design force provided by ASCE 7 and can be used for certain structures in lieu of a more complicated analysis. This is common practice in wood buildings that make use of horizontal diaphragms and shearwalls.

Seismic forces are generated by acceleration of the building mass. Typical practice is to consider that a lump of mass acts at each story level. This concept results in equivalent static forces being applied at each story level (i.e., at the roof and floors). Note that no such simplified forces are truly "equivalent" to the complicated combination of forces generated during an earthquake. For many buildings, however, it is felt that reasonable structural designs can be produced by designing to elastically resist the specified forces.

It should be realized that the forces given in ASCE 7 are at a strength level and must be multiplied by 0.7 for use in allowable stress design. This adjustment occurs in the required load combinations. See Sec. 2.17 for the required load combinations.

The empirical forces given in the equivalent lateral force procedure are considerably lower than would be expected in a major earthquake. In an allowable stress approach, the structure is designed to remain *elastic* under the ASCE 7 static forces. However, it is not expected that a structure will remain elastic in a major earthquake. The key in this philosophy is to *design* and *detail* the structure so that there is sufficient *system ductility* for the building to remain structurally safe when forced into the *inelastic* range in a major earthquake.

Therefore, in areas of high seismic risk, *detailing requirements* apply to all of the principal structural building materials (steel, concrete, masonry, and wood). The term "detailing" here refers to special connection design provisions and to a general tying together of the overall LFRS, so that there is a *continuous path* for the *transfer of lateral forces* from the top of the structure down into the foundation.

Anchorage is another term used to refer to the detailing of a structure so that it is adequately tied together for lateral forces. The basic seismic force requirements are covered in Chaps. 2 and 3, and the detailing and anchorage provisions as they apply to wood-frame structures are addressed in Chaps. 10, 15, and 16.

During an earthquake, vertical ground motion creates vertical forces in addition to the horizontal seismic forces discussed above. The vertical forces generated by earthquake ground motions are generally smaller than the horizontal forces. ASCE 7 directly incorporates vertical ground motions into the seismic force equations. The result of including the vertical component of ground motion in the equations is a slight increase in net uplift and downward forces when considering overturning.

The method used to calculate the horizontal story forces involves three parts. The first part is calculating the *base shear* (the horizontal force acting at the base of the building, V). The second part is assigning the appropriate percentages of this force to the various story levels throughout the height of the structure (story forces). The third part is to determine the forces on particular elements as a result of the story forces (element forces). As will be discussed later, there are several multiplying factors required to convert these *element forces* into design seismic forces for the elements at a story.

The *story forces* are given the symbol F_x (the force at level x). It should be clear that the sum of the F_x forces must equal the base shear V. See Fig. 2.13*a*. The formulas in ASCE 7 for calculating F_x are examined in detail in Sec. 2.15.

Before the expressions for these forces are reviewed, it should be noted that the story forces are shown to increase with increasing height above the base of the building. The magnitudes of the story forces depend on the mass (dead load) distribution throughout the height of the structure. The current vertical distribution provisions have put an exponent of between one and two on the height term in the vertical distribution. This exponent accounts for the increased top-story forces that can occur in buildings with longer periods. The exponent is taken as one for structures with a calculated approximate period of 0.5 sec or less, which is applicable for virtually all wood-frame buildings. With the exponent taken as one, a triangular distribution of story forces occurs.

The reason for this distribution is that ASCE 7 bases its forces on the fundamental mode of vibration of the structure. The fundamental mode is also known as the *first mode* of vibration, and it is the significant mode for most structures.

To develop a feel for the above force distribution, the dynamic model used to theoretically analyze buildings should briefly be discussed. See Fig. 2.13*b*. In this model, the mass (weight) tributary to each story is assigned to that level. In other words, the weight of the floor and the tributary wall loads halfway between adjacent floors is assumed to be concentrated or "lumped" at the floor level. In analytical studies, this model greatly simplifies the solution of the dynamic problem.

Now, with the term *lumped mass* defined, the concept of a mode shape can be explained. A *mode shape* is a simple displacement pattern that occurs as a

CODE SEISMIC FORCES

Figure 2.13*a* Seismic force distribution.

TRIB. DEAD LOADS   **LUMPED MASSES**

Figure 2.13b Seismic forces follow fundamental mode.

linear elastic structure moves when subjected to a dynamic force. The *first mode shape* is defined as the displacement pattern where all lumped masses are on one side of the reference axis. Higher mode shapes will show masses on both sides of the vertical reference axis. In a dynamic analysis, the complex motion of the complete structure is described by adding together the appropriate percentages of all of the modes of vibration. Again, ASCE 7 equivalent lateral forces are essentially based on the first mode.

The point of this discussion is to explain why the F_x story forces increase with increasing height above the base. To summarize, the fundamental or first mode is the critical displacement pattern (deflected shape). The first mode shape shows all masses on one side of the vertical reference axis. Greater displacements and accelerations occur higher in the structure, and the F_x story forces follow this distribution.

2.14 Seismic Forces

The notation E is used for seismic forces. Seismic forces are addressed in IBC Sec. 1613, which refers to ASCE 7 provisions for seismic design requirements. ASCE 7 information for determining the Seismic Design Category (SDC) is repeated in the IBC for the convenience of users. This book focuses on use of the ASCE 7 provisions for SDC; the IBC and ASCE 7 provisions should produce identical results. IBC Sec. 1613.6 contains two provisions that are noted as permitted alternatives to the relevant ASCE 7 provisions. The first addresses when a diaphragm may be considered flexible for purposes of seismic design. The second addresses seismically isolated structures, a topic beyond the scope of this book.

ASCE 7 provisions for determining seismic design criteria including the SDC and importance factor *I* are provided in ASCE 7 Chap. 11. Our discussion, however, will begin with ASCE 7 Sec. 12.4 seismic load effects and combinations. The seismic force E on an element of a structure can be defined as:

$$E = \rho Q_E \pm 0.2 S_{DS} D$$

in which ρ is a factor representing redundancy and reliability, Q_E is the horizontal seismic force component, S_{DS} is the design spectral response acceleration at short periods, and D is the dead load. The first term represents horizontal forces, while the second term represents forces acting vertically, reducing dead load for overturning resistance, and increasing downward vertical reactions.

Use of the redundancy factor, ρ, is addressed in ASCE 7 Sec. 12.3.4. Section 12.3.4 defines the redundancy factor and specifies where the value can be set to 1.0. ASCE 7 sets ρ equal to 1.0 for Seismic Design Categories A, B, and C, and for a number of other conditions. Seismic design categories will be introduced shortly in conjunction with the ASCE 7 base shear formula variables. To provide consistency in calculations, ρ will always be included, whether it defaults to 1.0 or has a higher value. The seismic force on an element will always be multiplied by ρ. It also needs to be kept in mind that these seismic forces are at a *strength level*. When these forces are included in the allowable stress design (ASD) *basic load combinations* (Sec. 2.17), E will be multiplied by 0.7 to adjust to an allowable stress design level.

Redundancy/reliability factor

The redundancy factor ρ is used to encourage the designer to provide a reasonable number and distribution of vertical LFRS elements. In wood structures this usually means providing a reasonable number of shearwalls, of reasonable length, well distributed through the building. The ρ factor is addressed in ASCE 7 Sec. 12.3.4. The approach used to determine ρ is different in ASCE 7-05 than it has been in previous standards and codes. ASCE 7 Sec. 12.3.4.1 contains a list of eight circumstances when ρ is permitted to be set equal to 1.0. Item one permits ρ to be 1.0 for all structures assigned to SDC A, B, or C. Item two permits ρ to be set to 1.0 for drift calculation and determining P-delta effects. The balance of the items apply to components and elements rather than the primary LFRS.

ASCE 7 Sec. 12.3.4.2 requires that ρ be 1.3 for all structures in SDC D, E, or F, unless the structure is qualified for a ρ of 1.0, using one of two possible methods. Both methods require further evaluation of each story that resists more than 35 percent of the base shear. In a three-story building, it would be anticipated that the bottom two stories require evaluation, but possibly not the top story.

In the first method (ASCE 7 Sec. 12.3.4.2, Item a), the user is asked to remove vertical resisting elements one at a time, and check to see if (1) the story strength is reduced by more than 33 percent or (2) if an extreme torsional irregularity is created with the element removed. If either of these conditions exist, a ρ of 1.3 must be used; if it does not, ρ may be taken as 1.0. For shearwall structures, it is only shearwalls with a length shorter than the wall height that need be investigated. If all shearwalls have a length at least equal to their height, the structure will automatically qualify for $\rho = 1.0$. Where removal of shearwall elements must be investigated, a rigid diaphragm analysis will be required, resulting in a significant analysis effort. Rigid diaphragm analyses are discussed in Chap. 9. Torsional irregularities are addressed in Chap. 16.

The second method (ASCE 7 Sec. 12.3.4.2, Item b) applies only to buildings that are regular in plan at all levels (i.e., no irregularities are triggered). It requires that there be two qualifying shearwalls in the building perimeter at all sides in each evaluated story. For wood-frame shearwall buildings, the length of each shearwall is to be not less than one-half the story height. For other wall types, the length of each wall is required to be not less than the height of the story. If the required perimeter shearwalls are provided, the structure will qualify for ρ of 1.0; otherwise ρ will need to be taken as 1.3. This approach is much easier to apply in most simple buildings.

Base shear calculation

The total horizontal base shear, V, is calculated from an expression which is essentially in the form:

$$F = Ma = \left(\frac{W}{g}\right)a = W\left(\frac{a}{g}\right)$$

where F = inertia force
    M = mass
    a = acceleration
    g = acceleration of gravity

The ASCE 7 form of this expression is somewhat modified. The (a/g) term is replaced by a "seismic base shear coefficient."

For the *equivalent lateral force procedure*, ASCE 7 Sec. 12.8.1 specifies the base shear formula as:

$$V = C_s W$$

V = base shear. The *strength level* horizontal seismic force acting at the base of the structure (Fig. 2.13*a*).

W = weight of structure. The total weight of the structure that is assumed to contribute to the development of seismic forces. For most structures, this weight is simply taken as the dead load. However, in structures where a large percentage of the live load is likely to be present at any given time, it is reasonable to include at least a portion of this live load in the value of W. ASCE 7 Sec. 12.7.2 lists four specific items that are to be included in the weight of the structure, W. For example, in storage warehouses W is to include at least 25 percent of the floor live load. Other live loads are not covered specifically by ASCE 7, and the designer must use judgment.

In offices and other buildings where the locations of *partitions* (nonbearing walls) are subject to relocation, ASCE 7 Chap. 4 requires that floors be designed for a *live load* of not less than 15 psf. However, this 15 psf value is to account for localized partition loads, and it is intended to be used only for gravity load design. For seismic design, it is recognized that the 15-psf loading does not occur at all locations at the same time. Consequently, an *average floor load* of 10 psf may be used for the weight of partitions in determining W for seismic design.

Roof live loads need not be included in the calculation of W, but ASCE 7 Sec. 12.7.2 does require that 20 percent of the snow load be included if it exceeds 30 psf.

C_s = the seismic response coefficient.   From ASCE 7 Sec. 12.8.1.1, the seismic response coefficient C_s is calculated as

$$C_s = \frac{S_{DS}}{R/I}$$

but need not be greater than

$$C_s = \frac{S_{D1}}{T(R/I)}$$

ASCE 7 includes another formula to be used where the structure fundamental period T is greater than T_L, a long period transition that is found in ASCE 7 Chap. 22. The third formula will not be applicable to common wood frame buildings, as the lowest mapped value of T_L is 4 sec, and structure periods for common wood structures will be significantly lower than 1 sec.

ASCE Sec. 12.8.1.1 specifies that C_s cannot be taken as less than 0.01, and in addition, where S_1 is equal to or greater than 0.6g, C_s shall not be less than

$$C_s = \frac{0.5S_1}{R/I}$$

where S_{DS} = short-period design spectral response acceleration
    S_{D1} = one-second design spectral response acceleration
    R = response modification factor
    I = importance factor
    T = building period
    S_1 = mapped one-second spectral acceleration

Design spectral response accelerations S_{DS} and S_{D1}

The design spectral response accelerations S_{DS} and S_{D1} are the primary variables defining the design response spectrum. The process for determining S_{DS} and S_{D1} is given in ASCE 7 Chap. 11. As previously noted this information is also provided in IBC Sec. 1613. The following discussion will refer to ASCE 7 section, figure, and table numbers.

The first step in defining S_{DS} and S_{D1} is to read maximum considered earthquake (MCE) spectral response accelerations from spectral response maps. There are two types of spectral response acceleration maps that need to be used. The mapped short-period (0.2 sec) spectral acceleration S_S is used in determining the acceleration-controlled portion of the design spectra, while S_1 the mapped one-second spectral acceleration determines the velocity-controlled portion. These maps are printed in ASCE 7 Fig. 22-1 through 22-14. It is more practical, however, to use computer CD maps distributed with the IBC, maps available on the United States Geological Survey (USGS) web site, or to consult with the building department. Alternatives for mapped values will be discussed in greater length later in this section.

ASCE 7 Sec. 12.8.1.3 permits the maximum value of S_S used in calculating C_S to be $1.5g$ for regular structures of five stories or less and having a period of 0.5 sec or less.

The variables S_S and S_1 are converted to maximum considered spectral response accelerations, S_{MS} and S_{M1}, by multiplying by site coefficients F_a and F_v, defined in ASCE 7 tables. F_a and F_v are a function of Site (soil) Classes A through F. Site classes can be assigned as a function of three different soil parameters: shear wave velocity, penetration resistance, or undrained shear strength. Most building designers would need input from a geotechnical engineer in order to determine site class. Site Class D is commonly assumed, provided that the building site is known to not have deep soft soils. Variables S_{MS} and S_{M1} are multiplied by $\frac{2}{3}$ to convert from maximum considered spectral response accelerations to design spectral response accelerations for the acceleration and velocity-controlled regions, S_{DS} and S_{D1}, respectively.

The maximum considered earthquake ground motion maps incorporated into ASCE 7 were developed through the National Seismic Hazard Mapping Project, conducted jointly by the USGS, the Building Seismic Safety Council (BSSC), and the Federal Emergency Management Agency (FEMA). As part of this process, significant effort went into collection of available data and into workshops to receive input on a regional level. The maps contain acceleration values obtained from a combination of probabilistic and deterministic methods. The commentary to the NEHRP provisions contains a detailed discussion of the basis of the maps. ASCE 7 mapping reflects the maximum considered earthquake (MCE), which is thought to represent for practical purposes the maximum earthquake that can be occur. These values are reduced to obtain design-level accelerations. Also of importance to the designer is that seismic hazard areas do not follow state or county lines. Mapped short-period and one-second spectral response accelerations are also available on a compact disk, prepared by the National Earthquake Hazard Mapping Project, conducted jointly by the USGS, FEMA, and BSSC. (Disks are available with the IBC.) With the compact disk and the USGS Website, the mapped spectral response accelerations can be accessed by location in degrees latitude and longitude.

The IBC design spectrum used for linear static design methods is a function of S_{DS} and S_{D1}/T (Fig. 2.17). In addition, S_{DS} and S_{D1}/T define the response

Figure 2.14   Period of vibration T is the time required for one cycle of free vibration. The heavy lines represent the tributary wall and roof dead load, which is assumed to be concentrated at the roof level.

spectrum that can be used for linear dynamic analysis (response spectrum analysis) methods. The following discussion will introduce the dynamic properties of a structure and define the general concept of a response spectrum.

The first and most basic dynamic property of a structure is its fundamental period of vibration. To define the period, first assume that a one-story building has its mass tributary to the roof level assigned or "lumped" at that level. See Fig. 2.14. The dynamic model then becomes a flexible column with a single, concentrated mass at its top. If the mass is given some horizontal displacement (point 1) and then released, it will oscillate back and forth (i.e., from 1 to 2 to 3). This movement, with no externally applied load, is termed *free vibration*. The *period of vibration, T*, of this structure is defined as the length of time (in seconds) that it takes for one complete cycle of free vibration. The period is a characteristic of the structure (a function of mass and stiffness), and it is a value that can be calculated from dynamic theory.

When the multistory building of Fig. 2.13 was discussed (Sec. 2.13), the concept of fundamental mode of vibration was defined. Characteristic periods are associated with all of the modes of vibration. The fundamental period can be defined as the length of time (in seconds) that it takes for the first or fundamental mode (deflection shape) to undergo one cycle of free vibration. The fundamental period can be calculated from theory, or the ASCE 7 simple, normally conservative, method for the approximate period can be used.

In this latter approach, Sec. 12.8.2.1 of ASCE 7 provides the following formula for the approximate period of vibration:

$$T_a = C_t h_n^x$$

where h_n = height of the highest (nth) level above the base, ft
  x = exponent dependent on structure type, from ASCE 7 Table 12.8-2

$\qquad$ = 0.80 for moment-resisting systems of steel
$\qquad$ = 0.90 for moment-resisting systems of concrete
$\qquad$ = 0.75 for eccentrically braced steel frames, and
$\qquad$ = 0.75 for all other structures
C_t = coefficient dependent on structure type, from ASCE 7 Table 12.8-2
$\qquad$ = 0.028 for moment-resisting systems of steel
$\qquad$ = 0.016 for moment-resisting systems of concrete
$\qquad$ = 0.030 for eccentrically braced steel frames, and
$\qquad$ = 0.020 for all other structures

ASCE 7 provides optional alternative definitions for the approximate period T_a in structures with steel or concrete moment frames and in structures with concrete or masonry shearwalls. However, for simplicity, $T_a = (0.020)(h_n)^{0.75}$ is used for all buildings in this text. The approximate period calculated using this formula is conservative for most structures. A conservative period is one that falls within the level plateau of the design spectrum.

Damping is another dynamic property of the structure that affects earthquake performance. Damping can be defined as the resistance to motion provided by the building materials. Damping mechanisms can include friction, metal yielding, and wood crushing as the structure moves during an earthquake. Damping will slowly reduce the free-vibration displacement of the structure, eventually bringing it to a stop.

With the concepts of period of vibration and damping now defined, the idea of a response spectrum can be introduced. A *response spectrum* is defined as a plot of the maximum response (acceleration, velocity, displacement, or equivalent static force) versus the period of vibration. See Example 2.12 and Fig. 2.15. In a study of structural dynamics, it has been found that structures with the same period and the same amount of damping have essentially the same response to a given earthquake acceleration record.

Earthquake records are obtained from strong-motion instruments known as *accelerographs,* which are triggered during an earthquake and record ground accelerations. Time histories of ground accelerations can then be used as the input for computer time-history analyses of a single degree of freedom model. Varying the period of the single degree of freedom model allows the development of a relationship between the period of vibration and the maximum response. A response spectrum can be determined from a single ground motion record or from a group of records. Once the response spectrum has been determined by analysis, it can be used to estimate the effect of the particular ground motion record, or group of records, on buildings. The information required to obtain values from a response spectrum is simply the fundamental or approximate period of the structure.

It should be pointed out that a number of earthquake ground acceleration records are available, and each record could be used to generate a unique response spectrum for a given damping level and soil condition. ASCE 7 simplifies this process, for design, by providing coefficients S_{DS} and S_{D1}/T for construction of a smoothed design response spectrum which is based on an assumed damping

EXAMPLE 2.12   Typical Theory Response Spectrum

The term *response spectrum* comes from the fact that *all building periods* are summarized on one graph (for a given earthquake record and a given percentage of critical damping). Figure 2.15 shows the complete *spectrum* of building periods. The curve shifts upward or downward for different amounts of damping.

Figure 2.15   Complete spectrum of building periods.

coefficient and results from a large number of individual ground motion records. The spectrum is specific to the mapped spectral accelerations, S_S and S_1, and the particular site class. The development of the ASCE 7 design spectra includes a modest amount of damping that is applicable to all building types. The additional damping capacity of particular seismic bracing systems is further considered in development of the R-factors.

Now that the basic dynamic properties (*period* and *damping*) of a building and the concept of a *response spectrum* have been introduced, the formulation for the response spectrum values S_{DS} and S_{D1}/T can be reviewed. It should be clear that S_{D1}/T will depend on the period of vibration, T. Experience in several earthquakes has shown that local soil conditions can have a significant effect on earthquake response. The 1985 Mexico earthquake shock is a prime example of earthquake ground motions being amplified by local soil conditions. See Example 2.13 and Fig. 2.16.

EXAMPLE 2.13   Effect of Local Soil Conditions

Soil-structure resonance is the term used to refer to the amplification of earthquake effects caused by local soil conditions (Fig. 2.16). The soil characteristics associated with a given building site (site specific) are incorporated into the definition of site coefficients F_a and F_v.

Figure 2.16   Geotechnical profile.

ASCE 7 establishes six *Site Classes* (Classes A through F, ASCE 7 Table 20.3-1), and different values of the site coefficients F_a and F_v are assigned to each class for each tabulated range of mapped spectral acceleration [ASCE 7 Table 11.4-1 and ASCE 7 Table 11.4-2]. If a structure is supported directly on hard rock (Site Class A), then F_a and F_v are 0.8 for all mapped spectral accelerations. However, if the structure rests on softer soil, the earthquake ground motion originating in the bedrock may be amplified. In the absence of a geotechnical evaluation, Site Class D is the default site class normally permitted for use in determining the site coefficients F_a and F_v. This is noted in ASCE 7 Sec. 20.1.

It is perhaps more difficult to visualize, but the soil layers beneath a structure have a period of vibration T_{soil} similar to the period of vibration of a building, T. Greater structural damage is likely to occur when the fundamental period of the structure is close to the period of the underlying soil. In these cases, a quasi-resonance effect between the structure and the underlying soil develops. The conditions at a specific site are classified into one of six soil profile types, designated as Site Classes A through F. Site class and site coefficients are determined in accordance with ASCE 7. ASCE 7 Chap. 20 and ASCE 7 Table 20.3-1 provide geotechnical definitions of site classes, which are then used to determine site coefficients F_a and F_v in accordance with ASCE 7 Tables 11.4-1 and 11.4-2.

Finally, based on the discussion of the last several pages, the ASCE 7 design response spectrum curve needs to be created. A generic design response spectrum curve is plotted in Fig. 2.17a. The response spectrum curve in Fig. 2.17b has been made specific to a location in California having $S_S = 1.5g$, $S_1 = 0.75g$, and Site Class D. From ASCE 7 Tables 11.4-1 and 11.4-2, $F_a = 1.0$ and $F_v = 1.5$. As a result, $S_{MS} = 1.5g$, $S_{M1} = 1.13g$, $S_{DS} = 1.0g$, and $S_{D1} = 0.75g$. S_{DS} determines the level plateau to the design response spectrum. At period $T_S = S_{D1}/S_{DS} = 0.75$ sec, the spectral acceleration level starts dropping in proportion to $1/T$. Below period $T_0 = 0.2T_S$, the spectral acceleration is reduced linearly from the plateau to $0.4\,S_{DS}$ at a zero period.

The graphs of Fig. 2.17 show the spectral accelerations for buildings of varying periods. The level plateau, defined by S_{DS} can be considered to apply to stiffer buildings. For periods above T_S, the downward trend of the curve shows that as buildings become more flexible, they tend to experience lower seismic forces. On the other hand, the more flexible buildings will experience greater deformations, and therefore damage to finishes and contents could become a problem. Because wood-framed buildings are almost always stiff enough to fall at the S_{DS} plateau, issues related to deformations of flexible buildings will not be discussed.

Figure 2.17a   ASCE 7 design response spectra. This generic curve will generate different specific values depending on seismic zone and soil type.

Figure 2.17b ASCE 7 design response spectrum using a California site with mapped spectral accelerations $S_S = 1.5$ and $S_1 = 0.75$, and site coefficients $F_a = 1.0$ and $F_v = 1.5$. This results in design spectral response accelerations of $S_{DS} = 1.0g$ and $S_{D1} = 0.75g$.

It was noted earlier that in a linear dynamic (response spectrum) analysis, the total response of a multidegree of freedom structure could be obtained by adding together appropriate percentages of spectral accelerations at a number of modes of vibration. However, the ASCE 7 design response spectrum, in addition to being a smoothed representation of multiple ground acceleration records, represents a multimode response spectrum envelope, modified to account for higher modes of vibration. These multimode effects are significant for relatively tall structures, which have correspondingly long periods. However, relatively low-rise structures are characterized by short periods of vibration. Consequently, it should be of little surprise that the flat plateau, defined by S_{DS} will apply to the buildings covered in this text. Numerical examples demonstrating this are given in Chap. 3.

Importance factor, I

An importance factor, I, was introduced into the seismic base shear formula as a result of failures which occurred in the 1971 San Fernando earthquake. ASCE 7 Sec. 11.5 addresses the assignment of an importance factor. The occupancy categories are found in ASCE 7 Table 1-1 or IBC Table 1604.5. This book will use the ASCE 7 table; however, the reader is reminded that the IBC table must be checked when conformance with the IBC is required. In general, facilities that house large groups of occupants, or occupants that have reduced mobility, are assigned to occupancy Category III and have a seismic importance factor of 1.25. In general, facilities needed for emergency response and facilities that house significant quantities of hazardous materials are assigned to Occupancy Category IV and have a seismic importance factor of 1.5. Other occupancy types generally fall under Occupancy Categories I and II, with a seismic importance factor of 1.0.

Seismic design category

The *Seismic Design Category* (SDC) is an indication of the relative seismic risk of a given structure, considering both the seismic demand in terms of the *design spectral response accelerations* (S_{DS} and S_{D1}) and the structure use in terms of *occupancy category*. As the SDC gets high, indicating higher relative risk, ASCE 7 requires the use of more ductile lateral-force-resisting systems, and imposes additional detailing requirements to help achieve the intended ductility.

The SDC is addressed in ASCE 7 Sec. 11.6. The *occupancy category* was introduced previously in relation to the *importance factor I*, and can be found in ASCE 7 Table 1-1. Once the design spectral response accelerations and importance factor are known, the SDC can be determined from the second paragraph of ASCE 7 Sec. 11.6. The assignment is as follows:

- SDC E is assigned to structures in Occupancy Categories I, II, or III located where the mapped one-second spectral response acceleration parameter S_1 is greater than or equal to 0.75.

- SDC F is assigned to structures in Occupancy Category IV located where the mapped 1-sec spectral response acceleration parameter S_1 is greater than or equal to 0.75.

- All other structures are assigned an SDC A through D based on the more critical of Tables 11.6-1 or 11.6-2.

Where a list of four criteria is met, ASCE 7 permits structures to be assigned an SDC based on Table 11.6-1 alone. This is particularly of interest to evaluate when a short-period building might otherwise have the SDC controlled by Table 11.6-2.

Response modification factor, *R*

The response modification factor (*R*-factor) is found in ASCE 7 Table 12.2-1. This factor reduces the design seismic forces as a function of the ductility and overstrength of the lateral force resisting system.

The premise of design procedures, as has been discussed previously, is that stresses in elements resulting from expected loading are required to be less than the strength of the elements. This generally results in element stresses remaining in the elastic range (stress proportional to strain) when subjected to design loads. If the premise of element stresses staying elastic were to be applied to seismic design, an *R* of approximately 1.0 would be used. This would mean that the full spectral acceleration plotted in the ASCE 7 design response spectrum would be used for design, resulting in many cases in design for seismic base shears in excess of 1.0g. Experience in past earthquakes, however, has demonstrated that buildings designed to a significantly lower base shear can adequately resist seismic forces without collapse. This experience provides the basis for use of the R factor. The reason for adequate performance at a lower base shear is thought to be the result of both extra or reserve strength in the structural system and nonstructural elements and finishes, and stable inelastic behavior of the structural elements. Based on reserve strength and inelastic behavior, the code allows use of R values significantly greater than 1.0, resulting in seismic base shears significantly lower than 1.0g.

The reserve strength in the structural system is called *overstrength*. The contribution of overstrength to the *response modification factor R* comes from several sources including *element overstrength* and *system overstrength*. The reader can visualize a wood structural panel shearwall. When the design seismic forces are applied, the wall stresses are well within the elastic range (stress nearly proportional to strain). More seismic force can be applied before the wall reaches what could be considered a yield stress (stresses no longer nearly proportional to strain). Yet more seismic force can usually be applied before the wall reaches its failure load and the strength starts decreasing. The difference between the initial design seismic force and the failure load is the *element overstrength*. The Ω_0 values tabulated in ASCE 7 Table 12.2-1 give an approximation of the expected *element overstrength* for each type of primary LFRS. The Ω_0 value and its use in ASCE 7 Sec. 12.4.3 will be discussed in Chaps. 9, 10, and 16.

The system overstrength comes from the practice of designing a group of elements for the forces on the most highly loaded elements. This results in the less highly loaded elements in that group having extra capacity. Because the capacity of elements must generally be exceeded at more than one location in a system before a system failure occurs, the result is reserve capacity or overstrength in the system.

The ability of structural elements to withstand stresses in the inelastic range is called *ductility*.

In a major earthquake a structure will not remain *elastic*, but will be forced into the *inelastic* range. Inelastic action absorbs significantly more energy from the system. Therefore, if a structure is properly detailed and constructed so that it can perform in a ductile manner (i.e., deform in the inelastic range), it can be designed for considerably smaller lateral forces (such as those given by the equivalent lateral force procedure).

Experience in previous earthquakes indicates that certain types of LFRSs perform better than others. This better performance can be attributed to the ductility (the ability to deform in the inelastic range without fracture) of the system. The *damping* characteristics of the various types of structures also affect seismic performance.

The expected ductility and overstrength of each LFRS is taken into account in the R factors. The R term in the denominator of the seismic base shear formula is the empirical factor that reduces the lateral seismic forces to an appropriate level for use in conventional design procedures. Numerical values of R are assigned to the various LFRSs in ASCE 7 Table 12.2-1 (*R*-factors for nonbuilding structures are given in ASCE 7 Chap. 15).

The five basic structural systems recognized by the ASCE 7 for conventional buildings are:

A. Bearing wall system

B. Building frame system

C. Moment-resisting frame system

D. Dual systems with special moment frames

E. Dual systems with intermediate moment frames

For these systems, *R*-factors range from 1 to 8. Because *R* appears in the denominator of the base shear coefficient, more ductile performance is expected of systems with larger *R*-factors. The range of the *R*-factors reflects the wide range of structural systems used regionally across the United States. An important feature in the *R*-factor table is the explicit listing of allowable seismic design categories and height limits for each listed structural system. The lowest *R*-factors correspond to *ordinary plain concrete shearwalls,* to *ordinary plain masonry shearwalls*, and to *ordinary steel moment frames* when occurring as cantilevered columns. As less ductile systems, all of these are prohibited in Seismic Design Category D (SDC D), and some are also prohibited in Seismic Design Category C.

In previous seismic design provisions the term *box system* was very descriptive of the LFRS used in typical wood-frame buildings with horizontal diaphragms and shearwalls. These structures are now classified as either a bearing wall system or a building frame system.

It is very common in a wood-frame building to have roof and floor beams resting on load-bearing stud walls. If a load-bearing stud wall is also a shearwall, the LFRS will be classified as a bearing wall system. For buildings with a bearing wall system, ASCE 7 Table 12.2-1 assigns the following values of *R*:

Bearing wall system	*R*
Light-framed walls sheathed with wood structural panels rated for shear resistance or steel sheets	$6^{1}/_{2}$
Light-framed walls with shear panels of all other materials	2
Special reinforced concrete walls (permitted in SDC D)	5
Special reinforced masonry walls (permitted in SDC D)	5

Past seismic design provisions included a modest difference between *R*-factors for structures with wood structural panel (plywood and oriented strand board) bracing, and structures braced by other materials. The *R*-factors for wood structural panel sheathing and other bracing materials (gypsum wallboard, stucco, etc.) are now different by a factor of 3. The very low *R*-factor is intended to reflect the perceived brittle nature of these materials and put their design on par with other brittle systems. Use of non-wood structural panel materials will significantly increase the design base shear, requiring not only additional bracing, but also additional fastening for shear transfer and overturning. The likely result is more extensive use of wood structural panels in order to qualify for a lower base shear. Special reinforced concrete and masonry shearwalls are included here because these systems are permitted in Seismic Design Category D. Additional system types are permitted in lower seismic design categories. It should be noted that ASCE 7, like the NEHRP provisions, has linked structural systems and structural detailing requirements. Each separate title used in the ASCE 7 table denotes a system with specific detailing requirements.

A *building frame system* may also use horizontal diaphragms and shearwalls to carry lateral forces, but in this case gravity loads are carried by what the ASCE 7 definitions term "an essentially complete space frame." For example, vertical loads could be supported entirely by a wood or steel frame, and lateral forces could be carried by a system of non-load-bearing shearwalls. The term *nonload-bearing* indicates that these walls carry no gravity loads (other than their own dead load). The term *shearwall* indicates that the wall is a lateral-force-resisting element.

The distinction between a bearing wall system and a building frame system is essentially this: In a bearing wall system, the walls serve a dual function in which both gravity loads and lateral forces are carried by the same structural element. Here, failure of an element in the LFRS during an earthquake could possibly compromise the ability of the system to support gravity loads. On the other hand, because of the separate vertical-load and lateral-force carrying elements in a building frame system, failure of a portion of the LFRS does not necessarily compromise the ability of the system to support gravity loads. Because of the expected better performance, slightly larger R-factors are assigned to building frame systems than to bearing wall systems:

Building frame system	R
Light-framed walls sheathed with wood structural panels rated for shear resistance or steel sheets	7
Light-framed walls with shear panels of all other materials	$2^{1}/_{2}$
Special reinforced concrete walls (permitted in SDC D)	6
Special reinforced masonry walls (permitted in SDC D)	$5^{1}/_{2}$

Building frame systems, however, are not extremely common in wood light-frame construction.

Each of the coefficients in the base shear formula has been reviewed, so the designer should have a solid understanding of these terms.

2.15 Seismic Forces—Primary System

Seismic forces are calculated and distributed throughout the structure in the reverse order used for most other forces. In evaluating wind forces, for example, the design pressures are calculated first. Later the shear at the base of the structure can be determined by summing forces in the horizontal direction. For earthquake forces, the process is just the reverse. The shear at the base of the structure is calculated first, using the base shear formula for V (Sec. 2.13). Then total *story forces F_x* are assigned to the roof and floor levels by distributing the base shear vertically over the height of the structure. Finally, individual story forces are distributed horizontally at each level in accordance with the mass distribution of that level.

The reasoning behind the vertical distribution of seismic forces was given in Sec. 2.13. The general distribution was described, and it was seen that the shape of the first mode of vibration serves as the basis for obtaining the story forces acting on the *primary* LFRS. When a part or portion of a building is considered, the seismic force F_p on the individual part may be larger than the seismic forces acting on the primary LFRS. Seismic forces on certain parts and elements of a structure are covered in Sec. 2.16. The methods used to calculate the distributed story forces on the primary LFRS are reviewed in the remaining portion of this section.

The primary LFRS is made up of both horizontal and vertical elements. In most wood-frame buildings, the horizontal elements are roof and floor systems that function as *horizontal diaphragms*, and the vertical elements are wall segments that function as *shearwalls*. A variety of other systems may be used (see Sec. 3.3 for a comparison of several types), but these alternative systems are more common in other kinds of structures (e.g., steel-frame buildings).

Another unique aspect of seismic force evaluation is that there are two different sets of story force distributions for the primary LFRS. One set of story forces is to be used in the design of the vertical elements in the LFRS, and the other set applies to the design of horizontal diaphragms. A different notation system is used to distinguish the two sets of story forces.

The forces for designing the vertical elements (i.e., the shearwalls) are given the symbol F_x, and the forces applied to the design of horizontal diaphragms are given the symbol F_{px}. Both F_x and F_{px} are horizontal *story forces* applied to level x in the structure. Thus, the horizontal forces are assumed to be concentrated at the story levels in much the same manner as the masses tributary to a level are "lumped" or assigned to a particular story height.

Initially it may seem strange that ASCE 7 would provide two different distributions (F_x and F_{px}) for designing the *primary LFRS*, but once the reasoning is understood, the concept makes sense. The rationale behind the F_x and F_{px} distributions has to do with the fact that the forces occurring during an earthquake change rapidly with time. Because of these rapidly changing forces and the different modes of vibration, it is likely that the maximum force on an individual horizontal diaphragm will not occur at the same instant in time as the maximum force on another horizontal diaphragm. Hence, the loading given by F_{px} is to account for the possible larger *instantaneous* forces that will occur on individual horizontal diaphragms. Therefore, the F_{px} story force is to be used in the design of individual horizontal diaphragms, diaphragm collectors (drag struts), and related connections. The design of horizontal diaphragms and the definition of terms (such as drag struts) are covered in details in Chap. 9.

On the other hand, when all of the story forces are considered to be acting on the structure *concurrently*, it is reasonable to use the somewhat smaller distribution of earthquake forces given by F_x. The simultaneous application of all of the F_x story forces does not affect the design of individual horizontal diaphragms. Thus, F_x is used to design the vertical elements (shearwalls) in the primary LFRS. The connections anchoring the shearwall to the foundation, and the foundation system itself, are also to be designed for the accumulated effects of the F_x forces. The design of shearwalls is covered in Chap. 10, and a brief introduction to foundation design for shearwalls is given in Chap. 16.

ASCE 7 Sec. 12.8.3 specifies vertical distribution of F_x story forces. The equation requires a vertical redistribution that will result in a triangular load distribution for most wood frame structures. ASCE 7 Sec. 12.10.1.1 specifies the distribution of F_{px} diaphragm story forces. These forces require a second vertical distribution that recognizes the higher instantaneous forces that can act on one story at a time. The formulas for story force F_x and diaphragm story force F_{px} are given in Example 2.14. In practice, the F_x story forces must be determined first because they are then used to evaluate the F_{px} story forces. The F_x story forces are to be applied simultaneously to all levels in the primary LFRS for designing the vertical elements in the system. In contrast, the F_{px} story forces are applied individually to each level x in the primary LFRS for designing the horizontal diaphragms.

Although the purpose of the F_x forces is to provide the design forces for the shearwalls, the F_x forces are applied to the shearwalls through the horizontal diaphragms. Thus, both F_{px} and F_x are shown as uniform forces on the horizontal diaphragms in Fig. 2.18. To indicate that the diaphragm design forces are applied individually, only one of the F_{px} forces is shown with solid lines. In comparison, the F_x forces act concurrently and are all shown with solid lines.

The formula for F_x will produce a triangular distribution of horizontal story forces if the masses (tributary weights) assigned to the various story levels are all equal (refer to Fig. 2.13a in Sec. 2.13). If the weights are not equal, some variation from the straight-line distribution will result, but the trend will follow the first-mode shape. Accelerations and, correspondingly, inertia forces ($F = Ma$) increase with increasing height above the base.

EXAMPLE 2.14 F_x and F_{px} Story Force Distributions

Two different distributions of seismic forces are used to define earthquake forces on the primary LFRS (Fig. 2.18). The *story forces* for the two major components of the primary LFRS are given by the following distributions.

F_x Distribution—Vertical Elements (Shearwalls)

All seismic design categories

$$F_x = C_{vx}V$$

and

$$C_{vx} = \frac{w_x h_x^k}{\sum\limits_{i=1}^{n} w_i h_i^k}$$

where C_{vx} = vertical distribution factor

$\quad\quad V$ = total base shear

$\quad w_i, w_x$ = tributary weights assigned to level i or x

$\quad h_i, h_x$ = height from the base of structure to level i or x, ft

$\quad\quad\; k$ = an exponent related to the structural period

$\quad\quad\quad$ = 1 for structures having a period of 0.5 sec or less

Figure 2.18 F_x and F_{px} story force distributions.

F_{px} *Distribution—Horizontal Elements (Diaphragms)*

$$F_{px} = \frac{\displaystyle\sum_{i=x}^{n} F_i}{\displaystyle\sum_{i=x}^{n} w_i}\, w_{px}$$

and $\quad\quad\quad\quad\quad\quad\quad\quad 0.2S_{\mathrm{DS}}Iw_{px} \le F_{px} \le 0.4S_{\mathrm{DS}}Iw_{px}$

where F_{px} = horizontal force on primary LFRS at story level x for designing horizontal elements

$\quad\; F_i$ = lateral force applied to level i (this is story force determined in accordance with formula for F_x)

$\quad w_{px}$ = weight of diaphragm and elements tributary to diaphragm at level x

Other terms are as defined for F_x.

ASCE 7 Sec. 12.10.1.1 specifies that the redundancy factor ρ is to be used in the design of diaphragms for structures assigned to SDC D, E, and F. Where the diaphragm is designed only for single-story forces, the redundancy factor ρ may be taken as 1.0. There are two circumstances where use of a higher ρ may be required. First, if an upper level shearwall were to be discontinued at lower levels, causing the shearwall force to be redistributed through a diaphragm, ASCE 7 Sec. 12.10.1.1 would require that the calculated redundancy factor ρ be included in forces for that diaphragm. Second, if a rigid diaphragm distribution of forces is being used, and changes in vertical element rigidity cause redistribution of forces through the diaphragm, the calculated redundancy factor ρ must be used for the redistributed forces.

In the formulas for distributing the seismic force over the height of the structure, the superscript k is to account for whip action in tall, slender buildings and to allow for the effects of the higher modes (i.e., other than the first mode) of vibration. When the period of vibration is less than 0.5 sec, there is no whipping effect.

It may not be evident at first glance, but the formulas for F_x and F_{px} can be simplified to a form that is similar to the base shear expression. In other words, the earthquake force can be written as the mass (weight) of the structure multiplied by a seismic coefficient. For example,

$$V = (\text{seismic coefficient})W$$

The seismic coefficient in the formula for V is known as the *base shear coefficient*. When all of the terms in the formulas for the story forces (F_x and F_{px}) are evaluated except the dead load w, seismic story *coefficients* are obtained. Obviously, since there are two formulas for story forces, there are two sets of seismic story coefficients.

The story coefficient used to define forces for designing shearwalls is referred to as the F_x *story coefficient*. It is obtained by factoring out the story weight from the formula for F_x:

$$F_x = C_{vx}V = \frac{w_x h_x^k}{\sum\limits_{i=1}^{n} w_i h_i^k} V$$

$$F_x = \left[\frac{V h_x^k}{\sum\limits_{i=1}^{n} w_i h_i^k} \right] w_x$$

$$= (F_x \text{ story coefficient})w_x$$

Likewise, the formula for F_{px} for use in diaphragm design can be viewed in terms of an F_{px} *story coefficient*. The formula for F_{px} is initially expressed in the format:

$$F_{px} = \left[\frac{\sum\limits_{i=x}^{n} F_i}{\sum\limits_{i=x}^{n} w_i} \right] w_{px}$$

$$= (F_{px} \text{ story coefficient}) w_{px}$$

It should be noted that a one-story building represents a special case for earthquake forces. In a one-story building, the diaphragm loads given by F_x and F_{px} are equal. In fact, the F_x and F_{px} story coefficients are the same as the base shear coefficient. In other words, for a *one-story buildings:*

Base shear coefficient = F_x story coefficient = F_{px} story coefficient

Having a single seismic coefficient for three forces greatly simplifies the calculation of seismic forces for one-story buildings. *Numerical examples* will greatly help to clarify the evaluation of lateral forces. Several one-story building examples are given in Chap. 3, and a comparison between the F_x and F_{px} force distributions for a two-story building is given in Example 3.10 in Sec. 3.6.

At this point, one final concept needs to be introduced concerning the distribution of seismic forces. After the story force has been determined, it is distributed at a given level in proportion to the mass (dead load, D) distribution of that level. See Example 2.15.

The purpose behind this distribution goes back to the idea of an inertial force. If it is visualized that each square foot of dead load has a corresponding inertial force generated by an earthquake, then the loading shown in the sketches becomes clear. If each square foot of area has the same D, the distributed seismic force is in proportion to the length of the roof or floor that is parallel to the direction of the force. Hence the magnitude of the distributed force is large where the dimension of the floor or roof parallel to the force is large, and it is small where the dimension parallel to the force is small.

EXAMPLE 2.15   Distribution of Seismic Force at Story Level x

Transverse and Longitudinal Directions Defined

A lateral force applied to a building may be described as being in the transverse or longitudinal direction. These terms are interpreted as follows:

Transverse lateral force is parallel to the short dimension of the building.
Longitudinal lateral force is parallel to the long dimension of the building.

Buildings are designed for seismic forces applied independently in both the transverse and longitudinal directions.

Figure 2.19*a* Distribution of story force in transverse direction.

TRANSVERSE FORCE

Each square foot of dead load, D, can be visualized as generating its own inertial force (Fig. 2.19*a*). If all of the inertial forces generated by these unit areas are summed in the transverse direction, the forces w_1 and w_2 are in proportion to lengths L_1 and L_2, respectively.

The sum of the distributed seismic forces w_1 and w_2 (i.e., the sum of their resultants) equals the transverse story force. For shearwall design the transverse story force is F_x, and for diaphragm design the transverse story force is F_{px}.

Figure 2.19*b* Distribution of story force in longitudinal direction.

LONGITUDINAL FORCE

In the longitudinal direction (Fig. 2.19*b*), L_3 and L_4 are measures of the distributed forces w_3 and w_4. The sum of these distributed seismic forces equals the story force in the longitudinal direction.

NOTE: The distribution of inertial forces generated by the dead load of the walls parallel to the direction of the earthquake is illustrated in Chap. 3.

The basic seismic forces acting on the primary LFRS of a *regular* structure have been described in this section. ASCE 7 requires that the designer consider the effects of structural *irregularities*. Section 12.3.2 of ASCE 7 identifies a number of these irregularities. In many cases, increased force levels and reduced stresses are required for the design of an irregular building.

It is important for the designer to be able to identify a structural irregularity and to understand the implications associated with the irregularity. However, a detailed study of these ASCE 7 provisions is beyond the scope of Chap. 2. In fact, the majority of this book is written as an *introduction* to the basic principles of engineered wood structures. To accomplish this, most of the structures considered are rather simple in nature. Structural irregularities may be common occurrences in daily practice, but they can be viewed as advanced topics at this point in the study of earthquake design.

It is felt the reader should first develop a good understanding of the design requirements for regular structures. Therefore, the provisions for irregular structures are postponed to Chap. 16, after the principles of structural design for regular buildings have been thoroughly covered.

The seismic forces required for the design of elements and components that are not part of the primary LFRS are given in Sec. 2.16.

2.16 Seismic Forces—Wall Components

The seismic forces which have been discussed up to this point are those assumed to be developed in the primary LFRS of a building as it responds to an earthquake. However, when individual elements of the structure are analyzed separately, it may be necessary to consider different seismic effects. One reason for this is that certain elements which are attached to the structure respond dynamically to the motion of the structure rather than to the motion of the ground. Resonance between the structure and the attached element may occur.

ASCE 7 Chap. 13 addresses the design of components attached to the structure. For these items, a component seismic force F_p is used. In ASCE 7, the F_p forces for exterior walls are specifically addressing exterior nonstructural "skin" walls with discreet attachments to the main building structure, rather than structural walls that would be integral. An exception to this is that seismic F_p forces for cantilevered wall parapets are included. It could be interpreted that these forces apply to parapets on "structural" walls.

The basic equation for out-of-plane seismic forces on structural walls can be found in ASCE 7 Sec. 12.11, which is applicable to Seismic Design Categories B and up. The specified force is $F_c = 0.4S_{DS}IW_c$, but not less than $0.10W_c$. S_{DS} is the design spectral response acceleration, I is the importance factor used for the main structure (as opposed to a component importance factor I_p), and W_c is the weight of the wall being anchored. This is a *strength level* force that can be multiplied by applicable load factors for allowable stress forces. Design will be addressed in a later chapter.

The seismic force for design of components F_p will be introduced for calculation of parapet forces. Seismic forces and design requirements for nonstructural components are addressed in ASCE 7 Sec. 13.3. The component forces also apply to a wide range of architectural, mechanical, and electrical components. Architectural components addressed include interior and exterior nonstructural walls and partitions, parapets, chimneys, veneer, ceilings, cabinets, etc. In fact, in higher SDCs virtually everything on or attached to the structure requires design per ASCE 7 Chap. 13. ASCE 7 Sec. 13.1.4 provides a list of five categories of components exempt from design:

- In SDC B, all architectural components having component importance factor I_p equal to 1.0 are exempt, except parapets supported by bearing walls or shearwalls.

- In SDC B, all mechanical and electrical components are exempt.

- In SDC C, all mechanical and electrical components having I_p equal to 1.0 are exempt.

- In SDC D, E, and F, mechanical and electrical components with I_p equal to 1.0 and meeting weight and/or installation restrictions are exempt.

The force on a portion of the structure is given by the following formula:

$$F_p = \frac{0.4a_p S_{DS} W_p}{R_p/I_p}\left(1 + 2\frac{z}{h}\right)$$

with F_p limited to the range:

$$0.3S_{DS}I_p W_p \le F_p \le 1.6S_{DS}I_p W_p$$

where F_p = component seismic design force centered at the component's center of gravity and distributed relative to the component's mass distribution

S_{DS} = short-period design spectral acceleration, discussed in Sec. 2.14

a_p = component amplification factor per ASCE 7 13.5-1 or 13.6-1

I_p = component importance factor per ASCE 7 Sec. 13.1.3

W_p = component operating weight

R_p = component response modification factor per ASCE 7 Table 13.5-1 or 13.6-1

z = height in structure of point of attachment of component with respect to the structure base. For items attached at or below the base, z/h need not exceed 1.0.

h = average roof height of structure with respect to the structure base

This equation for F_p was developed based on building acceleration data recorded during earthquakes. See the NEHRP commentary for further discussion. The term $1 + 2z/h$ allows F_p to vary from one value for components anchored at the ground level to 3 times that value for components anchored to the roof. This matches the general trend seen in recorded acceleration data. If the elevation at which the component will be anchored is not known, this term can default to 3.

The second paragraph of ASCE 7 Sec. 13.3.1 gives further direction on use of F_p component forces. Typically F_p will be considered in two perpendicular horizontal directions, combined with a vertical force component of plus or minus $0.2S_{DS}W_p$. Where the component under consideration is a vertically cantilevered system, F_p will be considered to act in any horizontal direction. It is specifically noted that the redundancy factor ρ will be taken as 1.0 for components, and that use of over-strength Ω_o forces in combination with F_p forces is not required. As with the seismic base shear equations, component forces F_p are at a strength level. Appropriate load factors are to be used in the load combination equations. Example 2.10 illustrates the use of F_p component forces for seismic forces normal to a wall.

EXAMPLE 2.16   Seismic Forces Normal to Wall

Determine the seismic design force normal to the wall for the building shown in Fig. 2.20. The wall spans vertically between the floor and the roof, which is 16 ft from ground level. The wall is constructed of reinforced brick masonry that weighs 90 psf. Known seismic information:

Mapped short- (0.2 sec) spectral acceleration, $S_S = 150\% \, g = 1.5g$

Mapped 1.0-sec spectral acceleration, $S_1 = 75\% \, g = 0.75g$

Site (soil) Class = D

Site coefficients F_a and F_v = 1.0 and 1.5, ASCE 7 Tables 11.4-1 and 11.4-2

Seismic Design Category D

Compare the seismic force to the wind force on components and cladding. Assume an effective wind area of 10 ft^2. Known wind information:

Wind speed, $V = 90$ mph

Exposure B conditions

Importance factor, $I = 1.0$ for standard occupancy

Topographical factor, $K_{zt} = 1.0$

Note that wind and seismic forces are not considered simultaneously.

Seismic Forces for Design of Wall Element

The equations for calculation of component force, F_p, and wall out-of-plane force, F_c, are based on the design spectral response acceleration S_{DS}. S_{DS} is calculated as follows:

$$S_{MS} = F_a S_S = 1.0 \times 1.50g = 1.50g$$

$$S_{DS} = \tfrac{2}{3} \times S_{MS} = \tfrac{2}{3} \times 1.50g = 1.0g$$

Figure 2.20 Seismic design force normal to the wall for a building.

Calculation of Wall Seismic Forces

Forces w_{u1} and w_{u2} act normal to the wall in either direction (i.e., inward or outward). These forces have the subscript u, denoting that the seismic forces are at a *strength* or *ultimate* level. Force w_{u1} will be calculated for wall out-of-plane design for portions of the wall other than the parapet. Force w_{u2} will be calculated for the parapet using the component force equations.

$$w_{u1} = 0.4S_{DS}Iw_c$$

Solving with $S_{DS} = 1.0$, $I = 1.0$, and $w_c = 90$ psf

$$w_{u1} = 0.4w_c$$

$$w_{u1} = 0.4 \times 90 \text{ psf} = 36 \text{ psf}$$

For the parapet

$$w_{u2} = \frac{0.4a_p S_{DS} w_c}{R_p/I_p}\left(1 + 2\frac{z}{h}\right)$$

Solving with

> $S_{DS} = 1.0$,
> $a_p = 1.0$, ASCE 7 Table 13.5-1
> $R_p = 2.5$, ASCE 7 Table 13.5-1
> $I_p = 1.0$, ASCE 7 Sec. 13.1.3
> $w_c = 90$ psf
> $z = h =$ roof height

$$w_{u2} = \frac{0.4w_c}{1.0}(1 + 2)$$

$$w_{u2} = 1.2w_c = 108 \text{ psf}$$

with F_p limited to the range:

$$0.3S_{DS}I_p w_c \leq F_p \leq 1.6S_{DS}I_p w_c$$

$$0.3w_c \leq F_p \leq 1.6w_c$$

Before comparing these forces with allowable stress design wind forces, several adjustment factors need to be considered. The basic equation for E, earthquake forces, is

$$\text{E} = \rho Q_E \pm 0.2S_{DS}D$$

The variable ρ is permitted to be taken as 1.0 for design of components. The component $0.2S_{DS}D$ is acting vertically, and would be combined with the wall dead load, to determine a worst-case condition for axial plus flexural forces on the wall component. In addition, the vertical component should be considered in checking anchorage of the component. For purposes of forces perpendicular to the wall surface, the vertical component has no effect.

Seismic forces calculated in accordance with ASCE 7 are at a strength or load and resistance factor design (LRFD) level, while the wind loads are at an allowable stress design (ASD) level. Direct comparison of these forces requires the use of load combinations, a topic that is discussed in the next section. In order to allow comparison in this example, seismic forces are multiplied by 0.7 to convert to an ASD force level. This results in $w_1 = 25$ psf and $w_2 = 76$ psf.

These forces can now be compared to the wind forces.

Wind Forces

Height and exposure factor:

$$\lambda = 1.0 \quad \text{for } 0 \leq h_{\text{mean}} \leq 30 \text{ ft in Exposure B}$$

Wall forces—Interior Zone 4 for effective wind area of 10 ft^2:

$$p_{\text{net30}} = \begin{cases} 14.6 \text{ psf} & \text{(inward pressure)} \\ -15.8 \text{ psf} & \text{(outward pressure)} \end{cases}$$

Design wind pressure:

$$w = p_{\text{net}} = \lambda K_{zt}Ip_{\text{net30}} = \begin{cases} 1.0(1.0)(1.0)(14.6) = 14.6 \text{ psf} & \text{(inward pressure)} \\ 1.0(1.0)(1.0)(-15.8) = -15.8 \text{ psf} & \text{(outward pressure)} \end{cases}$$

Wall forces—End Zone 5 for effective wind area of 10 ft^2:

$$p_{\text{net30}} = \begin{cases} 14.6 \text{ psf} & \text{(inward pressure)} \\ -19.5 \text{ psf} & \text{(outward pressure)} \end{cases}$$

Design wind pressure:

$$w = p_{net} = \lambda K_{zt} I p_{net30} = \begin{cases} 1.0(1.0)(1.0)(14.6) = 14.6 \text{ psf} & \text{(inward pressure)} \\ 1.0(1.0)(1.0)(-19.5) = -19.5 \text{ psf} & \text{(outward pressure)} \end{cases}$$

Wind $<$ seismic

$\therefore$ *seismic governs*

2.17 Load Combinations

Structural design criteria can be divided into two main groups (1) strength checks and (2) serviceability checks. Serviceability checks will be the focus of Sec. 2.18, while this section introduces load combinations to be used in strength checks. ASCE 7 and the IBC specify a number of load combinations that are to be considered in checking the strength of a structure. Two different groups of load combinations are addressed in this section, reflecting the two methods of design permitted by ASCE 7, the IBC, the AF&PA National Design Specification (NDS) [Ref. 2.1], and the AF&PA Special Design Provisions for Wind and Seismic (SDPWS) [Ref. 2.2]. The two methods are (1) ASD and (2) LRFD. Each of these methods of strength check involves comparing the demand (due to load) on a structure or element to the provided capacity (resistance or strength).

To date, the primary method used for design of wood structures has been *allowable stress design* (ASD). In this method, demand on the structure is calculated using loads that would be commonly anticipated to occur (also referred to as service level loads). In order to protect against failure, a factor of safety is incorporated into the capacity of the structure. The commonly anticipated (service level) load is compared to anywhere between one third and two thirds of the peak capacity of the structure. In this approach, factors of safety handed down from past practice have not been rationalized to the same extent as the newer LRFD method. Note that resulting structures have a good record of protecting life and providing serviceability.

More recently introduced is the *load and resistance factor* (LRFD) method (also referred to as strength design, and occasionally ultimate strength design). This method moves toward more rationally addressing factors of safety by specifically accounting for possible variations in demand (load), using a *load factor*, and possible variations in capacity (resistance), using a *resistance factor*. For example, a live load might be anticipated to be up to 60 percent greater than the minimum service level live load L required by ASCE 7 and IBC, giving a *load factor* of 1.6. This might be used in combination with a floor beam for which it is possible for the capacity to be only 90 percent of the peak capacity specified by the NDS, resulting in a *resistance factor* of 0.9. While LRFD methodologies have been available in the past, the 2005 edition of the NDS is the first time that this method has been presented in combination with the ASD method in the most widely used wood design standard. The determination of load factors to account for anticipated variations in loading is the purview of the ASCE 7 standard, while the establishment of resistance factors to account for variation in the material capacity is the purview of the AF&PA material standards—the NDS and SDPWS.

The ASD and LRFD methods have, for the most part, been calibrated so that they will produce similar results in common design situations, although there are likely to be some design cases in which one or the other method permits slightly higher demand for the same member size. For this reason, the LRFD approach is referred to as a "soft conversion," made available in wood primarily so those used to using the approach for other materials (concrete, steel, and masonry) will have the ability to use a parallel method in wood. This book presents both methodologies. Examples using LRFD are distinguished by shaded boxes.

Summary of load types

Before looking at the combination of loads, it is convenient to have the notations for different load types gathered in one location:

D = dead load
E = earthquake load
F = load due to fluids with well-defined pressures and maximum heights
F_a = flood load
H = load due to lateral earth pressure, ground water pressure, or pressure of bulk materials
L = live load
L_r = roof live load
R = rain load
S = snow load
T = self-straining force
W = wind load

Two additional load types are used in ASCE 7, but not used in the basic load combinations. They are D_i, weight of ice, and W_i, wind-on-ice. The reader is referred to ASCE for load and load combination information where these load types may be applicable.

ASD load combinations

The ASD basic load combinations are presented in ASCE 7 Sec. 2.4.1 and IBC Sec. 1605.3.1. ASCE 7 equation numbers precede the equations and IBC equation numbers follows.

1. $D + F$ (Equation 16-8)

2. $D + H + F + L + T$ (Equation 16-9)

3. $D + H + F + (L_r$ or S or $R)$ (Equation 16-10)

4. $D + H + F + 0.75 (L + T) + 0.75 (L_r$ or S or $R)$ (Equation 16-11)

5. $D + H + F + (W$ or $0.7E)$ (Equation 16-12)

6. $D + H + F + 0.75$ $(W$ or $0.7E) +$ (Equation 16-13)

     $0.75L + 0.75(L_r$ or S or $R)$

7. $0.6D + W + H$ (Equation 16-14)

8. $0.6D + 0.7E + H$ (Equation 16-15)

Instructions for including flood loads F_a in the above combinations are given in ASCE 7 Sec. 2.4.1, and vary by flood hazard zone.

A wood *load duration factor* C_D and other applicable adjustment factors from the NDS should be used with the above ASD load combinations, while *allowable stress increases* should not be used. These two terms will be described in detail following the load combinations.

An alternate set of ASD load combinations is published in the IBC. This set has been carried forward from previous code editions. This book focuses on the ASD basic load combinations shown above.

LRFD load combinations

The LRFD basic load combinations are presented in ASCE 7 Sec. 2.3.2 and IBC Sec. 1605.2.1. While the concept is the same, there are slight differences in presentation. The ASCE 7 load combinations are:

1. $1.4(D + F)$

2. $1.2(D + F + T) + 1.6(L + H) + 0.5(L_r$ or S or $R)$

3. $1.2D + 1.6(L_r$ or S or $R) + (L$ or $0.8W)$

4. $1.2D + 1.6W + L + 0.5(L_r$ or S or $R)$

5. $1.2D + 1.0E + L + 0.2S$

6. $0.9D + 1.6W + 1.6H$

7. $0.9D + 1.0E + 1.6H$

Three exceptions follow the ASCE load combinations. The first exception permits the load factor on L in combinations 3, 4 and 5 to be taken as 0.5 for all occupancies for which the tabulated minimum live load L_0 is less than or equal to 100 psf, excepting garages and places of public assembly. This also applies to the IBC load combinations, where it is more directly noted by using a variable load factor f_1. This will affect design for most wood structures.

The second exception gives direction regarding soil or groundwater pressure that might be counteracting other loads. Two concepts are important here. First, if a more critical condition exists with one or more loads not acting, this more critical condition must be considered in design. Second, if counteracting loads are relied on to provide resistance to loads, they need to be put on the resistance side of the equation and a resistance factor used.

The third exception deals with snow loads. Details of snow loads in the load combinations are different between ASCE 7 and IBC. The reader is cautioned to carefully consider all requirements.

The IBC LRFD load combinations are:

$1.4(D + F)$	(Equation 16-1)
$1.2(D + F + T) + 1.6(L + H) + 0.5(L_r$ or S or R)	(Equation 16-2)
$1.2D + 1.6(L_r$ or S or R) + (f_1L or 0.8W)	(Equation 16-3)
$1.2D + 1.6W + f_1L + 0.5(L_r$ or S or R)	(Equation 16-4)
$1.2D + 1.0E + f_1L + f_2S$	(Equation 16-5)
$0.9D + 1.6W + 1.6H$	(Equation 16-6)
$0.9D + 1.0E + 1.6H$	(Equation 16-7)

The variable load factor f_1 is set to 1 for garages, places of public assembly, and tabulated live loads over 100 psf. Variable f_1 is allowed to be 0.5 otherwise. This is consistent with ASCE 7. The variable f_2 is 0.7 for roof configurations that do not shed snow, and 0.2 otherwise. This is specific to the IBC.

Again ASCE 7 gives specific direction for inclusion of flood loads in the above combinations, varying based on flood hazard zone. Directions for D_i, weight of ice, and W_i, wind-on-ice, are also provided.

A *wood time effect factor* λ and other applicable adjustment factors from the NDS should be used with the above LRFD load combinations. This term will be described in detail following the load combinations.

Determining applicable loads

Both the ASD and LRFD load combinations include more types of loads than have previously been included in the IBC. This is done in an attempt to guide the designer toward systematic consideration of all applicable loads. It forces the designer to evaluate whether each of the noted load types is applicable to the structure under consideration. Discussion of each of the loads in earlier sections of this chapter will assist the designer in evaluating applicability. Most structures will need to be designed for dead load, live or roof live load, wind load, and seismic load as a minimum. Other loads that are not applicable to the structure under consideration can be crossed off in the appropriate load combination, thus simplifying the resulting design load combinations.

For the majority of wood structures, it would be unusual to resist fluid loads F and self-straining loads T. Flood hazard would most often be addressed by elevating the wood structure above the base flood elevation, thus avoiding design of the superstructure for flood loads F_a. Lateral soil pressure H would be applicable to a permanent wood foundation or to the design of a floor diaphragm supporting the top of concrete or masonry retaining walls. For above grade wood structures, however, H will generally not need to be considered. For (nearly) flat roofs, rain load R will likely need to be considered; however, most sloped roofs will not be able to accumulate rain water. Putting this information together, for most above grade wood structures with sloped roofs, the ASD load combinations can be simplified to:

1. D	(Equation 16-8)
2. D + L	(Equation 16-9)
3. D + (L$_r$ or S)	(Equation 16-10)
4. D + 0.75L + 0.75(L$_r$ or S)	(Equation 16-11)
5. D + (W or 0.7E)	(Equation 16-12)
6. D + 0.75(W or 0.7E) + 0.75L + 0.75(L$_r$ or S)	(Equation 16-13)
7. 0.6D + W	(Equation 16-14)
8. 0.6D + 0.7E	(Equation 16-15)

These load combinations are shown only as an illustration of possible simplification. The reader is encouraged to always start with the complete load combinations, and go through the process of considering each applicable load. The reader should find that after the first few times, this can be done quite quickly.

Simultaneous occurrence of loads

Each group of load combinations, ASD and LRFD, define what loads must be considered as acting simultaneously. It was noted earlier that dead loads are generally acting on the structure at all times. The magnitudes of most other load types tend to vary with time; the term *transient loads* is used for these varying loads. The probability of critical levels of some *transient loads* acting concurrently is very slight. For example, it would be unlikely that a design wind event would occur at the same time as a design earthquake event. Consequently, the load combinations include either wind load W or earthquake load E but not both at the same time. Of course, the loading that creates the most critical condition needs to be considered in design. Similarly, it is not necessary to design for roof live load and snow load at the same time.

In other cases, it is necessary to consider some portion of two transient loads occurring simultaneously. For example, ASD load combination 6 requires that 75 percent of wind load, 75 percent of floor live load, and 75 percent of roof live load be considered as acting simultaneously.

For more information on the development of the load combinations, the reader is referred to the ASCE 7 commentary.

Load duration factor, time effect factor, and allowable stress increases

An initial discussion of the load duration factor C_D, the time effect factor λ, and allowable stress increases is appropriate at this time because of the past practice of associating allowable stress increases with load combinations, including wind and seismic loads. The load duration factor C_D and time effect factor λ reflect the unique ability of wood to support higher stresses for short periods of time, as well as lower stresses for extended periods of time. The C_D and λ factors are not specific to wind or seismic loading, but are used as allowable stress or resistance modifications in all wood design calculations; C_D is used for ASD and λ is used for LRFD. In comparison, other materials such as structural steel and reinforced concrete exhibit very little variation in capacity for varying duration of load. The C_D and λ factors are discussed at length in Chap. 4.

Traditionally, the model building codes have permitted an allowable stress increase of one-third (i.e., allowable stresses may be multiplied by 1.33) for all materials when design includes wind or earthquake loads. The technical basis for the allowable stress increase is not clear. There are several theories regarding its origin. The first theory is that it accounts for the reduced probability that several transient load types will act simultaneously at the full design load level (i.e., full floor live load acting simultaneously with full design wind load). The second theory is that slightly higher stresses, and therefore lower factors of safety are acceptable when designing for wind or seismic loads due to their short duration. The exact justification for the allowable stress increase is not of great importance, as use of this factor is not permitted with either the ASD or LRFD basic load combinations. As a result, allowable stress increases will not be used for wood design in this book. If the reader were to use the IBC ASD alternate basic load combinations, use of an allowable stress increase might be permitted.

Consistent with the ASD and LRFD basic load combinations, design in this book uses load factors to account for the low probability of multiple transient loads acting simultaneously. Consistent with the NDS, ASD will use the load duration factor C_D and LRFD will use the time effect factor λ. These factors will be used for all designs, whether or not wind and seismic loads are included.

Load levels

Most of the loads defined by ASCE 7 and the IBC are specified at an allowable stress design (ASD) level. These were noted in the introduction to ASD as being at a commonly occurring or service level. The exception to this is seismic load, for which the load level defined by ASCE 7 is nominally at a strength or LRFD level. As a result, a load factor of 1.0 is used for LRFD and a load factor of 0.7 is used for ASD, except when further load factors for multiple transient loads apply.

Examples in this book vary between ASD load combinations, LRFD load combinations, and on occasion both. LRFD design examples are differentiated by a shaded box. In this book LRFD or strength level loads will be differentiated with a "u" subscript, while ASD or service level loads will not use this subscript.

2.18   Serviceability/Deflection Criteria

Serviceability criteria for structural elements are addressed briefly in ASCE 7 Sec. 1.3.2, and in more detail in IBC Sec. 1604.3. The broad statement made by ASCE 7 is "Structural systems, and members thereof, shall be designed to have adequate stiffness to limit deflections, lateral drift, vibration, or any other deformations that adversely affect the intended use and performance of the building." This broad statement could mean different things to different people, and so is not easily interpreted or enforced.

The IBC takes a more pragmatic, enforceable approach to minimum serviceability requirements. Deflection limits for structural members are specified in IBC Table 1604.3 for roof members, floor members, exterior wall members, partitions, farm buildings, and greenhouses. Maximum deflections are specified for live load, snow or wind load, and dead plus live load. While national design standard requirements are noted for some materials, wood defers to the IBC for serviceability requirements. This section will focus on use of the IBC deflection limits. Seismic drift limits will be addressed in later chapters.

While the IBC defines what steps as a minimum must be taken toward serviceability, there are many circumstances where it is prudent to take additional steps. Examples include long-span floor-framing systems where vibrations can lead to occupant discomfort, and brittle finish materials that may be damaged by standard deflections. In some cases, building component or finish manufacturers recommend or require use of serviceability limits more stringent than the IBC criteria.

The deflection criteria in IBC Table 1604.3 apply to roof members, floor members, and walls. These deflection limits are intended to ensure user comfort and to prevent excessive cracking of plaster ceilings and other interior finishes.

The question of user comfort is tied directly to the confidence that occupants have regarding the safety of a structure. It is possible for a structure to be safe with respect to satisfying stress limitations, but it may deflect under load to such an extent as to render it unsatisfactory.

Excessive deflections can occur under a variety of loading conditions. For example, user comfort is essentially related to deflection caused by live loads only. The IBC therefore requires that the deflection under live load be calculated. This deflection should typically be less than or equal to the span length divided by 360 ($\Delta_{\text{L or L}_r} \leq L/360$) for floor members and for roof members that support plaster ceilings.

Another loading condition that relates to the cracking of plaster and the creation of an unpleasant visual situation is that of deflection under dead load plus live load. For this case, the actual deflection is often controlled by the limit of the span divided by 240 ($\Delta_{K\text{D}+(\text{L or L}_r)} \leq L/240$). Although the IBC does not explicitly define the factor K, the footnotes to IBC Table 1604.3 explain that for structural wood members the dead load may be multiplied by a factor of either 0.5 or 1.0 when calculating deflection. The magnitude of the factor depends on the moisture conditions of the wood. Additional deflection limits are provided in IBC Table 1604.3 for various types of loads, including snow and wind.

Notice that in the second criterion above, the calculated deflection is to be under K times the dead load D plus the live load L or L_r. This K factor is an attempt to reflect the tendency of wood to creep under sustained load. Recall that when a beam or similar member is subjected to a load, there will be an *instantaneous deflection*. For certain materials and under certain conditions, additional deflection may occur under long-term loading, and this added deflection is known as *creep*. In practice, a portion of the live load on a floor may be a long-term or sustained load, but the IBC essentially treats the dead load as the only long-term load that must be considered.

Some structural materials are known to undergo creep and others do not. Furthermore, some materials may creep under certain conditions and not under others. The tendency of wood beams to creep is affected by the moisture content (see Chap. 4) of the member. The drier the member, the less the deflection under sustained load. Thus, for seasoned lumber, a K factor of 0.5 is used; for unseasoned wood, K is taken as 1.0. *Seasoned lumber* here is defined as wood having a moisture content of less than 16 percent at the time of construction, and it is further assumed that the wood will be subjected to dry conditions of use (as in most covered structures). Although the K factor of 0.5 is included in the IBC, many designers take a conservative approach and simply use the full dead load (that is, $K = 1.0$) in the check for deflection under D + L or L_r in wood beams. See Example 2.17.

EXAMPLE 2.17   Beam Deflection Limits

The deflection that occurs in a beam can be determined using the principles of strength of materials. For example, the maximum deflection due to bending in a simply supported beam with a uniformly distributed load over the entire span is

$$\Delta = \frac{5wL^4}{384EI}$$

There are several limits on the computed deflection which are not to be exceeded. See Fig. 2.21.

The IBC requires that the deflection of roof beams that support a rigid ceiling material (such as plaster) and floor beams be computed and checked against the following criteria:

1. Deflection under live load only shall not exceed the span length divided by 360:

$$\Delta_{(L \text{ or } L_r)} \leq \frac{L}{360}$$

2. Deflection under K times the dead load plus live load shall not exceed the span length divided by 240:

$$\Delta_{[KD + (L \text{ or } L_r)]} \leq \frac{L}{240}$$

The values of K may be used:

$$K = \begin{cases} 1.0 & \text{for } \textit{unseasoned} \text{ or green wood} \\ 0.5 & \text{for } \textit{seasoned} \text{ or dry wood} \end{cases}$$

As an alternative, the deflection limit for a wood member under total load (that is, $K = 1.0$) may be conservatively used:

$$\Delta_{TL} \leq \frac{L}{240}$$

Figure 2.21. *a.* Unloaded beam. *b.* Deflection under live load only. *c.* Deflection under *K* times dead load plus live load. *d.* Camber is curvature built into fabricated beams that opposes deflection due to gravity loading.

For members not covered by criteria given in the IBC, the designer may choose to use the deflection limits given in Fig. 2.22.

Fabricated wood members, such as glulam beams and wood trusses, may have curvature built into the member at the time of manufacture. This built-in curvature is known as *camber*, and it opposes the deflection under gravity loads to provide a more pleasing visual condition. See Example 6.15 in Sec. 6.6 for additional information. Solid sawn wood beams are not cambered.

Experience has shown that the IBC deflection criteria may not provide a sufficiently stiff wood floor system for certain types of buildings. In office buildings and other commercial structures, the designer may choose to use more restrictive deflection criteria than required by the IBC. The deflection criteria given in Fig. 2.22 are recommended by AITC (Ref. 2.4). These criteria include limitations for beams under *ordinary usage* (similar to the IBC criteria) and limitations for beams where *increased floor stiffness* is desired. These latter criteria depend on the type of beam (joist or girder), span length, and magnitude of floor live load. The added floor stiffness will probably result in increased user comfort and acceptance of wood floor systems.

Other deflection recommendations given in Fig. 2.22 can be used for guidance in the design of members not specifically covered by the IBC deflection criteria. In Fig. 2.22, the applied load is live load, snow load, wind load, and so on.

Recommended deflection limitations		
Use classification	Applied load only	Applied load + dead load
Roof beams		
Industrial	$L/180$	$L/120$
Commercial and institutional		
Without plaster ceiling	$L/240$	$L/180$
With plaster ceiling	$L/360$	$L/240$
Floor beams		
Ordinary usuage[*]	$L/360$	$L/240$
Highway bridge stringers	$L/300$	
Railway bridge stringers	$L/300$ to $L/400$	

[*]The ordinary usage classification is for floors intended for construction in which walking comfort and minimized plaster cracking are the main considerations. These recommended deflection limits may not eliminate all objections to vibrations such as in long spans approaching the maximum limits or for some office and institutional applications where increased floor stiffness is desired. For these usages the deflection limitations in the following table have been found to provide additional stiffness.

Deflection limitations for uses where increased floor stiffness is desired		
Use classification	Applied load only	Applied load + K (dead load)[*]
Floor beams		
Commercial, office and institutional		
Floor joists, spans to 26 ft†		
L ≤ 60 psf	$L/480$	$L/360$
60 psf < L < 80 psf	$L/480$	$L/360$
L ≥ 80 psf	$L/420$	$L/300$
Girders, spans to 36 ft†		
L ≤ 60 psf	$L/480$‡	$L/360$
60 psf < L < 80 psf	$L/420$‡	$L/300$
L ≥ 80 psf	$L/360$‡	$L/240$

[*]$K = 1.0$ except for seasoned members where $K = 0.5$. Seasoned members for this usage are defined as having a moisture content of less than 16 percent at the time of installation.
†For girder spans greater than 36 ft and joist spans greater than 26 ft, special design considerations may be required such as more restrictive deflection limits and vibration considerations that include the total mass of the floor.
‡Based on reduction of live load as permitted by the IBC.

Figure 2.22 Recommended beam deflection limitations from the TCM (Ref. 2.4). (*AITC.*)

The deflection of members in other possible critical situations should be evaluated by the designer. Members over large glazed areas and members that affect the alignment or operation of special equipment are examples of two such potential problems.

The NDS takes a somewhat different position regarding beam deflection from the IBC and the TCM. The NDS (Ref. 2.1) does not recommend deflection limits for designing beams or other components, and it essentially leaves these serviceability criteria to the designer or to the building code. However, the NDS recognizes the tendency of a wood member to creep under sustained loads in NDS Sec. 3.5.2 and NDS Appendix F.

According to the NDS, an unseasoned wood member will creep an amount approximately equal to the deflection under sustained load, and seasoned wood members will creep about half as much. With this approach, the total deflection of a wood member including the effects of creep can be computed.

For green lumber, or for seasoned lumber, glulam, and wood structural panels used in wet conditions:

$$\Delta_{\text{Total}} = 2.0(\Delta_{\text{long term}}) + \Delta_{\text{short term}}$$

For seasoned lumber, glulam, I-joists, and structural composite lumber used in dry conditions:

$$\Delta_{\text{Total}} = 1.5(\Delta_{\text{long term}}) + \Delta_{\text{short term}}$$

where $\Delta_{\text{long term}}$ = immediate deflection under long-term load. Long-term load is dead load plus an appropriate (long-term) portion of live load. Knowing the type of structure and nature of the live loads, the designer can estimate what portion of live load (if any) will be a long-term load.

$\Delta_{\text{short term}}$ = deflection under short-term portion of design load

The NDS thus provides a convenient method of estimating total deflection including creep. With this information, the designer can then make a judgment about the stiffness of a member. In other words, if the computed deflection is excessive, the design may be revised by selecting a member with a larger moment of inertia.

In recent years, there has been an increasing concern about the failure of roof systems associated with excessive deflections on *flat* roof structures caused by the entrapment of water. This type of failure is known as *ponding* failure, and it represents a progressive collapse caused by the accumulation of water on a flat roof. The initial beam deflection allows water to become trapped. This trapped water, in turn, causes additional deflection. A vicious cycle is generated which can lead to failure if the roof structure is too flexible.

Ponding failures may be prevented by proper design. The first and simplest method is to provide adequate drainage together with a positive slope (even on essentially flat roofs) so that an initial accumulation of water is simply not possible. IBC Sec. 1611 and ASCE 7 require that a roof have a minimum slope of one unit vertical to 48 units horizontal ($\frac{1}{4}$ in./ft) unless it is specifically designed for water accumulation. An adequate number and size of roof drains must be provided to carry off this water unless, of course, no obstructions are present. See ASCE 7 for additional requirements for roof drains and loads due to rain.

The second method is used in lieu of providing the minimum $\frac{1}{4}$-in./ft roof slope. Here, ponding can be prevented by designing a sufficiently stiff and strong roof structure so that water cannot accumulate in sufficient quantities to cause a progressive failure. This is accomplished by imposing additional deflection criteria for the framing members in the roof structure and by designing these members for increased stresses and deflections. The increased stresses are obtained by multiplying calculated actual stresses under service loads by a magnification factor. The magnification factor is a number greater than 1.0 and is a measure of the sensitivity of a roof structure to accumulate (pond) water. It is a function of the total design roof load ($D + L_r$) and the weight of ponding water.

Because the first method of preventing ponding is a more direct and less costly method, it is recommended for most typical designs. Where the minimum slope cannot be provided for drainage, the roof structure should be designed as described above for ponding. Because this latter approach is not the more common solution, the specific design criteria are not included here. The designer is referred to Ref. 2.4 for these criteria and a numerical example.

Several methods can be used in obtaining the recommended $\frac{1}{4}$-in./ft roof slope. The most obvious solution is to place the supports for framing members at different elevations. These support elevations (or the *top-of-sheathing*, abbreviated TS, elevations) should be clearly shown on the roof plan.

A second method which can be used in the case of glulam construction is to provide *additional* camber (see Chap. 5) so that the $\frac{1}{4}$-in./ft slope is built into supporting members. It should be emphasized that this slope camber is in addition to the camber provided to account for long-term (dead load) deflection.

2.19   References

[2.1] American Forest and Paper Association (AF&PA). 2005. *National Design Specification for Wood Construction* and *Supplement*, ANSI/AF&PA NDS-2005, AF&PA, Washington, DC.

[2.2] American Forest & Paper Association (AF&PA). 2005. *Special Design Provisions for Wind and Seismic* (SPDWS) Supplement to the NDS, 2005 ed., AF&PA, Washington, DC.

[2.3] American Institute of Steel Construction (AISC). 2005. *Steel Construction Manual,* 13th ed., AISC, Chicago, IL.

[2.4] American Institute of Timber Construction (AITC). 2005. *Timber Construction Manual*, 5th Edition, John Wiley & Sons, Hoboken, NJ.

[2.5] American Society of Civil Engineers (ASCE). 2006. *Minimum Design Loads for Buildings and Other Structures* (ASCE 7-05), ASCE, Reston, VA.

[2.6] American Society of Civil Engineers (ASCE). 2005 . *Flood Resistant Design and Construction* (ASCE 24-05), ASCE, Reston, VA.

[2.7] Building Officials and Code Administrators International (BOCA). 1999. *National Building Code*, 1999 ed., BOCA, Country Club Hills, IL.

[2.8] Building Seismic Safety Council (BSSC). 2003. *National Earthquake Hazards Reduction Program (NEHRP) Recommended Provisions for Seismic Regulations for New Buildings* and *Commentary*, 2003 ed., BSSC, Washington, DC.

[2.9] International Code Council (ICC). 2006. *International Building Code*, 2006 ed., ICC, Falls Church, VA.

[2.10] International Code Council (ICC). 2006. *International Residential Code*, 2006 ed., ICC, Falls Church, VA.

[2.11] International Conference of Building Officials (ICBO). 1997. *Uniform Building Code* (UBC), 1997 ed., ICBO, Whittier, CA.

[2.12] Mehta, K. C., and D. C. Perry. 2001. *Guide to the Use of the Wind Load Provisions of ASCE 7-02*, ASCE, Reston, VA.

[2.13] Southern Building Code Congress International (SBCCI). 1997. *Standard Building Code*, 1997 ed., SBCCI, Birmingham, AL.

[2.14] Structural Engineers Association of California (SEAOC). 1999. *Recommended Lateral Force Requirements and Commentary*, 7th ed., SEAOC, Sacramento, CA.

[2.15] Structural Engineers Association of Washington (SEAW). 2004. *SEAW Commentary on Wind Code Provisions* (ATC-60), Applied Technology Council, Redwood City, CA.

[2.16] Structural Engineers Association of Washington (SEAW). 2004. *SEAW's Handbook of a Rapid-Solutions Methodology for Wind Design* (SEAW RSM-03), Applied Technology Council, Redwood City, CA.

2.20   Problems

All problems are to be answered in accordance with the 2006 *International Building Code* (IBC) and ASCE 7-05. A number of tables are included in Appendix C.

2.1   *Given:*   The house framing section shown in Fig. 2.A

Figure 2.A

Find: a. Roof dead load D in psf on a horizontal plane
b. Wall D in psf of wall surface area
c. Wall D in lb/ft of wall
d. Basic (i.e., consider roof slope but not trib. area) unit roof live load, L_r, in psf
e. Basic unit roof L_r in psf if the slope is changed to $\frac{4}{12}$

2.2   *Given:*   The house framing section shown in Fig. 2.B. Note that a roofing square is equal to 100 ft^2.

Figure 2.B

Find: *a.* Roof dead load D in psf on a horizontal plane
            b. Ceiling dead load D in psf
            c. Basic (i.e., consider roof slope but not tributary area) unit roof live load L_r in psf

2.3   *Given:*   The building framing section shown in Fig. 2.C below.

       Find: *a.* Roof dead load D in psf
                   b. Second-floor dead load D in psf
                   c. Basic (i.e., consider roof slope but not tributary area) unit roof live load L_r in psf

SECTION

DETAIL A

Figure 2.C

2.4 *Given:* The roof framing plan of the industrial building shown in Fig. 2.D. Roof slope is $\frac{1}{4}$ in./ft. General construction:

Roofing—5-ply felt
Sheathing—$\frac{15}{32}$-in. plywood
Subpurlin—2 × 4 at 24 in. o.c.
Purlin—4 × 14 at 8 ft-0 in. o.c.
Girder—$6\frac{3}{4}$ × 33 DF-L glulam at 20 ft-0 in. o.c.

Assume loads are uniformly distributed on supporting members.

Find: *a.* Average dead load D of entire roof in psf
        b. Tributary dead load D to subpurlin in lb/ft
        c. Tributary dead load D to purlin in lb/ft
        d. Tributary dead load D to girder in lb/ft
        e. Tributary dead load D to column C1 in k
        f. Basic (i.e., consider roof slope but not tributary area) unit roof live load L_r in psf
        g. Rain load R in psf at roof low point, assuming that the height of water to the secondary drain system (overflow scupper) d_s is 5 in. and the hydraulic head d_h at the scupper at design flow is 0.5 in.

2.5 *Given:* Figure 2.A. The ridge beam spans 20 ft-0 in.

Find: *a.* Tributary area to the ridge beam
        b. Roof live load L_r in lb/ft

ROOF FRAMING PLAN

Figure 2.D

2.6 *Given:* A roof similar to Fig. 2.A with $^3\!/_{12}$ roof slope. The ridge beam spans 22 ft-0 in.

   Find: *a.* Tributary area to the ridge beam
   b. Roof live load L_r in lb/ft

2.7 *Given:* Figure 2.B, standard residential occupancy, heated building, a ground snow load of 70 psf, and Exposure C terrain with a fully exposed roof

   Find: Design snow load S in psf on a horizontal plane

2.8 *Given:* A roof similar to Fig. 2.B with $^8\!/_{12}$ slope and a ground snow load of 90 psf, Exposure B terrain with a sheltered roof, standard residential occupancy, and a heated building

   Find: Design snow load S in psf on a horizontal plane

2.9 *Given:* The roof structure in Fig. 2.D

   Find: *a.* Unit roof live load L_r in psf for
      1. 2 × 4 subpurlin
      2. 4 × 14 purlin
      3. $6^3\!/_4$ × 33 glulam beam
   b. Uniformly distributed roof live loads in lb/ft for each of the members, using the unit L_r from (*a*)

2.10 *Given:* The roof structure in Fig. 2.D and a 25-psf design snow load specified by the building official

   Find: *a.* Uniformly distributed snow load S in lb/ft for
      1. 2 × 4 subpurlin
      2. 4 × 14 purlin
      3. $6^3\!/_4$ × 33 glulam beam
   b. Tributary snow load S to column C1 in k

2.11 *Given:* The building in Fig. 2.C

   Find: Second-floor basic (i.e., consider occupancy but not tributary areas) unit floor live load L and concentrated loads for the following uses:
   a. Offices
   b. Light storage
   c. Retail store
   d. Apartments
   e. Hotel restrooms
   f. School classroom

2.12 *Given:* An interior column supports only loads from the second floor of an office building. The tributary area to the column is 240 ft^2, and the dead load is 35 psf.

   Find: *a.* Basic floor live load L_0 in psf
   b. Reduced floor live load L in psf
   c. Total load to column in k

2.13 *Given:* An interior beam supports the floor of a classroom in a school building. The beam spans 26 ft. and the tributary width is 16 ft. Dead load = 20 psf.

   Find: a. Basic floor live load L_0 in psf
   b. Reduced floor live load L in psf
   c. Uniformly distributed total load to the beam in lb/ft
   d. Compare the loading in part *c* with the alternate concentrated load required by the Code. Which loading is more critical for bending, shear, and deflection?

2.14 *Given:* IBC Table 1607.1

   Find: Four occupancies where the unit floor live load *L* cannot be reduced. List the occupancy and the corresponding unit floor live load *L*.

2.15 *Given:* IBC beam deflection criteria

   Find: The allowable deflection limits for the following members. Beams are unseasoned wood members
   a. Floor beam with 22-ft span
   b. Roof rafter that supports a plaster ceiling below. Span = 12 ft

2.16 *Given:* The *Timber Construction Manual* beam deflection recommendations in Fig. 2.22

   Find: The recommended limits for the following beams. Beams are seasoned wood members that remain dry in service.
   a. Roof rafter in a commercial building that supports a gypsum board ceiling below. Span = 16 ft.
   b. Roof girder in an office building supporting an acoustic suspended ceiling. Span = 40 ft.
   c. Floor joist in the second floor of a residential building to be designed for "ordinary usage." Span = 20 ft. Tributary width = 4 ft. Floor dead load D = 16 psf. (Give beam loads in lb/ft for each deflection limit.)
   d. Girder in the second floor of a retail sales building. Increased floor stiffness is desired to avoid public concern about perceived excessive floor deflections. Span = 32 ft. Tributary width = 10 ft. Floor dead load D = 20 psf. (Give beam loads in lb/ft for each deflection limit.)

2.17 *Given:* The ASCE 7 wind force provisions

   Find: a. The expression for calculating the design wind pressure
   b. The section in ASCE 7 where the terms of the expression are defined
   c. Distinguish between the following wind forces and the areas to which they are applied:
     1. Main wind-force-resisting systems
     2. Components and cladding away from discontinuities
     3. Components and cladding at or near discontinuities
   d. Describe the three wind exposure conditions.

2.18 *Given:* The ASCE 7 wind force provisions

   Find: a. The mean recurrence interval for the wind speeds given in ASCE 7 Fig 6-1.

        b. The approximate mean recurrence interval associated with the wind pressure for essential and hazardous facilities

        c. The height used to determine the height and exposure factor for the design wind pressure

2.19 *Given:* An enclosed building in Tampa, Florida, that is an essential facility. The roof is flat and is 30 ft above grade. Exposure B applies. $K_{zt} = 1.0$.

      Find: *a.* Basic wind speed V

          b. Importance factor I

          c. Height and exposure factor λ

          d. The design wind pressures p_s in each zone for main wind force-resisting systems

          e. The design wind pressures p_{net} for components and cladding in zones near discontinuities and zones away from discontinuities

2.20 *Given:* The enclosed building in Fig. 2.E is a two-story essential facility located near Denver, Colorado. Wind Exposure C applies.

      Find: The design wind pressures in both principal directions of the building for:

          a. Main wind force-resisting systems

          b. Design of individual structural elements having tributary areas of 20 ft^2 in the wall and roof systems away from discontinuities

          c. Design of individual structural elements in the roof near discontinuities having tributary areas of 50 ft^2.

Figure 2.E

2.21 *Given:* ASCE 7 seismic design force requirements

      Find: *a.* The formulas for the base shear. Give section reference.

          b. The mapped maximum considered spectral response accelerations

S_S and S_1 from seismic hazard maps. Cite reference. What is the physical significance of S_S and S_1?

c. The maximum tabulated Site Class coefficients F_a and F_v. Cite reference. Explain the purpose of these coefficients.

d. The maximum values of S_{MS}, S_{M1}, S_{DS}, and S_{D1}, based on previous values.

e. Briefly describe the purpose of the R-factor. What value of R is used for a building with wood-frame bearing walls that are sheathed with wood structural panel sheathing?

2.22 *Given:* ASCE 7 seismic design force requirements

Find: a. The definition of period of vibration and the methods for estimating the fundamental period.

b. How does period of vibration affect seismic forces?

c. Describe the effects of the interaction of the soil and structure on seismic forces.

d. What is damping, and how does it affect seismic forces? Do the ASCE 7 criteria take damping into account?

2.23 *Given:* ASCE 7 seismic design force requirements

Find: a. Briefly describe the general distribution of seismic forces over the height of a multistory building.

b. Describe differences in vertical distribution for vertical element and diaphragm forces between Seismic Design Categories B and D.

c. Describe forces for out-of-plane design of wall components. Cite Code reference.

2.24 *Given:* The $6\frac{3}{4} \times 33$ glulam beam in Fig. 2.D

Find: Identify and solve for unit vertical load in psf for all applicable ASCE 7 ASD load combinations given that the following are applicable unreduced vertical loads in psf acting downward:

D = 10 psf
L_r = 20 psf
S = 35 psf
R = 30 psf
W = 18 psf
E = 2 psf

2.25 Repeat problem 2.24 using ASCE 7 LRFD load combinations.

Chapter

3

Behavior of
Structures under
Loads and Forces

3.1 Introduction

The loads and forces required by the IBC (Ref. 3.2) and ASCE 7 (Ref. 3.1) for designing a building were described in Chap. 2. Chapter 3 deals primarily with the transfer of these from one member to another throughout the structure. The distribution
of *vertical loads* in a typical wood-frame building follows the traditional "post-and-
beam" concept. This subject is briefly covered at the beginning of the chapter.

The distribution of *lateral forces* may not be as evident as the distribution of
vertical loads. The majority of Chap. 3 deals with the transfer of lateral forces
from the point of origin, through the building, and into the foundation. This subject is introduced by reviewing the three basic types of lateral-force-resisting systems (LFRSs) commonly used in rectangular-type buildings.

Shearwalls and diaphragms make up the LFRS used in most wood-frame
buildings (or buildings with a combination of wood framing and concrete or
masonry walls). The chapter concludes with two detailed examples of lateral
force calculations for these types of buildings.

3.2 Structures Subject to Vertical Loads

The behavior of framing systems (post-and-beam type) under vertical loads is
relatively straightforward. Sheathing (decking) spans between the most closely
spaced beams; these short-span beams are given various names: rafters, joists,
subpurlins. The reactions of these members in turn cause loads on the next set
of beams in the framing system; these next beams may be referred to as beams,
joists, or purlins. Finally, reactions of the second set of beams impose loads on
the largest beams in the system. These large beams are known as girders. The
girders, in turn, are supported by columns. See Example 3.1 and Fig. 3.1.

EXAMPLE 3.1   Typical Post-and-Beam Framing

FRAMING PLAN

Figure 3.1

1. Sheathing spans between subpurlins.
2. Subpurlins span between purlins.
3. Purlins span between girders.
4. Girders span between columns.

Subpurlins and purlins are also supported by bearing walls. *Bearing walls* are defined as walls that support vertical loads in addition to their own weight.

When this framing system is used for a roof, it is often constructed as a *panelized* system. Panelized roofs typically use glulam girders spaced 18 to 40 ft on center, sawn lumber or glulam purlins at 8 ft on center, sawn lumber subpurlins at 24 in. on center, and plywood sheathing. The name of the system comes from the fact that 8-ft-wide roof *panels* are prefabricated and then lifted onto preset girders using forklifts. See Fig. 3.2. The speed of construction and erection makes panelized roof systems economical. Panelized roofs are widely used on large one-story commercial and industrial buildings. See Ref. 3.3 for more information on panelized roof structures.

Although the loads to successively larger beams are a result of reactions from lighter members, for structural design the loads on beams in this type of system

Figure 3.2 Panelized roof system installed with forklift. (*Photo by Mike Hausmann.*)

are often assumed to be uniformly distributed. To obtain a feel for whether this approach produces conservative values for shear and moment, it is suggested that a comparison be made between the values of shear and moment obtained by assuming a *uniformly distributed load* and those obtained by assuming *concentrated loads* from lighter beams. The actual loading probably falls somewhere between the two conditions described. See Example 3.2.

Regardless of the type of load distribution used, it should be remembered that it is the *tributary area of the member being designed* (Sec. 2.3) which is used in establishing unit live loads, rather than the tributary area of the lighter members that impose the load. This concept is often confusing when it is first encountered.

EXAMPLE 3.2   Beam Loading Diagrams

Figure 3.3 shows the girder from the building in Fig. 3.1. The load to the girder can be considered as a number of concentrated reaction loads from the purlins. However, a more common design practice is to assume that the load is uniformly distributed. The uniformly distributed load is calculated as the unit load times the tributary width to the girder. As the number of concentrated loads increases, the closer the loading approaches the uniform load case.

CONCENTRATED   **UNIFORM**

Figure 3.3   Loading assumptions for beam design.

As an example, consider the design load for the girder in Fig. 3.1. Confusion may occur when the *unit* live load for the girder (in psf based on a large tributary area) turns out to be less than the unit live load used in the design of the purlin. Obviously, the reaction of the purlin (using the higher live load) must be supported by the girder. Why is the lower live load used for design?

The reasoning is thus: The girder must be capable of supporting *individual* reactions from purlins calculated using the larger unit live load (obtained using the tributary area of the purlin). However, when the entire tributary area of the girder is considered loaded, the smaller unit live load may be used (this was discussed in detail in Chap. 2). Of course, each connection between the purlin and girder must be designed for the higher unit live load, but not all purlins are subjected to this higher load simultaneously.

The spacing of members and the span lengths depend on the function of the building. Closer spacing and shorter spans require smaller member sizes, but short spans require closely spaced columns or bearing walls. The need for clear, unobstructed space must be considered when the framing system is first established. Once the layout of the building has been determined, dimensions for framing should be chosen which result in the best utilization of materials. For example, the standard size of a sheet of plywood is 4 ft by 8 ft, and a joist spacing should be chosen which fits this basic module. Spacings of 16, 24, and 48 in. o.c. (o.c. = on center and c.c. = center to center) all provide supports at the edge of a sheet of plywood.

Certainly an unlimited number of framing systems can be used, and the choice of the framing layout should be based on a consideration of the requirements of a particular structure. Several other examples of framing arrangements are shown in Figs. 3.4*a*, *b*, and *c*. These are given to suggest possible arrangements and are not intended to be a comprehensive summary of framing systems.

It should be noted that in the framing plans, a break in a member represents a simple end connection. For example, in Fig. 3.4*a* the lines representing joists are broken at the girder. If a continuous joist is to be shown, a solid line with no break at the girder would be used. This is illustrated in Fig. 3.4*c* where the joist is continuous at the rear wall overhang. Such points may seem obvious, but a good deal of confusion results if they are not recognized.

Figure 3.4a Alternate post-and-beam framing.

Figure 3.4b Light frame trusses.

Figure 3.4c Roof framing with interior and exterior bearing walls.

3.3 Structures Subject to Lateral Forces

The behavior of structures under lateral forces usually requires some degree of explanation. In covering this subject, the various types of LFRSs commonly used in rectangular buildings should be clearly distinguished. See Example 3.3. These LFRSs include

1. Moment frame
2. Braced frame
3. Shearwall

EXAMPLE 3.3   Basic Lateral-Force-Resisting Systems

Moment Frame

Figure 3.5a   Moment frame system.

Resistance to lateral forces is provided by *bending* in the columns and girders of the moment frame.

Braced Frame

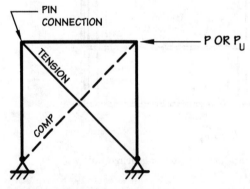

Figure 3.5b   Braced frame system.

Lateral forces develop *axial* forces in the braced frame.

Shearwall

Figure 3.5c Shearwall system.

Segments of walls can be designed to function as *shear-resisting* elements in carrying lateral forces. The deflected shape shows shear deformation rather than bending.

Note that two forces P and P_u are shown in Fig. 3.5a, b, and c. The designer needs to clearly understand the meaning of these two symbols. When the symbol for a force (or other property such as shear or moment) has the "u" subscript, the force (or other property) is at an LRFD or *strength level*. On the other hand, when the symbol for a force (or other property) does not have a "u" subscript, it is at an *ASD level*. This notation is necessary because the wind forces W calculated using ASCE 7 equations are at an ASD level, noted by P, whereas the seismic forces E are at a strength level, noted by P_u. Because strength and ASD force levels differ substantially, it is important that the reader make a clear distinction between the two. Throughout Chap. 3, and all chapters dealing with lateral force design, the reader should pay close attention to whether a quantity is at a *strength level* or at an *ASD level*. The use of a "u" subscript is the notation convention used in this book to distinguish between the two.

Also shown in Fig. 3.5c is the shearwall horizontal deflection. In this book, Δ_s is the symbol used to represent the horizontal deflection or drift of a shearwall. Shearwall drift is also closely related to the design story drift, Δ, the horizontal deflection of the full building, over the height of the story under consideration. The IBC places limits on the maximum acceptable story drift for seismic design, but does not limit story drift for wind design.

Moment frames, whether statically determinate or indeterminate, resist lateral forces by bending in the frame members. That is, the members have relatively small depths compared with their lengths, and the stresses induced as the structure deforms under lateral forces are essentially flexural. Some axial forces are also developed. In the United States, moment frames are often constructed of steel or reinforced concrete and are rarely constructed of wood. Earthquake experience with steel and reinforced concrete moment frames has shown that careful attention to detailing is required to achieve ductile behavior in these types of structures.

Braced frames are analyzed in a manner similar to horizontal trusses: connections are assumed to be pinned, and forces are assumed to be applied at the joints. They often take the form of cross or X-bracing. For the braced frame in Fig. 3.5*b*, both diagonal members are generally designed to function simultaneously (one in tension and the other in compression). For a hand calculation, the forces in the diagonals can be assumed equal due to the symmetry of the brace. A computer analysis would show a very small difference in member forces.

Code provisions for X-bracing (Fig. 3.5*b*) with tension-only bracing members have varied significantly in recent years. In the past, the use of X-bracing tension-only systems was not uncommon. Because of their slenderness, the brace members would buckle in compression, resulting in the full load being carried by the opposing tie rod in tension. More recent editions of the *Building Code* placed restrictions on the use of this type of system due to observed earthquake damage. The restrictions included limitation to one- and two-story buildings and increased design forces. ASCE 7 now recognizes the use of a new flat-strap tension-only system, used with special design requirements.

Shearwall structures make use of specially designed wall sections to resist lateral forces. A shearwall is essentially a vertical cantilever with the span of the cantilever equal to the height of the wall. The depth of these members (i.e., the length of the wall element parallel to the applied lateral force) is large in comparison with the depth of the structural members in the moment frame LFRS. For a member with such a large depth compared with its height, *shear* deformation replaces bending as the significant action (hence, the name "shearwall").

It should be mentioned that the LFRSs in Example 3.3 are the *vertical resisting elements* of the system (i.e., the vertical components). Because buildings are three-dimensional structures, some horizontal system must also be provided to carry the lateral forces to the vertical elements. See Example 3.4. A variety of horizontal systems can be employed:

1. Horizontal wall framing

2. Vertical wall framing with horizontal trusses at the story levels

3. Vertical wall framing with diaphragms at the story levels

EXAMPLE 3.4   Horizontal Elements in the LFRS

Horizontal Wall Framing

Horizontal wall members are known as *girts* and distribute the lateral force to the vertical LFRS. Dashed lines represent the deflected shape of the girts. See Fig. 3.6*a*. The lateral force to the shearwalls is distributed over the height of the wall.

Vertical Wall Framing

Vertical wall members are known as *studs*. The lateral force is carried by the studs to the roof level and the foundation. A horizontal truss in the plane of the roof distributes the lateral force to the transverse shearwalls. The diagonal members in the horizontal truss are sometimes steel rods which are designed to function in tension only.

Figure 3.6a   Wall horizontal framing (girts) spanning to supporting shearwalls.

Figure 3.6b   Wall vertical framing (studs) spanning to a supporting horizontal truss.

Vertical Wall Framing and Horizontal Diaphragm

In Fig. 3.6c the LFRS is similar to the system in Fig. 3.6b except that the horizontal truss in the plane of the roof is replaced with a *diaphragm*. The use of vertical wall framing and horizontal diaphragms is the most common system in wood-frame buildings because

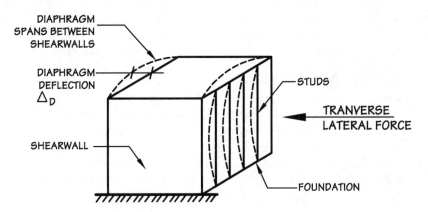

Figure 3.6c   Wall vertical framing (studs) spanning to a supporting diaphragm.

the roof sheathing can be designed economically to function as both a vertical-load- and a lateral-force-carrying element. The diaphragm is designed as a beam spanning between the shearwalls. The design requirements for diaphragms and shearwalls are given in Chaps. 9 and 10.

The first two framing systems in Example 3.4 are relatively easy to visualize. The third system is also easy to visualize once the concept of a diaphragm is understood. A diaphragm can be considered to be a large, thin structural element that is loaded in its plane. In Fig. 3.6c, the vertical wall members develop horizontal reactions at the roof level and at the foundation level (studs are assumed to span as simple beams between these two levels). The reaction of the studs at the roof level provides a force in the plane of the roof. The diaphragm acts as a large horizontal beam.

In wood buildings, or buildings with wood roof and floor systems and concrete or masonry walls, the roof or floor sheathing is designed and connected to the supporting framing members to function as a diaphragm. In buildings with concrete roof and floor slabs, the concrete slabs are designed to function as diaphragms.

The stiffness of a diaphragm refers to the amount of horizontal deflection that occurs in the diaphragm as a result of the in-plane lateral force (Fig. 3.6c shows the deflected shape). In this book, Δ_D is the symbol used to represent the deflection of the diaphragm. Because wood diaphragms are not nearly as stiff as concrete slabs, they have in the past been categorically considered *flexible* while concrete diaphragms have categorically been considered *rigid*. The designation of the diaphragm as rigid or flexible (or sometimes semi-rigid) is used to determine the method used to distribute seismic forces to the vertical resisting elements (shearwalls) and to determine required anchorage forces between concrete and masonry walls, and the diaphragm (see Chap. 15).

A diaphragm that is designated as flexible is generally modeled as spanning between shearwalls, similar to a single-span beam. A diaphragm that is designated as rigid is modeled to distribute seismic forces to the vertical elements in proportion to their stiffness, with consideration of redistribution due to torsion. A diaphragm designated as semi-rigid is likely to be modeled in a computer analysis using either plate or diagonal strut elements; load-deflection properties of the diaphragm will result in force distribution somewhere between the flexible and rigid models. All constructed diaphragms, in fact, fall somewhere between the flexible and rigid behavior; however, simplifications are required in order to keep design of structures reasonably simple. Study of the effect of diaphragm modeling choices on the earthquake performance of buildings is ongoing. ASCE 7 and the IBC provide requirements that must be used for design of new structures.

ASCE 7 Sec. 12.3 addresses designation of diaphragm flexibility. The general rule is that seismic analysis models consider semi-rigid diaphragm behavior; however, a series of simplifying assumptions are permitted:

Diaphragms of untopped steel decking or wood structural panels are permitted to be designated as flexible:

- When used in combination with concrete, masonry, steel, or composite shearwalls
- When used in combination with concrete, steel, or composite braced frames
- For one- and two-family residential buildings of light-frame construction (assumed to mean one- and two-family detached dwellings)

Diaphragms of concrete slabs or concrete filled metal deck (structural concrete, not insulating concrete) with aspect ratios not greater than 3:1 are permitted to be considered rigid.

Diaphragms of any construction are permitted to be considered flexible when demonstrated by calculation to quality as flexible per ASCE 7 Sec. 12.3.1.3.

Per these ASCE 7 requirements, semi-rigid diaphragm modeling will be required for a substantial portion of structures, including most wood shearwall buildings that are not one- and two-family detached dwellings. Because semi-rigid diaphragm behavior would give a distribution of forces somewhere between rigid and flexible diaphragm solutions, the ASCE 7 requirement can be met by using both rigid and flexible models and taking the highest forces from each.

IBC Sec. 1613.6.1 provides a permitted alternative for those that wish to use it. IBC permits diaphragms of untopped steel decking or wood structural panels to be designated as flexible when used in combination with light-frame shearwalls sheathed with wood structural panels or steel sheets, provided:

- Concrete toppings, if provided, are nonstructural and not thicker than $1\frac{1}{2}$ in., and
- Each line of vertical elements complies with story drift limitations, and
- Diaphragm cantilevers meet the requirements of IBC Chap. 23.

The IBC alternate method will permit a flexible diaphragm assumption for most wood shearwall buildings. In addition, the IBC alternate method reflects current design practice for wood shearwall buildings. The design requirements for diaphragms and shearwalls are given in Chaps. 9 and 10.

In each of the sketches in Example 3.4, the transverse lateral force is distributed horizontally to end shearwalls. The same horizontal systems shown in these sketches can be used to distribute lateral forces to the other basic vertical LFRSs (i.e., moment frames or vertical trusses). Any combination of *horizontal* and *vertical* LFRSs can be incorporated into a given building to resist lateral forces.

The discussion of lateral forces in Example 3.4 was limited to forces in the transverse direction. In addition, an LFRS must be provided for forces in the longitudinal direction. This LFRS will consist of both horizontal and vertical components similar to those used to resist forces in the transverse direction.

Different types of *vertical* elements can be used to resist transverse lateral forces and longitudinal lateral forces in the same building. For example, rigid frames can be used to resist lateral forces in the transverse direction, and shearwalls can be used to resist lateral forces in the other direction. The choice of LFRS in one direction does not necessarily limit the choice for the other direction.

In the case of the *horizontal* LFRS, it is unlikely that the horizontal system used in one direction will be different from the horizontal system used in the other direction. If the sheathing is designed to function as a diaphragm for lateral forces in one direction, it probably can be designed to function as a diaphragm for forces applied in the other direction. On the other hand, if the roof or floor sheathing is incapable of functioning as a diaphragm, a system of horizontal trusses spanning in both the transverse and longitudinal directions appears to be the likely solution.

The commonly used types of LFRSs have been summarized as a general introduction and overview. It should be emphasized that the large majority of wood-frame buildings, or buildings with wood roof and floor framing, and concrete or masonry walls use a combination of

1. Diaphragms

2. Shearwalls

to resist lateral forces. Because of its widespread use, only the design of this type of system is covered in this book.

3.4   Lateral Forces in Buildings with Diaphragms and Shearwalls

The majority of wood structures use the sheathing normally provided on floors, roofs, and walls, to form diaphragms and shearwalls that resist lateral wind and seismic forces. To function as a diaphragm or shearwall, the sheathing must be properly attached to the supporting members. The framing members must also be checked for stresses caused by the lateral forces. Furthermore, certain connections must be provided to transfer lateral forces down through the structure. The system must be tied together.

However, the economic advantage is clear. With the added attention to the framing and connection design, the usual sheathing material, which is required in any structure, can also be used to form the LFRS. In this way, one material serves two purposes (i.e., sheathing and lateral force resistance).

The following numerical examples illustrate how lateral forces are calculated and distributed in two different shearwall buildings. The first (Sec. 3.5) demonstrates the procedure for a one-story structure, and the second (Sec. 3.6) expands the system to cover a two-story building. The method used to calculate seismic forces in these two buildings is based on the same criteria, but the solution for the one-story structure is much more direct.

Before moving into design examples, an overview of seismic design forces is needed. There are six main types of seismic design forces which need to be considered for the primary LFRS. All six are at a *strength level*. These are:

1. Seismic base shear force *V*.

   This is the total shear force at the base of the structure. This is generally thought of as a seismic base shear coefficient times the building weight.

2. Seismic story (shearwall) force F_x.

This is a set of forces, with one force for each story level above ground. These forces are used for *design of the vertical elements* of the LFRS (the shearwalls). The shearwalls at any story are designed to resist the sum of the story forces above that level. For one-story buildings, the story force F_x is equal to the base shear force V. For buildings with more than one story, a vertical distribution procedure is used to assign a seismic story force at each story level. The vertical distribution is given by the ASCE 7 formula for F_x (Sec. 2.15). The subscript x in the notation is generally replaced with the specific story under consideration (i.e., F_r is the force at the roof level, F_3 is the force at the third-floor level). The vertical distribution procedure is illustrated in Sec. 3.6. The seismic story (shearwall) force F_x is most often thought of as an F_x story (shearwall) coefficient times the weight tributary to the story w_x. The seismic story (shearwall) force F_x applies for Seismic Design Categories B and up.

3. Seismic story (diaphragm) force F_{px}.

This is also a set of forces, one for each story level above ground. These forces are used for *design of the diaphragm*. The x subscript is again replaced with the specific story under consideration.

For a one-story building with wood diaphragms and wood shearwalls, the F_{px} force is equal to the base shear force V and the seismic story (shearwall) force F_x. For a building with more than one story, a vertical distribution procedure is used to assign the seismic story force at each level. The ASCE 7 formula for F_{px} was described in Sec. 2.15. The seismic story (diaphragm) force F_{px} is most often thought of as an F_{px} story (diaphragm) coefficient times the weight tributary to the story w_x.

Seismic force types 1 through 3 are used with ASCE 7 (Ref. 3.1) equations 12.4-1 through 12.4-4, giving:

$$E = \rho Q_E + 0.2 S_{DS} D$$

$$E = \rho Q_E - 0.2 S_{DS} D$$

Q_E is the horizontal force from 1, 2, or 3 above, and ρ is the reliability/redundancy factor. The term ρQ_E is the seismic force acting horizontally. S_{DS} is the design spectral response acceleration, and D is the vertical dead load. The term $0.2 S_{DS} D$ acts vertically. See Sec. 2.15. Both F_x and F_{px} forces are referred to as story forces, because there will be a force for each story in the building under design.

4. Wall component forces F_c.

These forces are used for structural walls subject to out-of-plane seismic forces. Shearwalls acting in-plane are considered part of the primary LFRS; whereas, when loaded out of plane, shearwalls and other walls are treated in a manner similar to components. The symbol F_c is used for wall out-of-plane forces in Seismic Design Categories B and higher. The force F_c will be greater than the base shear force level. The force F_c is most often thought of as an F_c seismic coefficient times the unit weight of the wall. It should be noted that there is an additional set of force requirements for anchorage of concrete and masonry walls to diaphragms.

5. Component forces F_p.

The force F_p, used for components anchored to the building (other than structural walls), was introduced in Sec. 2.16. The force F_p used for design of components will be greater than the base shear force level. The force F_p is most often thought of as an F_p seismic coefficient times the unit weight of the component.

The primary difference between F_c and F_p forces is that F_p forces are magnified dramatically as the component is anchored higher on the structure, as discussed in Sec. 2.15; whereas, F_c forces do not get magnified with height. The variables used to define the F_c and F_p forces are distinctly different, with F_p using component variables I_p, a_p, and R_p.

6. Special seismic force.

This force type, from ASCE 7 equations 12.4-5 through 12.4-7, defines a special magnified seismic force used for designing a limited number of structural elements, whose performance is viewed as critical to the performance of a building. The special seismic force is discussed in detail in Chaps. 9 and 16.

Designs using force types 1, 2, and 3 are illustrated in the remainder of this chapter. It is common for designers to think of these forces as a seismic coefficient (g factor) times the weight that would be acting under seismic loading:

1. V = Seismic base shear coefficient $\times W$

2. F_x = Seismic story (shearwall) coefficient $\times w_x$

3. F_{px} = Seismic story (diaphragm) coefficient $\times w_x$

Because there is significant repetition involved in the calculation of seismic coefficients, the examples in this book use the approach of calculating the seismic coefficient and multiplying it by the applicable weight acting on the element or portion under consideration.

For a *single-story building*, the seismic base shear force V, the story (shearwall) force F_x, and the story (diaphragm) force F_{px} are all equal, as are the corresponding coefficients:

$$\text{Base shear force } V = \text{Story (shearwall) force } F_x$$

$$= \text{Story (diaphragm) force } F_{px}$$

and

$$\text{Base shear coefficient} = F_x \text{ story coefficient} = F_{px} \text{ story coefficient.}$$

Wood-frame structures have traditionally been limited to relatively low-rise (one- and two-story) structures, and these are the primary focus of this book. It is interesting to note that there have been an increasing number of three- to five-story (and even taller) wood-frame buildings constructed in recent years. See Fig. 3.7. The method of analysis given for the two-story building design example

Figure 3.7   Four-story wood-frame building with diaphragms and shearwalls of plywood. (*Photo courtesy of APA—The Engineered Wood Association.*)

can be extended to handle the lateral force evaluation for taller multistory buildings.

Before proceeding with the one- and two-story building design examples, Example 3.5 will demonstrate the method used to compute some typical seismic base shear and story shear coefficients. Recall that the seismic base shear is the result of a fairly involved calculation. However, with a little practice it is fairly easy to determine the seismic coefficient for many common buildings which use diaphragms and shearwalls for the primary LFRS. These shearwall buildings have a low height-to-width ratio and are fairly rigid structures. As a result, they tend to have low fundamental periods. For this reason, it is common for low-rise shearwall buildings to use the maximum code response spectrum value of $S_{DS}I/R$ for the seismic coefficient, as demonstrated in Example 3.5.

The coefficients (g factors) summarized in the table in Example 3.5 can be used to determine the seismic base shear force V for both one-story and multistory buildings. In multistory buildings, the base shear force V is used to calculate the seismic story (shearwall) forces F_x and the F_x forces are then used to calculate the seismic story (diaphragm) force F_{px}. Example 3.5 evaluates base shear coefficients for two common types of buildings ($R = 5$ and $R = 6.5$).

EXAMPLE 3.5   Seismic Coefficient Calculation

Develop a table of seismic base shear coefficients that will apply to commonly encountered buildings. Many buildings will meet the following three conditions:

1. The building has an importance factor I of 1.0.

   This is the case for the great majority of wood-framed buildings which have residential, office, and commercial uses. Some uses where the importance factor might be greater than 1.0 include police stations, fire stations, and schools. See Sec. 2.14.

2. The building has a short fundamental period T so the seismic forces are controlled by the S_{DS} plateau of the ASCE 7 design spectrum (Fig. 2.17 in Sec. 2.14).

   This is the case for practically all buildings braced by wood shearwalls. This condition could potentially not be met, however, if tall slender shearwalls were to be used.

3. The site class is assumed to be D.

   Site Class D is often assumed in the absence of a geotechnical investigation.

For structures meeting all three of these conditions, the calculation of the *strength level* base shear coefficient can be simplified to S_{DS}/R. The two variables in this simplified expression for determining this seismic base shear force V are the design short-period spectral response acceleration S_{DS} and the *response modification factor R*. In the past, only two R-factors were needed to address the great majority of wood-frame construction. In ASCE 7, due to the incorporation of systems used nationally, the number of systems and R-factors has increased. R-factors of interest for wood-frame construction include:

$R = 6.5$ is used for wood and metal light-frame shearwalls in bearing wall systems that are braced with wood structural panel sheathing.

$R = 2$ is used for wood and metal light-frame shearwalls in bearing wall systems that are braced with shear panels other than wood structural panels.

$R = 5$ is used for special reinforced concrete and masonry shearwalls in bearing wall systems, which is permitted in all seismic design categories.

$R = 4$ is used for ordinary reinforced concrete shearwalls in bearing wall systems, which is permitted for Seismic Design Categories B and C.

$R = 3.5$ is used for intermediate reinforced masonry shearwalls in bearing wall systems, which also is permitted for Seismic Design Categories B and C.

$R = 2$ is used for ordinary reinforced masonry shearwalls in bearing wall systems and is permitted in Seismic Design Categories B and C.

Other types of concrete and masonry shearwall systems are permitted in Seismic Design Category B. The terms *special, intermediate, ordinary,* and *plain,* when applied to concrete or masonry shearwalls, refer to the extent of seismic detailing required. Additional reinforcing and detailing is required for the higher Seismic Design Categories. Examples in this text will focus on $R = 6.5$ (for light-frame bearing wall buildings) and $R = 5$ (for concrete and masonry shearwall buildings), because they can be used across all seismic design categories. Methods used to calculate seismic base shear forces do not change with seismic design category, so the method could be applied equally to a system with any R-factor.

Sample Calculation

The following calculation is for a building in Hayward, California. From the maximum considered earthquake ground motion maps, at 37.7 degrees latitude and 122 degrees longitude, the mapped short-period spectral acceleration S_s, is 2.05g. The mapped one-second spectral acceleration S_1, is 0.91g.

The occupancy type is commercial, which is assigned Occupancy Category II in ASCE 7 Table 1-1.

The site class is assumed to be D. Based on the mapped short-period spectral acceleration and the site class, ASCE 7 Table 11.4-1 assigns a site coefficient F_a of 1.0. From this, the design short-period spectral response acceleration S_{DS} can be calculated as $S_{DS} = S_{MS} \times \frac{2}{3} = F_a S_s \times \frac{2}{3} = 1.37g$.

ASCE 7 Sec. 11.6 would identify the Seismic Design Category as E. The structural system is light-frame bearing walls with wood structural panel shear walls, using an R-factor of 6.5. This system is permitted in all SDCs per ASCE 7 Table 12.2-1, as is bracing with steel sheets. Note that light-frame walls sheathed with other materials (the next line down in Table 12.2-1) would not be permitted for the example building in SDC E. In addition, IBC Secs. 2304.6.3 and 2304.6.4 prohibit use of particleboard and fiberboard in SDC D.

The importance factor I is assumed to be 1.0, and is, therefore, dropped from the equation.

$$\text{Seismic base shear coefficient}$$

$$= S_{DS}/R$$

$$= 1.37g/6.5$$

$$= 0.21g$$

This *strength level* base shear coefficient is entered into the table. Additional values are computed and entered in a similar way, using the S_s ranges from ASCE 7 Table 11.6-1. While the tabulated values provide an overview, the base shear coefficient will need to be calculated for virtually every building site, because mapped S_s and S_1 values vary for every site.

Example Seismic Coefficients (1)(2)

S_S (g)	Site Class D F_a	Site Class D S_{DS} (g)	Seismic base shear coefficient (g)	
			$R = 6.5$	$R = 5$
0.25	1.6	0.27	0.041	0.053
0.50	1.4	0.47	0.072	0.093
0.75	1.2	0.60	0.092	0.120
1.00	1.1	0.73	0.113	0.147
1.25	1.0	0.83	0.128	0.167
2.05	1.0	1.37	0.210	0.273

(1) The seismic coefficients in this table apply only to buildings that meet the three conditions listed above.
(2) To use this table, enter with the S_S value and read either S_{DS} or base shear coefficient. Base shear coefficients are at a *strength level*.

The coefficients in this table can be viewed as seismic base shear force V coefficients for *both* one-story and multistory buildings.

For *one-story structures*, the base shear force V coefficient is also equal to the F_x story (shearwall) coefficient and the F_{px} story (diaphragm) coefficient.

In *multistory structures*, the base shear force V must be determined first. However, in a multistory building the F_x story (shearwall) coefficients and the F_{px} story (diaphragm) coefficients vary from one-story level to another. These seismic coefficients must be evaluated using the appropriate Code equations (Sec. 2.14).

After the *strength level* forces are determined using the coefficients, two additional steps are required. 1. The seismic force must be multiplied by the redundancy/reliability factor ρ, and 2. The seismic force must be combined with other applicable loads in ASD or LRFD load combinations (Sec. 2.17), using designated load factors.

3.5 Design Problem: Lateral Forces on One-Story Building

In this section, a rectangular one-story building with a wood roof system and masonry walls is analyzed to determine both wind and seismic forces. The building chosen for this example has been purposely simplified so that the basic procedure is demonstrated without "complications." The structure is essentially defined in Fig. 3.8 with a plan view of the diaphragm and a typical transverse cross section.

The example is limited to the consideration of the lateral force in the transverse direction. This force is shown in both plan and section views. In the plan view, it is seen as a uniformly distributed force to the diaphragm, and the diaphragm spans between the shearwalls. In the section view, the lateral force is shown at the point where the walls tie to the diaphragm. This height is taken as the *reference* location for evaluating the tributary heights to the diaphragm for both wind and seismic forces.

Figure 3.8 One-story building subjected to lateral force in transverse direction.

The critical lateral force for the diaphragm will be taken as the larger of the two tributary forces: wind or seismic. Although the IBC requires the effects of the horizontal and vertical wind or seismic forces to be considered simultaneously, only the horizontal component of the force affects the unit shear in the diaphragm. The possible effects of the vertical wind pressure (uplift) and seismic forces are not addressed in this example.

Wind.   The wind force in this problem is determined using the simplified wind load method of ASCE 7 Sec. 6.4. The force to the roof diaphragm is obtained by multiplying the design wind pressures by the respective wall areas tributary to the reference point. See Example 3.6. One method for calculating the force to the roof diaphragm is to assume that the wall spans vertically between the roof diaphragm and the foundation. Thus, the tributary height below the diaphragm is simply one-half of the wall height. Above the reference point, the tributary height to the diaphragm is taken as the cantilever height of the parapet wall plus the projected height of the roof above the top of the parapet.

EXAMPLE 3.6   Wind Force Calculation

Determine the horizontal component of the wind force tributary to the roof diaphragm. The basic wind speed is given as 85 mph. Standard occupancy and Exposure B apply. $K_{zt} = 1.0$.

Figure 3.9   Wind pressures tributary to roof diaphragm.

Building Geometry

Roof slope = 7:25 = 15.64°

$$h_{mean} = \text{height to reference point} + \tfrac{1}{2}(\text{projected height of roof})$$

$$= 14 \text{ ft} + \tfrac{1}{2}(7 \text{ ft})$$

$$= 17.5 \text{ ft}$$

$$0.4h_{mean} = 0.4(17.5 \text{ ft}) = 7 \text{ ft}$$

Least horizontal building dimension = b = 50 ft

$$0.1b = 0.1(50 \text{ ft}) = 5 \text{ ft}$$

End zone dimension:

$$a = \text{lesser of } \{0.4h_{mean} \text{ or } 0.1b\}$$

$$= 5 \text{ ft}$$

Length of end zone = $2a$ = 10 ft

Wind Pressures

Wind pressure formula:

$$p_s = \lambda K_{zt} I p_{s30}$$

$$I = 1.0 \qquad \text{ASCE 7 Table 1-1}$$

$$\lambda = 1.0 \qquad \text{for 0 to 30 ft ASCE 7 Fig. 6-2}$$

Basic wind pressures p_{s30} from ASCE 7 Fig. 6-2 with K_{zt} taken as 1.0

	Roof slope		Linear interpolation for 15.64° slope	$p_s = \lambda K_{zt} I p_{s30}$
	15°	20°		
Zone A	14.4 psf	15.9 psf	p_{s30} = 14.6 psf	14.6 psf
Zone B	−4.8 psf	−4.2 psf	p_{s30} = −4.7 psf	−4.7 psf
Zone C	9.6 psf	10.6 psf	p_{s30} = 9.7 psf	9.7 psf
Zone D	−2.7 psf	−2.3 psf	p_{s30} = −2.6 psf	−2.6 psf

These wind pressures are shown in the section view in Fig. 3.9.

Load to Diaphragm

$$\left\{ \begin{matrix} \text{Tributary height} \\ \text{to roof diaphragm} \end{matrix} \right\} = \left\{ \begin{matrix} \text{Tributary wall height} \\ \text{below reference point} \end{matrix} \right\} + \left\{ \begin{matrix} \text{Height of} \\ \text{parapet wall} \end{matrix} \right\} + \left\{ \begin{matrix} \text{Projected roof} \\ \text{height above} \\ \text{parapet wall} \end{matrix} \right\}$$

Tributary wall height below reference point = $\tfrac{1}{2}(14 \text{ ft})$ = 7 ft

Height of parapet wall = 4 ft

Projected roof height above parapet wall = 7 ft − 4 ft = 3 ft

∴ Tributary height to roof diaphragm = 7 ft + 4 ft + 3 ft = 14 ft

Since wind pressures are negative (outward) on the projected roof area above the parapet wall, the total wind force to the roof diaphragm will be greater if it is assumed that $p_s = 0$ psf in roof Zones B and D (per ASCE 7 Fig. 6-2.). Thus,

$$w_{end} = 14.6 \text{ psf } (7 \text{ ft} + 4 \text{ ft}) + 0 \text{ psf } (3 \text{ ft}) = 160.6 \text{ lb/ft}$$

$$w_{int} = 9.7 \text{ psf } (7 \text{ ft} + 4 \text{ ft}) + 0 \text{ psf } (3 \text{ ft}) = 106.7 \text{ lb/ft}$$

$$W = w_{end}(2a) + w_{int}(L - 2a)$$

$$= 160.6(10) + 106.7(110 - 10)$$

$$= 12{,}276 \text{ lb}$$

Use static equilibrium equations to determine the larger reaction force (R_A or R_B) at either end of the roof diaphragm.

Summing moments about the reaction at B:

$$R_A = w_{end}(2a)(L - a)\left(\frac{1}{L}\right) + w_{int}(L - 2a)\left(\frac{L}{2} - a\right)\left(\frac{1}{L}\right)$$

$$= (160.6)(10)(110 - 5)\left(\frac{1}{110}\right) + (106.7)(110 - 10)\left(\frac{110}{2} - 5\right)\left(\frac{1}{110}\right)$$

$$= 6383 \text{ lb}$$

Summing forces:

$$R_B = W - R_A = 12{,}276 - 6383 = 5893 \text{ lb}$$

Check minimum load to diaphragm based on $p_s = 10$ psf throughout Zones A, B, C, and D.

$$w_{min} = p_s \text{ (Tributary height to roof diaphragm)} = 10 \text{ psf } (14 \text{ ft}) = 140 \text{ lb/ft}$$

$$W_{min} = w_{min}(L) = 140(110) = 15{,}400 \text{ lb}$$

$$R_A = R_B = (\tfrac{1}{2})(15{,}400) = 7700 \text{ lb}$$

∴ minimum wind pressure of 10 psf governs

Reaction forces due to wind load to diaphragm:

$$\boxed{R = 7700 \text{ lb}}$$

Seismic.   Compared with the seismic analysis for multistory buildings, the calculation of forces on a one-story structure is greatly simplified. This is because no vertical distribution of the seismic forces occurs with a single-story building. For the masonry wall building in this example, the seismic coefficients for the base shear force V, the story (shearwall) force F_x, and the story (diaphragm) force F_{px} will be the same.

The direct evaluation of the uniform force on the diaphragm requires a clear understanding of the way inertial forces are distributed. To see how the earthquake forces work their way down through the structure, it is helpful to make use of the weight of a 1-ft-wide strip of dead load W_1 taken parallel to the direction of the earthquake being considered. For example, in the case of lateral forces in the transverse direction, the weight of a 1-ft-wide strip of dead load parallel to the short side of the building is used. See the 1-ft strip in Fig. 3.10. Only the dead loads tributary to the roof level are included in W_1.

The weight of this 1-ft-wide strip includes the *roof dead load* and the weight of the *walls* that are *perpendicular* to the direction of the earthquake force being considered. Thus, for seismic forces in the transverse direction, the tributary dead load D of the longitudinal walls is included in W_1. In the example being considered, there are only two walls perpendicular to the direction of the seismic force. The weight of interior partition walls (both parallel and perpendicular to the seismic force) as well as other nonstructural items including mechanical equipment and ornamentation need to be considered in the weight of a 1-ft strip. These additional weights have not been included in this example for simplicity.

When shearwalls are wood framed, it is common to include the weight of the parallel as well as perpendicular shearwalls in the calculated roof weight. This makes the roof forces a bit conservative, but greatly streamlines the design calculations.

The forces determined in this manner satisfy the requirement that the seismic force be applied "in accordance with the mass distribution" of the level. The 1-ft strip of dead load can be viewed as the mass that causes inertial (seismic) forces to develop in the diaphragm. The weight of the transverse walls does not contribute to the seismic force in the diaphragm. The forces in the transverse shearwalls are handled in a later part of the example.

This example demonstrates the complete evaluation of the seismic coefficients for a particular structure. However, it will be recalled that Example 3.5 developed a table of example seismic coefficients applying to buildings that meet the three criteria listed for a "typical" structure. Example 3.7 confirms the seismic coefficients for the table in Example 3.5.

The distribution of forces to the primary LFRS in Example 3.7 assumes that the longitudinal walls span between the roof diaphragm and the foundation. A similar loading for the seismic force on elements and components was shown in Fig. 2.20 (Sec. 2.16).

EXAMPLE 3.7   Seismic Force Calculation

Determine the story (diaphragm) force F_{px} and the unit shear (lb/ft) in the diaphragm. See Fig. 3.10. The "u" subscripts in Fig. 3.10 are a reminder that ASCE 7 seismic forces are at a *strength level*, as was discussed in Sec. 3.3. Keep in mind that the seismic forces

UNIFORM SEISMIC
FORCE = W_U

LONGIT. WALL D

ROOF D

W_1

50'

1'

LONGIT. WALL D

PLAN

ROOF D = 10 psf
(HORIZ. PLANE – INCLUDES TRUSS)

11'

4'

14/2 = 7'

WALL D TRIB.
TO ROOF
WALL D = 60 psf

w_u

TRANSVERSE SECTION

Figure 3.10 Plan view shows a typical 1-ft-wide strip of dead load D in transverse direction. Weight of this strip W_1 generates a uniform seismic force on the diaphragm. Section view has mass of walls tributary to roof level indicated by cross-hatching. Both views show the force acting on the diaphragm.

used for design of the diaphragm (F_{px}, Sec. 2.15) are different from those used to design the shearwalls (in-plane walls F_x, Sec. 2.15) and different from those used for wall out-of-plane design (F_c, Sec. 2.16).

As in Example 3.5, the building is located in Hayward, California, with a short-period spectral acceleration S_S of 2.05g, as determined from the maximum considered earthquake ground motion maps. Site Class D is assigned, and Seismic Design Category E is assigned. The structure is a bearing wall system with masonry shearwalls. Per ASCE 7 Table 12.2-1, these walls are required to be specially reinforced masonry shearwalls in Seismic Design Category E, resulting in an R-factor of 5.

The roof dead load has been determined by prior analysis. The roof dead load D of 10 psf has been converted to the load on a horizontal plane. The masonry walls are 8-in. medium-weight concrete block units with cells grouted at 16 in. o.c. For this construction, the wall dead load D is 60 psf.

Building Period T_a and Design Spectral Response Accelerations S_{DS} and S_{D1}

To start, the building fundamental period will be estimated in accordance with the approximate method introduced in Sec. 2.13:

$$T_a = C_t h_n^x$$

$$T_a = 0.020(14)^{0.75} = 0.145 \text{ sec}$$

As in Example 3.5, for the building location, from the maximum considered earthquake ground motion maps, the mapped short-period spectral acceleration S_S is $2.05g$, and the mapped one-second spectral acceleration S_1 is $0.91g$.

For Site Class D, ASCE 7 Tables 11.4-1 and 11.4-2 assign values of 1.0 and 1.5 for site coefficients F_a and F_v, respectively. Using this information, the maximum considered spectral response accelerations S_{MS} and S_{M1}, and the design spectral response accelerations S_{DS} and S_{D1} can be calculated:

$$S_{MS} = S_S \times F_a \qquad S_{M1} = S_1 \times F_v$$

$$= 2.05g \times 1.0 \qquad = 0.91g \times 1.5$$

$$= 2.05g \qquad = 1.37g$$

$$S_{DS} = \tfrac{2}{3} \times S_{MS} \qquad S_{D1} = \tfrac{2}{3} \times S_{M1}$$

$$= \tfrac{2}{3} \times 2.05g \qquad = \tfrac{2}{3} \times 1.37g$$

$$= 1.37g \qquad = 0.91g$$

The value of S_{DS} can be confirmed against the Example 3.5 table. In addition, from the S_{DS} and S_{D1} values, T_S can be calculated, and it can be verified that the approximate period T_a falls on the level plateau of the design response spectrum.

$$T_S = \frac{S_{D1}}{S_{DS}} = \frac{0.91g}{1.37g} = 0.66 \text{ sec}$$

Because T_a is less than T_S (the period at which the response spectrum plateau ends and the spectral acceleration starts dropping), the period is confirmed to fall on the plateau. As a result, the design short-period spectral response accelerations S_{DS} will define the seismic design forces.

Base Shear Force V, Story (Shearwall) Force F_x, and Story (Diaphragm) Force F_{px}

For this one-story building in Seismic Design Category E, the base shear force, story (shearwall) force, and story (diaphragm) force will all use the same seismic force coefficient. The seismic base shear can be calculated as:

$$V = C_s W$$

and, for short-period buildings

$$C_s = \frac{S_{DS} I}{R}$$

For this building, S_{DS} has just been calculated as 1.37g. In this case, the importance factor I is taken as 1.0, and R is taken as 5 for a bearing wall system with special reinforced masonry shearwalls. This results in

$$C_s = \frac{1.37g \times 1.0}{5}$$

$$= 0.274g$$

This is the *strength level* seismic coefficient for the base shear, the story (shearwall) forces, and the story (diaphragm) forces.

Seismic Force

For a one-story building, the uniform force to the diaphragm can be obtained by multiplying the seismic coefficient by the weight of a 1-ft-wide strip of dead load (W_1) tributary to the roof level.

$$
\begin{array}{llr}
W_1 = \text{roof dead load } D = 10 \text{ psf} \times 50 \text{ ft} & = 500 \text{ lb/ft} \\
+ \text{ wall dead load } D = 60 \text{ psf} \times 11 \text{ ft} \times 2 \text{ walls} & = 1320 \\
\hline
& W_1 = 1820 \text{ lb/ft}
\end{array}
$$

$$W_u = 0.274W_1 = 0.274(1820)$$

$$W_u = 499 \text{ lb/ft} \quad (500 \text{ lb/ft will be used})$$

Redundancy/Reliability Factor ρ

The ASCE 7 equations for seismic force are:

$$\text{E} = \rho\text{Q}_\text{E} + 0.2\text{S}_{DS}\text{D}$$

and

$$\text{E} = \rho\text{Q}_\text{E} - 0.2\text{S}_{DS}\text{D}$$

where ρQ_E is the force acting horizontally, and $0.2\text{S}_{DS}\text{D}$ is a vertical force component. The redundancy/reliability factor ρ is a multiplier on the horizontal forces, including the base shear force, the story (shearwall) force, and the story (diaphragm) force.

ASCE 7 Sec. 12.3.4.2 requires that ρ be 1.3 for all structures in SDC D, E, or F, unless the structure is qualified for a ρ of 1.0 using one of two possible methods. Both methods require further evaluation of each story that resists more than 35 percent of the base shear force. ASCE 7 Sec. 12.3.4.2 Item b applies only to buildings that are regular in plan at all levels (no irregularities are triggered). It requires that there be two qualifying shearwalls in the building perimeter at all sides in each evaluated story. For wood frame shearwall buildings, the length of each shearwall is to be not less than one-half the story height. For other wall types, the length of each wall is required to be not less than the height of the story. If the required perimeter shearwalls are provided, the structure will qualify for ρ of 1.0.

Our example building has been assumed to have a regular configuration. Assuming the building is 110 ft long and 50 ft wide and has 20 ft of door or window openings in each wall, this leaves 90 ft of solid shearwall in the longitudinal walls and 30 ft in the transverse walls. The story height is approximately 14 ft, per Fig. 3.10. To qualify for ρ of 1.0 by Item b, a minimum shearwall length of 2×14 ft or 28 ft needs to be provided at each exterior wall. This 28 ft will need to include two sections not less than 14 ft each. It will be assumed that this criterion is met, and so ρ will be taken as 1.0.

Wind and Seismic Forces in Load Combinations

Wind and seismic forces must be used in ASD or LRFD load combinations with appropriate load factors. The load combinations are simplified by the fact that often only wind or seismic forces will be causing load in the plane of the shearwalls and diaphragms.

ASD load combinations with E or W:

$D + H + F + $ (W or 0.7E)

$D + H + F + 0.75$(W or 0.7E) $+ 0.75L + 0.75(L_r$ or S or R)

$0.6D + W + H$

$0.6D + 0.7E + H$

With only E or W acting in the plane of the diaphragm, the forces are simplified to 0.7E and 1.0W for ASD design of the diaphragm shear capacity.

LRFD load combinations with E or W:

$1.2D + 1.0E + L + 0.2S$

$1.2D + 1.6W + L + 0.5(L_r$ or S or R)

$0.9D + 1.6W + 1.6H$

$0.9D + 1.0E + 1.6H$

With only E or W acting in the plane of the diaphragm, the forces are simplified to 1.0E or 1.6W for LRFD design of the diaphragm shear capacity.

The story (diaphragm) F_{px} forces are compared with wind loading as follows:

ASD Design: 0.7E with $\rho = 1.0$ gives 0.7E $= 0.7(1.0)(500$ plf$) = 350$ plf

$R = 350$ plf $(110$ ft$/2) = 19,250$ lb

1.0W

$R = 7700$ lb

Therefore seismic controls.

LRFD Design: 1.0E $= 500$ plf

$R_u = 500$ plf $(110$ ft$/2) = 27,500$ lb

1.6 W

$R_u = 1.6$ (7700 lb) $= 12,300$ lb

Seismic again controls.

Note that the ratio of wind to earthquake reaction is 7700/19,250 = 0.40 for ASD forces and 12,300/27,500 = 0.45 for LRFD forces. This slightly favors use of ASD load combinations where wind controls. The reader is strongly cautioned against mixing of ASD and LRFD design methods and load combinations. One or the other should be chosen and used for the full building design.

One of the criteria used to design diaphragms and shearwalls is the unit shear. Although the design of diaphragms and shearwalls is covered in Chaps. 9 and 10, the calculation of unit shear is illustrated here. See Example 3.8 for the unit shear in the roof diaphragm.

For the building in this example subjected to lateral forces in the transverse direction, the only shearwalls are the exterior end walls. As discussed in Sec. 3.3, ASCE 7 Sec. 12.3 permits wood structural panel diaphragms used in combination with concrete or masonry shearwalls to be modeled as flexible diaphragms for analysis and seismic force distribution. The deflected shape of the roof diaphragm is again shown in Fig. 3.11. The reaction of the diaphragm on the transverse end walls is the reaction of a uniformly loaded simple beam with a span length equal to the distance between shearwalls. The shear diagram for a simple beam shows that the maximum internal shear is equal to the external reaction. The maximum total shear is converted to a unit shear by distributing it along the width of the diaphragm available for resisting the shear.

The reader is again cautioned to pay close attention to the subscripts for seismic loads. The IBC equations define *strength level* (LRFD) seismic forces, which are noted in this book with a "u" subscripts (e.g., w_u = 500 lb/ft). Seismic forces which have been reduced for use in ASD are shown in this book without the "u" subscript (e.g., w = 350 lb/ft). Other symbols with and without "u" subscripts have similar meanings in this book. For example, v_u indicates a unit shear calculated with a strength level seismic force, while v indicates an ASD unit shear.

EXAMPLE 3.8   Unit Shear in Roof Diaphragm

The simple beam strength (LRFD) level loading diagram in Fig. 3.11 is a repeat of the loading on the diaphragm. The simple beam reactions R_u are shown along with the shear diagram.

The free-body diagram at the bottom of Fig. 3.11 is cut through the diaphragm a small distance away from the transverse shearwalls. The unit shear (lb/ft) is obtained by dividing the maximum total shear from the shear diagram by the width of the diaphragm b.

Diaphragm reaction:

$$R_u = \frac{w_u(L)}{2} = \frac{500(110)}{2} = 27,500 \text{ lb} = 27.50 \text{ k}$$

For a simple beam, the shear equals the reaction:

$$V_u = R_u = 27.50 \text{ k}$$

Figure 3.11 Diaphragm unit shear. For simplicity, the calculations for unit shear are shown using the nominal length L and diaphragm width b (i.e., wall thickness is ignored).

The unit shear distributes the total shear over the width of the diaphragm. The conventional symbol for *total shear* is V_u, and the *unit shear* in the diaphragm is assigned the symbol v_u.

$$v_{u,\text{roof}} = \frac{V_u}{b} = \frac{27{,}500}{50} = \boxed{550 \text{ lb/ft}}$$

Seismic Forces in Load Combinations

As was discussed in Example 3.7, seismic forces must be used in ASD or LRFD load combinations with appropriate load factors. The same load combinations used for Example 3.7 are applicable.

ASD Load Combinations with E:

D + H + F +(W or 0.7E)

D + H + F + 0.75(W or 0.7E) + 0.75L + 0.75(L$_r$ or S or R)

0.6D + 0.7E + H

LRFD Load Combinations with E:

1.2D + 1.0E + L + 0.2S

0.9D + 1.0E + 1.6H

And again, with only E acting in the plane of the diaphragm, the forces are simplified to 0.7E for ASD design and 1.0E for LRFD design of the diaphragm shear capacity.

The redundancy factor ρ is taken as 1.0, not only because ρ was determined to be 1.0 in Example 3.7, but also because it can always be taken as 1.0 for diaphragm F_{px} forces. The LRFD unit shear in the diaphragm is $1.0E = 1.0(\rho)v_u = 550$ plf. The ASD unit shear in the diaphragm is $0.7E = 0.7(\rho)v_u = 0.7(550) = 385$ plf.

The next step in the lateral force design process is to consider similar quantities in the shearwalls. In determining the uniform force to the diaphragm, it will be recalled that seismic forces due to only the dead load D of the roof and the longitudinal walls was included in the seismic force. The inertial force generated in the transverse walls was not included in the load to the roof diaphragm. The reason is that the shearwalls carry directly their own seismic force parallel to the wall. These forces do not, therefore, contribute to the force or shear in the diaphragm.

Several approaches are used by designers to compute wall seismic forces. The ASCE 7 requirement is that the wall be designed at the most critical shear location. In the more common method, the unit shear in the shearwall is evaluated at the *midheight* of the wall. See Example 3.9. This convention developed because the length of the shearwall b is often a minimum at this location. If the wall openings were different than shown, some location other than midheight might result in critical seismic forces. However, any openings in the wall (both doors and windows) are typically intersected by a horizontal line drawn at the midheight of the wall. In addition, using the midheight is consistent with the lumped-mass seismic model presented in Chap. 2.

The seismic force generated by the top half of the wall is given the symbol R_1. It can be computed as the dead load D of the top portion of the wall times the seismic coefficient. The total shear at the midheight of the wall is the sum of all forces above this level. For a one-story building, these forces include the reaction from the roof diaphragm R plus the wall seismic force R_1. The unit shear v may be computed once the total shear has been obtained.

The reader should note that, consistent with earlier examples, Example 3.9 computes the seismic forces to the wall at a strength level. At the end of the calculations, the strength level force is multiplied by ρ and used in ASD and LRFD load combinations.

EXAMPLE 3.9   Unit Shear in Shearwall

Determine the total shear and unit shear at the midheight of the shearwall in Fig. 3.12. For simplicity, ignore the reduction in wall dead load due to the opening (conservative).

END ELEVATION

Figure 3.12   Shearwall unit shear. Maximum unit shear occurs at midheight of wall.

Roof Diaphragm Seismic Force

Compute the F_x story (shearwall) coefficient:

$$F_x \text{ story coefficient} = \frac{S_{DS}I}{R}$$

$$= \frac{1.37(1.0)}{5}$$

$$= 0.274g$$

Alternatively, this F_x coefficient could have been obtained from the seismic table in Example 3.5.

From Example 3.7, the weight of a 1-ft wide strip of dead load (W_1) tributary to the roof level is 1820 lb/ft. The strength level uniform seismic load to the diaphragm and resulting diaphragm reaction are:

$$w_u = 0.274W_1 = 0.274(1820) = 500 \text{ lb/ft}$$

$$R_u = w_u L/2 = 500(110 \text{ ft})/2 = 27{,}500 \text{ lb}$$

Wall Seismic Force

Seismic force generated by the top half of the wall (see Fig. 3.12):

$$\text{Wall area} = (11 \times 50) + \tfrac{1}{2}(3 \times 50) = 625 \text{ ft}^2$$

$$\text{Wall D} = 625 \text{ ft}^2 \times 60 \text{ psf} = 37{,}500 \text{ lb} \qquad \text{(neglect window reduction)}$$

$$R_{u1} = 0.274\text{W}$$

$$= 0.274(37{,}500)$$

$$= 10{,}300 \text{ lb}$$

Wall Shear

$$\left\{ \begin{array}{l} \text{Total shear at} \\ \text{midheight of wall} \end{array} \right\} = \left\{ \begin{array}{l} \text{sum of all forces on free-body diagram} \\ \text{of shearwall above midheight} \end{array} \right\}$$

$$V_u = R_u + R_{u1} = 27{,}500 + 10{,}300 = 37{,}800 \text{ lb}$$

$$\text{Unit wall shear} = v_u = \frac{V_u}{b} = \frac{37{,}800}{15 + 15}$$

$$\boxed{{}^\dagger v_{u \text{ wall}} = 1260 \text{ lb/ft}}$$

Seismic Forces in Load Combinations

As was discussed in Examples 3.7 and 3.8, seismic forces must be used in ASD or LRFD load combinations with appropriate load factors. The same load combinations used for Examples 3.7 and 3.8 are applicable.

ASD load combinations with E:

D + H + F +(W or 0.7E)

D + H + F + 0 .75(W or 0.7E) + 0.75L + 0.75(L$_r$ or S or R)

0.6D + 0.7E + H

LRFD load combinations with E:

1.2D + 1.0E + L + 0.2S

0.9D + 1.0E + 1.6H

---

†For wall openings not symmetrically located, see Sec. 9.6.

And again, with only E acting in the plane of the shearwall, the forces are simplified to 0.7E for ASD design and 1.0E for LRFD design of the shearwall shear capacity.

The redundancy factor ρ is again taken as 1.0, per Example 3.7. The LRFD unit shear in the shearwall is $1.0E = 1.0\ (\rho)\ v_u = 1260$ plf. The ASD unit shear in the shearwall is $0.7E = 0.7\ (\rho)\ v_u = 0.7(1260) = 882$ plf.

As mentioned earlier, the unit shear in the roof diaphragm and in the shearwall constitute one of the main parameters in the design of these elements. There are additional design factors that must be considered, and these are covered in subsequent chapters.

These examples have dealt only with the transverse lateral forces, and a similar analysis is used for the longitudinal direction. Roof diaphragm shears are usually critical in the transverse direction, but both directions should be analyzed. Shearwalls may be critical in either the transverse or longitudinal directions depending on the sizes of the wall openings.

3.6   Design Problem: Lateral Forces on Two-Story Building

A multistory building has a more involved analysis of seismic forces than a one-story structure. Once the seismic base shear force V has been determined, the forces are distributed to the story levels in accordance with the ASCE 7 formulas for F_x and F_{px}. These seismic forces were reviewed in Secs. 2.14 and 2.15. There it was noted that all three of the seismic forces on the primary LFRS (V, F_x, and F_{px}) could be viewed as a seismic coefficient times the appropriate mass or dead load of the structure:

$$V = \text{Base shear coefficient} \times W$$

$$F_x = F_x \text{ story coefficient} \times w_x$$

$$F_{px} = F_{px} \text{ story coefficient} \times w_x$$

The purpose of the two-story building problem in Examples 3.10 and 3.11 is to compare the maximum wind and seismic forces and to evaluate the unit shears in the diaphragms and shearwalls. The base shear force V, the F_x story (shearwall) forces, and the unit shears in shearwalls are covered in Example 3.10. The F_{px} story (diaphragm) forces and unit shears in diaphragms will be covered in Example 3.11. The wind pressures are slightly different from those in the previous one-story building of Example 3.5, due to the different wall height. Although the final objective of the earthquake analysis is to obtain numerical values of the design forces, it is important to see the overall process. To do this, the calculations emphasize the determination of the various seismic coefficients (g forces). Once the seismic coefficients have been determined, it is a simple matter to obtain the numerical values.

The one-story building example in Sec. 3.5 was divided into a number of separate problems. The two-story structure in Examples 3.10 and 3.11 is organized into two sets of design calculations which are more representative of what might be done in practice. However, sufficient explanation is provided to describe the process required for multistory structures.

EXAMPLE 3.10   Two-Story Lateral Force Calculation, Base Shear and Shearwalls

Determine the lateral wind and seismic forces in the transverse direction for the two-story office building in Fig. 3.13a. For the critical loading, determine the unit shear in the transverse shearwalls at the midheight of the first- and second-story walls. Assume that there are no openings in the masonry walls.

Figure 3.13a   Wind pressures and tributary heights to roof and second-floor diaphragms.

Wind forces are to be based on the simplified wind load provisions of ASCE 7 Chap. 6. The basic wind speed is 85 mph. Standard occupancy and Exposure B apply. $K_{zt} = 1.0$.

The building is located in Pullman, Washington ($S_S = 0.286g$, $S_1 = 0.089g$). Site Class D is assumed and $I = 1.0$. The following dead loads have been determined in a prior analysis: roof dead load D = 20 psf, floor dead load D = 12 psf, additional floor dead load D to account for the weight of interior wall partitions = 10 psf, and exterior wall dead load D = 60 psf.

NOTE: In buildings where the location of nonbearing walls and partitions is subject to change, the ASCE 7 requires a partition live load L of 15 psf for designing individual floor members for vertical loads. However, for evaluation of seismic design forces, an average partition weight of not less than 10 psf is required (ASCE 7 Sec. 12.7.2 definition of W).

Wind Forces

Building Geometry

$$h_{\text{mean}} = 19 \text{ ft}$$

$$0.4h_{\text{mean}} = 0.4(19 \text{ ft}) = 7.6 \text{ ft}$$

$$\text{Least horizontal building dimension} = b = 32 \text{ ft}$$

$$0.1b = 0.1(32 \text{ ft}) = 3.2 \text{ ft}$$

End zone dimension:

$$a = \text{lesser of } 0.4h_{\text{mean}} \text{ or } 0.1b$$

$$= 3.2 \text{ ft}$$

Length of end zone = $2a$ = 6.4 ft

Wind Pressures

Wind pressure formula:

$$p_s = \lambda K_{zt} I \, p_{s30}$$

$$I = 1.0 \quad \text{(ASCE 7 Table 6-1)}$$

$$\lambda = 1.0 \quad \text{for 0 to 30 ft (ASCE 7 Fig. 6-2)}$$

$$K_{zt} = 1.0$$

Basic wind pressures p_{s30} from ASCE 7 Fig. 6-2 for a flat roof:

$$\text{Zone A} \quad p_{s30} = p_s = 11.5 \text{ psf}$$

$$\text{Zone C} \quad p_{s30} = p_s = 7.6 \text{ psf}$$

These wind pressures are shown in the section view in Fig. 3.13a.

Load to Diaphragms

Roof:

$$\left\{ \begin{array}{l} \text{Tributary height to} \\ \text{roof diaphragm} \end{array} \right\} = \left\{ \begin{array}{l} \text{Tributary wall height} \\ \text{below roof} \end{array} \right\} + \left\{ \begin{array}{l} \text{Height of} \\ \text{parapet wall} \end{array} \right\}$$

$$= \tfrac{1}{2}(9 \text{ ft}) + 2 \text{ ft} = 6.5 \text{ ft}$$

$$w_{\text{end}} = 11.5 \text{ psf}(6.5 \text{ ft}) = 69.0 \text{ lb/ft}$$

$$w_{\text{int}} = 7.6 \text{ psf}(6.5 \text{ ft}) = 49.4 \text{ lb/ft}$$

$$W_r = w_{\text{end}}(2a) + w_{\text{int}}(L - 2a)$$

$$= 69.0(6.4) + 49.4(60 - 6.4)$$

$$= 3089 \text{ lb}$$

Use static equilibrium equations to determine the larger reaction force (R_{rA} or R_{rB}) at either end of the roof diaphragm.

Summing moments about the reaction at B:

$$R_{rA} = w_{\text{end}}(2a)(L - a)\left(\frac{1}{L}\right) + w_{\text{int}}(L - 2a)\left(\frac{L}{2} - a\right)\left(\frac{1}{L}\right)$$

$$= (69.0)(6.4)(60 - 3.2)\left(\frac{1}{60}\right) + (49.4)(60 - 6.4)\left(\frac{60}{2} - 3.2\right)\left(\frac{1}{60}\right)$$

$$= 1601 \text{ lb}$$

Summing forces:

$$R_{rB} = W_r - R_{rA} = 3089 - 1601 = 1488 \text{ lb}$$

Check minimum load to roof diaphragm based on $p_s = 10$ psf throughout Zones A and C.

$$w_{\text{min}} = p_s \text{ (Tributary height to roof diaphragm)} = 10 \text{ psf}(6.5 \text{ ft}) = 65 \text{ lb/ft}$$

$$W_{\text{min}} = w_{\text{min}}(L) = 65(60) = 3900 \text{ lb}$$

$$R_{rA} = R_{rB} = (\tfrac{1}{2})(3900) = 1950 \text{ lb}$$

$$\therefore \text{ minimum wind pressure of 10 psf governs}$$

Reaction forces due to wind load to roof diaphragm:

$$\boxed{R_r = 1950 \text{ lb}}$$

Second floor:

$$\left\{\begin{matrix}\text{Tributary height to} \\ \text{2nd floor diaphragm}\end{matrix}\right\} = \left\{\begin{matrix}\text{Tributary wall height} \\ \text{above 2nd floor}\end{matrix}\right\} + \left\{\begin{matrix}\text{Tributary wall height} \\ \text{below 2nd floor}\end{matrix}\right\}$$

$$= \tfrac{1}{2}(9 \text{ ft}) + \tfrac{1}{2}(10 \text{ ft}) = 9.5 \text{ ft}$$

$$w_{\text{end}} = 11.5 \text{ psf}(9.5 \text{ ft}) = 109.25 \text{ lb/ft}$$

$$w_{\text{int}} = 7.6 \text{ psf}(9.5 \text{ ft}) = 72.2 \text{ lb/ft}$$

$$W_2 = w_{\text{end}}(2a) + w_{\text{int}}(L - 2a)$$

$$= 109.25(6.4) + 72.2(60 - 6.4)$$

$$= 4569 \text{ lb}$$

Summing moments about the reaction at B:

$$R_{2A} = w_{end}(2a)(L - a)\left(\frac{1}{L}\right) + w_{int}(L - 2a)\left(\frac{L}{2} - a\right)\left(\frac{1}{L}\right)$$

$$= (109.25)(6.4)(60 - 3.2)\left(\frac{1}{60}\right) + (72.2)(60 - 6.4)\left(\frac{60}{2} - 3.2\right)\left(\frac{1}{60}\right)$$

$$= 2390 \text{ lb}$$

Summing forces:

$$R_{2B} = W_2 - R_{2A} = 4569 - 2390 = 2179 \text{ lb}$$

Check minimum load to second-floor diaphragm based on $p_s = 10$ psf throughout Zones A and C.

$$w_{min} = p_s \text{ (Tributary height to roof diaphragm)} = 10 \text{ psf(9.5 ft)} = 95 \text{ lb/ft}$$

$$W_{min} = w_{min}(L) = 95(60) = 5700 \text{ lb}$$

$$R_{2A} = R_{2B} = (\tfrac{1}{2})(5700) = 2850 \text{ lb}$$

$$\therefore \text{ minimum wind pressure of 10 psf governs}$$

Reaction forces due to wind load to second-floor diaphragm:

$$\boxed{R_2 = 2850 \text{ lb}}$$

Seismic Forces

Building Period T_a and Design Spectral Response Accelerations S_{DS} and S_{D1}

To start, the building fundamental period will be estimated in accordance with the approximate method introduced in Sec. 2.13:

$$T_a = C_t h_n^x$$

$$T_a = 0.020(19)^{0.75} = 0.182 \text{ sec}$$

For the building location in Pullman, Washington, from the maximum considered earthquake ground motion maps, the mapped short-period spectral acceleration S_S is $0.286g$, and the mapped one-second spectral acceleration S_1 is $0.089g$.

For Site Class D, ASCE 7 Tables 11.4-1 and 11.4-2 assign values of 1.57 and 2.4 for site coefficients F_a and F_v, respectively. The F_a value of 1.57 is found by interpolating between the tabulated values of 1.6 and 1.4. Using the information, the maximum considered spectral response accelerations S_{MS} and S_{M1}, and the design spectral response accelerations S_{DS} and S_{D1} can be calculated:

$$S_{MS} = S_S \times F_a \qquad\qquad S_{M1} = S_1 \times F_v$$

$$= 0.286g \times 1.57 \qquad\qquad = 0.089g \times 2.4$$

$$= 0.449g \qquad\qquad\qquad = 0.214g$$

$$S_{DS} = \tfrac{2}{3} \times S_{MS} \qquad\qquad S_{D1} = \tfrac{2}{3} \times S_{M1}$$

$$= \tfrac{2}{3} \times 0.449g \qquad\qquad = \tfrac{2}{3} \times 0.214g$$

$$= 0.299g \qquad\qquad\qquad = 0.143g$$

From the S_{DS} and S_{D1} values, T_S can be calculated, and it can be verified that the approximate period T_a falls on the level plateau of the design response spectrum.

$$T_S = \frac{S_{D1}}{S_{DS}} = \frac{0.143g}{0.299g} = 0.48\,\text{sec}$$

Because T_a is less than T_S (the period at which the response spectrum plateau ends and spectral acceleration starts dropping), the period is confirmed to fall on the plateau. As a result, the design short-period spectral response acceleration S_{DS} will define the seismic design forces.

Redundancy Factor ρ

For this example, there will be a number of element forces that will need to be adjusted using the ρ factor. Assume that the building is 60 ft long and 32 ft wide, has no door or window openings in the transverse walls, and has 10 ft of opening in each longitudinal wall. As was done in Example 3.7, ASCE 7 Sec. 12.3.4.2 Item b will be checked first. For each story resisting more than 35 percent of the base shear, there must be two qualifying shearwalls in the building perimeter at all sides in each evaluated story. For wall types other than wood frame, the length of each wall is required to be not less than the height of the story. If the required perimeter shearwalls are provided, the structure will qualify for ρ of 1.0.

Our example building has been assumed to have a regular configuration. The first story height is 10 ft, and the second story height is 9 ft. In the first story, each exterior wall must have two shearwall segments not less than 10 ft long and in the second story 9 ft. This can be easily accommodated with solid transverse walls, and longitudinal walls having 50 ft without openings, so ρ will be taken as 1.0. Note that it could be difficult to meet this criterion if openings were introduced in the transverse walls.

Seismic Base Shear Coefficient

The seismic base shear force V will be used to calculate the story (shearwall) forces F_x, which are used to design the vertical elements of the LFRS. The seismic base shear can be calculated as:

$$V = C_s W$$

where

$$C_s = \frac{S_{D1}I}{RT} \le \frac{S_{DS}I}{R}$$

For this building, S_{DS} has just been calculated as $0.299g$, and S_{D1} as $0.143g$. A previous calculation identified the period T_S up to which the plateau defined by S_{DS} controls. Because the approximate period for this building T_a is less than T_S, calculation of C_S can be simplified to:

$$C_s = \frac{S_{DS}I}{R}$$

In this case I is taken as 1.0, and R is taken as 5 for a bearing wall system with special reinforced masonry shearwalls. This results in:

$$C_s = \frac{0.299g \times 1.0}{5}$$

$$C_s = 0.060g$$

This is the *strength length* seismic coefficient for the base shear. This example again demonstrates the complete calculation of the seismic coefficient. With this low calculated value of C_s, it is a good idea to check against minimum permitted C_s values. ASCE 7 Sec. 12.8.1.1 has two minimum requirements. First, C_s cannot be less than 0.01g. The second applies where S_1 is $\geq 0.6g$, and is not applicable to the example building since S_1 for the example site is 0.089g. Therefore, the calculated C_s value of 0.06g may be used.

Tributary Roof Dead Loads

The total dead load for the structure is all that is required to complete computation of the base shear force V. However, in the process of developing the total dead load, it is beneficial to summarize the weight tributary to the roof diaphragm and second-floor diaphragm using the idea of a 1-ft-wide strip. Recall from Example 3.7 that W_1 represents the mass or weight that will cause a uniform seismic force to be developed in a horizontal diaphragm.

The values of W_1, tributary to the roof and second floor, will eventually be use to determine the distributed story forces. It is recommended that the reader sketch the 1-ft-wide strip on the plan view in Fig. 3.13a. The tributary wall heights are shown on the section view.

Weight of 1-ft-wide strip tributary to roof:

Roof dead load D	$= (20 \text{ psf})(32 \text{ ft})$	$= 640 \text{ lb/ft}$
+ Wall dead load D (2 longitudinal walls)	$= 2(60 \text{ psf})\left(\frac{9}{2} + 2\right)$	$= 780$
Dead load D of 1-ft strip at roof	$= W_1$	$= 1420 \text{ lb/ft}$

The mass that generates the entire seismic force in the roof diaphragm is given the symbol W'_r. It is the sum of all the W_1 values at the roof level.

$$W'_r = \Sigma W_1 = 1420 \text{ lb/ft}(60 \text{ ft}) = 85.2 \text{ k}$$

To obtain the total mass tributary to the roof level, the weight of the top half of the transverse shearwalls is added to W'_r. The total dead load D tributary to the roof level is given the symbol W_r.

$$\text{Dead load D of 2 end walls} = 2(60 \text{ psf})(32)\left(\frac{9}{2} + 2\right) = 25.0 \text{ k}$$

$$\text{Total dead load D tributary to roof} = W'_r = 85.2 + 25.0 = 110.2 \text{ k}$$

Similar quantities are now computed for the second floor.

Tributary Second-Floor Dead Loads

Weight of 1-ft-wide strip tributary to second floor:

Second-floor dead load D	$= (12 \text{ psf})(32 \text{ ft})$	$= 384 \text{ lb/ft}$
+ Partition load	$= (10 \text{ psf}) (32 \text{ ft})$	$= 320$
+ Wall dead load D (2 longitudinal walls)	$= 2(60 \text{ psf})\left(\dfrac{9}{2} + \dfrac{10}{2}\right)$	$= 1140$
Deal load D of 1-ft strip at second floor	$= W_1$	$= 1844 \text{ lb/ft}$

The mass that generates the entire seismic force in the second floor diaphragm is W_2'.

$$W_2' = \Sigma W_1 = 1844 \text{ lb/ft}(60 \text{ ft}) = 110.6 \text{ k}$$

The total mass tributary to the second-floor level is the sum of W_2' and the tributary weight of the transverse shearwalls. The total dead load D tributary to the second-floor level is given the symbol W_2.

$$\text{Dead load D of 2 end walls} = 2(60 \text{ psf})(32)\left(\frac{9}{2} + \frac{10}{2}\right) = 36.5 \text{ k}$$

$$\text{Total dead load D tributary to second floor} = W_2 = 110.6 + 36.5 = 147.1 \text{ k}$$

Seismic Tables

Calculations of seismic forces for multistory buildings are conveniently carried out in tables. Tables are not only convenient for bookkeeping, but also provide a comparison of the F_x and F_{px} story coefficients.

Tables and the necessary formulas can easily be stored in equation solving computer software. Once stored on a computer, tables serve as a template for future problems. In this way, the computer can be used to handle repetitive calculations and problem formatting (e.g., setting up the table), and the designer can concentrate on the best way to solve the problem at hand. Tables can be expanded to take into account taller buildings and to include items such as overturning moments.

The balance of Example 3.10 will look at the F_x story (shearwall) forces and the resulting unit shears in the shearwalls. The F_{px} story (diaphragm) forces and resulting unit shears in the diaphragms are calculated in Example 3.11.

The F_x table below is shown completely filled out. However, at this point in the solution of the problem, only the first four columns can be completed. Columns 1, 2, and 3 simply list the story levels, heights, and masses (dead loads). The values in column 4 are the products of the respective values in columns 2 and 3. The sum of the story masses at the bottom of column 3, Σw_x, is the weight of the structure W to be used in the calculation of base shear.

The steps necessary to complete the remaining columns in the F_x table are given in the two sections immediately following the table.

F_x Story (Shearwall) Force Table—$R = 5$

1	2	3	4	5	6	7
Story	Height h_x	Weight w_x	$w_x h_x$	Story force $F_x = .0043 h_x w_x$	F_x Coef.	Story shear V_x
R	19	110.2	2094	$F_r = 9.00$ k	0.0817	9.00 k
2	10	147.1	1471	$F_2 = 6.33$ k	0.0430	15.33 k
1	0					
Sum		257.3 k	3565 k-ft	$V = .060W = 15.4$ k		

Base Shear

The *strength level* seismic base shear coefficient for the F_x forces was determined previously to be 0.060g. The *strength level* base shear for F_{px} forces will be calculated in Example 3.11. The total base shear for the building is 0.060 times the total weight from column 3.

$$V = 0.060(257.3) = 15.4 \text{ k}$$

The story coefficients for distributing the seismic force over the height of the structure can now be determined. The distribution of forces to the vertical elements in the primary LFRS is given by the ASCE 7 formula for F_x.

F_x Story (Shearwall) Coefficients

In Chap. 2 it was noted that the formula for F_x can be written as an F_x story coefficient times the mass tributary to level x, w_x:

$$F_x = C_{vx}V = \left[\frac{w_x h_x^k}{\sum\limits_{i=1}^{n} w_i h_i^k} \right] V$$

$$F_x = (F_x \text{ story coefficient})w_x = \left[\frac{V h_x^k}{\sum\limits_{i=1}^{n} w_i h_i^k} \right] w_x$$

The exponent k is related to the building period, and can be taken as 1.0 for buildings with an estimated period of less than 0.5 sec. The period has been estimated to be 0.182 sec for this building, so the story force coefficient can be simplified to:

$$F_x = (F_x \text{ story coefficient})w_x = \left[\frac{V h_x}{\sum\limits_{i=1}^{n} w_i h_i} \right] w_x$$

The *strength level* F_x story coefficients will now be evaluated. The base shear force V is known. The summation term in the denominator is obtained as the last item in column 4 of the seismic table.

$$F_x = \left[\frac{V h_x}{\sum\limits_{i=1}^{n} w_i h_i} \right] w_x = \left[\frac{(15.4)h_x}{3565} \right] w_x$$

This general formula for F_x is entered at the top of column 5 as

$$F_x = (0.0043h_x)w_x$$

Individual F_x story coefficients follow this entry. At the roof level F_x is given the symbol F_r, and at the second-floor level the symbol is F_2.

Roof:

$$F_r = (0.0043h_r)w_r = (0.0043)(19)w_r = 0.0817w_r$$

The numerical value for the *strength level* seismic force at the roof level is added to column 5 next to the F_x story coefficient:

$$F_r = 0.0817w_r = 0.0817(110.2 \text{ k}) = 9.00 \text{ k}$$

Second floor:

$$F_2 = (0.0043h_2)w_2 = (0.0043)(10)w_2 = 0.043w_2$$

The numerical value for the seismic force at the second-floor level is also added to column 5.

$$F_2 = 0.043w_2 = 0.043(147.1 \text{ k}) = 6.33 \text{ k}$$

The summation at the bottom of column 5 serves as a check on the numerical values. The *sum* of all the F_x story forces must equal the total base shear.

$$V = \Sigma F_x = F_r + F_2 = 9.00 + 6.33 = 15.33 \text{ k} \qquad OK$$

The values in column 7 of the seismic table represent the total *strength level* story shears between the various levels in the structure. The story shear can be obtained as the sum of all the F_x story forces above a given section.

In a simple structure of this nature, the story shears from column 7 may be used directly in the design of the vertical elements (i.e., the shearwalls). However, as the structure becomes more complicated, a more progressive distribution of seismic forces from the diaphragms to the vertical elements may be necessary. Both approaches are illustrated in this example.

With all of the F_x story coefficients determined, the individual distributed forces for designing the shearwalls can be evaluated.

Uniform Forces to Diaphragms Using F_x Story Coefficients

For shearwall design, the forces to the diaphragms are based on the F_x story coefficients. See Fig. 3.13b. These uniformly distributed forces will be used to compute the forces in the shearwalls following the progressive distribution in Method 1 (described later in this example).

Load to roof diaphragm:

The load to the roof diaphragm that is used for design of the shearwalls needs to be based on the F_x story forces from the F_x Story (Shearwall) Force table. The strength level roof diaphragm reaction can be calculated as follows:

$$w_{ur} = 0.0817(1420) = 116 \text{ lb/ft}$$

$$R_{ur} = \frac{w_{ur}L}{2} = \frac{116(60)}{2} = 3.48 \text{ k}$$

Figure 3.13b   Seismic forces to roof diaphragm w_{ur} and second-floor diaphragm w_{u2} are for designing the vertical elements in the LFRS. Concentrated forces on shear-walls are diaphragm reactions R_{ur} and R_{u2}.

Load from second-floor diaphragm:

The uniform force on the second-floor diaphragm is also determined using the F_x story coefficient from column 6 of the F_x Story Force table.

$$w_{u2} = 0.0430W_1 = 0.0430(1844)$$

$$= 79 \text{ lb/ft}$$

The reaction of the second-floor diaphragm on the shearwall is

$$R_{u2} = \frac{w_{u2}L}{2} = \frac{79(60)}{2} = 2.37 \text{ k}$$

The diaphragm reactions due to wind and seismic force F_x can be summarized as:

The wind load diaphragm reactions are

$$R_r = 1950 \text{ lb}$$

$$R_2 = 2850 \text{ lb}$$

The seismic F_{px} diaphragm reactions are

$$R_{ur} = 3480 \text{ lb}$$

$$R_{u2} = 2370 \text{ lb}$$

Here it is important to note that these forces cannot yet be used to compare wind and seismic forces for design of the shearwalls. There is no additional loading to be considered for wind design. The second-story shearwall wind load would match the diaphragm wind reaction at the roof. The first-story shearwall wind load would be the sum of the roof and second-floor diaphragm wind reactions. The seismic design force for the second-story shearwall, however, must include both the seismic reaction at the roof diaphragm and the seismic force due to the weight of the second-story shearwall (the wall parallel to the direction of seismic load). The seismic design force for the first-story shearwall must include the seismic reactions of the roof and second-floor diaphragms plus seismic force due to the weight of the wall at the first and second stories.

Shear at Midheight of Second-Story Walls (Using F_x Story Shearwall Forces)

Two methods for evaluating the shear in the shearwalls are illustrated. The first method demonstrates the progressive distribution of the forces from the horizontal diaphragms to the shearwalls. Understanding Method 1 is essential to the proper use of the F_x story forces for more complicated shearwall arrangements. Method 2 can be applied to simple structures where the distribution of seismic forces to the shearwalls can readily be seen.

METHOD 1

For the shear between the second floor and the roof, the free-body diagram (FBD) of the wall includes two seismic forces. See Fig. 3.13c. One force is the reaction from the roof diaphragm (from Fig. 3.13b), and the other is the inertial force developed by the mass of the top half of the shearwall.

Figure 3.13c FBD of shearwall cut midway between second-floor and roof levels.

Force from top half of shearwall:

The seismic force generated by the top half of the second-story shearwall is given the symbol R_{u1}. This force is obtained by multiplying the dead load of the wall by the F_x story coefficient for the *roof* level.

$$R_{u1} = 0.0817w_u = 0.0817\left[(60 \text{ psf})\left(\frac{9}{2} + 2\right)(32 \text{ ft})\right]$$

$$= 1.02 \text{ k}$$

The shear in the wall between the second floor and the roof is given the symbol V_{u2r}, and it is obtained by summing forces in the x direction.

$$\Sigma F_x = 0$$

$$V_{u2r} = R_{ur} + R_{u1} = 3.48 + 1.02 = 4.50 \text{ k}$$

METHOD 2

For this simple rectangular building with two equal-length transverse shearwalls, the shear in one wall V_{u2r} can be obtained as one-half of the total story shear from column 7 of the F_x Story (Shearwall) Force table.

$$\text{Wall shear } V_{u2r} = \tfrac{1}{2}(\text{story shear } V_{u2r})$$

$$= \tfrac{1}{2}(9.00) = 4.50 \text{ k} \qquad (\text{same as Method 1})$$

For other shearwall arrangements, including interior shearwalls, the progressive distribution of forces using Method 1 is required.

Comparison of Wind and Seismic Forces in Load Combinations

In order to determine whether wind or seismic forces govern shearwall design, wind and seismic forces will be used in ASD or LRFD load combinations with appropriate load factors. The load combinations are simplified by the fact that often only wind or seismic forces will be causing loading in the plane of the shearwalls and diaphragms.
    ASD load combinations with E or W:

D + H + F + (W or 0.7E)
D + H + F + 0.75(W or 0.7E) + 0.75L + 0.75(L$_r$ or S or R)
0.6D + W + H
0.6D + 0.7E + H

With only E or W acting in the plane of the shearwall, the forces are simplified to 0.7E or 1.0W for ASD design of the shearwall shear capacity.
    LRFD load combinations with E or W:

1.2D + 1.0E + L + 0.2S
1.2D + 1.6W + L + 0.5(L$_r$ or S or R)
0.9D + 1.6W + 1.6H
0.9D + 1.0E + 1.6H

With only E or W acting in the plane of the shearwall, the forces are simplified to 1.0E or 1.6W for LRFD design of the shearwalls.

The shearwall wind load is $W = R_r = 1950 \text{ lb}$

The shearwall seismic force is $Q_E = V_{u2r} = 4.5 \text{ k} = 4500 \text{ lb}$

The redundancy factor ρ was determined to be 1.0. Using applicable load factors, the story F_x forces are compared with wind loading as follows:

ASD design: Second story $1.0W = 1.0(1950)$ $= 1950$ lb

$$0.7E = 0.7(\rho)\, Q_E$$

$$= 0.7(1.0)(4500) = 3150 \text{ lb} - \text{seismic controls}$$

LRFD design: Second story $1.6W = 1.6(1950)$ $= 3120$ lb

$$1.0E = 1.0(\rho)\, Q_E$$

$$= 1.0(1.0)(4500) = 4500 \text{ lb} - \text{seismic controls}$$

Dividing by the 32 ft length of the transverse shearwalls, the controlling design forces can be converted to unit shears in lb/ft as follows:

ASD $v_{2r} = (3150 \text{ lb})/32 \text{ ft} = 98 \text{ lb/ft}$

LRFD $v_{u2r} = (4500 \text{ lb})/32 \text{ ft} = 140 \text{ lb/ft}$

Shear at Midheight of First-Story Walls [Using F_x Story (Shearwall) Forces]

METHOD 1

The shear in the walls between the first and second floors is obtained from the FBD in Fig. 3.13*d*. The two forces on the top are the forces from Fig. 3.13*c*. The load R_{u2} is the reaction from the second-floor diaphragm (from Fig. 3.13*b*). The final seismic force is the second force labeled R_{u1}. This represents the inertial force generated by the mass of the shearwall tributary to the second floor.

Figure 3.13*d*   FBD of shearwall cut midway between first-floor and second-floor levels.

Force from wall mass tributary to second-floor level:

The R_{u1} force for the middle portion of the shearwall uses the F_x story coefficient for the second-floor level:

$$R_{u1} = 0.0430w_u = 0.0430\left[(60 \text{ psf})\left(\frac{9}{2} + \frac{10}{2}\right)(32 \text{ ft})\right]$$

$$= 0.78 \text{ k}$$

The shear between the first- and second-floor levels is given the symbol V_{12}. It is obtained by summing forces in the x direction (Fig. 3.13d):

$$\Sigma F_x = 0$$
$$V_{u12} = R_{ur} + R_{u1} + R_{u2} + R_{u1}$$
$$= 3.48 + 1.02 + 2.37 + 0.78$$
$$= 7.65 \text{ k}$$

METHOD 2

Again for a simple rectangular building with two exterior equal-length shearwalls, the total shear in a wall can be determined as one-half of the story shear from column 7 of the F_x Story (Shearwall) Force Table.

$$\text{Wall shear } V_{u12} = \tfrac{1}{2} \text{ (story shear } V_{u12})$$
$$= \tfrac{1}{2}(15.33) = 7.67 \text{ k} \qquad \text{(same as Method 1)}$$

Comparison of Wind and Seismic Forces in Load Combinations

In order to determine whether wind or seismic forces govern shearwall design, wind and seismic forces will be used in ASD or LRFD load combinations with appropriate load factors. The load combinations are simplified by the fact that often only wind or seismic forces will be causing loading in the plane of the shearwalls and diaphragms.

With only E or W acting in the plane of the shearwall, the forces are simplified to 0.7E or 1.0W for ASD design of the shearwall shear capacity.

With only E or W acting in the plane of the shearwall, the forces are simplified to 1.0E or 1.6W for LRFD design of the shearwalls.

The shearwall wind load is $W = R_r + R_2 = 1950 + 2850 \text{ lb} = 4800 \text{ lb}$

The shearwall seismic force is $Q_E = V_{u12} = 7.65 \text{ k} = 7650 \text{ lb}$

The redundancy factor ρ was determined to be 1.0. Using applicable load factors, the story F_x forces are compared with wind loading as follows:

ASD design: First story $1.0W = 1.0(4800) \qquad = 4800 \text{ lb}$

$0.7E = 0.7(\rho) Q_E$

$= 0.7(1.0)(7650) = 5360 \text{ lb} - \text{seismic controls}$

LRFD design: First story $1.6W = 1.6(4800) = 7680 \text{ lb}$

$1.0E = 1.0(\rho) Q_E$

$= 1.0(1.0)(7650) = 7650 \text{ lb} - \text{seismic and wind}$
are approximately equal

Here it can be seen that the LRFD load combinations create somewhat larger wind demands than the ASD. While this might lead the reader toward use of ASD for wind design, the reader is strongly discouraged from mixing and matching ASD and LRFD design methods. One or the other should be chosen and used consistently for the entire structure design.

Dividing by the 32 ft length of the transverse shearwalls, the controlling design forces can be converted to unit shears in lb/ft as follows:

ASD $v_{2r} = (5360 \text{ lb})/32 \text{ ft} = 168 \text{ lb/ft}$

LRFD $v_{u2r} = (7680 \text{ lb})/32 \text{ ft} = 240 \text{ lb/ft}$

The above analysis is for the lateral forces in the transverse direction. A similar analysis is required in the longitudinal direction.

Example 3.11 continues design calculations for the two-story building from Example 3.10. In Example 3.10 the applied seismic and wind forces used for design of *shearwalls* were calculated and compared. The shearwall unit shears were then calculated based on the more critical seismic forces, determined using F_x story (shearwall) coefficients.

Example 3.11 shifts the focus from *shearwall* design forces to *diaphragm* design forces. This is done by calculating the F_{px} story (diaphragm) forces and diaphragm unit shears due to seismic forces and comparing them to the diaphragm unit shears from wind forces.

EXAMPLE 3.11   Two-Story Lateral Force Calculation, Diaphragm Forces

Determine the wind and seismic diaphragm forces for the two-story office building from Example 3.10. Evaluate the unit shears in the roof and second-floor diaphragms. The wind and seismic criteria remain unchanged from Example 3.10.

F_{px} Story (Diaphragm) Coefficient

The F_{px} coefficients will be used to calculate the forces for design of the diaphragms. The seismic story (diaphragm) forces F_{px} were introduced in Sec. 2.15. The F_{px} forces can be calculated as:

$$F_{px} = \frac{\sum\limits_{i=x}^{n} F_i}{\sum\limits_{i=x}^{n} w_i} w_{px}$$

and

$$0.2 S_{DS} I w_{px} \leq F_{px} \leq 0.4 S_{DS} I w_{px}$$

This calculation makes use of information in the story (shearwall) force coefficient table, so it makes sense to repeat this information and add the F_{px} forces.

F_{px} Story (Diaphragm) Table

1	2	3	4	5	6	7	8
Story	Height h_x	Weight w_x	Story (shearwall) force F_x	$\sum\limits_{i=x}^{n} F_i$	$\sum\limits_{i=x}^{n} w_i$	Story (diaphragm) force F_{px}	Story (diaphragm) force F_{px} coefficient
R	19	110.2	9.00	9.00	110.20	9.00	0.0817
2	10	147.1	6.33	15.33	257.30	8.76	0.0598
1	0						
Sum		257.3 k	15.33 k				

The story (diaphragm) force in column 6 is calculated as:

$$F_{px} = \frac{\sum\limits_{i=x}^{n} F_i}{\sum\limits_{i=x}^{n} w_i} w_{px}$$

At the roof diaphragm, the sum of $F_x = 9.00$, and the sum of $w_x = 110.2 = w_{px}$, giving:

$$F_{pr} = \frac{9.00}{110.2}110.2 = 9.00 \text{ k}$$

Note that the sums of F_x and w_x are calculated from the topmost level down to the level being considered. The upper and lower limits for F_{pr} also need to be checked:

$$0.2S_{DS}Iw_{px} \leq F_{px} \leq 0.4S_{DS}Iw_{px}$$

$$0.2(0.299)1.00(110.2) \leq F_{pr} \leq 0.4(0.299)1.00(110.2)$$

$$6.58 \text{ k} \leq F_{pr} \leq 13.18 \text{ k}$$

F_{pr} of 9.00 k falls between these limits, so 9.00 is the controlling value. In column 7, this value is converted to a coefficient by dividing by $w_r = 110.2$ k.

At the second-floor diaphragm, summing from the top level down, the sum of $F_x = 9.00 + 6.33 = 15.33$, and the sum of $w_x = 110.2 + 147.1 = 257.3$, giving:

$$F_{p2} = \frac{15.33}{257.3}147.1 = 8.76 \text{ k}$$

The upper and lower limits for F_{p2} need to be checked:

$$0.2S_{DS}Iw_{px} \leq F_{px} \leq 0.4S_{DS}Iw_{px}$$

$$0.2(0.299)1.00(147.1) \leq F_{pr} \leq 0.4(0.299)1.00(147.1)$$

$$8.79 \text{ k} \leq F_{p2} \leq 17.59 \text{ k}$$

F_{p2} of 8.76 k falls just below the lower limit of 8.79 k, so 8.79 k is the controlling value. In column 7, this value is converted to a coefficient by dividing by $w_2 = 147.1$ k.

Again, it is appropriate to make some observations regarding the base shear coefficient, the story (shearwall) force coefficients F_x, and the story (diaphragm) force coefficients F_{px}.

1. The coefficients F_x and F_{px} are the same at the roof level.
2. The maximum story coefficients (at the roof level) exceed the magnitude of the base shear coefficient.
3. The minimum value for the F_x story (shearwall) coefficient (at the second-floor level) is less than the base shear coefficient.
4. The minimum value for the F_{px} story (diaphragm) coefficient (at the second-floor level) is equal to the magnitude of the seismic base shear coefficient.

These rules are not limited to two-story structures, and they hold true for multistory buildings in general.

With all of the *strength level* F_{px} story coefficients determined, the individual distributed forces for designing the diaphragms can be evaluated. The forces are considered in the following order: roof diaphragm (using F_{px}) and second-floor diaphragm (using F_{px}).

Shear in Roof Diaphragm Using F_{px} Forces

Compare the F_{px} seismic force at the roof level with the wind force to determine which is critical. The uniformly distributed *strength level* seismic force is determined by multiplying the F_{px} story coefficient at the roof level by the weight of a 1-ft-wide strip of roof dead load D. The weight W_1 at the roof level was determined in Example 3.10.

$$w_{upr} = 0.0817W_1 = 0.0817(1420) = 116 \text{ lb/ft}$$

The roof diaphragm is treated as a simple beam spanning between transverse end shearwalls. See Fig. 3.14a. For a simple span, the shear is equal to the beam reaction. The unit shear in the roof diaphragm is the total shear in the diaphragm divided by the width of the diaphragm.

$$V_{ur} = R_{ur} = \frac{w_{upr}L}{2} = \frac{116(60)}{2} = 3480 \text{ lb}$$

$$v_{ur} = \frac{V_{ur}}{b} = \frac{3480}{32} = \boxed{109 \text{ lb/ft}}$$

This unit shear may be used with the information in Chap. 9 to design the roof diaphragm.

Shear in Second-Floor Diaphragm Using F_{px} Forces

The second-floor diaphragm is analyzed in a similar manner. See Fig. 3.14b. The seismic force is again obtained by multiplying the F_{px} story coefficient from column 7 by the dead load D of a 1-ft-wide strip. The weight W_1 comes from Example 3.10.

$$w_{up2} = 0.0596W_1 = 0.0596(1844)$$

$$= 110 \text{ lb/ft}$$

Figure 3.14a   Roof diaphragm strength level design force w_{upr} and the corresponding unit shear in the roof diaphragm v_{ur}.

Figure 3.14b   Second-floor strength level diaphragm design force w_{up2} and the corresponding unit shear in the second-floor diaphragm v_{u2}.

$$V_{u2} = R_{u2} = \frac{w_{up2}L}{2} = \frac{110(60)}{2} = 3300 \text{ lb}$$

$$v_{u2} = \frac{V_{u2}}{b} = \frac{3300}{32} = \boxed{103 \text{ lb/ft}}$$

This unit shear may be used with the information in Chap. 9 to design the second-floor diaphragm.

As was done with the F_x story forces in Example 3.10, we can now compare diaphragm loads due to wind with seismic F_{px} story (diaphragm) forces. The diaphragm reactions due to wind and seismic forces can be summarized as:

Wind load diaphragm reactions	$R_r = 1950$ lb
	$R_2 = 2850$ lb
Seismic F_{px} diaphragm reactions	$R_{ur} = 3480$ lb
	$R_{u2} = 3300$ lb

Unlike the comparison of shearwall forces, where diaphragm and shearwall contributions had to be summed, for diaphragm design it is only the forces directly acting on the diaphragm that need to be considered.

Comparison of Wind and Seismic Forces in Load Combinations

In order to determine whether wind or seismic forces govern diaphragm design, wind and seismic reaction forces will be used in ASD or LRFD load combinations with appropriate load factors. We will again start by looking at a full set of applicable load combinations. Again, the load combinations are simplified by the fact that generally, only wind or seismic forces will be causing loading in the plane of the shearwalls and diaphragms.

ASD load combinations with E or W:

D + H + F + (W or 0.7E)
D + H + F + 0.75(W or 0.7E) + 0.75L + 0.75(L$_r$ or S or R)
0.6D + W + H
0.6D + 0.7E + H

With only E or W acting in the plane of the diaphragm, the forces are simplified to 0.7E or 1.0W for ASD design of the diaphragm shear capacity.

LRFD load combinations with E or W:

1.2D + 1.0E + L + 0.2S
1.2D + 1.6W + L + 0.5(L$_r$ or S or R)
0.9D + 1.6W + 1.6H
0.9D + 1.0E + 1.6H

With only E or W acting in the plane of the diaphragm, the forces are simplified to 1.0E or 1.6W for LRFD design of the diaphragms.

The redundancy factor ρ is always taken as 1.0 when determining F_{px} forces. Using applicable load factors, the story F_{px} forces are compared with wind loading as follows:

ASD design: Roof $1.0W = 1.0(1950)$ $= 1950$ lb

$0.7E = 0.7(\rho)\, Q_E$

$= 0.7(1.0)(3480) = 2440$ lb – seismic controls

Second story $1.0W = 1.0(2850)$ $= 2850$ lb – wind controls

$0.7E = 0.7(\rho)\, Q_E$

$= 0.7(1.0)(3300) = 2310$ lb

LRFD design: Roof $1.6W = 1.6(1950)$ $= 3120$ lb

$1.0E = 1.0(\rho)\, Q_E$

$= 1.0(1.0)(3480) = 3480$ lb – seismic controls

Second story $1.6W = 1.6(2850)$ $= 4560$ lb – wind controls

$1.0E = 1.0(\rho)\, Q_E$

$= 1.0(1.0)(3300) = 3300$ lb

Dividing by the 32 ft length of the transverse shearwalls, the controlling design forces can be converted to unit shears in lb/ft as follows:

ASD $v_r = (2440 \text{ lb})/32 \text{ ft} = 76$ lb/ft

$v_2 = (2850 \text{ lb})/32 \text{ ft} = 89$ lb/ft

LRFD $v_{ur} = (3480 \text{ lb})/32 \text{ ft} = 109$ lb/ft

$v_{u2} = (4560 \text{ lb})/32 \text{ ft} = 143$ lb/ft

3.7   References

[3.1] American Society of Civil Engineers (ASCE). 2006. *Minimum Design Loads for Buildings and Other Structures (ASCE 7-05)*, ASCE, Reston, VA.
[3.2] International Codes Council (ICC). 2006. *International Building Code*, 2006 ed., ICC, Falls Church, VA.
[3.3] Rood, Roy. 1991. "Panelized Roof Structures," *Wood Design Focus*, Vol. 2, No. 3, Forest Products Society, Madison, WI.

3.8   Problems

All problems are to be answered in accordance with ASCE 7-05. Assume $K_{zt} = 1.0$ for wind loads.

3.1   The purpose of this problem is to compare the design values of shear and moment for a girder with different assumed load configurations (see Fig. 3.3 in Example 3.2).

Given: The roof framing plan in Fig. 3.A with girders G1, G2, and G3 supporting loads from purlin P1. Roof dead load D = 13 psf. Roof live load L_r is
to be obtained from IBC Sec. 1607.11.

Find: *a.* Draw the shear and moment diagrams for girder G1(D + L_r), assuming
1. A series of concentrated reaction loads from the purlin P1.
2. A uniformly distributed load over the entire span (unit load times
the tributary width).
b. Rework part *a* for girder G2.
c. Rework part *a* for girder G3.

3.2   This problem is the same as Prob. 3.1 except that the roof dead load D = 23 psf.

ROOF FRAMING PLAN

LONGIT. SECT. IS SIMILAR

TRANSVERSE SECTION

Figure 3.A

3.3   *Given:*   IBC Chap. 16 lateral force requirements

     Find: The definition of
       a. Building frame system
       b. LFRS
       c. Shearwall
       d. Braced frame
       e. Bearing wall system

3.4   *Given:*   The plan and section of the building in Fig. 3.B. The basic wind speed is 100 mph, and Exposure B applies. The building is enclosed and has a standard occupancy classification. Roof dead load D = 15 psf on a horizontal plane. Wind forces to the primary LFRS are to be in accordance with ASCE 7.

Figure 3.B

     Find: *a.* The wind force on the roof diaphragm in the transverse direction. Draw the loading diagram.
          b. The wind force distribution on the roof diaphragm in the longitudinal direction. Draw the loading diagram.

c. The total diaphragm shear and the unit diaphragm shear at line 1
d. The total diaphragm shear and the unit diaphragm shear at line 4

3.5 *Given:* The plan and section of the building in Fig. 3.B. Roof dead load D = 15 psf on a horizontal plane, and wall dead load D = 12 psf. The seismic diaphragm force F_{px} coefficient has been calculated as 0.200.

Find: *a.* Uniform seismic force on the roof diaphragm in the transverse direction. Draw the loading diagram.
b. The seismic force distribution on the roof diaphragm in the longitudinal direction. Draw the loading diagram noting the lower force at the overhang.
c. The total diaphragm shear and the unit diaphragm shear at line 1
d. The total diaphragm shear and the unit diaphragm shear at line 4

3.6 Repeat Prob. 3.4 except that the wind forces are for Exposure C.

3.7 *Given:* The plan and section of the building in Fig. 3.B. Roof dead load D = 10 psf on a horizontal plane, and wall dead load D = 8 psf. The seismic base shear coefficient and seismic diaphragm force coefficient have been calculated as 0.200.

Find: *a.* Uniform seismic force on the roof diaphragm in the transverse direction. Draw the loading diagram.
b. The seismic force distribution on the roof diaphragm in the longitudinal direction. Draw the loading diagram, noting the lower force at the overhang.
c. The total diaphragm shear and the unit diaphragm shear at line 1
d. The total diaphragm shear and the unit diaphragm shear at line 4

3.8 *Given:* The plan and section of the building in Fig. 3.A. The basic wind speed is 85 mph, and Exposure C applies. The building is an enclosed structure with a standard occupancy classification. Roof dead load D = 13 psf. Wind forces to the primary LFRS are to be in accordance with ASCE 7.

Find: *a.* The tributary wind force to the roof diaphragm. Draw the loading diagram
b. The total diaphragm shear and the unit diaphragm shear at the 60-ft transverse end walls
c. The total diaphragm shear and the unit diaphragm shear at the 96-ft longitudinal side walls

3.9 Repeat Prob. 3.8 except that the wind forces are for 120 mph.

3.10 *Given:* The plan and section of the building in Fig. 3.A. The basic wind speed is 85 mph, and Exposure C applies. The building is an enclosed structure with an essential occupancy classification. Roof dead load D = 13 psf.

Find: *a.* The wind pressure (psf) for designing components and cladding in the roof system away from discontinuities

   b. The tributary wind force to a typical purlin using the load from part
      a. Draw the loading diagram
   c. The wind pressure (psf) for designing an element in the roof system
      near an eave

3.11 *Given:* The plan and section of the building in Fig. 3.A. Roof dead load D = 10 psf,
      and the walls are 7½-in.-thick concrete. The building has a bearing wall
      system, braced with special reinforced concrete shearwalls. The building
      location is Charleston, South Carolina, where the mapped short-period
      spectral acceleration S_S is 1.50g, and the mapped one-second spectral
      acceleration S_1 is 0.42g. Site Class D, Occupancy Category II should be
      assumed.

   Find: For the transverse direction:
   a. The seismic base shear coefficient and the seismic diaphragm force
      coefficient.
   b. The uniform force to the roof diaphragm in lb/ft. Draw the loading
      diagram.
   c. The total diaphragm shear and the unit diaphragm shear adjusted to
      an ASD level adjacent to the transverse walls.
   d. The total shear and the unit shear at the midheight of the transverse
      shearwalls using ASD and LRFD load combinations.

3.12 Repeat Prob. 3.11 except that the longitudinal direction is to be considered.

3.13 *Given:* The plan and section of the building in Fig. 3.A. Roof dead load D = 12 psf,
      and the walls are 6-in.-thick concrete. The building has a bearing wall
      system, braced with special reinforced concrete shearwalls. The build-
      ing location is Memphis, Tennessee, where the mapped short-period
      spectral acceleration S_S is 1.26g, and the mapped one-second spectral
      acceleration S_1 is 0.38g. Site Class D, Occupancy Category IV should be
      assumed.

   Find: *a.* The seismic base shear and diaphragm force coefficients
   b. The uniform force to the roof diaphragm in lb/ft. Draw the loading
      diagram
   c. The total diaphragm shear and the unit diaphragm shear adjacent
      to the transverse walls using ASD and LRFD load combinations
   d. The total shear and the unit shear at the midheight of the transverse
      shearwalls using ASD and LRFD load combinations

3.14 Repeat Prob. 3.13 except that the longitudinal direction is to be considered.

3.15 *Given:* The elevation of the end shearwall of a building as shown in Fig. 3.C.
      The force from the roof diaphragm to the shearwall is 10 k. The wall dead
      load D = 20 psf, and the seismic coefficient is 0.200.

   Find: The total shear and the unit shear at the midheight of the wall adjusted
      to an ASD level

ELEVATION

Figure 3.C

3.16 *Given:* The elevation of the side shearwall of a building as shown in Fig. 3.D. The force from the roof diaphragm to the shearwall is 50 k. The wall dead load D = 65 psf, and the seismic base shear coefficient is 0.244.

   Find: The total shear and the unit shear at the midheight of the wall using ASD and LRFD load combinations.

Figure 3.D

3.17 *Given:* The elevation of the side shearwall of a building as shown in Fig. 3.D. The force from the roof diaphragm to the shearwall is 43 k. The wall dead load D = 75 psf, and the seismic base shear coefficient is 0.244.

   Find: The total shear and the unit shear at the midheight of the wall using ASD and LRFD load combinations.

3.18 *Given:* The plan and section of the building in Fig. 3.E. Roof dead load D = 15 psf, floor dead load D = 20 psf (includes an allowance for interior walls), exterior wall dead load D = 53 psf. Basic wind speed = 100 mph. Exposure C and ASCE 7 are specified. The building has a bearing wall system, braced with special reinforced masonry shearwalls. The building is located in Sacramento, California with a mapped short-period spectral acceleration of 0.56*g* and a mapped one-second spectral acceleration of 0.22*g*. Site Class B, Occupancy Category IV should be assumed.

Find: For the transverse direction using ASD load combinations:
   a. The unit shear in the roof diaphragm
   b. The unit shear in the floor diaphragm
   c. The unit shear in the second-floor shearwall
   d. The unit shear in the first-floor shearwall

3.19 Repeat Prob. 3.18 except that LRFD load combinations are to be used.

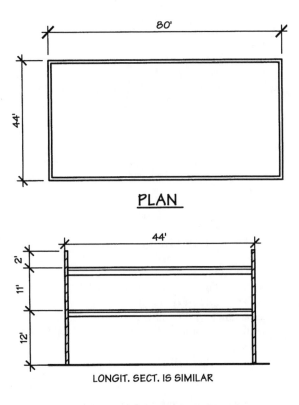

PLAN

LONGIT. SECT. IS SIMILAR

TRANSVERSE SECTION

Figure 3.E

3.20 *Given:* The plan and section of the building in Fig. 3.E. Roof dead load D = 10 psf, floor dead load D = 18 psf plus 10 psf for interior partitions, exterior wall dead load D = 16 psf. Basic wind speed = 85 mph. Exposure C and ASCE 7 are specified. Enclosed bearing wall structure has standard occupancy classification. The building has a bearing wall system, braced with wood structural panel shearwalls. The building is located in Sacramento, California, with a mapped short-period spectral acceleration of 0.56g and a mapped one-second spectral acceleration of 0.22g. Site Class B, Occupancy Category IV, should be assumed. Neglect any wall openings.

Find: For the transverse direction using ASD load combinations:
   a. The unit shear in the roof diaphragm
   b. The unit shear in the floor diaphragm
   c. The unit shear in the second-floor shearwall
   d. The unit shear in the first-floor shearwall

3.21 Repeat Prob. 3.20 except that the longitudinal direction is to be considered.

3.22 Use a microcomputer spreadsheet to set up the solution of seismic forces for the primary lateral-force-resisting system for a multistory building up to four stories.

The LFRS to be considered consists of horizontal diaphragms and shearwalls. The structural systems may be limited to bearing wall systems with wood-frame roof, floor, and wall construction or wood-frame roof and floor construction and masonry or concrete walls. Thus, an R of special reinforced 5 or 6.5 (ASCE 7 Table 9.5.2.2 applies).

The spreadsheet is to handle structures without "complications." For example, buildings will be limited to structures that are seismically *regular*. In addition, only the exterior walls will be used for shearwalls, and openings in the horizontal diaphragms and shearwalls may be ignored in this assignment. Wind forces are not part of this problem.

The following is to be used for input:

   Short-period spectral acceleration S_S
   One-second spectral acceleration S_1
   Site Class
   Seismic importance factor I
   Type of bearing wall system (used to establish R)
   Plan dimensions of rectangular building
   Story heights and parapet wall height (if any)
   Roof, floor, and wall dead loads; interior wall dead loads may be handled by increasing the floor dead loads.

The spreadsheet is to do the following:
   a. Evaluate the seismic base shear coefficient and numerical value of base shear.
   b. Evaluate the seismic diaphragm force coefficient, if different from the base shear coefficient.
   c. Generate the *seismic tables* summarizing the F_x and F_{px} story coefficients.
   d. Compute the w_{ux} and w_{upx} uniformly distributed seismic forces to the horizontal diaphragms.
   e. Determine the design unit shears in the horizontal diaphragms using ASD and LRFD load combinations.
   f. Determine the total shear and unit shear using ASD and LRFD load combinations in the exterior shearwalls between each story level.

Properties of Wood and Lumber Grades

4.1 Introduction

The designer should have a basic understanding of the characteristics of wood, especially as they relate to the functioning of structural members. The terms *sawn lumber* and *solid sawn lumber* are often used to refer to wood members that have been manufactured by cutting a member directly from a log. Other structural members may start as lumber and then undergo additional fabrication processes. For example, small pieces of lumber can be graded into laminating stock then glued and laid up to form larger wood members, known as *glued laminated timbers*, or *glulams.*

Many other wood-based products are available for use in structural applications. Some examples include solid members such as wood poles and timber piles; fabricated components such as trusses, wood I-joists, and box beams; and other manufactured products such as wood structural panels (e.g., plywood and oriented strand board) and structural composite lumber (e.g., laminated veneer lumber and parallel strand lumber). A number of these products are recent developments in the wood industry. They are the result of new technology and the economic need to make use of different species and smaller trees that cannot be used to produce solid sawn lumber.

This chapter introduces many of the important physical and mechanical properties of wood. In addition, the sizes and grades of sawn lumber are covered. A number of other wood products are addressed later in this book. For example, glulam is covered in Chap. 5, and the properties and grades of plywood and other wood structural panels are reviewed in Chap. 8. Additionally, structural composite lumber and several types of manufactured components are described, in part, in Chap. 6.

4.2   Design Specification

The *2005 National Design Specification for Wood Construction* (NDS)(Ref. 4.3) is the basic specification in the United States for the design of wood structures. All or part of the NDS is usually incorporated into the IBC.

Traditionally, the NDS has been updated on a 3- to 5-year cycle. Although there have been significant changes from time to time, the usual revisions often involved minor changes and the clarification of certain design principles. The reader should have a copy of the 2005 NDS to follow the discussion in this book. Having a copy of the NDS in order to learn timber design is analogous to having a copy of the steel manual in order to learn structural steel design. One can read about the subject, but it is difficult to truly develop a feel for the material without both an appropriate text *and* the basic industry design specification. In the case of wood design, the NDS is this basic industry document.

The NDS is actually the formal design section of what is a series of interrelated design documents known as the *2005 Wood Design Package.* There are two primary companion documents to the NDS that support and complete the dual-format *National Design Specification for Wood Construction.* The first companion document is the *NDS Supplement: Design Values for Wood Construction,* which is often referred to simply as the *Supplement* or the *NDS Supplement.* The NDS Supplement provides basic design values for wood construction, including structural lumber and glued laminated timber. The second companion design document to the NDS is the *NDS Supplement: Special Design Provisions for Wind and Seismic (SDPWS),* also called the *Wind and Seismic Supplement.* The Wind and Seismic Supplement is the newest supplement and is maintained as a separate document due to the unique requirements related to wind- and seismic-resistant design. Included in the SDPWS are design values for shearwalls and diaphragms, which comprise the primary lateral-force-resisting system (LFRS) in most wood structures.

In addition to the NDS and its supplements, two other associated documents are part of the *2005 Wood Design Package.* The *NDS Commentary* (Ref. 4.2) provides additional guidance and other supporting information for the design provisions included in the NDS. Prior to the 2005 edition, the Commentary was more of a historic account of the NDS than a true design commentary with supporting documentation.

The final component of the *2005 Wood Design Package* is the *ASD/LRFD Manual for Engineered Wood Construction* (Ref. 4.1). The ASD/LRFD Manual contains supporting information for both LRFD and ASD, and is organized to parallel the NDS.

NDS

The *NDS* covers the basic principles of wood engineering that apply to all products and species groups. The design section is written and published by the American Forest & Paper Association (AF&PA) with input from the wood industry, government agencies, universities, and the structural engineering profession.

Many of the chapters in *this* book deal with the provisions in the design section of the NDS, including procedures for beams, columns, members with combined stress, and connections. The study of each of these topics should be accompanied by a review of the corresponding section in the NDS. To facilitate this, a number of sections and tables in the 2005 NDS are referenced throughout the discussion in this book.

The designer should be aware of changes and additions for any new editions of a design specification or code. The 2005 NDS contains chapters specific to sawn lumber, glued-laminated timber, poles and piles, wood I-joists, structural composite lumber, wood structural panels, mechanical connections, dowel-types fasteners, split ring and shear plate connectors, timber rivets, shearwalls and diaphragms, special loading conditions, and fire design. While some of the basic provisions have been changed and some new provisions added, the major advancement for the 2005 NDS is the integration of new LRFD provisions with the traditional ASD provisions. As a result of the addition of LRFD provisions, the presentation of many of the provisions has changed in order to provide clear presentation of both design philosophies in a single specification.

NDS Supplement

The second part of the NDS is traditionally referred to simply as the *NDS Supplement,* even though there are now two formal supplements to the NDS. The NDS Supplement contains the numerical values of design stresses for the various species groupings of structural lumber and glued-laminated timber. Although the Supplement is also published by AF&PA, the mechanical properties for sawn lumber are obtained from the agencies that write the grading rules for structural lumber. See Fig. 4.1. There are currently seven rules-writing agencies for visually graded lumber that are certified by the American Lumber Standards Committee (ALSC). The design values in the NDS Supplement are reviewed and approved by the ALSC.

The NDS Supplement provides *reference design values* for the following mechanical properties:

Bending stress F_b

Tension stress parallel to grain F_t

Shear stress F_v

Compression stress parallel to grain F_c

Compression stress perpendicular to grain $F_{c\perp}$

Modulus of elasticity E and $E_{\min}$ (some publications use the notation (MOE)

An important part of timber design is being able to locate the proper design value in the tables. The reader is encouraged to verify the numerical values in the examples given throughout this book. As part of this process, it is suggested that tabulated values be checked against the NDS Supplement. Some of the design value

1. Northeastern Lumber Manufacturers Association (NELMA)
2. Northern Softwood Lumber Bureau (NSLB)
3. Redwood Inspection Service (RIS)
4. Southern Pine Inspection Bureau (SPIB)
5. West Coast Lumber Inspection Bureau (WCLIB)
6. Western Wood Products Association (WWPA)
7. National Lumber Grades Authority (NLGA), a Canadian agency

Figure 4.1   List of rules-writing agencies for visually graded structural lumber. The addresses of the American Lumber Standards Committee (ALSC) and the seven rules-writing agencies are listed in the NDS Supplement.

adjustments are found in the NDS Supplement, and others are in the NDS design section. An active review of the numerical examples will require use of both.

The material in the 2005 NDS represents the latest in wood design principles for both ASD and LRFD. Likewise, the design values in the 2005 NDS Supplement are the most recent structural properties. It should be understood that these two sections of the NDS are an integrated package. In other words, both parts of the 2005 NDS should be used together, and the user should not mix the design section from one edition with the supplement from another edition.

Wind and Seismic Supplement

The *NDS Supplement: Special Design Provisions for Wind and Seismic (SDPWS)* is also called the *Wind and Seismic Supplement*. The *Wind and Seismic Supplement* is maintained as a separate part of the *Wood Design Package* due to the unique requirements related to wind- and seismic-resistant design. The SDPWS includes numerical design values for shearwalls and diaphragms considering both wind and seismic loading. Shearwalls and diaphragms comprise the primary LFRS in most structures. The SDPWS will be referenced extensively in Chaps. 9 and 10, where the design of diaphragms and shearwalls is presented.

NDS Commentary

First introduced in 1993 for the 1991 NDS, the Commentary has traditionally provided a historic account of the development of NDS provisions. The 2005 *NDS Commentary* is now similar to the commentaries of other design specifications and codes, providing additional guidance and other supporting information for the design provisions included in the NDS. Some provisions that are not yet adopted in the NDS are also included in the Commentary.

Design Manual

The *2005 ASD/LRFD Manual for Engineered Wood Construction* is significantly rewritten from that of the previous *2001 ASD Manual*. The 2001 ASD Manual included several additional supplements and guidelines focused entirely on allowable stres design. The 2005 ASD/LRFD Manual contains supporting information for both LRFD and ASD, including nonmandatory design information such as capacity tables and fire resistive assembly details. The Manual is organized, chapter by chapter, to parallel the NDS.

The reader is cautioned that the NDS and supporting documents represent *recommended* design practice by the wood industry, and it does not have *legal outhority* unless it becomes part of a local building code. The code change process can be lengthy, and some codes may not accept all of the industry recommendations. Consequently, it is recommended that the designer verify local code acceptance before using the 2005 NDS.

4.3 Methods of Grading Structural Lumber

The majority of sawn lumber is graded by visual inspection, and material graded in this way is known as *visually graded* structural lumber. As the lumber comes out of the mill, a person familiar with the lumber grading rules examines each piece and assigns a grade by stamping the member. The *grade stamp* includes the grade, the species or species group, and other pertinent information. See. Fig. 4.2*a*. If the lumber grade has recognized mechanical properties for use in structural design, it is often referred to as a *stress grade*.

The lumber grading rules establish limits on the size and number of growth (strength-reducing) characteristics that are permitted in the various stress grades. A number of the growth characteristics found in full-size pieces of lumber and their effect on strength are discussed later in this chapter.

The term *resawn lumber* is applied to smaller pieces of wood that are cut from a larger member. The resawing of previously graded structural lumber invalidates the initial grade stamp. The reason for this is that the acceptable size of a defect (e.g., a knot) in the original large member may not be permitted in the same grade for the smaller resawn size. The primary example is the changing of a centerline knot into an edge knot by resawing. Restrictions on edge knots are more severe than those on centerline knots. Thus, if used in a structural application, resawn lumber must be *regraded*.

(a)

Figure 4.2*a* Typical grade stamps for visually graded structural lumber. Elements in the grade stamp include (*a*) lumber grading agency (e.g., WWPA), (*b*) mill number (e.g., 12), (*c*) lumber grade (e.g., Select Structural and No.3), (*d*) commercial lumber species (e.g., Douglas Fir-Larch and Western Woods), (*e*) moisture content at time of surfacing (e.g., S-GRN, and S-DRY), (*Courtesy of WWPA*)

```
┌─────────────────────────────┐
│  MACHINE RATED              │
│  ⓌⓌⓅ®  12   ┌─────┐        │
│          S-DRY │HEM  │       │
│                │FIR  │       │
│                └─────┘        │
│  1650 Fb 1.5E               │
└─────────────────────────────┘
```

(b)

Figure 4.2b Typical grade stamp for machine stress-rated (MSR) lumber. Elements in the grade stamp include (a) MSR marking (e.g., MACHINE RATED), (b) lumber grading agency (e.g., WWPA), (c) mill number (e.g., 12), (d) reference bending design value (e.g., 1650 psi) and modulus of elasticity (e.g., 1.5×10^6 psi), (e) commercial lumber species (e.g., Hem-Fir), (f) moisture content at time of surfacing (e.g., S-DRY). (*Courtesy of WWPA*)

The designer should be aware that more than one set of grading rules can be used to grade some commercial species groups. For example, Douglas Fir-Larch can be graded under Western Wood Products Association (WWPA) rules or under West Coast Lumber Inspection Bureau (WCLIB) rules. There are some differences in reference design values between the two sets of rules. The tables of design properties in the NDS Supplement have the grading rules clearly identified (e.g., WWPA and/or WCLIB). The differences in reference design values occur only in large-size members known as *Timbers,* and reference design values are the same under both sets of grading rules for *Dimension lumber.* The sizes of Timbers and Dimension lumber are covered later in this chapter.

Because the designer usually does not have control over which set of grading rules will be used, the lower reference design value should be used in design when conflicting values are listed in NDS tables. The higher reference design value is justified only if a grade stamp associated with the higher design value actually appears on a member. This situation could arise in reviewing the capacity of an existing member.

Although most lumber is visually graded, a small percentage of lumber is *machine stress rated* by subjecting each piece of wood to a nondestructive test. The nondestructive test is typically highly automated, and the process takes little time. As lumber comes out of the mill, it passes through a series of rollers. In this process, a bending load is applied about the minor axis of the cross section, and the modulus of elasticity of each piece is measured. In addition to the nondestructive test, machine stress rated lumber is subjected to a visual check. Because of the testing procedure, machine stress rating (MSR) is limited to thin material (2 in. or less in thickness). Lumber graded in this manner is typically known as *MSR lumber.*

Each piece of MSR lumber is stamped with a grade stamp that allows it to be fully identified, and the grade stamp for MSR lumber differs from the stamp for visually graded lumber. The grade stamp for MSR lumber includes a numerical value of reference bending design value F_b and modulus of elasticity E. A typical grade stamp for MSR lumber is shown in Fig. 4.2b. Tabulated design values for MSR lumber are given in NDS Supplement Table 4C.

MSR lumber has less variability in some mechanical properties than visually graded lumber. Consequently, MSR lumber is often used to fabricate engineered wood products. For example, MSR lumber is used for laminating stock for some glulam beams. Another application is in the production of wood components such as light frame trusses and wood I-joists.

A more recent development in the sorting of lumber by nondestructive measurement of its properties is known as *machine evaluated lumber* (MEL).This process typically employs radiographic (x-ray) inspection to measure density, and allows a greater mix of F_b/E combinations than is permitted in MSR lumber. As with MSR lumber, MEL is subjected to a supplementary visual check. Design values for machine evaluated lumber are also listed in NDS Supplement Table 4C.

4.4 In-Grade Versus Clear Wood Design Values

Design values tabulated in the 2005 NDS are based in part on *clear-wood test procedures* and in part on full-size lumber *in-grade test methods*. There are two broad size classifications of sawn lumber:

1. Dimension lumber
2. Timbers

Dimension lumber is the smaller (thinner) sizes of structural lumber. Dimension lumber usually ranges in size from 2 × 2 through 4 × 16. In other words, dimension lumber constitutes any material that has nominal *thickness* of 2 to 4 in. Note that in lumber grading terminology *thickness* refers to the smaller cross-sectional dimension of a piece of wood and *width* refers to the larger dimension. The availability of lumber in the wider widths varies with species, and not all sizes are available in all species.

Timbers are the larger sizes and have a 5-in. minimum nominal dimension. Thus, practically speaking, the smallest size timber is a 6 × 6, and any member larger than a 6 × 6 is classified as a timber. There are additional size categories within both Dimension lumber and Timbers, and the further subdivision of sizes is covered later in this chapter.

In the 2005 NDS Supplement, the design properties for visually graded sawn lumber are based on two different sets of American Society for Testing and Materials (ASTM) standards:

1. *In-grade* procedures (ASTM D 1990), applied to Dimension lumber (Ref. 4.13)
2. *Clear wood* procedures (ASTM D 2555 and D 245), applied to Timbers (Ref. 4.11 and 4.12)

The method of establishing reference design values for visually graded sawn lumber for *Timbers* is based on the *clear wood strength* of the various species and species combinations. The clear wood strength is determined by testing small, clear, straight-grained specimens of a given species. For example, the clear wood bending strength test is conducted on a specimen that measures 2 × 2 × 30 in. The testing methods to be used on small, clear wood specimens are given in ASTM D 143 (Ref. 4.14). The unit strength (stress) of a small, clear, straight-grained piece of wood is much greater than the unit strength of a full-size member.

After the clear wood strength properties for the species have been determined, the effects of the natural growth characteristics that are permitted in the different grades of full-size members are taken into account. This is accomplished by multiplying the clear wood values by a reduction factor known as a *strength ratio*. In other words, the strength ratio takes into account the various strength-reducing defects (e.g., knots) that may be present.

As noted, the procedure for establishing reference design values using the clear wood method is set forth in ASTM Standards D 245 and D 2555 (Refs. 4.11 and 4.12). Briefly, the process involves the following:

1. A statistical analysis is made of a large number of clear wood strength values for the various commercial species. With the exception of $F_{c\perp}$ and E, the 5 *percent exclusion value* serves as the starting point for the development of reference design values. The 5 percent exclusion value represents a strength property (e.g., bending strength). Out of 100 clear wood specimens, 95 could be expected to fail at or above the 5 percent exclusion value, and 5 could be expected to fail below this value.

2. For Dimension lumber, the 5 percent exclusion value for an unseasoned specimen is then increased by an appropriate seasoning adjustment factor to a moisture content of 19 percent or less. This step is not applicable to Timbers since reference design values for Timbers are for unseasoned conditions.

3. Strength ratios are used to adjust the clear wood values to account for the strength-reducing defects permitted in a given stress grade.

4. The stresses are further reduced by a general adjustment factor which accounts for the duration of the test used to establish the initial clear wood values, a manufacture and use adjustment, and several other factors.

The combined effect of these adjustments is to provide an average factor of safety on the order of 2.5. Because of the large number of variables in a wood member, the factor of safety for a given member may be considerably larger or smaller than the average. However, for 99 out of 100 pieces, the factor of safety will be greater than 1.25, and for 1 out of 100, the factor of safety will exceed 5. References 4.18, 4.19, and 4.23 give more details on the development of mechanical properties using the clear wood strength method.

The *In-Grade Testing Program* was undertaken jointly between the lumber industry and the U.S. Forest Products Laboratory (FPL). The purpose of the In-Grade Program was to test full-size Dimension lumber that had been graded in the usual way. The grading rules for the various species did not change, and as the name "In-Grade" implies, the members tested were representative of lumber available in the market place. Approximately 73,000 pieces of full-size *Dimension lumber* were tested in bending, tension, and compression parallel to grain in accordance with ASTM D 4761 (Ref. 4.15). Relationships were also developed between mechanical properties and moisture content, grade, and size.

The objective of the In-Grade Program was to verify the published design values that had been determined using the clear-wood strength method. Although some of the values from the In-Grade Testing Program were close, there were enough differences between the In-Grade results and the clear wood strength values that a new method of determining reference design values was developed. These procedures are given in ASTM Standard D 1990 (Ref. 4.13).*

This brief summary explains why the design values for *Dimension lumber* and *Timbers* are published in separate tables in the NDS Supplement. As a practical matter, the designer does not need the ASTM standards to design a wood structure. The ASTM standards simply document the methods used by the rules-writing agencies to develop the tabulated reference design values listed in the NDS Supplement.

4.5   Species and Species Groups

A large number of *species* of trees can be used to produce structural lumber. As a general rule, a number of species are grown, harvested, manufactured, and marketed together. From a practical standpoint, the structural designer uses lumber from a commercial *species group* rather than a specific individual species. The same grading rules, reference design values, and grade stamps are applied to all species in the species group. Reference design values for a species group were derived using statistical procedures that ensure conservative values for all species in the group.

In some cases, the mark of one or more *individual* species may be included in the grade stamp. When one or more species from a species *group* are identified in the grade stamp, the reference design values for the species group are the appropriate values for use in structural design. In other cases, the grade stamp on a piece of lumber may reflect only the name of the *species group*, and the actual species of a given piece will not be known. Special knowledge in wood identification would be required to determine the individual species.

The 2005 NDS Supplement contains a complete list of the species groups along with a summary of the various individual species of trees that may be included in each group. Examples of several commonly used species groups are shown in Fig. 4.3. Individual species as well as the species groups are shown.

It should be noted that there are a number of species groups that have similar names [e.g., Douglas Fir-Larch *and* Douglas Fir-Larch (N), Hem-Fir *and* Hem-Fir (N), and Spruce-Pine-Fir *and* Spruce-Pine-Fir (S)]. It is important to understand that each is a separate and distinct species group, and there are different sets of reference design values for each group. Different properties may be the result of the trees being grown in different geographical locations. However, there may also be different individual species included in combinations with similar names.

The choice of species for use in design is typically a matter of economics. For a given location, only a few species groups will be available, and a check with local lumber distributors or a wood products agency will narrow the selection considerably. Although the table in Fig. 4.3 identifies only a small number of the

---

*ASTM D 1990 does not cover shear and compression perpendicular to grain. Therefore, values of F_v and $F_{c\perp}$ for Dimension lumber are obtained from ASTM D 2555 and D 245.

Species Group Name and Group Name Mark that may appear in grade stamp	Individual Species that may be included in the Species Group and Individual Species Mark that may appear in grade stamp	Notes
Douglas Fir-Larch [DOUG. FIR-L]	Douglas Fir [D FIR] Western Larch [L]	Individual species mark for Douglas Fir may also appear as "DOUGLAS FIR" or "D. FIR"
Douglas Fir-Larch (N) **D. FIR(N)**	Douglas Fir[1] Western Larch[1]	(N) - indicates a Canadian species group
Douglas Fir South [D FIR S]	Douglas Fir South [D FIR S]	South indicates Douglas Fir grown in Arizona, Colorado, Nevada, New Mexico, and Utah
Hem-Fir [HEM FIR]	California Red Fir[1] Grand-Fir[1] Noble Fir[1] Pacific Silver Fir[1] Western Hemlock [HEM] White Fir[1]	
Hem-Fir (N) **HEM-FIR-N**	Amabilis Fir Western Hemlock	(N) - indicates a Canadian species group
Southern Pine **SYP**	Loblolly Pine[1] Longleaf Pine[1] Shortleaf Pine[1] Slash Pine[1]	Group mark is not used when graded under Southern Pine Inspection Bureau - grade stamp will show: [SPIB]
Spruce-Pine-Fir **S-P-F**	Alpine Fir[1] Balsam Fir[1] Black Spruce[1] Englemann Spruce[1] Jack Pine[1] Lodgepole Pine[1] Red Spruce[1] White Spruce[1]	Canadian species group
Spruce-Pine-Fir (S) **SPF** [S]	Balsam Fir[2] Eastern Spruce[2] Englemann Spruce [ES] Jack Pine[1] Lodgepole Pine [LP] Red Pine[2] Sitka Spruce [SS]	(S) - indicates USA species group (established 1991). Eastern Spruce is any combination of Black Spruce, Red Spruce, and White Spruce.

Figure 4.3 Typical species groups of structural lumber. These and a number of additional species groups are given in the NDS Supplement. The species groups listed here account for a large percentage of the structural lumber sold in the United States. Also shown are the individual species that may be included in a given species group.

commercial lumber species, those listed account for much of the total volume of structural lumber in North America.

The species of trees used for structural lumber are classified as hardwoods and softwoods. These terms are not necessarily a description of the physical properties of the wood, but are rather classifications of trees. *Hardwoods* are broadleafed deciduous trees. *Softwoods,* on the other hand, have narrow, needle-like leaves, are generally evergreen, and are known as conifers. By far, the large majority of structural lumber comes from the softwood category.

For example, Douglas Fir-Larch and Southern Pine are two species groups that are widely used in structural applications. Although these contain species that are all classified as softwoods, they are relatively dense and have structural properties that exceed those of many hardwoods.

It has been noted that the lumber grading rules establish the limits on the strength-reducing characteristics permitted in the various lumber grades. Before discussing the various lumber grades, a number of the natural growth characteristics found in lumber will be described.

4.6   Cellular Makeup

As a biological material, wood represents a unique structural material because its supply can be renewed by growing new trees in forests which have been harvested. Proper forest management is necessary to ensure a continuing supply of lumber.

Wood is composed of elongated, round, or rectangular tubelike cells. These cells are much longer than they are wide, and the length of the cells is essentially parallel with the length of the tree. The cell walls are made up of *cellulose,* and the cells are bound together by material known as *lignin.*

If the cross section of a log is examined, concentric rings are seen. One ring represents the amount of wood material deposited on the outside of the tree during one growing season. One ring then is termed an *annual ring.* See Fig. 4.4.

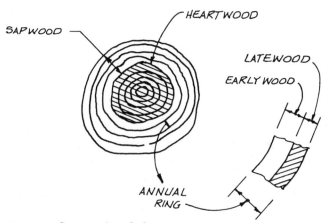

Figure 4.4   Cross section of a log.

The annual rings develop because of differences in the wood cells that are formed in the early portion of the growing season compared with those formed toward the end of the growing season. Large, thin-walled cells are formed at the beginning of the growing season. These are known as *earlywood* or *spring wood* cells. The cells deposited on the outside of the annual ring toward the end of the growing season are smaller, have thicker walls, and are known as *latewood* or *summerwood* cells. It should be noted that annual rings occur only in trees that are located in climate zones which have distinct growing seasons. In tropical zones, trees produce wood cells which are essentially uniform throughout the entire year.

Because summerwood is denser than springwood, it is stronger (the more solid material per unit volume, the greater the strength of the wood). The annual rings, therefore, provide one of the *visual* means by which the strength of a piece of wood may be evaluated. The more summerwood in relation to the amount of springwood (other factors being equal), the stronger the piece of lumber. This comparison is normally made by counting the number of growth rings per unit width of cross section.

In addition to annual rings, two different colors of wood may be noticed in the cross section of the log. The darker center portion of the log is known as *heartwood*. The lighter portion of the wood near the exterior of the log is known as *sapwood*. The relative amount of heartwood compared with sapwood varies with the species of tree. Heartwood, because it occurs at the center of the tree, is obviously much older than sapwood, and, in fact, heartwood represents woods cells which are inactive. These cells, however, provide strength and support to the tree. Sapwood, on the other hand, represents both living and inactive wood cells. Sapwood is used to store food and transport water. The strength of heartwood and sapwood is essentially the same. Heartwood is somewhat more decay resistant than sapwood, but sapwood more readily accepts penetration by wood-preserving chemicals.

4.7   Moisture Content and Shrinkage

The *solid* portion of wood is made of a complex cellulose-lignin compound. The cellulose comprises the framework of the cell walls, and the lignin cements and binds the cells together.

In addition to the solid material, wood contains moisture. The moisture content (MC) is measured as the percentage of water to the oven dry weight of the wood:

$$\text{MC} = \frac{\text{moist weight} - \text{oven dry weight}}{\text{oven dry weight}} \times 100 \text{ percent}$$

The moisture content in a living tree can be as high as 200 percent (i.e., in some species the weight of water contained in the tree can be 2 times the weight of the solid material in the tree). However, the moisture content of structural lumber in service is much less. The average moisture content that lumber assumes in service is known as the *equilibrium moisture content* (EMC). Depending on atmospheric conditions, the EMC of structural framing lumber in a covered structure (dry conditions) will range somewhere between 7 and 14 percent. In most cases, the MC at the time of construction will be higher than the EMC of a building (perhaps 2 times higher). See Example 4.1.

Moisture is held within wood in two ways. Water contained in the cell cavity is known as *free water*. Water contained within the cell walls is known as *bound water*. As wood dries, the first water to be driven off is the free water. The moisture content that corresponds to a complete loss of free water (with 100 percent of the bound water remaining) is known as the *fiber saturation point* (FSP). No loss of bound water occurs as lumber dries above the fiber saturation point. In addition, no volume changes or changes in other structural properties are associated with changes in moisture content above the fiber saturation point.

However, with moisture content changes below the fiber saturation point, bound water is lost and volume changes occur. If moisture is lost, wood shrinks; if moisture is gained, wood swells. Decreases in moisture content below the fiber

---

EXAMPLE 4.1   Bar Chart Showing Different MC Conditions

Figure 4.5 Bar chart showing different MC conditions.

Figure 4.5 shows the moisture content in lumber in comparison with its solid weight. The values indicate that the lumber was manufactured (point 1) at an MC below the fiber saturation point. Some additional drying occurred before the lumber was used in construction (point 2). The EMC is shown to be less than the MC at the time of construction. This is typical for most buildings saturation point are accompanied by increases in strength properties. Prior to the In-Grade Testing Program, it was generally believed that the more lumber dried, the greater would be the increase in strength. However, results from the In-Grade Program show that strength properties peak at around 10 to 15 percent MC. For a moisture content below this, member strength capacities remain about constant.

The fiber saturation point varies with species, but the *Wood Handbook* (Ref. 4.20) indicates that 30 percent is average. Individual species may differ from the average. The drying of lumber in order to increase its structural properties is known as *seasoning.* As noted, the MC of lumber in a building typically decreases after construction until the EMC is reached. Although this drying in service can be called seasoning, the term *seasoning* often refers to a controlled drying process. Controlled drying can be performed by air or kiln drying (KD), and both increase the cost of lumber.

From this discussion it can be seen that there are opposing forces occurring as wood dries below the fiber saturation point. On one hand, shrinkage decreases the size of the cross section with a corresponding reduction in section properties. On the other hand, a reduction in MC down to approximately 15 percent increases most structural properties. The net effect of a decrease in moisture content in the 10 to 30 percent range is an overall increase in structural capacity.

Shrinkage can also cause cracks to form in lumber. As lumber dries, the material near the surface of the member loses moisture and shrinks before the wood at the inner core. Longitudinal cracks, known as *seasoning checks,* may occur near the neutral axis (middle of wide dimension) of the member as a result of this nonuniform drying process. See Fig. 4.6*a* in Example 4.2. Cracking of this nature causes a reduction in shear strength which is taken into account in the lumber grading rules and reference design values. This type of behavior is more common in thicker members.

EXAMPLE 4.2   Shrinkage of Lumber

The *Wood Handbook* (Ref. 4.20) lists average clear wood shrinkage percentages for many individual species of wood. Tangential shrinkage is greatest. Radial shrinkage is on the order of one-half of the tangential value, but is still significant. Longitudinal shrinkage is small and is usually disregarded.

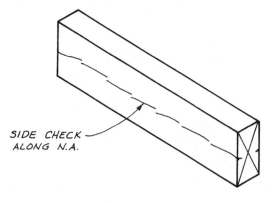

SIDE CHECK
ALONG N.A.

Figure 4.6*a* Seasoning checks may occur in the wide side of a member at or near the neutral axis. These cracks form because wood near the surface dries and shrinks first. In larger pieces of lumber, the inner core of the member loses moisture and shrinks much slower. Checking relieves the stresses caused by nonuniform drying.

LUMBER   SECTION

Figure 4.6b Tangential shrinkage is greater than radial shrinkage. This promotes the formation of radial cracks known as *end checks*.

Figure 4.6c Tangential, radial, and longitudinal shrinkage.

Another point should be noted about the volume changes associated with shrinkage. The dimensional changes as the result of drying are not uniform. Greater shrinkage occurs parallel (tangent) to the annual ring than normal (radial) to it. See Fig. 4.6b. These nonuniform dimensional changes may cause radial checks.

In practice, the orientation of growth rings in a member will be arbitrary. In other words, the case shown in Fig. 4.6c is a rather unique situation with the annual rings essentially parallel and perpendicular to the sides of the member. Annual rings can be at any angle with respect to the sides of the member.

It is occasionally necessary for the designer to estimate the amount of shrinkage or swelling that may occur in a structure. The more common case involves shrinkage of lumber as it dries in service. Several approaches may be used to estimate shrinkage.

One method comes from the *Wood Handbook*. Values of tangential, radial, and volumetric shrinkage from clearwood samples are listed for many individual species. The shrinkage percentages are assumed to take place from no shrinkage at a nominal FSP of 30 percent to full shrinkage at zero MC. A linear interpolation is used for shrinkage at intermediate MC values. See Fig. 4.6c. The maximum shrinkage can be estimated using the tangential shrinkage, and the minimum can be evaluated with the radial value. Thus, the method from the *Wood Handbook* can be used to bracket the probable shrinkage.

A second approach to shrinkage calculations is given in Ref. 4.21. It provides formulas for calculating the percentage of shrinkage for the *width* and *thickness* of a piece of lumber. This method was used for shrinkage adjustments for the In-Grade Test data and is included in the appendix to ASTM D 1990 (Ref. 4.13).

In structural design, there are several reasons why it may be more appropriate to apply a simpler method for estimating the shrinkage than either of the two methods just described:

1. Shrinkage is a variable property. The shrinkage that occurs in a given member may be considerably different from those values obtained using the published *average* radial and tangential values.

2. Orientation of the annual rings in a real piece of lumber is unknown. The sides of a member are probably not parallel or perpendicular to the growth rings.

3. The designer will probably know only the *species group,* and the *individual species* of a member will probably not be known.

For these and perhaps other reasons, a simple method of estimating shrinkage in structural lumber is recommended in Ref. 4.26. In this third approach, a constant shrinkage value of 6 percent is used for both the width and the thickness of a member. The shrinkage is taken as zero at an FSP of 30 percent, and the full 6 percent shrinkage is assumed to occur at an MC of zero. A linear relationship is used for MC values between 30 and 0. See Example 4.3.

Although Ref. 4.26 deals specifically with western species lumber, the recommended general shrinkage coefficient should give reasonable estimates of shrinkage in most species.

To carry out the type of shrinkage estimate illustrated in Example 4.3, the designer must be able to establish reasonable values for the initial and final MC for the lumber. The *initial moisture content* is defined to some extent by the specification for the lumber for a particular job. The general MC range at the time of manufacture is shown in the grade stamp, and this value needs to be reflected in the lumber specification for a job.

EXAMPLE 4.3   Simplified Method of Estimating Shrinkage

Estimate the shrinkage that will occur in a four-story wood-frame wall that uses Hem-Fir framing lumber. Consider a decrease in moisture content from 15 to 8 percent.

Framing is typical platform construction with 2×12 floor joints resting on bearing walls. Wall framing is conventional $2 \times$ studs with a typical single $2 \times$ bottom plate and double $2 \times$ top plates. See Fig. 4.7.

Figure 4.7   Details for estimating shrinkage in four-story building.

The species group of Hem-Fir is given. The list in Fig. 4.3 indicates that any one of six species may be grade marked with the group name of Hem-Fir. It the individual species is known, the shrinkage coefficients from the *Wood Handbook* (Ref. 4.20) could be used to bracket the total shrinkage, using tangential and radial values. However, for practical design purposes, the simplified approach from Ref. 4.26 is used to develop a design estimate of the shrinkage.

A shrinkage of 6 percent of the member dimension is assumed to occur between MC = 30 percent and MC = 0 percent. Linear interpolation allows the shrinkage value (SV) per unit (percent) change in moisture content to be calculated as

$$\text{Shrinkage value SV} = \tfrac{6}{30} = 0.2 \text{ percent per 1 percent change in MC}$$

$$= 0.002 \text{ in./in. per 1 percent change in MC}$$

The shrinkage S that occurs in the dimension d of a piece is calculated as the shrinkage value times the dimension times the change in moisture content:

$$\text{Shrinkage } S = \text{SV} \times d \times \Delta_{\text{MC}}$$

$$= 0.002 \times d \times \Delta_{\text{MC}}$$

Shrinkage in the depth of one 2 × 12 floor joist:

$$S_{\text{floor}} = 0.002 \times d \times \Delta_{\text{MC}}$$

$$= 0.002 \times 11.25 \times (15 - 8) = 0.158 \text{ in.}$$

Shrinkage in the thickness[*] of one 2 × wall plate.

$$S_{\text{plate}} = 0.002 \times d \times \Delta_{\text{MC}}$$

$$= 0.002 \times 1.5 \times (15 - 8) = 0.021 \text{ in.}$$

Shrinkage in the length of a stud: The longitudinal shrinkage of a piece of lumber is small.

$$S_{\text{stud}} \approx 0$$

The first floor is a concrete slab. The second, third, and fourth floors each use 2 × 12 floor joists (three total). There is a 2 × bottom plate on the first, second, third, and fourth floors (four total). There is a double 2 × plate on top of the first-, second-, third-, and fourth-floor wall studs (a total of eight 2 × top plates).

$$\text{Total } S = \sum S = 3(S_{\text{floor}}) + 12(S_{\text{plate}})$$

$$= 3(0.158) + (4 + 8)(0.021)$$

$$\boxed{\text{Total } S = 0.725 \text{ in.} \approx \tfrac{3}{4} \text{ in.}}$$

The *grade stamp* on a piece of lumber will contain one of three MC designations, which indicates the condition of the lumber at the time of manufacture. *Dry lumber* is defined as lumber having a moisture content of 19 percent or less. Material with a moisture content of over 19 percent is defined as *unseasoned* or *green lumber.*

When unseasoned lumber is grade stamped, the term "S-GRN" (surfaced green) will appear. "S-DRY" (surfaced dry) or "KD" (kiln dried) indicates that the lumber

---

[*]In lumber terminology, the larger cross-sectional dimension of a piece of wood is known as the width, and the smaller is the thickness.

was manufactured with an MC of 19 percent or less. Refer to the sample grade stamps in Fig. 4.2*a* for examples of these markings. Some smaller lumber sizes may be seasoned to 15 percent or less in moisture content and marked "MC 15," or "KD 15." It should be understood that larger-size wood members (i.e., Timbers) are not produced in a dry condition. The large cross-sectional dimensions of these members would require an excessive amount of time for seasoning.

In addition to the MC range reflected in the grade stamp, the initial moisture content of lumber in place in a structure is affected by a number of variables including the size of the members, time in transit to the job site, construction delays, and time for construction. Reference 4.26 recommends that in practical situations the following assumptions can reasonably be made:

Moisture designation in grade stamp	Initial moisture content assumed in service
S-GRN (MC greater than 19 percent at time of manufacture)	19 percent
S-DRY or KD (MC of 19 percent or less at time of manufacture)	15 percent

It should be noted that these recommendations are appropriate for relatively thin material (e.g., the 2 × floor joists and wall plates in Example 4.3). However, larger-size members will dry slower. The designer should take this and other possible factors into consideration when estimating the initial moisture content for shrinkage calculations.

The *final moisture content* can be taken as the equilibrium moisture content (EMC) of the wood. Various surveys of the moisture content in existing buildings have been conducted, and it was previously noted that the EMC in most buildings ranges between 7 and 14 percent.

Reference 4.7 gives typical EMC values of several broad atmospheric zones. The average EMC for framing lumber in the "dry southwestern states" (eastern California, Nevada, southern Oregon, southwest Idaho, Utah, and western Arizona) is given as 9 percent. The MC in most covered structures in this area is expected to range between 7 and 12 percent. For the remainder of the United States, the average EMC is given as 12 percent with an expected range of 9 to 14 percent. These values basically agree with the 8 to 12 percent MC suggested in Ref. 4.26.

The average values and the MC ranges can be used to estimate the EMC for typical buildings. Special conditions must be analyzed individually. As an alternative, the moisture content of wood in an existing structure can be measured with a portable, hand-held moisture meter.

The discussion of moisture content and shrinkage again leads to an important conclusion that was mentioned earlier: Wood is a unique structural material, and its behavior must be understood if it is to be used successfully. Wood is not a static material, and significant changes in dimensions can result because of atmospheric conditions.

Even if shrinkage calculations are not performed, the designer should allow for the movement (shrinkage or swelling) that may occur. This may be necessary in a number of cases. A primary concern in structural design is the potential splitting of wood members.

Wood is very weak in tension perpendicular to grain. The majority of shrinkage occurs across the grain, and connection details must accommodate this movement. If a connection does not allow lumber to shrink freely, tension stresses perpendicular to grain may develop. Splitting of the member will be the likely result. The proper detailing of connections to avoid built-in stresses due to changes in MC is covered in Chaps. 13 and 14.

In addition to the structural failures that may result from cross-grain tension, there are a number of other practical shrinkage considerations. Although these may not affect structural safety, they may be crucial to the proper functioning of a building. Consider the shrinkage evaluated in the multistory structure in Example 4.3. Several types of problems could occur.

For example, consider the effect of ceiling joists, trusses, or roof beams supported by a four-story wood-frame wall on one end *and* a concrete or masonry wall on the other end. Shrinkage will occur in the wood wall but not in the concrete or masonry. Thus, one end of the member in the top level will eventually be $\frac{3}{4}$ in. lower than the other end. This problem occurs as a result of *differential* movement.

Even greater differential movement problems can occur. Consider an all-wood-frame building again, of the type in Example 4.3. In all-wood construction, the shrinkage will be uniform throughout. However, consider the effect of adding a short-length concrete masonry block wall (say, a stair tower enclosure) in the middle of one of the wood-frame walls. The differential movement between the wood-frame wall and the masonry wall now takes place in a short distance. Distress of ceiling, floor, and wall sheathing will likely develop.

Other potential problems include the possible buckling of finish wall siding. Even if the shrinkage is uniform throughout a wall, there must be sufficient clearance in wall covering details to accommodate the movement. This may require slip-type architectural details (for example, *Z* flashing). In addition, plumbing, piping, and electrical and mechanical systems must allow for movement due to shrinkage. This can be accomplished by providing adequate clearance or by making the utilities flexible enough to accommodate the movement without distress. See Ref. 4.26 for additional information.

4.8   Effect of Moisture Content on Lumber Sizes

The moisture content of a piece of lumber obviously affects the cross-sectional dimensions. The width and depth of member are used to calculate the section properties used in structural design. These include area A, moment of inertia I, and section modulus S.

Fortunately for the designer, it is not necessary to compute section properties based on a consideration of the initial MC and EMC and the resulting shrinkage (or swelling) that occurs in the member. Grading practices for *Dimension lumber* have established the *dry* size (MC $\leq$ 19 percent) of a member as the basis for structural calculations. This means that only one set of cross-sectional properties needs to be considered in design.

This is made possible by manufacturing lumber to different cross-sectional dimensions based on the MC of the wood at the time of manufacture. Therefore, lumber which is produced from green wood will be somewhat larger at the time of manufacture. However, when this wood reaches a dry moisture content condition, the cross-sectional dimensions will closely coincide with those for lumber produced in the dry condition. Again, this discussion has been based on the manufacturing practices for *Dimension lumber*.

Because of their large cross-sectional dimensions, *Timbers* are not produced in a dry condition since an excessive amount of time would be required to season these members. For this reason, cross-sectional dimensions that correspond to a *green* (MC > 19 percent) condition have been established as the basis for design calculations for these members. In addition, reference design values have been adjusted to account for the higher MC of *Timbers*.

4.9   Durability of Wood and the Need for Pressure Treatment

The discussion of the MC of lumber often leads to concerns about the durability of wood structures and the potential for decay. However, the record is clear. If wood is used properly, it can be a permanent building material. If wood is used incorrectly, major problems can develop, sometimes rapidly. Again, understanding the material is the key to its proper use. The performance of many classic wood structures (Ref. 4.9) is testimony to the durability of wood in properly designed structures.

Generally, if it is protected (i.e., not exposed to the weather or not in contact with the ground) and is used at a relatively low moisture content (as in most covered structures), wood performs satisfactorily without chemical treatment. Wood is also durable when *continuously* submerged in *fresh* water. However, if the MC is high and varies with time, or if wood is in contact with the ground, the use of an appropriate preservative treatment should be considered.

High MC values can occur in wood roof systems over swimming pools and in processing plants with high-humidity conditions. *High* MC is generally defined as exceeding 19 percent in sawn lumber and as being 16 percent or greater in glulam. Problems involving high MC can also occur in geographic locations with high humidity.

In some cases, moisture can become entrapped in roof systems that have *below-roof insulation*. This type of insulation can create dead-air spaces, and moisture from condensation or other sources may lead to decay. Moisture-related problems have occurred in some flat or nearly flat roofs. To create air movement, a minimum roof slope of $\frac{1}{2}$ in./ft is now recommended for panelized roofs that use below-roof insulation. A number of other recommendations have been developed by the industry to minimize these types of problems. See Refs. 4.6, 4.16, and 4.25.

The issue of mold in buildings, particularly wood-frame buildings, has received considerable attention in recent years. Mold and mildew are often present in buildings where there is excessive moisture. Moist, dark, or low-light environments with stagnant airflow contribute to active mold and mildew growth. Such conditions are common in foundations and basements, but also susceptible are exterior walls and roof or attic areas. Industry associations such as APA—The Engineered Wood Association have produced considerable technical and nontechnical literature for seeking to mitigate mold-and mildew-related problems (See Ref. 4.17).

When required for new construction, chemicals can be impregnated into lumber and other wood products by a pressure treating process. The chemical preservatives prevent or effectively retard the destruction of wood. Pressure treating usually takes place in a large steel cylinder. The wood to be treated is transported into the cylinder on a tram, and the cylinder is closed and filled with a preservative. The cylinder is then subjected to pressure which forces the chemical into the wood.

The chemical does not saturate the complete cross section of the member. Therefore, field cutting and drilling of holes for connections after treating should be minimized. It is desirable to carry out as much fabrication of structural members as possible before the members are treated. The depth of penetration is known as the *treated zone*. The retention of the chemical treatment is measured in lb/ft^3 in the treated zone. The required retention amounts vary with the end use and type of treatment.

Many species, most notably the southern pines, readily accept preservative treatments. Other species, however, do not accept pressure treatments as well and require *incising* to make the treatment effective. In effect, incised lumber has small cuts, or incisions, made into all four sides along its length. The incisions create more surface area for the chemicals to penetrate the wood member, thereby increasing the effectiveness of the pressure treating. Incising, while increasing the effectiveness of preservative treatment, adversely affects many mechanical properties. When incised lumber is used, modification of modulus of elasticity and bending, tension and compression parallel to grain design values must be made. See Sec. 4.20.

Rather than focusing only on moisture content, a more complete overview of the question of long-term performance and durability recognizes that several instruments can destroy wood. The major ones are

1. Decay

2. Termites

3. Marine borers

4. Fire

Each of these is addressed briefly in this section, but a comprehensive review of these subjects is beyond the scope of this book. Detailed information is available in Refs. 4.18, 4.20, and 4.24. In the case of an existing wood structure that has been exposed to some form of destruction, guidelines are available for its evaluation, maintenance, and upgrading (Ref. 4.8).

Decay is caused by fungi which feed on the cellulose or lignin of the wood. These fungi must have food, moisture (MC greater than approximately 20 percent), air, and favorable temperatures. All of these items are required for decay to occur (even so-called dry rot requires moisture).

If any of the requirements is not present, decay will not occur. Thus, untreated wood that is continuously dry (MC < 20 percent, as in most covered structures), or continuously wet (submerged in fresh water—no air), will not decay. Exposure to the weather (alternate wetting and drying) can set up the conditions necessary for decay to develop. Pressure treatment introduces chemicals that poison the food supply of the fungi.

Termites can be found in most areas of the United States, but they are more of a problem in the warmer-climate areas. Subterranean termites are the most common, but drywood and dampwood species also exist. Subterranean termites nest in the ground and enter wood which is near or in contact with damp ground. The cellulose forms the food supply for termites.

The IBC (Sec. 2304.11.2.1) requires a minimum clearance of 18 in. between the bottom of unprotected floor joists (12 in. for girders) and grade. Good ventilation of crawl spaces and proper drainage also aid in preventing termite attack. Lumber which is near or in contact with the ground, and wall plates on concrete ground-floor slabs and footings, must be pressure treated to prevent termite attack. (Foundation-grade redwood has a natural resistance and can be used for wall plates.) The same pressure treatments provide protection against decay and termites.

Marine borers are found in salt waters, and they present a problem in the design of marine piles. Pressure treated piles have an extensive record in resisting attack by marine borers.

A brief introduction to the fire-resistive requirements for buildings was given in Chap. 1. Where necessary to meet building code requirements, or where the designer decides that an extra measure of fire protection is desirable, *fire-retardant treated wood* may be used. This type of treatment involves the use of chemicals in formulations that have fire-retardant properties. Some of the types of chemicals used are preservatives and thus also provide decay and termite protection. Fire-retardant treatment, however, requires higher concentrations of chemicals in the treated zone than normal preservative treatments.

The reference design values in the NDS Supplement apply to both untreated and pressure-preservative treated lumber. In other words, there is no required design value modification for preservative treatments. The exception to this is if the lumber is incised to increase the penetration of the preservatives and thereby increasing the effectiveness of the pressure treatment. Incising effectively decreases the strength and stiffness and must be accounted for in design when incised lumber is used. *Preservative treatments* are those that guard against decay, termites, and marine borers. The high concentrations of chemicals used in fire-retardant treated lumber will probably require that reference design values be reduced. However, the reduction coefficients vary with the treating process, and the NDS refers the designer to the company providing the fire-retardant treatment and redrying service for the appropriate factors.

The three basic types of pressure preservatives are

1. Creosote and creosote solutions

2. Oilborne treatments (pentachlorophenol and others dissolved in one of four hydrocarbon solvents)

3. Waterborne oxides

There are a number of variations in each of these categories. The choice of the preservative treatment and the required retentions depend on the application. Detailed information on pressure treatments and their use can be obtained from the American Wood Preservers Association (AWPA). For the address of AWPA, see the list of organizations in the Nomenclature section.

An introduction to pressure treatments is given in Ref. 4.7. This reference covers fire-resistive requirements and fire hazards as well as preservative and fire-retardant treatments. Reference 4.5 provides a concise summary of preservative treatments. See Ref. 4.24 for additional information on the use of wood in adverse environments.

4.10   Growth Characteristics of Wood

Some of the more important growth characteristics that affect the structural properties of wood are density, moisture content, knots, checks, shakes, splits, slope of grain, reaction wood, and decay. The effects of density, and how it can be measured visually by the annual rings, were described previously. Likewise, moisture content and its effects have been discussed at some length. The remaining natural growth characteristics also affect the strength of lumber, and limits are placed on the size and number of these structural defects permitted in a given stress grade. These items are briefly discussed here.

Knots constitute that portion of a branch or limb that has been incorporated into the main body of the tree. See Fig. 4.8. In lumber, knots are classified by form, size, quality, and occurrence. Knots decrease the mechanical properties of the wood because the knot displaces *clear wood* and because the slope of the grain is forced to deviate around the knot. In addition, stress concentrations

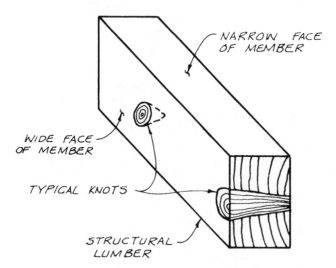

Figure 4.8 Examples of knots. Lumber grading rules for the commercial species have different limits for knots occurring in the wide and narrow faces of the member.

occur because the knot interrupts wood fibers. Checking also may occur around the knot in the drying process. Knots have an effect on both tension and compression capacity, but the effect in the tension zone is greater. Lumber grading rules for different species of wood describe the size, type, and distribution (i.e., location and number) of knots allowed in each stress grade.

Checks, shakes and *splits* all constitute separations of wood fibers. See Fig. 4.9. Checks have been discussed earlier and are radial cracks caused by nonuniform volume changes as the moisture content of wood decreases (Sec. 4.7). Recall that the outer portion of a member shrinks first, which may cause longitudinal cracks. In addition, more shrinkage occurs tangentially to the annual ring than radially.

Figure 4.9   Checks, shakes, and splits.

Figure 4.10 Slope of grain.

Checks therefore are seasoning defects. Shakes, on the other hand, are cracks that are usually parallel to the annual ring and develop in the standing tree. Splits represent complete separations of the wood fibers through the thickness of a member. A split may result from a shake or seasoning or both. Splits are measured as the penetration of the split from the end of the member parallel to its length. Again, lumber grading rules provide limits on these types of defects.

The term *slope of grain* is used to describe the deviation of the wood fibers from a line that is parallel to the edge of a piece of lumber. Slope of grain is expressed as a ratio (for example, 1:8, 1:15, etc.). See Fig. 4.10. In structural lumber, the slope of grain is measured over a sufficient length and area to be representative of the general slope of wood fibers. Local deviations, such as around knots, are disregarded in the general slope measurement. Slope of grain has a marked effect on the structural capacity of a wood member. Lumber grading rules provide limits on the slope of grain that can be tolerated in the various stress grades.

Reaction wood (known as compression wood in softwood species) is abnormal wood that forms on the underside of leaning and crooked trees. It is hard and brittle, and its presence denotes an unbalanced structure in the wood. Compression wood is not permitted in readily identifiable and damaging form in stress grades of lumber.

Decay is a degradation of the wood caused by the action of fungi. Grading rules establish limits on the decay allowed in stress-grade lumber. Section 4.9 describes the methods of preserving lumber against decay attack.

4.11   Sizes of Structural Lumber

Structural calculations are based on the *standard net size* of a piece of lumber. The effects of moisture content on the size of lumber are discussed in Sec. 4.8.

The designer may have to allow for shrinkage when detailing connections, but standard dimensions are accepted for stress calculations.

Most structural lumber is *dressed lumber*. In other words, the lumber is *surfaced* to the standard net size, which is less than the *nominal* (stated) size. See Example 4.4. Lumber is dressed on a planing machine for the purpose of obtaining smooth surfaces and uniform sizes. Typically lumber will be S4S (surfaced four sides), but other finishes can be obtained (for example , S2S1E indicates surfaced two sides and one edge).

Dressed lumber is used in many structural applications, but large timbers are commonly *rough sawn* to dimensions that are close to the standard net sizes. The textured surface of rough-sawn lumber may be desired for architectural purposes and may be specially ordered in smaller sizes. The cross-sectional dimensions of rough-sawn lumber are approximately $\frac{1}{8}$ in larger than the standard dressed size. A less common method of obtaining a rough surface is to specify *full-sawn* lumber. In this case, the actual size of the lumber should be the same as the specified size. Cross-sectional properties for rough-sawn and full-sawn lumber are not included in the NDS because of their relatively infrequent use.

EXAMPLE 4.4   Dressed, Rough-Sawn, and Full-Sawn Lumber

Figure 4.11 Actual and dressed sizes for dressed, rough-sawn,and full-sawn lumber.

Consider an 8 × 12 member (nominal size = 8 in. × 12 in.).

1. *Dressed lumber.* Standard net size = $7\frac{1}{2}$ in. × $11\frac{1}{2}$ in. Refer to NDS Supplement Tables 1A and 1B for dressed lumber sizes.

2. *Rough-sawn lumber.* Approximate size = $7\frac{5}{8}$ in. × $11\frac{5}{8}$ in. Rough size is approximately $\frac{1}{8}$ in. larger than the dressed size.

3. *Full-sawn lumber.* Minimum size 8 in. × 12 in. Full-sawn lumber is not generally available.

The terminology in the wood industry that is applied to the dimensions of a piece of lumber differs from the terminology normally used in structural calculations. The grading rules refer to the thickness and width of a piece of lumber. It was previously stated that the *thickness* is the smaller cross-sectional dimension, and the *width* is the larger.

However, in the familiar case of a beam, design calculations usually refer to the *width* and *depth* of a member. The width is parallel to the neutral axis of the cross section, and the depth is perpendicular. In most beam problems, the member is loaded about the strong or *x* axis of the cross section. Therefore, the width of a beam is usually the smaller cross-sectional dimension, and the depth is the larger. Naturally, the strong axis has larger values of section modulus and moment of inertia. Loading a beam about the strong axis is also described as having the *load applied to the narrow face of the beam*.

Another type of beam loading is less common. If the bending stress is about the weak axis or *y* axis, the section modulus and moment of inertia are much smaller. *Decking* is an obvious application where a beam will have the *load applied to the wide face of the member*. In this case the width is the larger cross-sectional dimension, and the depth is the smaller. As with all structural materials, the objective is to make the most efficient use of materials. Thus, a wood beam is used in bending about the strong axis whenever possible.

The dimensions of sawn lumber are given in the 2005 NDS Supplement Table 1A, *Nominal and Minimum Dressed Sizes of Sawn Lumber*. However, a more useful table for design is the list of cross-sectional properties in the NDS Supplement Table 1B, *Section Properties of Standard Dressed (S4S) Sawn Lumber*. The properties include nominal and dressed dimensions, area, section modulus and moment of inertia for both the *x* and *y* axes. The section properties for a typical sawn lumber member are verified in Example 4.5. The weight per linear foot for various densities of wood is also given in Table 1B.

EXAMPLE 4.5   Section Properties for Dressed Lumber

Show calculations for the section properties of a 2×8 sawn lumber member. Use standard net sizes for dressed (S4S) lumber, and verify the section properties in NDS Table 1B.

WIDTH = $1\frac{1}{2}''$

DEPTH = $7\frac{1}{4}''$

STRONG AXIS

Figure 4.12a Dimensions for section properties about strong x axis of 2×8.

2×8 *SECTION*

Section Properties for x Axis

$$A = bd = 1\tfrac{1}{2} \times 7\tfrac{1}{4} = 10.875 \text{ in.}^2$$

$$S_x = \frac{bd^2}{6} = \frac{1.5(7.25)^2}{6} = 13.14 \text{ in.}^3$$

$$I_x = \frac{bd^3}{12} = \frac{1.5(7.25)^3}{12} = 47.63 \text{ in.}^4$$

The section properties for the x axis agree with those listed in the NDS Supplement.

Figure 4.12b Dimensions for section properties about weak y axis of 2×8.

Section Properties for y Axis

$$S_y = \frac{bd^2}{6} = \frac{7.25(1.5)^2}{6} = 2.719 \text{ in.}^3$$

$$I_y = \frac{bd^3}{12} = \frac{7.25(1.5)^3}{12} = 2.039 \text{ in.}^4$$

The section properties for the y axis agree with those listed in the NDS Supplement.

4.12   Size Categories and Commercial Grades

The lumber grading rules which establish reference design values for use in structural design have been developed over many years. In this development process, the relative size of a piece of wood was used as a guide in anticipating the application or "use" that a member would receive in the field. For example, pieces of lumber with rectangular cross sections make more efficient beams than members with square (or approximately square) cross sections. Thus, if the final application of a piece of wood were known, the commercial grading rules would take into account the primary function (e.g., axial strength or bending strength) of the member. See Example 4.6.

EXAMPLE 4.6   Size and Use Categories

There are three main *size categories* of lumber. The categories and nominal size ranges are:

Boards	$\frac{3}{4}$ to $1\frac{1}{2}$ in. thick 2 in. and wider
Dimension lumber	2 to 4 in. thick 2 in. and wider
Timbers	5 in. and thicker 5 in. and wider

A number of additional subdivisions are available within the main size categories. Each represents a size and use category in the lumber grading rules. The primary *size and use categories* for commercial graded (structural) lumber are as follows:

Boards

  Stress-Rated Board (SRB)

Dimension lumber

  Structural Light Framing (SLF)

  Light Framing (LF)

  Studs

  Structural Joists and Planks (SJ&P)

  Decking

Timbers

  Beams and Stringers (B&S)

  Posts and Timbers (P&T)

Stress-Rated Boards may be used in structural applications. However, because they are relatively thin pieces of lumber, Stress-Rated Boards are not commonly used for structural framing. Therefore, the remaining discussion is limited to a consideration of Dimension lumber and Timbers. Sizes in the seven basic subcategories of structural lumber are summarized in the following table.

		Nominal dimensions		
Symbol	Name	Thickness	Width	Examples of sizes
LF	Light Framing and	2 to 4 in.	2 to 4 in.	$2 \times 2, 2 \times 4, 4 \times 4$
SLF	Structural Light Framing			
SJ&P	Structural Joist and Plank	2 to 4 in.	5 in. and wider	$2 \times 6, 2 \times 14, 4 \times 10$
	Stud	2 to 4 in	2 in. and wider	$2 \times 4, 2 \times 6, 4 \times 6$ (lengths limited to 10 ft and shorter)
	Decking*	2 to 4 in.	4 in. and wider	$2 \times 4, 2 \times 8, 4 \times 6$
B&S	Beams and Stringers	5 in. and thicker	More than 2 in. greater than thickness	$6 \times 10, 6 \times 14, 12 \times 16$
P&T	Posts and Timbers	5 in. and thicker	Not more than 2 in. greater than thickness	$6 \times 6, 6 \times 8, 12 \times 14$

*Decking is normally stressed about its minor axis. In this book, all other bending members are assumed to be stressed about the major axis of the cross section, unless otherwise noted.

It has been noted that size and use are related. However, in the process of determining the reference design values for a member, the structural designer needs to place emphasis on understanding the *size* classifications. The reason is that *different* design values apply to the *same grade name* in the *different size categories.* For example, Select Structural (a commercial grade) is available in SLF, SJ&P, B&S, and P&T size categories. Reference design values for a given commercial species of lumber are generally different for Select Structural in all of these size categories. See Example 4.7.

Several important points should be made about the *size and use categories* given in Example 4.6 *and* the *commercial grades* listed in Example 4.7:

1. *Decking* is normally stressed in bending about the minor axis of the cross section, and reference design values for Decking are listed in a separate table. See NDS Supplement Table 4E, *Reference Design Values for Visually Graded Decking.*

EXAMPLE 4.7   Commercial Grades

Typical commercial grades vary within the various size and use categories. The grades shown are for Douglas Fir-Larch.

1. *Structural Light Framing* (SLF)
   Select Structural
   No. 1 and Better
   No. 1
   No. 2
   No. 3

2. *Light Framing* (LF)
   Construction
   Standard
   Utility

3. *Structural Joist and Plank* (SJ&P)
   Select Structural
   No. 1 and Better
   No. 1
   No. 2
   No. 3

4. *Stud*
   Stud

5. *Decking*
   Select Decking
   Commercial Decking

6. *Beams and Stringers* (B&S)
   Dense Select Structural
   Select Structural
   Dense No. 1
   No. 1
   Dense No. 2
   No. 2

7. *Posts and Timber* (P&T)
   Dense Select Structural
   Select Structural
   Dense No. 1
   No. 1
   Dense No. 2
   No. 2

NOTE: The grades listed are intended to be representative, and they are not available in all species groups. For example, *No. 1 and Better* is available only in DF-L, Hem-Fir, DF-L(N), and Hem-Fir(N). Southern Pine has a number of additional *dense* and *nondense* stress grades.

---

2. Reference design values for Dimension lumber (except Decking) are given in a number of separate tables. In these tables, the commercial grades are grouped together regardless of the size and use subcategory.

Reference design values for *Dimension lumber* are listed in the following tables in the 2005 NDS Supplement:

Table 4A. *Reference Design Values for Visually Graded Dimension Lumber (2–4 in. thick) (All Species except Southern Pine)*

Table 4B. *Reference Design Values for Visually Graded Southern Pine Dimension Lumber (2–4 in. thick)*

Table 4C. *Reference Values for Mechanically Graded Dimension Lumber*

Table 4F. *Reference Design Values for Non-North American Visually Graded Dimension Lumber (2–4 in. thick)*

The simplification of the reference design values in Tables 4A, 4B, and 4F requires the use of several adjustment factors (Sec. 4.13).

3. Reference design values for Beams and Stringers (B&S) and Posts and Timbers (P&T) are given in NDS Supplement Table 4D, *Reference Design Values for Visually Graded Timbers (5 × 5 in. and larger)*. Table 4D covers all species groups including Southern Pine.

Reference design values for B&S are generally different from the reference design values for P&T. This requires a complete listing of values for all of the commercial grades for both of these size categories. Furthermore, it should be noted that there are two sets of design values for both B&S and P&T in two species groups: Douglas Fir-Larch and Sitka Spruce. This is the result of differences in grading rules from two agencies.

As noted, the lumber grading rules reflect the *anticipated use* of a wood member based on its size, but no such restriction exists for the *actual use* of the member by the designer. In other words, lumber that falls into the B&S size category was originally anticipated to be used as a bending member. As a rectangular member, a B&S bending about its strong axis is a more efficient beam (because of its larger section modulus) than a square (or essentially square) member such as a P&T. However, reference design values are tabulated for tension, compression, and bending for *all size categories*. The designer may, therefore, use a B&S in any of these applications.

Although size and use are related, it must be emphasized again that the *reference design values depend on the size* of a member rather than its *use*. Thus, a member in The P&T size category is always graded as a P&T even though it could possibly be used as a beam. Therefore, if a 6 × 8 is used as a beam, the reference bending design value for a P&T applies. Similarly, if a 6 × 10 is used as a column, the compression value for a B&S must be used.

The general notation used in the design of wood structures is introduced in the next section. This is followed by a review of a number of the adjustment factors required in wood design.

4.13   General Notation

The 2005 NDS provides guidelines for the design of wood structures following the principles of both the *allowable stress design* (ASD) and the *load and resistance factor design* (LRFD) methods. ASD has been the traditional basis for the NDS, with the LRFD provisions being new for 2005. While it is expected that ASD will continue to be the popular method for wood design in the near future, the wood design profession is in a transition period when both methods may be applied in practice. It is expected that eventually the LRFD method will become the primary design technique.

The notation system used in the NDS for both ASD and LRFD is similar to that used in the design of steel and concrete structures. However, wood is a unique material and its proper use may require a number of adjustment factors. Although the basic concepts of wood design are fairly straightforward, the many possible adjustment factors can make wood design cumbersome in the beginning. Most of the adjustment factors are common between ASD and LRFD. However, some factors and adjustment are specific or applied to one design method versus the other. It is important for the designer to understand all the adjustment factors used in wood design, with additional attention to those adjustment that may apply uniquely or differently to either ASD or LRFD.

In ASD, *working* or *service-level* stresses in a member subjected to a set of Code-required loads cannot exceed the allowable design values for the member. For LRFD, the design is based on *nominal* strengths and *strength-level* (factored) loads versus allowable design values and working stresses. That is, the service loads are *factored* up to a strength level and the resulting member forces cannot exceed the nominal capacity of the member adjusted for end-use conditions.

Generally speaking, the forces and stresses in wood structures are computed according to principles of engineering mechanics and strength of materials, and simple linear elastic theory is applied in the design of wood members with the unique properties and behavior of wood usually taken into account with adjustment factors. Because of this, the dual format NDS is possible. One set of *reference design values* is provided for use in both ASD and LRFD. These reference design values are then adjusted for use in either method. This is similar to what is done for loads (structural demand). ASCE 7 specifies basic loads for use in either ASD or LRFD (Ref. 4.10). The Code also prescribes load combinations that are specified for each method, effectively adjusting the loads for appropriate use in either ASD or LRFD.

For consistency, whether working with ASD or LRFD, it is highly recommended that the adjustment factors for wood design be kept as multiplying factors on the design values. An alternative approach of using the design value adjustments to modify the design loads can lead to confusion and error. The modification of design loads with wood design adjustment factors is not recommended.

The general notation system for use in ASD for wood structures is summarized in Example 4.8. Example 4.9 provides a summary of the general notation system for use in LRFD.

EXAMPLE 4.8   Symbols for Design Values and Adjustment Factors Using ASD

Symbols for use in ASD for wood are standardized in the NDS.

Actual Stresses

Actual member stresses are calculated from prescribed loads and member sizes. These stresses are given the symbol of lowercase f, and a subscript is added to indicate the type of stress. For example, the axial tension stress in a member is calculated as the force divided by the cross-sectional area. The notation is

$$f_t = \frac{P}{A}$$

Reference Design Values

The design values listed in the tables in the NDS Supplement are referred to as *reference design values*. All the reference design values (except modulus of elasticity) include reductions for safety. The values of modulus of elasticity listed in the tables are average values and do not include reductions for safety. Reference design values are given the symbol of an uppercase F, and a subscript is added to indicate the type of stress. For example, F_t represents the reference tension design value parallel to grain. The modulus of elasticity is assigned the traditional symbol E.

Adjusted ASD Design Values

Reference design values for wood simply represent a starting point in the determination of the allowable stress for a particular design. *Adjusted ASD design values* are determined by multiplying the reference values by the appropriate adjustment factors.

It is highly desirable to have a notation system that permits the designer to readily determine whether a design value in a set of calculations is a *reference* or an *adjusted* property. A prime is simply added to the symbol for the reference value to indicate that the necessary adjustments have been applied to obtain the adjusted design value. For example, the adjusted ASD tension design value is obtained by multiplying the reference value for tension by the appropriate adjustment factors:

$$F_t' = F_t \times \text{(product of adjustment factors)}$$

For a design to be acceptable, the actual stress must be less than or equal to the adjusted design value:

$$f_t \leq F_t'$$

On the other hand, if the actual stress exceeds the adjusted design value, the design needs to be revised.

The following design values are included in the NDS Supplement:

Design value	Symbol for reference design value	Symbol for adjusted ASD design value
Bending stress	F_b	F_b'
Tension stress parallel to grain	F_t	F_t'
Shear stress parallel to grain	F_v	F_v'
Compression stress perpendicular to grain	$F_{c\perp}$	$F_{c\perp}'$
Compression stress parallel to grain	F_c	F_c'
Modulus of elasticity	E	E'
Modulus of elasticity for stability calculations	E_{min}	F_{min}'

Adjustment Factors

The adjustment factors in wood design are usually given the symbol of an upper-case *C,* and one or more subscripts are added to indicate the purpose of the adjustment. Some of the subscripts are uppercase letters, and others are lowercase. Therefore, it is important to pay close attention to the form of the subscript, because simply changing from an uppercase to a lowercase subscript can change the meaning of the adjustment factor. Some of the possible factors for use in determining adjusted design values are

$$C_D = \text{load duration factor } (ASD \ only)$$

$$C_M = \text{wet service factor}$$

$$C_F = \text{size factor}$$

$$C_{fu} = \text{flat use factor}$$

$$C_i = \text{incising factor}$$

$$C_t = \text{temperature factor}$$

$$C_r = \text{repetitive member factor}$$

These adjustment factors do not apply to all reference design values. In addition, other adjustments may be necessary in certain types of problems. For example, the column stability factor C_P is required in the design of wood columns, and the beam stability factor C_L is used in the design of many wood beams. The factors listed here are simply representative, and the additional adjustment factors are covered in the chapters where they are needed.

The load duration factor, C_D, is a unique adjustment factor as it is limited to use only in ASD. Duration of load effects are handled differently in LRFD. Additional discussion of load duration adjustments and the difference in how they are addressed in ASD and LRFD is provided in Secs. 4.15 and 4.16. All other adjustment factors listed here are applicable to both ASD and LRFD.

Example 4.8 provided a summary of the general notation used in the NDS for ASD. The same basic notation is used in the NDS for LRFD, which allows both design formats to be presented in the NDS but may lead to some confusion. In ASD, the adjusted design values for a member are checked against *working* or *service-level* stresses calculated for the member. For LRFD, the design is based on adjusted *nominal* strength and *strengths level* loads. Example 4.8 is repeated using the LRFD format in Example 4.9

EXAMPLE 4.9   Symbols for Design Values and Adjustment Factors Using LRFD

Symbols for use in LRFD for wood are standardized in the 2005 NDS.

Actual Forces

Actual member forces are calculated from prescribed loads and member sizes. These internal member forces are determined using Code-specified factored load combinations and are given a subscript u to indicate the force is due to factored loads. For example, the bending moment in a member due to factored loads is M_u and the axial tension in a member is T_u.

Reference Design Values

The design values listed in the tables in the NDS Supplement are referred to as *reference design values.* All the tabulated design values (except the average modulus of elasticity E) include reductions for safety and are primarily intended for direct application in ASD. The values of E listed in the tables are average values and do not include reductions for safety. The reference design values provided in the NDS Supplement are given the symbol of an uppercase F, and a subscript is added to indicate the type of stress. For example, F_t represents the reference tension design value parallel to grain. The modulus of elasticity is assigned the traditional symbol E.

Reference design values published in the NDS Supplement simply represent a starting point for the determination of the adjusted resistances for a particular design. To establish *nominal design values* for use in LRFD, reference design values tabulated in the NDS Supplement are multiplied by a *format conversion factor, K_F.* For example, the nominal design value for tension is obtained by multiplying the tabulated value by K_F for tension:

$$F_{tn} = F_t \times K_F$$

The subscript n has been added to differentiate this as a *nominal* reference design value for use in LRFD.[†] Format conversion factors are provided for different applications and properties in Table N1 of NDS Appendix N.

Adjusted LRFD Design Values

The nominal design value must be adjusted for various end-use factors applicable for a particular design. *Adjusted LRFD design values* are determined by multiplying the nominal design values by the appropriate adjustment factors. It is highly desirable to have a notation system that permits the designer to readily determine whether a design value in a set of calculations is a *nominal* or an *adjusted* property. A prime is simply added to the symbol for the nominal design value to indicate that the necessary adjustments have been applied. For example, the adjusted LRFD design value for tension is obtained by multiplying the nominal design value by the appropriate adjustment factors:

$$F'_{tn} = F_{tn} \times \text{(product of adjustment factors)}$$

[†]It is also becoming customary to use ksi as the base stress unit when using LRFD, versus psi for ASD, to help easily identify a set of design calculations as following the LRFD format.

Reference and Adjusted Resistances

Typically in LRFD, designs are performed using member forces and moments versus, member stresses in ASD. Accordingly, the nominal design value should be scaled to a nominal member *resistance,* sometimes also referred to as *nominal capacity.* For example, the nominal axial tension resistance of a member, denoted T_n, is determined by multiplying the nominal design value by the cross-sectional area:

$$T_n = F_{tn} \times A = (F_t \times K_F) \times A$$

Other resistances are similarly identified by a single uppercase symbol, including M_n for moment resistance, V_n for shear resistance, and P_n for compression resistance. *Adjusted LRFD resistances,* which are used in design checking equations, are determined in a similar manner. For example, the adjusted LRFD axial tension resistance of a member T_n' is determined by multiplying adjusted LRFD the design value by the cross-sectional area:

$$T_n' = F_{tn}' \times A$$

For a design to be acceptable, the factored member force must be less than or equal to the adjusted LRFD resistance:

$$T_u \le \phi T_n'$$

On the other hand, if the factored member force exceeds the adjusted LRFD resistance, the design needs to be revised.

The following design values are provided in the NDS Supplement:

Design values	Symbol for reference design value from NDS supplement	Symbol for *nominal* design value	Symbol for adjusted LRFD design value	Symbol for adjusted LRFD resistance
Bending moment	F_b	F_{bn}	F_{bn}'	M_n'
Tension parallel to grain	F_t	F_{tn}	F_{tn}'	T_n'
Shear parallel to grain	F_v	F_{vn}	F_{vn}'	V_n'
Compression perpendicular to grain	$F_{c\perp}$	$F_{c\perp n}$	$F_{c\perp n}'$	$P_{\perp n}'$
Compression parallel to grain	F_c	F_{cn}	F_{cn}'	P_n'
Modulus of elasticity*	E	E	E'	
Modulus of elasticity for stability calculations	E_{min}	E_{min-n}	E_{min-n}'	

*Modulus of elasticity is an average value for both ASD and LRFD.

Adjustment Factors

The adjustment factors in wood design are usually given the symbol of an uppercase C, and one or more subscripts are added to indicate the purpose of the adjustment. Some of the subscripts are uppercase letters, and others are lowercase.

Therefore, it is important to pay close attention to the form of the subscript, because simply changing from an uppercase to a lowercase subscript can change the meaning of the adjustment factor. Additionally, there are several new adjustment factors specific for use in LRFD. One such factor is the *format conversion factor, K_F,* which adjusts the published reference design values to a nominal level appropriate for LRFD. Some of the possible adjustment factors for use in determining the adjusted resistance are

$$\lambda = \text{time effect factor (} LRFD \ only\text{)}$$

$$C_M = \text{wet service factor}$$

$$C_F = \text{size factor}$$

$$C_{fu} = \text{flat use factor}$$

$$C_i = \text{incising factor}$$

$$C_t = \text{temperature factor}$$

$$C_r = \text{repetitive member factor}$$

These adjustment factors do not apply to all nominal resistances. In addition, other adjustments may be necessary in certain types of problems. For example, the column stability factor C_P is required in the design of wood columns, and the beam stability factor C_L is used in the design of many wood beams. The factors listed here are simply representative, and the additional adjustment factors are covered in the chapters where they are needed.

The time effect factor, λ, is a unique adjustment factor as it is limited to use only in LRFD and is linked to specific load combinations. It is similar to the traditional load duration factor, C_D, which is for use only in ASD. Additional discussion of load duration effects and how they are addressed in ASD and LRFD is provided in Secs. 4.15 and 4.16.

Another factor the NDS lists as an adjustment factor is more commonly referred to in LRFD as a resistance factor and is applicable only to LRFD. *Resistance factors, ϕ,* are provided for different applications and properties in Table N2 of NDS Appendix N. For example, the resistance factor for bending is 0.85, and it is 0.80 for tension. All other adjustment factors are applicable for both ASD and LRFD.

Care must be taken not to mix formats in a single design. Either ASD or LRFD may be used in a design, but not both. The designer must choose which method to use.

Tables summarizing the adjustment factors for various products are given in specific tables in the NDS for both ASD and LRFD. For example, NDS Table 4.3.1 provides a summary of the *Applicability of Adjustment Factors for Sawn Lumber.* Other NDS tables provide similar information for glued laminated timber (Table 5.3.1), round timber poles and piles (Table 6.3.1), wood I-joists (Table 7.3.1), structural composite lumber (Table 8.3.1), and wood structural panels (Table 9.3.1). A summary of the factors for use in the design of connections is given in NDS Table 10.3.1, *Applicability of Adjustment Factors for Connections.*

The large number of factors is an attempt to remind the designer not to over-look something that can affect the performance of a structure. However, in many practical design situations, a number of adjustment factors may have a value of 1.0. In such a case, the adjustment is said to *default to unity.* Thus, in many common designs, the problem will not be as complex as the long list of adjustment factors would make it appear.

Some of the adjustment factors will cause the reference design values to decrease, and others will cause the value to increase. When factors that reduce strength are considered, a larger member size will be required to support a given load. On the other hand, when circumstances exist that produce increased strength, smaller, more economical members can result if these factors are taken into consideration. The point here is that a number of items can affect the strength of wood. These items *must* be considered in design when they result in a reduction of member capacity. Factors that increase the calculated strength of a member *may* be considered in the design.

This discussion emphasizes that a *conservative* approach (i.e., in the direction of greater safety) in structural design is the general rule. Factors that cause member sizes to increase *must* be considered. Factors that cause them to decrease *may* be considered or ignored. The question of whether the latter should be ignored has to do with economics. It may not be practical to ignore reductions in member sizes that result from a beneficial set of conditions.

Most adjustments for wood design are handled as a string of multiplying fac-tors that are used to convert reference values to adjusted values for a given set of design circumstances. However, to avoid an excessive number of coefficients, often only those coefficients which have an effect on the final design are shown in calculations. In other words, if an adjustment has no effect on a design value (i.e., it defaults to $C = 1.0$), the factor is often omitted from design calculations.

In this book, the adjustment factors will generally be shown, including those with values of unity. The adjustment factors mentioned in Examples 4.8 and 4.9 are described in the remainder of this chapter. Others are covered in the chap-ters that deal with specific problems.

4.14 Wet Service Factor C_M

The moisture content (MC) of wood and its relationship to strength were described in Sec. 4.7. Reference design values in the NDS Supplement gener-ally apply to wood that is used in a *dry condition,* as in most covered structures.

For sawn lumber, the reference values apply to members with an equilibrium moisture content (EMC) of 19 percent of less. Values apply whether the lumber is manufactured S-DRY, KD, or S-GRN. If the MC in service exceeds 19 per-cent for an extended period of time, the reference values are to be multiplied by an appropriate *wet service factor* C_M. Note that the subscript M refers to moisture.

For *sawn lumber,* the appropriate values of C_M are obtained from the sum-mary of *adjustment factors* at the beginning of each table in the NDS

Supplement (i.e., at the beginning of Tables 4A to 4F). In most cases, C_M is less than 1.0 when the moisture content exceeds 19 percent. The exceptions are noted in the tables for C_M. For lumber used at an MC of 19 percent or less, the default value of $C_M = 1.0$ applies.*

For *connection design,* the moisture content at the time of fabrication of the connection *and* the moisture content in service are both used to evaluate C_M. Values of C_M for connection design are summarized in NDS Table 10.3.3.

For *glulam members* (Chap. 5), reference design values apply to MC values of less than 16 percent (that is, $C_M = 1.0$). For an MC of 16 percent or greater, use of C_M less than 1.0 is required. Values of C_M for softwood glulam members are given in the summary of *adjustment factors* preceding the NDS Supplement Tables 5A and 5B, and in Tables 5C and 5D for hardwood glulam members.

4.15 Load Duration Factor C_D (ASD Only)

Wood has a unique structural property. It can support higher stresses if the loads are applied for a short period of time. This is particularly significant when one realizes that if an overload occurs, it is probably the result of a temporary load.

All reference design values for wood members and connections apply to *normal* duration loading. In fact, the tables in the NDS generally remind the designer that the published values apply to "normal-load duration and dry-service conditions." This should highlight the need for the designer to account for other conditions. In ASD, the load duration factor, C_D, is the adjustment factor used to convert reference design values to adjusted ASD values based on the expected duration of full design load.

In other words, C_D converts values for normal duration to design values for other durations of loading. Normal duration is taken as 10 years, and floor live loads are conservatively associated with this time of loading. Because reference design values apply directly to floor live loads, $C_D = 1.0$ for this type of loading. For other loads, the load duration factor lies in the range $0.9 \le C_D \le 2.0$. It should be noted that C_D applies to all reference design values except compression perpendicular to grain $F_{c\perp}$ and modulus of elasticity E. In the case of pressure-preservative treated and fire-retardant treated wood, the NDS limits the load duration factor to a maximum of 1.6 ($C_D \le 1.6$). This is due to a tendency for treated material to become less resistant to impact loading.

The historical basis for the load duration factor is the curve shown in Fig. 4.13. See Example 4.10. The load duration factor is plotted on the vertical axis versus the accumulated duration of load on the horizontal axis. This graph appears in the *Wood Handbook* (Ref. 4.20) and in the NDS Appendix B. Over the years this plot has become known as the *Madison Curve* (the FPL is located in Madison,

---

*Prior to the 1991 NDS, lumber grade marked MC15 was permitted use of a C_M greater than 1.0, but as a result of the In-Grade Program, this has been deleted. For some grades of Southern Pine, the wet service factor has been incorporated into the tabulated values, and for these cases the use of an additional C_M is not appropriate.

EXAMPLE 4.10   Load Duration Factor (ASD Only)

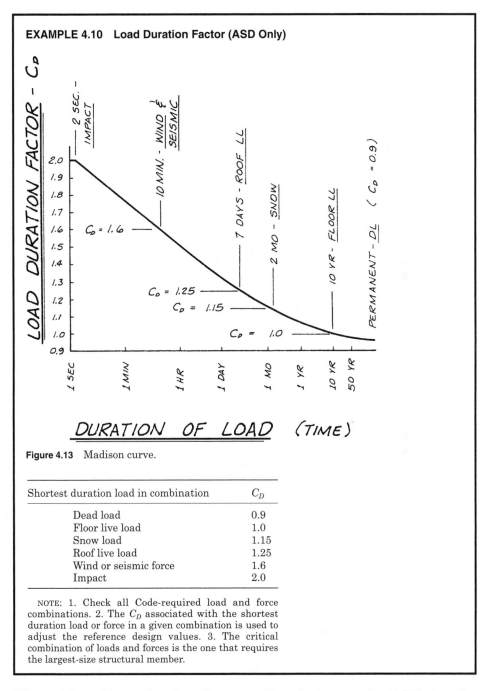

Figure 4.13   Madison curve.

Shortest duration load in combination	C_D
Dead load	0.9
Floor live load	1.0
Snow load	1.15
Roof live load	1.25
Wind or seismic force	1.6
Impact	2.0

NOTE: 1. Check all Code-required load and force combinations. 2. The C_D associated with the shortest duration load or force in a given combination is used to adjust the reference design values. 3. The critical combination of loads and forces is the one that requires the largest-size structural member.

Wisconsin), and its use has been integrated into design practice (ASD) since the 1940s. The durations associated with the various design loads are shown on the graph and in the summary below the graph.

The term "duration of load" refers to the total *accumulated* length of time that the full design load is applied during the life of a structure. Furthermore, in considering duration, it is indeed the *full design load* that is of concern, and not the length of time over which a portion of the load may be applied. For example, it is obvious that some wind or air movement is almost always present. However, in assigning C_D for wind, the duration is taken as the total length of time over which the design *maximum* wind force will occur.

It was previously recommended that adjustment factors, including C_D, be applied as multiplying factors for adjusting reference design values. Modifications of actual stresses or modifications of applied loads should not be used to account for duration of load. A consistent approach in the application of C_D to the reference design value will avoid confusion.

The stresses that occur in a structure are usually not the result of a single applied load (see Sec. 2.16 for a discussion of ASCE-7-required load combinations). Quite to the contrary, they are normally caused by a combination of loads and forces that act simultaneously. The question then arises about which load duration factor should be applied when checking a stress caused by a given combination. It should be noted that the load duration factor applies to the *entire combination of loads* and not just to that portion of the stress caused by a load of a particular duration. The C_D to be used is the one associated with the *shortest duration* load or force in a given combination.

For example, consider the possible load combinations on a floor beam that also carries a column load from the roof. What are the appropriate load duration factors for the various load combinations? If stresses under the dead load alone are checked, $C_D = 0.9$. If stresses under (D + L) are checked, the shortest duration load in the combination is floor live load, and $C_D = 1.0$. For (D + L + S), $C_D = 1.15$. If the structure is located in an area where snow loads do not occur, the last combination becomes (D + L + L_r), and $C_D = 1.25$.

In this manner, it is possible for a smaller load of longer duration (with a small C_D) to be more critical than a larger load of shorter duration (with a large C_D). Whichever combination of loads, together with the appropriate load duration factor, produces the largest required member size is the one that must be used in the design of the structure. It may be necessary, therefore, to check several different combinations of loads to determine which combination governs the design. With some practice, the designer can often tell by inspection which combinations need to be checked. In many cases, only one or possibly two combinations need be checked. See Example 4.11.

EXAMPLE 4.11   Comparison of Load Combinations Using ASD

Determine the design loads and the critical load combination for the roof beam in Fig. 4.14. The tributary width to the beam and the design unit loads are given. The beam is fully braced to prevent lateral-torsional buckling.

$$\text{Tributary width} = 10 \text{ ft}$$
$$D = 20 \text{ psf}$$
$$L_r = 16 \text{ psf}$$

Figure 4.14

Part a

Load combination 1 (D alone):
$$w_D = 20 \times 10 = 200 \text{ lb/ft}$$
$$C_D = 0.90$$

Load combination 2 (D + L_r):
$$w_{TL} = (20 + 16)10 = 360 \text{ lb/ft}$$
$$C_D = 1.25$$

The reference design values for the beam are to be multiplied by 0.90 for load combination 1 and 1.25 for load combination 2. Theoretically, both load combinations must be considered. However, with some practice, the designer will be able to tell from the relative magnitude of the loads which combination is critical. For example, 360 lb/ft is so large in comparison with 200 lb/ft that load combination 2 will be critical. Therefore, calculations for load combination 1 are not required. If it cannot be determined by inspection which loading is critical, calculations for both load causes should be performed.

In some cases, calculations for two or more cases *must* be performed. Often this occurs in members with combined axial and bending loads. These types of problems are considered in Chap. 7

Part b

Show calculations which verify the critical load case for the fully braced beam in part *a* without complete stress calculations. Remove "duration" by dividing the design loads by the appropriate C_D factors.[*]

Load combination 1:
$$\frac{w_D}{C_D} = \frac{200}{0.9} = 222 \text{ lb/ft}^\dagger$$

Load combination 2:
$$\frac{w_{TL}}{C_D} = \frac{360}{1.25} = 288 \text{ lb/ft}^\dagger$$

$$222 < 288$$

$$\therefore \text{ load combination 2 governs}$$

---

[*] This method is only appropriate for tension members, connections, and beams with full lateral support. For columns and beams susceptible to buckling failure, this method is not appropriate and could lead to design errors.

[†] These modified loads are used to determine the critical load combination only. Actual design loads (for example, $w = 360$ lb/ft) should be used in calculations, and $C_D = 1.25$ should be applied to reference design values.

When designers first encounter the adjustment for duration of load, they like to have a *system* for determining the critical loading combination. See Example 4.11 part *b*. Essentially, the system involves removing the question of load duration from the problem. If the sum of the loads in a given combination is divided by the C_D for the combination, duration is removed from the load. If this is done for each required load combination, the resulting loads can be compared. The largest modified load represents the critical combination.

This method is not foolproof. The C_D has full effect for tension members, connections, and beams that are braced on the compression edge to prevent lateral-torsional buckling failure. However, since C_D does not apply for modulus of elasticity E, this method is not appropriate for structural members susceptible to buckling failure. Inappropriate use of this method for columns or laterally unsupported beams could lead to errors in design.

There is a second objection to the system just described. It runs counter to the recommendation that adjustment factors be applied to the reference design values and not to the design loads. Thus, if this analysis is used, the calculations should be done separately (perhaps on scrap paper). Once the critical combination is known, the actual design loads (not modified) can be used in formal calculations, and C_D can be applied to the reference design values in the usual manner.

4.16   Time Effect Factor λ (LRFD Only)

As mentioned in the previous section, wood has the unique property that it can support higher stresses if the loads are applied for a short period of time. This is particularly significant when one recognizes that overload, if it occurs, is probably the result of a temporary or short duration load. The approach used to account for this unique property differs significantly between ASD and LRFD. In Sec. 4.15, the development and use of the ASD load duration factor, C_D, was presented. Basically, in ASD, the load duration factor is determined based on the shortest duration load in a given combination.

In LRFD, the load duration effect is accounted for using the time effect factor, λ, which adjusts the nominal resistance based on the given LRFD load combination. A reliability-based approach was used in the development of the time effect factor for LRFD. The reliabilities resulting from specific load combinations were compared considering both short-term and long-term effects. Due to the load duration effect and depending on the temporal nature of the loads, the resulting reliabilities considering long-term loads were sometimes lower than the reliabilities considering only short-term loading. The time effect factor is, therefore, used to adjust the design resistance to ensure that consistent reliability is achieved for various load duration effects. This is similar to the resistance factor, ϕ, which is used to ensure that appropriate reliability is achieved for various limit states (failure modes).

It is important to note that, in addition to linking the time effect factor to specific load combination, the nominal resistance is based on short-duration (10 min) loading versus normal (10-year) duration as used in ASD. The format conversion factor, K_F, adjusts the reference design value from *normal duration* loading to a *nominal* design value for *short duration* loading. Therefore, values of the time effect

EXAMPLE 4.12   Time Effect Factor (LRFD Only)

Load combination	Time effect factor λ
1.4(D + F)	0.6
1.2(D + F + T) + 1.6(H) + 0.5 (L$_r$ or S or R)	0.6
1.2(D + F + T) + 1.6(L + H) +0.5 (L$_r$ or S or R)	0.7 when L is from storage
	0.8 when L is from occupancy
	1.25 when L is from impact
1.2D + 1.6(L$_r$ or S or R) + (L or 0.8W)	0.8
1.2D + 1.6W + L + 0.5 (L$_r$ or S or R)	1.0
1.2D + 1.0E + L + 0.2S	1.0
0.9D + 1.6W + 1.6H	1.0
0.9D + 1.0E + 1.6H	1.0

The load combinations and load factors given above are from ASCE 7-05 (Ref. 4.10). However, the relationship between the specific load combinations and time effect factors may be taken less directly than the table implies. Load combinations and load factors will continue to evolve and change based on improved understanding of loads. As they change and are adopted in the Code, the following relationships between the time effect factor and load combinations should be recognized: When the load combination is dominated by permanent loads such as D, F, or H, the time effect factor is taken as 0.6. Conversely, when the load combination is governed by short-duration actions such as W or E, the time effect factor of 1.0 is used. In the case of load combinations dominated by sustained live loads such as S, R, L$_r$, or L from storage or occupancy, then the time effect factor is either 0.7 or 0.8. When L is the result of impact, the time effect factor is taken as 1.25.

Finally, due to recognized behavior of connections and certain treated materials under impact-type loads, time effect factors greater than 1.0 (impact) are not allowed for connections or structural members that are pressure treated with waterborne preservatives or fire-retardant chemicals.

factor range from 0.6 for load combinations dominated by dead loads to 1.0 for load combinations dominated by short-duration loads such as wind and seismic actions. Unlike the load duration factor used in ASD, the time effect factor cannot be represented graphically. Rather, specific time effect factors are given for various load combinations. See Example 4.12.

4.17   Size Factor C_F

It has been known for some time that the size of a wood member has an effect on its unit strength (stress). This behavior is taken into account by the *size factor* C_F (NDS Sec. 4.3.6). The size factors are based on the size classification.

Visually graded Dimension lumber.   The size factors C_F for most species of visually graded Dimension lumber are summarized in the *Adjustment Factor* section that precedes NDS Supplement Tables 4A and 4F. The size factors for F_b, F_t, and F_c

are given in a table which depends on the *grade* and the *width* (depth) of the piece of lumber. For bending, the *thickness* of the member also affects the size factor.

The size factors for Southern Pine Dimension lumber are handled somewhat differently. For Southern Pine, a number of the size factors have been incorporated into the reference design values given in NDS Supplement Table 4B. Thus, the tabulated values for Southern Pine are said to be "size-specific." Unfortunately, the size-specific tables for Southern Pine do not completely avoid the use of a C_F multiplier. Bending values in Table 4B apply to lumber that has a nominal thickness of 2 in. A size factor of $C_F = 1.1$ is provided for F_b if the lumber being considered has a nominal thickness of 4 in. instead of 2 in. In addition, a size factor of $C_F = 0.9$ is provided for F_b, F_t, and F_c for Dimension lumber that has a width greater than 12 in.

Timbers. For Timbers, the size factor C_F applies only to F_b. Essentially, the size factor reflects the fact that as the *depth* of a beam increases, the *unit strength* (and correspondingly the bending design value) decreases. When the depth d of a timber exceeds 12 in., the size factor is defined by the expression.

$$C_F = \left(\frac{12}{d}\right)^{1/9}$$

For members that are less than 12 in. deep, the size factor defaults to unity: $C_F = 1.0$.

At one time this size factor was also used for glulam beams. However, the size factor for glulams has been replaced with the volume factor C_V (see Chap. 5).

4.18 Repetitive Member Factor C_r

Many wood structures have a series of closely spaced parallel members. The members are often connected together by sheathing or decking. In this arrangement, the performance of the *system* does not depend solely on the capacity of the weakest individual member. This can be contrasted to an engineered wood structure with relatively large structural members spaced a greater distance apart. The failure of one large member would essentially be a failure of the system.

The system performance of a series of small, closely spaced wood members is recognized in the NDS by providing either a 15 percent increase or a 4 percent increase in the reference design value for bending F_b. This increase is provided by the *repetitive member factor* C_r (NDS Sec. 4.3.9). It applies only to F_b for Dimension lumber or structural composite lumber used in a repetitive system. A *repetitive member system* is defined as one that has

1. Three or more parallel members of Dimension lumber or structural composite lumber

2. Members spaced not more than 24 in. o.c.

3. Members connected together by a load-distributing element such as roof, floor, or wall sheathing

For a repetitive member system, the reference F_b may be multiplied by $C_r = 1.15$. For all other framing systems and design values $C_r = 1.0$.

The repetitive member factor recognizes system performance. If one member should become overloaded, parallel members come into play. The load is distributed by sheathing to adjacent members, and the load is shared by a number of beams. The repetitive member factor is not applied to the larger sizes of wood members (i.e., Timbers and glulams) because these large members are not normally spaced closely enough together to qualify as repetitive members.

When a concentrated load is supported by a deck which distributes the load to parallel beams, the entire concentrated load need not be assumed to be supported by one member. NDS Sec. 15.1 provides a method for the *Lateral Distribution of a Concentrated Load* to adjacent parallel beams. According to Ref. 4.22, the single-member bending stress (that is, $C_r = 1.0$) applies if the load distribution in NDS Sec. 15.1 is used.

4.19   Flat Use Factor C_{fu}

Except for decking, reference design values for Dimension lumber apply to wood members that are stressed in flexure about the strong axis (major axis) of the cross section. The NDS refers to this conventional type of beam loading as *edge-wise* use or *load applied to narrow face* of the member.

In a limited number of situations, Dimension lumber may be loaded in bending about the minor axis (weak axis) of the cross section. The terms *flatwise use* and *load applied to wide face* describe this application. When members are loaded in bending about the weak axis, the reference value of F_b may be increased by multiplying by the *flat-use factor* C_{fu} (NDS Sec. 4.3.7). Numerical values for C_{fu} are given in the *adjustment factor* sections of NDS Supplement Tables 4A, 4B, 4C, and 4F.

Reference bending design values for Beams and Stringers also apply to the usual case of bending about the strong axis of the cross section. The NDS does not provide a flat-use factor for bending about the weak axis. However, size factors are applicable to both F_b and E for Beams and Stringers with bending about the weak axis. See NDS Supplement Table 4D. It is recommended that the designer contact the appropriate rules-writing agency for assistance if this situation is encountered. For example, WWPA and WCLIB provide adjustment factors for Beams and Stringers for Western Species for both bending strength and modulus of elasticity.

The reference bending design value for a glulam beam that is stressed in bending about the weak axis is given the symbol F_{byy}. Values of F_{byy} apply to glulams that have a cross-sectional dimension *parallel to the wide face of the laminations* of at least 12 in. For beams that are less than 12 in. wide, the value of F_{byy} may be increased by a flat-use factor (NDS Sec. 5.3.7). For values of C_{fu} for glulams, see NDS Supplement Tables 5A, 5B, 5C, and 5D.

4.20   Temperature Factor C_t

The strength of a member is affected by the temperature of the wood in service. Strength is increased as the temperature cools below the normal temperature range found in most buildings. On the other hand, the strength decreases as temperatures are increased. The *temperature factor* C_t is the multiplier that is used to reduce reference design values if prolonged exposure to higher than normal temperatures are encountered in a design situation.

Reference design values apply to wood used in the ordinary temperature range and perhaps *occasionally* heated up to 150°F. Prolonged exposure to temperatures above 150°F may result in a *permanent* loss of strength. Reductions in strength caused by heating below 150°F are generally reversible. In other words, strength is recovered as the temperature is reduced to the normal range.

Values of C_t are given NDS Sec. 2.3.3. The first temperature range that requires a reduction in design values is $100°F < T < 125°F$. At first, this seems to be a rather low temperature range. After all, temperatures in many of the warmer areas of the country often exceed 100°F. In these locations, is it necessary to reduce the reference design values for members in a wood roof system? The answer is that it is generally *not* considered necessary. Members in roof structures subjected to *temporarily* elevated temperatures are not usually subjected to the *full* design load under these conditions. For example, snow loads will not be present at these elevated temperatures, and roof live loads occur infrequently. Furthermore, any loss in strength should be regained when the temperature returns to normal.

For these and other reasons, $C_t = 1.0$ is normally used in the design of ordinary wood frame buildings. However, in an industrial plant there may be operations that cause temperatures to be consistently elevated. Structural members in these types of situations may require use of a temperature factor less than 1. Additional information is given in NDS Appendix C.

4.21   Incising Factor C_i

Many species, most notably the Southern Pines, readily accept preservative treatments, while other species do not accept pressure treatments as well. For species that are not easily treated, incising is often used to make the treatment effective. See Sec. 4.9. If incising is used to increase the penetration of the preservatives, some design values for Dimension lumber in the NDS Supplement must be adjusted (NDS Sec. 4.3.8). For the modulus of elasticity, $C_i = 0.95$, and for bending, tension, shear, and compression parallel to grain, $C_i = 0.80$. For compression perpendicular to grain as well as for non-incised treated lumber, the incising factor is taken as 1.0.

4.22   Resistance Factor ϕ (LRFD Only)

In LRFD a *resistance factor* ϕ, sometimes also referred to as a strength reduction factor, is used to allow for the possibility that the resistance may be less than computed. The resistance factor accounts for a variety of situations. In addition to variability of the strengths of the material, the resistance factor may account for differences between as-built dimensions and those used in design, effects of simplifying design assumptions, and the potential consequences of failure. Some

EXAMPLE 4.13   Resistance Factor (LRFD Only)

Property	Symbol	Value
Bending F_b	ϕ_b	0.85
Tension F_t	ϕ_t	0.80
Shear F_v	ϕ_v	0.75
Compression F_c, $F_{c\perp}$	ϕ_c	0.90
Stability $E_{\min}$	ϕ_s	0.85

Shear is a highly variable strength value and may produce sudden, brittle failures. Conversely, compression is a very ductile mode of failure. Accordingly, the resistance factor for shear is significantly lower than that for compression.

modes of failure can be potentially sudden or brittle, while other modes may be more ductile. Accordingly, the resistance factor ϕ is a function of the application and material property.

The NDS provides resistance factors in Appendix N, Table N2. For member design, the values of ϕ range from 0.75 for shear to 0.90 for compression. See Example 4.13.

4.23   Format Conversion Factor K_F (LRFD Only)

Reference design values published in the NDS Supplement include safety adjustments appropriate for ASD, as specified in ASTM standards (Refs. 4.12 and 4.13). These ASD safety adjustments must be removed to obtain nominal design values for LRFD. The format conversion factor K_F converts tabulated reference design values to nominal design values for LRFD. For example, the nominal design value for bending is obtained by multiplying the reference design value by K_F for bending:

$$F_{bn} = F_b \times K_F$$

The subscript n has been added to differentiate this as a *nominal* design value for use in LRFD.

It is important to note that the reference resistance used in LRFD is based on short duration (10-min) loading versus the normal (10-year) duration used in ASD. The format conversion factor K_F, also adjusts the reference design value for *normal duration loading* to a nominal design value for *short duration loading*.

Format conversion factors are provided for different applications and properties in Table N1 of NDS Appendix N. For member design, the values of K_F range from 1765 for $E_{\min}$ to 2.88 for shear. See Example 4.14.

EXAMPLE 4.14 **Format Conversion Factor (LRFD Only)**

Property	Value
Bending F_b	$2.16/\phi_b = 2.16/0.85 = 2.54$
Tension F_t	$2.16/\phi_t = 2.16/0.8 = 2.70$
Shear F_v,	$2.16/\phi_v = 2.16/0.75 = 2.88$
Compression parallel to the grain F_c	$2.16/\phi_c = 2.16/0.9 = 2.40$
Compression perpendicular to the grain $F_{c\perp}$	$1.875/\phi_c = 1.875/0.9 = 2.083$
Stability E_{min}	$1.5/\phi_s = 1.5/0.85 = 1.765$

All conversions are based on the prescribed resistance factor ϕ, and all except compression perpendicular to grain and modulus of elasticity used for stability have the same coefficient in the numerator. In ASD, compression perpendicular to grain is not adjusted for duration of load; that is, C_D is not applicable to $F_{c\perp}$. However, in LRFD, the time effect factor λ is applied to $F_{c\perp}$. E_{min} is not adjusted for load duration or time effect in either ASD or LRFD. Therefore, the format conversions for $F_{c\perp}$ and E_{min} differ from those for other properties.

4.24 Design Problem: Adjusted Design Values

If one examines NDS Tables 4.3.1, 5.3.1, 6.3.1, 7.3.1, 8.3.1, and 9.3.1, *Applicability of Adjustment Factors,* it will be noticed that the temperature factor C_t applies to all design values. However, it should be realized that this table shows what adjustments *may* be required under certain conditions. It simply serves as a reminder to the designer not to overlook a necessary adjustment. The table says nothing about the *frequency of use* of an adjustment factor.

It was observed in Sec. 4.20 that the temperatures in most wood buildings do not require a reduction in design values. Thus, C_t is a factor that is rarely used in the design of typical wood buildings, and the default value of $C_t = 1.0$ often applies. On the other hand, the load duration factor C_D (ASD) and time effect factor λ (LRFD) are common adjustment factors that are used in the design of practically all wood structures.

Several examples are given to illustrate the use of the NDS tables and the required adjustment factors. ASD and LRFD are presented together for side-by-side comparison. Complete design problems are given later in this book, and the current examples simply emphasize obtaining the correct *adjusted* design value. The first requirement is to obtain the correct *reference* value from the NDS Supplement for the given size, grade, and species group. The second step is to apply the appropriate *adjustment* factors.

Example 4.15 deals with four different sizes of the same grade (No. 1) in a single species group (Douglas Fir-Larch). Dimensions are obtained from NDS Supplement Tables 1A and 1B. The example clearly shows the effect of a number of variables. Several different loading conditions and adjustment factors are illustrated. The reader is encouraged to verify the reference design values and the adjustment factors from the NDS.

Some adjustment factors in NDS Table 4.3.1 for sawn lumber are not shown in this example. These factors do not apply to the given problem. Other factors may have a default value of unity and are shown for information purposes.

EXAMPLE 4.15   Determination of Adjusted Design Values

Determine the *adjusted* design values in both ASD and LRFD for the four members given below. All members are No. 1 DF-L. Bending loads will be about the strong axis of the cross section (load applied to narrow face). Bracing conditions are such that buckling is not a concern ($C_L = 1.0$). Consider dry-service conditions (EMC $\leq$ 19 percent) unless otherwise indicated. Normal temperature conditions apply.

For each member, a *single* load duration factor C_D or time effect factor λ will be used to adjust the design values for the given load combination. In practice, a number of loading conditions must be checked, and each load case will have an appropriate C_D or λ. Limiting each member to a single-load case is done for simplicity in this example.

Part a

Roof rafters are 2 × 8 at 24 in. o.c., and they directly support the roof sheathing. Loads are (D + L_r) for ASD and (1.2D + 1.6L_r) for LRFD. See Fig 4.15*a*.

Figure 4.15*a*   A 2 × 8 is a Dimension lumber size.

Reference design values of visually graded DF-L *Dimension lumber* are obtained from NDS Supplement Table 4A. The framing arrangement qualifies for the 15 percent increase in bending stress for repetitive members. Dimension lumber requires a size factor for F_b, F_t, and F_c.

ASD.   The load duration factor is 1.25 for the combination of (D + L_r).

$$F'_b = F_b(C_D \times C_M \times C_t \times C_F \times C_r \times C_i)$$
$$= 1000(1.25 \times 1.0 \times 1.0 \times 1.2 \times 1.15 \times 1.0)$$
$$= 1725 \text{ psi}$$
$$F'_t = F_t(C_D \times C_M \times C_t \times C_F \times C_i)$$
$$= 675(1.25 \times 1.0 \times 1.0 \times 1.2 \times 1.0) = 1012 \text{ psi}$$
$$F'_v = F_v(C_D \times C_M \times C_t \times C_i)$$
$$= 180(1.25 \times 1.0 \times 1.0 \times 1.0) = 225 \text{ psi}$$

$$F'_c = F_c(C_D \times C_M \times C_t \times C_F \times C_i)$$

$$= 1500(1.25 \times 1.0 \times 1.0 \times 1.05 \times 1.0) = 1968 \text{ psi}$$

$$E' = E(C_M \times C_t \times C_i)$$

$$= 1,700,000(1.0 \times 1.0 \times 1.0)$$

$$= 1,700,000 \text{ psi}$$

$$E'_{\min} = E_{\min}(C_M \times C_t \times C_i)$$

$$= 620,000(1.0 \times 1.0 \times 1.0) = 620,000 \text{ psi}$$

This value would be used in all beam and column stability calculations.

LRFD. The time effect factor λ is 0.8 for the combination $(1.2D + 1.6L_r)$. Also, LRFD is typically done using resistances instead of stresses. Section properties (S_x and A) are obtained from NDS Supplement Table 1B.
Bending:

$$F_{bn} = F_b \times K_F$$

$$= 1000(2.16/\phi_b) = 1000(2.16/0.85) = 2541 \text{ psi} = 2.54 \text{ ksi}$$

$$F'_{bn} = F_{bn}(\phi_b \times \lambda \times C_M \times C_t \times C_F \times C_r \times C_i)$$

$$= 2.54(0.85 \times 0.8 \times 1.0 \times 1.0 \times 1.2 \times 1.15 \times 1.0) = 2.38 \text{ ksi}$$

$$M'_n = F'_{bn} \times S_x$$

$$= 2.38 \times 13.14 = 31.3 \text{ kip-in.}$$

Tension:

$$F_{tn} = F_t \times K_F$$

$$= 675(2.16/\phi_t) = 675(2.16/0.80) = 1822 \text{ psi} = 1.82 \text{ ksi}$$

$$F'_{tn} = F_{tn}(\phi_t \times \lambda \times C_M \times C_t \times C_F \times C_i)$$

$$= 1.82(0.80 \times 0.8 \times 1.0 \times 1.0 \times 1.2 \times 1.0) = 1.40 \text{ ksi}$$

$$T'_n = F'_{tn} \times A$$

$$= 1.40 \times 10.88 = 15.2 \text{ k}$$

Shear:

$$F_{vn} = F_v \times K_F$$

$$= 180(2.16/\phi_v) = 180(2.16/0.75) = 518 \text{ psi} = 0.518 \text{ ksi}$$

$$F'_{vn} = F_{vn}(\phi_v \times \lambda \times C_M \times C_t \times C_i)$$

$$= 0.518(0.75 \times 0.8 \times 1.0 \times 1.0 \times 1.0) = 0.311 \text{ ksi}$$

$$V'_n = F'_{vn} \times (2/3 \times A)^*$$

$$= 0.311 \times (2/3 \times 10.88) = 2.26 \text{ k}$$

---

*For rectangular sections only.

Compression parallel to grain:

$$F_{cn} = F_c \times K_F$$
$$= 1500(2.16/\phi_c) = 1500(2.16/0.90) = 3600 \text{ psi} = 3.60 \text{ ksi}$$
$$F'_{cn} = F_{cn}(\phi_c \times \lambda \times C_M \times C_t \times C_F \times C_i)$$
$$= 3.60(0.90 \times 0.8 \times 1.0 \times 1.0 \times 1.05 \times 1.0) = 2.72 \text{ ksi}$$
$$P'_n = F'_{cn} \times A$$
$$= 2.72 \times 10.88 = 29.6 \text{ k}$$

Modulus of elasticity:

$$E' = E(C_M \times C_t \times C_i)^{\dagger}$$
$$= 1,700,000(1.0 \times 1.0 \times 1.0) = 1,700,000 \text{ psi} = 1700 \text{ ksi}$$

Modulus of Elasticity for stability calculations:

$$E_{\min \text{-}n} = E_{\min} \times K_F$$
$$= 620,000(1.5/\phi_s) = 620,000(1.5/0.85) = 1,094,000 \text{ psi} = 1094 \text{ ksi}$$
$$E'_{\min \text{-}n} = E_{\min \text{-}n}(\phi_s \times C_M \times C_t \times C_i)$$
$$= 1094(0.85 \times 1.0 \times 1.0 \times 1.0) = 930 \text{ ksi}$$

This value would be used in all beam and column stability calculations.

Part b

Roof beams are 4 × 10 at 4 ft-0 in. o.c. Loads are (D + S) for ASD and (1.2D + 1.6S) for LRFD. See Fig. 4.15b.

Figure 4.15b A 4 × 10 is a Dimension lumber size.

Reference design values for visually graded DF-L *Dimension lumber* are again obtained from NDS Supplement Table 4A. A 4-ft framing module exceeds the 24-in. spacing limit for repetitive members, and $C_r = 1.0$.

---

†There is no difference between the E' used in ASD and the E' used in LRFD.

ASD.   The load duration factor is 1.15 for the combination of (D + S).

$$F'_b = F_b(C_D \times C_M \times C_t \times C_F \times C_r \times C_i)$$

$$= 1000(1.15 \times 1.0 \times 1.0 \times 1.2 \times 1.0 \times 1.0) = 1380 \text{ psi}$$

$$F'_t = F_t(C_D \times C_M \times C_t \times C_F \times C_i)$$

$$= 675(1.15 \times 1.0 \times 1.0 \times 1.1 \times 1.0) = 854 \text{ psi}$$

$$F'_v = F_v(C_D \times C_M \times C_t \times C_i)$$

$$= 180(1.15 \times 1.0 \times 1.0 \times 1.0) = 207 \text{ psi}$$

$$F'_c = F_c(C_D \times C_M \times C_t \times C_F \times C_i)$$

$$= 1500(1.15 \times 1.0 \times 1.0 \times 1.05 \times 1.0) = 1725 \text{ psi}$$

$$E' = E(C_M \times C_t \times C_i)$$

$$= 1,700,000(1.0 \times 1.0 \times 1.0) = 1,700,000 \text{ psi}$$

$$E'_{min} = E_{min}(C_M \times C_t \times C_i)$$

$$= 620,000(1.0 \times 1.0 \times 1.0) = 620,000 \text{ psi}$$

This value would be used in all beam and column stability calculations.

LRFD.   The time effect factor λ is 0.8 for the combination (1.2D + 1.6S). LRFD is typically done using resistances instead of stresses, and section properties (S_x and A) are obtained from NDS Supplement Table 1B.
Bending:

$$F_{bn} = F_b \times K_F$$

$$= 1000(2.16/\phi_b) = 1000(2.16/0.85) = 2541 \text{ psi} = 2.54 \text{ ksi}$$

$$F'_{bn} = F_{bn}(\phi_b \times \lambda \times C_M \times C_t \times C_F \times C_r \times C_i)$$

$$= 2.54(0.85 \times 0.8 \times 1.0 \times 1.0 \times 1.2 \times 1.0 \times 1.0) = 2.07 \text{ ksi}$$

$$M'_n = F'_{bn} \times S_x$$

$$= 2.07 \times 49.91 = 103.5 \text{ kip-in.}$$

Tension:

$$F_{tn} = F_t \times K_F$$

$$= 675(2.16/\phi_t) = 675(2.16/0.80) = 1822 \text{ psi} = 1.82 \text{ ksi}$$

$$F'_{tn} = F_{tn}(\phi_t \times \lambda \times C_M \times C_t \times C_F \times C_i)$$

$$= 1.82(0.80 \times 0.8 \times 1.0 \times 1.0 \times 1.1 \times 1.0) = 1.28 \text{ ksi}$$

$$T'_n = F'_{tn} \times A$$

$$= 1.28 \times 32.38 = 41.5 \text{ k}$$

Shear:

$$F_{vn} = F_v \times K_F$$

$$= 180(2.16/\phi_v) = 180(2.16/0.75) = 518 \text{ psi} = 0.518 \text{ ksi}$$

$$F'_{vn} = F_{vn}(\phi_v \times \lambda \times C_M \times C_t \times C_i)$$

$$= 0.518(0.75 \times 0.8 \times 1.0 \times 1.0 \times 1.0) = 0.311 \text{ ksi}$$

$$V'_n = F'_{vn} \times (2/3 \times A)^*$$

$$= 0.311 \times (2/3 \times 32.38) = 6.71 \text{ k}$$

Compression parallel to grain:

$$F_{cn} = F_c \times K_F$$

$$= 1500(2.16/\phi_c) = 1500(2.16/0.90) = 3600 \text{ psi} = 3.60 \text{ ksi}$$

$$F'_{cn} = F_{cn}(\phi_c \times \lambda \times C_M \times C_t \times C_F \times C_i)$$

$$= 3.60(0.90 \times 0.8 \times 1.0 \times 1.0 \times 1.05 \times 1.0) = 2.72 \text{ ksi}$$

$$P'_n = F'_{cn} \times A$$

$$= 2.72 \times 32.38 = 88.1 \text{ k}$$

Modulus of Elasticity:

$$E' = E(C_M \times C_t \times C_i)^\dagger$$

$$= 1{,}700{,}000(1.0 \times 1.0 \times 1.0) = 1{,}700{,}000 \text{ psi} = 1700 \text{ ksi}$$

Modulus of Elasticity for stability calculations:

$$E_{\min\text{-}n} = E_{\min} \times K_F$$

$$= 620{,}000(1.5/\phi_s) = 620{,}000(1.5/0.85) = 1{,}094{,}000 \text{ psi} = 1094 \text{ ksi}$$

$$E'_{\min\text{-}n} = E_{\min\text{-}n}(\phi_s \times C_M \times C_t \times C_i)$$

$$= 1094(0.85 \times 1.0 \times 1.0 \times 1.0) = 930 \text{ ksi}$$

This value would be used in all beam and column stability calculations.

Part c

A 6 × 16 floor beam supports loads from both the floor and the roof. Several load combinations have been studied, and the critical loading for ASD is (D + L + L_r). For LRFD the critical load combination is (1.2D + 1.6L_r + L). See Fig. 4.15c.

---

*For rectangular sections only.

†There is no difference between the E′ used in ASD and the E′ used in LRFD.

Figure 4.15c   A 6 × 16 is a Beams and Stringers size.

A B&S has a minimum cross-sectional dimension of 5 in., and the width is more than 2 in. larger than the thickness. *Beams and Stringers* sizes are described in Example 4.6 (Sec. 4.12). Reference design values are obtained from NDS Supplement Table 4D. To be conservative, take the smaller tabulated design values listed for the two sets of grading rules (WCLIB and WWPA). In this problem, the values are the same for both.

Unlike Dimension lumber, large members have one size factor, and it applies to the bending design value only. When the depth of a Timber exceeds 12 in., the size factor is given by the following expression

$$C_F = \left(\frac{12}{d}\right)^{1/9} = \left(\frac{12}{15.5}\right)^{1/9} = 0.972$$

ASD.   The load duration factor for the combination of loads is based on the shortest duration load in the combination. Therefore, $C_D = 1.25$.

$$F'_b = F_b(C_D \times C_M \times C_t \times C_F)$$

$$= 1350(1.25 \times 1.0 \times 1.0 \times 0.972) = 1640 \text{ psi}$$

$$F'_t = F_t(C_D \times C_M \times C_t)$$

$$= 675(1.25 \times 1.0 \times 1.0) = 844 \text{ psi}$$

$$F'_v = F_v(C_D \times C_M \times C_t)$$

$$= 170(1.25 \times 1.0 \times 1.0) = 212 \text{ psi}$$

$$F'_c = F_c(C_D \times C_M \times C_t)$$

$$= 925(1.25 \times 1.0 \times 1.0) = 1156 \text{ psi}$$

$$E' = E(C_M \times C_t)$$

$$= 1,600,000(1.0 \times 1.0) = 1,600,000 \text{ psi}$$

$$E'_{min} = E_{min}(C_M \times C_t)$$

$$= 580,000(1.0 \times 1.0) = 580,000 \text{ psi}$$

This value would be used in all beam and column stability calculations.

LRFD.   The time effect factor λ is 0.8 for the combination ($1.2D + 1.6L_r + L$). Section properties to determine resistances values for LRFD are obtained from NDS Supplement Table 1B.

Bending:

$$F_{bn} = F_b \times K_F$$
$$= 1350(2.16/\phi_b) = 1350(2.16/0.85) = 3431 \text{ psi} = 3.43 \text{ ksi}$$
$$F'_{bn} = F_{bn}(\phi_b \times \lambda \times C_M \times C_t \times C_F)$$
$$= 3.43(0.85 \times 0.8 \times 1.0 \times 1.0 \times 0.972) = 2.27 \text{ ksi}$$
$$M'_n = F'_{bn} \times S_x$$
$$= 2.27 \times 220.2 = 499 \text{ kip-in.}$$

Tension:

$$F_{tn} = F_t \times K_F$$
$$= 675(2.16/\phi_t) = 675(2.16/0.80) = 1822 \text{ psi} = 1.82 \text{ ksi}$$
$$F'_{tn} = F_{tn}(\phi_t \times \lambda \times C_M \times C_t)$$
$$= 1.82(0.80 \times 0.8 \times 1.0 \times 1.0) = 1.17 \text{ ksi}$$
$$T'_n = F'_{tn} \times A$$
$$= 1.17 \times 85.25 = 99.4 \text{ k}$$

Shear:

$$F_{vn} = F_v \times K_F$$
$$= 170(2.16/\phi_v) = 170(2.16/0.75) = 490 \text{ psi} = 0.490 \text{ ksi}$$
$$F'_{vn} = F_{vn}(\phi_v \times \lambda \times C_M \times C_t)$$
$$= 0.490(0.75 \times 0.8 \times 1.0 \times 1.0) = 0.294 \text{ ksi}$$
$$V' = F'_{vn} \times (2/3 \times A)^*$$
$$= 0.294 \times (2/3 \times 85.25) = 16.7 \text{ k}$$

Compression parallel to grain:

$$F_{cn} = F_c \times K_F$$
$$= 925(2.16/\phi_c) = 925(2.16/0.90) = 2220 \text{ psi} = 2.22 \text{ ksi}$$
$$F'_{cn} = F_{cn}(\phi_c \times \lambda \times C_M \times C_t)$$
$$= 2.22(0.90 \times 0.8 \times 1.0 \times 1.0) = 1.60 \text{ ksi}$$
$$P'_n = F'_{cn} \times A$$
$$= 1.60 \times 85.25 = 136 \text{ k}$$

*For rectangular sections only.

Modulus of Elasticity:

$$E' = E(C_M \times C_t)^\dagger$$

$$= 1,600,000(1.0 \times 1.0) = 1,600,000 \text{ psi} = 1600 \text{ ksi}$$

Modulus of Elasticity for stability calculations:

$$E_{\min\text{-}n} = E_{\min} \times K_F$$

$$= 580,000(1.5/\phi_s) = 580,000(1.5/0.85) = 1,024,000 \text{ psi} = 1024 \text{ ksi}$$

$$E'_{\min\text{-}n} = E_{\min\text{-}n}(\phi_s \times C_M \times C_t)$$

$$= 1,024(0.85 \times 1.0 \times 1.0) = 870 \text{ ksi}$$

This value would be used in all beam and column stability calculations.

Part d

A 6 × 8 is used as a column to support a roof. It also supports tributary wind forces, and the critical loading condition has been determined for ASD to be D + 0.75 (L + W) and for LRFD to be (1.2D + 1.6W + L). High humidity conditions exist, and the moisture content of this member may exceed 19 percent. See Fig. 4.15*d*.

A P&T has a minimum cross-sectional dimension of 5 in., and the width is not more than 2 in. larger than the thickness. *Posts and Timbers* sizes are described in Example 4.6 (Sec. 4.12). Reference design values are obtained from NDS Supplement Table 4D. To be conservative, take the smaller tabulated design values listed for the two sets of grading rules (WCLIB and WWPA). In this problem, the values are the same for both.

The depth of this member is less than 12 in., and C_F defaults to unity. Recall that for Timbers the size factor applies only to F_b.

ASD.   The load duration factor for the combination of loads is based on the shortest-duration load in the combination. Therefore, $C_D = 1.6$.

$$F'_b = F_b(C_D \times C_M \times C_t \times C_F)$$

$$= 1200(1.6 \times 1.0 \times 1.0 \times 1.0) = 1920 \text{ psi}$$

$$F'_t = F_t(C_D \times C_M \times C_t)$$

$$= 825(1.6 \times 1.0 \times 1.0) = 1320 \text{ psi}$$

$$F'_v = F_v(C_D \times C_M \times C_t)$$

$$= 170(1.6 \times 1.0 \times 1.0) = 272 \text{ psi}$$

$$^*F'_c = F_c(C_D \times C_M \times C_t)$$

$$= 1000(1.6 \times 0.91 \times 1.0) = 1456 \text{ psi}$$

---

†There is no difference between the E′ used in ASD and the E′ used in LRFD.

* The compression parallel to grain design value would also need to be adjusted for buckling using the column stability factor C_P. See Sec. 7.4.

Figure 4.15d A 6 × 8 is a Posts and Timbers size.

$$E' = E(C_M \times C_t)$$
$$= 1{,}600{,}000(1.0 \times 1.0) = 1{,}600{,}000 \text{ psi}$$
$$E'_{\min} = E_{\min}(C_M \times C_t)$$
$$= 580{,}000(1.0 \times 1.0) = 580{,}000 \text{ psi}$$

This value would be used in all beam and column stability calculations.

LRFD. The time effect factor λ is 1.0 for the combination (1.2D + 1W + L). LRFD is typically done using resistances instead of stresses, and section properties are obtained from NDS Supplement Table 1B.
Bending:

$$F_{bn} = F_b \times K_F$$
$$= 1200(2.16/\phi_b) = 1200(2.16/0.85) = 3049 \text{ psi} = 3.05 \text{ ksi}$$
$$F'_{bn} = F_{bn}(\phi_b \times \lambda \times C_M \times C_t \times C_F)$$
$$= 3.05(0.85 \times 1.0 \times 1.0 \times 1.0 \times 1.0) = 2.59 \text{ ksi}$$
$$M'_n = F'_{bn} \times S_x$$
$$= 2.59 \times 51.56 = 134 \text{ kip-in.}$$

Tension:

$$F_{tn} = F_t \times K_F$$
$$= 825(2.16/\phi_t) = 825(2.16/0.80) = 2227 \text{ psi} = 2.23 \text{ ksi}$$
$$F'_{tn} = F_{tn}(\phi_t \times \lambda \times C_M \times C_t)$$
$$= 2.23(0.80 \times 1.0 \times 1.0 \times 1.0) = 1.78 \text{ ksi}$$
$$T'_n = F'_{tn} \times A$$
$$= 1.78 \times 41.25 = 73.5 \text{ k}$$

Shear:

$$F_{vn} = F_v \times K_F$$
$$= 170(2.16/\phi_v) = 170(2.16/0.75) = 490 \text{ psi} = 0.490 \text{ ksi}$$
$$F'_{vn} = F_{vn}(\phi_v \times \lambda \times C_M \times C_t)$$
$$= 0.490(0.75 \times 1.0 \times 1.0 \times 1.0) = 0.367 \text{ ksi}$$

$$V'_n = F'_{vn} \times (2/3 \times A)^*$$
$$= 0.367 \times (2/3 \times 41.25) = 10.1 \text{ k}$$

Compression parallel to grain:

$$F_{cn} = F_c \times K_F$$
$$= 1000(2.16/\phi_c) = 1000(2.16/0.90) = 2400 \text{ psi} = 2.40 \text{ ksi}$$
$$F'_{cn} = F_{cn}(\phi_c \times \lambda \times C_M \times C_t)$$
$$= 2.40(0.90 \times 1.0 \times 0.91 \times 1.0) = 1.97 \text{ ksi}$$

$$^{**}P'_n = F'_{cn} \times A$$
$$= 1.97 \times 41.25 = 81.1 \text{ k}$$

Modulus of Elasticity:

$$E' = E(C_M \times C_t)^\dagger$$
$$= 1,600,000(1.0 \times 1.0) = 1,600,000 \text{ psi} = 1600 \text{ ksi}$$

Modulus of Elasticity for stability calculations:

$$E_{\min\text{-}n} = E_{\min} \times K_F$$
$$= 580,000(1.5/\phi_s) = 580,000(1.5/0.85) = 1,024,000 \text{ psi} = 1024 \text{ ksi}$$
$$E'_{\min\text{-}n} = E_{\min\text{-}n}(\phi_s \times C_M \times C_t)$$
$$= 1024(0.85 \times 1.0 \times 1.0) = 870 \text{ ksi}$$

This value would be used in all beam and column stability calcaulations.

4.25 Future Directions in Wood Design

The wood industry is not a static business. If there is a better way of doing something, a better way of doing it will be found. In this book, the question of "better" generally refers to developing more accurate methods of structural design, but there is an underlying economic force that drives the system.

It was noted at the beginning of Chap. 4 that many wood-based products that are in widespread use today were unavailable only a few years ago. These include a number of structural-use panels, wood I-joists, resawn glulam beams, laminated veneer lumber (LVL), and more recently parallel strand lumber (PSL). A number of these developments, especially in the area of reconstituted wood products, are the result of new technology, and they represent an economic response to environmental concerns and resource constraints. With these products, the move is plainly in the direction of *engineered* wood construction.

---

*For rectangular sections only.

†There is no difference between the E' used in ASD and the E' used in LRFD.

** The compression parallel to grain capacity (resistence) would also need to be adjusted for buckling using the column stability factor C_P. See Sec. 7.4.

The design profession is caught in the middle of this development spiral. Anyone who is at all familiar with previous editions of the NDS will testify to the broad changes to the wood design criteria. Relatively recent changes included new lumber values originating from the In-Grade Test Program, new column and laterally unbraced beam formulas, new interaction equation for members with combined stresses, an engineering mechanics approach to the design of wood connections, and the inclusion of various wood product lines and systems formerly not covered by the NDS, including shearwalls, diaphragms, structural wood panels, structural composite lumber, I-joists, and metal plate connected wood trusses.

Traditionally, the NDS has been based on a deterministic method known as *allowable stress design* (ASD). Some argue that the method should be referred to more appropriately as *working stress design* (WSD) because the stresses that are computed are based on working or service loads. Both names have been used in the past, but ASD seems to be the term most widely used today.

Recent trends in structural design are based on reliability theory. As with ASD, various terms are used to refer to this alternative system. These include *reliability-based design, probability-based design, limit states design,* and *load and resistance factor design* (LRFD). Of these, LRFD is the approach that is generally agreed upon by the profession as the appropriate technique for use in structural design in the United States. Reinforced concrete has operated under this general design philosophy in the "strength" method for quite some time. The structural-steel industry has recently adopted a combined ASD and LRFD format similar to the NDS (Ref. 4.4).

As evidenced by the inclusion of LRFD in the new NDS, the wood industry is in the process of moving to an LRFD format. In mid-1991, the wood industry completed a 3-year project to develop a *draft LRFD Specification for Engineered Wood Construction.* This document was developed by a team from the wood industry, university faculty, and the design profession, and was subjected to trial use by professionals knowledgeable in the area of timber design. The first edition of the LRFD specification was published by the American Society of Civil Engineers (ASCE). Following this effort, the AF&PA committee responsible for maintaining the NDS developed the new version of the NDS that presents both design methods, ASD and LRFD, side by side in a single design specification. This was possible, since both the ASD and LRFD design specifications are based on the same behavioral equations. Already, all model building codes, including the IBC, recognize the LRFD specification as an alternate design procedure.

4.26   References

[4.1]   American Forest and Paper Association (AF&PA). 2006. ASD/LRFD *Manual for Engineered Wood Construction,* 2005 ed., AF&PA, Washington, DC.

[4.2]   American Forest and Paper Association (AF&PA). 2006. *Commentary on the National Design Specification for Wood Construction,* 2005 ed., AF&PA, Washington DC.

[4.3]   American Forest and Paper Association (AF&PA). 2005. *National Design Specification for Wood Construction and Supplement.* ANSI/AF&PA NDS-2005, AF&PA, Washington DC.

[4.4]   American Institute of Steel Construction (AISC). 2005. *Steel Construction Manual—13th* ed., AISC, Chicago, IL.

[4.5]   American Institute of Timber Construction (AITC). 1998. *Standard for Preservative Treatment of Structural Glued Laminated Timber, AITC 109-98,* AITC, Englewood, CO.

[4.6]   American Institute of Timber Construction (AITC). 1992. *Guidelines to Minimize Moisture Entrapment in Panelized Wood Roof Systems, AITC Technical Note 20,* AITC, Englewood, CO.

[4.7]   American Society of Civil Engineers (ASCE). 1975. *Wood Structures: A Design Guide and Commentary,* ASCE, Reston, VA.

[4.8]   American Society of Civil Engineers (ASCE). 1982. *Evaluation, Maintenance, and Upgrading of Wood Structures,* ASCE, Reston, VA.

[4.9]   American Society of Civil Engineers (ASCE). 1989. *Classic Wood Structures,* ASCE, Reston, VA.

[4.10] American Society of Civil Engineers (ASCE). 2006. *Minimum Design Loads for Buildings and Other Structures, ASCE 7-05,* ASCE, Reston, VA.

[4.11] American Society for Testing and Materials (ASTM). 2006. "Standard Practice for Establishing Clear Wood Strength Values," ASTM D2555-06, *Annual Book of Standards, Vol. 04.10 Wood,* ASTM, Philadelphia, PA.

[4.12] American Society for Testing and Materials (ASTM). 2006. "Standard Practice for Establishing Structural Grades and Related Allowable Properties for Visually Graded Lumber," ASTM D245-06, *Annual Book of Standards, Vol. 04.10 Wood,* ASTM, Philadelphia, PA.

[4.13] American Society for Testing and Materials (ASTM). 2006. "Standard Practice for Establishing Allowable Properties for Visually-Graded Dimension Lumber from In-Grade Tests of Full-Size Specimens," ASTM D1990-00 (reapproved 2002), *Annual Book of Standards, Vol. 04.10 Wood,* ASTM, Philadelphia, PA.

[4.14] American Society for Testing and Materials (ASTM). 2006. "Standard Test Methods for Small Clear Specimens of Timber," ASTM D143-94 (reapproved 2000), *Annual Book of Standards, Vol. 04.10 Wood,* ASTM, Philadelphia, PA.

[4.15] American Society for Testing and Materials (ASTM). 2006. "Standard Test Methods for Mechanical Properties of Lumber and Wood-Base Structural Material," ASTM D4761-05, *Annual Book of Standards, Vol. 04.10 Wood,* ASTM, Philadelphia, PA.

[4.16] APA—The Engineered Wood Association. 1999. *Moisture Control in Low Slope Roofs, Technical Note EWS R525B,* APA—The Engineered Wood Association, Engineered Wood Systems, Tacoma, WA.

[4.17] APA—The Engineered Wood Association. 2003. *Controlling Mold and Mildew. Form A525A.* APA—The Engineered Wood Association, Tacoma, WA.

[4.18] Dietz, A.G., Schaffer, E.L., and Gromala, D.S. (eds.). 1982. *Wood as a Structural Material,* Clark C. Heritage Memorial Series, vol. 2, Pennsylvania State University, University Park, PA.

[4.19] Faherty, K.F., and Williamson, T.G. (eds.). 1995. *Wood Engineering and Construction Handbook,* 2nd ed., McGraw-Hill, New York, NY.

[4.20] Forest Products Laboratory (FPL). 1999. *Wood Handbook: Wood as an Engineering Material, Technical Report 113,* FPL, Forest Service, U.S.D.A., Madison, WI.

[4.21] Green, D.W. 1989. "Moisture Content and the Shrinkage of Lumber," Research Paper FPL-RP-489, Forest Products Laboratory, Forest Service, U.S.D.A., Madison, WI.

[4.22] Gurfinkel, G. 1981. *Wood Engineering,* 2nd ed., Kendall/Hunt Publishing (available through Southern Forest Products Association, Kenner, LA).

[4.23] Hoyle, R.J., and Woeste, F.E. 1989. *Wood Technology in the Design of Structures,* 5th ed., Iowa State University Press, Ames, IA.

[4.24] Meyer, R.W., and Kellogg, R.M. (eds.). 1982. *Structural Use of Wood in Adverse Environments,* Van Nostrand Reinhold, New York, NY.

[4.25] Pneuman, F.C. 1991. "Inspection of a Wood-Framed-Warehouse-Type Structure," *Wood Design Focus,* vol. 2, no. 2.

[4.26] Rummelhart, R., and Fantozzi, J.A. 1992. "Multistory Wood-Frame Structures: Shrinkage Considerations and Calculations," *Proceedings of the 1992 ASCE Structures Congress,* American Society of Civil Engineers, Reston, VA.

4.27   Problems

Design values and adjustment factors in the following problems are to be taken from the 2005 NDS. Assume wood will be used in dry-service conditions and at normal temperatures unless otherwise noted.

4.1 *a.* Describe softwoods.
  b. Describe hardwoods.
  c. What types of trees are used for most structural lumber?

4.2 Sketch the cross section of a log. Label and define the following items:
  a. Annual ring
  b. The two types of wood cells
  c. Heartwood and sapwood

4.3 Define the following terms:
  a. Moisture content
  b. Fiber saturation point
  c. Equilibrium moisture content

4.4 Give the moisture content ranges for:
  a. Dry lumber
  b. Green lumber

4.5 What is the average EMC for an enclosed building in southern California? Cite reference.

4.6 List the components of the 2005 *Wood Design Package.*

4.7 *a.* List and describe the *Supplements* of the NDS.
  b. Why is the Supplement for *special design provisions for wind and seismic* separated from the NDS and the other Supplement?

4.8 Determine the dressed size, area, moment of inertia, and section modulus for the following members. Give values for both axes. Tables may be used (cite reference).
  a. 2 × 4
  b. 8 × 8
  c. 4 × 10
  d. 6 × 16

4.9 *a.* Give the range of sizes of lumber in Dimension lumber.
  b. Give the range of sizes of lumber in Timbers.
  c. Briefly summarize why the design values in the NDS Supplement for members in these broad categories are given in separate tables. What tables apply to Dimension lumber and what tables apply to Timbers?

4.10 Give the range of sizes for the following size and use subcategories. In addition, indicate whether these categories are under the general classification of *Dimension lumber* or *Timbers.*
  a. Beams and Stringers
  b. Structural Light Framing
  c. Decking
  d. Structural Joists and Planks
  e. Posts and Timbers
  f. Light Framing
  g. Stud

4.11   Briefly describe what is meant by the terms "visually graded sawn lumber" and "machine stress-rated (MSR) lumber." What tables in the NDS Supplement give design values for each? Are there any size distinctions? Explain.

4.12   Assume that the following members are visually graded lumber from a species group other than Southern Pine. Indicate whether the members are a size of Dimension lumber, Beams and Stringers (B&S), or Posts and Timbers (P&T). Also give the appropriate table in the NDS Supplement for obtaining design values. The list does not include material that is graded as Decking.
  a. 10×12 e. 2×12
  b. 14×14 f. 6×12
  c. 4×8 g. 8×12
  d. 4×4 h. 8×10

4.13   Repeat Prob. 4.12 except the material is Southern Pine.

4.14   What grades are listed in the NDS Supplement for visually graded Hem-Fir in the following size categories? Give table reference.
  a. Dimension lumber
  b. Beams and Stringers (B&S)
  c. Posts and Timbers (P&T)

4.15   What grades are listed in the NDS Supplement for visually graded Southern Pine in the following size categories? Give table reference.
  a. Dimension lumber
  b. Beams and Stringers (B&S)
  c. Posts and Timbers (P&T)

4.16   Give the reference design values for No.1 DF-L for the following sizes. List values for F_b, F_t, F_v, $F_{c\perp}$, F_c, E, and E_{min}. Give table reference.
  a. 10×10 e. 2×10
  b. 12×14 f. 6×12
  c. 4×16 g. 6×8
  d. 4×4 h. 10×14

4.17   Using the format conversion factor K_F, determine the nominal design values for LRFD for Prob. 4.16. What resistance factor ϕ would be used for each value?

4.18   Give the notation for the following adjustment factors. In addition, list the design properties (that is, F_b, F_t, F_v, $F_{c\perp}$, F_c, E, or E_{min}) that may require adjustment (NDS Table 4.3.1) by the respective factors.
  a. Size factor   e. Temperature factor
  b. Time effect factor   f. Wet service factor
  c. Load duration factor   g. Flat-use factor
  d. Repetitive member factor

4.19   Briefly describe the following adjustments. To what design values do they apply? Give NDS reference for numerical values of adjustment factors.

    a. Load duration factor
    b. Time effect factor
    c. Wet service factor
    d. Size factor
    e. Repetitive member factor

4.20   Distinguish between the *load duration factor* and *time effect factor.*

4.21   Give the load duration factor C_D associated with the following loads:
    a. Snow
    b. Wind
    c. Floor live load
    d. Roof live load
    e. Dead load

4.22   Give the time effect factor λ for the following load combinations:
    a. 1.2D + 1.6S + L
    b. 1.2D + 1.6W + L + 0.5S
    c. 1.2D + 1.6L + 0.5S with L from occupancy
    d. 1.2D + 1.6L$_r$ + L
    e. 1.4D

4.23   What reference design values for wood, if any, are not subject to adjustment for duration of loading (ASD)? For time effect (LRFD)?

4.24   Above what moisture content is it necessary to reduce the reference design values for most species of (*a*) sawn lumber and (*b*) glulam?

4.25   Under what conditions is it necessary to adjust the reference design values in wood design for temperature effects? Cite NDS reference for the temperature modification factors.

4.26   Distinguish between pressure-preservative treated wood and fire-retardant treated wood. Under what conditions is it necessary to adjust reference design values in wood design for the effects of pressure-impregnated chemicals? Where are adjustment factors obtained?

4.27   Should lumber be pressure-treated if it is to be used in an application where it will be continuously submerged in fresh water? Salt water? Explain.

4.28   *Given:* A fully braced structural member in a building is subjected to several different loads, including roof loads of D = 3 k and L$_r$ = 5 k; floor loads of D = 6 k and occupancy L = 10 k; and W = 10 k (resulting from overturning forces on the lateral-force-resisting system). Consider the ASD load combinations described in Sec. 2.16.

       Find: The critical ASD combination of loads.

4.29   *Given:* A fully braced structural member in a building is subjected to several different loads, including roof loads of D = 5 k and L$_r$ = 7 k; floor loads of

D = 6 k and occupancy L = 15 k; S = 18 k; and W = 10 k and E = 12 k (resulting from overturning forces on the lateral-force-resisting system). Consider the ASD load combinations described in Sec. 2.16.

Find: The critical ASD combination of loads.

4.30   Determine the ASD reference and adjusted design values for the following members and loading conditions. All members are No. 2 Hem-Fir and are fully braced to prevent lateral-torsional buckling. Bending occurs about the strong axis.
a. Roof joists are 2 × 10 at 16 in. o.c. which directly support the roof sheathing. Loads are (D + S).
b. A 6 × 14 carries an equipment load that can be considered a permanent load.
c. Purlins in a roof are 4 × 14 at 8 ft o.c. Loads are (D + L_r).
d. Floor beams are 4 × 6 at 4 ft o.c. Loads are (D + L). High-humidity conditions exist, and the moisture content may exceed 19 percent.

4.31   Determine the LRFD nominal and adjusted design values and adjusted design resistances for the following members and loading conditions. All members are No. 2 Hem-Fir and are fully braced to prevent lateral-torsional buckling. Bending occurs about the strong axis.
a. Roof joists are 2 × 10 at 16 in. o.c. which directly support the roof sheathing. Loads are 1.2D + 1.6S.
b. A 6 × 16 carries an equipment load that can be considered storage using 1.2D + 1.6L.
c. Purlins in a roof are 4 × 14 at 8 ft o.c. Loads are 1.2D + 1.6L_r.
d. Floor beams are 4 × 6 at 4 ft o.c. Loads are 1.2D + 1.6L with L from occupancy. High-humidity conditions exist, and the moisture content may exceed 19 percent.

4.32   Determine the ASD reference and adjusted design values for the following members and loading conditions. All members are Select Structural Southern Pine and are fully braced to prevent lateral-torsional buckling. Bending occurs about the strong axis.
a. Roof joists are 2 × 6 at 24 in. o.c. which directly support the roof sheathing. Loads are (D + S).
b. A 4 × 12 supports (D + L + L_r).
c. Purlins in a roof are 2 × 10 at 4 ft o.c. Loads are (D + L_r).
d. Floor beams are 4 × 10 at 4 ft o.c. Loads are (D + L + W).

4.33   Determine the LRFD nominal and adjusted design values and adjusted design resistances for the following members and loading conditions. All members are Select Structural Southern Pine and are fully braced to prevent lateral-torsional buckling. Bending occurs about the strong axis.
a. Roof joists are 2 × 6 at 24 in. o.c. which directly support the roof sheathing. Loads are 1.2D + 1.6S.
b. A 4 × 12 that supports 1.2D + 1.6L + 0.5L_r with L from occupancy.
c. Purlins in a roof are 2 × 10 at 4 ft o.c. Loads are 1.2D + 1.6L_r.
d. Floor beams are 4 × 10 at 4 ft o.c. Loads are 1.2D + 1.6W + L.

4.34   Estimate the amount of shrinkage that will occur in the depth of the beam in Fig. 4.A. Use the simplified shrinkage approach recommended in Ref. 4.26. Assume an initial moisture content of 19 percent and a final MC of 10 percent.

Figure 4.A Top of roof beams set higher for shrinkage.

NOTE: The top of the wood beams should be set higher than the top of the girder by an amount equal to the estimated shrinkage. After shrinkage, the roof sheathing will be supported by beams and girders that are all at the same elevation. Without this allowance for shrinkage, a wave or bump may be created in the sheathing where it passes over the girder.

4.35 Estimate the total shrinkage that will occur in a four-story building similar to the one in Example 4.3. Floor joists are 2×10's instead of 2×12's. The initial moisture content can be taken as 19 percent, and the final moisture content is assumed to be 9 percent. All other information is the same as in Example 4.3.

4.36 Use a personal computer spreadsheet or a database to input the reference design values for one or more species (as assigned) of sawn lumber. Include values for both Dimension lumber, Beams and Stringers, and Posts and Timbers sizes.

The purpose of the spreadsheet is to list reference and nominal design values (F_b, F_t, F_v, $F_{c\perp}$, F_c, E, or $E_{\min}$) as output for a specific problem with the following input being provided by the user:

a. Species (if values for more than one species are in spreadsheet or database)
b. Grade of lumber (e.g., Select Structural, No. 1, etc.)
c. Nominal size of member (for example, 2×4, 6×12, 6×6, and so on)
d. ASD or LRFD

4.37 Expand or modify the spreadsheet from Prob. 4.36 to develop adjusted ASD or LRFD design values. The spreadsheet should be capable of applying all the adjustment factors introduced in Chap. 4. The input should be expanded to provide sufficient information to the spreadsheet template so that the appropriate adjustment factors can be computed or drawn from a database or table.

Output should include a summary of the adjustment factors and the final design values F'_b, F'_t, F'_v, $F'_{c\perp}$, F'_c, E', or $E_{\min}$. Default values of unity may be listed for any adjustment factor that does not apply.

Structural
Glued Laminated
Timber

5.1 Introduction

Sawn lumber is manufactured in a large number of sizes and grades (Chap. 4) and is used for a wide variety of structural members. However, the cross-sectional dimensions and lengths of these members are limited by the size of the trees available to produce this type of lumber.

When the span becomes long or when the loads become large, the use of sawn lumber may become impractical. In these circumstances (and possibly for architectural reasons), structural glued laminated timber (*glulam*) can be used.

Glulam members are fabricated from relatively thin laminations (nominal 1 and 2 in.) of wood. These laminations can be end-jointed and glued together in such a way to produce wood members of practically any size and length. Lengths of glulam members are limited by handling systems and length restrictions imposed by highway transportation systems rather than by the size of the tree.

This chapter provides an introduction to glulam timber and its design characteristics. The similarities and differences between glulam and sawn wood members are also noted.

5.2 Sizes of Glulam Members

The specifications for glulam permit the fabrication of a member of any width and depth. However, standard practice has resulted in commonly accepted widths and thicknesses of laminations (see Ref. 5.6). The generally accepted dimensions for glulams fabricated from the Western Species are slightly different from those for Southern Pine glulams as given in NDS Table 5.1.3 (Ref. 5.1).

See Fig. 5.1. Because of surfacing requirements, Southern Pine laminations are usually thinner and narrower, although they can be manufactured to the same net sizes as Western Species, if necessary. The dimensions given in Fig. 5.1 are net sizes, and the total depth of a member will be a multiple of the lamination thickness.

Straight or slightly curved glulams will be fabricated with $1\frac{1}{2}$-in. (or $1\frac{3}{8}$-in.) laminations. If a member is sharply curved, thinner ($\frac{3}{4}$-in. or less) laminations should be used in the fabrication because smaller built-in, or residual, stresses

Figure 5.1   Structural glued laminated timber (glulam).

USUAL CASE   UNUSUAL CASE

Figure 5.2 The orientation of the x and y axes for glulams is actually related to the orientation of the laminations, not the strong and weak directions. The usual case is for a rectangular beam with its depth significantly greater than its width, and thus the x axis is parallel to the laminations and is the strong axis of the section. The unusual case is where the depth is less than the width. In this unusual situation, the x axis for the glulam, being parallel to the laminations, is not the strong axis of the section.

will result. These thinner laminations are not used for straight or slightly curved glulams because cost is heavily influenced by the number of glue lines in a member. Only the design of straight and slightly curved rectangular members is included in this text. The design of tapered members and curved members (including arches) is covered in the *Timber Construction Manual* (TCM, Ref. 5.7).

The sizes of glulam members are called out on plans by giving their net dimensions (unlike sawn lumber which uses "nominal" sizes). Cross-sectional properties for glulams are listed in the 2005 NDS Supplement Table 1C, Section Properties of *Western Species* Glued Laminated Timber, and Table 1D, Section Properties of *Southern Pine* Glued Laminated Timber. Section properties include

1. Cross-sectional area A (in.^{2})

2. Moment of inertia about the strong axis I_x (in.^{4})

3. Section modulus about the strong axis S_x (in.^{3})

4. Radius of gyration about the strong axis r_x (in.)

5. Moment of inertia about the weak axis I_y (in.^{4})

6. Section modulus about the weak axis S_y (in.^{3})

7. Radius of gyration about the weak axis r_y (in.)

Strictly speaking, the x and y axes are not always the strong and weak axes as indicated in the above listing. For glulams, the x and y orientation is actually related to the orientation of the laminations, not the strong and weak axes of the section. See Fig. 5.2. For the vast majority of glulams produced and used in structural applications, the *usual* case orientation (Fig. 5.2), of the x axis being parallel to the laminations *and* being the strong axis, holds true. Therefore, the above definitions for the section properties will be used throughout this book.

Section properties for glulam members are determined using the same basic principles illustrated in Example 4.5 (Sec. 4.11). The approximate weight per linear foot for a given size glulam can be obtained by converting the cross-sectional area in NDS Supplement Table 1C or 1D from in.^{2} to ft^2 and multiplying by the following unit weights:

Type of glulam	Unit weight
Southern Pine	36 pcf
Western Species	
Douglas Fir-Larch	35 pcf
Alaska Cedar	35 pcf
Hem-Fir and California Redwood	27 pcf

5.3   Resawn Glulam

In addition to the standard sizes of glulams shown in Fig. 5.1, NDS Supplement Tables 1C and 1D give section properties for a narrower width glulam. Glulams that are $2\frac{1}{2}$ in. wide are obtained by ripping a glulam manufactured from nominal 2×6 laminations into two pieces. The relatively narrow beams that are produced in this way are known as *resawn glulams*. See Fig. 5.3. Although section properties are listed only for $2\frac{1}{2}$ in.-wide beams, wider resawn members can be produced from glulams manufactured from wider laminations.

The resawing of a glulam to produce two narrower members introduces some additional manufacturing controls that are not required in the production of a normal-width member which is not going to be resawn. For example, certain strength-reducing characteristics (such as knot size or location) may be permitted in a $5\frac{1}{8}$-in. glulam that is not going to be resawn. If the member is going to be resawn, a more restrictive set of grading limitations apply.

Resawn glulams are a fairly recent development in the glulam industry. These members can have large depths. With a narrow width and a large depth, resawn glulams produce beams with efficient cross sections. In other words, the section modulus and moment of inertia for the strong axis (that is, S_x and I_x)

Figure 5.3   Resawn glulams are obtained by longitudinally cutting standard-width glulams to form narrower members. For example, a $2\frac{1}{2}$-in.-wide member is obtained by resawing a glulam manufactured from nominal 2×6 laminations. A resawn glulam is essentially an industrial-use (i.e., not architectural) beam because three sides of the member are finished and one side is sawn.

are large for the amount of material used in the production of the member. On the negative side, narrow members are weak about the minor axis (that is, S_y and I_y are small). The relatively thin nature of these members requires that they be handled properly in the field to ensure that they are not damaged during construction. In addition, it is especially important that the compression edge of a deep, narrow beam be properly braced so that the member does not buckle when a load is applied. Bracing of the tension edge, other than at the supports, is not necessary except in cases where moment reversal is anticipated.

Resawn glulams are used as an alternative to certain sizes of sawn lumber. They also provide an alternative to wood I-joists in some applications. Resawn beams are normally used where appearance is not a major concern.

5.4 Fabrication of Glulams

Specifications and guidelines covering the design and fabrication of glulam members (Refs. 5.4, 5.8, 5.10, 5.11), are published by the American Institute of Timber Construction (AITC) and Engineered Wood Systems (EWS), a related corporation of APA—The Engineered Wood Association. AITC and EWS are technical trade associations representing the structural glued laminated timber industry. AITC also publishes the *Timber Construction Manual* (TCM, Ref. 5.7), which was introduced in Chap. 1 and is referenced throughout this book. The TCM is a wood engineering handbook that can be considered the basic reference on glulam (for convenience, it also includes information on other structural wood products such as sawn lumber).

Most structural glulam members are produced using Douglas Fir or Southern Pine. Hem-Fir, Spruce-Pine-Fir, Alaska Cedar, and various other species including hardwood species can also be used. Quality control standards ensure the production of a reliable product. In fact, the structural properties of glulam members in most cases exceed the structural properties of sawn lumber.

The reason that the structural properties for glulam are so high is that the material included in the member can be selected from relatively high-quality laminating stock. The growth characteristics that limit the structural capacity of a large solid sawn wood member can simply be excluded in the fabrication of a glulam member.

In addition, laminating optimizes material use by dispersing the strength-reducing defects in the laminating material throughout the member. For example, consider the laminations that are produced from a sawn member with a knot that completely penetrates the member at one section. See Fig. 5.4. If this member is used to produce laminating stock which is later reassembled in a glulam member, it is unlikely that the knot defect will be reassembled in all the laminations at exactly the same location in the glulam member. Therefore, the reduction in cross-sectional properties at any section consists only of a portion of the original knot. The remainder of the knot is distributed to other locations in the member.

Figure 5.4 Dispersion of growth defects in glulam. Growth characteristics found in sawn lumber can be eliminated or (as shown in this sketch) dispersed throughout the member to reduce the effect at a given cross section.

Besides dispersing strength-reducing characteristics, the fabrication of glulam members makes efficient use of available structural materials in another way. High-quality laminations are located in the portions of the cross section which are more highly stressed. For example, in a typical glulam beam, wood of superior quality is located in the outer tension and compression zones. This coincides with the location of maximum bending stresses under typical loading. See Example 5.1. Although the maximum bending compressive and tensile stresses are equal, research has demonstrated that the outer laminations in the tension zone are the most critical laminations in a beam. For this reason, additional grading requirements are used for the outer tension laminations.

The different grades of laminations over the depth of the cross section really make a glulam a composite beam. Recall from strength of materials that a composite member is made up of more than one material with different values of modulus of elasticity. Composite members are analyzed using the transformed section method. The most obvious example of a composite member in building construction is a reinforced concrete beam, but a glulam is also a composite member because the different grades of laminations have different E's.

EXAMPLE 5.1   Distribution of Laminations is Glulam Beams

Figure 5.5   Distribution of high-and lower-quality laminations in glulam beams.

Bending stress calculation:

$$\text{Arbitrary point} \quad f_b = \frac{My}{I}$$

$$\text{Maximum stress} \quad f_b = \frac{Mc}{I}$$

In glulam beams, high-quality laminations are located in areas of high stress (i.e., near the top and bottom of the beam). Lower-quality wood is placed near the neutral axis where the stresses are lower. The outer tension laminations are critical and require the highest-grade stock.

However, from a designer's point of view, a glulam beam can be treated as a homogeneous material with a rectangular cross section. Design values have been determined in accordance with ASTM D 3737 (Ref. 5.9) using transformed sections. All glulam design values have been mathematically transformed to allow the use of apparent rectangular section properties. Thus, except for differences in design values and section properties, a glulam design is carried out in much the same manner as the design of a solid sawn beam.

Glulam beams are usually loaded in bending about the strong axis of the cross section. Large section properties and the distribution of laminations over the depth of the cross section make this an efficient use of materials. This is the loading condition assumed in Example 5.1, and bending about the strong axis should be assumed unless otherwise noted. In the tables for glulam design values, bending about the *strong axis* is described as the transverse load being applied *perpendicular to the wide face of the laminations.* See Example 5.2.

EXAMPLE 5.2   Bending of Glulams

Figure 5.6   Major and minor axis bending of glulam beams.

Bending can occur about either the x or y axis of a glulam, or both. In section 1, the load is perpendicular to the wide faces of the laminations, and bending occurs about the major axis of the member. This is the more common situation. In section 2, the load is parallel to the wide faces of the laminations, and bending occurs about the weak or minor axis of the member.

Loading about the minor axis is also possible, but it is much less common. One common example of glulam beams loaded about the minor axis is bridge decks. Different tabulated stresses apply to members loaded about the x and y axes.

Laminations are selected and dried to a moisture content (MC) of 16 percent or less before gluing. Differences in moisture content for the laminations in a member are not permitted to exceed 5 percent in order to minimize internal stresses and checking. Because of the relatively low MC of glulam members at the time of fabrication, the change in moisture content in service (i.e., the initial MC minus the EMC) is generally much smaller for glulam than it is for sawn lumber. Thus, glulams are viewed as being more dimensionally stable. Even though the percent change in MC is normally less, the depth of a glulam is usually much larger than that of a sawn lumber member. Thus, the possible effects of shrinkage need to be considered in glulam design. See Sec. 4.7 for more discussion of shrinkage and Chap. 14 for recommendations about how to avoid shrinkage-related problems in connections.

Traditionally, two types of glue have been permitted in the fabrication of glulam members: (1) dry-use adhesives (casein glue) and (2) wet-use adhesives (usually phenolresorcinol-base, resorcinol-base, or melamine-base adhesives). While both types of glue are capable of producing joints which have horizontal shear capabilities in excess of the capacity of the wood itself, today only wet-use adhesives are permitted in glulam manufacturing, according to ANSI/AITC 190.1-2002 (Ref. 5.3). Wet-use adhesives have been used almost exclusively for a number of years and only recently has their use become required. The increased, and now required, use of wet-use adhesives was made possible with the development of room-temperature-setting glues for exterior use. Wet-use adhesives, as the name implies, can withstand severe conditions of exposure.

The laminations run parallel to the length of a glulam member. The efficient use of materials and the long length of many glulam members require that effective end splices be developed in a given lamination. While several different configurations of lamination *end-joint* splices are possible, including *finger* and *scarf* joints (see Fig. 5.7), virtually all glued laminated timber produced in North America uses some form of finger joint.

Finger joints produce high-strength joints when the fingers have relatively flat slopes. The fingers have small blunt tips to ensure adequate bearing pressure. Finger joints also make efficient use of laminating stock because the lengths of the fingers are usually short in comparison with the lengths of scarf joints. With scarf joints, the flatter the slope of the joint, the greater the strength of the connection. Scarf slopes of 1 in 5 or flatter for compression and 1 in 10 or flatter for tension are recommended (Ref. 5.12).

If the width of the laminating stock is insufficient to produce the required net width of glulam, more than one piece of stock can be used for a lamination. The

Figure 5.7   End-joint splices in laminating stock. Most glulam fabricators use either the vertical or horizontal finger joints for end-joint splices. In addition, proof loading of joints is common, and in this case the location of end joints is not restricted.

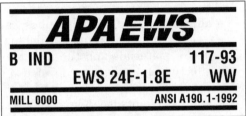

Figure 5.8 Typical grade stamps for glulam. (*Courtesy of AITC and EWS.*)

edge joints in a lamination can be glued. However, the vast majority of glulam producers do not edge-glue laminates. Rather, the edge joints are staggered in adjacent laminations.

Although one should be aware of the basic fabrication procedures and concepts outlined in this section, the building designer does not have to be concerned about designing the individual laminations, splices, and so on. The manufacturing standards for glulam are based on ANSI/AITC A190.1, *Structural Glued Laminated Timber* (Ref. 5.3), and implementation is ensured through a quality control system. Quality assurance involves the inspection and testing of glulam production by a qualified agency. The majority of glulam produced in the United States is inspected by two agencies: AITC Inspection Bureau and Engineered Wood Systems (EWS), a related corporation of APA. Each glulam is grade-stamped for identification purposes. See Fig. 5.8. In addition, because of the importance of the tension laminations, the top of a glulam bending member using an unbalanced layup is also marked with a stamp. This identification allows construction personnel in the field to orient the member properly in the structure (i.e., get it right side up). If a glulam were inadvertently turned upside down, the compression laminations would be stressed in tension and the strength of the member could be greatly reduced. For applications such as continuous or cantilevered beams, the designer should specify a balanced layup that has high-quality tension laminations on both the top and bottom of the member, and therefore has equivalent positive and negative moment capacities. Some of the items in the grade stamp include

1. Quality control agency (e.g., American Institute of Timber Construction or Engineered Wood Systems)

2. Structural use (possible symbols: *B*, simple span bending member; *C*, compression member; *T*, tension member; and CB, continuous or cantilever bending member)

3. Appearance grade (FRAMING, framing; IND, industrial; ARCH, architectural; PREM, premium)

4. Plant or mill number (e.g., 143 and 0000 shown)

5. Standard for structural glued laminated timber (i.e., ANSI/AITC A190.1-2002)

6. Laminating specification and combination symbol (e.g., 117-1993 24F-1.8E)

7. Species/species group of the glulam

8. Proof-loaded end joints, if used during the manufacturing process

A complete list of all required markings is provided in ANSI/AITC A190.1 (Ref. 5.3).

5.5 Grades of Glulam Members

For strength, grades of glulam members traditionally have been given as combinations of laminations. The two main types are *bending combinations* and *axial combinations*. The 2005 NDS uses the *Stress Class System* for softwood glulam bending combinations. The Stress Class System is recommended by the glulam industry for specifying beams, as it will greatly benefit designers and manufactures alike. Softwood glulam bending combinations with similar properties have been grouped into stress classes, markedly reducing the complexity in selecting an appropriate grade combination. With the Stress Class System, the number of tabulated values has been reduced to simplify things for the designer.

In addition to grading for strength, glulam members are graded for appearance. One of the four appearance grades (framing, industrial, architectural, and premium) should be specified along with the strength requirements to ensure that the member furnished is appropriate for the intended use. It is important to understand that the selection of an appearance grade does not affect the strength of a glulam (see Ref. 5.2 for additional information), but can increase costs due to the required additional finishing to achieve the higher-quality appearance.

Members that are stressed principally in bending and loaded in the normal manner (i.e., with the applied load perpendicular to the wide faces of the laminations) are produced from the bending combinations. Bending combinations are defined by a *combination symbol* and the *species* of the laminating stock. The combination symbol is made up of two parts. The first is the reference bending design value for the grade in hundreds of psi followed by the latter F. For example, 24F indicates a bending combination with a reference design value in bending of 2400 psi for normal duration of loading and dry-service conditions. Bending combinations that are available include 16F, 20F, 22F, 24F, and 26F flexural stress levels for most species. Additionally, 28F and 30F combinations are available for Southern Pine glulams.

It should be noted that a number of combinations of laminations can be used to produce a given bending stress level. Therefore, there is an abbreviation that follows the bending stress level which gives the distribution of laminating stock to be used in the fabrication of a member. Two basic abbreviations are used in defining the combinations: one is for visually graded laminating stock (e.g., 24F-V3), and the other is for laminating stock that is mechanically graded, or E-rated, for stiffness (e.g., 22F-E5).

In addition to the combination symbol, the species of wood is required to define the grade. The symbols for the species are DF for Douglas Fir-Larch, DFS for Douglas Fir South, HF for Hem-Fir, SW for softwoods,* and SP for Southern Pine. Section 5.4 indicated that higher-quality laminating stock is located at the outer faces of a glulam bending combination, and lower-quality stock is used for the less highly stressed inner zone. In a similar manner, the laminating specifications allow the mixing of more than one species of wood in certain combinations. The idea is again to make efficient use of raw materials by allowing the use of a strong species for the outer laminations and a weaker species for the center core.

If more than one species of wood is used in a member, both species are specified (e.g., DF/HF indicates DF outer laminations and HF inner core laminations). If only one species is used throughout the member, the species symbol is repeated (e.g., DF/DF). Although the laminating specification allows the mixing of more than one species, most glulam production currently uses laminating stock from only one species of wood for a given member.

As mentioned, the *Stress Class System* for softwood glulam bending combinations is used in the 2005 NDS. The basic premise of the Stress Class System is to simplify the designer's choices and give the manufacturer more flexibility to meet the needs of the designer. The glulam industry has had a longstanding position that designers should specify by required design values, rather than by combination symbol. This was intended to give manufacturers flexibility in choosing combinations to fit their resource. Unfortunately, the design community never broadly followed this practice. Designers have been trained to choose a material and then size the member appropriately, so engineers continue to specify by combination symbol, leaving the manufacturer with no flexibility. This old system of specifying combinations and species was also far from simple for a new designer. In looking at the glulam design tables in the NDS, it was easy to become overwhelmed with the choices. The Stress Class System improves the design process for the designer and the ability of the manufacturer to meet the design requirements.

Softwood bending combinations with similar properties are now grouped into stress classes. All of the design values in a higher stress class equal or exceed those from a lower stress class. This allows designers and manufacturers the ability to substitute higher grades based on availability. The number of reference values has also been reduced for design simplicity. Since glulam combinations are derived from broad ranges of species groups, some anomalies are bound to be present. Accordingly, unusual cases have been footnoted in the design tables instead of being separately tabulated. This is intended to further simplify the design process.

As with the bending combinations, stress classes are defined by a two part *stress class symbol*. The first part of the symbol is the reference bending design

---

*Formerly, the symbol WW was used for Western Species. This was changed in AITC 117-2001 to SW for softwoods, because the species allowed for this glulam combination do not match the lumber grading agencies' definitions of "Western Woods."

value for the class in hundreds of psi followed by the letter F. For example, 24F indicates a stress class with a reference bending design value of 2400 psi for normal duration of loading and dry-service conditions. Stress classes that are readily available are 16F, 20F, and 24F. Higher stress classes of 26F, 28F, and 30F may be available from some manufacturers.

The second part of the stress class symbol also provides information about a design value of the glulam. Recall that for a traditional bending combination, the second part of the combination symbol indicates the distribution of laminating stock used in the fabrication of the member (e.g., V3 for visually-graded laminating stock or E5 for E-rated laminating stock). With the Stress Class System, the second part of the symbol is the bending modulus of elasticity in millions of psi. For example, 24F-1.8E indicates a stress class with a reference bending design value of 2400 psi and modulus of elasticity of 1.8×10^6 psi.

The 2005 NDS provides the stress classes in Table 5A, and the individual combinations in each stress class are shown in Table 5A Expanded. It is the intent of the glulam industry to transition from NDS Table 5A Expanded to NDS Table 5A. However, AITC and APA-EWS will likely continue to list the design values for the individual combinations after this transition, because there may be special cases where use of a particular combination is desirable.

Members that are principally axial-load-carrying members are identified with a numbered combination symbol such as 1, 2, 3, and so on. The new Stress Class System applies only to softwood glulam members, used primarily in bending. Members stressed primarily in axial tension or compression are designed using numbered combinations. See NDS Supplement Table 5B. Because axial load members are assumed to be uniformly stressed throughout the cross section, the distribution of lamination grades is uniform through the member section, compared with the distribution of lamination quality used for beams.

Glulam combinations are, in one respect, similar to the "use" categories of sawn lumber. The *bending combinations* anticipate that the member will be used as a beam, and the *axial combinations* assume that the member will be loaded axially.

Bending combinations are often fabricated with higher-quality laminating stock at the outer fibers, and consequently they make efficient beams. This fact, however, does not mean that a bending combination cannot be loaded axially. Likewise, an axial combination can be designed for a bending moment. The combinations have to do with efficiency, but they do not limit the use of a member. The ultimate use is determined by stress calculations.

Design values for glulams are listed in the following tables in the 2005 NDS Supplement:

Table 5A *Reference Design Values for Structural Glued Laminated Softwood Timber (Members stressed primarily in bending)*. These are the bending stress classes.

Table 5A Expanded *Reference Design Values for Structural Glued Laminated Softwood Timber (Members stressed primarily in bending)*. These are the bending combinations that meet the requirements of each stress class in Table 5A.

Table 5B *Reference Design Values for Structural Glued Laminated Softwood Timber (Members stressed primarily in axial tension and compression).* These are the axial load combinations.

Table 5C *Reference Design Values for Structural Glued Laminated Hardwood Timber (Members stressed primarily in bending).* These are the bending combinations.

Table 5D *Reference Design Values for Structural Glued Laminated Hardwood Timber (Members stressed primarily in axial tension and compression).* These are the axial load combinations.

Reference design values for glulam members are also available in a number of other publications including Refs. 5.4, 5.7, 5.8, 5.10, and 5.11. The tables include the following reference design values:

Bending F_{bx} and F_{by}

Tension parallel to grain F_t

Shear parallel to grain F_{vx} and F_{vy}

Compression parallel to grain F_c

Compression perpendicular to grain $F_{c\perp x}$ and $F_{c\perp y}$

Modulus of elasticity E_x, $E_{x\,min}$, E_y, $E_{y\,min}$, and E_{axial}

Reference design values for glulam are the same basic properties that are listed for solid sawn lumber, but the glulam tables are more complex. The reason for this is the way glulams are manufactured with different grades of laminations. As a result, different design properties apply for bending loads about both axes, and a third set of properties is provided for axial loading. It is suggested that the reader accompany this summary with a review of NDS Supplement Tables 5A, 5B, 5C, and 5D.

For softwood glulam members, the design values for the more common application of a glulam member are listed first in the tables. For example, a *member stressed primarily in bending* will normally be used as a beam loaded about the strong axis, and reference design values for loading about the *x* axis are the first values given in NDS Table 5A and Table 5A Expanded. These are followed by values for loading about the *y* axis and for axial loading.

For loading about the x axis, two values of F_{bx} are listed. The first value represents the more efficient use of a glulam, and consequently it is the more frequently used value in design. F_{bx}^{+} indicates that the high-quality tension laminations are stressed in tension (i.e., tension zone stressed in tension). If, for example, a glulam were installed upside down, the second value of F_{bx} would apply. In other words, F_{bx}^{-} indicates that the lower-quality compression laminations are stressed in tension (i.e., compression zone stressed in tension).

The real purpose for listing F_{bx}^- is not to analyze beams that are installed improperly (although that is one possible use). An application of this design value in a continuous beam that is properly installed is given in Chap. 6. The reason for mentioning the case of a beam being installed upside down is to simply illustrate why the two values for F_{bx} can be so different.

Five values of modulus of elasticity are given in Tables 5A and 5A Expanded: E_x, $E_{x\,\mathrm{min}}$, E_y, $E_{y\,\mathrm{min}}$, and E_{axial}. Values of E_x and E_y are for use in beam deflection calculations about the x and y axes, respectively. Values of $E_{x\,\mathrm{min}}$ and $E_{y\,\mathrm{min}}$ are used in stability calculations for columns and laterally unbraced beams. On the other hand, E_{axial} is used for deformation calculations in members subjected to axial loads, such as the shortening of a column or the elongation of a tension member.

Two design values for compression perpendicular to grain $F_{c\perp x}$ are listed in NDS Table 5A Expanded. One value applies to bearing on the face of the outer tension lamination, and the other applies to bearing on the compression face. The reference bearing value may be larger for the tension face because of the higher-quality laminations in the tension zone. Under the Stress Class System, a single design value for compression perpendicular to grain is listed in NDS Table 5A. The value for $F_{c\perp x}$ listed in Table 5A is taken as the minimum design value for the group of combinations comprising the stress class. Because compression perpendicular to grain rarely governs a design, this conservative approach simplifies the design value table with minimal impact.

Design values listed in NDS Table 5B are for *axial combinations* of glulams, and therefore the properties for axial loading are given first in the table. The distribution of laminations for the axial combinations does not follow the distribution for beams given in Example 5.1. Consequently, values for F_{bx}^+ and F_{bx}^- do not apply to axial combinations. The design values for a glulam from an axial combination depend on the number of laminations in a particular member.

Design values for hardwood glulam members are provided in Tables 5C and 5D of the NDS Supplement. These tables are identical in format to Tables 5A Expanded and 5B, which provide design values for softwood glulam members. Prior to the 2001 NDS, the design of hardwood glulam members was significantly different from that of softwood glulam members. The reason for this past difference was that softwood glulam members were more popular and were used to such an extent that establishing various combinations (and now stress classes) was warranted. However, for hardwood glulam members, the use was not as extensive, and the design of the lamination layup was less standardized. Now the use of hardwood glulam members has increased, and the design approach has been unified with that of softwood glulam members. Regardless, the remainder of this chapter focuses on the design of softwood glulam members.

All of the tables for glulam have an extensive set of footnotes, which should be consulted for possible modification of design values. Additional information on ordering and specifying glulam members can be obtained from AITC or APA-EWS.

5.6   Adjustment Factors for Glulam

The notation for reference design values, adjustment factors, and adjusted design values is essentially the same for glulam and for sawn lumber. Refer to Sec. 4.13 for a review of the notation used in wood design. The basic system involves the determination of adjusted design values by multiplying the reference design values by a series of adjustment factors.

$$F' = F \times \text{(product of } C \text{ factors)}$$

The reference design values for glulams are generally larger than similar properties for sawn lumber. This is essentially a result of the selective placement of laminations and the dispersion of imperfections. However, glulams are a wood product, and they are subject to many of the adjustments described in Chap. 4 for sawn lumber. Some of the adjustment factors are numerically the same for glulam and sawn lumber, and others are different. In addition, some adjustments apply only to sawn lumber, and several other factors are unique to glulam design.

The general summary of adjustment factors for use in glulam design is given in NDS Table 5.3.1, *Applicability of Adjustment Factors for Glued Laminated Timber.* Several adjustment factors for glulam were previously described for sawn lumber. A brief description of the similarities and differences for glulam and sawn lumber is given here. Where appropriate, the reader is referred to Chap. 4 for further information. The TCM (Ref. 5.7) also includes adjustment factors and design procedures for glulam not covered in the NDS, such as tapered beams which are quite common with glulam.

Wet service factor (C_M)

Reference design values for glulam are for dry conditions of service. For glulam, dry is defined as MC < 16 percent. For moisture contents of 16 percent or greater, reference design values are multiplied by C_M. Values of C_M for glulam are given in the Adjustment Factors section preceding NDS Supplement Tables 5A through 5D. When a glulam member is used in high moisture conditions, the need for pressure treatment (Sec. 4.9) should be considered.

Load duration factor (C_D)—ASD only

Reference design values for glulam are for normal duration of load. Normal duration is defined as 10 years and is associated with floor live loads. Loads and load combinations of other durations are taken into account by multiplying reference design values by C_D. The same load duration factors are used for both glulam and sawn lumber. See Sec. 4.15 for a complete discussion.

Time effect factor (λ)—LRFD only

Similar to the load duration factor for ASD, the time effect factor λ is used in LRFD to adjust nominal design values for the expected duration of loading. The same time effect factors are used for both glulam and sawn lumber. See Sec. 4.16 for further detail regarding λ.

Temperature factor (C_t)

Reference design values for glulam are for use at normal temperatures. Section 4.20 discussed adjustment factors for other temperature ranges.

Flat-use factor (C_{fu})

The flat-use factor is somewhat different for sawn lumber and for glulam. For sawn lumber, tabulated values for F_b apply to bending about the x axis. When bending occurs about the y axis, tabulated values of F_b are multiplied by C_{fu} (Sec. 4.19) to convert the value to a property for the y axis.

On the other hand, glulam members have tabulated bending values for both the x and y axes (that is, F_{bx} and F_{by} are both listed). When the depth of the member for bending about the y axis (i.e., the cross-sectional dimension parallel to the wide faces of the laminations) is less then 12 in., the tabulated value of F_{by} may be increased by multiplying by C_{fu}. Values of C_{fu} for glulam are found in the Adjustment Factors section preceding NDS Supplement Tables 5A through 5D.

Because most beams are stressed about the x axis rather than the y axis, the flat-use factor is not a commonly applied adjustment factor. In addition, C_{fu} exceeds unity, and it can conservatively be ignored.

Volume factor (C_v)

It has been noted that the reference design values in a wood member is affected by the relative size of the member. This general behavior is termed *size effect*. In sawn lumber, the size effect is taken into account by the size factor C_F.

In the past, the same size factor was applied to F_b for glulam that is currently applied to the F_b for sawn lumber in the Timber sizes. However, full-scale test data indicate that the size effect in glulam is related to the *volume* of the member rather than to only its depth.

Therefore, the volume factor C_V replaces the size factor C_F for use in glulam design. Note that C_V applies only to bending stress. Reference values of F_b apply to a standard-size glulam beam with the following base dimensions: width = $5\frac{1}{8}$ in., depth = 12 in., length = 21 ft. The volume factor C_V is used to obtain the adjusted bending design value for other sizes of glulams. See Example 5.3. It has been shown that the volume effect is less significant for Southern Pine than for other species, and the volume factor is thus species-dependent.

EXAMPLE 5.3 Volume Factor C_V for Glulam

Reference values of F_b apply to a glulam with the dimensions shown in Fig. 5.9.

Figure 5.9 *Base dimensions* for reference bending design value in glulam.

The reference bending design value for a glulam of another size is obtained by multiplying the design value by the volume factor.

$$F_b' = F_b \times C_V \times \cdots$$

For Western Species of glulam

$$C_V = \left(\frac{21\,\text{ft}}{L}\right)^{1/10}\left(\frac{12\,\text{in.}}{d}\right)^{1/10}\left(\frac{5.125\,\text{in.}}{b}\right)^{1/10} \leq 1.0$$

or

$$C_V = \left(\frac{15{,}498\,\text{in.}^3}{V}\right)^{1/10} \leq 1.0$$

For Southern Pine glulam

$$C_V = \left(\frac{21\,\text{ft}}{L}\right)^{1/20}\left(\frac{12\,\text{in.}}{d}\right)^{1/20}\left(\frac{5.125\,\text{in.}}{b}\right)^{1/20} \leq 1.0$$

or

$$C_V = \left(\frac{15{,}498\,\text{in.}^3}{V}\right)^{1/20} \leq 1.0$$

where L = length of beam between points of zero moment, ft
$\quad\quad d$ = depth of beam, in.
$\quad\quad b$ = width of beam, in. (*Note:* For laminations that consist of more than one piece, b is the width of widest piece in layup.)
$\quad\quad V$ = volume of beam between points of zero moment, in.^{3}
$\quad\quad\quad$ = $(L \times 12\,\text{in./ft}) \times d \times b$

The application of the volume-effect factor is shown in Sec. 5.7. Other modification factors for glulam design are introduced as they are needed.

Resistance factor (ϕ)—LRFD only

In LRFD a *resistance factor* ϕ, sometimes also referred to as a strength reduction factor, is used to allow for the possibility that the resistance may be less than computed. Accordingly, the resistance factor ϕ is a function of the application and material property. The NDS provides resistance factors in Appendix N, Table N2. See Sec. 4.22 for additional discussion of the resistance factor ϕ.

Format conversion factor (K_F)—LRFD only

Reference design values for glulam are for direct use in ASD for normal duration of load. For LRFD, the reference design values must be converted to nominal design values using the format conversion factor K_F. The same format conversion factors are used for glulam and sawn lumber. Format conversion factors are provided for different applications and properties in Table N1 of NDS, Appendix N. See Sec. 4.23 for further information regarding K_F.

5.7 Design Problem: Adjusted Design Values

The adjusted design values for a glulam member are evaluated in Examples 5.4 and 5.5 for ASD and LRFD, respectively. As with sawn lumber, the first step is to obtain the correct reference design values from the NDS Supplement. The second step is to apply the appropriate adjustment factors. A primary difference between a glulam problem and sawn lumber problem is the use of the volume factor instead of the size factor.

In these examples, a single load combination is given. It is recognized that in practice as number of different loading combinations must be considered (Sec. 2.16), and the same load duration factor or time effect factor may not apply to all load cases. Appropriate loading combinations are considered in more complete problems later in this book. A single C_D (ASD) is used in Example 5.4 and a single (LRFD) is used in Example 5.5 for simplicity.

EXAMPLE 5.4   Determination of Adjusted Design Values for a Glulam (ASD only)

A glulam beam is shown in Fig. 5.10. The member is Douglas Fir from the 24F-1.7E Stress Class. From the sketch, the bending load is about the x axis of the cross section. Loads are [D + 0.75(S + W)]. Use a single C_D based on the shortest duration load in the combination. Bracing conditions are such that buckling is not a concern. Consider dry-service application (EMC < 16 percent). Normal temperature conditions apply.

Determine the following adjusted ASD values:

Positive bending design value about the strong axis $F_{bx}^{-\prime}$
Negative bending design value about the strong axis $F_{bx}^{-\prime}$
Tension design value parallel to grain $F_t^{\prime}$
Compression design value parallel to grain $F_c^{\prime}$
Compression design value perpendicular to grain under concentrated load $F_{c\perp}^{\prime}$ on compression face

Figure 5.10   Load, shear, and moment diagrams for glulam beam.

Compression design value perpendicular to grain at support reactions $F'_{c\perp}$ on tension face

Shear design value parallel to grain F'_{vx} (with bending about the strong axis)

Modulus of elasticity for deflection calculations (beam loaded about strong axis) E_x'

Douglas Fir is a Western Species glulam. The Stress Class 24F-1.7E is recognized as a *bending combination*. Reference values are taken from NDS Supplement Table 5A. Bending is about the strong axis of the member. The member is properly installed (top-side up), and the tension laminations are on the bottom of the beam. The moment diagram is positive throughout the span, and bending tension stresses are on the bottom of the member. It is thus confirmed that the normally used bending stress is appropriate (that is, the adjusted ASD "tension zone stressed in tension" value $F_{bx}^{+\prime}$ applies to the problem at hand; but for illustrative purposes, the "compression zone stressed in tension" design value $F_{bx}^{-\prime}$ will also be determined in this example).

The shortest duration load in the combination is wind, and $C_D = 1.6$. Any design value adjustment factors in NDS Table 5.3.1 that are not shown in this example do not apply to the given problem or have a default value of unity. Recall that C_D does not apply to $F_{c\perp}$ or to E.

Volume Factor C_V

The dimensions of the given member do not agree with the base dimensions for the standard-size glulam in Example 5.3. Therefore, the bending design value will be multiplied by C_V. The length L in the formula is the distance between points of zero moment, which in this case is the span length of 48 ft.

$$C_V = \left(\frac{21}{L}\right)^{1/10}\left(\frac{12}{d}\right)^{1/10}\left(\frac{5.125}{b}\right)^{1/10}$$

$$= \left(\frac{21}{48}\right)^{0.1}\left(\frac{12}{37.5}\right)^{0.1}\left(\frac{5.125}{6.75}\right)^{0.1} = 0.799$$

Adjusted Design Values

$$F_{bx}^{+\prime} = F_{bx}(C_D \times C_M \times C_t \times C_V) = 2400(1.6 \times 1.0 \times 1.0 \times 0.799) = 3068\,\text{psi}$$

$$F_{bx}^{-\prime} = F_{bx}^{-}(C_D \times C_M \times C_t \times C_V) = 1450(1.6 \times 1.0 \times 1.0 \times 0.799) = 1854\,\text{psi}$$

(NOTE: If a balanced section were required (that is, $F_{bx}^{-} = F_{bx}^{+}$), the designer must specify that a balanced layup is required. See footnote 1 of NDS Supplement Table 5A.)

$$F_t' = F_t(C_D \times C_M \times C_t) = 775(1.6 \times 1.0 \times 1.0) = 1240\,\text{psi}$$

$$F_c' = F_c(C_D \times C_M \times C_t) = 1000(1.6 \times 1.0 \times 1.0) = 1600\,\text{psi}$$

$$F_{c\perp}' = F_{c\perp}(C_M \times C_t) = 500(1.0 \times 1.0) = 500\,\text{psi}$$

$$F_{vx}' = F_{vx}(C_D \times C_M \times C_t) = 210(1.6 \times 1.0 \times 1.0) = 336\,\text{psi}$$

$$E_x' = E(C_M \times C_t) = 1{,}700{,}000(1.0 \times 1.0) = 1{,}700{,}000\,\text{psi}$$

Example 5.4 is repeated for LRFD is Example 5.5. As introduced in Chap. 4, the subscript n is used to indicate a *nominal* design value appropriate for use in LRFD.

EXAMPLE 5.5   Determination of Adjusted Design Values for a Glulam (LRFD only).

Example 5.4 is now repeated for selected design values in Example 5.5 using LRFD provisions. For LRFD, the load combination is 1.2D + 1.6W + 0.5S. Therefore, a time effect factor λ of 1.0 will be used. The volume effect factor is the same for both ASD and LRFD. Therefore, from Example 5.4, $C_V = 0.799$.

Positive Bending:

$$F_{bn}^{+} = F_b^{+} \times K_F$$

$$= 2400(2.16/\phi_b) = 2400(2.16/0.85) = 6099\,\text{psi} = 6.10\,\text{ksi}$$

$$F_{bn}^{+\prime} = F_{bn}^{+}(\phi_b \times \lambda \times C_M \times C_t \times C_V)$$

$$= 6.10(0.85 \times 1.0 \times 1.0 \times 1.0 \times 0.799) = 4.14\,\text{ksi}$$

LRFD is typically done using resistances instead of stresses. Section properties are obtained from NDS Supplement Table 1C.

$$M_n^{+\prime} = F_{bn}^{+\prime} \times S_x$$

$$= 4.14 \times 1582 = 6550\,\text{kip-in.}$$

Negative Bending:

$$F^-_{bn} = F^-_b \times K_F$$
$$= 1450(2.16/\phi_b) = 1450(2.16/0.85) = 3685 \text{ psi} = 3.68 \text{ ksi}$$
$$F^{-\prime}_{bn} = F^-_{bn}(\phi_b \times \lambda \times C_M \times C_t \times C_V)$$
$$= 3.68(0.85 \times 1.0 \times 1.0 \times 1.0 \times 0.799) = 2.50 \text{ ksi}$$
$$M^{-\prime}_n = F^{-\prime}_{bn} \times S_x$$
$$= 2.50 \times 1582 = 3960 \text{ kip-in.}$$

Tension:

$$F_{tn} = F_t \times K_F$$
$$= 775(2.16/\phi_t) = 775(2.16/0.80) = 2092 \text{ psi} = 2.09 \text{ ksi}$$
$$F'_{tn} = F_{tn}(\phi_t \times \lambda \times C_M \times C_t)$$
$$= 2.09(0.80 \times 1.0 \times 1.0 \times 1.0) = 1.67 \text{ ksi}$$
$$T'_n = F'_{tn} \times A$$
$$= 1.67 \times 253.1 = 424 \text{ k}$$

Compression Parallel to Grain:

$$F_{cn} = F_c \times K_F$$
$$= 1000(2.16/\phi_c) = 1000(2.16/0.90) = 2400 \text{ psi} = 2.40 \text{ ksi}$$
$$F'_{cn} = F_{cn}(\phi_c \times \lambda \times C_M \times C_t)$$
$$= 2.40(0.90 \times 1.0 \times 1.0 \times 1.0) = 2.16 \text{ ksi}$$
$$P'_n = F'_{cn} \times A$$
$$= 2.16 \times 253.1 = 547 \text{ k}$$

Modulus of Elasticity:

$$E' = E(\phi \times C_M \times C_t)^\dagger$$
$$= 1{,}700{,}000(1.0 \times 1.0 \times 1.0) = 1{,}700{,}000 \text{ psi} = 1700 \text{ ksi}$$

[†] There is no difference between the E' used in ASD and the E' used in LRFD for deflection.

5.8   References

[5.1]   American Forest and Paper Association (AF&PA). 2005. *National Design Specification for Wood Construction and Supplement.* ANSI/AF&PA NDS-2005, AF&PA, Washington, DC.

[5.2]   American Institute of Timber Construction (AITC). 2001. *Standard Appearance Grades for Structural Glued Laminated Timber, AITC 110-2001,* AITC, Englewood, CO.

[5.3]   American Institute of Timber Construction (AITC).2002. *Structural Glued Laminated Timber, ANSI/AITC Standard 190.1-2002, AITC,* Englewood, CO.

[5.4]   American Institute of Timber Construction (AITC). 2004. *Standard Specifications for Structural Glued Laminated Timber of Softwood Species, Design and Manufacturing AITC 117-2004,* AITC, Englewood, CO.

[5.5]   American Institute of Timber Construction (AITC). 2005. *Use of a Volume Effect Factor in the Design of Glued Laminated Timber Beams, AITC TN-21,* AITC, Englewood, Co.

[5.6]   American Institute of Timber Construction (AITC). 2001. *Standard Dimensions for Structural Glued Laminated Timber, AITC 113-2001,* AITC, Englewood, CO.

[5.7]   American Institute of Timber Construction (AITC). 2005. *Timber Construction Manual, 5th* ed., AITC, Englewood, CO.

[5.8]   American Institute of Timber Construction (AITC). 1996. *Standard Specifications for Structural Glued Laminated Timber of Hardwood Species, AITC 119–96,* AITC, Englewood, CO.

[5.9]   American Society for Testing and Materials (ASTM). 2001. "Standard Practice for Establishing Stresses for Structural Glued Laminated Timber (Glulam)," ASTM D3737-01a, *Annual Book of Standards, Vol. 04.10 Wood,* ASTM, Philadelphia, PA.

[5.10] APA—The Engineered Wood Association. 2001. *Data file: Glued Laminated Beam Design Tables, EWS S475,* APA—The Engineered Wood Association, Engineered Wood Systems, Tacoma, WA.

[5.11] APA—The Engineered Wood Association. 2003. *Glulam Product Guide, EWS, 440,* APA—The Engineered Wood Association, Engineered Wood Systems, Tacoma, WA.

[5.12] Forest Products Laboratory (FPL). 1999. *Wood Handbook: Wood as an Engineering Material, Technical Report 113,* FPL, Forest Service, U.S.D.A., Madison, WI.

5.9   Problems

Design values and adjustment factors in the following problems are to be taken from the 2005 NDS. Assume that glulams will be used in dry-service conditions and at normal temperatures unless otherwise noted.

5.1   What is the usual thickness of laminations used to fabricate glulam members from
  a. Western Species?
  b. Southern Pine?
  c. Under what conditions would thinner laminations be used?

5.2   What are the usual widths of glulam members fabricated from
  a. Western Species?
  b. Southern Pine?

5.3   How are the *stress grades* denoted for a glulam that is
  a. Primarily a bending member fabricated with visually graded laminations?
  b. Primarily a bending member fabricated with *E*-rated laminations?
  c. Primarily an axial-load-carrying member?
  d. What are the *appearance grades* of glulam members, and how do they affect the grading for strength?

5.4   Briefly describe what is meant by resawn glulam. What range of sizes is listed in NDS Supplement Tables 1C and 1D for resawn glulam?

5.5   What is the most common type of lamination end-joint splice used in glulam members? Sketch the splice.

5.6   If the width of a lamination in a glulam beam is made up of more than one piece of wood, must the edge joint between the pieces be glued?

5.7   Describe the distribution of laminations used in the fabrication of a glulam to be used principally as an axial load member.

5.8   Describe the distribution of laminations used in the fabrication of a glulam member that is used principally as a bending member.

5.9   Briefly describe the meaning of the following glulam designations:
  a. Stress Class 20F-1.5E
  b. Stress Class 24F-1.8E
  c. Stress Class 30F-2.1E SP
  d. Combination 20F-V7 DF/DF
  e. Combination 24F-V5 DF/HF
  f. Combination 24F-E11 HF/HF
  g. Combination 26F-V2 SP/SP
  h. Combination 5 DF
  i. Combination 32 DF
  j. Combination 48 SP

5.10 Reference values of F_{bx} for a member stressed primarily in bending apply to a glulam of a "standard" size. What are the dimensions of this hypothetical beam? Describe the adjustment that is required if a member of another size is used.

5.11 *Given:* $5^1/_8 \times 28.5$ 24F-1.8E Douglas Fir glulam is used to span 32 ft, carrying an ASD load combination of $(D + S)$. The load is a uniform load over a simple span, and the beam is supported so that buckling is prevented.

  Find: a. Sketch the beam and the cross section. Show calculations to verify the section properties S_x and I_x for the member, and compare with values in NDS Supplement Table 1C.
  b. Determine the adjusted ASD values associated with the section properties in part *a*. These include $F_{bx}^{+\prime}$ $F_{ux}^{\prime}$ and $F_x^{\prime}$
  c. Repeat part b except the moisture content of the member may exceed 16 percent.

5.12 Repeat Prob. 5.11 using LRFD. The applicable load combination is $(1.2D + 1.6S)$.

5.13 Repeat Prob. 5.11 except the member is a 24F-V4 Douglas Fir-Larch glulam.

5.14 Repeat Prob. 5.13 using LRFD.

5.15 *Given:* Assume that the member in Prob. 5.11 may also be loaded about the minor axis.

Find: *a.* Show calculations to verify the section properties S_y and I_y for the member. Compare with values in NDS Supplement Table 1C.

      b. Determine the adjusted ASD values associated with the section properties in part *a.* These include F'_{by}, F'_{vy}, and E'_y.

      c. Repeat part *b,* except the moisture content of the member may exceed 16 percent.

5.16   Repeat Prob. 5.15 using LRFD.

5.17   *Given:* Assume that the member in Prob. 5.11 may also be subjected to an axial tension or compression load.

      Find: *a.* Show calculations to verify the cross-sectional area A for the member. Compare with the value in NDS Supplement Table 1C.

         b. Determine the adjusted ASD values associated with the section properties in part *a.* These include F'_t, F'_c, and E'_{axial}.

         c. Repeat part *b,* except the moisture content of the member may exceed 16 percent.

5.18   Repeat Prob. 5.17 using LRFD.

5.19   Repeat Prob. 5.11 except the member is a 5 × 33 26F-1.9E Southern Pine glulam.

5.20   Repeat Prob. 5.19 using LRFD.

5.21   Explain why the design value tables for glulam bending combinations (NDS Supplement Table 5A Expanded) list two values of compression perpendicular to grain for loads normal to the x axis ($F_{c\perp x}$).

5.22   Over what moisture content are the reference design values in glulam to be reduced by a wet-service factor C_M?

5.23   List the wet-service factors C_M to be used for designing glulam beams with high moisture contents.

Beam Design

6.1 Introduction

The design of rectangular sawn wood beams and straight or slightly curved rectangular glulam beams is covered in this chapter. Glulam members may be somewhat different than sawn lumber beams, and the special design procedures that apply only to glulam design are noted. Where no distinction is made, it may be assumed that essentially the same procedures apply to both sawn lumber and glulam design.

Glulam beams are sometimes tapered and/or curved for architectural considerations, to improve roof drainage, or to lower wall heights. The design of these types of members requires additional considerations beyond the information presented in this book. For the additional design considerations of these advanced subjects, see the *Timber Construction Manual* (TCM) (Ref. 6.5).

The design of wood beams follows the same basic overall procedure used in the design of beams of other structural materials. The criteria that need to be considered are

1. Bending (including lateral stability)

2. Shear

3. Deflection

4. Bearing

The first three items can govern the size of a wood member. The fourth item must be considered in the design of the supports. In many beams, the bending stress is a critical design item. For this reason, a trial size is often obtained from bending stress calculations. The remaining items are then simply checked using the trial size. If the trial size proves inadequate in any of the checks, the design is revised.

Computer solutions to these problems can greatly speed up the design process, and with the use of the computer, much more thorough beam deflection studies are possible. However, the basic design process needs to be fully understood first.

Designers are cautioned about using programs in a *blackbox* approach. Any program used should be adequately documented and sufficient output should be available so that results can be verified by hand solutions. The emphasis throughout this book is on understanding the design criteria. Modern spreadsheet or equation-solving software can be effective tools in design. With such an application program, the user can tailor the solution to meet a variety of goals. With little computer training, the designer can develop a *template* to solve a basic problem. A basic template can serve as the starting point for more sophisticated solutions.

6.2   Bending

In discussing the strength of a wood beam, it is important to understand that the bending stresses are parallel to the length of the member and are thus parallel to the grain of the wood. This is a common beam design problem (Fig. 6.1*a*), and it is the general subject of this section. See Example 6.1.

Occasionally, however, bending stresses across the grain (Fig. 6.1*b*) are developed, and the designer needs to recognize this situation. It has been noted previously that wood is relatively weak in *tension perpendicular to grain*. This is true whether the cross-grain tension stress is caused by a direct tension force perpendicular to grain or by loading that causes cross-grain bending. Cross-grain tension should generally be avoided.

EXAMPLE 6.1   Bending in Wood Members

Longitudinal Bending Stresses—Parallel to Grain

Ordinarily, the bending stress in a wood beam is parallel to the grain. The free-body diagram (FBD) in Fig. 6.1*a* shows a *typical beam* cut at an arbitrary point. The internal force and moment *V* and *M* are required for equilibrium. The bending stress diagram indicates that the stresses developed by the moment are longitudinal stresses, and they are, therefore, *parallel to grain*. Bending is shown about the strong or *x* axis of the member.

Cross-Grain Bending—Not Allowed

Section 1 in Fig. 6.1*b* shows a concrete wall connected to a wood horizontal diaphragm. The lateral force is transferred from the wall through the *wood ledger* by means of anchor bolts and nailing.

Section 2 indicates that the ledger cantilevers from the anchor bolt to the diaphragm level. Section 3 is an FBD showing the internal forces at the anchor bolt and the bending stresses that are developed in the ledger. The bending stresses in the ledger are *across the grain* (as opposed to being parallel to the grain). Wood is very weak in cross-grain bending and tension. This connection is introduced at this point to illustrate the cross-grain bending problem. Reference bending design values for wood design apply to longitudinal bending stresses only.

Because of failures in some ledger connections of this type, cross-grain bending and cross-grain tension are not permitted by the IBC for the anchorage of seismic forces. Even

FBD of **BEAM**   **BENDING STRESS**   **BEAM SECTION**

$$f_b = \frac{Mc}{I}$$

Figure 6.1a Bending stress is parallel to grain in the usual beam design problem.

SECTION ①   ②   ③

Figure 6.1b Cross-grain bending in a wood member should be avoided.

for other loading conditions, designs should generally avoid stressing wood in bending or tension across the grain.

It should be noted that the use of a wood ledger in a building with concrete or masonry walls is still a common connection. However, additional anchorage hardware is required to prevent the ledger from being stressed across the grain. Anchorage for this type of connection is covered in detail in Chap. 15.

The design moment in a wood beam is obtained using ordinary elastic theory. Most examples in this book use the nominal span length for evaluating the shear and moment in a beam. This is done to simplify the design calculations. However, in some problems it may be advantageous to take into account the technical definition of *span length* given in NDS Sec. 3.2.1 (Ref. 6.2).

Practically speaking, the span length is usually taken as the distance from the center of one support to the center of the other support. However, in most cases the *furnished* bearing length at a support will exceed the *required* bearing length. Thus, the NDS permits the designer to consider the span to be the *clear* distance between supports *plus* one-half of the *required* bearing length at each end. The required bearing length is a function of the compression design value perpendicular to grain $F_{c\perp}$ (Sec. 6.8).

The critical location for shear in a uniformly loaded wood beam is at a distance d from the face of the beam support (a similar practice is followed in reinforced-concrete design). The span length for bending and the critical loading condition for shear are shown later in this chapter in Fig. 6.13 (Sec. 6.5). Again, for hand calculations the shear and moment in a beam are often determined using a nominal span length. The added effort to obtain the more technical definition of span length is normally justified only in cases where the member appears to be overstressed using the nominal center-to-center span length.

The check for bending stress in a wood beam uses the familiar formula from strength of materials

$$f_b = \frac{Mc}{I} = \frac{M}{S} \le F_b'$$

where f_b = actual (computed) bending stress
$\quad\ M$ = moment in beam
$\quad\quad c$ = distance from neutral axis to extreme fibers
$\quad\quad I$ = moment of inertia of beam cross section about axis of bending
$\quad\quad S = I/c$
$\quad\quad\ $ = section modulus of beam cross section about axis of bending
$\quad\ F_b'$ = adjusted bending design value

According to allowable stress design (ASD) principles, this formula says that the *actual* (computed) bending stress must be less than or equal to the adjusted ASD bending design value. Similarly, for load and resistance factor design (LRFD), this formula says that the *actual* bending stress computed using *factored* loads must be less than or equal to the adjusted LRFD bending design value. However, LRFD is typically cast in terms of forces and moments rather than stresses, or

$$M_u \le M_n'$$

$$\le F_{bn}' \cdot S$$

where M_u = moment in beam due to factored loads
$\quad\ M_n'$ = adjusted LRFD moment capacity

F'_{bn} = adjusted LRFD bending design value

S = section modulus of beam cross section

Regardless, the adjusted bending design value for ASD (F'_b) or LRFD (F'_{bn}) takes into account the necessary adjustment factors to reference design values that may be required for a wood member.

Most wood beams are used in an efficient manner. In other words, the *moment is applied about the strong axis (x axis) of the cross section.* From an engineering point of view, this seems to be the most appropriate description of the common loading situation. However, other terms are also used in the wood industry to refer to bending about the strong axis. For solid *sawn* lumber of rectangular cross section, the terms *loaded edgewise, edgewise bending,* and *load applied to the narrow face* of the member all refer to bending about the *x* axis. For *glulam* beams, the term *load applied perpendicular to the wide face* of the laminations is commonly used.

As wood structures become more highly engineered, there is a need to generalize the design expressions to handle a greater variety of situations. In a general approach to beam design, the moment can occur about either the *x* or *y* axis of the beam cross section. See Example 6.2. For sawn lumber, the case of bending about the weak axis (*y* axis) is described as loaded *flatwise, flatwise bending,* and *load applied to the wide face of the member.* For glulam, it is referred to as *load applied parallel to the wide face of the laminations.* In engineering terms, *weak-axis bending* and *bending about the y axis* are probably better descriptions.

Throughout this book the common case of bending about the strong axis is assumed, unless otherwise noted. Therefore, the symbols f_b and F'_b imply bending about the *x* axis and thus represent the values f_{bx} and F'_{bx}. Where needed, the more complete notation of f_{bx} and F'_{bx} is used for clarity. (An exception to the general rule of bending about the strong axis is Decking, which is normally stressed about the *y* axis.)

EXAMPLE 6.2   Strong- and Weak-Axis Bending

The large majority of wood beams are rectangular in cross section and are loaded as efficient bending members. See Fig. 6.2*a*. This common condition is assumed, unless otherwise noted.

The bending stress in a beam about the strong axis (Fig. 6.2*a*) is

$$f_{bx} = \frac{M_x}{S_x} = \frac{M_x}{bd^2/6}$$

$$\leq F'_{bx}$$

A less efficient (and therefore less common) type of loading is to stress the member in bending about the minor axis. See Fig. 6.2*b*. Although it is not common, a structural member may occasionally be loaded in this manner.

Figure 6.2a   Most wood beams have bending about the strong axis. For sawn lumber, *loaded edgewise*. For glulam, *load perpendicular to wide face of laminations*.

Figure 6.2b   Occasionally beams have bending about the weak axis. For sawn lumber, *loaded flatwise*. For glulam, *load parallel to wide face of laminations*.

The bending stress in a beam loaded about the weak axis (Fig. 6.2b) is

$$f_{by} = \frac{M_y}{S_y} = \frac{M_y}{bd^2/6}$$

$$\leq F'_{by}$$

The designer must be able to recognize and handle either bending application.

The formula from engineering mechanics for bending stress f_b was developed for an ideal material. Such a material is defined as a solid, homogeneous, and isotropic (having the same properties in all directions) material. In addition, plane sections before bending are assumed to remain plane during bending, and stress is assumed to be linearly proportional to strain.

From the discussion of some of the properties of wood in Chap. 4, it should be clear that wood does not fully satisfy these assumptions. Wood is made up of hollow cells, which generally run parallel to the length of a member. In addition, there are a number of growth characteristics and service conditions such as annual rings, knots, slope of grain, and moisture content (MC). However, adequate beam designs are obtained by applying the ordinary bending formula and adjusting the reference design value to account for the unique characteristics of wood beams.

The starting point is to obtain the correct *reference* bending design value for the appropriate species and grade of member. Values of F_b are listed in NDS Supplement Tables 4A to 4F for sawn lumber and NDS Supplement Tables 5A to 5D for glulam. NDS Table 4.3.1, *Applicability of Adjustment Factors for Sawn Lumber* and Table 5.3.1, *Applicability of Adjustment Factors for Glued Laminated Timber*, then provide a string of multiplying factors to obtain the adjusted bending design values once the reference value is known.

For ASD, the adjusted bending design value is defined as

$$F'_b = F_b(C_D)(C_M)(C_t)(C_L)(C_F \text{ or } C_V)(C_{fu})(C_c)(C_i)(C_r)$$

and for LRFD, the adjusted bending design value is

$$F'_{bn} = F_{bn}(\phi_b)(\lambda)(C_M)(C_t)(C_L)(C_F \text{ or } C_V)(C_{fu})(C_c)(C_i)(C_r)$$

where F'_b = adjusted ASD bending value

$\quad F'_{bn}$ = adjusted LRFD bending value

$\quad F_b$ = reference bending design value

$\quad F_{bn}$ = nominal bending design value—LRFD

$\quad\quad = F_b \cdot K_F$

$\quad K_F$ = format conversion factor (Sec. 4.23) = 2.54—LRFD only

$\quad \phi_b$ = resistance factor for bending (See. 4.22) = 0.85—LRFD only

$\quad C_D$ = load duration factor (Sec. 4.15)—ASD only

$\quad \lambda$ = time effect factor (Sec. 4.16)—LRFD only

$\quad C_M$ = wet-service factor (Sec. 4.14—note that subscript M stands for moisture)

C_t = temperature factor (Sec. 4.20)

C_L = beam stability factor (consider when lateral support to compression side of beam may permit beam to buckle laterally— Sec. 6.3)

C_F = size factor (Sec. 4.17)

C_V = volume factor (Sec. 5.6)

C_{fu} = flat-use factor (Sec. 4.19)

C_c = curvature factor [Apply only to curved glulam beams; C_c = 1.0 for straight and cambered (slightly curved) glulams. The design of curved beams is beyond the scope of this book.]

C_i = incising factor for sawn lumber (Sec. 4.21)

C_r = repetitive member factor (Sec. 4.18)

The reader is referred to the appropriate sections in Chaps. 4 and 5 for background on the adjustment factors discussed previously. Lateral stability is an important consideration in the design of a beam. The beam stability factor C_L is an adjustment factor that takes into account a reduced moment capacity if lateral torsional buckling can occur. Initially it is assumed that buckling is prevented, and C_L defaults to unity. See Example 6.3.

It should be realized that the long list of adjustment factors for determining F_b' or F_{bn}' is basically provided as a reminder that a number of special conditions may require an adjustment of the tabulated design value. However, in many practical design situations, a number of the possible adjustment factors will default to 1.0. In addition, not all of the possible adjustments apply to all types of wood beams. Section 6.4 shows how the string of adjustment factors can be greatly reduced for practical beam design.

EXAMPLE 6.3   Full Lateral Support a Beam

The analysis of bending stresses is usually introduced by assuming that lateral torsional buckling of the beam is prevented. Continuous support of the compression edge of a beam essentially prevents the member from buckling (Fig. 6.3a).

Figure 6.3a   Direct attachment of roof or floor sheathing provides full lateral support to top edge of a beam. When subjected to bending loads, a beam with full lateral support is stable, and it will deflect only in its plane of loading.

Figure 6.3b When plywood sheathing is properly attached to framing, a diaphragm is formed that provides stability to beams. (*Photo courtesy of APA*—The Engineered Wood Association.)

A beam with positive moment everywhere has *compressive* bending stresses on the top edge of the member throughout its length. An effective connection (proper nailing) of roof or floor sheathing to the top of such a beam reduces the unbraced length to zero ($l_u = 0$). Technically, the unbraced length is the *spacing of the nails* through the sheathing and into the compression edge of the beam. For most practical diaphragm construction and most practical beam sizes, the unbraced length can be taken as zero.

Many practical wood structures have *full* or *continuous* lateral support as part of their normal construction. See Fig. 6.3b. Closely spaced beams in a repetitive framing arrangement are shown. However, a roof or floor diaphragm can also be used to provide lateral support to larger beams and girders, and the concept is not limited to closely spaced members.

With an unbraced length of zero, lateral buckling is eliminated, and the beam stability factor C_L defaults to unity. For other conditions of lateral support, C_L may be less than 1.0. The stability of laterally unbraced beams is covered in detail in Sec. 6.3.

Several points should be mentioned concerning the reference bending design values for different kinds of wood beams. Unlike glulam, the tables for sawn lumber do not list separate design properties for bending about the x and y axes. Therefore, it is important to understand which axis is associated with the tabulated values.

Reference bending design values F_b for *visually graded sawn lumber* apply to both the x and y axes *except* for Beams and Stringers (B&S) and Decking. Because decking is graded with the intent that the member will be used flatwise (i.e., weak-axis bending), the reference value in the NDS Supplement is F_{by}. A flat-use factor C_{fu} has already been incorporated into the reference value, and the designer should not apply C_{fu} to Decking. The use of Decking is mentioned only briefly, and it is not a major subject in this book.

For members in the B&S size category, the reference bending design values apply to the x axis only (i.e., $F_b = F_{bx}$). However, for other members including Dimension lumber and Posts and Timbers sizes, the tabulated bending design value applies to both axes (i.e., $F_b = F_{bx} = F_{by}$). To obtain the bending design value for the y axis, F_b must be multiplied by the appropriate flat-use factor C_{fu}.

In the infrequent case that a member in the B&S size category is loaded in bending about the minor axis, the designer should use the size adjustment factors provided with Table 4D in the NDS Supplement to determine the adjusted bending design value for the y axis.

Another point needs to be understood about the reference bending design values in the B&S size category. ASTM D 245 allows the application of a less restrictive set of grading criteria to the outer thirds of the member length. This practice anticipates that the member will be used in a *simple beam* application. It further assumes that the length of the member will not be reduced substantially by sawing the member into shorter lengths.

Therefore, if a B&S is used in some other application where the maximum bending stress does not occur in the middle third of the original member length (e.g., a cantilever beam or a continuous beam), the designer should specify that the grading provisions applicable to the middle third of the length shall be applied to the entire length. See Example 6.4.

EXAMPLE 6.4   Bending Design Values for Beams and Stringers

Reference bending design values for B&S sizes are for bending about the x axis of the cross section (See Fig. 6.4). Lumber grading agencies may apply less restrictive grading

Figure 6.4

rules to the outer thirds of the member length. This assumes that the maximum moment will be located in the middle third of the member length. The common uniformly loaded simply supported beam is the type of loading anticipated by this grading practice.

If the loading or support conditions result in a moment diagram which does not agree with the assumed distribution, the designer should specify that the grading rules normally applied to the middle third shall be applied to the entire length.

A similar problem develops if a long B&S member is ordered and then cut into shorter lengths (see NDS Sec. 4.1.7). A note on the plans should prohibit cutting beams of this type, *or* full-length grading should be specified.

NOTE: A way to reduce the length of a B&S without affecting its stress grade is to cut approximately equal lengths from *both* ends.

A brief introduction to reference bending design values for glulams was given in Chap. 5. Recall that *two* values of F_{bx} are listed for the softwood glulam bending combinations in NDS Supplement Table 5A, along with a value of F_{by}. It should be clear that F_{by} is for the case of bending about the weak axis of the member, but the two values for F_{bx} require further explanation. Although the computed compression and tension bending stresses in a rectangular beam are equal at the extreme fibers, tests have shown that the outer *tension laminations* are critical. Therefore, high-grade tension laminations are placed in the outer tension zone of the beam. The *top* of a glulam beam is marked in the laminating plant so that the member can be identified at the job site and oriented properly in the structure.

If the beam is loaded so that the tension laminations are stressed in tension, the appropriate bending value is F_{bx} *tension zone stressed in tension.* In this book, the following notation is used to indicate this value: F_{bx}^{+}. In most cases a glulam beam is used in an efficient manner, and F_{bx} is normally F_{bx}^{+}. In other words, F_{bx} is assumed to be F_{bx}^{+} unless otherwise indicated. On the other hand, if the member is loaded in such a way that the compression laminations are stressed in bending tension, the reference value known as F_{bx} *compression zone stressed in tension* (F_{bx}^{-}) is the corresponding reference bending design value.

A review of the NDS Supplement for glulam shows that the two reference design values for F_{bx} just described can differ by a factor of 2. Accordingly, depending on the combination, the calculated bending strength of a member could be 50 percent less than expected if the beam were inadvertently installed upside down. Thus, it is important that the member be installed properly in the field.

Simply supported beams under gravity loads have positive moment throughout, and bending tensile stresses are located on the bottom edge of the member. Here the designer is just concerned with F_{bx} *tension zone stressed in tension.* See Example 6.5.

EXAMPLE 6.5 F_{bx} in Glulam Bending Combinations

Some glulam beams are fabricated so that the reference bending tensile design value is the same for both faces of the member. Others are laid up in such a manner that the

Figure 6.5a Glulam with positive moment everywhere.

Figure 6.5b Glulam with positive and negative moments.

reference bending tensile design values are not the same for both faces of the beam. Two different reference bending design values are listed in the glulam tables:

1. F_{bx} tension zone stressed in tension F_{bx}^+
2. F_{bx} compression zone stressed in tension F_{bx}^- (this value never exceeds F_{bx}^+, and it may be much less)

In Fig. 6.5a, the designer needs to consider only F_{bx}^+ because there is tension everywhere on the bottom edge of the beam.

However, when positive and negative moments occur (Fig. 6.5b), both values of F_{bx} need to be considered.

In Figure 6.5b F_{bx}^+ applies to M_1, and F_{bx}^- is used for M_2. If M_1 and M_2 are equal, a bending combination can be used that has equal values for the two F_{bx} values.

Figure 6.5c Large glulam beam in manufacturing plant undergoing finishing operation. A stamp is applied to the "top" of a glulam so that the field crews will install the member right side up. (*Photo courtesy of FPL.*)

In the design of beams with both positive and negative moments, both values of F_{bx} need to be considered. In areas of negative moment (tension on the top edge of the beam), the value of F_{bx} *compression zone stressed in tension* applies. When the negative moment is small, the reduced reference bending design value for the compression zone stressed in tension may be satisfactory. Small negative moments may occur, for example, in beams with relatively small cantilever spans.

On the other hand, when the negative moment is large, the designer is not limited to a small value of F_{bx}^-. The designer can specify that tension zone grade requirements, including end-joint spacing, must be applied to both edges of the member. In this case, the higher reference bending design value F_{bx}^+ may be used to design for both positive and negative moments. Large negative moments often occur in cantilever beam systems (Sec. 6.16). For additional information regarding F_{bx} tension zone stressed in tension and F_{bx} compression zone stressed in tension, see Refs. 6.4 and 6.5.

A final general point should be made about the strength of a wood beam. The *notching of structural members* to accommodate plumbing or mechanical systems is the subject of considerable concern in the wood industry. The notching and cutting of members in residential construction is fairly common practice. Although this may not be a major concern for members in repetitive systems which are lightly loaded, it can cause serious problems in other situations. Therefore, a note on the building plans should prohibit the cutting or notching of any structural member unless it is specifically detailed on the structural plans.

The effects of notching in areas of *bending stresses* are often addressed separately from the effects of notching on the *shear capacity* near supports. The discussion in the remaining portion of this section deals primarily with the effects of a notch where a bending moment exists. For shear considerations see Sec. 6.5.

The effect of a notch on the bending strength of a beam is not fully understood, and convenient methods of analyzing the bending stress at a notch are not currently available. However, it is known that the critical location of a notch is in the bending tension zone of a beam. Besides reducing the depth available for resisting the moment, stress concentrations are developed. Stress concentrations are especially large for the typical square-cut notch.

To limit the effect in sawn lumber, NDS Sec. 4.4.3 limits the maximum depth of a notch to one-sixth the depth of the member and states that the notch shall not be located in the middle third of the span. Although not stated, it is apparent that this latter criterion applies to simply supported beams because of the high-bending stresses in this region of the span. The NDS further limits notches at ends of the member for bearing over a support to no more than one-fourth the depth. Except for notches at the ends of a member, the NDS prohibits the notching of the tension edge of beams when the nominal width of the member is 4 in. or greater.

Notches are especially critical in glulams because of the high-quality laminations at the outer fibers. Again, the tension laminations are the most critical and are located on the bottom of a beam that is subjected to a positive moment. For a glulam beam, NDS Sec. 5.4.4 prohibits notching in the tension face, except at the ends of the beam for bearing over a support. Even where allowed at the ends, the NDS limits the maximum depth of a notch in the tension face to one-tenth the depth of the glulam or 3 in. The NDS further prohibits notching in the compression face in the middle third of the span and limits the depth of a notch in the compression face at the end of a beam to two-fifths the depth of the member.

These negative statements about the use of notches in wood beams should serve as a warning about the potential hazard that can be created by stress concentrations due to reentrant corners. Failures have occurred in beams with notches located at some distance from the point of maximum bending and at a load considerably less than the design load.

The problem is best handled by avoiding notches. In the case of an existing notch, some strengthening of the member at the notch may be advisable.

6.3   Lateral Stability

When a member functions as a beam, a portion of the cross section is stressed in compression and the remaining portion is stressed in tension. If the compression zone of the beam is not braced to prevent lateral movement, the member may buckle at a bending stress that is less than the adjusted design value defined in Sec. 6.2. The adjusted bending design value described in Sec. 6.2 assumed that lateral torsional buckling was prevented by the presence of adequate bracing.

The bending compressive stress can be thought of as creating an equivalent column buckling problem in the compressive half of the cross section. Buckling in the plane of loading is prevented by the presence of the stable tension portion of the cross section. Therefore, if buckling of the compression edge occurs, movement will take place *laterally* between points of lateral support. See Example 6.6.

EXAMPLE 6.6   Lateral Buckling of Bending Member

Unlike the beam in Example 6.3, the girder in Fig. 6.6 does not have full lateral support.

Figure 6.6   Bending member with span length L and unbraced length l_u.

1. The distance between points of lateral support to the compressive edge of a bending member is known as the *unbraced length* l_u of the beam. The beams that frame into the girder in Fig. 6.6 provide lateral support of the compression (top) edge of the girder at a spacing of $l_u = L/2$.

2. It is important to realize that the *span* of a beam and the *unbraced length* of a beam are two different items. They may be equal, but they may also be quite different. The span is used to calculate moments, stresses, and deflections. The unbraced length, together with the cross-sectional dimensions, is used to analyze the stability of a bending member.

    In other words, the span L gives the actual bending stress f_b, and the unbraced length l_u defines the adjusted bending design values F'_{bx} or F'_{bxn}.

3. The section view in Fig. 6.6 shows several possible conditions:
    a. The unloaded position of the girder.
    b. The deflected position of the girder under a vertical load with no instability. Vertical deflection occurs if the girder remains stable.

    c. The buckled position. If the unbraced length is excessive, the compression edge of the member may buckle laterally in a manner similar to a slender column. Buckling takes place between points of lateral support. This buckled position is also shown in the plan view.

4. When the top of a beam is always in compression (positive moment everywhere) and when roof or floor sheathing is effectively connected directly to the beam, the unbraced length approaches zero. Such a member is said to have full lateral support. When lateral buckling is prevented, the strength of the beam depends on the bending design value of the material and not on stability considerations.

In many practical situations, the concern of lateral instability is simply eliminated by providing lateral support to the compression edge of the beam at close intervals. It has been noted that an effective connection (proper nailing) of roof or floor sheathing to the compression edge of a beam causes the unbraced length to approach zero ($l_u = 0$), and lateral instability is prevented by *full* or *continuous* lateral support.

In the case of laterally unbraced *steel beams* (W shapes), the problem of stability is amplified because cross-sectional dimensions are such that relatively *slender* elements are stressed in compression. Slender elements have large width-to-thickness (b/t) ratios, and these elements are particularly susceptible to buckling.

In the case of *rectangular wood beams*, the dimensions of the cross section are such that the depth-to-thickness ratios (d/b) are relatively small. Common framing conditions and cross-sectional dimensions cause large reductions in adjusted bending design value to be the exception rather than the rule. Procedures are available; however, for taking lateral stability into account and these are outlined in the remainder of this section.

Two methods for handling the lateral stability of beams are currently in use. One method is based on *rules of thumb* that have been developed over time. These rules are applied to the design of sawn lumber beams. In this approach, the required type of lateral support is specified on the basis of the depth-to-thickness ratio (d/b) of the member.

These rules are outlined in NDS Sec. 4.4.1, *Stability of Bending Members* for sawn lumber. As an example, the rules state that if $d/b = 6$, bridging, full-depth solid blocking, or diagonal cross-bracing is required at intervals of 8 ft- 0 in. maximum, full lateral support must be provided for the compression edge, and the beam must be supported at bearing points such that rotation is prevented. The requirement for bridging, blocking, or cross-bracing can be omitted if both edges are held in line for their entire length. See Example 6.7.

EXAMPLE 6.7   Lateral Support of Beams—Approximate Method

When $d/b = 6$, lateral support can be provided by full-depth solid blocking, diagonal cross bracing or by bridging spaced at 8-ft-0 in. maximum. *Solid blocking* must be the same depth as the beams. Adjacent blocks may be staggered to facilitate construction (i.e., end nail-ing through beam). *Bridging* is cross-bracing made from wood (typically 1×3 or 1×4) or light-gauge steel (available prefabricated from manufacturers of hardware for wood construction). See Fig. 6.7.

Figure 6.7   Solid (full depth) blocking or bridging for lateral stability based on traditional rules involving (d/b) ratio of beam.

These requirements for lateral support are *approximate* because only the proportions of the cross section (i.e., the d/b ratio) are considered. The second, more accurate method of accounting for lateral stability uses the *slenderness ratio* R_B of the beam. See Example 6.8. The slenderness ratio considers the unbraced length (distance between points of lateral support to the compression edge of the beam) in addition to the dimensions of the cross section. This method was developed for large glulam beams, but it applies equally well to sawn lumber beams.

EXAMPLE 6.8   Slenderness Ratio for Bending Members

BEAM
SECTION

Figure 6.8   Cross section of a glulam bending member.

The *slenderness ratio* for a beam measures the tendency of the member to buckle laterally between points of lateral support to the *compression edge* of the beam. Dimensions are in inches.

$$R_B = \sqrt{\frac{l_e d}{b^2}}$$

where R_B = slenderness ratio for a bending member
   b = beam width
   d = beam depth
   l_u = unbraced length of beam (distance between points of lateral support as in Fig. 6.6)
   l_e = effective unbraced length

The effective unbraced length is a function of the type of span, loading condition, and l_u/d ratio of the member. Several definitions of l_u are given here for common beam configurations. NDS Table 3.3.3., *Effective Length, l_e, for Bending Members*, summarizes these and a number of additional loading conditions involving multiple concentrated loads.

Cantilever Beam		
Type of load	When $l_u/d < 7$	When $l_u/d \geq 7$
Uniformly distributed load	$l_e = 1.33l_u$	$l_e = 0.90l_u + 3d$
Concentrated load at free end	$l_e = 1.87l_u$	$l_e = 1.44l_u + 3d$

Single-Span Beam		
Type of load	When $l_u/d < 7$	When $l_u/d \geq 7$
Uniformly distributed load	$l_e = 2.06l_u$	$l_e = 1.63l_u + 3d$
Concentrated load at center with no lateral support at center	$l_e = 1.80l_u$	$l_e = 1.37l_u + 3d$
Concentrated load at center with lateral support at center	$l_e = 1.11l_u$	
Two equal concentrated loads at one-third points and lateral support at one-third points	$l_e = 1.68l_u$	

NOTE: For a cantilever or single-span beam with any loading, the following values of l_e may conservatively be used:

$$
l_e = \begin{cases}
2.06l_u & \text{when } l_u/d < 7 \\
1.63l_u + 3d & \text{when } 7 \leq l_u/d \leq 14.3 \\
1.84l_u & \text{when } l_u/d > 14.3
\end{cases}
$$

In calculating the beam slenderness ratio R_B, the *effective unbraced length* is defined in a manner similar to the effective length of a column (Chap. 7). For a beam, the effective length l_e depends on the end conditions (span type) and type of loading. In addition, the ratio of the unbraced length to the beam depth l_u/d may affect the definition of effective unbraced length.

Once the slenderness ratio of a beam is known, the effect of lateral stability on the bending design value may be determined. For large slenderness ratios, the adjusted bending design values are reduced greatly, and for small slenderness ratios, lateral stability has little effect. At a slenderness ratio of zero, the beam can be considered to have full lateral support, and the adjusted bending design value is as defined in Sec. 6.2 with $C_L = 1.0$.

The acceptable range for beam slenderness is $0 \leq R_B \leq 50$. The effect of lateral stability on the bending strength of a beam is best described on a graph of the adjusted bending design value F'_{bx} plotted against the beam slenderness ratio R_B. See Example 6.9. The NDS formula for evaluating the effect of lateral stability on beam capacity gives a continuous curve for F'_{bx} over the entire range of beam slenderness ratios.

EXAMPLE 6.9   Adjusted Bending Design Value Considering Lateral Stability

The NDS has a continuous curve for evaluating the effects of lateral torsional buckling on the bending strength of a beam. See Fig. 6.9a. Lateral torsional buckling may occur between points of lateral support to the compression edge of a beam as the member is stressed in bending about the x axis of the cross section.

The tendency for a beam to buckle is eliminated if the moment occurs about the weak axis of the member. Therefore, the adjusted design value reduction given by the curve in Fig. 6.9a is limited to bending about the x axis, and the adjusted bending design value is labeled F'_{bx}. However, the x subscript is often omitted, and it is understood that the reduced bending design value is about the x axis (i.e., $F'_b = F'_{bx}$).

Figure 6.9a   Typical plot of adjusted bending design value about the x axis F'_{bx} versus beam slenderness ratio R_B.

Adjusted Bending Design Value

The adjusted bending design value curve in Fig. 6.9a is obtained by multiplying the reference bending design value by the beam stability factor C_L and all other appropriate adjustment factors.

For ASD:

$$F'_{bx} = F_{bx}(C_L) \times \cdots$$

For LRFD:

$$F'_{bxn} = F_{bxn}(C_L) \times \cdots$$

where F'_{bx} = adjusted ASD bending value about x axis
   F'_{bn} = adjusted LRFD bending value about x axis
   F_{bx} = reference ASD bending value x axis
   F_{bn} = nominal bending value about x axis for LRFD
   C_L = beam stability factor (defined below)
   $\times \cdots$ = product of other appropriate adjustment factors

Beam Stability Factor C_L
For ASD:

$$C_L = \frac{1 + F_{bE}/F^*_{bx}}{1.9} - \sqrt{\left(\frac{1 + F_{bE}/F^*_{bx}}{1.9}\right)^2 - \frac{F_{bE}/F^*_{bx}}{0.95}}$$

For LRFD:

$$C_L = \frac{1 + F_{bEn}/F_{bxn}^*}{1.9} - \sqrt{\left(\frac{1 + F_{bEn}/F_{bxn}^*}{1.9}\right)^2 - \frac{F_{bEn}/F_{bxn}^*}{0.95}}$$

where F_{bE} = Euler-based ASD critical buckling value for bending members

$$= \frac{1.20E'_{min}}{R_B^2}$$

F_{bEn} = Euler-based LRFD critical buckling value for bending members

$$= \frac{1.20E'_{min\,n}}{R_B^2}$$

F_{bx}^* = reference ASD bending value about x axis multiplied by certain adjustment factors

$= F_{bx} \times$ (product of all adjustment factors except C_{fu}, C_V, and C_L)

F_{bxn}^* = nominal LRFD bending value about x axis multiplied by certain adjustment factors

$= F_{bxn} \times$ (product of all adjustment factors except C_{fu}, C_V, and C_L)

E'_{min} = adjusted ASD modulus of elasticity associated with lateral torsional buckling

$E'_{min\text{-}n}$ = adjusted LRFD modulus of elasticity associated with lateral torsional buckling

R_B = slenderness ratio for bending member (Example 6.8)

Figure 6.9b   Effect of load duration factor on F'_{bx} governed by lateral stability.

In lateral torsional buckling, the bending stress is about the x axis. With this mode of buckling, instability is related to the y axis, and the modulus of elasticity about the y axis is used to evaluate F_{bE}. For sawn lumber, the modulus of elasticity about the y axis is the same as it is about the x axis. For glulam, however, $E_{y\,min}$ and $E_{x\,min}$ may be different. Also, recall that C_D (ASD) and λ (LRFD) do not apply to E_{min}.

A similar plot could be generated for LRFD considering the effect of λ.

The load duration factor C_D or time effect factor λ have full effect on the adjusted bending design value in a beam that has full lateral support. On the other hand, C_D and λ have no influence on the adjusted bending design value when instability predominates. A transition between C_D or λ having full effect at a slenderness ratio of 0 and C_D or λ having no effect at a slenderness of 50 is automatically provided in the definition of C_L. This relationship is demonstrated for ASD with C_D in Fig. 6.9b.

The form of the expression for the *beam stability factor* C_L is the same as the form of the *column stability factor* C_P. The column stability factor is presented in Sec. 7.4 on column design. Both expressions serve to reduce the reference design value based on the tendency of the member to buckle. For a beam, C_L measures the effects of lateral torsional buckling, and for a member subjected to axial compression, C_P evaluates column buckling.

The general form of the beam and column buckling expressions is the result of column studies by Ylinen. They were confirmed by work done at the Forest Products Laboratory (FPL) as part of a unified treatment of combined axial and bending loads for wood members (Ref. 6.13). The beam stability factor and the column stability factor provide continuous curves for adjusted design values.

The expressions for C_L and C_P both make use of an elastic buckling stress. The Euler critical buckling stress is the basis of the elastic buckling stress

$$F_{\text{Euler}} = F_E = \frac{\pi^2 E}{(\text{slenderness ratio})^2}$$

Recall that two values of modulus of elasticity are listed in the NDS Supplement, E representing the average value and E_{min} for use in stability calculations. Using E_{min} is equivalent to using a factor of safety in the F_E formula.

For beam design, the elastic buckling stress is stated as

$$F_{bE} = \frac{1.20E'_{min}}{R_B^2}$$

In the lateral torsional buckling analysis of beams, the bending stress about the x axis is the concern. However, instability with this mode of buckling is associated with the y axis (see the section view in Fig. 6.6), and $E_{y\,min}$ is used to compute F_E. For glulams, the values of $E_{x\,min}$ and $E_{y\,min}$ may not be equal, and the designer should use $E_{y\,min}$ from the glulam tables to evaluate the Euler-based critical buckling value for beams.

For the beam and column stability factors, the elastic buckling value F_E is divided by a materials strength property to form a ratio that is used repeatedly in the formulas for C_L and C_P. For beams, the material strength property is given the notation F_b^*. In this book subscript x is sometimes added to this notation. This is a reminder that F_b^* is the reference bending design value for the x axis

F_{bx} multiplied by certain adjustment factors. Again, the ratio F_{bE}/F_b^* is used a number of times in the Ylinen formula.

From strength of materials it is known that the Euler formula defines the critical buckling stress in long slender members. The effect of beam and column stability factors (C_L and C_P) is to define an adjusted value curve that converges on the Euler curve for large slenderness ratios.

Several numerical examples given later in this chapter demonstrate the application of C_L for laterally unbraced beams. Section 6.4 summarizes the adjustments of bending design values for different types of beam problems.

6.4   Adjusted Bending Design Value Summary

The comprehensive listing of adjustment factors given in Sec. 6.2 for determining F_b' or F_{bn}' is a general summary and not all of the factors apply to all beams. The purpose of this section is to identify the adjustment factors required for specific applications. In addition, a number of the adjustment factors that frequently default to unity are noted.

Some repetition of material naturally occurs in a summary of this nature. The basic goal, however, is to simplify the long list of possible adjustment factors and provide a concise summary of the factors relevant to a particular type of beam problem. The objective is to have a complete outline of the design criteria without making the problem appear overly complicated. Knowing what adjustment factors default to unity for frequently encountered design problems should help in the process.

The adjusted bending design values for sawn lumber are given in Example 6.10. The example covers *visually graded sawn lumber*, and bending values apply to all size categories except Decking. The grading rules for Decking presume that loading will be about the minor axis, and published values are F_{by}. The flat-use factor C_{fu} has already been applied to the tabulated F_b for Decking.

EXAMPLE 6.10   Adjusted Bending Design Value—Visually Graded Sawn Lumber

The adjusted bending design values for sawn lumber beams of rectangular cross section are summarized in this example. The common case of bending about the strong axis is covered first. See Fig. 6.10a. The appropriate adjustment factors are listed, and a brief comment is given as a reminder about each factor. Certain common default values are suggested (e.g., dry-service conditions and normal temperatures, as found in most covered structures).

Figure 6.10a   Sawn lumber beam with moment about strong axis.

Adjusted Bending Design Value for Strong Axis

For ASD:

$$F'_{bx} = F_{bx}(C_D)(C_M)(C_t)(C_L)(C_F)(C_r)(C_i)$$

For LRFD:

$$F'_{bxn} = F_{bxn}(\phi_b)(\lambda)(C_M)(C_t)(C_L)(C_F)(C_r)(C_i)$$

where F'_{bx} = adjusted ASD bending value about x axis
$\quad\quad F'_{bxn}$ = adjusted LRFD bending value about x axis
$\quad\quad F_{bx} = F_b$ = reference bending design value. Recall that for sawn lumber, reference bending design values apply to x axis (except Decking). Values are listed in NDS Supplement Tables 4A, 4B, 4C, and 4F for Dimension lumber and in Table 4D for Timbers.
$\quad\quad F_{bxn} = F_{bn}$ = nominal LRFD bending design value
$\quad\quad\quad\quad = F_b \cdot K_F$
$\quad\quad K_F$ = format conversion factor for bending (see Example 4.9)—LRFD only
$\quad\quad\quad\quad = 2.16/\phi_b = 2.54$
$\quad\quad \phi_b$ = resistance factor for bending (See. 4.22) = 0.85—LRFD only
$\quad\quad C_D$ = load duration factor (Sec. 4.15)—ASD only
$\quad\quad \lambda$ = time effect factor (See. 4.16)—LRFD only
$\quad\quad C_M$ = wet-service factor (Sec. 4.14)
$\quad\quad\quad\quad = 1.0$ for MC ≤ 19 percent (as in most covered structures)
$\quad\quad C_t$ = temperature factor (Sec. 4.20)
$\quad\quad\quad\quad = 1.0$ for normal temperature conditions
$\quad\quad C_L$ = beam stability factor
$\quad\quad\quad\quad = 1.0$ for continuous lateral support of compression face of beam. For other conditions compute C_L in accordance with Sec. 6.3.
$\quad\quad C_F$ = size factor (Sec. 4.17). Obtain values from Adjustment Factors section of NDS Supplement Tables 4A, 4B and 4F for Dimension lumber and in Table 4D for Timbers.
$\quad\quad C_r$ = repetitive member factor (Sec. 4.18)
$\quad\quad\quad\quad = 1.15$ for Dimension lumber applications that meet the definition of a repetitive member
$\quad\quad\quad\quad = 1.0$ for all other conditions
$\quad\quad C_i$ = incising factor (Sec. 4.21)
$\quad\quad\quad\quad = 0.8$ for incised Dimension lumber
$\quad\quad\quad\quad = 1.0$ for Dimension lumber not incised (whether the member is treated or untreated)
$\quad\quad\quad\quad = 1.0$ for Timbers (whether the member is incised or not)

Although bending about the strong axis is the common bending application, the designer should also be able to handle problems when the loading is about the weak axis. See Fig. 6.10b.

Figure 6.10*b* Sawn lumber beam with moment about weak axis.

Adjusted Bending Design Value for Weak Axis

For ASD:

$$F'_{by} = F_{by}(C_D)(C_M)(C_t)(C_F)(C_{fu})(C_i)$$

For LRFD:

$$F'_{byn} = F_{byn}(\phi_b)(\lambda)(C_M)(C_t)(C_F)(C_{fu})(C_i)$$

where F'_{by} = adjusted ASD bending value about y axis

F'_{byn} = adjusted LRFD bending value about y axis

F_{by} = F_b = reference bending design value. Recall that reference values of bending stress apply to y axis for all sizes of sawn lumber except Beams and Stringers. Values of F_b are listed in NDS Supplement Tables 4A, 4B, 4C, and 4F for Dimension lumber and in Table 4D for Timbers. For F_{by} in a B&S size, a size factor is provided for bending about the y axis.

F_{byn} = F_{bn} = nominal LRFD bending design value

 = $F_b \cdot K_F$

K_F = format conversion factor for bending (see Example 4.9)—LRFD only

 = $2.16/\phi_b = 2.54$

ϕ_b = resistance factor for bending (See. 4.22) = 0.85—LRFD only

C_D = load duration factor (Sec. 4.15)—ASD only

λ = time effect factor (Sec. 4.16)—LRFD only

C_M = wet-service factor (Sec. 4.14)

 = 1.0 for MC ≤ 19 percent (as in most covered structures)

C_t = temperature factor (Sec. 4.20)

 = 1.0 for normal temperature conditions

C_F = size factor (Sec. 4.17). Obtain values from Adjustment Factors section of NDS Supplement Tables 4A, 4B, and 4F for Dimension lumber and in Table 4D for Timbers.

C_{fu} = flat-use factor (Sec. 4.19). Obtain values from Adjustment Factors section of NDS Supplement Tables 4A, 4B, 4C, and 4F for dimension lumber.

C_i = incising factor (Sec. 4.21)

 = 0.8 for incised Dimension lumber

 = 1.0 for Dimension lumber not incised (whether the member is treated or untreated)

 = 1.0 for Timbers

A summary of the appropriate adjustment factors for a glulam beam is given in Example 6.11. Note that the size factor C_F that is used for sawn lumber beams is replaced by the volume factor C_V in glulams. However, in glulams the volume factor C_V is not applied simultaneously with the beam stability factor C_L. The industry position is that volume factor C_V is a bending value coefficient that adjusts for strength in the tension zone of a beam. Therefore, it is not applied concurrently with the beam stability factor C_L, which is an adjustment related to the bending strength in the compression zone of the beam.

EXAMPLE 6.11   Adjusted Bending Design Value—Glulam

The adjusted bending design values for straight or slightly curved glulam beams of rectangular cross section are summarized in this example. The common case of bending about the strong axis with the tension laminations stressed in tension is covered first. See Fig. 6.11*a*. This summary is then revised to cover the case of the compression laminations stressed in tension. As with the sawn lumber example, the appropriate adjustment factors are listed along with a brief comment.

Figure 6.11*a*   Glulam beam with moment about strong axis.

Adjusted Bending Design Value for Strong Axis

A glulam beam *bending combination* is normally stressed about the *x* axis. The usual case is with the tension laminations stressed in tension. The notation F_{bx} typically refers to this loading situation. The adjusted bending design value is taken as the smaller of two values, one considering C_L and the other considering C_V:

For ASD:

$$F'_{bx} = F^{+'}_{bx} = F_{bx}(C_D)(C_M)(C_t)(C_L)$$

or

$$F'_{bx} = F^{+'}_{bx} = F^+_{bx}(C_D)(C_M)(C_t)(C_V)$$

For LRFD:

$$F'_{bxn} = F^+_{bxn} = F^+_{bxn}(\phi_b)(\lambda)(C_M)(C_t)(C_L)$$

or

$$F'_{bxn} = F^{+'}_{bxn} = F^+_{bxn}(\phi_b)(\lambda)(C_M)(C_t)(C_V)$$

where $F'_{bx} = F^{+'}_{bx} =$ adjusted ASD bending value about *x* axis with high-quality tension laminations stressed in tension

$F'_{bxn} = F^{+'}_{bxn} =$ adjusted LRFD bending value about *x* axis with high-quality tension laminations stressed in tension

$F_{bx} = F^+_{bx} =$ reference bending design value about *x* axis with tension zone stressed in tension. Values are listed in NDS Supplement Table 5A for softwood glulam.

$F_{bxn} = F^+_{bxn} =$ nominal LRFD bending value about *x* axis with tension zone stressed in tension
$= F^+_{bx} \cdot K_F$

$K_F =$ format conversion factor for bending (see Example 4.9)—LRFD only
$= 2.16/\phi_b = 2.54$

ϕ_b = resistance factor for bending (See 4.22) = 0.85—LRFD only
C_D = load duration factor (Sec. 4.15)—ASD only
λ = time effect factor (Sec. 4.16)—LRFD only
C_M = wet-service factor (Sec. 4.14)
= 1.0 for MC < 16 percent (as in most covered structures)
C_t = temperature factor (Sec. 4.20)
= 1.0 for normal temperature conditions (as in most covered structures)
C_L = beam stability factor
= 1.0 for continuous lateral support of compression face of beam. For other conditions of lateral support C_L is evaluated in accordance with Sec. 6.3.
C_V = volume factor (Sec. 5.6)

Glulam beams are sometimes loaded in bending about the x axis with the compression laminations stressed in tension. The typical application for this case is in a beam with a relatively short cantilever (Fig. 6.5b). The adjusted bending design value is taken as the smaller of two values, one considering C_L and the other considering C_V:

For ASD:

$$F_{bx}^{-\prime} = F_{bx}^{-}(C_D)(C_M)(C_t)(C_L)$$

or

$$F_{bx}^{-\prime} = F_{bx}^{-}(C_D)(C_M)(C_t)(C_V)$$

For LRFD:

$$F_{bxn}^{-\prime} = F_{bxn}^{-}(\phi_b)(\lambda)(C_M)(C_t)(C_L)$$

or

$$F_{bxn}^{-\prime} = F_{bxn}^{-}(\phi_b)(\lambda)(C_M)(C_t)(C_V)$$

where $F_{bx}^{-\prime}$ = adjusted ASD bending value about x axis with compression laminations stressed in tension
$F_{bxn}^{-\prime}$ = adjusted LRFD bending value about x axis with compression laminations stressed in tension
F_{bx}^{-} = reference bending design value about x axis with compression zone stressed in tension. Values are listed in NDS Supplement Table 5A for softwood glulam.
F_{bxn}^{-} = nominal LRFD bending value about x axis with compression zone stressed in tension
= $F_{bx}^{-} \cdot K_F$

Other terms are as defined above.
Although loading about the strong axis is the common application for a bending combination, the designer may occasionally be required to handle problems with bending about the weak axis See Fig. 6.11b.

Figure 6.11b Glulam beam with moment about weak axis.

Adjusted Bending Design Value For Weak Axis

For ASD:

$$F'_{by} = F_{by}(C_D)(C_M)(C_t)(C_{fu})$$

For LRFD:

$$F'_{byn} = F_{byn}(\phi_b)(\lambda)(C_M)(C_t)(C_{fu})$$

where F'_{by} = adjusted ASD bending value about y axis
$\quad$ F_{by} = reference bending value about y axis. Values are listed in NDS
$\qquad$ Supplement Table 5A for softwood glulam.
$\quad$ F'_{byn} = adjusted LRFD bending value about y axis
$\quad$ F_{byn} = nominal LRFD bending value about y axis
$\qquad$ = $F_{by} \cdot K_F$
$\quad$ K_F = format conversion factor for bending (see Example 4.9)—LRFD only
$\qquad$ = $2.16/\phi_b$ = 2.54
$\quad$ ϕ_b = resistance factor for bending (See. 4.22) = 0.85—LRFD only
$\quad$ C_D = load duration factor (Sec. 4.15) —ASD only
$\quad$ λ = time effect factor (Sec. 4.16)—LRFD only
$\quad$ C_M = wet-service factor (Sec. 4.14)
$\qquad$ = 1.0 for MC < 16 percent (as in most covered structures)
$\quad$ C_t = temperature factor (Sec. 4.20)
$\qquad$ = 1.0 for normal temperature conditions
$\quad$ C_{fu} = flat-use factor (Sec. 4.19). Obtain values from Adjustment Factors section
$\qquad$ of NDS Supplement Table 5A for softwood glulam. Flat-use factor may
$\qquad$ conservatively be taken equal to 1.0.

Example 6.11 deals with the most common type of glulam which is a softwood bending combination. A glulam constructed from an axial load combination does not have the distribution of laminations that is used in a bending combination. Therefore, only one value of F_{bx} is provided for axial combination glulams, and the distinction between F_{bx}^+ and F_{bx}^- is not required. Other considerations for the adjusted bending design value in an axial combination glulam are similar to those in Example 6.11. Reference values and adjustment factors for axial combination softwood glulams are given in NDS Supplement Table 5B. Design values for all hardwood glulam combinations are given in

NDS Supplement Tables 5C and 5D for bending and axial combinations, respectively.

The designer should not be overwhelmed by the fairly extensive summary of adjusted bending design values. Most sawn lumber and glulam beam applications involve bending about the strong axis, and most glulams have the tension zone stressed in tension. The other definitions of adjusted bending design values are simply provided to complete the summary and to serve as a reference in the cases when they may be needed.

Numerical examples later in this chapter will demonstrate the evaluation of bending design values for both visually graded sawn lumber and glulams.

6.5   Shear

The shear stress in a beam is often referred to as *horizontal* shear. From strength of materials it will be recalled that the shear stress at any point in the cross section of a beam can be computed by the formula

$$f_v = \frac{VQ}{Ib}$$

Recall also that through equalibrium the horizontal and vertical shear stresses at a given point are equal. The shear strength of wood parallel to the grain is much less than the shear strength across the grain, and in a wood beam the grain is parallel with the longitudinal axis. In a typical horizontal beam, then, the horizontal shear is critical.

It may be helpful to compare the shear stress distribution given by VQ/Ib for a typical steel beam and a typical wood beam. See Example 6.12. Theoretically, the formula applies to the calculation of shear stresses in both types of members. However, in design practice the shear stress in a steel W shape is approximated by a nominal (average web) shear calculation.

The average shear stress calculation gives reasonable results in typical steel I-beams, but it does not apply to rectangular wood beams. The maximum shear in a rectangular beam is 1.5 times the average shear stress. This difference is significant and cannot be disregarded.

EXAMPLE 6.12   Horizontal Shear Stress Distribution

Steel Beam

For a steel W shape (see Fig. 6.12a), a nominal check on shear is made by dividing the total shear by the cross-sectional area of the web.

$$\text{Avg. } f_v = \frac{V}{A_{\text{web}}} = \frac{A}{dt_w} \approx \text{max. } f_v$$

WIDE FLANGE
STEEL BEAM   SHEAR STRESS

Figure 6.12a   Theoretical shear and average web shear in a steel I-beam.

Wood Beam

For rectangular beams, the theoretical *maximum* "horizontal" shear *must* be used. The following development shows that the maximum shear is 1.5 times the average. See Fig. 6.12b.

$$\text{Avg. } f_v = \frac{V}{A}$$

$$\text{Max. } f_v = \frac{VQ}{Ib} = \frac{VA\,\bar{y}}{Ib} = \frac{V(bd/2)(d/4)}{(bd^3/12) \times b}$$

$$= \frac{3v}{2bd} = 1.5\,\frac{V}{A} = 1.5\,(\text{avg. } f_v)$$

RECT BEAM   SHEAR STRESS

Figure 6.12b   Shear stress distribution in a typical wood beam section).

A convenient formula for horizontal shear stresses in a rectangular beam is developed in Example 6.12. For wood cross sections of other configurations, the distribution of shear stresses will be different, and it will be necessary to use the basic shear stress formula or some other appropriate check, depending on the type of member involved. The check on shear for a rectangular wood beam is

$$f_v = \frac{1.5V}{A}$$

$$\le F_v'$$

where f_v = actual (computed) shear stress in beam
V = maximum design shear in beam
A = cross-sectional area of beam
F_v' = adjusted ASD shear design value

According to ASD principles, this formula says that the *actual* computed working shear stress must be less than or equal to the adjusted ASD shear design value. Similarly for LRFD, this formula says that the *actual* shear stress computed using *factored* loads must be less than or equal to the adjusted LRFD shear design value. However, LRFD is typically put in terms of forces rather stress, or

$$V_u \le \frac{2}{3}F_{vn}'A$$

where V_u = shear force in a beam due to factored loads
F_{vn}' = adjusted LRFD shear value

For ASD:

$$F_v' = F_v(C_D)(C_M)(C_t)(C_i)$$

For LRFD:

$$F_{vn}' = F_{vn}(\phi_v)(\lambda)(C_M)(C_t)(C_i)$$

where F_v = reference shear design value
F_{vn} = nominal LRFD shear value
$= F_v \cdot K_F$
K_F = format conversion factor for shear (see Example 4.9)—LRFD only
$= 2.16/\phi_v = 2.88$
ϕ_v = resistance factor for shear (See. 4.22) = 0.75—LRFD only

The factors used to evaluate the adjusted shear design value were introduced in Chap. 4. The adjustment factors and typical values for frequently encountered conditions are

C_D = load duration factor (Sec. 4.15)—ASD only

λ = time effect factor (Sec. 4.16)—LRFD only

C_M = wet-service factor (Sec. 4.14)

$= 1.0$ for dry-service conditions, as in most covered structures. Dry-service conditions are defined as

MC $\leq$ 19 percent for sawn lumber

MC $<$ 16 percent for glulam

$C_t =$ temperature factor (Sec. 4.20)

$= 1.0$ for normal temperature conditions

$C_i =$ incising factor (Sec. 4.21)

$= 0.8$ for incised Dimension lumber

$= 1.0$ for Dimension lumber not incised (Note: The incising factor is only applicable to Dimension lumber and is *not* applicable to glulam or Timbers.)

In beams that are not likely to be critical in shear, the value of V (ASD) or V_u (LRFD) used in the shear design checking formula is often taken as the maximum shear from the shear diagram. However, the NDS Sec. 3.4.3.1 permits the maximum design shear to be reduced. To take this reduction into account, the load must be applied to one face of the beam, with the support reactions on the opposite face. This is the usual type of loading for a beam. The reduction does not apply, for example, to the case where the loads are hung or suspended from the bottom face of the beam.

The reduction in shear is accomplished by neglecting or removing all uniformly distributed loads within a distance d (equal to the depth of the beam) from the face of the beam supports. Concentrated loads within a distance d from the face of the supports *cannot* be simply neglected. Rather, concentrated loads can be reduced by a factor x/d, where x is the distance from the face of the beam support to the load. Therefore, a concentrated load *can* be ignored if located at the face of the support, but must be fully considered if located a distance d from the face of the support, and be considered with a linear reduction in magnitude if located anywhere else within a distance d from the support. See Example 6.13. In the case of a single moving concentrated load, the design shear may be obtained by locating the moving load at a distance d from the support (rather than placing it directly at the support). These reductions in computed shear stress can be applied in the design of both glulam and sawn lumber beams.

EXAMPLE 6.13   Reduction in Loads for Horizontal Shear Calculations

1. The maximum design shear may be reduced by:
   a. Omitting uniformly distributed loads within a distance d (the depth of the beam) from the face of the support, and
   b. Reducing the magnitude of a concentrated load by a factor x/d, where x is the distance from the face of the beam support to the load. Concentrated loads can be ignored ($x/d = 0$) if located at the face of the support ($x = 0$); must be fully considered ($x/d = 1$) if located a distance $x = d$ from the face of the support; and be considered with a linear reduction in magnitude, if located anywhere else within a distance d from the support.

Figure 6.13 Permitted reduction in shear for calculating f_v.

2. The modified loads are *only for horizontal shear stress* calculations in wood (sawn lumber and glulam) beams. The full design loads must be used for other design criteria.

3. The concept of omitting or reducing loads within d from the support is based on an assumption that the loads are applied to one side of the beam (usually the top) and the member is supported by bearing on the opposite side (usually the bottom). In this way the loads within d from the support are transmitted to the supports by diagonal compression. A similar type of adjustment for shear is used in reinforced concrete design.

The *span length for bending* is defined in Sec. 6.2 and is shown in Fig. 6.13 for information. It is taken as the clear span plus one-half of the required bearing length at each end (NDS Sec. 3.2.1). Although this definition is permitted, it is probably more common (and conservative) in practice to use the distance between the centers of bearing. The span for bending is normally the length used to construct the shear and moment diagrams.

When the details of the beam support conditions are fully known, the designer may choose to calculate the shear stress at a distance d from the *face* of the support. However, in this book many of the examples do not have the support details completely defined. Consequently, if the reduction for shear is used in an example, the loads are conservatively considered within a distance d from the reaction point in the "span for bending." The designer should realize that the point of reference is technically the face of the support and a somewhat greater reduction in calculated shear may be obtained.

Since a higher actual shear stress will be calculated without this modification, it is conservative not to apply it. It is convenient in calculations to adopt a notation which indicates whether the reduced shear from Example 6.13 is being used. Here V represents a shear which is not modified, and V^* is used in this book to indicate a shear which has been reduced. Similarly, f_v is the shear stress calculated using V, and f_v^* is the shear stress based on V^*.

$$f_v^* = \frac{1.5V^*}{A}$$

$$\leq F_v'$$

Other terms are as previously defined. The modified load diagram is to be used for horizontal shear stress calculations only. Reactions and moments are to be calculated using the full design loads.

It was noted earlier that bending stresses often govern the size of a beam, but secondary items, such as shear, *can* control the size under certain circumstances. It will be helpful if the designer can learn to recognize the type of beam in which shear is critical. As a general guide, shear is critical on relatively *short, heavily loaded spans*. With some experience, the designer will be able to identify by inspection what probably constitutes a "short, heavily loaded" beam. In such a case, the design would start by obtaining a trial beam size that satisfies the horizontal shear formula. Other items, such as bending and deflection, would then be checked.

If a beam is notched at a support, the shear at the notch must be checked (NDS Sec. 3.4.3.2). To do this, the theoretical formula for horizontal shear is applied with the actual depth at the notch d_n used in place of the total beam depth d. See Example 6.14. For square-cut notches in the tension face, the calculated stress must be increased by a stress concentration factor which is taken as the square of the ratio of the total beam depth to the net depth at the notch $(d/d_n)^2$. Notches of other configurations which tend to relieve stress concentrations will have lower stress concentration factors. *The notching of a beam in areas of bending tensile stresses is not recommended* (see Sec. 6.2 for additional comments).

EXAMPLE 6.14   Shear in Notched Beams

Figure 6.14a   Notch at supported end.

For square-cut notches at the *end* of a beam on the tension side, the NDS provides an expression for the design shear, V_r', which must be greater than or equal to the calculate shear force on the member, V or V^* (see Example 6.13):

For ASD:
$$V_r' = \frac{2}{3} F_v' \, bd_n \left(\frac{d_n}{d}\right)^2$$
$$\geq V \quad \text{or} \quad V^*$$

For LRFD:
$$V_{rn}' = \frac{2}{3} F_{vn}' \, bd_n \left(\frac{d_n}{d}\right)^2$$
$$\geq V_u \text{ or } V_u^*$$

The shear force must then be less than the design shear, as determined from the above equation. The approach outlined here, where the design shear is compared to the applied shear force, is not typical in ASD. It is, however, quite common when the LRFD method is used. For ASD, it is much more common to compare the adjusted design values to the calculated or working shear stress. In this case, the above equation can be rewritten in terms of stresses as follows:

$$f_v = \frac{1.5V}{bd_n}\left(\frac{d}{d_n}\right)^2$$
$$\leq F_v'$$

Notches in the tension face of a beam induce tension stresses perpendicular to the grain. These interact with horizontal shear to cause a splitting tendency at the notch. Tapered notches can be used to relieve stress concentrations (dashed lines in Fig. 6.14a).

Mechanical reinforcement such as the fully threaded lag bolt in Fig. 6.14b can be used to resist splitting.

Figure 6.14b   Mechanical reinforcement at notched end.

When designing a notched glulam beam, the reference shear design value must be multiplied by a reduction factor of 0.72. This reduction is specified as a footnote in Tables 5A and 5B of the NDS Supplement and is applicable only to softwood glulam members.

Notches at the end of a beam in the compression face are less critical than notches in the tension side. NDS Sec. 3.4.3.2(e) provides a method for analyzing the effects of reduced stress concentrations for notches in the compression side. Additional provisions for horizontal shear at bolted connections in beams are covered in Sec. 13.9.

6.6 Deflection

The deflection limits for wood beams required by the IBC and the additional deflection limits recommended by AITC are discussed in Sec. 2.7. Actual deflections for a trial beam size are calculated for a known span length, support conditions, and applied loads. Deflections may be determined from a traditional deflection analysis, from standard beam formulas, or from a computer analysis. The actual (calculated) deflections should be less than or equal to the deflections limits given in Chap. 2. ASD and LRFD approach deflection design in an identical manner. See Example 6.15.

EXAMPLES 6.15   Beam Deflection Criteria and Camber

Figure 6.15a   Deflected shape of beam.

Actual Deflection

The maximum deflection is a function of the loads, type of span, moment of inertia and modulus of elasticity:

$$\text{Max. } \Delta = f\left(\frac{P, w, L}{I, E'}\right)$$

where E' = adjusted modulus of elasticity
$= E(C_M)(C_t)(C_i)$

Other terms for beam deflection analysis are as normally otherwise defined.

The adjustment factors for evaluating the adjusted modulus of elasticity were introduced in Chap. 4. The factors and typical values for frequently encountered conditions are

E = reference modulus of elasticity
$= E_x$ for usual case of bending about strong axis

C_M = wet service factor (Sec. 4.14)

= 1.0 for dry-service conditions as in most covered structures.

Dry-service conditions are defined as

MC ≤ 19 percent for sawn lumber

MC < 16 percent for glulam

C_t = temperature factor (Sec. 4.20)

= 1.0 for normal temperature conditions

C_i = incising factor (Sec. 4.21)

= 0.95 for incised Dimension lumber

= 1.0 for Dimension lumber not incised (whether the member is treated or untreated).

(Note: The incising factor is *not* applicable to glulam or Timbers.)

Deflections are often checked under live load alone Δ_L and under total load Δ_{TL} (dead load plus live load). Recall that in the total load deflection check, the dead load may be reduced by a factor of 0.5 if the wood member has an MC of less than 16 percent at the time of installation and is used under dry conditions (IBC Table 1604.3 footnote d, Ref. 6.10). See Sec. 2.7 for additional information.

Deflection Criteria

$$\text{Max. } \Delta_L \leq \text{allow. } \Delta_L$$
$$\text{Max. } \Delta_{TL} \leq \text{allow. } \Delta_{TL}$$

If these criteria are not satisfied, a new trial beam size is selected using the moment of inertia and the allowable deflection limit as a guide.

Camber

Camber is initial curvature built into a member which is opposite to the deflection under gravity loads. See Fig. 6.15*b*

CAMBER = 1.5 Δ_D

Figure 6.15*b* Typical camber built into glulam beam is 1.5 times dead load deflection.

The material property that is used to evaluate beam deflection is the adjusted modulus of elasticity E'. The modulus of elasticity has relatively few adjustment factors. The few adjustments that technically apply to E default to unity for many common beam applications. Note that the load duration factor C_D and time effect factor λ do not apply to modulus of elasticity (Sec. 4.15 and 4.16).

It will be recalled that the reference modulus of elasticity is an average value. It is common design practice in both ASD and LRFD to evaluate deflections using the average E. However, in certain cases deflection may be a critical consideration, and NDS Appendix F may be used to convert the average E to a lower-percentile modulus of elasticity. Depending on the required need, the average modulus of elasticity can be converted to a value that will be exceeded by either 84 or 95 percent of the individual pieces of lumber. These values are given the symbols $E_{0.16}$ and $E_{0.05}$ and are known as the 16 and 5 percent lower exclusion values, respectively. See NDS Appendix F for additional information.

In the design of glulam beams and wood trusses, it is common practice to call for a certain amount of camber to be built into the member. *Camber* is defined as an initial curvature or reverse deflection which is built into the member when it is fabricated. In glulam design, the typical camber is $1.5 \Delta_D$. This amount of camber should produce a nearly level member under long-term deflection, including creep. Additional camber may be required to improve appearance or to obtain adequate roof slope to prevent ponding (see Ref. 6.5).

See Chap. 2 for more information on deflection, specifically Fig. 2.8.

6.7   Design Summary

One of the three design criteria discussed in the previous sections (bending, shear, and deflection) will determine the required size of a wood beam. In addition, consideration must be given to the type of lateral support that will be provided to prevent lateral instability. If necessary, the bending analysis will be expanded to take the question of lateral stability into account. With some practice, the structural designer may be able to tell which of the criteria will be critical by inspection.

The sequence of the calculations used to design a beam has been described in the above sections. It is repeated here in summary.

For many beams, the bending stress is the critical design item. Therefore, a trial beam size is often developed from the bending stress formula

$$\text{Req'd } S = \frac{M}{F'_b} \quad \text{or} \quad \frac{M_u}{F'_{bn}}$$

A trial member is chosen which provides a furnished section modulus S that is greater than the required value. Because the magnitude of the size factor C_F or the volume factor C_V is not definitely known until the size of the beam has been chosen, it may be helpful to summarize the actual versus adjusted bending design value after a size has been established:

For ASD:

$$f_b = \frac{M}{S}$$

$$\leq F'_b$$

For LRFD:

$$M_u \leq M'_n$$

$$\leq F'_{bn} \cdot S$$

After a trial size has been established, the remaining items (shear and deflection) should be checked. For a rectangular beam, the shear is checked by the expression

For ASD:

$$f_v = \frac{1.5V}{A}$$

$$\le F_v'$$

For LRFD:

$$V_u \le V_n'$$

$$\le \frac{2}{3} F_{vn}' A$$

In this calculation, a reduced shear V^* or V_u^* can be substituted for V or V_u, and f_v^* becomes the computed shear in place of f_v in ASD. If this check proves unsatisfactory, the size of the trial beam is revised to provide a sufficient area A so that the shear strength is adequate.

The deflection is checked by calculating the actual deflection using the moment of inertia for the trial beam. The actual deflection is then compared with the allowable deflection limit:

$$\Delta \le \text{allow. } \Delta$$

If this check proves unsatisfactory, the size of the trial beam is revised to provide a sufficient moment of inertia I so that the deflection criteria are satisfied.

It is possible to develop a trial member size by starting with something other than bending stress. For example, for a beam with a short, heavily loaded span, it is reasonable to establish a trial size using the shear calculation

$$\text{Req'd } A = \frac{1.5V}{F_v'} \quad \text{or} \quad \frac{1.5V_u}{F_{vn}'}$$

The trial member should provide an area A which is greater than the required area.

If the structural properties of wood are compared with the properties of other materials, it is noted that the modulus of elasticity for wood is relatively low. For this reason, in fairly long span members, deflections can control the design. Obviously, if this case is recognized, or if more restrictive deflection criteria are being used in design, the trial member size should be based on satisfying deflection limits. Then the remaining criteria of bending and shear can be checked.

The section properties such as section modulus and moment of inertia increase rapidly with an increase in depth. Consequently, narrow and deep cross sections are more efficient beams.

In lieu of other criteria, the most economical beam for a given grade of lumber is the one that satisfies all stress and deflection criteria with the minimum cross-sectional area. Sawn lumber is purchased by the board foot (a board foot is a volume of wood based on nominal dimensions that corresponds to a 1 × 12 piece of wood 1 ft long). The number of board feet for a given member is obviously directly proportional to the cross-sectional area of the member.

A number of factors besides minimum cross-sectional area can affect the final choice of a member size. First, there are detailing considerations in which a member size must be chosen that fits in the structure and accommodates other members and their connections. Second, a member size may be selected that is uniform with the size of members used elsewhere in the structure. This may be convenient from a structural detailing point of view, and it also can simplify material ordering and construction. Third, the availability of lumber sizes and grades must also be considered. However, these other factors can be considered only with knowledge about a specific job, and the general practice in this book is to select the beam with the least cross-sectional area.

The design summary given above is essentially an outline of the process that may be used in a hand solution. Computer solutions can be used to automate the process. Generally, computer designs will be more direct in that the required section properties for bending, shear, and deflection (that is, S, A, and I) will be computed directly with less work done by trial and error. However, even with computer solutions, wood design often involves iteration to some extent in order to obtain a final design.

The designer is encouraged to start using the computer by developing simple spreadsheet or equation-solving software templates for beam design calculations. If a dedicated computer program is used, the designer should ensure that sufficient output and documentation are available for verifying the results by hand.

6.8 Bearing at Supports

Bearing perpendicular to the grain of wood occurs at beam supports or where loads from other members frame into the beam. See Example 6.16. The actual bearing stress is calculated by dividing the load or reaction by the contact area between the members or between the member and the connection bearing plate. The actual stress must be less than the adjusted bearing design value.

$$f_{c\perp} = \frac{P}{A}$$

$$\leq F'_{c\perp}$$

The adjusted compressive design value perpendicular to grain is obtained by multiplying the reference design value by a series of adjustment factors.

For ASD:

$$F'_{c\perp} = F_{c\perp}(C_M)(C_t)(C_i)(C_b)$$

For LRFD:

$$F'_{c\perp n} = F_{c\perp n}(\phi_c)(\lambda)(C_M)(C_t)(C_i)(C_b)$$

where $F'_{c\perp}$ = adjusted ASD compressive (bearing) value perpendicular to grain

$F_{c\perp}$ = reference compressive (bearing) design value perpendicular to grain

$F'_{c\perp n}$ = adjusted LRFD compressive (bearing) value perpendicular to grain

$F_{c\perp n}$ = nominal compressive (bearing) design value perpendicular to grain (LRFD)

$\quad$ = $F_{c\perp} \cdot K_F$

K_F = format conversion factor for bearing (see Example 4.9)—LRFD only

$\quad$ = $1.875/\phi_c = 2.08$

ϕ_c = resistance factor for compression (See. 4.22) = 0.9—LRFD only

λ = time effect factor (see Sec. 4.16)—LRFD only

C_M = wet service factor (Sec. 4.14)

$\quad$ = 1.0 for dry-service conditions, as in most covered structures. Dry-service conditions are defined as

$\quad$ MC $\leq$ 19 percent for sawn lumber

$\quad$ MC $<$ 16 percent for glulam

C_t = temperature factor (Sec. 4.20)

$\quad$ = 1.0 for normal temperature conditions

C_i = incising factor (Sec. 4.21)

$\quad$ = 1.0 for sawn lumber. (*Note:* The incising factor is not applicable to glulam or Timbers.)

C_b = bearing area factor (defined below)

$\quad$ = 1.0 is conservative for all cases

For *sawn lumber*, a single value of $F_{c\perp}$ is listed for individual stress grades in the NDS Supplement. For *glulams*, a number of different reference values of $F_{c\perp}$ are listed. For a glulam bending combination stressed about the x axis, the value of $F_{c\perp x}$ depends on whether the bearing occurs on the *compression* laminations or on the higher-quality *tension* laminations. For the common case of a beam with a positive moment, the compression laminations are on the top side of the member, and the tension laminations are on the bottom.

The bearing area factor C_b is used to account for an effective increase in bearing length. The bearing length l_b (in.) is defined as the dimension of the contact area measured *parallel* to the grain. The bearing area factor C_b may be used to account for additional wood fibers beyond the actual bearing length l_b that develop normal resisting force components. Under the conditions shown in Fig. 6.16b, a value of C_b greater than 1.0 is obtained by adding $3/8$ in. to the actual bearing length.

Note that C_b is always greater than or equal to 1.0. It is, therefore, conservative to disregard the bearing area factor (i.e., use a default value of unity). Values of C_b may be read from NDS Table 3.10.4, or they may be calculated as illustrated in Example 6.16.

Compression perpendicular to grain is generally not considered to be a matter of life safety. Instead, it relates to the amount of deformation that is acceptable in a structure. Currently published values of bearing perpendicular to grain

$F_{c\perp}$ are *average* values which are based on a *deformation limit* of 0.04 in. when tested in accordance with ASTM D 143 (Ref. 6.6). This deformation limit has been found to provide adequate service in typical wood-frame construction.

A significant difference between ASD and LRFD is noted for bearing. In LRFD, the time effect factor λ(Sec. 4.16) is used to adjust the bearing design value based on the load combination being considered. However, in ASD the load duration factor C_D (Sec. 4.15) is not applied to compression perpendicular to grain design values. The justification for not applying C_D in ASD is the recognition that $F_{c\perp}$ is a deformation based limit state or mode of failure. In LRFD, since this is a capacity-level design check, λ is applied.

In addition, reference values of $F_{c\perp}$ are generally lower for glulam than for sawn lumber of the same deformation limit. (For a discussion of these differences see Ref. 6.5.)

EXAMPLE 6.16   Bearing Perpendicular to Grain

Figure 6.16a   Compression perpendicular to grain.

For ASD the bearing stress calculation is

$$f_{c\perp} = \frac{P}{A}$$

$$\leq F'_{c\perp}$$

where $f_{c\perp}$ = actual (computed) bearing stress perpendicular to grain
$\quad\quad\; P$ = applied load or reaction (force P_1 or P_2 in Fig. 6.16a)
$\quad\quad\; A$ = contact area
$\quad\quad F'_{c\perp}$ = adjusted ASD bearing value perpendicular to grain

For LRFD the bearing check is

$$P_u \leq P'_{c\perp n}$$

$$\leq F'_{c\perp n} \cdot A$$

where P_u = applied *factored* load or reaction
$P'_{c\perp n}$ = adjusted LRFD bearing resistance perpendicular to grain
$F'_{c\perp n}$ = adjusted LRFD bearing value perpendicular to grain

Adjustment Based on Bearing Length

When the bearing length l_b (Fig. 6.16b) is less than 6 in. *and* when the distance from the end of the beam to the contact area is more than 3 in., the adjusted bearing design value may be increased (multiplied) by the bearing area factor C_b.

Essentially, C_b increases the effective bearing length by $\frac{3}{8}$ in. This accounts for the additional wood fibers that resist the applied load after the beam becomes slightly indented.

Figure 6.16b   Required conditions to use C_b greater than 1.0.

Bearing area factor:

$$C_b = \frac{l_b + 0.375}{l_b}$$

In design applications where deformation may be critical, a reduced value of $F_{c\perp}$ may be considered. The following expressions are recommended when a deformation limit of 0.02 in. (one-half of the limit associated with the "reference value") is desired:

$$F_{c\perp 0.02} = 0.73 F_{c\perp}$$

where $F_{c\perp 0.02}$ = reduced compressive design value perpendicular to grain at
deformation limit of 0.02 in.

$F_{c\perp}$ = reference compressive design value perpendicular to grain
(deformation limit of 0.04 in.)

The other adjustments described previously for $F_{c\perp}$ also apply to $F_{c\perp 0.02}$.

The bearing stress discussed thus far has been perpendicular to the grain in the wood member. A second type of bearing stress is known as the bearing stress parallel to grain (NDS Sec. 3.10.1). It applies to the bearing that occurs on the end of a member, and it is not to be confused with the compressive stress parallel to the grain that occurs away from the end (e.g., column stress in Sec. 7.4). The bearing stress parallel to the grain assumes that the member is adequately braced and that buckling does not occur. The actual bearing stress parallel to the grain is not to exceed the adjusted design value

For ASD:

$$f_c = \frac{P}{A} \le F_c^*$$

For LRFD:

$$P_u \le P_{cn}^*$$

$$\le (F_{cn}^* \cdot A)$$

where f_c = actual (computed) bearing stress parallel to grain
P = load parallel to grain on end of wood member (ASD)
P_u = factored load parallel to grain on end of wood member (LRFD)
A = net bearing area
F_c^* = adjusted ASD compressive (bearing) design value parallel to grain on end of wood member including all adjustments *except* column stability
= $F_c(C_D)(C_M)(C_t)(C_F)(C_i)$
F_c = reference compressive (bearing) design value parallel to grain
P_{cn}^* = adjusted LRFD compressive (bearing) resistance parallel to grain on end of wood member including all adjustments *except* column stability
F_{cn}^* = adjusted LRFD compressive (bearing) value parallel to grain on end of wood member including all adjustments *except* column stability
= $F_{cn}(\phi_c)(\lambda)(C_M)(C_t)(C_F)(C_i)$
F_{cn} = $F_c \cdot K_F$
K_F = format conversion factor for compression (see Example 4.9)—LRFD only
= $2.16/\phi_c$ = 2.40
ϕ_c = compression resistance factor = 0.90—LRFD only
C_D = load duration factor (Sec. 4.15)—ASD only
λ = time effect factor (Sec. 4.16)—LRFD only

C_M = wet-service factor (Sec. 4.14)

= 1.0 for dry-service conditions, as in most covered structures. Dry-service conditions are defined as

MC ≤ 19 percent for sawn lumber

MC < 16 percent for glulam

C_t = temperature factor (Sec. 4.20)

= 1.0 for normal temperature conditions

C_F = size factor (Sec. 4.17). Obtain values from Adjustment Factors section of NDS Supplement Tables 4A, 4B, and 4F for Dimension lumber and in Table 4D for Timbers.

C_i = incising factor (Sec. 4.21)

= 0.8 for incised Dimension lumber

= 1.0 for Dimension lumber not incised, whether the member is treated or untreated (*Note*: The incising factor is not applicable to glulam or Timbers.)

Bearing parallel to grain applies to two wood members bearing end to end as well as end bearing on other surfaces. Member ends are assumed to be accurately cut square. When f_c exceeds $0.75F_c^*$, bearing is to be on a steel plate or other appropriate rigid bearing surface. When required for end-to-end bearing of two wood members, the rigid insert shall be at least a 20-gage metal plate with a snug fit between abutting ends.

A comparison of the reference bearing design values parallel to grain F_c and perpendicular to grain $F_{c\perp}$ shows that the values differ substantially. To make this comparison, refer to NDS Supplement Tables 4A to 4D and 4F.

It is also possible for bearing in wood members to occur at some angle other than 0 to 90 degrees with respect to the direction of the grain. In this case, an adjusted bearing design value somewhere between F_c' and $F_{c\perp}'$ is determined from the Hankinson formula (NDS Sec. 3.10.3). See Example 6.17.

EXAMPLE 6.17   Bearing at an Angle to Grain

Bearing at some angle to grain (Fig. 6.17) other than 0 to 90 degrees:

For ASD:

$$f_\theta = \frac{P}{A}$$

$$\leq F_\theta'$$

For LRFD:

$$P_u \leq P_{\theta n}'$$

$$\leq F_{\theta n}'A$$

where f_θ = actual (unfactored) bearing stress at angle to grain θ

P = applied load or reaction

P_u = factored applied load or reaction
A = contact area
F'_θ = adjusted ASD bearing value at angle to grain θ
$F'_{\theta n}$ = adjusted LRFD bearing value at angle to grain θ
$P'_{\theta n}$ = adjusted LRFD resistance at angle to grain θ

Hankinson Formula

The adjusted value at an angle to grain θ is given by the Hankinson formula (NDS Eq. 3.10-1)

For ASD:
$$F'_\theta = \frac{F_c^* F'_{c\perp}}{F_c^* \sin^2\theta + F'_{c\perp}\cos^2\theta}$$

For LRFD:
$$F'_{\theta n} = \frac{F_{cn}^* F'_{c\perp n}}{F_{cn}^* \sin^2\theta + F'_{c\perp n}\cos^2\theta}$$

where F_c^* = adjusted ASD bearing value parallel to grain excluding column stability factor
$F'_{c\perp}$ = adjusted ASD bearing value perpendicular to grain
F_{cn}^* = adjusted LRFD bearing value parallel to grain excluding column stability factor
$F'_{c\perp n}$ = adjusted LRFD bearing value perpendicular to grain

RAFTER CONNECTION

Figure 6.17 Bearing stress in two wood members. Bearing in rafter is at an angle to grain θ. Bearing in the supporting beam or header is perpendicular to grain.

This formula can probably best be solved mathematically, but the graphical solution in NDS Appendix J, *Solution of Hankinson Formula*, may be useful in visualizing the effects of angle of load to grain.

NOTE: The connection in Fig. 6.17 is given to illustrate bearing at an angle θ. For the condition shown, bearing may be governed by compression perpendicular to the grain $f_{c\perp}$ in the beam supporting the rafter, rather than by f_θ in the rafter. If $f_{c\perp}$ in the beam is excessive, a bearing plate between the rafter and the beam can be used to reduce the bearing stress in the beam. The bearing stress in the rafter would not be relieved by use of a bearing plate.

As indicated in Example 6.17, adjustments are applied individually to F_c and $F_{c\perp}$ *before* F_θ' or $F_{\theta n}'$ is computed using the Hankinson formula.

A number of examples are now given to illustrate the design procedures for beams. A variety of sawn lumber and glulam beams are considered with different support conditions and types of loading.

6.9   Design Problem: Sawn Beam

In this beam example and those that follow, the span lengths for bending and shear are, for simplicity, taken to be the same length. However, the designer may choose to determine the *design moment* based on the clear span plus one-half the required bearing length at each end (Sec. 6.2) and the *design shear* at a distance d from the support (Sec. 6.5). These different span length considerations are described in Example 6.13 (Sec. 6.5) for a simply supported beam.

In Example 6.18, a typical sawn lumber beam is designed for a roof that is essentially flat. Minimum slope is provided to prevent ponding. An initial trial beam size is determined from bending stress calculations. The extensive list of possible adjustment factors for bending design values is reduced to seven for the case of a visually graded sawn lumber beam with bending about the strong axis using ASD procedures (see Example 6.10 in Sec. 6.4).

The beam in this problem is used in dry service conditions and at normal temperatures, and C_M and C_t both default to unity. In addition, the roof sheathing provides continuous lateral support to the compression side of the beam. Consequently, there is no reduction in moment capacity due to lateral stability, and C_L is unity. The beam is not incised for pressure treatment since it is protected from exposure to moisture in service. Accordingly, C_i also defaults to unity. Therefore, the potential number of adjustment factors for bending is reduced to three in this typical problem. The bending is affected by the load duration factor C_D (ASD) or the time effect factor λ (LRFD), the size factor for Dimension lumber C_F, and the repetitive-member factor C_r.

After selection of a trial size, shear and deflection are checked. The shear stress is not critical, but the second deflection check indicates that deflection under (D + L$_r$) is slightly over the recommended allowable deflection limit. The decision of whether to accept this deflection is a matter of judgment. In this case it was decided to accept the deflection, and the trial size was retained for the final design.

EXAMPLE 6.18   Sawn Beam Design Using ASD

Design the roof beam in Fig. 6.18 to support the given loads. Beams are spaced 16 in. o.c. (1.33 ft), and sufficient roof slope is provided to prevent ponding. The ceiling is gypsum wallboard. Plywood roof sheathing prevents lateral buckling. Material is No. 1 Douglas Fir-Larch (DF-L).

D = 14 psf and L_r = 20 psf. The MC ≤ 19 percent, and normal temperature conditions apply. Reference design values and section properties are to be taken from the NDS Supplement.

Loads

Uniform loads are obtained by multiplying the given design loads by the tributary width.

$$w_D = 14 \text{ psf} \times 1.33 \text{ ft} \quad = 18.67 \text{ lb/ft}$$

$$w_L = 20 \times 1.33 \quad\quad\quad = 26.67$$

$$\overline{\text{Total load } w_{TL} \quad\quad = 45.33 \text{ lb/ft}}$$

Figure 6.18   Trial size from bending calculations is 2 × 6 (Dimension lumber size).

The required load combinations are D alone with C_D = 0.9, and (D + L_r) with C_D = 1.25. By a comparison of the loads and the load duration factors (Example 4.10 in Sec. 4.15), it is determined that the critical load combination is D + L_r (i.e., total load governs).

Determine a trial size based on bending, and then check other criteria.

Bending

The span length and load for this beam are fairly small. It is assumed that the required beam size is from the range of sizes known as Dimension lumber. Reference design values are found in NDS Supplement Table 4A. The beam qualifies for the repetitive member increase of 15 percent. A size factor of $C_F = 1.2$ is initially assumed, and the true size factor is confirmed after a trial beam is developed. C_M, C_t, C_L, and C_i default to 1.0.

$$F_b' = F_{bx}' = F_{bx}(C_D)(C_M)(C_t)(C_L)(C_F)(C_r)(C_i)$$

$$= 1000(1.25)(1.0)(1.0)(1.0)(1.2)(1.15)(1.0)$$

$$= 1725 \text{ psi}$$

$$\text{Req'd } S = \frac{M}{F_b'} = \frac{12{,}390}{1725} = 7.18 \text{ in.}^3$$

A trial beam size is obtained by reviewing the available sizes in NDS Supplement Table 1B. The general objective is to choose the member with the least area that furnishes a section modulus greater than that required. However, certain realities must also be considered. For example, a 1-in. nominal board would not be used for this type of beam application.

$$\text{Try} \quad 2 \times 6 \quad S = 7.56 \text{ in.}^3 > 7.18 \text{ in.}^3 \quad OK$$

The trial size of a 2×6 was determined using an assumed value for C_F. The size factor can now be verified in the Adjustment Factors section of NDS Supplement Table 4A:

$$C_F = 1.3 > 1.2 \quad OK$$

At this point, the member has been shown to be adequate for bending. However, it is often convenient to compare the actual stress and the adjusted design value in a summary.

$$f_b = \frac{M}{S} = \frac{12{,}390}{7.56} = 1640 \text{ psi}$$

$$F_b' = F_b(C_D)(C_M)(C_t)(C_L)(C_F)(C_r)(C_i)$$

$$= 1000(1.25)(1.0)(1.0)(1.0)(1.3)(1.15)(1.0)$$

$$= 1870 \text{ psi} > 1640 \text{ psi}$$

$$\therefore \text{ Bending} \quad OK$$

NOTE: A 2×5 can be checked with $C_F = 1.4$, but the reduced section modulus causes f_b to exceed F_b'. Furthermore, 2×5s are not commonly available.

Shear

Because it is judged that the shear stress for this beam is likely not to be critical, the maximum shear from the shear diagram is used without modification. C_M, C_t, and C_i are all set equal to 1.0.

$$f_v = \frac{1.5V}{A} = \frac{1.5(306)}{8.25} = 55.6 \text{ psi}$$

$$F_v' = F_v(C_D)(C_M)(C_t)(C_i)$$

$$= 180(1.25)(1.0)(1.0)(1.0)$$

$$= 225 \text{ psi} > 55.6 \text{ psi}$$

$$\therefore \text{Shear} \quad OK$$

Deflection

The IBC specifies deflection criteria for roof beams (IBC Table 1604.3, Ref. 6.10). The calculations below use these recommended deflection criteria.

Recall that the modulus of elasticity for a wood member is not subject to adjustment for load duration, and the buckling stiffness factor C_T does not apply to deflection calculations. The adjustment factors for E in this problem all default to unity.

$$E' = E(C_M)(C_t)(C_i)$$

$$= 1,700,000(1.0)(1.0)(1.0) = 1,700,000 \text{ psi}$$

$$\Delta_L = \frac{5w_L L^4}{384E'I} = \frac{5(26.7)(13.5)^4(1728 \text{ in.}^3/\text{ft}^3)}{384(1,700,000)(20.8)} = 0.56 \text{ in.}$$

$$\text{Allow, } \Delta_L = \frac{L}{240} = \frac{13.5 \times 12}{240} = 0.67 \text{ in.} > 0.56 \text{ in.} \quad OK$$

Deflection under total load can be calculated using the same beam deflection formula, or it can be figured by proportion.

$$\Delta_{TL} = \Delta_L \left(\frac{w_{TL}}{w_L}\right) = 0.56\left(\frac{45.3}{26.7}\right) = 0.95 \text{ in.}$$

$$\text{Allow. } \Delta = \frac{L}{180} = \frac{13.5 \times 12}{180} = 0.90 \text{ in.} < 0.95 \text{ in.}$$

In the second deflection check, the actual deflection is slightly over the allowable limit.

NOTE: In accordance with the IBC, the full live load was included since the moisture content was specified as MC ≤ 19 percent. If the member was specified as less than 16 percent moisture content at the time of installation and maintained in a dry condition, then only half of the dead load would be required when checking against the $L/180$ limit. Using 0.5D + L, the $\Delta_{TL} = 0.76$ in. < 0.90 in.

After consideration of the facts, either select a higher grade or larger size, which is necessary. Assume No. 1& Better with an E = 1,800,000 psi. The $\Delta_{TL} = 0.90$ in. = allow Δ.

> Use 2×6 No. 1 & Btr DF-L
> MC ≤ 19 percent

Bearing

Evaluation of bearing requires knowledge of the support conditions. Without such information, the minimum bearing length will simply be determined. Recall that C_D does not apply to $F_{c\perp}$.

$$F'_{c\perp} = F_{c\perp}(C_M)(C_t)(C_b) = 625(1.0)(1.0)(1.0) = 625 \text{ psi}$$

$$\text{Req'd } A = \frac{R}{F'_{c\perp}} = \frac{306}{625} = 0.49 \text{ in.}^2$$

$$\text{Req'd } l_b = \frac{A}{b} = \frac{0.49}{1.5} = 0.33 \text{ in.}$$

All practical support conditions provide bearing lengths in excess of this minimum value.

Example 6.18 demonstrated the design of a typical sawn lumber beam using ASD. The design is now repeated using LRFD in Example 6.19.

EXAMPLE 6.19   Sawn Beam Design Using LRFD

Repeat the design of the roof beam in Example 6.18 using LRFD. Beams are spaced 16 in. o.c. (1.33 ft), and sufficient roof slope is provided to prevent ponding. The ceiling is gypsum wallboard. Plywood roof sheathing prevents lateral buckling. Material is No. 1 Douglas Fir-Larch (DF-L).

Unfactored D = 14 psf, and L_r = 20 psf. The MC ≤ 19 percent, and normal temperature conditions apply. Reference design values and section properties are to be taken from the NDS Supplement.

Loads

Unfactored uniform loads are obtained by multiplying the given design loads by the tributary width.

$$w_D = 14 \text{ psf} \times 1.33 \text{ ft} \qquad = 18.67 \text{ lb/ft}$$

$$w_L = 20 \times 1.33 \qquad\qquad = 26.67$$

The LRFD load combination are

$$1.4D = 1.4 (18.67) = 26.14 \text{ lb/ft}$$

$$1.2D + 1.6\,L_r = 1.2 (18.67) + 1.6 (26.67) = 65.08 \text{ lb/ft}$$

The time effect factor for 1.4D is λ = 0.6 and for $1.2D + 1.6L_r$ is λ = 0.8. The critical load combination is $1.2D + 1.6L_r$.

$$M_u = \frac{w_u L^2}{8} = \frac{(65.08)(13.5)^2}{8} = 1482 \text{ ft-lb}$$

$$= 17.79 \text{ in. k}$$

Determine a trial size based on bending, and then check other criteria.

Bending

The span length and load for this beam are fairly small. It is assumed that the required beam size is from the range of sizes known as Dimension lumber. Reference design values are found in NDS Supplement Table 4A. The beam qualifies for the repetitive member stress increase of 15 percent. A size factor of C_F = 1.2 is initially assumed, and the true size factor is confirmed after a trial beam is developed. C_M, C_t, C_L, and C_i default to 1.0. The resistance factor for bending is $\phi_b = 0.85$ and the format conversion factor for bending is $K_F = 2.16/\phi_b = 2.54$.

$$F_{bxn} = F_{bx}(K_F) = 1000(2.54) = 2540 \text{ psi}$$

$$F'_{bn} = F'_{bxn} = F_{bxn}(\phi_b)(\lambda)(C_M)(C_t)(C_L)(C_F)(C_r)(C_i)$$

$$= (2540)(0.85)(0.8)(1.0)(1.0)(1.2)(1.15)(1.0)$$

$$= 2384 \text{ psi} = 2.38 \text{ ksi}$$

$$\text{Req'd } S = \frac{M_u}{F'_{bn}} = \frac{17.79}{2.38} = 7.47 \text{ in.}^3$$

A trial beam size is obtained by reviewing the available sizes in NDS Supplement Table 1B. The general objective is to choose the member with the least area that furnishes a section modulus greater than that required. However, certain realities must also be considered. For example, a 1-in. nominal board would not be used for this type of beam application.

$$\textit{Try} \quad 2 \times 6 \quad S = 7.56 \text{ in.}^3 > 7.47 \text{ in.}^3 \quad \textit{OK}$$

The trial size of a 2×6 was determined using an assumed value for C_F. The size factor can how be verified in the Adjustment Factors section of NDS Supplement Table 4A:

$$C_F = 1.3 > 1.2 \quad \textit{OK}$$

$$F'_{bn} = F'_{bxn} = 2.38(1.3/1.2) = 2.58 \text{ ksi}$$

At this point the member has been shown to be adequate for bending. However, it is often convenient to compare the actual factored moment and the adjusted LRFD moment capacity in a summary.

$$M'_n = F'_{bxn} \cdot S$$

$$= (2.58)(7.56)$$

$$= 19.5 \text{ k-in.} > 17.79 \text{ k-in.}$$

$$\therefore \text{Bending} \quad \textit{OK}$$

Shear

Because it is judged that the shear for this beam is likely not to be critical, the maximum shear from the shear diagram is used without modification. C_M, C_t, and C_i are all set equal to 1.0. The resistance factor for shear is $\phi_v = 0.75$ and the format conversion factor for shear is $K_P = 2.16/\phi_v = 2.88$.

$$F_{vn} = F_v \cdot K_F = 180(2.88) = 518 \text{ psi}$$

$$F'_{vn} = F_{vn}(\phi_v)(\lambda)(C_M)(C_t)(C_i)$$

$$= (518)(0.75)(0.8)(1.0)(1.0)(1.0)$$

$$= 311 \text{ psi} = 0.311 \text{ ksi}$$

$$V'_n = \left(\frac{2}{3}F'_{vn}A\right) = \left(\frac{2}{3}\right)(0.311)(8.25) = 1.71 \text{ k}$$

$$V_u = \frac{w_u L}{2} = \frac{(65.08)(13.5)}{2}$$

$$= 439 \text{ lb} = .44 \text{ k} < 1.71 \text{ k}$$

$$\therefore \text{ Shear} \qquad OK$$

Deflection

The deflection check for LRFD is identical to the deflection check for ASD. See Example 6.18.

> Use   2 × 6 No. 1 & Btr DF-L
> MC ≤ 19 percent

Bearing

Evaluation of bearing requires knowledge of the support conditions. Without such information, the minimum bearing length will simply be determined. Recall that λ applies to $F_{c\perp}$. For bearing, $\phi_c = 0.90$ and $K_F = 1.875/\phi_c = 2.08$.

$$F_{c\perp n} = F_{c\perp} \cdot K_F = 625(2.08) = 1300 \text{ psi}$$

$$F'_{c\perp n} = F_{c\perp n}(\phi_c)(\lambda)(C_M)(C_t)(C_b)$$

$$= (1300)(0.90)(0.8)(1.0)(1.0)(1.0)$$

$$= 938 \text{ psi} = 0.938 \text{ ksi}$$

$$\text{Req'd } A = \frac{R_u}{F'_{c\perp n}} = \frac{.44}{(0.938)} = .47 \text{ in.}^2$$

$$\text{Req'd } l_b = \frac{A}{b} = \frac{0.47}{1.5} = 0.31 \text{ in.}$$

All practical support conditions provide bearing lengths in excess of this minimum value.

Comparing Example 6.18 and 6.19 allows similarities and differences between ASD and LRFD to be recognized. Both procedures result in a 2 × 6 No. 1 & Btr DF-L to be selected and the details of the design are quite similar. It is important to recognize that the approach used in ASD and LRFD for the deflection check are the same. It is only with respect to the strength design that the format approaches differ.

6.10   Design Problem: Rough-Sawn Beam Using ASD

In this ASD example, a large rough-sawn beam with a fairly short span is analyzed. The cross-sectional properties for dressed lumber (S4S) are smaller than those for rough-sawn lumber, and it would be conservative to use S4S section properties for this problem. However, the larger section properties obtained using the rough-sawn dimensions are used in this example. Refer to Sec. 4.11 for information on lumber sizes.

In this problem, the basic shear adjustment of neglecting loads within a distance d from the support is used. See Example 6.20.

The importance of understanding the size categories for sawn lumber is again emphasized. The member in this problem is a Beams and Stringers size, and reference design values are taken from NDS Supplement Table 4D.

EXAMPLE 6.20   Rough-Sawn Beam Using ASD

Determine if the 6 × 14 rough-sawn beam in Fig. 6.19a is adequate to support the given loads. The member is Select Structural DF-L. The load is a combination of (D + L). Lateral buckling is prevented. The beam is used in dry-service conditions (MC ≤ 19 percent) and at normal temperatures. Reference design values are taken from the NDS Supplement Table 4D.

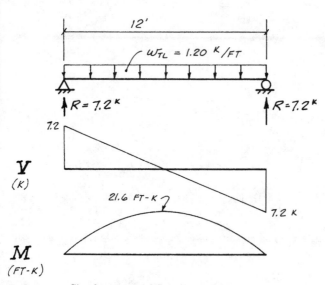

Figure 6.19a   Simply supported floor beam.

Section Properties

The dimensions of rough-sawn members are approximately ⅛ in. larger than standard dressed sizes.

$5\frac{5}{8}''$

$13\frac{5}{8}''$

Figure 6.19b Rough-sawn 6 × 14. For a member in the *Beams and Stringers* size category, the smaller cross-sectional dimension (i.e., the thickness) is 5 in. or larger, and the width is more than 2 in. greater than the thickness.

$$A = bd = (5\tfrac{5}{8})(13\tfrac{5}{8}) = 76.64 \text{ in.}^2$$

$$S = \frac{bd^2}{6} = \frac{(5\tfrac{5}{8})(13\tfrac{5}{8})^2}{6} = 174.0 \text{ in.}^3$$

$$I = \frac{bd^3}{12} = \frac{(5\tfrac{5}{8})(13\tfrac{5}{8})^3}{12} = 1185 \text{ in.}^4$$

Bending

$$M = 21.6 \times 12 = 259 \text{ in.-k}$$

$$f_b = \frac{M}{S} = \frac{259,000}{174.0} = 1488 \text{ psi}$$

The size factor for a sawn member in the Beams and Stringers category is given by the formula

$$C_F = \left(\frac{12}{d}\right)^{1/9} = \left(\frac{12}{13.625}\right)^{1/9} = 0.986$$

The load duration factor for the combination of (D + L) is 1.0. All of the adjustment factors for bending design values default to unity except C_F.

$$F'_b = F_b(C_D)(C_M)(C_t)(C_L)(C_F)(C_r)$$

$$= 1600(1.0)(1.0)(1.0)(1.0)(0.986)(1.0)$$

$$= 1578 \text{ psi} > 1488 \text{ psi} \quad OK$$

Shear

Consider the shear diagram given in Fig. 6.19a.

$$f'_v = \frac{1.5V}{A} = \frac{1.5(7200)}{76.64} = 140.9 \text{ psi}$$

$$F'_v = F_v(C_D)(C_M)(C_t)$$

$$= 170(1.0)(1.0)(1.0)$$

$$= 170 \text{ psi} > 140.9 \text{ psi} \quad OK$$

If the calculated actual shear stress had exceeded the adjusted shear design value, the beam may not have been adequate. Recall that the NDS permits the uniform load within a distance d from the face of the support to be ignored for shear calculations (see Sec. 6.5). Therefore, a modified shear V^* could be determined and used to compute a reduced shear stress f_v^*.

Deflection

Because the percentages of D and L were not given, only the total load deflection is calculated.

$$E' = E(C_M)(C_t)$$

$$= 1,600,000(1.0)(1.0) = 1,600,000 \text{ psi}$$

$$\Delta_{TL} = \frac{5w_{TL}L^4}{384E'I} = \frac{5(1200)(12)^4(12 \text{ in./ft})^3}{384(1,600,000)(1185)} = 0.30 \text{ in.}$$

$$\text{Allow. } \Delta_{TL} = \frac{L}{240} = \frac{12(12)}{240} = 0.60 \text{ in.} > 0.30 \text{ in.} \quad OK$$

Bearing

$$F'_{c\perp} = F_{c\perp}(C_M)(C_t)(C_b) = 625(1.0)(1.0)(1.0) = 625 \text{ psi}$$

$$\text{Req'd } A_b = \frac{R}{F'_{c\perp}} = \frac{7200}{625} = 11.52 \text{ in.}^2$$

$$\text{Req'd } l_b = \frac{A_b}{b} = \frac{11.52}{5.625} = 2.05 \text{ in. minimum}$$

> 6 × 14 rough-sawn
> Sel. Str. DF-L beam is   *OK*

The importance of understanding the size categories of sawn lumber can be seen by comparing the adjusted design value shown above for No. 1 DF-L Beams and Stringers (B&S) with those in Example 6.18 for No. 1 DF-L Dimension lumber. For a given grade, the adjusted values depend on the size category.

6.11   Design Problem: Notched Beam

Example 6.20 will be reevaluated assuming that the ends of a deeper section are notched at the supports to provide the same member depth over the support. In Example 6.20, the adequacy of a 6 × 14 rough-sawn beam was determined for the given loads. In this example, a notched 6 × 18 rough-sawn member is used as the beam in Fig. 6.19a. Since the notches are at the supports only, they will only impact the shear calculations.

EXAMPLE 6.21   Notched Beam Using ASD

Determine if a 6 × 18 rough-sawn beam with notched supports is adequate, considering shear only, to support the given loads. The notches are 4-in. deep and sufficiently long to provide adequate bearing length at the support. The notch does not extend past the face of the support. See Fig. 6.20. The depth of the notch allows the remaining depth of the member to be compatible with the 6 × 14 beams supported on the same level of the structure.

Figure 6.20   Notch detail at support for rough-sawn beam.

Notch Size

NDS Sec. 4.4.3 limits the size of notches at the ends of members for bearing over a support. The depth of such notches cannot exceed one-fourth of the total depth d of the beam. In this example, $d = 17\frac{5}{8}$ in. The maximum notch depth is $4\frac{13}{32}$ in., which is greater than the 4-in. notch specified in Fig. 6.20. The notch is permissible by the NDS.

Shear Check

$$f_v = \frac{1.5V}{bd_n}\left(\frac{d}{d_n}\right)^2 = \frac{1.5(7200)}{(5.625)(13.625)}\left(\frac{17.625}{13.625}\right)^2 = 235.8 \text{ psi}$$

$$F_v' = F_v(C_D)(C_M)(C_t)$$

$$= 170(1.0)(1.0)(1.0)$$

$$= 170 \text{ psi} < 235.8 \text{ psi} \quad \text{No Good}$$

From Example 6.20, the shear stress for the unnotched 6 × 14 section was 140.9 psi. This illustrates the effect of notching and the resulting stress concentration.

NOTE: For notched sections, the NDS *requires* that all loads, including a uniform load within a distance d from the face of the support, are to be considered for shear calculations (see NDS Sec. 3.4.3.2). Therefore, a modified shear V^* *cannot* be used to compute a reduced shear stress f_v^* for notched sections.

Sometimes notching is required to maintain uniform depth above a support for a series of members. As seen by this example, the effect of notching is significant on the shear strength of a beam and can cause shear to control the design. Consider the general case of a maximum depth notch with $d_n = 0.75d$. The stress increase caused by this maximum notch depth at a support is $(d/d_n)^2 = 1.78$.

6.12   Design Problem: Sawn-Beam Analysis

The three previous examples have involved beams in the Dimension lumber and Beams and Stringers size categories. Examples 6.22 and 6.23 are provided to give additional practice in determining adjusted design values. The member is again Dimension lumber, but the load duration and time effect factors, the wet service factor, and the size factor are all different from those in previous problems. Wet service factors and size factors are obtained from the Adjustment Factors section in the NDS Supplement and are identical for both ASD and LRFD.

EXAMPLE 6.22   Sawn-Beam Analysis Using ASD

Determine if the 4 × 16 beam given in Fig. 6.21 is adequate for a dead load of 70 lb/ft and a snow load of 180 lb/ft. Lumber is grade No. 1 & Better, and the species group is Hem-Fir. The member is not incised. Adequate bracing is provided so that lateral stability is not a concern.

Figure 6.21   A 4 × 16 beam is in the *Dimension lumber* size category.

This beam is used in a factory where the EMC will exceed 19 percent,* but temperatures are in the normal range. Beams are 4 ft-0 in. o.c. The minimum roof slope for drainage is provided so that ponding need not be considered. Allowable deflection limits for this design are assumed to be $L/360$ for snow load and $L/240$ for total load.

Design values and section properties are to be in accordance with the NDS.

---

*The need for pressure treated lumber to prevent decay should be considered (Sec. 4.9).

Bending

Section properties for a 4 × 16 are listed in NDS Supplement Table 1B.

$$f_b = \frac{M}{S} = \frac{150{,}000}{135.7} = 1105 \text{ psi}$$

The load duration factor is $C_D = 1.15$ for the load combination of (D + S). Beam spacing does not qualify for the repetitive member increase, and $C_r = 1.0$. Lateral stability is not a consideration, and $C_L = 1.0$.

For a 4 × 16, the size factor is read from NDS Table 4A:

$$C_F = 1.0$$

Also from Table 4A, the wet-service factor for bending is given as

$$C_M = 0.85$$

Except that, when $F_b(C_F) \le 1150$ psi, $C_M = 1.0$.

In the case of 4 × 16 No. 1 & Better Hem-Fir:

$$F_b\ (C_F) = 1100(1.0) < 1150 \text{ psi}$$

$$\therefore C_M = 1.0$$

The coefficients for determining F'_{bx} for a sawn lumber beam are obtained from the summary in Example 6.10 in Sec. 6.4 (see also NDS Table 4.3.1). In the bending design value summary given below, most of the adjustment factors default to unity. However, it is important for the designer to follow the steps leading to this conclusion.

$$F'_b = F_b(C_D)(C_M)(C_t)(C_L)(C_F)(C_r)(C_i)$$

$$= 1100(1.15)(1.0)(1.0)(1.0)(1.0)(1.0)(1.0)$$

$$= 1265 \text{ psi} > 1105 \text{ psi} \quad OK$$

Shear

$$f_v = \frac{1.5V}{A} = \frac{1.5(2500)}{53.375} = 70.3 \text{ psi}$$

$$F'_v = F_v(C_D)(C_M)(C_t)$$

$$= 150(1.15)(0.97)(1.0)$$

$$= 167.3 \text{ psi} > 70.3 \text{ psi}^\dagger \quad OK$$

[†]If f_v had exceeded F'_v, the design shear could have been reduced in accordance with Sec. 6.5.

Deflection

$$E' = E(C_M)(C_t)(C_i) = 1,500,000(0.9)(1.0)(1.0) = 1,350,000 \text{ psi}$$

$$\Delta_S = \frac{5wL^4}{384E'I}$$

$$= \frac{5(180)(20)^4(1728)}{384(1,350,000)(1034)} = 0.46 \text{ in.}$$

Allow. $\Delta_S = \dfrac{L}{360} = \dfrac{20 \times 12}{360} = 0.67 \text{ in.} > 0.46 \text{ in.}$ *OK*

By proportion,

$$\Delta_{TL} = \left(\frac{250}{180}\right) 0.46 = 0.64 \text{ in.}$$

Allow. $\Delta_{TL} = \dfrac{L}{240} = \dfrac{20 \times 12}{240} = 1.00 \text{ in.} > 0.64 \text{ in.}$ *OK*

> 4 × 16 No. 1 & Btr
> Hem-Fir beam   *OK*

Example 6.22 is repeated using LRFD. The load duration factor is replaced by the time effect factor. However, the wet service factor and the size factor are applied in LRFD as they were in ASD.

EXAMPLE 6.23   Sawn-Beam Analysis Using LRFD

Repeat Example 6.22 using LRFD. Consider the 1.2D + 1.6S load combination with a dead load of 70 lb/ft and a snow load of 180 lb/ft. Determine if the 4 × 16 No. 1 and Better Hem-Fir shown in Fig. 6.21 and spaced 4 ft o.c. is adequate.

$$w_u = 1.2D + 1.6S = 1.2(70) + 1.6(180) = 372 \text{ lb/ft}$$

$$M_u = \frac{w_u L^2}{8} = \frac{(372)(20)^2}{8} = 18600 \text{ ft-lb} = 223 \text{ in-k}$$

$$V_u = \frac{w_u L}{2} = \frac{(372)(20)}{2} = 3720 \text{ lb} = 3.72 \text{ k}$$

Bending

The time effect factor is $\lambda = 0.8$ for the load combination 1.2D + 1.6S.
Beam spacing does not qualify for the repetitive member factor, so $C_r = 1.0$.
   For a 4 × 16, the size factor is determined from Table 4A:

$$C_F = 1.0$$

Also from Table 4A the wet service factor for bending is given us

$$C_M = 0.85$$

Except when $F_b(C_F) \le 1150 \text{ psi}$, $C_M = 1.0$.

In the case of 4×16 No. 1 & Better Hem-Fir:

$$F_b(C_F) = 1100(1.0) < 1150 \text{ psi}$$

$$\therefore \ C_M = 1.0$$

The coefficients for determining F'_{bn} for a sawn lumber beam are obtained from the summary in Example 6.10 in Sec. 6.4 (see also NDS Table 4.3.1). The resistance factor $\phi_b = 0.85$ and the format conversion factor $K_F = 2.16/\phi_b = 2.54$. In the bending stress summary given below, most adjustment factors default to unity. However, it is important for the designer to follow the steps leading to this conclusion.

$$F_{bn} = F_b \cdot K_F = 1100(2.54) = 2794 \text{ psi}$$

$$F'_{bn} = F_{bn}(\phi_b)(\lambda)(C_M)(C_t)(C_L)(C_F)(C_r)(C_i)$$

$$= (2794)(0.85)(0.8)(1.0)(1.0)(1.0)(1.0)(1.0)(1.0)$$

$$= 1901 \text{ psi} = 1.90 \text{ ksi}$$

Section properties for a 4×16 are listed in NDS Supplement Table 1B.

$$
\begin{aligned}
M'_n &= F'_{bn} S \\
&= (1.90)(135.7) \\
&= 258 \text{ in.-k} > 223 \text{ in.-k} \quad \textit{OK}
\end{aligned}
$$

Shear

For shear, $\phi_v = 0.75$ and $K_F = 2.16/\phi_v = 2.88$.

$$F_{vn} = F_v \cdot K_F = 150(2.88) = 432 \text{ psi}$$

$$F'_{vn} = F_{vn}(\phi_v)(\lambda)(C_M)(C_t)(C_i)$$

$$= (432)(0.75)(0.8)(1.0)(1.0)(1.0)$$

$$= 259 \text{ psi} = 0.259 \text{ ksi}$$

$$V'_n = \left(\frac{2}{3} F'_{vn}\, A\right)$$

$$= \left(\frac{2}{3}\right)(0.259)(53.375)$$

$$= 9.22 \text{ k} > 3.72 \text{ k} \quad \textit{OK}^*$$

Deflection

The deflection checks for $L/360$ and $L/240$ are the same in LRFD as in ASD. See Example 6.22. Deflection is *OK*

> 4×16 No. 1 & Better
> Hem-Fir beam   *OK*

---

*If V_u had exceeded V'_n the factored shear could have been reduced in accordance with Sec. 6.5.

6.13 Design Problem: Glulam Beam with Full Lateral Support

The examples in Secs. 6.13, 6.14, and 6.15 all deal with the design of the same glulam beam, but different conditions of lateral support for the beam are considered in each problem. These examples illustrate the application of the beam stability factor, which is the same in both ASD and LRFD. The first example deals with the design of a beam that has *full lateral support* to the compression side of the member, and lateral stability is simply not a concern. See Example 6.24.

EXAMPLE 6.24   Glulam Beam—Full Lateral Support Using ASD

Determine the required size of a 24F-1.8E DF stress class glulam, using ASD for the simple-span roof beam shown in Fig. 6.22. Assume dry service conditions and normal temperature range. D = 200 lb/ft, and S = 800 lb/ft. Use the IBC-required deflection limits for a roof beam in a commercial building supporting a nonplaster ceiling. By inspection, the critical load combination is

$$D + S = 200 + 800 = 1000 \text{ lb/ft}$$

A number of adjustment factors for determining adjusted design values can be determined directly from the problem statement. For example, the load duration factor is $C_D = 1.15$ for the combination of (D + S). In addition, the wet service factor is $C_M = 1.0$ for a glulam with MC < 16 percent, and the temperature factor is $C_t = 1.0$ for members used at normal temperatures.

Bending

The glulam beam will be loaded such that the tension laminations will be stressed in tension, and the reference design value F_{bx}^+ applies. The summary for glulam beams in Example 6.11 in Sec. 6.4 (and NDS Table 5.3.1) indicates that there are two possible definitions of the adjusted ASD bending value. One considers the effects of lateral stability as measured by the beam stability factor C_L. The other evaluates the effect of beam width, depth, and length as given by the volume factor C_V.

Lateral stability: $F'_{bx} = F_{bx}(C_D)(C_M)(C_t)(C_L)$

Volume effect: $F'_{bx} = F_{bx}(C_D)(C_M)(C_t)(C_V)$

The sketch of the beam cross section shows that the compression edge of the beam (positive moment places the top side in compression) is restrained from lateral movement by an effective connection to the roof diaphragm. The unbraced length is zero, and the beam slenderness ratio is zero. Lateral buckling is thus prevented, and the beam stability factor is $C_L = 1.0$. In this case, only the adjusted bending design value using the volume factor needs to be considered.

Before the volume factor can be evaluated, a trial beam size must be established. This is done by assuming a value for C_V which will be verified later. Assume $C_V = 0.82$. Reference design values are obtained from NDS Supplement Table 5A.

Figure 6.22 Glulam beam with span of 48 ft and full lateral support to the compression edge of the member provided by the roof diaphragm.

$$F'_{bx} = F_{bx}(C_D)(C_M)(C_t)(C_V)$$

$$= 2400(1.15)(1.0)(1.0)(0.82)$$

$$= 2263 \text{ psi}$$

$$\text{Req'd } S = \frac{M}{F'_b} = \frac{3{,}456{,}000}{2263} = 1527 \text{ in.}^3$$

As in most beam designs, the objective is to select the member with the least cross-sectional area that provides a section modulus grater than that required. This can be done using the S_x column for Western Species glulams in NDS Supplement Table 1C.

Try $6\frac{3}{4} \times 37\frac{1}{2}$ (twenty-five $1\frac{1}{2}$-in. laminations)

$A = 253.1 \text{ in.}^2$

$S = 1582 \text{ in.}^3 > 1527 \text{ in.}^3$ *OK*

$I = 29{,}660 \text{ in.}^4$

The trial size was based on an assumed volume factor. Determine the actual C_V (see Sec. 5.6 for a review of C_V), and revise the trial size if necessary.

$$C_V = \left(\frac{21}{L}\right)^{1/10}\left(\frac{12}{d}\right)^{1/10}\left(\frac{5.125}{b}\right)^{1/10}$$

$$= \left(\frac{21}{48}\right)^{0.1}\left(\frac{12}{37.5}\right)^{0.1}\left(\frac{5.125}{6.75}\right)^{0.1}$$

$$= 0.799 < 0.82$$

Because the assumed value of C_V was not conservative, the actual bending stress and adjusted bending design value will be compared in order to determine if the trial size is adequate.

$$f_b = \frac{M}{S} = \frac{3{,}456{,}000}{1582} = 2185 \text{ psi}$$

$$F_b' = F_b(C_D)(C_M)(C_t)(C_V)$$

$$= 2400(1.15)(1.0)(1.0)(0.799)$$

$$= 2205 \text{ psi} > 2185 \text{ psi}   OK$$

Shear

Ignore the reduction of shear given by V^* (conservative).

$$f_v = \frac{1.5V}{A} = \frac{1.5(24{,}000)}{253.1} = 142 \text{ psi}$$

$$F_{vx}' = F_v(C_D)(C_M)(C_t) = 265(1.15)(1.0)(1.0)$$

$$= 305 \text{ psi} > 142 \text{ psi}   OK$$

Deflection

$$E_x' = E_x(C_M)(C_t)$$

$$= 1{,}800{,}000(1.0)(1.0)$$

$$= 1{,}800{,}000 \text{ psi}$$

$$\Delta_{TL} = \frac{5w_{TL}L^4}{384E'I} = \frac{5(1000)(48)^4(12 \text{ in./ft})^3}{384(1{,}800{,}000)(29{,}660)} = 2.24 \text{ in.}$$

$$\frac{\Delta_{TL}}{L} = \frac{2.24}{48 \times 12} = \frac{1}{257} < \frac{1}{180}   OK$$

NOTE: The total load was conservatively used for this deflection check. According to footnote d in the IBC Table 1604.3, the dead load component may be reduced by a factor of 0.5. If this deflection check had not been satisfied, then the deflection caused by 0.5D + S could be checked against $L/180$.

By proportion,

$$\Delta_S = \left(\frac{800}{1000}\right)\Delta_{TL} = 0.8(2.24) = 1.79 \text{ in.}$$

$$\frac{\Delta_S}{L} = \frac{1.79}{48 \times 12} = \frac{1}{321} < \frac{1}{240} \quad OK$$

$$\text{Camber} = 1.5\Delta_D = 1.5\left(\frac{200}{1000}\right)(2.24) = 0.67 \text{ in.}$$

Bearing

The support conditions are unknown, so the required bearing length will simply be determined. Use $F'_{c\perp}$ for bearing on the tension face of a glulam bending about the x axis. Recall that C_D does not apply to $F_{c\perp}$.

$$F'_{c\perp} = F_{c\perp}(C_M)(C_t)(C_b)$$
$$= 650(1.0)(1.0)(1.0) = 650 \text{ psi}$$

$$\text{Req'd } A = \frac{R}{F'_{c\perp}} = \frac{24,000}{650} = 36.9 \text{ in.}^2$$

$$\text{Req'd } l_b = \frac{36.9}{6.75} = 5.47 \text{ in.} \quad \text{Say } l_b = 5\frac{1}{2} \text{ in. min.}$$

> Use $6\frac{3}{4} \times 37\frac{1}{2}$ (twenty-five $1\frac{1}{2}$ -in. laminations 24F-1.8E glulam—camber 0.67 in.

Example 6.24 followed ASD procedures. The LRFD approach to this same problem is provided in Example 6.25.

EXAMPLE 6.25   Glulam Beam—Full Lateral Support Using LRFD

Using LRFD, determine the required size of the 24F-1.8E DF stress class glulam from Example 6.24. Assume dry-services conditions and normal temperature range. Unfactored loads are D = 200 lb/ft, and S = 800 lb/ft. Use the IBC required deflection limits for a roof beam in a commercial building supporting a nonplastar ceiling. By inspection, the critical LRFD load combination is

$$1.2D + 1.6S = 1.2(200) + 1.6(800) = 1520 \text{ lb/ft}$$

A number of adjustment factors for determining adjusted LRFD design values can be determined directly from the problem statement. For example, the time effect factor is $\lambda = 0.8$ for the combination of (1.2D + 1.6S). In addition, the wet service factor is $C_M = 1.0$ for a glulam with MC < 16 percent, and the temperature factor is $C_t = 1.0$ for members used at normal temperatures.

Bending

Since the beam is simply supported with a uniformly distributed load (see Fig. 6.22), the factored moment, M_u, is

$$M_u = w_u L^2/8$$

$$= (1520)(48^2)/8$$

$$= 437,760 \text{ ft-lb} = 5253 \text{ in.-k}$$

The glulam beam will be loaded such that the tension laminations will be stressed in tension, and the reference design value F_{bx}^+ applies. The summary for glulam beams in Example 6.11 in Sec. 6.4 (and NDS Table 5.3.1) indicates that there are two possible definitions of the adjusted LRFD design value. One considers the effects of lateral stability as measured by the beam stability factor C_L. The other evaluates the effect of beam width, depth, and length as given by the volume factor C_V.

Lateral stability: $F'_{bxn} = F_{bxn}(\phi_b)(\lambda)(C_M)(C_t)(C_L)$

Volume effect: $F'_{bxn} = F_{bxn}(\phi_b)(\lambda)(C_M)(C_t)(C_V)$

The sketch of the beam cross section (see Fig. 6.22) shows that the compression side of the beam (positive moment places the top side in compression) is restrained from lateral movement by an effective connection to the roof diaphragm. The unbraced length is zero, and the beam slenderness ratio is zero. Lateral buckling is thus prevented, and the beam stability factor is $C_L = 1.0$. In this case, only the adjusted bending design value using the volume factor needs to be considered.

Before the volume factor can be evaluated, a trial beam size must be established. This is done by assuming a value for C_V which will be verified later. Assume $C_V = 0.82$, $\phi_b = 0.85$ and $K_F = 2.16/\phi_b = 2.54$. Reference design values are obtained from NDS Supplement Table 5A.

$$F_{bxn} = F_{bx} \cdot K_F = 2400(2.54) = 6096$$

$$F'_{bxn} = F_{bxn}(\phi_b)(\lambda)(C_M)(C_t)(C_V)$$

$$= (6096)(0.85)(0.8)(1.0)(1.0)(0.82)$$

$$= 3400 \text{ psi} = 3.40 \text{ ksi}$$

$$\text{Req'd } S = \frac{M_u}{F'_{bn}} = \frac{5253}{3.40} = 1545 \text{ in.}^3$$

As in most beam designs, the objective is to select the member with the least cross-sectional area that provides a section modulus greater than that required. This can be done using the S_x column for Western Species glulams in NDS Supplement Table 1C.

Try $7\frac{3}{4} \times 34\frac{1}{2}$ (twenty-five $1\frac{1}{2}$-in. lams)

$$A = 253.1 \text{ in.}^2$$

$$S = 1582 \text{ in.}^3 > 1545 \text{ in.}^3 \quad OK$$

$$I = 29,660 \text{ in.}^4$$

$$C_V = \left(\frac{21}{L}\right)^{1/10}\left(\frac{12}{d}\right)^{1/10}\left(\frac{5.125}{b}\right)^{1/10}$$

$$= \left(\frac{21}{48}\right)^{0.1}\left(\frac{12}{37.5}\right)^{0.1}\left(\frac{5.125}{6.75}\right)^{0.1}$$

$$= 0.799 < 0.82$$

$$F'_{bxn} = (F_{bxn})(\phi_b)(\lambda)(C_M)(C_t)(C_V)$$

$$= (6096)(0.85)(0.8)(1.0)(1.0)(0.799)$$

$$= 3316 \text{ psi} = 3.32 \text{ ksi}$$

$$M'_n = F'_{bxn}\cdot S$$

$$= (3.32)(1582)$$

$$= 5252 \text{ in.-k} < 5253 \text{ in.-k}$$

Considering moment, the $6^{3/4} \times 37^{1/2}$ is *OK*

Shear

Ignore the reduction of shear given by V^* (conservative). $\phi_v = 0.75$ and $K_F = 2.16/\phi_v = 2.88$.

$$V_u = w_u L/2$$

$$= (1520)(48)/2$$

$$= 36{,}480 \text{ lb} = 36.5 \text{ k}$$

$$F_{vn} = F_v\cdot K_F = 265(2.88) = 763 \text{ psi}$$

$$F'_{vn} = F_{vn}(\phi_v)(\lambda)(C_M)(C_t)$$

$$= (763)(0.75)(0.8)(1.0)(1.0)$$

$$= 458 \text{ psi} = 0.458 \text{ ksi}$$

$$V'_n = \left(\frac{2}{3}F'_{vn}A\right)$$

$$= \left(\frac{2}{3}\right)(0.458)(253.1)$$

$$= 77.3 \text{ k} > 36.5 \text{ k} \qquad OK$$

Deflection

The deflection checks for $L/180$ and $L/240$ are the same in LRFD as in ASD. See Example 6.24. Deflection is *OK*.

---

*If V_u had exceeded V'_n the factored shear could have been reduced in accordance with Sec. 6.5.

Bearing

The support conditions are unknown, so the required bearing length will be determined. Use $F'_{c\perp n}$ for bearing on the tension face of a glulam bending about the x axis. Unlike ASD where C_D does not apply to compression perpendicular to the grain, the time effect factor λ does apply in LRFD. $\phi_c = 0.90$ and $K_F = 1.875/\phi_c = 2.08$.

$$F_{c\perp n} = F_{c\perp} \cdot K_F = 650 \cdot 2.08 = 1352 \text{ psi}$$

$$F'_{c\perp n} = F_{c\perp n}(\phi_c)(\lambda)(C_M)(C_t)(C_b)$$

$$= (1352)(0.90)(0.8)(1.0)(1.0)(1.0)$$

$$= 973 \text{ psi} = 0.973 \text{ ksi}$$

$$\text{Assuming } R_u = V_u = 36.5 \text{ k}$$

$$\text{Req'd } A = \frac{R_u}{F'_{c\perp n}} = \frac{36.5}{(0.973)} = 37.5 \text{ in.}^2$$

$$\text{Req'd } l_b = \frac{A}{b} = \frac{37.5}{6.75} = 5.56 \text{ in.} \quad \text{Say } l_b = 5\tfrac{5}{8} \text{ in. min.}$$

> *Use* $6 \times 37\tfrac{1}{2}$ (twenty-five $1\tfrac{1}{2}$-in. laminations)
> 24F-1.8E DF glulam – camber 0.67 in.

In Sec. 6.14 this example is reworked with lateral supports at 8 ft-0 in. o.c. This spacing is obtained when the purlins are attached to the top of the glulam. With this arrangement the sheathing is separated from the beam, and the distance between points of lateral support becomes the spacing of the purlins.

In Sec. 6.15, the beam is analyzed for an unbraced length of 48 ft-0 in. In other words, only the ends of the beam are braced against translation and rotation. This condition would exist if no diaphragm action is developed in the sheathing, or if no sheathing or effective bracing is present along the beam. Fortunately, this situation is not common in ordinary building design.

6.14   Design Problem: Glulam Beam with Lateral Support at 8 ft-0 in.

In order to design a beam with an unbraced compression zone, it is necessary to check both *lateral stability* and *volume effect*. To check lateral stability, a trial beam size is required so that the beam slenderness ratio R_B can be computed. This is similar to column design, where a trial size is required before the column slenderness ratio and the strength of the column can be evaluated.

All criteria except unbraced length are the same for this example and the previous problem (Sec. 6.13). Therefore, the initial trial beam size is taken from Examples 6.24 and 6.25. The $6\tfrac{3}{4} \times 37\tfrac{1}{2}$ trial represents the minimum beam size based on the volume-effect criterion. Because all other factors are the same, only the lateral stability criteria are considered in this example. Both ASD and LRFD approach as presented. See Example 6.26. The calculations for C_L indicate that lateral stability is less critical than the volume effect for this problem. The trial size, then, is adequate.

EXAMPLE 6.26   Glulam Beam—Lateral Support at 8 ft-0 in. Using both ASD and LRFD

Rework Examples 6.24 and 6.25, using the modified lateral support condition shown in the beam section view in Fig. 6.23. All other criteria are the same.

Bending

The adjusted design values for a glulam beam bending about the x axis are summarized in Example 6.11. Separate design bending design values are provided for lateral stability and volume effect.

For ASD:

$$\text{Lateral stability:} \quad F'_{bx} = F_{bx}(C_D)(C_M)(C_t)(C_L)$$

$$\text{Volume effect:} \quad F'_{bx} = F_{bx}(C_D)(C_M)(C_t)(C_V)$$

For LRFD:

$$\text{Lateral stability:} \quad F'_{bxn} = F_{bxn}(\phi_b)(\lambda)(C_D)(C_M)(C_t)(C_L)$$

$$\text{Volume effect:} \quad F'_{bxn} = F_{bxn}(\phi_b)(\lambda)(C_M)(C_t)(C_V)$$

See Examples 6.24 (ASD) and 6.25 (LRFD) for the development of a trial size based on the volume effect. This trial size will now be analyzed for the effects of lateral stability using an unbraced length of 8 ft-0 in.

$$\text{Try} \quad 6\tfrac{3}{4} \times 37\tfrac{1}{2} \ 24\text{F-}1.8\text{E glulam}$$

Figure 6.23   Beam from Examples 6.24 (ASD) and 6.25 (LRFD) with revised lateral support conditions.

Slenderness ratio for bending member R_B

Unbraced length l_u = 8 ft = 96 in.

Effective unbraced lengths are given in Example 6.8 and in NDS Table 3.3.3. For a single-span beam with a uniformly distributed load, the definition of l_e depends on the l_u/d ratio

$$\frac{l_u}{d} = \frac{96}{37.5} = 2.56 < 7$$

$$\therefore\ l_e = 2.06 l_u = 2.06(96) = 198 \text{ in.}$$

$$R_B = \sqrt{\frac{l_e d}{b^2}} = \sqrt{\frac{198(37.5)}{(6.75)^2}} = 12.76$$

Coefficients for computing beam stability factor C_L

A beam subject to lateral torsional buckling is governed by stability about the y axis, and the reference modulus of elasticity for use in determining the beam stability factor is $E_{y\ \min}$.

For ASD:

$$E'_{y\ \min} = E_{y\ \min}(C_M)(C_t) = 830{,}000(1.0)(1.0)$$

$$= 830{,}000 \text{ psi}$$

$$F_{bE} = \frac{1.20\,E'_{y\ \min}}{R_B^2} = \frac{1.20(830{,}000)}{12.76^2} = 6117 \text{ psi}$$

The reference bending design value about the x axis modified by all factors except C_V and C_L is given the notation F_b^*

$$F_{bx}^* = F_{bx}(C_D)(C_M)(C_t)$$

$$= 2400(1.15)(1.0)(1.0) = 2760 \text{ psi}$$

$$\frac{F_{bE}}{F_{bx}^*} = \frac{6117}{2760} = 2.216$$

$$\frac{1 + F_{bE}/F_{bx}^*}{1.9} = \frac{1 + 2.216}{1.9} = 1.693$$

Beam stability factor

$$C_L = \frac{1 + F_{bE}/F_{bx}^*}{1.9} - \sqrt{\left(\frac{1 + F_{bE}/F_{bx}^*}{1.9}\right)^2 - \frac{F_{bE}/F_{bx}^*}{0.95}}$$

$$= 1.693 - \sqrt{1.693 - 2.216/0.95}$$

$$= 0.963$$

From Example 6.24, the volume factor for this beam is

$$C_V = 0.799$$

$$< C_L$$

$$\therefore \text{ Volume effect governs over lateral stability.}$$

The adjusted ASD bending design value for the beam with lateral support to the compression edge at 8 ft-0 in. is the same as that for the beam in Example 6.24:

$$F'_b = 2205 \text{ psi} > 2185 \text{ psi} \quad OK$$

For ASD:

> *Use* $6\frac{3}{4} \times 37\frac{1}{2}$
> 24F-1.8E DF glulam

For LRFD: $\phi_s = 0.85$ and $K_F = 1.50/\phi_s = 1.76$.

$$E_{y \min n} = E_{y \min} \cdot K_F = 830{,}000(1.76) = 1{,}460{,}800 \text{ psi}$$
$$E'_{y \min n} = E_{y \min n}(\phi_s)(C_M)(C_t)$$
$$= (1{,}460{,}800)(0.85)(1.0)(1.0)$$
$$= 1{,}245{,}000 \text{ psi} = 1245 \text{ ksi}$$

$$F_{bEn} = \frac{1.20 E'_{y \min n}}{R_B^2} = \frac{1.20(1245)}{12.76^2} = 9.18 \text{ ksi}$$

The reference bending design value about the x axis modified by all factors except C_V and C_L is given the notation F_b^*. $\phi_b = 0.85$ and $K_F = 2.16/\phi_b = 2.54$.

$$F_{bn} = F_b \cdot K_F = 2400(2.54) = 6096 \text{ psi}$$

$$F_{bn}^* = F_{bn}(\phi_b)(\lambda)(C_M)(C_t)$$
$$= (6096)(0.85)(0.8)(1.0)(1.0)$$
$$= 4145 \text{ psi} = 4.14 \text{ ksi}$$

$$\frac{F_{bEn}}{F_{bxn}^*} = \frac{9.15}{4.14} = 2.210$$

$$\frac{1 + F_{bEn}/F_{bxn}^*}{1.9} = \frac{1 + 2.210}{1.9} = 1.689$$

Beam stability factor:

$$C_L = \frac{1 + F_{bEn}/F_{bxn}^*}{1.9} - \sqrt{\left(\frac{1 + F_{bEn}/F_{bxn}^*}{1.9}\right)^2 - \frac{F_{bEn}/F_{bxn}^*}{0.95}}$$

$$= 1.689 - \sqrt{1.689^2 - 2.210/0.95}$$

$$= 0.963$$

From Example 6.25, the volume factor for this beam is
$$C_V = 0.806$$
$$< C_L$$

∴ Volume effect governs over lateral stability.

The adjusted LRFD moment resistance, M'_n, for the beam with lateral support to the compression edge at 8 ft-0 in. is the same as that for the beam in Example 6.25.

$$M'_n = 5252 \text{ in.-k} \approx 5253 \text{ in.-k} \quad OK$$

Use $6 \times 37\frac{1}{2}$
24F-1.8E DF glulam

The beam in Example 6.26 is seen to be unaffected by an unbraced length of 8 ft. The beam slenderness ratio R_B is the principal measure of lateral stability, and R_B is a function of the unbraced length, beam depth, and beam width. The slenderness ratio is especially sensitive to beam *width* because of the square in the denominator.

A large slenderness ratio is obtained in Example 6.27 by increasing the unbraced length from 8 to 48 ft.

6.15 Design Problem: Glulam Beam with Lateral Support at 48 ft-0 in.

The purpose of this brief example is to illustrate the impact of a long unbraced length and a correspondingly large beam slenderness ratio. See Example 6.27. As with the previous example, the initial trial size is taken from Examples 6.24 and 6.25 for ASD and LRFD, respectively, because a trial size is required in order to calculate the beam slenderness ratio.

This example, using both ASD and LRFD, illustrates why it is desirable to have at least some intermediate lateral bracing. The long unbraced length causes the trial size to be considerably overstressed, and a new trial beam size is required.

The problem is not carried beyond the point of checking the initial trial beam because the purpose of the example is simply to demonstrate the effect of lateral buckling. A larger trial size would be evaluated in a similar manner.

EXAMPLE 6.27   Glulam Beam—Lateral Support at 48 ft-0 in. Using both ASD and LRFD

Rework the beam design problem in Examples 6.24 through 6.26 with lateral supports at the ends of the span only. See Fig. 6.22.

For ASD:

Try $6\frac{3}{4} \times 37\frac{1}{2}$ 24F-1.8E Stress Class glulam

Slenderness ratio for bending member R_B

Unbraced length l_u = 48 ft = 576 in.

Effective unbraced lengths are given in Example 6.8 and in NDS Table 3.3.3. For a single-span beam with a uniformly distributed load, the definition of l_e depends on the l_u/d ratio

$$\frac{l_u}{d} = \frac{576}{37.5} = 15.36 > 7$$

$$\therefore \ l_e = 1.63l_u + 3d = 1.63(576) + 3(37.5) = 1051 \text{ in.}$$

$$R_B = \sqrt{\frac{l_e d}{b^2}} = \sqrt{\frac{1051(37.5)}{(6.75)^2}} = 29.42$$

Coefficients for computing beam stability factor C_L

$$F_{bE} = \frac{1.20 E'_{y \, min}}{R_B^2} = \frac{1.2(830,000)}{(29.42)^2} = 1151 \text{ psi}$$

$$F_{bx}^* = F_{bx}(C_D)(C_M)(C_t)$$

$$= 2400(1.15)(1.0)(1.0) = 2760 \text{ psi}$$

$$\frac{F_{bE}}{F_{bx}^*} = \frac{1151}{2760} = 0.417$$

$$\frac{1 + F_{bE}/F_{bx}^*}{1.9} = \frac{1 + 0.417}{1.9} = 0.746$$

Beam stability factor

$$C_L = \frac{1 + F_{bE}/F_{bx}^*}{1.9} - \sqrt{\left(\frac{1 + F_{bE}/F_{bx}^*}{1.9}\right)^2 - \frac{F_{bE}/F_{bx}^*}{0.95}}$$

$$= 0.746 - \sqrt{0.746^2 - 0.417/0.95}$$

$$= 0.398$$

From Example 6.24, the volume factor for this beam is

$$C_V = 0.799$$

$$> C_L$$

$\therefore$ Lateral stability governs over the volume factor.

The adjusted bending design value for the beam with lateral support to the compression edge at 48 ft-0 in. is

$$F_b' = F_b(C_D)(C_M)(C_t)(C_L)$$

$$= 2400(1.15)(1.0)(1.0)(0.398)$$

$$= 1098 \text{ psi}$$

$$f_b = 2185 \text{ psi}$$

$$> 1098 \text{ psi} \quad NG$$

For LRFD:

Slenderness ratio for bending member R_B

The effective length and slenderness ratio of a member are dependent on the support conditions, type of loading, and geometry of the section. Therefore, the determination of R_B is the same for both ASD and LRFD.

$$R_B = 29.42$$

Coefficients for computing beam stability factor C_L

$$F_{bEn} = \frac{1.20 E'_{y\,min\,n}}{R_B^2} = \frac{1.20(1245 \text{ ksi})}{29.42^2} = 1.72 \text{ ksi}$$

$$F_{bn}^* = F_{bn}(\phi_b)(\lambda)(C_M)(C_t)$$

$$= (6096)(0.85)(0.8)(1.0)(1.0)$$

$$= 4145 \text{ psi} = 4.14 \text{ ksi}$$

$$\frac{F_{bEn}}{F_{bxn}^*} = \frac{1.72}{4.14} = 0.415$$

$$\frac{1 + F_{bEn}/F_{bxn}^*}{1.9} = \frac{1 + 0.416}{1.9} = 0.745$$

Beam stability factor:

$$C_L = \frac{1 + F_{bEn}/F_{bxn}^*}{1.9} - \sqrt{\left(\frac{1 + F_{bEn}/F_{bxn}^*}{1.9}\right)^2 - \frac{F_{bEn}/F_{bxn}^*}{0.95}}$$

$$= 0.745 - \sqrt{0.745^2 - 0.416/0.95}$$

$$= 0.403$$

From Example 6.25, the volume factor for this beam is

$$C_V = 0.799$$

$$> C_L$$

∴ Lateral stability governs over the volume factor.

The adjusted LRFD moment resistance, M'_n, for the beam with lateral support to the compression side at 48 ft-0 in. is

$$F'_{bn} = F_{bn}(\phi_b)(\lambda)(C_M)(C_t)(C_L)$$

$$= (6096)(0.85)(0.8)(1.0)(1.0)(0.403)$$

$$= 1671 \text{ psi} = 1.67 \text{ ksi}$$

$$M'_n = F'_{bn}S$$

$$= (1.67)(1582)$$

$$= 2642 \text{ in-k} < 5253 \text{ in.-k} \qquad NG$$

Using either ASD or LRFD the trial size of a $6\frac{3}{4} \times 37\frac{1}{2}$ is considerably overstressed in bending and is no good (*NG*). A revised trial size is thus required and is left as an exercise for the reader.

6.16 Design Problem: Glulam with Compression Zone Stressed in Tension

Some glulam beams have *balanced* combinations of laminations. These have the same bending design value on the top and bottom faces of the member. Other combinations have tension lamination requirements only on one side of the beam. For this latter case there are two different values for bending:

1. F_{bx} tension zone stressed in tension (F_{bx}^{+})
2. F_{bx} compression zone stressed in tension (F_{bx}^{-})

The beam in Example 6.28 has a large positive moment and a small negative moment. The member in this example involves a combination that is not balanced. The beam is first designed for the large positive moment using F_{bx}^{+}. The bending stress that results from the negative moment is then checked against the smaller adjusted bending design value F_{bx}^{-}. This example is provided using only ASD. For LRFD, the approach would be similar except that the bending design values would be converted to nominal values F_{bxn}^{+} and F_{bxn}^{-}, and associated nominal resistances $M_n^{+\prime}$ and $M_n^{-\prime}$ would be checked against the factored moments M_u^{+} and M_u^{-}.

EXAMPLE 6.28   Compression Zone Stressed in Tension Using ASD

The roof beam in Fig. 6.24 is a 24F-1.8E DF glulam. The design load includes a concentrated load and a uniformly distributed load. Loads are a combination of (D + L$_r$). Lateral support is provided to the top face of the beam by the roof sheathing. However, the bottom face is laterally unsupported in the area of negative moment except at the reaction point. The beam is used in dry service conditions and at normal temperatures. The minimum roof slope is provided so that ponding need not be considered. For this problem consider bending stresses only.

Positive Moment (Tension Zone Stressed in Tension)

In the area of positive bending moment, the adjusted ASD bending value is F_{bx}^{+}.

Lateral stability: $F_b' = F_{bx}^{+\prime} = F_{bx}^{+}(C_D)(C_M)(C_t)(C_L)$

Volume effect: $F_b' = F_{bx}^{+} = F_{bx}^{+}(C_D)(C_M)(C_t)(C_V)$

Also in the area of positive bending moment, the unbraced length is $l_u = 0$ because continuous lateral support is provided to the top side of the beam by the roof sheathing. Therefore, C_L defaults to unity, and lateral stability *does not govern* (DNG).

Develop a trial beam size, using an assumed value for the volume factor, and check the actual C_V later. The load duration factor is $C_D = 1.25$ for the combination of (D + L$_r$). Both C_M and C_t default to unity. Reference design values are given in NDS Supplement Table 5A.

Figure 6.24 Glulam beam with small cantilever.

Assume $C_V = 0.90$:

$$F'_b = F^{+'}_{bx} = F^+_{bx}(C_D)(C_M)(C_t)(C_V)$$

$$= 2400(1.25)(1.0)(1.0)(0.90) = 2700 \text{ psi}$$

Max. $M = 108 \text{ ft-k} = 1295 \text{ in.-k}$ (from Fig. 6.24)

$$\text{Req'd } S = \frac{M}{F'_b} = \frac{1,295,000}{2700} = 480 \text{ in.}^3$$

Refer to NDS Supplement Table 1C, and choose the smallest Western Species glulam size that furnishes a section modulus greater than the required.

$$\textit{Try}\quad 5\tfrac{1}{8} \times 24 \quad 24\text{F-}1.8\text{E glulam}$$

$$S = 492 \text{ in.}^3 > 480 \text{ in.}^3 \qquad \textit{OK}$$

Verify C_V:

The volume factor is a function of the length, depth, and width of a beam. The length is to be taken as the distance between points of zero moment in Fig. 6.24

($L = 36 - 2.11 = 33.89$ ft). However, it is simple and conservative to use the full span length of 36 ft.

$$C_V = K_L \left(\frac{21}{L}\right)^{1/10}\left(\frac{12}{d}\right)^{1/10}\left(\frac{5.125}{b}\right)^{1/10} \leq 1.0$$

$$= 1.0\left(\frac{21}{36}\right)^{0.1}\left(\frac{12}{24}\right)^{0.1}\left(\frac{5.125}{5.125}\right)^{0.1}$$

$$= 0.884 < 1.0$$

The assumed value of $C_V = 0.90$ was not conservative. Therefore, compare the actual and adjusted bending design values in a summary:

$$f_b = \frac{M}{S} = \frac{1,295,000}{492} = 2630 \text{ psi}$$

$$F_b' = F_{bx}^{+'} = F_{bx}(C_D)(C_M)(C_t)(C_V)$$

$$= 2400(1.25)(1.0)(1.0)(0.884)$$

$$= 2650 \text{ psi} > 2630 \text{ psi}$$

$$\therefore \text{ Positive moment} \qquad OK$$

Negative Moment (Compression Zone Stressed in Tension)

The trial beam size remains the same, and the computed bending stress is

$$\text{Neg. } M = 28.5 \text{ ft-k} = 342 \text{ in.-k}$$

$$f_b = \frac{M}{S} = \frac{342,000}{492} = 695 \text{ psi}$$

In the area of negative bending moment, the reference bending design value is $F_{bx}^- = 1450$ psi. The adjusted design value is the smaller value determined from the two criteria

Lateral stability: $\qquad\qquad F_b' = F_{bx}^{-'} = F_{bx}^-(C_D)(C_M)(C_t)(C_L)$

Volume effect: $\qquad\qquad F_b' = F_{bx}^{-'} = F_{bx}^-(C_D)(C_M)(C_t)(C_V)$

Lateral stability

The possibility of lateral buckling needs to be considered because the bottom edge of the beam does not have continuous lateral support.

Slenderness ratio for beam R_b:

To the left of the support:

$$l_u = 6 \text{ ft} = 72 \text{ in.}$$

To the right of the support to the inflection point (IP):

$$l_u = 2.11 \text{ ft} = 25.3 \text{ in.} \qquad \text{(not critical)}$$

Effective unbraced lengths are given in Example 6.8 (Sec. 6.3) and in NDS Table 3.3.3. For a cantilever beam with any loading, the definition of l_e depends on the l_u/d ratio

$$\frac{l_u}{d} = \frac{72}{24} = 3.0 < 7$$

$$\therefore l_e = 2.06 l_e = 2.06(72) = 148 \text{ in.}$$

$$R_B = \sqrt{\frac{l_e d}{b^2}} = \sqrt{\frac{148(24)}{(5.125)^2}} = 11.64$$

Coefficients for computing beam stability factor C_L:

$$E'_{y \min} = E_{y \min}(C_M)(C_t) = 830{,}000(1.0)(1.0)$$

$$= 830{,}000 \text{ psi}$$

$$F_{bE} = \frac{1.20 E'_{y \min}}{R_B^2} = \frac{1.20(830{,}000)}{11.64^2} = 7351 \text{ psi}$$

$$F_{bx}^* = F_{bx}(C_D)(C_M)(C_t)$$

$$= 1450(1.25)(1.0)(1.0) = 1812 \text{ psi}$$

$$\frac{F_{bE}}{F_{bx}^*} = \frac{7351}{1812} = 4.057$$

$$\frac{1 + F_{bE}/F_{bx}^*}{1.9} = \frac{1 + 4.057}{1.9} = 2.662$$

Beam stability factor

$$C_L = \frac{1 + F_{bE}/F_{bx}^*}{1.9} - \sqrt{\left(\frac{1 + F_{bE}/F_{bx}^*}{1.9}\right)^2 - \frac{F_{bE}/F_{bx}^*}{0.95}}$$

$$= 2.662 - \sqrt{2.662^2 - 4.057/0.95}$$

$$= 0.984$$

Volume effect

The length to compute the volume factor is defined as the distance between points of zero moment ($L = 6 + 2.11 = 8.11$ ft).

$$C_V = K_L \left(\frac{21}{L}\right)^{1/10} \left(\frac{12}{d}\right)^{1/10} \left(\frac{5.125}{b}\right)^{1/10} \leq 1.0$$

$$= 1.0 \left(\frac{21}{8.11}\right)^{0.1} \left(\frac{12}{24}\right)^{0.1} \left(\frac{5.125}{5.125}\right)^{0.1}$$

$$= 1.026 > 1.0$$

$$\therefore \quad C_V = 1.0$$

The lateral stability factor governs over the volume factor.

$$F'_b = \overline{F'_{bx}}(C_D)(C_M)(C_t)(C_L)$$

$$= 1450(1.25)(1.0)(1.0)(0.984)$$

$$= 1784 \text{ psi}$$

$$f_b = 695 \text{ psi} < 1784 \text{ psi} \qquad OK$$

$5\frac{1}{8} \times 24$ 24F-1.8E DF glulam   *OK*   for bending

6.17 Cantilever Beam Systems

Cantilever beam systems that have an internal hinge connection are often used in glulam construction. The reason for this is that a smaller-size beam can generally be used with a cantilever system compared with a series of simply supported beams. The cantilever length L_c in the cantilever beam system is an important variable. See Example 6.29. A cantilever length can be established for which an optimum beam size can be obtained.

EXAMPLE 6.29   Cantilever Beam Systems

The *bending strength* of a cantilever beam system can be optimized by choosing the cantilever length L_c so that the local maximum moments M_1, M_2, and M_3 will all be equal. For the two-equal-span cantilever system shown in Fig. 6.25, with a constant uniform load on both spans, the cantilever length

$$L_c = 0.172L$$

gives equal local maximum moments

$$M_1 = M_2 = M_3 = 0.086wL^2$$

This maximum moment is considerably less than the maximum moment for a uniformly loaded simple beam:

$$M = \frac{wL^2}{8} = 0.125wL^2$$

Recommended cantilever lengths for a number of cantilever beam systems are given in the TCM (Ref. 6.5).

Figure 6.25 Two-span cantilever beam system.

Cantilever beam systems are not recommended for floors. Proper cambering is difficult, and cantilever beam systems in floors may transmit vibrations from one span to another. AITC recommends the use of simply supported beams for floors.

For the design of both roofs and floors, IBC Chap. 16 requires that the case of dead load on all spans plus roof live load on alternate spans (unbalanced L_r) must be considered in addition to full ($D + L_r$) on all spans. See Example 6.30.

EXAMPLE 6.30   Load Cases for a Two-Span Cantilever Beam System

Load Case 1: **(D + L or 1.2D + 1.6L on All Spans)**

This load constitutes the maximum total load and can produce the critical design moment, shear, and deflection. See Fig. 6.26.

ASD	LRFD
$w = D + L$	$w_u = 1.2D + 1.6L$
$w_1 = D + L$	$w_{u1} = 1.2D + 1.6L$
$w_2 = D$ only	$w_{u2} = 1.2D$ only
$w_1 = D$ only	$w_{u1} = 1.2D$ only
$w_2 = D + L$	$w_{u2} = 1.2D + 1.6L$

Figure 6.26   When $L_r < 20$ psf, *full* and *unbalanced* live load analyses are required.

Load Case 2: **(D + Unbalanced L or 1.2D + 1.6L on Left Span)**

When unbalanced live load is required, this load will produce the critical positive moment in the left span.

Load Case 3: **(D + Unbalanced L or 1.2D + 1.6L on Right Span)**

This load case will produce the same maximum negative moment as load case 1. It will also produce the maximum length from the interior support to the inflection point on the moment diagram for the left span. Depending on bracing conditions this length could be critical for lateral stability. In addition, this load case will produce the minimum reaction at the left support. For a large live load and a long cantilever length, it is possible to develop an uplift reaction at this support.

The case of unbalanced live loads can complicate the design of cantilever beam systems. This is particularly true if deflections are considered. When unbalanced live loads are required, the optimum cantilever span length L_c will be different from those established for the same uniform load on all spans.

In a cantilever system, the compression edge of the member is not always on the top of the beam. This will require a *lateral stability* analysis of bending stresses even though the top of the girder may be connected to the horizontal diaphragm. See Example 6.31.

EXAMPLE 6.31   Lateral Stability of Cantilever Systems

Moment diagram sign convention:

Positive moment = compression on top of beam

Negative moment = compression on bottom of beam

In areas of negative moment (Fig. 6.27a), the horizontal diaphragm is connected to the tension edge of the beam, and this does not provide lateral support to the compression edge of the member. If the lower face of the beam is braced (Fig. 6.27b) at the interior column, the unbraced length l_u for evaluating lateral stability is the cantilever length L_c, or it is the distance from the column to the inflection point (IP). For the given beam these unbraced lengths are equal (Fig. 6.27a) under balanced loading. If lateral stability considerations cause a large reduction in the adjusted bending design value, additional diagonal braces from the diaphragm to the bottom face of the beam may be required.

Several types of knee braces can be used to brace the bottom edge of the beam. A *prefabricated metal knee brace* and a *lumber knee brace* are shown in Fig. 6.27b. The distance between knee braces, or the distance between a brace and a point of zero moment, is the unbraced length. For additional information on unbraced lengths, see Ref. 6.5.

In order to avoid the use of diagonal braces for aesthetic reasons, some designers use a beam-to-column connection which is designed to provide lateral support to the bottom

Figure 6.27a Unbraced length considerations for negative moment.

FRAMING   SHEATHING

METAL
KNEE BRACE   GLULAM BEAM

FRAMING   SHEATHING

LUMBER
KNEE BRACE   METAL HANGER

GLULAM BEAM

TYP. BRACES FOR
NEGATIVE MOMENT

Figure 6.27b   Methods of bracing bottom edge of beam.

face of the beam. Considerable care and engineering judgment must be exercised in the design of this type of connection to ensure effective lateral restraint.

6.18   Lumber Roof and Floor Decking

Lumber sheathing (1-in. nominal thickness) can be used to span between closely spaced roof or floor beams. However, plywood and other wood structural panels are more often used for this application. Plywood and other panel products are covered in Chap. 8.

Timber Decking is used for longer spans. It is available as *solid decking* or *laminated decking*. Solid decking is made from dry lumber and is available in several grades in a number of commercial wood species. Common sizes are 2×6, 3×6, and 4×6 (nominal sizes). Various types of edge configurations are available, but tongue-and-groove (T&G) edges are probably the most common. See Fig. 6.28. A single T&G is used on 2-in.-nominal decking, and a double T&G is used on the larger thicknesses.

Glued laminated decking is fabricated from three or more individual laminations. Laminated decking also has T&G edge patterns.

Figure 6.28   Solid lumber decking. Decking can be obtained with various surface patterns if the bottom side of the decking is architecturally exposed. These sketches show a V-joint pattern.

Decking essentially functions as a series of parallel beams that span between the roof or floor framing. Bending stresses (or capacities) or deflection criteria usually govern the maximum loads on decking. Spans range from 3 to 20 ft and more depending on the load, span type, grade, and thickness of decking. The *layup* of decking affects the load capacity. See Example 6.32. It has been noted elsewhere that decking is graded for bending about the minor axis of the cross section.

EXAMPLE 6.32   Layup of Decking

Layup refers to the arrangement of end joints in decking. Five different layups are defined in Ref. 6.5, and three of these are shown in Fig. 6.29. Controlled random layup is economical and simply requires that end joints in adjacent courses be well staggered. Minimum end-joint spacing is 2 ft for 2-in. nominal decking and 4 ft for 3- and 4-in. nominal decking. In addition, end joints that occur on the same transverse line must be separated by at least two courses. End joints may be mechanically interlocked by matched T&G ends or by wood or metal splines to aid in load transfer. For other requirements see Ref. 6.5.

Figure 6.29 Three forms of layup for decking.

The TCM gives bending and deflection coefficients for the various types of layups. These can be used to calculate the required thickness of decking. However, the designer can often refer to allowable spans and load tables for decking requirements. IBC Table 2308.10.9 gives the allowable spans for 2-in. T&G decking. Reference 6.5 includes allowable load tables for simple span and controlled random layups for a variety of thicknesses.

6.19 Fabricated Wood Components

Several fabricated wood products are covered in considerable detail in this book. These include glulam (Chap. 5), plywood, and other wood structural panel products (Chap. 8). In addition to these, a number of other fabricated wood elements can be used as beams in a roof or floor system. Many of these components are produced as proprietary products, and consequently design criteria and material properties may vary from manufacturer to manufacturer.

The purpose of this section is simply to describe some of the wood components that may be used in typical wood-frame buildings. The structural design of some of these products may be performed by the manufacturer. For example, the design engineer for a building may decide to use a certain system in a roof application. After the spacing of members has been established and the loading has been determined, the engineering staff of the supplier may design the components to perform in the specified manner.

For other components, the project engineer may use certain information supplied by the manufacturer to determine the size of the required structural member. The information provided by the manufacturer can take the form of load/span tables or reference design values and effective section properties. Cooperation between the designer and the supplier is recommended in the early planning stages. The designer should also verify local building code recognition of the product and the corresponding design criteria.

The fabricated wood components covered in this section are

1. Structural composite lumber (SCL)
   a. Laminated veneer lumber (LVL)
   b. Parallel strand lumber (PSL)

2. Prefabricated wood I-joists

3. Light-frame wood trusses

4. Fiber-reinforced glulam

Structural composite lumber (SCL) refers to engineered lumber that is produced in a manufacturing plant. Although glulam was described in Chap. 5 as a composite material, the term *structural composite lumber* generally refers to a reconstituted wood product made from much smaller pieces of wood. It is fabricated by gluing together thin pieces of wood that are dried to a low moisture content. The glue is a waterproof adhesive. As a result of the manufacturing process, SCL is dimensionally stable and has less variability than sawn lumber.

The design values for glulams are generally higher than those for solid sawn lumber, and design values for structural composite lumber are higher than those for glulam. Reference bending values F_b for SCL range from 2300 to 3200 psi, and reference shear values F_v range from 150 to 290 psi. Current practice involves production of two general types of SCL which are known as laminated veneer lumber and parallel or oriented strand lumber. The basic design process for SCL is covered in the *ASD/LRFD Manual for Engineered Wood Construction* (Ref. 6.1).

Laminated veneer lumber (LVL) is similar in certain respects to glulam and plywood. It is fabricated from veneer similar to that used in plywood. The veneer typically ranges in thickness between $\frac{1}{10}$ and $\frac{1}{6}$ in. and is obtained from the rotary cutting process illustrated in Fig. 8.3. Laminated veneer lumber generally makes use of the same species of wood used in the production of structural plywood (i.e., Douglas Fir-Larch and Southern Pine).

Unlike plywood which is cross-laminated, the veneers in LVL are laid up with the wood fibers all running in one direction (i.e., parallel to the length of the member). The parallel orientation of the wood fiber is one reason for the high reference design values for LVL. Selective grading of veneer and the dispersion of defects as part of the manufacturing process (similar to the dispersion of defects in glulam, see Fig. 5.3) are additional reasons for the higher design values. The layup of veneer for LVL can also follow a specific pattern similar to glulam to meet strength requirements.

LVL is produced in either a continuous-length manufacturing operation or in fixed lengths. Fixed lengths are a function of the press size in a manufacturing plant. However, any desired length can be obtained by end jointing members of fixed lengths. Laminated veneer lumber is produced in boards or billets that can range from $3/4$ to $3\,1/2$ in. thick and may be 4 ft wide and 80 ft long. A billet is then sawn into sizes as required for specific applications. See Example 6.33.

EXAMPLE 6.33   Laminated Veneer Lumber

Laminated veneer lumber is fabricated from sheets of veneer that are glued into panels called billets. Unlike the cross-lamination of veneers in plywood (Sec. 8.3), LVL has the direction of the wood grain in all veneers running parallel with the length of the billet. Pieces of LVL are trimmed from the billet for use in a variety of applications. See Figs. 6.30a and 6.30b.

Figure 6.30a   Billet of laminated veneer lumber.

Figure 6.30b   Typical uses of LVL

Uses of laminated veneer lumber include beams, joists, headers, and scaffold planking. Beams and headers may require multiple thicknesses of LVL to obtain the necessary member width. LVL can also be used for the higher-quality tension laminations in glulams. Additional applications include flanges of prefabricated wood I-joists and chords of trusses. Two LVL beams are shown in Fig. 6.30c.

Figure 6.30c   Laminated veneer lumber beams. (*Photo courtesy of Truss Joist—A Weyerhaeuser Company.*)

The use of LVL is economical where the added expense of the material is offset by its increased strength and greater reliability.

There are two types of *parallel strand lumber* (PSL) currently in production. One type is made from the same species of wood used for plywood (i.e., Douglas Fir-Larch and Southern Pine). It starts with a sheet of veneer, which is clipped into narrow *strands* that are approximately $\frac{1}{2}$ in. wide and up to 8-ft long. The strands are dried, coated with a waterproof adhesive, and bonded together under pressure and heat. The strands are aligned so that the wood grain is parallel to the length of the member (hence the name).

The second type of PSL is also known as *oriented strand lumber* (OSL) and is made from small-diameter trees of Aspen that previously could not be used in structural applications because of size. Flaking machines are used for small-diameter logs (instead of veneer peelers) to produce wood flakes that are approximately $\frac{1}{2}$ in. wide, 0.03 in. thick, and 1 ft long. The flakes are also glued and bonded together under pressure and heat.

Both forms of parallel strand lumber (i.e., the types made from *strands* or *flakes*) result in a final piece called a billet. Billets of PSL are similar to those of LVL (Fig. 6.30*a*), but the sizes are different. Billets of PSL can be as large as 12 in. wide, 17 in. deep, and 60 ft long. Again, final sizes for field applications are obtained by sawing the billet. Parallel strand lumber may be used alone as high-grade structural lumber of beams and columns. See Example 6.34. It may also be used in the fabrication of other structural components similar to the LVL applications in Example 6.33.

EXAMPLE 6.34   Parallel Strand Lumber

Parallel strand lumber is manufactured from strands or flakes of wood with the grain parallel to the length of the member. High-quality wood members in large sizes are possible with this form of SCL. See Fig. 6.31*a*. Applications include beams and columns which can be left architecturally exposed. See Fig. 6.31*b*.

Figure 6.31*a* Parallel strand lumber. (*Photo courtesy of Trus Joist—A Weyerhaeuser Company.*)

Figure 6.31*b* Beams and columns of PSL. (*Photo courtesy of Trus Joist—A Weyerhaeuser Company.*)

The use of prefabricated wood components has increased substantially in recent years. The most widely used form of these composite members is the wood I-Joist. See Example 6.35. Wood I-joists are efficient bending members for two reasons. First, the cross section is an efficient shape. The most popular steel beams (W shapes and S shapes) have a similar configuration. The large flange areas are located away from the neutral axis of the cross section, thus increasing the moment of inertia and section modulus. In other words, the shape is efficient because the flanges are placed at the point in the cross section where the material does the most good: at the point of maximum flexural stress. The relatively thin web is satisfactory as long as it has adequate shear strength.

Second, wood I-joists are efficient from a material usage standpoint. The flanges are stressed primarily in tension and compression as a result of the flexural stresses in the member. The material used for the flanges in wood I-joists has high tensile and compressive strengths. Some manufacturers use sawn lumber flanges, but laminated veneer lumber flanges are common.

Although the bending moment is primarily carried by the flanges, it is noted that the shear in I-beams is essentially carried by the web. Wood I-joists also gain efficiency by using web materials that are strong in shear. Plywood and oriented strand board panels are used in other high-shear applications such as horizontal diaphragms (Chap. 9) and shearwalls (Chap. 10), in addition to being used as the web material in fabricated wood I-beams.

Additional information on the design of wood composite I-joists is provided in the *ASD/LRFD Manual* (Ref. 6.1). Supplemental design considerations (specific to wood I-joists) and recommended installation details are also provided in this Manual.

EXAMPLE 6.35   Prefabricated Wood I-Joists

Initially, prefabricated wood I-joists were constructed with solid sawn lumber flanges and plywood webs. However, more recently I-joists are produced from some of the newer wood products. For example, laminated veneer lumber (LVL) is used for flanges and oriented strand board (OSB) for web material (Fig. 6.32a).

Figure 6.32a Typical prefabricated wood I-joists. (*Photo courtesy of Louisiana-Pacific Corporation.*)

Prefabricated wood I-joists have gained wide acceptance in many areas of the country for repetitive framing applications (Figs. 6.32*b* and 6.32*c*). Web stiffeners for wood I-joists may be required to transfer concentrated loads or reactions in bearing through the flange and into the web. Prefabricated metal hardware is available for a variety of connection applications. Because of the slender cross section of I-joists, particular attention must be paid to stabilizing the members against translation and rotation. The manufacturer's recommendations for bracing and blocking should be followed in providing stability for these members.

Figure 6.32*b* Wood I-joists supported on LVL header. (*Photo courtesy of Trus Joist—A Weyerhaeuser Company.*)

Figure 6.32*c* Wood I-joists as part of a wood roof system in a building with masonry walls. (*Photo courtesy of Trus Joist—A Weyerhaeuser Company.*)

Wood I-joists make efficient use of materials, and because of this they are relatively lightweight and easy to handle by crews in the field. In addition to strength, the depth of the cross section provides members that are relatively stiff for the amount of material used. Wood I-joists can be used to span up to 40 or 50 ft, but many applications are for shorter spans. Wood I-joists can be deep and slender, and care should be taken in the installation of these members to ensure adequate stability. Information on the design of lumber and plywood beams is available in Refs. 6.8 and 6.9. Additional information on the design and installation of wood I-joists is available from individual manufacturers.

Wood trusses represent another common type of fabricated wood component. Heavy wood trusses have a long history of performance, but light wood trusses are more popular today. The majority of residential wood structures, and many commercial and industrial buildings, use some form of closely spaced light wood trusses in roof and floor systems. Common spans for these trusses range up to 75 ft, but larger spans are possible. The spacing of trusses is on the order of 16 to 24 in. o.c. for floors and up to 8 ft o.c. in roof systems.

Some manufacturers produce trusses that have wood top and bottom chords and steel web members. However, the majority of truss manufacturers use light-gage toothed metal plates to connect wood chords and wood web members. See Example 6.36. The metal plates have teeth that are produced by stamping the metal plates. The metal plates are placed over the members to be connected together and the teeth are pressed into the wood.

EXAMPLE 6.36   Light-Frame Wood Trusses

Trusses can be manufactured with sawn lumber or LVL chords and steel web members (Fig. 6.33a). Trusses can be supported in a variety of ways. The top or bottom chord of a truss can bear on wood walls or beams, on steel beams, or on concrete or masonry walls. Another method is to suspend the truss from a ledger attached to a concrete or masonry wall.

Figure 6.33a Wood trusses with tubular steel webs. Trusses in the foreground are supported on a wood member attached to a steel W-shape beam. In the background, trusses rest on glulam. (*Photo courtesy of Trus Joist—A Weyerhaeuser Company*).

Metal plate connected trusses (Fig. 6.33*b*) use toothed or barbed plates to connect the truss members. See Fig. 11.3 in Sec. 11.2 for a photograph of metal plate connectors. Typically, the metal plates are assigned a unit load capacity (lb/in.^{2} of contact area). Thus, the required plate size is determined by dividing the forces to be transferred through the connection by the unit load capacity of the metal plate.

Figure 6.33*b*   Metal plate connected trusses being placed in roof system supported on masonry walls. (*Photo courtesy of Alpine Engineered Products, Inc.*)

Light wood trusses are rather limber elements perpendicular to their intended plane of loading. Because of this flexibility, proper handling procedures are required in the field to avoid damage to the truss during erection. The use of strongbacks with a sufficient number of pickup points for lifting the trusses into place will avoid buckling of the truss about its weak axis during installation. Once a truss is properly positioned, it must be braced temporarily until the sheathing and permanent bracing are in place (Ref. 6.11). Trusses which are not adequately braced can easily buckle or rotate. However, once the bracing is in place, the trusses provide a strong, stiff, and economical wood framing system.

The Truss Plate Institute (TPI) is the technical trade association of the metal plate truss industry. TPI also publishes the comprehensive "National Design Standard for Metal Plate Connected Wood Trusses" (Ref. 6.12), which provides detailed design requirements for these trusses. Among other considerations, this standard requires that the continuity of the chords at the joints in the truss be taken into account. In addition to its design standard, TPI publishes other pertinent literature such as the "Commentary and Recommendations for Handling, Installing, and Bracing Metal Plate Connected Wood Trusses" (Ref. 6.11). Additional information on wood trusses and bracing can be obtained from individual truss manufacturers.

Fiber-reinforced glulam is one of the newer products to be developed. While, in general, many different materials can be used to reinforce glulam, the use of *fiber-reinforced polymers* (FRPs) has proven to be the most effective reinforcement. FRPs consist of synthetic fibers (including glass, carbon, and graphite) and a thermoplastic polymer that serves as a binder, holding the fibers together.

Fiber-reinforced glulam is actually a modification to traditional glulam where FRP sheets are placed between laminations, particularly in the tension region, to improve performance. See Example 6.37. Specifically, the advantages of integrating FRP sheets into a glulam layup include increased strength and stiffness, increased ductility, reduced creep, and reduced overall variability.

EXAMPLE 6.37   Fiber-Reinforced Glulam

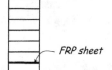
FRP sheet

Figure 6.34 Fiber-reinforced glulam with FRP sheet between tension laminations.

As discussed in Chap. 5, glulam is an engineered wood composite that allows more efficient use of materials and provides larger size members. With the addition of FRP sheets, an increase in strength of the section is possible, but the primary advantages are increased stiffness, increased ductility, reduced creep, and reduced overall variability. Since deflections can oftentimes control the size of a member, the increase in stiffness and reduced creep allow for smaller sections to be required. The reduction in variability can result in higher design values as well. In addition, the increase in ductility provides for a potentially safer failure mechanism. These same attributes can also allow the utilization of low-quality wood in glulam without reduction in overall design values. That is, even though the use of low-quality wood in a glulam may translate to a reduced set of design values, this reduction may be offset by the improved performance resulting from the addition of FRP.

Fiber-reinforced glulam is a proprietary product that is not covered in the NDS. Additional information on the use and design of fiber-reinforced glulam can be obtained from APA—The Engineered Wood Association, from AITC or from the manufacturers.

The use of FRPs to improve the performance of wood materials is not limited to glulam. The combination of FRPs with SCL, I-joists, and wood panels have all been used with varying degrees of success.

6.20   References

[6.1]   American Forest and Paper Association (AF&PA). 2006. ASD/LRFD *Manual For Engineered Wood Construction* 2005 ed., AF&PA, Washington, DC.

[6.2]   American Forest and Paper Association (AF&PA). 2005. *National Design Specification for Wood Construction and Supplement.* ANSI/AF&PA NDS-2005, AF&PA, Washington DC.

[6.3]   American Institute of Steel Construction (AISC). 2005. *Steel Construction Manual,* 13th ed., AISC, Chicago, IL.

[6.4]   American Institute of Timber Construction (AITC). 2004. *Standard Specifications for Structural Glued Laminated Timber of Softwood Species. Design and Manufacturing Requirements. AITC 117-*2004, AITC, Englewood, CO.

[6.5]   American Institute of Timber Construction (AITC). 2005. *Timber Construction Manual,* 5th ed., John Wiley & Sons Inc., Hoboken, NJ.

[6.6]   American Society for Testing and Materials (ASTM). 2006. "Standard Test Methods for Small Clear Specimens of Timber," ASTM D143-94 (Reapproved 2000), *Annual Book of Standards, Vol. 04.10 Wood*, ASTM, Philadelphia, PA.

[6.7]   American Society for Testing and Materials (ASTM). 2006. "Standard Practice for Establishing Structural Grades and Related Allowable Properties for Visually Graded Lumber," ASTM D245-06, *Annual Book of Standards, Vol. 04.10 Wood,* ASTM, Philadelphia, PA.

[6.8]   APA—The Engineered Wood Association. 2004. *Panel Design Specification, APA Form D510A*, APA—The Engineered Wood Association, Engineered Wood Systems, Tacoma, WA.

[6.9]   APA—The Engineered Wood Association. 1998. *Plywood Design Specification and Supplements 1-5, APA Form No. Y510T*, APA—The Engineered Wood Association, Tacoma, WA.

[6.10] International Codes Council (ICC). 2006. *International Building Code,* 2006 ed., ICC, Falls Church, VA.

[6.11] Truss Plate Institute (TPI). 1991. *Commentary and Recommendations for Handling, Installing and Bracing Metal Plate Connected Wood Trusses, HIB-91,* TPI, Madison, WI.

[6.12] Truss Plate Institute (TPI). 2002. *National Design Standard for Metal Plate Connected Wood Truss Construction, ANSI/TPI 1-2002,* TPI, Madison, WI.

[6.13] Zahn, J.J. 1991. "Biaxial Beam-Column Equation for Wood Members," *Proceedings of Structures Congress '91*, American Society of Civil Engineers, Reston, Virginia pp. 56–59.

6.21   Problems

Reference design values and section properties for the following problems are to be in accordance with the 2005 NDS. Dry-service conditions, normal temperatures, and bending about the strong axis apply unless otherwise indicated.

Some problems require the use of spreadsheet or equation-solving software. Problems that are solved on a spreadsheet or equation-solving software can be saved and used as a *template* for other similar problems. Templates can have many degrees of sophistication. Initially, a template may only be a hand (i.e., calculator) solution worked on a computer. In a simple template of this nature, the user will be required to provide many of the lookup functions for such items as

Reference design values

Lumber dimensions

Load duration factor (ASD), time effect factor (LRFD)

Wet service factor

Size factor

Volume factor

Format conversion factor (LRFD)

Resistance factor (LRFD)

As the user gains experience with the chosen software, the template can be expanded to perform lookup and decision-making functions that were previously done manually.

Advanced computer programming skills are not required to create effective spreadsheet or equation-solving templates. Valuable templates can be created by designers who normally do only hand solutions. However, some programming techniques are helpful in automating *lookup* and *decision-making* steps.

The first requirement is that a template operate correctly (i.e., calculate correct values). Another major consideration is that the input and output be structured in an orderly manner. A sufficient number of intermediate answers should be displayed and labeled so that the solution can be verified by hand.

6.1 *Given:* The roof beam in Fig. 6.A with the following information:

Load:	P_D = 400 lb
	P_{Lr} = 1600 lb
Load combination:	$D + L_r$ (ASD)
	$1.2D + 1.6L_r$ (LRFD)
Span:	L = 8 ft
Member size:	4 × 8
Stress grade and species:	No. 1 DF-L
Unbraced length:	$l_u = 0$
Moisture content:	MC ≤ 19 percent
Live load deflection limit:	Allow. $\Delta \le L/360$

Find: *a.* Size category (Dimension lumber, B&S, or P&T)
   b. Reference design values: F_b, F_v and E
   c. Adjusted ASD values: F'_b, F'_v and E'
   d. Actual stresses (ASD) and deflection: f_b, f_v, and Δ
   e. Compare the actual stresses and adjusted design values, and determine if the member is adequate (ASD).
   f. Nominal LRFD values: F_{bn}, F_{vn}, and E
   g. Adjusted LRFD values: F'_{bn}, F'_{vn}, and E'
   h. Adjusted LRFD moment and shear resistances: M'_n and V'_n
   i. Factored moment and shear (LRFD), and actual deflection: M_u V_u, and Δ
   j. Determine if the member is adequate (LRFD).

Figure 6.A

6.2 Repeat Prob. 6.1 except the moisture content exceeds 19 percent.

6.3 Repeat Prob. 6.1 except the unbraced length is $l_u = L/2 = 4$ ft. $C_M = 1.0$.

6.4 Use the hand solution to Probs. 6.1 and 6.3 as a guide to develop a personal computer template to solve similar problems using ASD and LRFD.
   a. Consider only the specific criteria given in Probs. 6.1 and 6.3.
   b. Expand the template to handle any

      Span length L
      Magnitude load P or P_u

Unbraced length l_u
Sawn lumber trial member size

The template is to include a list (i.e., database) of reference design values for all size categories (Dimension lumber B&S, P&T) of No. 1 DF-L.

c. Expand the database in part b to include all grades of DF-L from No. 2 through Select Structural.

6.5   Repeat Prob. 6.1 except the unbraced length is $l_u = L = 8$ ft. $C_M = 1.0$.

6.6   *Given:*   The roof beam in Fig. 6.A with the following information:

Load:	$P_D = 500$ lb
	$P_S = 1000$ lb
Load combination:	D + S (ASD)
	1.2D + 1.6S (LRFD)
Span:	$L = 24$ ft
Member size:	$3^1/_8 \times 21$
Bending Stress Class glulam:	24F-1.8E
Unbraced length:	$l_u = 0$
Moisture content:	MC <16 percent
Snow load (alone) deflection limit:	Allow. $\Delta \leq L/240$

Find: a. Reference design values: F_b, F_v, E_x, and E_y
b. Adjusted ASD values: F'_b, F'_v, E'_x, and E'_y
c. Actual stresses (ASD) and deflection: f_b, f_v, and Δ
d. Compare the actual stresses and adjusted design values and determine if the member is adequate (ASD). How much camber should be provided?
e. Nominal LRFD values: F_{bn}, F_{vn}, E_x, and E_y
f. Adjusted LRFD values: F'_{bn}, F'_{vn}, E'_x, and E'_y
g. Adjusted LRFD moment and shear resistances: M'_n and V'_n
h. Factored moment and shear (LRFD), and actual deflection: M_u, V_u, and Δ
i. Determine if the member is adequate (LRFD). How much camber should be provided?

6.7   Repeat Prob. 6.6 except the moisture content exceeds 16 percent.

6.8   Repeat Prob. 6.6 except the unbraced length is $l_u = L/2 = 12$ ft. $C_M = 1.0$.

6.9   Use the hand solution to Probs. 6.6 and 6.8 as a guide to develop a personal computer template to solve similar problems using ASD and LRFD.
a. Consider only the specific criteria given in Probs. 6.6 and 6.8.
b. Expand the template to handle any

Span length L
Magnitude load P or P_u
Unbraced length l_u
Size Western Species bending combination glulam

The template is to include a list (i.e., database) of reference design values for glulam bending Stress Classes 16F-1.3E, 20F-1.5E, and 24F-1.8E.

6.10   *Given:*   The roof beam in Fig. 6.B with the following information:

Load:	w_D = 200 lb/ft
	w_{L_r} = 250 lb/ft
Load combination:	D + L_r (ASD)
	1.2D + 1.6L_r (LRFD)
Span:	L = 10 ft
Member size:	4 × 10
Stress grade and species:	Sel. Str. Hem-Fir
Unbraced length:	l_u = 0
Moisture content:	MC ≤ 19 percent
Deflection limit:	Allow. Δ_L ≤ L/360
	Allow. $\Delta_{(D + L)}$ ≤ L/240

Find: *a.* Size category (Dimension lumber, B&S, or P&T)
b. Reference design values: F_b, F_v, and E
c. Adjusted ASD values: F'_b, F'_v, and E'
d. Actual stresses (ASD) and deflection: f_b, f_v, and Δ
e. Compare the actual stresses and adjusted design values, and determine if the member is adequate (ASD).
f. Nominal LRFD values: F_{bn}, F_{vn}, and E
g. Adjusted LRFD values: F'_{bn}, F''_{vn}, and E'
h. Adjusted LRFD moment and shear resistances: M'_n and V'_n
i. Factored moment and shear (LRFD), and actual deflection: M_u, V_u, and Δ
j. Determine if the member is adequate (LRFD).

6.11   Repeat Prob. 6.10 except the moisture content exceeds 19 percent.

Figure 6.B

6.12   Repeat Prob. 6.10 except the unbraced length is l_u = L/2 = 5 ft.

6.13   Use the hand solution to Probs. 6.10 and 6.12 as a guide to develop a personal computer template to solve similar problems using ASD and LRFD.
a. Consider only the specific criteria given in Probs. 6.10 and 6.12.
b. Expand the template to handle any

Span length L
Magnitude load w or w_u
Unbraced length l_u
Sawn lumber trial member size

The template is to include a list (i.e., database) of reference design values for all size categories (Dimension lumber, B&S, P&T) of Sel. Str. Hem-Fir.
c. Expand the database in part *b* to include all grades of Hem-Fir from No. 2 through Select Structural.

6.14   *Given:*   The roof beam in Fig. 6.B with the following information:

Load:	w_D = 200 lb/ft
	w_S = 300 lb/ft
Load combination:	D + S (ASD)
	1.2D + 1.6S (LRFD)
Span:	L = 20 ft
Member size:	5 × 19$^{1/4}$
Glulam bending Stress Class:	24F-1.7E SP
Unbraced length:	l_u = 0
Moisture content:	MC < 16 percent
Deflection limit:	Allow. $\Delta_S \leq L/360$
	Allow. $\Delta_{(D + S)} \leq L/240$

   Find: a. Reference design values: F_b, F_v, E_x, and E_y
   b. Adjusted ASD values: F'_b, F'_v, E'_x, and E'_y
   c. Actual stresses (ASD) and deflection: f_b, f_v, and Δ
   d. Compare the actual stresses and adjusted design values, and determine if the member is adequate (ASD). How much camber should be provided?
   e. Nominal LRFD values: F_{bn}, F_{vn}, E_x, and E_y
   f. Adjusted LRFD values: F'_{bn}, F'_{vn}, E'_x, and E'_y
   g. Adjusted LRFD moment and shear resistances: M'_n and V'_n
   h. Factored moment and shear (LRFD), and actual deflection: M_u, V_u, and Δ
   i. Determine if the member is adequate (LRFD). How much camber should be provided?

6.15   Repeat Prob. 6.14 except the moisture content exceeds 16 percent.

6.16   Repeat Prob. 6.14 except the unbraced length is $l_u = L/2 = 10$ ft $C_M = 1.0$.

6.17   Use the hand solution to Probs. 6.14 and 6.16 as a guide to develop a personal computer template to solve similar problems using ASD and LRFD.
   a. Consider only the specific criteria given in Probs. 6.14 and 6.16.
   b. Expand the template to handle any

   Span length L
   Magnitude load w or w_u
   Unbraced length l_u
   Size Southern Pine bending combination glulam

   The template is to include a list (i.e., database) of reference design values for glulam bending Stress Classes 16F-1.3E, 20F-1.5E, and 24F-1.8E SP.

6.18   *Given:*   The floor beam in Fig. 6.C with the following information:

Load:	P_D = 400 lb
	P_L = 1600 lb
Load combination:	D + L (ASD)
	1.2D + 1.6L (LRFD)
Span:	L_1 = 8 ft
	L_2 = 4 ft
Member size:	4 × 12

Stress grade and species: Sel. Str.   SP
Unbraced length: $l_u = 0$
Moisture content: $MC \leq 19$ percent
Deflection limit: Allow. $\Delta_{\text{free end}} \leq L_2/180$
                            Allow. $\Delta_{\text{between supports}} \leq L_1/360$

Find: *a.* Size category (Dimension lumber, B&S, or P&T)
 b. Reference design values: F_b, F_v, and E
 c. Adjusted ASD values: F'_b, F'_v, and E'
 d. Actual stresses (ASD) and deflection: f_b, f_v, and Δ
 e. Compare the actual stresses and adjusted design values, and determine if the member is adequate (ASD).
 f. Nominal LRFD values: F_{bn}, F_{vn}, and E
 g. Adjusted LRFD values: F'_{bn}, F'_{vn}, and E'
 h. Adjusted LRFD moment and shear resistances: M'_n and V'_n
 i. Factored moment and shear (LRFD), and actual deflection: M_u, V_u, and Δ
 j. Determine if the member is adequate (LRFD).

6.19 Repeat Prob. 6.18 except the moisture content exceeds 19 percent.

Figure 6.C

6.20 Repeat Prob. 6.18 except lateral support is provided at the vertical supports and at the free end.

6.21 Use the hand solution to Probs. 6.18 and 6.20 as a guide to develop a personal computer template to solve similar problems using ASD and LRFD.
a. Consider only the specific criteria given in Probs. 6.18 and 6.20.
b. Expand the template to handle any

    Span lengths L_1 and L_2
    Magnitude load P or P_u
    Unbraced length l_u
    Sawn lumber trial member size

    The template is to include a list (i.e., database) of reference design values for all size categories (Dimension lumber, B&S, P&T) of Sel. Str. SP.
c. Expand the database in part *b* to include the stress grades of No. 2, No. 1, and Select Structural Southern Pine.

6.22 A series of closely spaced floor beams is to be designed. Loading is similar to Fig. 6.B. The following information is known:

Load: $w_D = 18$ psf

                                                   $w_L = 50$ psf

Load combination: D + L (ASD)

                                                  1.2D + 1.6L (LRFD)

Span: $L = 14$ ft

Member spacing: Trib. width = b = 16 in. o.c.

Stress grade and species: No. 1 Hem-Fir

Unbraced length: $l_u = 0$

Moisture content: MC $\leq$ 19 percent

Deflection limit: Allow. $\Delta_L \leq L/360$

                                           Allow. $\Delta_{(D + L)} \leq L/240$

Find: Minimum beam size. As part of the solution also give

       a. Size category (Dimension lumber, B&S, or P&T)

       b. Reference design values: F_v, F_b, and E

       c. Adjusted ASD values: F'_b, F'_v, and E'

       d. Actual stresses (ASD) and deflection: f_b, f_v, and Δ

       e. Nominal LRFD values: F_{bn}, F_{vn}, and E

       f. Adjusted LRFD values: F'_{bn}, F'_{vn}, and E'

       g. Adjusted LRFD moment and shear resistances: M'_n and V'_n

       h. Factored moment and shear (LRFD), and actual deflection: M_u, V_u, and Δ

6.23   For the beam designed in Prob. 6.22, determine the size of notches allowed by the NDS on both the tension and compression side at (a) the supports or ends of the member and (b) in the interior of the span. Determine the shear capacity at the support, if a 1-in. deep notch were assumed at the support. Use ASD procedures.

6.24   If the member designed in Prob. 6.22 were a glulam, determine the size of notches allowed by the NDS on both the tension and compression side at (a) the supports or ends of the member and (b) in the interior of the span.

6.25   *Given:*   The roof rafters in Fig. 6.D are 24 in o.c. Roof dead load is 12 psf along the roof, and roof live load is in accordance with the IBC. See Example 2.4. Calculate design shear and moment, using the horizontal plane method of Example 2.6 (see Fig. 2.6*b*). Lumber is No. 2 DF-L. Lateral stability is not a problem. Disregard deflection. $C_M = 1.0$ and $C_t = 1.0$.

      Find: Minimum rafter size. As part of the solution also give

         a. Size category (Dimension lumber, B&S, or P&T)

         b. Reference design values: F_b and F_v

         c. Adjusted ASD values: F'_b and F'_v

         d. Actual stresses (ASD): f_b and f_v

         e. Nominal LRFD values: F_{bn} and F_{vn}

         f. Adjusted LRFD values (LRFD): F'_{bn} and F'_{vn}

         g. Adjusted LRFD moment and shear resistances: M'_n and V'_n

         h. Factored moment and shear (LRFD): M_u and V_u

6.26   Repeat Prob. 6.25 except that the rafters are spaced 6 ft-0 in. o.c.

6.27 *Given:* The roof rafters in Fig. 6.D are 24 in. o.c. The roof dead load is 15 psf along the roof, and the design snow load is 50 psf. Calculate the design shear and moment, using the horizontal plane method of Example 2.6 (see Fig. 2.6*b*). Disregard deflection. Lateral stability is not a problem. Lumber is No. 1 DF-L. $C_M = 1.0$ and $C_t = 1.0$.

Find: The Minimum rafter size. As part of the solution also give
    a. Size category (Dimension lumber, B&S, or P&T)
    b. Reference design values: F_b and F_v
    c. Adjusted ASD values: F'_b and F'_v
    d. Actual stresses (ASD): f_b and f_v
    e. Nominal LRFD values: F_{bn} and F_{vn}
    f. Adjusted LRFD values: F'_{bn} and F'_{vn}
    g. Adjusted LRFD moment and shear resistances: M'_n and V'_n
    h. Factored moment and shear (LRFD): M_u and V_u

SECTION

Figure 6.D

6.28 *Given:* The beam in Fig. 6.E has the compression side of the member supported laterally at the ends and midspan only. The span length is $L = 25$ ft. The member is a $3^{1}/_{8} \times 18$ DF glulam Stress Class 24F-1.8E. The load is a combination of D + L (ASD) or 1.2D + 1.6L (LRFD). $C_M = 1.0$ and $C_t = 1.0$.

Find: *a.* The maximum allowable bending moment in ft-k and the corresponding maximum allowable load P in k (ASD).
    b. The maximum factored moment M_u in ft-k and the corresponding maximum factored load P_u in k (LRFD).

Figure 6.E

6.29 *Given:* The beam in Fig. 6.E has the compression side of the member supported laterally at the ends and the quarter points. The span length is $L = 24$ ft. The member is a resawn glulam $2^{1}/_{2} \times 19^{1}/_{2}$ DF 24F-1.8E. The load is a combination of D + L (ASD) or 1.2D + 1.6L (LRFD). $C_M = 1.0$ and $C_t = 1.0$.

   Find: a. The maximum allowable bending moment in ft-k and the corresponding maximum allowable load P in k (ASD).
   b. The maximum factored moment M_u in ft-k and the corresponding maximum factored load P_u in k (LRFD).

6.30 Repeat Prob. 6.28 except that the member is a $3 \times 17^{7}/_{8}$ + Southern Pine glulam 24F-1.8E, and the load is a combination of D + S (ASD) or 1.2D + 1.6S (LRFD).

6.31 *Given:* The rafter connection in Fig. 6.F. The load is a combination of $P_D = 140$ lb and $P_S = 560$ lb. Consider the D + S (ASD) or 1.2D + 1.6S (LRFD) load combination. Lumber is No. 1 Spruce-Pine-Fir (South). $C_M = 1.0$ and $C_t = 1.0$.

   Find: a. The actual bearing stress $f_{c\perp}$ in the rafter and in the top plate of the wall (ASD).
   b. The adjusted ASD bearing value $F'_{c\perp}$ in the top plate.
   c. The adjusted ASD bearing value $F'_{c\theta}$ in the rafter.

Figure 6.F

        d. Factored bearing load P_u in the rafter and in the top plate of the wall (LRFD).

        e. The adjusted LRFD bearing resistance $P'_{n\perp}$ in the top plate.

        f. The adjusted LRFD bearing resistance $P'_{n\theta}$ in the rafter.

6.32 *Given:* The rafter connection in Fig. 6.F with the slope changed to $^{12}/_{12}$. The load is a combination of $P_D = 140$ lb and $P_{L_r} = 560$ lb. Consider the D + L_r (ASD) or 1.2D + 1.6 L_r (LRFD) load combination. Lumber is No. 2 DF-L that is used in a high-moisture-content condition (MC > 19 percent). $C_t = 1.0$.

      Find: *a.* The actual bearing stress $f_{c\perp}$ in the rafter and in the top plate of the wall (ASD).

        b. The adjusted ASD bearing value $F'_{c\perp}$ in the top plate.

        c. The adjusted ASD bearing value $F'_{c\theta}$ in the rafter.

        d. Factored bearing load P_u in the rafter and in the top plate of the wall (LRFD).

        e. The adjusted LRFD bearing resistance $P'_{n\perp}$ in the top plate.

        f. The adjusted LRFD bearing resistance $P'_{n\theta}$ in the rafter.

6.33 *Given:* The beam-to-column connection in Fig. 6.G. The gravity reaction from the simply supported beam is transferred to the column by bearing (not by the bolts). Assume the column and the metal bracket have adequate strength to carry the load. $C_t = 1.0$.

      Find: Using ASD, the maximum allowable beam reaction governed by bearing stresses for the following conditions:

        a. The beam is a 4 × 12 No. 1 DF-L. MC ≤ 19 percent, and the dimensions are $A = 12$ in. and $B = 5$ in. Loads are (D + S).

        b. The beam is a $5^{1}/_{8}$ × 33 DF glulam Combination 24F-V4. MC = 18 percent, and the dimensions are $A = 0$ and $B = 12$ in. Loads are (D + S).

        c. The beam is a 6 × 16 No. 1 DF-L. MC = 20 percent, and the dimensions are $A = 8$ in. and $B = 10$ in. Loads are (D + L_r).

        d. What deformation limit is associated with the bearing stresses used in parts *a* to *c*?

Figure 6.G

6.34 Repeat Prob. 6.33 using LRFD and determine the maximum factored beam reaction governed by bearing for each of the conditions.

6.35 *Given:* The beam in Fig. 6.H is a $5^{1}/_{8} \times 19.5$ DF glulam 16F-1.3E. The ASD load is (D + S). MC < 16 percent. Lateral support is provided to the top side of the member by roof sheathing. $C_t = 1.0$.

Find: Check the given member for bending and shear using ASD procedures.

Figure 6.H

6.36 Repeat Prob. 6.35 using LRFD. The LRFD load combination is 1.2D + 1.6S with the factored uniformly distributed load $w_u = 500$ lb/ft and the factored concentrated load $P_u = 3000$ lb.

6.37 *Given:* The roof framing plan of the commercial building in Fig. 6.I. There is no ceiling. The *total* dead loads to the members are

$$\text{Subpurlin}(2 \times 4 \text{ at } 24 \text{ in. o.c.}) = 7.0 \text{ psf}$$
$$\text{Purlins } (4 \times 14 \text{ at } 8 \text{ ft} - 0 \text{ in. o.c.}) = 8.5 \text{ psf}$$
$$\text{Girder} = 10.0 \text{ psf}$$

ROOF FRAMING PLAN

Figure 6.I

The roof is flat except for a minimum slope of $\frac{1}{4}$ in./ft to prevent ponding. Roof live loads are to be in accordance with the IBC. See Example 2.4. The roof diaphragm provides continuous lateral support to the top side of all beams. $C_M = 1.0$ and $C_t = 1.0$.

Find: *a.* Using ASD, check the subpurlins, using No. 1 & Better DF-L. Are the AITC-recommended deflection criteria satisfied?

   b. Using ASD, check the purlins, using No. 1 & Better DF-L. Are the AITC deflection limits met?

   c. Using ASD, design the girder, using 24F-1.8E DF glulam. Determine the minimum size, considering both strength and stiffness.

   d. Repeat parts a, b, and c using LRFD.

6.38 *Given:* The girder in the roof framing plan in Fig. 6.J is to be designed using the optimum cantilever length L_c. The girder is 20F-1.5E DF glulam. D = 16 psf. The top of the girder is laterally supported by the roof sheathing. Deflection need not be checked, but camber requirements are to be determined. The roof is flat except for a minimum slope of $\frac{1}{4}$ in./ft to prevent ponding. $C_M = 1.0$ and $C_t = 1.0$.

ROOF FRAMING PLAN

Figure 6.J

Find: *a.* Using ASD, the minimum required beam size if the girder is designed for a 20-psf roof live load with no reduction for tributary area.

   b. Using ASD, the minimum required beam size if the girder is designed for a basic 20-psf roof live load that is to be adjusted for tributary area. Roof live load is to be determined in accordance with the IBC. See Example 2.4. For the roof live load reduction, consider the tributary area of the suspended portion of the cantilever system.

   c. Repeat part a using LRFD.

   d. Repeat part b using LRFD.

Axial Forces and Combined Bending and Axial Forces

7.1 Introduction

An axial force member has the load applied parallel to the longitudinal axis through the centroid of the cross section. The axial force may be either tension or compression. Because of the need to carry vertical gravity loads down through the structure into the foundation, columns are more often encountered than tension members. Both types of members, however, see widespread use in structural design in such items as trusses and diaphragms.

In addition to the design of axial force members, this chapter covers the design of members with a more complicated loading condition. These include members with bending (beam action) occurring simultaneously with axial forces (tension or compression). This type of member is often referred to as a *combined stress* member. A combination of loadings is definitely more critical than the case of the same forces being applied individually. The case of compression combined with bending is probably encountered more often than tension plus bending, but both types of members are found in typical wood-frame buildings.

To summarize, the design of the following types of members is covered in this chapter:

1. Axial tension

2. Axial compression

3. Combined bending and tension

4. Combined bending and compression

The design of axial force tension members is relatively straightforward, and the required size of a member can be solved for directly. For the other three types of members, however, a trial-and-error solution is the typical design approach. Trial-and-error solutions may seem awkward in the beginning. However, with a little practice the designer will be able to pick an initial trial size which will be relatively close to the required size. The final selection can often be made with few trials. Several examples will illustrate the procedure used in design.

As noted, the most common axial load member is probably the column, and the most common combined stress member is the beam-column (combined bending and compression). See Fig. 7.1. In this example, an axial load is assumed to be applied to the interior column by the girder. For the exterior column, there is both a lateral force that causes bending and a vertical load that causes axial compression.

The magnitude of the lateral force (wind or seismic) to the column depends on the design load and how the wall is framed. The wall may be framed horizontally to span between columns, or it may be framed vertically to span between story levels (Example 3.4 in Sec. 3.3).

Numerous other examples of axial force members and combined load members can be cited. However, the examples given here are representative, and they adequately define the type of members and loadings that are considered in this chapter.

7.2 Axial Tension Members

Wood members are stressed in tension in a number of structural applications. For example, trusses have numerous axial force members, and roughly half of these are in tension. It should be noted that unless the loads frame directly into the joints in the truss and unless the joints are pinned, bending stresses will be developed in addition to the axial stresses obtained in the standard truss analysis.

Axial tension members also occur in the chords of shearwalls and diaphragms. In addition, tension members are used in diaphragm design when the length of the diaphragm is greater than the length of the shearwall to which it is attached. This type of member is known as a *collector* or *drag strut,* and it is considered in Chap. 9.

According to allowable stress design (ASD) principles, the check for the axial tension stress in a member of known size uses the formula

$$f_t = \frac{P}{A_n}$$
$$\leq F_t'$$

where f_t = actual (computed) tension stress parallel to grain
$\quad\quad P$ = axial tension force in member
$\quad\quad A_n$ = net cross-sectional area
$\quad\quad\quad$ = $A_g - \Sigma A_h$
$\quad\quad A_g$ = gross cross-sectional area
$\quad\quad \Sigma A_h$ = sum of projected area of holes at critical section
$\quad\quad F_t'$ = adjusted ASD tension design value parallel to grain

Figure 7.1 Example of columns and beam-columns. Both of the members in this example are vertical, but horizontal or inclined members with this type of loading are also common.

According to load and resistance factor design (LRFD) principles, the axial check for a tension member is typically cast in terms of factored resistance and loads using the formula

$$T_u \leq T'_n$$
$$\leq F'_{tn} \cdot A_n$$

where T_u = tension force in the member due to factored loads
T'_n = adjusted LRFD tension resistance
F'_{tn} = adjusted LRFD tension design value
A_n = net cross-sectional area
$\quad\ = A_g - \Sigma A_h$
A_g = gross cross-sectional area
ΣA_h = sum of projected area of holes at critical section

Regardless of whether ASD or LRFD is used, the design check for tension is usu-ally referred to as an analysis expression. In other words, an *analysis* problem involves *checking* the adequacy of a member of known size. In a *design* situa-tion, the size of the member is unknown, and the usual objective is to establish the minimum required member size.

In a tension member design, the axial checking formula can be rewritten to solve directly for the required area. Although certain assumptions may be involved, this can be described as a *direct solution*. The member size deter-mined in this way is usually close to the final solution.

It was previously noted that the design of axial tension members is the only type of problem covered in Chap. 7 that can be accomplished by direct solution. Columns, for example, involve design by trial and error because the adjusted design value depends on the column slenderness ratio. The slenderness ratio, in turn, depends on the size of the column, and it is necessary to first establish a trial size. The adequacy of the trial size is evaluated by performing an analy-sis. Depending on the results of the analysis, the trial size is accepted or adjusted up or down. Members with combined stresses are handled in a similar way.

It should be emphasized that the tension member problems addressed in this chapter are for *parallel-to-grain loading*. The weak nature of wood in tension perpendicular to grain is noted throughout this book, and the general recom-mendation is to avoid stressing wood in tension across the grain.

There are a variety of fasteners that can be used to connect wood members. The projected area of holes or grooves for the installation of fasteners is to be deducted from the *gross area* to obtain the *net area*. Some frequently used fasteners in wood connections include nails, bolts, lag bolts, split rings, and shear plates, and the design procedures for these are covered in Chaps. 11 through 13.

In determining the net area of a tension member, the projected area for nails is usually disregarded. The projected area of a bolt hole is a rectangle. A split ring or shear plate connector involves a dap or groove cut in the face of the wood member plus the projected area of the hole for the bolt (or lag bolt) that holds the assembly together. See Example 7.1. The projected area removed from the cross section for the installation of a lag bolt is determined by the shank diam-eter and the diameter of the lead or pilot hole for the threads.

Perhaps the most common situation that requires a reduction of area for ten-sion member design is a bolted connection. The NDS (Ref. 7.1) requires that the hole diameter be $\frac{1}{32}$ to $\frac{1}{16}$ in. larger than the bolt diameter. It also recommends

EXAMPLE 7.1   Net Areas at Connections

The gross area of a wood member is the width of the member times its depth:

$$A_g = b \times d$$

The standard net dimensions and the gross cross-sectional areas for sawn lumber are given in NDS Supplement Table 1B. Similar properties for glulam members are listed in NDS Supplement Tables 1C and 1D.

The projected areas for fasteners to be deducted from the gross area are as follows:

Nail holes—disregarded.

Bolt holes—computed as the hole diameter times the width of the wood member. See Fig. 7.2. The hole diameter is between $\frac{1}{32}$ and $\frac{1}{16}$ in. larger than the bolt diameter (NDS Sec. 11.1.2.2). In this book the bolt hole, for strength calculation purposes, is taken as the bolt diameter plus $\frac{1}{16}$ in.

Lag bolt holes—a function of the connection details. See NDS Appendix L for lag bolt dimensions. Drill diameters for lead holes and shank holes are given in NDS Sec. 11.1.3.

MEMBER WITH BOLT

MEMBER WITH CONNECTOR IN ONE FACE

Figure 7.2 *Net section* through two wood members. One member is shown cut at a bolt hole. The other is at a joint with a split ring or shear plate connector in one face plus the projected area of a bolt. The bolt is required to hold the entire assembly (wood members and connectors) together. Photographs of split ring and shear plate connectors are included in Chap. 13 (Fig. 13.26*a* and *b*).

Split ring and shear plate connectors—a function of the connection details. See NDS Appendix K for the projected areas of split ring and shear plate connectors. See Fig. 7.2.

If more than one fastener is used, the sum of the projected areas of all the fasteners *at the critical section* is subtracted from the gross area. For staggered fastener patterns, see NDS Sec. 3.1.2.

---

against tight-fitting installations that require forcible driving of the bolt. In *ideal* conditions it is appropriate to take the hole diameter for calculation purposes equal to the actual hole diameter.

In practice, ideal installation procedures are often viewed as goals. There are many field conditions that may cause the actual installation to be less than perfect. For example, a common bolt connection is through two steel side plates with the wood member between the two metal plates. Holes in the steel plates are usually punched in the shop, and holes in the wood member are drilled in the field.

It is difficult to accurately drill the hole in the wood member from one side (through a hole in one of the steel plates) and have it align perfectly with the hole in the steel plate on the opposite side. The hole will probably be drilled partially from both sides with some misalignment where they meet. Some oversizing of the bolt hole typically occurs as the two holes are reamed to correct alignment for the installation of the bolt. This is one example of a practical field problem, and a number of others can be cited.

In this book, the hole diameter for net area calculations will be taken as the bolt diameter plus $\frac{1}{16}$ in. as specified in the NDS. See Chap. 13 for more information on bolted connections.

The adjusted tension design value in a wood member is determined by multiplying the reference tension design value by the appropriate adjustment factors.

For ASD, the adjusted tension design value is defined as

$$F'_t = F_t(C_D)(C_M)(C_t)(C_F)(C_i)$$

For LRFD, the nominal and adjusted tension design values are

$$F_{tn} = F_t \cdot K_F$$

$$F'_{tn} = F_{tn}(\phi_t)(\lambda)(C_M)(C_t)(C_F)(C_i)$$

where F'_t = adjusted ASD tension design value parallel to grain
F'_{tn} = adjusted LRFD tension design value parallel to grain
F_t = reference tension design value parallel to grain
F_{tn} = nominal LRFD tension design value parallel to grain
K_F = format conversion factor (Sec. 4.23)—LRFD only
ϕ_t = resistence factor for tension = 0.8—LRFD only
C_D = load duration factor (Sec. 4.15)—ASD only
λ = time effect factor (Sec. 4.16)—LRFD only
C_M = wet service factor (Sec. 4.14)
= 1.0 for dry service conditions (as in most covered structures).

Dry service is defined as

MC $\leq$ 19 percent for sawn lumber

MC $<$ 16 percent for glulam

C_t = temperature factor (sec. 4.20)

= 1.0 for normal temperature conditions

C_F = size factor (Sec. 4.17) for sawn lumber in tension. Obtain values for visually-graded Dimension lumber from the Adjustment Factors section of NDS Supplement Tables 4A, 4B, and 4F.

= 1.0 for sawn lumber in B&S and P&T sizes and MSR lumber (Note: The size factor is not applicable for glulam.)

C_i = incising factor for sawn lumber (Sec. 4.21)

= 0.80 for incised Dimension lumber

= 1.0 for sawn lumber not incised (whether the member is treated or untreated)

(Note: The incising factor is not applicable for glulam or Timbers.)

It can be seen that the usual adjustment factors for load duration (ASD), time effect (LRFD), moisture content, temperature, size effect, and incising apply to tension design values parallel to grain. Values of C_M, C_t, and C_i frequently default to unity, but the designer should be aware of conditions that may require an adjustment. The size factor for tension applies to visually-graded Dimension lumber only, and values are obtained from NDS Supplement Tables 4A, 4B, and 4F.

7.3 Design Problem: Tension Member

In this example, the required size for the lower chord of a truss is determined. The loads are assumed to be applied to the top chord of the truss only. See Example 7.2 for ASD and Example 7.3 for LRFD.

To determine the axial forces in the members using simple truss analysis techniques, it is useful to assume the loads to be applied to the joints. Loads for the truss analysis are obtained by taking the tributary width to one joint times the uniform load. Because the actual loads are applied uniformly to the top chord, these members will have combined stresses. Other members in the truss will have only axial forces if the joints are pinned.

The tension force in the bottom chord is obtained through a standard truss analysis (method of joints). The member size is determined by calculating the required net area and adding to it the area removed by the bolt hole.

EXAMPLE 7.2 Tension Chord Using ASD

Determine the required size of the lower (tension) chord in the truss in Fig. 7.3a using ASD. The loads are (D + S), and the effects of roof slope on the magnitude of the snow load have already been taken into account. Joints are assumed to be pinned.

Figure 7.3a Uniform load on top chord converted to concentrated joint loads.

Connections will be made with a single row of $\frac{3}{4}$-in.-diameter bolts. Trusses are 4 ft-0 in. o.c. Lumber is No. 1 Spruce-Pine-Fir (South) [abbreviated S-P-F(S)]. MC ≤ 19 percent, and normal temperatures apply. Reference design values and cross-sectional properties are to be taken from the NDS Supplement.

Loads

$$D = 14 \text{ psf} \qquad \text{horizontal plane}$$

$$\underline{S = 30 \text{ psf}}$$

$$TL = 44 \text{ psf}$$

$$w_{TL} = 44 \times 4 = 176 \text{ lb/ft}$$

For truss analysis (load to joint),

$$P = 176 \times 7.5 = 1320 \text{ lb}$$

Force in Lower Chord

Use method of joints (Fig. 7.3b).

Figure 7.3b Free body diagram of joint A.

Determine Required Size of Tension Member

The relatively small tension force will require a Dimension lumber member size. The reference tension design value is obtained from NDS Supplement Table 4A for No. 1 S-P-F(S). A value for the size factor for tension parallel to grain will be assumed and checked later.

Assume $C_F = 1.3$.

$$F'_t = F_t(C_D)(C_M)(C_t)(C_F)(C_i)$$

$$= 400(1.15)(1.0)(1.0)(1.3)(1.0) = 598 \text{ psi}$$

$$\text{Req'd } A_n = \frac{P}{F'_t} = \frac{3960}{598} = 6.62 \text{ in.}^2$$

Figure 7.3c   Net section of tension member.

The actual hole diameter is to be $\frac{1}{32}$ to $\frac{1}{16}$ in. larger than the bolt size. For net area calculations, assume that the bolt hole is $\frac{1}{16}$ in. larger than the bolt (for member sizing calculations only). See Fig. 7.3c. Select a trial size from NDS Supplement Table 1B.

$$\text{Req'd } A_g = A_n + A_h$$

$$= 6.62 + 1.5(\frac{3}{4} + \frac{1}{16}) = 7.84 \text{ in.}^2$$

Try   2 × 6:

$$A = 8.25 \text{ in.}^2 > 7.84 \text{ in.}^2 \quad OK$$

Verify the size factor for tension in NDS Supplement Table 4A for a 6-in. nominal width:

$$C_F = 1.3 \quad \text{(same as assumed)} \quad OK$$

> *Use*   2 × 6   No. 1   S-P-F(S)

NOTE: The simplified truss analysis used in this example applies only to trusses with pinned joints. If some form of toothed metal plate connector (see Sec. 11.2) is used for the connections, the design should conform to Ref. 7.6 or applicable building code standards.

Example 7.2 is repeated using LRFD. See Example 7.3. The loads are factored using the combination 1.2D + 1.6S, and the load duration factor is replaced by the time effect factor. The resistance factor for tension ϕ_t and format conversion factor K_F are used in LRFD calculations. All other adjustment factors are applied in LRFD as they were in ASD.

EXAMPLE 7.3   Tension Chord Using LRFD

Repeat Example 7.2 using LRFD. Determine the required size of the lower (tension) chord in the truss in Fig. 7.3a. The load combination is 1.2D + 1.6S, and the effects of roof slope on the magnitude of the snow load have already been taken into account. Joints are assumed pinned.

Loads

$$1.2D + 1.6S = 1.2(14) + 1.6(30) = 65 \text{ psf}$$

$$w_u = 65 \times 4 = 260 \text{ lb/ft}$$

For truss analysis (load to joint),

$$P_u = 260 \text{ lb/ft} \times 7.5 \text{ ft} = 1950 \text{ lb} = 1.95 \text{ k}$$

Force in Lower Chord

Using methods of joints, the factored tension for member AC is

$$T_{\text{AC-}u} = 5.85 \text{ k}$$

Determine Required Size of Tension Member

The relatively small factored tension force will require a Dimension lumber member size. The reference design tension design value is obtained from NDS Supplement Table 4A for No. 1 S-P-F(S). A value for the size factor for tension parallel to grain will be assumed and checked later. Assume $C_F = 1.3$. The time effect factor is $\lambda = 0.8$ for the load combination 1.2D + 1.6S. $\phi_t = 0.8$ and $K_F = 2.70$.

$$F_{tn} = F_t \cdot K_F = 400(2.70) = 1080 \text{ psi}$$

$$F'_{tn} = F_{tn}(\phi_t)(\lambda)(C_M)(C_t)(C_F)(C_i)$$

$$= 1080(0.80)(0.8)(1.0)(1.0)(1.3)(1.0)$$

$$= 898 \text{ psi} = 0.898 \text{ ksi}$$

$$\text{Req'd } A_n = \frac{T_{\text{AC-}u}}{F'_{tn}} = \frac{5.85}{0.898} = 6.52 \text{ in.}^2$$

$$\text{Req'd } A_g = A_n - A_h$$

$$= 6.52 + 1.5(\tfrac{3}{4} + \tfrac{1}{16}) = 7.74 \text{ in.}^2$$

Try 2 × 6:

$$A = 8.25 \text{ in.}^2 > 7.74 \text{ in.}^2 \qquad OK$$

Verify the size factor for tension in NDS Supplement Table 4a for a 6-in. nominal width:

$$C_F = 1.3 \qquad \text{(same as assumed)} \qquad OK$$

$$Use \quad 2 \times 6 \quad No. \quad 1 \text{ S-P-F(S)}$$

NOTE: See note at end of Example 7.2 regarding simplified truss analysis assumptions.

7.4 Columns

In addition to being a compression member, a *column* generally is sufficiently long that the possibility of buckling needs to be considered. On the other hand, the term *short column* usually implies that a compression member will not buckle, and its strength is related to the crushing capacity of the material.

In order to evaluate the tendency of a column to buckle, it is necessary to know the size of the member. Thus, in a design situation, a trial size is first established. With a known size or a trial size, it is possible to compare the actual stress with the adjusted design value. Based on this comparison, the member size will be accepted or rejected.

Following ASD procedures, the check on the capacity of an axially loaded wood column of known size uses the formula

$$f_c = \frac{P}{A}$$
$$\leq F'_c$$

where f_c = actual (computed) compressive stress parallel to grain
   P = axial compressive force in member
   A = cross-sectional area
   F'_c = adjusted ASD compressive design value parallel to grain

According to LRFD principles, the axial check for a column is typically cast in terms of factored resistance and loads,

$$P_u \leq P'_n$$
$$\leq F'_{cn} \cdot A$$

where P_u = axial compressive force in the member due to factored loads
   P'_n = adjusted LRFD compressive resistance parallel to grain
   F'_{cn} = adjusted LRFD compressive design value parallel to grain
   A = cross-sectional area

Regardless of whether ASD or LRFD is used, the cross-sectional area to be used will be either the gross area A_g of the column or the net area A_n at some hole in the member. The area to be used depends on the location of the hole along the length of the member and the tendency of the member at that point to buckle laterally. If the hole is located at a point which is braced, the gross area of the member may be used in the check for column stability between brace locations.

EXAMPLE 7.4   Actual Stresses in a Column

Figure 7.4   Pinned end column.

Actual Stresses

In Fig. 7.4, it is assumed that there are no holes in the column except at the supports (connections). Check the following column stresses:

1. Away from the supports
   ASD:

$$f_c = \frac{P}{A_g} \le F'_c \qquad \text{as determined by column stability (using } C_P\text{)}$$

   LRFD:

$$P_u \le P'_n = F'_{cn} \cdot A_g \qquad \text{as determined by column stability (using } C_P\text{)}$$

2. At the connection where buckling is not a factor
   ASD:

$$f_c = \frac{P}{A_n} \le F'_c \qquad \text{as determined for a } short \text{ column (without } C_P\text{)}$$

   LRFD:

$$P_u \le P'_n = F'_{cn} \cdot A_n \qquad \text{as determined for a } short \text{ column (without } C_P\text{)}$$

Another check at the reduced cross section (using the net area) should be compared with the adjusted compressive design value or capacity for a short column with no reduction for stability at the braced location. See Example 7.4. The other possibility is that some reduction of column area occurs in the laterally unbraced portion of the column. In the latter case, the net area is used directly in the stability check.

The adjusted design value in a column reflects many of the familiar adjustment factors in addition to column stability. For ASD, the adjusted compressive design value is defined as

$$F'_c = F_c(C_D)(C_M)(C_t)(C_F)(C_P)(C_i)$$

for LRFD, the nominal and adjusted compressive design values are

$$F_{cn} = F_c \cdot K_F$$
$$F'_{cn} = F_{cn}(\phi_c)(\lambda)(C_M)(C_t)(C_F)(C_P)(C_i)$$

where F'_c = adjusted ASD compressive design value parallel to grain

F'_{cn} = adjusted LRFD compressive design value parallel to grain

F_c = reference compressive design value parallel to grain

F_{cn} = nominal LRFD compressive design value parallel to grain

C_D = load duration factor (Sec. 4.15)—ASD only

K_F = format conversion factor (Sec. 4.23)—LRFD only

ϕ_c = resistence factor for compression—LRFD only

$\quad$ = 0.9

λ = time effect factor (Sec. 4.16)—LRFD only

C_M = wet service factor (Sec. 4.14)

$\quad$ = 1.0 for dry service conditions as in most covered structures. Dry service conditions are defined as

$\quad$ MC $\leq$ 19 percent for sawn lumber

$\quad$ MC $<$ 16 percent for glulam

C_t = temperature factor (Sec. 4.20)

$\quad$ = 1.0 for normal temperature conditions

C_F = size factor (Sec. 4.17). Obtain values for visually graded Dimension lumber from Adjustment Factors section of NDS Supplement Tables 4A, 4B, and 4F.

$\quad$ = 1.0 for Timbers

$\quad$ = 1.0 for MSR and MEL lumber

$\quad$ (Note: The size factor is not applicable for glulam.)

C_P = column stability factor

$\quad$ = 1.0 for fully supported column

C_i = incising factor for sawn lumber (Sec. 4.21)

$\quad$ = 0.80 for incised Dimension lumber

$\quad$ = 1.0 for sawn lumber not incised (whether the member is treated or untreated)

$\quad$ (Note: The incising factor is not applicable for glulam or for Timbers.)

The size factor for compression applies only to Dimension lumber sizes, and C_F defaults to 1.0 for other members. The column stability factor takes buckling into account, and the slenderness ratio is the primary measure of buckling. The column slenderness ratio and C_P are the subjects of the remainder of this section.

In its traditional form, the slenderness ratio is expressed as the effective unbraced length of a column divided by the least radius of gyration, l_e/r. For the design of rectangular wood columns, however, the slenderness ratio is modified to a form that is somewhat easier to apply. Here the slenderness ratio is the effective unbraced length of the column divided by the least dimension of the cross section, l_e/d.

Use of this modified slenderness ratio is possible because the radius of gyration r can be expressed as a direct function of the width of a rectangular column. See Example 7.5. The constant in the conversion of the modified slenderness ratio is simply incorporated into the column design formulas.

EXAMPLE 7.5   Column Slenderness Ratio—Introduction

Figure 7.5   Typical wood column with rectangular cross section.

Consider a simple column as shown in Fig. 7.5. Column stability is measured by the slenderness ratio.

$$\text{General slenderness ratio} = \frac{l_e}{r}$$

where l_e = effective unbraced length of column
$\quad\quad r = r_y$ = least radius of gyration of column cross section

$$\text{Slenderness ratio for } \textit{rectangular columns} = \frac{l_e}{d}$$

where l_e = effective unbraced length of column
$\quad\quad d$ = least cross-sectional dimension of column*

For a rectangular cross section, the dimension d is directly proportional to the radius of gyration.

$$r_y = \sqrt{\frac{I_y}{A}} = \sqrt{\frac{bd^3/12}{bd}} = \sqrt{\frac{d^2}{12}} = d\sqrt{\frac{1}{12}}$$
$$\therefore r \propto d$$

A more complete review of the column slenderness ratio is given in Sec. 7.5.

---

*In beam design, d is normally associated with the strong axis.

Most wood columns have rectangular cross sections, and the design formula given in this chapter is for this common type of column. However, a column of nonrectangular cross section may be analyzed by substituting $r\sqrt{12}$ in place of d in the formula for rectangular columns. For round columns, see NDS Sec. 3.7.3. A much more detailed analysis of the slenderness ratio for columns is given in the next section.

The column stability factor C_P was shown previously as an adjustment factor for obtaining the adjusted compressive design value for a column. The treatment of C_P as another multiplying factor is convenient from an organizational or bookkeeping point of view. However, the expression for C_P times F_c (ASD) or F_{cn} (LRFD) essentially defines the *column curve* or *column equation* for wood design. The other coefficients in the expression for F_c' or F_{cn}' are more in keeping with the general concept of adjustment factors, and the column equation as given by $F_c \times C_P$ or $F_{cn} \times C_P$ is probably more basic to column behavior than the term *adjustment factor* implies.

The column equation in the NDS provides a continuous curve over the full range of slenderness ratios. See Example 7.6. The column expression in the NDS was originally developed by Ylinen and was verified by studies at the Forest Products Laboratory. The Ylinen formula also serves as the basis for the expression used for laterally unsupported beams in Sec. 6.3.

Zahn (Ref. 7.8) explained that the *behavior of wood columns* as given by the Ylinen formula is the result of the *interaction* of two modes of failure: *buckling and crushing*. Pure *buckling* is defined by the Euler critical buckling stress

$$F_{\mathrm{cr}} = \frac{\pi^2 E}{(l_e/r)^2}$$

For use in design, the Euler stress is expressed in the NDS as

$$F_{cE} = \frac{0.822 E_{\min}'}{(l_e/d)^2}$$

The coefficient 0.822 equals $\pi^2/12$, which recognizes the relationship $r^2 = d^2/12$ for rectangular sections. $E_{\min}$ is used since stability is a strength limit state. The Euler column buckling stress F_{cE} is graphed in Fig. 7.6a. It will be noted that the Ylinen formula converges to the Euler-based formula for columns with large slenderness ratios.

The second mode of failure is crushing of the wood fibers. When a compression member fails by pure crushing, there is no column buckling. Therefore, *crushing* is measured by the reference compressive design value parallel to grain multiplied by all applicable adjustment factors except C_P. This value is given the symbol F_c^* and is defined mathematically as

ASD:

$$F_c^* = F_c(C_D)(C_M)(C_t)(C_F)(C_i)$$

LRFD:

$$F_{cn} = F_c \cdot K_F$$

$$F_{cn}^* = F_{cn}(\phi_c)(\lambda)(C_M)(C_t)(C_F)(C_i)$$

The value of F_c^* (ASD) or F_{cn}^* (LRFD) is the upper limit of adjusted column design values for a slenderness ratio of zero.

Again, column behavior is defined by the *interaction* of the buckling and crushing modes of failure, and the ratio F_{cE}/F_c^* (ASD) or F_{cEn}/F_{cn}^* (LRFD) appears several times in the Ylinen formula. The coefficient c in the Ylinen formula is viewed by Zahn as a generalized interaction parameter. The value of c lies in the range

$$0 \leq c \leq 1.0$$

A value of $c = 1.0$ is an upper bound of column behavior and can only be met by an ideal material, loaded under ideal conditions. Because practical columns do not satisfy these idealizations, the value of c for wood compression members is less than one. The more a wood column deviates from the ideal situation, the smaller c becomes. Glulam members are generally thought to be straighter and more homogeneous than sawn lumber, and consequently glulam is assigned a larger value for c. The effects of different values of c are shown in Fig. 7.6*b*.

As the slenderness ratio increases, the column expression makes a transition from a design value based on the crushing strength of wood (at a zero slenderness ratio) to a design value based on the Euler curve (for large slenderness ratios). The Ylinen column curve closely fits the results of column tests. Compared with previously used column formulas, the Ylinen equation gives slightly more conservative values for compressive members with intermediate slenderness ratios.

EXAMPLE 7.6   Ylinen Column Equation

The NDS uses a continuous curve for evaluating the effects of column buckling. The adjusted column stress given by the Ylinen equation is plotted versus column slenderness ratio in Fig. 7.6*a*.

Figure 7.6*a*   Ylinen column curve: plot of F_c' versus l_e/d.

Adjusted Column Design Value

In general terms, the column stress curve in Fig. 7.6*a* is obtained by multiplying the reference compressive design value parallel to grain F_c by the column stability factor C_P and all other appropriate factors.

$$F'_c = F_c(C_P) \times \cdots$$

where F'_c = adjusted compressive design value in a column
$\quad F_c$ = reference compressive design value parallel to grain
$\quad C_P$ = column stability factor (defined below)
$\quad \times \cdots$ = product of other appropriate adjustment factors

Column Stability Factor

The column stability factor is determined using the same basic formula for both ASD and LRFD by substituting nominal design values for LRFD.

ASD:

$$C_P = \frac{1 + F_{cE}/F_c^*}{2c} - \sqrt{\left(\frac{1 + F_{cE}/F_c^*}{2c}\right)^2 - \frac{F_{cE}/F_c^*}{c}}$$

LRFD:

$$C_P = \frac{1 + F_{cEn}/F_{cn}^*}{2c} - \sqrt{\left(\frac{1 + F_{cEn}/F_{cn}^*}{2c}\right)^2 - \frac{F_{cEn}/F_{cn}^*}{c}}$$

where F_{cE} = Euler critical buckling stress for columns—ASD
$\qquad = \dfrac{0.822E'_{\min}}{(l_e/d)^2}$
$\quad F_{cEn}$ = nominal Euler critical buckling stress for columns—LRFD
$\qquad = \dfrac{0.822E'_{\min n}}{(l_e/d)^2}$
$\quad F_c^*$ = limiting ASD compressive design value in column at zero slenderness ratio
$\qquad$ = reference compressive design value parallel to grain multiplied by all adjustment factors except C_P
$\qquad = F_c(C_D)(C_M)(C_t)(C_F)(C_i)$
$\quad F_{cn}^*$ = limiting LRFD compressive design value in column at zero slenderness ratio
$\qquad$ = nominal LRFD compressive design value parallel to grain multiplied by all adjustment factors except C_P
$\qquad = F_{cn}(\phi_c)(\lambda)(C_M)(C_t)(C_F)(C_i)$
$\qquad = (F_c \cdot K_F)(\phi_c)(\lambda)(C_M)(C_t)(C_F)(C_i)$
$\quad E'_{\min}$ = adjusted ASD modulus of elasticity for column buckling. Recall that C_D does not apply to $E_{\min}$. For sawn lumber, $E_{\min x} = E_{\min y}$. For glulam, $E_{\min x}$ and $E_{\min y}$ may be different.
$\qquad = E(C_M)(C_t)(C_T)(C_i)$
$\quad E'_{\min n}$ = adjusted LRFD modulus of elasticity for column buckling. Recall that λ does not apply to $E_{\min n}$.

$$= E_{\min n}(\phi_s)(C_M)(C_t)(C_T)(C_i)$$

$$E_{\min n} = E_{\min} \cdot K_F$$

K_F = format conversion factor—LRFD

ϕ_s = resistance factor for stability—LRFD

$\quad = 0.85$

ϕ_c = resistance factor for compression—LRFD

$\quad = 0.90$

c = buckling and crushing interaction factor for columns

$\quad = 0.8$ for sawn lumber columns

$\quad = 0.85$ for round timber poles and piles

$\quad = 0.9$ for glulam or structural composite lumber columns

C_T = buckling stiffness factor for *2 × 4 and smaller compression chords in trusses* with $\frac{3}{8}$-in. or thicker plywood nailed to narrow face of member

$\quad = 1.0$ for all other members

C_F = size factor (Sec. 4.17) for compression. Obtain values for visually graded Dimension lumber from the Adjustment Factors section of NDS Supplement Tables 4A, 4B, and 4F.

$\quad = 1.0$ for sawn lumber in B&S and P&T sizes and MSR and MEL lumber (Note: The size factor is not applicable for glulam.)

Other factors are as previously defined.

The interaction between column buckling and crushing of wood fibers in a compression member is measured by parameter c. The effect of several different values of c is illustrated in Fig. 7.6b.

Figure 7.6b   Plot of F_c' versus l_e/d showing the effect of different values of c. The parameter c measures the interaction between crushing and buckling in wood columns.

A value of $c = 1.0$ applies to idealized column conditions. Practical wood columns have $c < 1.0$:

For sawn lumber: $c = 0.8$

For glulam and structural composite lumber: $c = 0.9$

Figure 7.6c Plot of F_c' versus l_e/d showing the effect of load duration on adjusted ASD compressive design value.

The effect of the load duration factor (ASD) varies depending on the mode of column failure that predominates. See Fig. 7.6c.

The load duration factor C_D has full effect on compressive design values when crushing controls (i.e., at a slenderness ratio of zero). On the other hand, C_D has no influence on the compressive design value when instability predominates. A transition between C_D having full effect at $l_e/d = 0$, and C_D having no effect at the maximum slenderness ratio of 50, is automatically provided in the definition of C_P.

A similar plot to that shown in Fig. 7.6c for C_D and ASD could be developed for λ and LRFD. The important point, though, is that the time effect factor λ, like C_D, has full effect when crushing controls and no influence when stability predominates. A transition between these cases is provided by the stability factor C_P.

7.5 Detailed Analysis of Slenderness Ratio

The concept of the slenderness ratio was briefly introduced in Sec. 7.4. There it was stated that the *least* radius of gyration is used in the l_e/r ratio and that the *least* dimension of the column cross section is used in the l_e/d ratio.

These statements assume that the unbraced length of the column is the same for both the x and the y axes. In this case the column, if loaded to failure, would buckle about the weak y axis. See Fig. 7.7a. Note that if buckling occurs about the y axis, the column moves in the x direction. Figure 7.7a illustrates this straightforward case of column buckling. Only the slenderness ratio about the weak axis needs to be calculated.

Although this concept of column buckling applies in many situations, the designer should have a deeper insight into the concept of the slenderness ratio. Conditions can exist under which the column may actually buckle about the *strong* axis of the cross section rather than the weak axis.

Figure 7.7a   By inspection the slenderness ratio about the *y* axis is larger and is therefore critical.

In this more general sense, the column can be viewed as having two slenderness ratios. One slenderness ratio would evaluate the tendency of the column to buckle about the *strong* axis of the cross section. The other would measure the tendency of buckling about the *weak* axis. For a rectangular column, these slenderness ratios would be written

$$\left(\frac{l_e}{d}\right)_x \quad \text{for buckling about strong } x \text{ axis (column movement in } y \text{ direction)}$$

$$\left(\frac{l_e}{d}\right)_y \quad \text{for buckling about weak } y \text{ axis (column movement in } x \text{ direction)}$$

If the column is loaded to failure, buckling will occur about the axis that has the larger slenderness ratio. In design, the larger slenderness ratio is used to calculate the adjusted compressive design value. (Note that it is conceivable in a glulam column with different values for E_x and E_y that a slightly smaller l_e/d could produce the critical F_c'.)

The reason that the strong axis can be critical can be understood from a consideration of the *bracing* and *end conditions* of the column. The *effective unbraced length* is the length to be used in the calculation of the slenderness ratio. It is possible to have a column with different unbraced lengths for the *x* and *y* axes. See Fig. 7.7*b*. In this example, the unbraced length for the *x* axis is twice as long as the unbraced length for the *y* axis. In practice, bracing can occur at any interval. The effect of column end conditions is explained later.

Another case where the unbraced lengths for the *x* and *y* axes are different occurs when sheathing is attached to a column. If the sheathing is attached to the column with an effective connection, buckling about an axis that is perpendicular to the sheathing is prevented. See Fig. 7.8. The most common example

Figure 7.7b Different unbraced lengths for both axes. Because of the intermediate bracing for the *y* axis, the critical slenderness ratio cannot be determined by inspection. Both slenderness ratios must be calculated, and the larger value is used to determine F_c'.

of this type of column is a stud in a bearing wall. The wall sheathing can prevent column buckling about the weak axis of the stud, and only the slenderness ratio about the strong axis of the member needs to be evaluated.

The final item regarding the slenderness ratio is the effect of column *end conditions*. The length *l* used in the slenderness ratio is theoretically the unbraced length of a *pinned-end* column. For columns with other end conditions, the length is taken as the distance between inflection points (IPs) on a sketch of the buckled column. An inflection point corresponds to a point of reverse curvature on the deflected shape of the column and represents a point of zero moment. For this

Figure 7.8 Column braced by sheathing. Sheathing attached to stud prevents column buckling about the weak (*y*) axis of the stud. Therefore, consider buckling about the *x* axis only.

reason, the inflection point is considered as a pinned end for purposes of column analysis. The *effective unbraced length* is taken as the distance between inflection points. When only one inflection point is on the sketch of the buckled column, the mirror image of the column is drawn to give a second inflection point.

Typically, six "ideal" column end conditions are identified in various fields of structural design. See Fig. 7.9*a*. The recommended effective length factors for use in the design of wood columns are given in Fig. 7.9*b*. The effective unbraced

Figure 7.9*a* Six typical idealized columns showing buckled shapes and theoretical effective lengths.

Column No.	1	2	3	4	5	6
Theoretical effective length factors	0.5	0.7	1.0	1.0	2.0	2.0
Recommended design K_e	0.65	0.8	1.0	1.2	2.1	2.4

Figure 7.9b   Table of theoretical and recommended effective length factors K_e. Values of recommended K_e are from NDS Appendix G.

length can be determined by multiplying the effective length factor K_e times the actual unbraced length.

$$\text{Effective length} = \text{distance between inflection points}$$

$$= \text{effective length factor} \times \text{unbraced length}$$

$$l_e = K_e \times l$$

The effective lengths shown on the column sketches in Fig. 7.9a are *theoretical* effective lengths, and practical field column end conditions can only approximate the ideal pinned and fixed column end conditions. The recommended *design* effective length factors from NDS Appendix G are to be used for practical field end conditions.

In practice, the designer must determine which "ideal" column most closely approximates the actual end conditions for a given columns. Some degree of judgment is required for this evaluation, but several key items should be considered in making this comparison.

First, note that three of the ideal columns undergo sidesway, and the other three do not. *Sidesway* means that the top of the column is relatively free to displace laterally with respect to the bottom of the column. The designer must be able to identify which columns will undergo sidesway and, on the other hand, what constitutes restraint against sidesway. In general, the answer to this depends on the type of lateral-force-resisting system (LFRS) used (Sec. 3.3).

Usually, if the column is part of a system in which lateral forces are resisted by *bracing* or by *shearwalls*, sidesway will be prevented. These types of LFRSs are relatively rigid, and the movement of one end of the column with respect to the other end is restricted. If an overload occurs in this case, column buckling will be symmetric. See Fig. 7.10. However, if the column is part of a *rigid frame* type of LFRS, the system is relatively flexible, and sidesway can occur. Typical columns for the types of buildings considered in this text will have sidesway prevented.

It should also be noted that columns with *sidesway prevented* have an effective length that is *less than* or *equal to* the actual unbraced length ($K_e \leq 1.0$). A common and conservative practice is to consider the effective length equal to the unbraced length for these columns ($K_e = 1.0$).

Figure 7.10 Columns with and without sidesway. (*a*) Braced frames or buildings with shearwalls limit the displacement of the top end of the column so that sidesway does not occur. (*b*) Columns in rigid frames (without bracing) will undergo sidesway if the columns buckle.

For columns where *sidesway* can occur, the effective length is *greater than* the actual unbraced length ($K_e > 1.0$). For these types of columns, the larger slenderness ratio causes the maximum axial load to be considerably less than the maximum load on a column with both ends pinned and braced against sidesway.

In addition to answering the question of sidesway, the comparison of an actual column to an ideal column should evaluate the effectiveness of the column connections. Practically all wood columns have *square-cut ends*. For structural design purposes, this type of column end condition is normally assumed to be pinned. Square-cut column ends do offer some restraint against column end rotation. However, most practical column ends are not exactly square, and some *accidental eccentricity* may be present due to nonuniform bearing. These effects are often assumed to be compensating. Therefore, columns in a typical wood-frame building with shearwalls are usually assumed to be type 3 in Fig. 7.9a, and the effective length factor is taken to be unity.

It is possible to design moment-resisting connections in wood members, but they are the exception rather than the rule. As noted, the majority of connections in ordinary wood buildings are "simple" connections, and it is generally conservative to take the effective length equal to the unbraced length. However, the designer should examine the actual *bracing conditions* and *end conditions* for a given column and determine whether or not a larger effective length should be used.

7.6 Design Problem: Axially Loaded Column

The design of a column is a trial-and-error process because, in order to determine the adjusted column design value F'_c, it is first necessary to know the slenderness ratio l_e/d. In the following examples, only two trials are required to determine the size of the column. See Examples 7.7 (ASD) and 7.8 (LRFD).

Several items in the solution should be emphasized. First, the importance of the size category should be noted. The initial trial is a Dimension lumber size, and the second trial is a Posts and Timbers size. Reference design values are different for these two size categories.

The second item concerns the load duration (ASD) and time effect (LRFD) factors. The ASD example (7.7) involves (D + L$_r$). It will be remembered that roof live load is an arbitrary minimum load required by the Code, and C_D for this combination is 1.25. For many areas of the country the design load for a roof will be (D + S), and the corresponding C_D for snow would be 1.15. The designer is cautioned that Example 7.8 is illustrative in that the critical load combination (D + L$_r$) was predetermined. To be complete, the designer should also check D only with the corresponding C_D for permanent load of 0.9.

In Example 7.8 (LRFD), the corresponding load combination $1.2D + 1.6L_r$ with a time effect factor $\lambda = 0.8$ is used. If snow governed over the roof live load, the combination $1.2D + 1.6S$ would be used with the same time effect factor $\lambda = 0.8$. Thus, in LRFD, there is no change in time effect factor when considering roof live load versus snow load. To be complete, the designer should consider dead-only load using 1.4D and $\lambda = 0.6$.

EXAMPLE 7.7   Sawn Lumber Column using ASD

Design the column in Fig. 7.11a, using No. 1 Douglas Fir-Larch and ASD procedures. Bracing conditions are the same for buckling about the x and y axes. The load is combined dead load and roof live load. MC $\leq$ 19 percent ($C_M = 1.0$), the member is not treated or incised ($C_i = 1.0$) and normal temperatures apply ($C_t = 1.0$). Adjusted design values are to be in accordance with the NDS.

Figure 7.11a Elevation view of column.

Trial 1

Try 4 × 6 (Dimension lumber size category). See Fig. 7.11b. From NDS Supplement Table 4A,

$$F_c = 1500 \text{ psi}$$

$$C_F = 1.1 \qquad \text{size factor for compression}$$

$$E_{\min} = 620,000 \text{ psi}$$

$$A = 19.25 \text{ in.}^2$$

Figure 7.11b Cross section of 4 × 6 trial column.

Determine capacity using Ylinen column equation (Example 7.6):

$$\left(\frac{l_e}{d}\right)_{\max} = \left(\frac{K_e l}{d}\right)_y = \frac{1 \times 10 \text{ ft} \times 12 \text{ in./ft}}{3.5 \text{ in.}} = 34.3$$

$$E'_{\min} = E_{\min}(C_M)(C_t)(C_T)(C_i) = 620,000(1.0)(1.0)(1.0)(1.0)$$

$$= 620,000 \text{ psi}$$

For visually graded sawn lumber, $c = 0.8$

$$F_{cE} = \frac{0.822 E'_{\min}}{(l_e/d)^2} = \frac{0.822(620,000)}{(34.3)^2} = 433 \text{ psi}$$

$$F_c^* = F_c(C_D)(C_M)(C_t)(C_F)(C_i)$$

$$= 1500(1.25)(1.0)(1.0)(1.1)(1.0) = 2062 \text{ psi}$$

$$\frac{F_{cE}}{F_c^*} = \frac{433}{2062} = 0.210$$

$$\frac{1 + F_{cE}/F_c^*}{2c} = \frac{1 + 0.210}{2(0.8)} = 0.756$$

$$C_P = \frac{1 + F_{cE}/F_c^*}{2c} - \sqrt{\left(\frac{1 + F_{cE}/F_c^*}{2c}\right)^2 - \frac{F_{cE}/F_c^*}{c}}$$

$$= 0.756 - \sqrt{(0.756)^2 - 0.210/0.8} = 0.200$$

$$F_c' = F_c(C_D)(C_M)(C_t)(C_F)(C_P)(C_i)$$

$$= 1500(1.25)(1.0)(1.0)(1.1)(0.200)(1.0)$$

$$= 412 \text{ psi}$$

Allow. $P = F_c'A = 0.412(19.25) = 7.94 \text{ k} < 15 \text{ k}$ *NG*

Trial 2

Try 6×6 (P&T size category). Values from NDS Supplement Table 4D:

$$F_c = 1000 \text{ psi}$$

$$E_{\min} = 580,000 \text{ psi}$$

Size factor for compression defaults to unity for all sizes except Dimension lumber $(C_F = 1.0)$.

$$d_x = d_y = 5.5 \text{ in.}$$

$$A = 30.25 \text{ in.}^2$$

Determine column capacity and compare with the given design load.

$$\left(\frac{l_e}{d}\right)_{\max} = \frac{1.0(10 \text{ ft} \times 12 \text{ in./ft})}{5.5 \text{ in.}} = 21.8$$

$$E'_{\min} = E_{\min}(C_M)(C_t)(C_i) = 580,000(1.0)(1.0)(1.0)$$

$$= 580,000 \text{ psi}$$

$$c = 0.8$$

$$F_{cE} = \frac{0.822E'_{\min}}{(l_e/d)^2} = \frac{0.822(580,000)}{(21.8)^2} = 1003 \text{ psi}$$

$$F_c^* = F_c(C_D)(C_M)(C_t)(C_F)(C_i)$$

$$= 1000(1.25)(1.0)(1.0)(1.0)(1.0) = 1250 \text{ psi}$$

$$\frac{F_{cE}}{F_c^*} = \frac{1003}{1250} = 0.802$$

$$\frac{1 + F_{cE}/F_c^*}{2c} = \frac{1 + 0.802}{2(0.8)} = 1.126$$

$$C_P = \frac{1 + F_{cE}/F_c^*}{2c} - \sqrt{\left(\frac{1 + F_{cE}/F_c^*}{2c}\right)^2 - \frac{F_{cE}/F_c^*}{c}}$$

$$= 1.126 - \sqrt{(1.126)^2 - 0.802/0.8} = 0.611$$

$$F_c' = F_c(C_D)(C_M)(C_t)(C_F)(C_P)(C_i)$$

$$= 1000(1.25)(1.0)(1.0)(1.0)(0.611)(1.0)$$

$$= 764 \text{ psi}$$

Allow. $P = F_c'A = 764(30.25) = 23.1 \text{ k} > 15 \text{ k}$ *OK*

> *Use* 6×6 column   No. 1   DF-L

In this example, the load combination of (D + L$_r$) was predetermined to be critical.

Example 7.7 is now repeated using LRFD. Again, the load combination 1.2D + 1.6L$_r$ has been predetermined to govern the design.

EXAMPLE 7.8   Sawn Lumber Column Using LRFD

Design the column illustrated in Fig. 7.11a, using No. 1 Douglas Fir-Larch and LRFD procedures. Bracing conditions are the same for buckling about the x and y axes. The factored load P_u = 25 k is combined dead load and roof live load (1.2D + 1.6L$_r$). MC $\leq$ 19 percent (C_M = 1.0), the member is not treated or incised (C_i = 1.0) and normal temperatures apply (C_t = 1.0). Adjusted design values are to be in accordance with the NDS. ϕ_c = 0.90 and ϕ_s = 0.85 are used for LRFD. For compression K_F = 2.16/ϕ_c = 2.40, and for stability K_F = 1.5/ϕ_s = 1.765.

Trial 1

Try 4 $\times$ 6 (Dimension lumber size category). From NDS Supplement Table 4A

$$F_{cn} = F_c \cdot K_F$$

$$= 1500(2.40) = 3600 \text{ psi} = 3.6 \text{ ksi}$$

$$C_F = 1.1 \quad \text{size factor for compression}$$

$$E_{\min n} = E_{\min} \cdot K_F$$

$$= 620{,}000(1.765) = 1{,}094{,}000 \text{ psi} = 1094 \text{ ksi}$$

$$A = 19.25 \text{ in.}^2$$

Determine capacity using Ylinen column equation (Example 7.6):

$$\left(\frac{l_e}{d}\right)_{max} = \left(\frac{K_e l}{d}\right)_y = \frac{1 \times 10 \text{ ft} \times 12 \text{ in./ft}}{3.5 \text{ in.}} = 34.3$$

$$E'_{min\,n} = E_{min\,n}(\phi_s)(C_M)(C_t)(C_T)(C_i) = 1094(0.85)(1.0)(1.0)(1.0)(1.0)$$
$$= 930 \text{ ksi}$$

For visually graded sawn lumber,

$$c = 0.8$$

$$F_{cEn} = \frac{0.822E'_{min\,n}}{(l_e/d)^2} = \frac{0.822(930)}{(34.3)^2} = 0.650 \text{ ksi}$$

$$F_{cn}^* = F_{cn}(\phi_c)(\lambda)(C_M)(C_t)(C_F)(C_i)$$
$$= 3.6(0.90)(0.8)(1.0)(1.0)(1.1)(1.0) = 2.85 \text{ ksi}$$

$$\frac{F_{cEn}}{F_{cn}^*} = \frac{(0.650)}{(2.85)} = 0.228$$

$$\frac{1 + F_{cEn}/F_{cn}^*}{2c} = \frac{1 + 0.228}{2(0.8)} = 0.768$$

$$C_P = \frac{1 + F_{cEn}/F_{cn}^*}{2c} - \sqrt{\left(\frac{1 + F_{cEn}/F_{cn}^*}{2c}\right)^2 - \frac{F_{cEn}/F_{cn}^*}{c}}$$
$$= 0.768 - \sqrt{(0.768)^2 - 0.228/0.8} = 0.216$$

$$F'_{cn} = F_{cn}(\phi_c)(\lambda)(C_M)(C_t)(C_F)(C_P)(C_i)$$
$$= 3.6(0.90)(0.8)(1.0)(1.0)(1.1)(0.216)(1.0)$$
$$= 0.616 \text{ ksi}$$

$$P_n = F'_{cn}A = (0.616)(19.25) = 11.9 \text{ k} < 25 \text{ k} \qquad NG$$

Trial 2

Try 6 × 6 (P&T size category). Values from NDS Supplement Table 4D:

$$F_{cn} = F_c \cdot K_F$$
$$= 1000(2.40) = 2400 \text{ psi} = 2.40 \text{ ksi}$$

$$E_{min\,n} = E_{min} \cdot K_F$$
$$= 580,000(1.765) = 1,024,000 \text{ psi} = 1024 \text{ ksi}$$

Size factor for compression defaults to unity for all sizes except Dimension lumber ($C_F = 1.0$).

$$d_x = d_y = 5.5 \text{ in.}$$
$$A = 30.25 \text{ in.}^2$$

Determine column capacity and compare with the given design load.

$$\left(\frac{l_e}{d}\right)_{max} = \frac{1.0(10 \text{ ft} \times 12 \text{ in./ft})}{5.5 \text{ in.}} = 21.8$$

$$E'_{\min n} = E_{\min n}(\phi_s)(C_M)(C_t)(C_i) = 1024(0.85)(1.0)(1.0)(1.0)$$

$$= 870 \text{ ksi}$$

$$c = 0.8$$

$$F_{cEn} = \frac{0.822 E'_{\min n}}{(l_e/d)^2} = \frac{0.822(870)}{(21.8)^2} = 1.50 \text{ ksi}$$

$$F^*_{cn} = F_{cn}(\phi_c)(\lambda)(C_M)(C_t)(C_F)(C_i)$$

$$= 2.40(0.90)(0.8)(1.0)(1.0)(1.0)(1.0) = 1.73 \text{ ksi}$$

$$\frac{F_{cEn}}{F^*_{cn}} = \frac{(1.50)}{(1.73)} = 0.869$$

$$\frac{1 + F_{cEn}/F^*_{cn}}{2c} = \frac{1 + 0.869}{2(0.8)} = 1.168$$

$$C_P = \frac{1 + F_{cEn}/F^*_{cn}}{2c} - \sqrt{\left(\frac{1 + F_{cEn}/F^*_{cn}}{2c}\right)^2 - \frac{F_{cEn}/F^*_{cn}}{c}}$$

$$= 1.168 - \sqrt{(1.168)^2 - 0.869/0.8} = 0.641$$

$$F'_{cn} = F_{cn}(\phi_c)(\lambda)(C_M)(C_t)(C_F)(C_P)(C_i)$$

$$= 2.40(0.90)(0.8)(1.0)(1.0)(1.0)(0.641)(1.0)$$

$$= 1.11 \text{ ksi}$$

$$P_u \le F'_{cn}A = (1.11)(30.25) = 33.5 \text{ k} > 25 \text{ k} \qquad OK$$

$$\boxed{Use \quad 6 \times 6 \text{ column} \quad \text{No. 1} \quad \text{DF-L}}$$

In this example, the load combination of $1.2D + 1.6L_r$ was predetermined to be critical.

7.7   Design Problem: Capacity of a Glulam Column

This next example determines the axial load capacity of a glulam column that is fabricated from a bending combination. See Examples 7.9 (ASD) and 7.10 (LRFD). Usually, if a glulam member has an axial force only, an axial combination (rather than a bending combination) will be used. However, this example demonstrates the proper selection of $E_{x \min}$ and $E_{y \min}$ in the evaluation of a column. (Note that for an axial combination $E_{x \min} = E_{y \min} = E_{\text{axial min}}$ and the proper selection of modulus of elasticity is simplified.)

To make the use of a glulam bending combination a practical problem, one might consider the given loading to be one possible load case. Another load case could involve the *axial* force *plus* a transverse *bending* load. This second load case would require a combined stress analysis (Sec. 7.12). Two different unbraced lengths are involved in this problem, and the designer must determine the slenderness ratio for the x and y axes.

EXAMPLE 7.9   Capacity of a Glulam Column Using ASD

Determine the axial compression load capacity of the glulam column in Fig. 7.12 using ASD. The column is a $6\frac{3}{4} \times 11$ 24F-1.7E Southern Pine glulam. It is used in an industrial plant where the MC will exceed 16 percent. Normal temperatures apply. Loads are (D + S). The designer is cautioned that the critical load combination (D + S) was predetermined. To be complete, the designer should also check D alone with the corresponding C_D for permanent load of 0.9.

Glulam properties are from the NDS Supplement.

Reference design values from NDS Supplement Table 5A:

When the MC of glulam is 16 percent or greater, a wet use factor C_M less than one is required and the need for pressure treatment should be considered.

Figure 7.12   Front and side elevation views of glulam column showing different bracing conditions for column buckling about x and y axes. Also shown are section views above the respective elevations.

$$F_c = 1000 \text{ psi} \qquad C_M = 0.73$$

$$E_{x\,min} = 880,000 \text{ psi} \qquad C_M = 0.833$$

$$E_{y\,min} = 670,000 \text{ psi} \qquad C_M = 0.833$$

NOTE: E_x and E_y are used in beam deflection calculations $E_{x\,min}$ and $E_{y\,min}$ are used for stability analysis, and E_{axial} is used for axial deformation computations.

In this problem there are different unbraced lengths about the x and y axes. Therefore, the effects of column buckling about both axes of the cross section are evaluated. In a member with $E_{x\,min}$ equal to $E_{y\,min}$, this analysis would simply require the comparison of the slenderness ratios for the x and y axes [that is, $(l_e/d)_x$ and $(l_e/d)_y$]. The load capacity of the column would then be evaluated using the larger slenderness ratio $(l_e/d)_{max}$. However, in the current example there are different material properties for the x and y axes, and a full evaluation of the column stability factor is given for both axes.

Analyze Column Buckling About x Axis

$$\left(\frac{l_e}{d}\right)_x = \frac{1.0(22 \text{ ft} \times 12 \text{ in./ft})}{11 \text{ in.}} = 24.0$$

Use $E_{x\,min}$ to analyze buckling about the x axis.

$$E'_{x\,min} = E_{x\,min}(C_M)(C_t) = 880,000(0.833)(1.0)$$

$$= 733,000 \text{ psi}$$

Column stability factor x axis:

$$c = 0.9 \qquad \text{for glulam}$$

$$F_{cE} = \frac{0.822 E'_{x\,min}}{[(l_e/d)_x]^2} = \frac{0.822(733,000)}{(24.0)^2} = 1046 \text{ psi}$$

$$F_c^* = F_c(C_D)(C_M)(C_t)$$

$$= 1000(1.15)(0.73)(1.0) = 840 \text{ psi}$$

$$\frac{F_{cE}}{F_c^*} = \frac{1046}{840} = 1.25$$

$$\frac{1 + F_{cE}/F_c^*}{2c} = \frac{1 + 1.25}{2(0.9)} = 1.248$$

$$\text{For } x \text{ axis}, \quad C_P = \frac{1 + F_{cE}/F_c^*}{2c} - \sqrt{\left(\frac{1 + F_{cE}/F_c^*}{2c}\right)^2 - \frac{F_{cE}/F_c^*}{c}}$$

$$= 1.248 - \sqrt{(1.248)^2 - 1.25/0.9} = \boxed{0.832}$$

Analyze Column Buckling About y Axis

$$\left(\frac{l_e}{d}\right)_y = \frac{1.0(11 \text{ ft} \times 12 \text{ in./ft})}{6.75 \text{ in.}} = 19.6$$

Use $E_{y\,min}$ to analyze buckling about the y axis.

$$E'_{y\,min} = E_{y\,min}(C_M)(C_t) = 670{,}000(0.833)(1.0)$$

$$= 558{,}000 \text{ psi}$$

$$F_{cE} = \frac{0.822E'_{y\,min}}{[(l_e/d)_y]^2} = \frac{0.822(558{,}000)}{(19.6)^2} = 1200 \text{ psi}$$

$$F_c^* = F_c(C_D)(C_M)(C_t)$$

$$= 1000(1.15)(0.73)(1.0) = 840 \text{ psi}$$

$$\frac{F_{cE}}{F_c^*} = \frac{1194}{840} = 1.429$$

$$\frac{1 + F_{cE}/F_c^*}{2c} = \frac{1 + 1.429}{2(0.9)} = 1.349$$

$$\text{For } y \text{ axis,} \quad C_P = \frac{1 + F_{cE}/F_c^*}{2c} - \sqrt{\left(\frac{1 + F_{cE}/F_c^*}{2c}\right)^2 - \frac{F_{cE}/F_c^*}{c}}$$

$$= 1.349 - \sqrt{(1.349)^2 - 1.429/0.9} = \boxed{0.867}$$

The x axis produces the smaller value of the column stability factor, and the x axis is critical for column buckling.

$$F'_c = F_c(C_D)(C_M)(C_t)(C_P)$$

$$= 1000(1.15)(0.73)(1.0)(0.832)$$

$$= 699 \text{ psi}$$

$$P = F'_c A = 699(74.25) = 51{,}900 \text{ lb}$$

$$\boxed{P = 51.9 \text{ k (ASD)}}$$

Example 7.9 is repeated using LRFD in Example 7.10. The LRFD load combination of 1.2D + 1.6S is predetermined as controlling the design. To be complete, the designer should check all possible load combinations, including 1.4D for dead load alone.

EXAMPLE 7.10   Capacity of a Glulam Column Using LRFD

Determine the axial compression load capacity of the glulam column in Fig. 7.12 using LRFD. The column is a $6\frac{3}{4} \times 11$ 24F-1.7E Southern Pine glulam. It is used in an industrial plant where the MC will exceed 16 percent. Normal temperatures apply. Loads are 1.2D + 1.6S with $\lambda = 0.8$. $\phi_c = 0.90$ and. $\phi_s = 0.85$. For compression $K_F = 0.216/\phi_c = 2.40$, and for stability $K_F = 1.5/\phi_s = 1.765$.

Glulam properties are from the NDS Supplement.

Reference design values from NDS Supplement Table 5A:

When the MC of glulam is 16 percent or greater, a wet use factor C_M less than one is required and the need for pressure treatment should be considered.

$$F_{cn} = F_c \cdot K_F$$

$$= 1000(2.40) = 2.40 \text{ ksi} \qquad\qquad C_M = 0.73$$

$$E_{x \min n} = E_{x \min} \cdot K_F$$

$$= 880,000(1.765) = 1553 \text{ ksi} \qquad\qquad C_M = 0.833$$

$$E_{y \min n} = E_{y \min} \cdot K_F$$

$$= 670,000(1.765) = 1182 \text{ ksi} \qquad\qquad C_M = 0.833$$

NOTE: E_x and E_y are used in beam deflection calculations $E_{x \min n}$ and $E_{y \min n}$ are used for stability analysis, and E_{axial} is used for axial deformation computations.

In this problem there are different unbraced lengths about the x and y axes. Therefore, the effects of column buckling about both axes of the cross section are evaluated. In a member with $E_{x \min n}$ equal to $E_{y \min n}$, this analysis would simply require the comparison of the slenderness ratios for the x and y axes [that is, $(l_e/d)_x$ and $(l_e/d)_y$]. The load capacity of the column would then be evaluated using the larger slenderness ratio $(l_e/d)_{\max}$. However, in the current example there are different material properties for the x and y axes, and a full evaluation of the column stability factor is given for both axes.

Analyze Column Buckling About x Axis

$$\left(\frac{l_e}{d}\right)_x = \frac{1.0(22 \text{ ft} \times 12 \text{ in./ft})}{11 \text{ in.}} = 24.0$$

Use $E_{x \min n}$ to analyze buckling about the x axis.

$$E'_{x \min} = E_{x \min n}(\phi_s)(C_M)(C_t) = 1553(0.85)(0.833)(1.0)$$

$$= 1100 \text{ ksi}$$

Column stability factor x axis:

$$c = 0.9 \qquad \text{for glulam}$$

$$F_{cEn} = \frac{0.822E'_{x \min n}}{[(l_e/d)_x]^2} = \frac{0.822(1100)}{(24.0)^2} = 1.57 \text{ ksi}$$

$$F^*_{cn} = F_{cn}(\phi_c)(\lambda)(C_M)(C_t)$$

$$= 2.40(0.90)(0.8)(0.73)(1.0) = 1.26 \text{ ksi}$$

$$\frac{F_{cEn}}{F_{cn}^*} = \frac{(1.57)}{(1.26)} = 1.24$$

$$\frac{1 + F_{cEn}/F_{cn}^*}{2c} = \frac{1 + 1.24}{2(0.9)} = 1.247$$

$$\text{For } x \text{ axis, } C_P = \frac{1 + F_{cEn}/F_{cn}^*}{2c} - \sqrt{\left(\frac{1 + F_{cEn}/F_{cn}^*}{2c}\right)^2 - \frac{F_{cEn}/F_{cn}^*}{c}}$$

$$= 1.247 - \sqrt{(1.247)^2 - 1.24/0.9} = \boxed{0.832}$$

Analyze Column Buckling About y Axis

$$\left(\frac{l_e}{d}\right) = \frac{1.0(11 \text{ ft} \times 12 \text{ in./ft})}{6.75 \text{ in.}} = 19.6$$

Use $E_{y\,\min\,n}$ to analyze buckling about the y axis.

$$E'_{y\,\min\,n} = E_{y\,\min\,n}(\phi_s)(C_M)(C_t) = 1182(0.85)(0.833)(1.0)$$

$$= 837 \text{ ksi}$$

$$F_{cEn} = \frac{0.822E'_{y\,\min\,n}}{[(l_e/d)_y]^2} = \frac{0.822(837)}{(19.6)^2} = 1.80 \text{ ksi}$$

$$F_{cn}^* = F_{cn}(\phi_c)(\lambda)(C_M)(C_t)$$

$$= 2.40(0.90)(0.8)(0.73)(1.0) = 1.26 \text{ ksi}$$

$$\frac{F_{cEn}}{F_{cn}^*} = \frac{(1.80)}{(1.26)} = 1.427$$

$$\frac{1 + F_{cEn}/F_{cn}^*}{2c} = \frac{1 + 1.427}{2(0.9)} = 1.348$$

$$\text{For } y \text{ axis, } C_P = \frac{1 + F_{cEn}/F_{cn}^*}{2c} - \sqrt{\left(\frac{1 + F_{cEn}/F_{cn}^*}{2c}\right)^2 - \frac{F_{cEn}/F_{cn}^*}{c}}$$

$$= 1.348 - \sqrt{(1.348)^2 - 1.427/0.9} = \boxed{0.866}$$

The x axis produces the smaller value of the column stability factor, and the x axis is critical for column buckling.

$$F'_{cn} = F_{cn}(\phi_c)(\lambda)(C_M)(C_t)(C_P)$$

$$= 2.40(0.90)(0.8)(0.73)(1.0)(0.832)$$

$$= 1.05 \text{ ksi}$$

$$P_u \leq F'_{cn}A = (1.05)(74.25) = 77.9 \text{ k}$$

$$\boxed{P_u \leq 77.9 \text{ k (LRFD)}}$$

7.8   Design Problem: Capacity of a Bearing Wall

An axial compressive load may be applied to a wood-frame wall. For example, an interior bearing wall may support the reactions of floor or roof joists (Fig. 3.4c). Exterior bearing walls also carry reactions from joists and rafters, but in addition, exterior walls usually must be designed to carry lateral wind forces.

In Example 7.11, the vertical load capacity of a wood-frame wall is determined using ASD. As noted in the previous examples, there are few differences in column design between ASD and LRFD. Therefore, only ASD is presented for this problem. Two main factors should be noted about this problem, however. The first relates to the column capacity of a stud in a wood-frame wall. Because sheathing is attached to the stud throughout its height, continuous lateral support is provided in the x direction. Therefore, the possibility of buckling about the weak y axis is prevented. The column capacity is evaluated by the slenderness ratio about the strong axis of the stud, $(l_e/d)_x$.

The second factor to consider is the bearing capacity of the top and bottom wall plates. It is possible that the vertical load capacity of a bearing wall may be governed by compression perpendicular to the grain on the wall plates rather than by the column capacity of the studs. This is typically not a problem with major columns in a building because steel bearing plates can be used to distribute the load perpendicular to the grain on supporting members. However, in a standard wood frame wall, the stud bears *directly* on the horizontal wall plates.

EXAMPLE 7.11   Capacity of a Stud Wall Using ASD

Determine the vertical load capacity of the stud shown in Fig. 7.13a. There is no bending. Express the maximum load in pounds per lineal foot of wall. Lumber is Standard-grade Hem-Fir. Load is (D + S). $C_M = 1.0$, $C_t = 1.0$, and $C_i = 1.0$.

Figure 7.13a   Sheathing provides lateral support about y axis of stud.

Wall studs are 2 × 4 (Dimension lumber size category). Values from NDS Supplement Table 4A:

$$F_c = 1300 \text{ psi}$$

$$C_F = 1.0 \quad \text{for compression}$$

$$E_{min} = 440{,}000 \text{ psi}$$

$$F_{c\perp} = 405 \text{ psi} \quad \text{bearing capacity of wall plate}$$

Column Capacity of Stud

Buckling about the weak axis of the stud is prevented by the sheathing, and the only slenderness ratio required is for the x axis. The length of the stud is the height of the wall less the thicknesses of the top and bottom plates.

$$l_e = (9.5 \text{ ft} \times 12 \text{ in./ft}) - 3 \times 1.5 \text{ in.} = 109.5 \text{ in.}$$

$$\left(\frac{l_e}{d}\right)_x = \frac{1.0(109.5)}{3.5 \text{ in.}} = 31.3$$

$$E'_{min} = E_{min}(C_M)(C_t)(C_T)(C_i) = 440{,}000(1.0)(1.0)(1.0)(1.0)$$

$$= 440{,}000 \text{ psi}$$

For visually graded sawn lumber, $c = 0.8$

$$F_{cE} = \frac{0.822E'_{min}}{[(l_e/d)_x]^2} = \frac{0.822(440{,}000)}{(31.3)^2} = 370 \text{ psi}$$

$$F_c^* = F_c(C_D)(C_M)(C_t)(C_F)(C_i)$$

$$= 1300(1.15)(1.0)(1.0)(1.0)(1.0) = 1495 \text{ psi}$$

$$\frac{F_{cE}}{F_c^*} = \frac{370}{1495} = 0.247$$

$$\frac{1 + F_{cE}/F_c^*}{2c} = \frac{1 + 0.247}{2(0.8)} = 0.779$$

$$C_P = \frac{1 + F_{cE}/F_c^*}{2c} - \sqrt{\left(\frac{1 + F_{cE}/F_c^*}{2c}\right)^2 - \frac{F_{cE}/F_c^*}{c}}$$

$$= 0.779 - \sqrt{(0.779)^2 - 0.247/0.8} = 0.233$$

$$F'_c = F_c(C_D)(C_M)(C_t)(C_F)(C_P)(C_i)$$

$$= 1300(1.15)(1.0)(1.0)(1.0)(0.233)(1.0)$$

$$= 348 \text{ psi}$$

$$P = F'_c A = 348(5.25) = 1829 \text{ lb}$$

$$\text{Max. } w = \frac{1829 \text{ lb}}{1.33 \text{ ft}} = \boxed{1371 \text{ lb/ft}}$$

Bearing Capacity of Wall Plates (Fig. 7.13b)

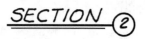

Figure 7.13b Bearing on bottom wall plate.

The conditions necessary to apply the bearing area factor are summarized in Fig. 6.16b (Sec. 6.8). Since the bottom plate will typically be composed of multiple pieces of sawn lumber placed end-to-end, it is possible that some studs will be located within 3 in. of the cut end of the wall plate. Therefore, the bearing area factor conservatively defaults to 1.0. Recall that C_D does not apply to $F_{c\perp}$.

$$F'_{c\perp} = F_{c\perp}(C_M)(C_t)(C_b)$$

$$= 405(1.0)(1.0)(1.0)$$

$$= 405 \text{ psi} > 348 \text{ psi}$$

$$F'_{c\perp} > F'_c$$

∴ column capacity governs over bearing perpendicular to the grain.

While the design of columns, including load bearing walls, is identical for both ASD and LRFD, it is important to note that a difference does arise when considering bearing capacity of wall plates. As shown in Example 7.11, the design value for compression perpendicular to grain $F_{c\perp}$ is not adjusted for load duration when using ASD. However, when using LRFD, the nominal compression perpendicular to grain design value $F_{c\perp n}$ is adjusted for time effects. For Example 7.11 using LRFD, $K_F = 1.875/\phi_c = 2.083$, and

$$F_{c\perp n} = F_\perp \cdot K_F = 405(2.083) = 844 \text{ psi} = 0.844 \text{ ksi}$$

$$F'_{c\perp n} = F_{c\perp n}(\phi_c)(\lambda)(C_M)(C_t)(C_b)$$

$$= (0.844)(0.90)(0.8)(1.0)(1.0)(1.0) = 0.608 \text{ ksi}$$

The column capacity of the studs would then need to be less than 0.608 ksi for the bearing capacity of the wall plates to govern in LRFD.

7.9   Built-Up Columns

A *built-up* column is constructed from several parallel wood members that are nailed or bolted together to function as a composite column. These are distinguished from *spaced columns,* which have specially designed timber connectors to transfer shear between the separate parallel members.

The NDS includes criteria for designing spaced columns (NDS Sec. 15.2). Spaced columns can be used to increase the maximum load capacity in compression members in heavy wood trusses. They are, however, relatively expensive to fabricate and are not often used in ordinary wood buildings. For this reason, spaced columns are not covered in this text.

Built-up columns see wider use because they are fairly easy to fabricate. Their design is briefly covered here. The combination of several members in a built-up column results in a member with a larger cross-sectional dimension d and, correspondingly, a smaller slenderness ratio l_e/d.

With a smaller slenderness ratio, a larger column capacity can be achieved. However, the fasteners connecting the members do not fully transfer the shear between the various pieces, and the capacity of a *built-up* column is less than the capacity of a *solid* sawn or glulam column of the same size and grade.

The capacity of a built-up column is determined by first calculating the column capacity of an equivalent solid column. This value is then reduced by an adjustment factor K_f that depends on whether the built-up column is fabricated with nails or bolts. This procedure is demonstrated in Example 7.12 using LRFD.

Recall that, in general, a column has two slenderness ratios: one for possible buckling about the x axis, and another for buckling about the y axis. In a typical problem, the adjusted column design value is simply evaluated using the larger of the two slenderness ratios.

For a built-up column this may or may not be the controlling condition. Because the reduction factor K_f measures the effectiveness of the shear transfer between the individual laminations, the K_f factor applies only to the column slenderness ratio for the *axis parallel to the weak axis of individual laminations.* In other words, the column slenderness ratio parallel to the strong axis of the individual laminations does not require the K_f reduction. Depending on the relative magnitude of $(l_e/d)_x$ and $(l_e/d)_y$, the evaluation of *two* column design values may be required.

The design of built-up columns is covered in NDS Sec. 15.3. The procedure applies to columns that are fabricated from two to five full-length parallel members that are nailed or bolted together. Details for the nailing or bolting of the members in order to qualify as a built-up column are given in the NDS.

Research has been conducted at the Forest Products Laboratory (FPL) (Ref. 7.3) on the effect of built-up columns fabricated with members that are not continuous over the full length of the column. Information regarding design recommendations for nail-laminated posts with butt joints may be obtained from the FPL.

EXAMPLE 7.12   Strength of Built-Up Column Using LRFD

Determine the maximum factored axial load on the built-up column in Fig. 7.14. Lumber is No. 1 DF-L. Column length is 13 ft-0 in., $\lambda = 0.8$, $C_M = 1.0$, $C_t = 1.0$, and $C_i = 1.0$. $\phi_c = 0.90$ and $\phi_s = 0.85$. For compression $K_F = 2.16/\phi_c = 2.40$, and for stability $K_F = 1.5/\phi_s = 1.765$.

The 2 × 6s comprising the built-up column are in the Dimension lumber size category. Design values are obtained from NDS Supplement Table 4A:

$$F_{cn} = F_c \cdot K_F$$

$$= 1500(2.40) = 3.60 \text{ ksi}$$

$$C_F = 1.1 \qquad \text{for compression}$$

$$E_{\min n} = E_{\min} \cdot K_F$$

$$= 620{,}000(1.765) = 1094 \text{ ksi}$$

3 FULL LENGTH 2x6 LAMINATIONS NAILED TOGETHER

Figure 7.14 Cross section of nailed built-up column. For nailing requirements see NDS Sec. 15.3.3.

BUILT-UP COLUMN

Column Capacity

$$\left(\frac{l_e}{d}\right)_x = \frac{1.0(13.0 \text{ ft} \times 12 \text{ in./ft})}{5.5} = 28.4$$

$$\left(\frac{l_e}{d}\right)_y = \frac{1.0(13.0 \text{ ft} \times 12 \text{ in./ft})}{4.5 \text{ in.}} = 34.7$$

$$\left(\frac{l_e}{d}\right)_{\max} = \left(\frac{l_e}{d}\right)_y$$

The adjustment factor K_f applies to the design value for the column axis that is parallel to the weak axis of the individual laminations. Therefore, K_f applies to the column design value based on $(l_e/d)_y$. In this problem, the y axis is critical for both column buckling as well as the reduction for built-up columns, and only one adjusted column design value needs to be evaluated.

$$E'_{\min n} = E_{\min n}(\phi_s)(C_M)(C_t)(C_T)(C_i) = 1094(0.85)(1.0)(1.0)(1.0)(1.0)$$

$$= 930 \text{ ksi}$$

For visually graded sawn lumber, $c = 0.8$

$$F_{cEn} = \frac{0.822E'_{\min n}}{(l_e/d)^2} = \frac{0.822(930)}{(34.7)^2} = 0.636 \text{ ksi}$$

$$F^*_{cn} = F_{cn}(\phi_c)(\lambda)(C_M)(C_t)(C_F)(C_i)$$

$$= 3.60(0.90)(0.8)(0.1)(1.0)(1.1)(1.0) = 2.85 \text{ ksi}$$

$$\frac{F_{cEn}}{F^*_{cn}} = \frac{(0.636)}{(2.85)} = 0.223$$

$$\frac{1 + F_{cEn}/F^*_{cn}}{2c} = \frac{1 + 0.223}{2(0.8)} = 0.764$$

For nailed built-up columns (see NDS Sec. 15.3.2),

$$K_f = 0.6$$

$$C_P = K_f \left[\frac{1 + F_{cEn}/F^*_{cn}}{2c} - \sqrt{\left(\frac{1 + F_{cEn}/F^*_{cn}}{2c} \right)^2 - \frac{F_{cEn}/F^*_{cn}}{c}} \right]$$

$$= 0.6[0.764 - \sqrt{(0.764)^2 - 0.233/0.8}]$$

$$= 0.6(0.212) = 0.127$$

$$F'_{cn} = F_{cn}(\phi_c)(\lambda)(C_M)(C_t)(C_F)(C_P)(C_i)$$

$$= 3.60(0.90)(0.8)(1.0)(1.0)(1.1)(0.127)(1.0)$$

$$= 0.362 \text{ ksi}$$

Capacity of a nailed built-up (three 2 × 6s) column:

$$P_u \le F'_{cn}A = (0.362 \text{ ksi})(3 \times 8.25 \text{ in.}^2)$$

$$\boxed{P_u \le 8.96 \text{ k (LRFD)}}$$

For the nailing requirements of a mechanically laminated built-up column, see NDS Sec. 15.3.3.

In Example 7.12, the y axis gave the maximum slenderness ratio, and it also required the use of the reduction factor K_f. In such a problem, only one adjusted column design value needs to be evaluated. However, another situation could require the evaluation of a second adjusted column design value. For example, if an additional lamination is added to the column in Fig. 7.14, the maximum slenderness ratio would become $(l_e/d)_x$, and an adjusted column design value would be determined using $(l_e/d)_x$ *without* K_f. Another design value would be evaluated using $(l_e/d)_y$ *with* K_f. The smaller of the two adjusted design values would then govern the capacity of the built-up member.

7.10 Combined Bending and Tension

When a bending moment occurs simultaneously with an axial tension force, the effects of combined stresses must be taken into account. The distribution of axial tensile stresses and bending stresses can be plotted over the depth of the cross section of a member. See Example 7.13.

From the plots of combined stress (Fig. 7.15a), it can be seen that on one face of the member the axial tensile stresses and the bending tensile stresses add. On the opposite face, the axial tensile stresses and bending compressive stresses cancel. Depending on the magnitude of the stresses involved, the resultant stress on this face can be *either* tension or compression.

Therefore, the capacity of a wood member with this type of combined loading can be governed by either a combined *tension* criterion or a net *compression* criterion. These criteria are given in NDS Sec. 3.9.1, *Bending and Axial Tension*.

Combined axial tension and bending tension

First, the combined *tensile* stresses are analyzed in an *interaction equation*. In this case, the interaction equation is a straight-line expression (Fig. 7.15b) which is made up of two terms known as *stress ratios*. The first term measures the effects of axial tension, and the second term evaluates the effects of bending. In each case, the actual (working or factored) stress is divided by the corresponding adjusted design value. The adjusted design values in the denominators are determined in the usual way with one exception. Because the actual (working or factored) bending stress is the *bending tensile* stress, the adjusted bending design value does *not* include the lateral stability factor C_L. In other words, F_b^* is F_b' or F_{bn}^* is F_{bn}' determined with C_L set equal to unity.

It may be convenient to think of the stress ratios in the interaction equation as percentages or fractions of total member capacity. For example, the ratio of actual tension stress to adjusted tension design value, f_t/F_t' or f_{tu}/F_{tn}', can be viewed as the fraction of total member capacity that is used to resist axial tension.[†] The ratio of actual bending stress to adjusted bending design value f_b/F_b^* or f_{bu}/F_{bn}^* then represents the fraction of total member capacity used to resist bending.[†] The sum of these fractions must be less than the total member capacity, 1.0.

Net compressive stress

Second, the combined stresses on the opposite face of the member are analyzed. If the sense of the combined stress on this face is tension, no additional work is required. However, if the combined stress at this point is *compressive*, a bending analysis is required. The adjusted bending design value in the denominator of the second combined stress check must reflect the lateral stability of the compression edge of the member. This is done by including the lateral stability factor C_L in evaluating F_b' or F_{bn}'. Refer to Sec. 6.3 for information on C_L.

[†]It is common when using LRFD to express these ratios in terms of forces and moments, or T_u/T_n' and M_u/M_n'.

EXAMPLE 7.13   Criteria for Combined Bending and Tension

The axial tensile stress and bending stress distributions are shown in Fig. 7.15a. If the individual stresses are added algebraically, one of two possible combined stress distributions will result. If the axial tensile stress is larger than the bending stress, a *trapezoidal* combined stress diagram results in tension everywhere throughout the depth of the member (combined stress diagram 1). If the axial tensile stress is smaller than the bending stress, the resultant combined stress diagram is *triangular* (combined stress diagram 2).

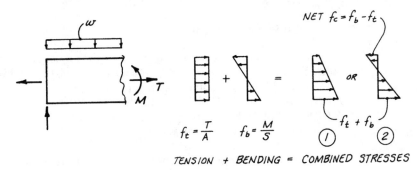

Figure 7.15a   Combined bending and axial tension stresses.

Theoretically the following stresses in Fig. 7.15a are to be analyzed:

1. *Combined axial tension and bending tension stress.* This is done by using the straight-line interaction equation described below. These combined stresses are shown at the bottom face of the member in stress diagrams 1 and 2.

2. *Net compressive stress.* This stress is shown on the top surface of the member in stress diagram 2.

Combined Axial Tension and Bending Tension

A basic straight-line interaction equation is used for combined axial tensile and bending tensile stresses (NDS Eq. 3.9-1). The two stress ratios define a point on the graph in Fig. 7.15b. If the point lies on or below the line representing 100 percent of member strength, the interaction equation is satisfied.

INTERACTION EQUATION

$$\frac{f_t}{F'_t} + \frac{f_b}{F^*_b} \leq 1.0 \qquad \text{(ASD)}$$

$$\frac{f_{tu}}{F'_{tn}} + \frac{f_{bu}}{F^*_{bn}} \leq 1.0 \qquad \text{(LRFD)}$$

where f_t = actual working (ASD) tensile stress parallel to grain
    $= T/A$
  f_{tu} = factored (LRFD) tensile stress parallel to grain
    $= T_u/A$
  F'_t = adjusted ASD tensile design value (Sec. 7.2)

$\qquad = F_t(C_D)(C_M)(C_t)(C_F)(C_i)$ for sawn lumber

$\qquad = F_t(C_D)(C_M)(C_t)$ for glulam

F_{tn} = nominal LRFD tension design value

$\qquad = F_t \cdot K_F$

F'_{tn} = adjusted LRFD tension design value

$\qquad = F_{tn}(\phi_t)(\lambda)(C_M)(C_t)(C_F)(C_i)$ for sawn lumber

$\qquad = F_{tn}(\phi_t)(\lambda)(C_M)(C_t)$ for glulam

f_b = actual working (ASD) bending stress. For usual case of bending about x axis, this is f_{bx}.

$\qquad = M/S$

f_{bu} = factored (LRFD) bending stress

$\qquad = M_u/S$

F_b^* = adjusted ASD bending design value without the adjustment for lateral stability. For the usual case of bending about the x axis of a rectangular cross section, the adjusted bending design value is

$\qquad = F_b(C_D)(C_M)(C_t)(C_F)(C_r)(C_i)$ for sawn lumber

$\qquad = F_b(C_D)(C_M)(C_t)(C_V)$ for glulam

F_{bn} = nominal LRFD bending design value

$\qquad = F_b K_F$

F_{bn}^* = adjusted LRFD bending design value without the adjustment for lateral stability

$\qquad = F_b^* \cdot K_F$

$\qquad = F_{bn}(\phi_b)(\lambda)(C_M)(C_t)(C_F)(C_r)(C_i)$ for sawn lumber

$\qquad = F_{bn}(\phi_b)(\lambda)(C_M)(C_t)(C_V)$ for glulam

ϕ_t = resistance factor for tension (LRFD)

$\qquad = 0.80$

ϕ_b = resistance factor for bending (LRFD)

$\qquad = 0.85$

Figure 7.15b Interaction curve for axial tension plus bending tension.

Net Compressive Stress

When the bending compressive stress exceeds the axial tensile stress, the following stability check is given by NDS Eq. 3.9-2.

$$\frac{Net\ f_c}{F'_b} = \frac{f_b - f_t}{F^{**}_b} \le 1.0 \quad (ASD)$$

$$\frac{Net\ f_{cu}}{F'_{bn}} = \frac{f_{bu} - f_{tu}}{F^{**}_{bn}} \le 1.0 \quad (LRFD)$$

where Net f_c = net compressive stress (ASD)

Net f_{cu} = factored net compression stress (LRFD)

f_b = actual working (ASD) bending stress. For usual case of bending about x axis, this is f_{bx}.

   = M/S (ASD/LRFD)

f_{bu} = factored (LRFD) bending stress

   = M_u/S

f_t = actual working (ASD) axial tensile stress parallel to grain

   = T/A (ASD/LRFD)

f_{tu} = factored (LRFD) axial tensile stress parallel to grain

   = T_u/A

F^{**}_b = F'_b adjusted ASD bending design value (Sec. 6.3). The beam stability factor C_L applies, but the volume factor C_V does not. For usual case of bending about x axis of rectangular cross section, adjusted bending design value is

   = $F_b(C_D)(C_M)(C_t)(C_L)(C_F)(C_r)(C_i)$ for sawn lumber

   = $F_b(C_D)(C_M)(C_t)(C_L)$ for glulam

F^{**}_{bn} = adjusted LRFD bending design value, including the beam stability factor C_L but not the volume factor C_V.

   = $F_{bn}(\phi_b)(\lambda)(C_M)(C_t)(C_L)(C_F)(C_r)(C_i)$ for sawn lumber

   = $F_{bn}(\phi_b)(\lambda)(C_M)(C_t)(C_L)$ for glulam

ϕ_b = resistance factor for bending (LRFD)

   = 0.85

The designer should use a certain degree of caution in applying the criterion for the *net compressive stress*. Often, combined stresses of this nature are the result of different loadings, and the maximum bending compressive stress may occur *with* or *without* the axial tensile stress.

For example, the bottom chord of the truss in Fig. 7.16*a* will always have the dead load moment present regardless of the loads applied to the top chord. Thus, f_b is a constant in this example. However, the tension force in the member varies depending on the load applied to the top chord. The tensile stress f_t will be small if dead load alone is considered, and it will be much larger under dead load plus snow load. To properly check the net compressive stress, the designer must determine the *minimum* f_t that will occur simultaneously with the bending stress f_b.

On the other hand, a simple and conservative approach for evaluating bending compressive stresses is to ignore the reduction in bending stress provided by the axial tensile stress. In this case, the compressive stress ratio becomes

$$\frac{Gross\ f_c}{F'_b} = \frac{f_b}{F^{**}_b} \le 1.0$$

This check on the *gross* bending compressive stress can be rewritten as

$$f_b \leq F'_b = F_b^{**}$$

where $F'_b = F_b^{**}$ is the adjusted bending design value considering the effects of lateral stability and other applicable adjustments. This more conservative approach is used for the examples in this book. Under certain circumstances, however, the designer may wish to consider the *net* compressive stress. For example, the designer may be overly conservative in using the gross bending compressive stress rather than the net compressive stress when tension and bending are caused by the same load, such as in the design of wall studs for wind-induced bending and uplift in high wind zones.

It should be noted that NDS Sec. 3.9.1 introduces special notation for F'_b for use in the combined bending plus axial tension interaction equations. The symbols F_b^* and F^{**} are adjusted bending design values obtained with certain adjustment factors deleted. These are noted in Example 7.13.

The design expressions given in Example 7.13 are adequate for most combined bending and axial tension problems. However, as wood structures become more highly engineered, there may be the need to handle problems involving axial tension plus bending about both the x and y axes. In this case, the expanded criteria in Example 7.14 may be used.

EXAMPLE 7.14   Generalized Criteria for Combined Bending and Tension

The problem of biaxial bending plus axial tension has not been studied extensively. However, the following interaction equations are extensions of the NDS criteria for the general axial tension plus bending problem.

Combined Axial Tension and Bending Tension

$$\frac{f_t}{F'_t} + \frac{f_{bx}}{F_{bx}^*} + \frac{f_{by}}{F'_{by}} \leq 1.0 \qquad \text{(ASD)}$$

$$\frac{f_{tu}}{F'_{tn}} + \frac{f_{bxu}}{F_{bxn}^*} + \frac{f_{byu}}{F'_{by}} \leq 1.0 \qquad \text{(LRFD)}$$

The three terms all have the same sign. The adjusted design values include applicable adjustments (the adjustment for lateral stability C_L does not apply to bending tension).

Net Compressive Stress

For the check on *net* bending compressive stress, the tension term is negative:

$$-\frac{f_t}{F'_t} + \frac{f_{bx}}{F'_{bx}} + \frac{f_{by}}{F'_{by}} \leq 1.0 \qquad \text{(ASD)}$$

$$-\frac{f_{tu}}{F'_{tn}} + \frac{f_{bxu}}{F'_{bxn}} + \frac{f_{byu}}{F'_{byn}} \leq 1.0 \qquad \text{(LRFD)}$$

Depending on the magnitudes of the stresses involved, or for reasons of simplicity, the designer may prefer to omit the negative term in the expression.

An even more conservative approach is to apply the general interaction formula for the net compressive stress in Example 7.17 (Sec. 7.12) with the axial component of the expression set equal to zero. This provides a more conservative biaxial bending (i.e., bending about the x and y axes) interaction formula. The adjusted bending design values in the biaxial interaction equation would include all appropriate adjustment factors. Because the focus is on compression, the effect of lateral torsional buckling is to be taken into account with the beam stability factor C_L, but the volume factor C_V does not apply.

7.11 Design Problem: Combined Bending and Tension

The truss in Example 7.15 is similar to the truss in Example 7.2. The difference is that in the current example, an additional load is applied to the bottom chord. This load is uniformly distributed and represents the weight of a ceiling supported by the bottom chord of the truss.

The first part of the example deals with the calculation of the axial force in member AC. In order to analyze a truss using the method of joints it is necessary for the loads to be resolved into joint loads. The tributary width to the three joints along the top chord is 7.5 ft, and the tributary width to the joint at the midspan of the truss on the bottom chord is 15 ft. The remaining loads (both top and bottom chord loads) are tributary to the joints at each support.

The design of a combined stress member is a trial-and-error procedure. In this example, a 2 × 8 bottom chord is the initial trial, and it proves satisfactory. *Independent* checks on the tension and bending stresses are first completed. The independent check on the bending stress automatically satisfies the check on the *gross* bending compression discussed in Sec. 7.10. Finally, the combined effects of tension and bending are evaluated.

It should be noted that the load duration factor used for the independent check for axial tension is $C_D = 1.15$ for combined (D + S). Dead load plus snow causes the axial force of 4.44 k. The independent check of bending uses $C_D = 0.9$ for dead load, because only the dead load of the ceiling causes the bending moment of 10.8 in.-k in the bottom chord. Dead load bending alone is a simplified and conservative assumption since dead load of the upper chords would add tension to the number.

In the combined stress check, however, $C_D = 1.15$ applies to *both* the axial and the bending portions of the interaction formula. Recall that the C_D to be used in checking stresses caused by a *combination* of loads is the one associated with the shortest-duration load in the combination. For combined stresses, then, the same C_D applies to both terms.

EXAMPLE 7.15   Combined Bending and Tension Using ASD

Design the lower chord of the truss in Fig. 7.16a using ASD. Use No. 1 and Better Hem-Fir. MC ≤ 19 percent, and normal temperature conditions apply. Connections will be made with a single row of $\tfrac{3}{4}$-in.-diameter bolts. Connections are assumed to be pinned. Trusses are 4 ft-0 in. o.c. Loads are applied to both the top and bottom chords. Assume that the ends of the member are held in position (not allowed to rotate laterally) at each joint. Design values are from the NDS Supplement.

Figure 7.16a Loading diagram for truss. The uniformly distributed loads between the truss joints cause bending stresses in top and bottom chords, in addition to axial truss forces. Bottom chord has combined bending and axial tension.

Loads

TOP CHORD:

$$D = 14 \text{ psf} \qquad \text{(horizontal plane)}$$
$$\underline{S = 30 \text{ psf} \qquad \text{(reduced snow load based on roof slope)}}$$
$$TL = 44 \text{ psf}$$
$$w_{TL} = 44 \times 4 = 176 \text{ lb/ft to truss}$$

Load to joint for truss analysis (Fig. 7.16b):

$$P_T = 176 \times 7.5 = 1320 \text{ lb/joint}$$

BOTTOM CHORD:

$$\text{Ceiling } D = 8 \text{ psf}$$
$$w_D = 8 \times 4 = 32 \text{ lb/ft}$$

Load to joint C for truss analysis:

$$P_B = 32 \times 15 = 480 \text{ lb/joint}$$

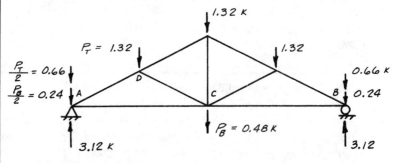

Figure 7.16b Loading diagram for truss. The distributed loads to the top and bottom chords are converted to concentrated joint forces for conventional truss analysis.

Force in lower chord (method of joints) (Fig. 7.16c):

Figure 7.16c   Free-body diagram of joint A.

Load diagram for tension chord AC (Fig. 7.16d):

Figure 7.16d   Loading diagram for member AC. The tension force is obtained from the truss analysis, and the bending moment is the result of the transverse load applied between joints A and C.

Member Design

Try 2 × 8 No. 1 & Better Hem-Fir (Dimension lumber size category).
  Values from NDS Supplement Table 4A:

$$F_b = 1100 \text{ psi}$$

$$F_t = 725 \text{ psi}$$

Size factors:

$$C_F = 1.2 \text{ for bending}$$

$$C_F = 1.2 \text{ for tension}$$

Section properties:

$$A_g = 10.875 \text{ in.}^2$$

$$S = 13.14 \text{ in.}^3$$

AXIAL TENSION:

1. Check axial tension at the net section Fig. 7.16e. Because this truss is assumed to have pinned connections, the bending moment is theoretically zero at this point. Assume the hole diameter is $\frac{1}{16}$ in. larger than the bolt diameter (for stress calculations only).

Figure 7.16e Net section for tension member.

$$A_n = 1.5[7.25 - (0.75 + 0.0625)] = 9.66 \text{ in.}^2$$

$$f_t = \frac{T}{A_n} = \frac{4440}{9.66} = 460 \text{ psi}$$

$$F_t' = F_t(C_D)(C_M)(C_t)(C_F)(C_i)$$

$$= 725(1.15)(1.0)(1.0)(1.2)(1.0) = 1000 \text{ psi}$$

$$1000 \text{ psi} > 460 \text{ psi} \quad OK$$

2. Determine tension stress at the point of maximum bending stress (midspan) for use in the interaction formula.

$$f_t = \frac{T}{A_g} = \frac{4440}{10.875} = 408 \text{ psi} < 1000 \text{ psi} \quad OK$$

BENDING:

For a simple beam with a uniform load,

$$M = \frac{wL^2}{8} = \frac{32(15)^2}{8} = 900 \text{ ft-lb} = 10,800 \text{ in.-lb}$$

$$f_b = \frac{M}{S} = \frac{10,800}{13.14} = 822 \text{ psi}$$

The problem statement indicates that rotation of the ends of the member at the truss joints was restrained. In addition, the truss spacing exceeds the limit for repetitive members. Therefore, the beam stability factor C_L, according to NDS Section 4.4.1.2, and the repetitive-member factor C_r are both 1.0.

The bending stress of 822 psi is caused by dead load alone. Therefore, use $C_D = 0.9$ for an independent check on bending. Later use $C_D = 1.15$ for the combined stress check in the interaction formula.

$$F_b' = F_b(C_D)(C_M)(C_t)(C_L)(C_F)(C_r)(C_i)$$

$$= 1100(0.9)(1.0)(1.0)(1.0)(1.2)(1.0)(1.0)$$

$$= 1188 \text{ psi} > 822 \text{ psi} \quad OK$$

In terms of NDS notation, this value of F_b' is F_b^{**}.

COMBINED STRESSES:

1. Axial tension plus bending. Two load cases should be considered for axial tension plus bending: D-only and D + S. It has been predetermined that the D + S load combination with a $C_D = 1.15$ governs over dead load acting alone with a $C_D = 0.9$.

$$F_b^* = F_b' = F_b(C_D)(C_M)(C_t)(C_F)(C_r)(C_i)$$

$$= 1100(1.15)(1.0)(1.0)(1.2)(1.0)(1.0)$$

$$= 1518 \text{ psi}$$

$$\frac{f_t}{F_t'} + \frac{f_{bx}}{F_{bx}^*} = \frac{408}{1000} + \frac{822}{1518} = 0.95$$

$$0.95 < 1.0 \qquad \text{OK}$$

NOTE: It can be determined that $f_t = 160$ psi results from D-only and $F_t' = 783$ psi with $C_D = 0.9$. The f_b and F_b' values were determined previously for D-only bending. The D-only axial tension plus bending interaction results in a value of 0.90 vs. 0.95 for D + S. If the specified snow load had been slightly smaller, then D acting alone would have governed the design of the lower chord.

2. Net bending compressive stress. The *gross* bending compressive stress was shown to be not critical in the independent check on bending ($f_b = 822$ psi $< F_b' = 1188$ psi). Therefore, the *net* bending compressive stress is automatically *OK*. The simpler, more conservative check on compression is recommended in this book. However, the more accurate interaction approach should be used if the gross bend compressive check indicated the member is inadquate.

The 2 × 8 No. 1 & Btr Hem-Fir member is seen to pass the combined stress check. The combined stress ratio of 0.95 indicates that the member is roughly overdesigned by 5 percent (1.0 corresponds to full member capacity or 100 percent of member strength).

$$(1.0 - 0.95)100 = 5\% \text{ overdesign}$$

Because of the inaccuracy involved in estimating design loads and because of variations in material properties, a calculated overstress of 1 or 2 percent (i.e., combined stress ratio of 1.01 or 1.02) is considered by many designers to still be within the "spirit" of the design specifications. Judgments of this nature must be made individually with knowledge of the factors relating to a particular problem. However, in this problem the combined stress ratio is less than 1.0, and the trial member size is acceptable.

$$\boxed{Use \quad 2 \times 8 \quad \text{No. 1 \& Btr} \quad \text{Hem-Fir}}$$

NOTE: The simplified analysis used in this example applies only to trusses with pinned joints. For a truss connected with toothed metal plate connectors, a design approach should be used which takes the continuity of the joints into account (Ref. 7.6).

While Example 7.15 is presented using ASD, the approach using LRFD would be the same. It should be noted, however, that the time effect factor used for the independent check for axial tension is $\lambda = 0.8$ for the 1.2D + 1.6S load combination. The independent check for bending would use $\lambda = 0.6$ for the 1.4D load combination. In the combined stress check, the time effect factor $\lambda = 0.8$ is used for both the axial and bending portions of the interaction formula since the load combination creating both the tension and bending is 1.2D + 1.6S.

7.12   Combined Bending and Compression

Structural members that are stressed simultaneously in bending and compression are known as *beam-columns*. These members occur frequently in wood buildings, and the designer should have the ability to handle these types of problems. In order to do this, it is first necessary to have a working knowledge of laterally unsupported beams (Sec. 6.3) and axially loaded columns (Secs. 7.4 and 7.5). The interaction formulas presented in this section can then be used to handle the combination of these stresses.

The straight-line interaction equation was introduced in Fig. 7.15*b* (Sec. 7.10) for combined bending and axial tension. At one time a similar straight-line equation was also used for the analysis of beam-columns. More recent editions of the NDS use a modified version of the interaction equation.

There are many variables that affect the strength of a beam-column. The NDS interaction equation for the analysis of beam-columns was developed by Zahn (Ref. 7.7). It represents a unified treatment of

1. Column buckling

2. Lateral torsional buckling of beams

3. Beam-column interaction

The Ylinen buckling formula was introduced in Sec. 7.4 for column buckling and in Sec. 6.3 for the lateral buckling of beams. The adjusted design values F'_c and F'_{bx} determined in accordance with these previous sections are used in the Zahn interaction formula to account for the first two items. The added considerations for the *simultaneous* application of beam and column loading can be described as beam-column interaction. These factors are addressed in this section.

When a bending moment occurs simultaneously with an axial compressive force, a more critical combined stress problem exists in comparison with combined bending and tension. In a beam-column, an additional bending stress is created which is known as the *P-Δ effect*. The *P-Δ* effect can be described in this way. First consider a member without an axial load. The bending moment developed by the *transverse* loading causes a deflection Δ. When the axial force *P* is then applied to the member, an *additional* bending moment equal to $P \times \Delta$ is generated. See Example 7.16. The *P-Δ* moment is known as a *second-order effect* because the added bending stress is not calculated directly. Instead, it is taken into account by *amplifying* the computed bending stress in the interaction equation.

The most common beam-column problem involves axial compression combined with a bending moment about the *strong* axis of the cross section. In this case, the actual bending stress f_{bx} is multiplied by an amplification factor that reflects the magnitude of the load P and the deflection Δ. This concept should be familiar to designers who also do structural steel design. The amplification factor in the NDS is similar to the one used for beam-columns in the AISC steel specification (Ref. 7.2).

The amplification factor is a number greater than 1.0 given by the following expression:

$$\text{Amplification factor for } f_{bx} = \left(\frac{1}{1 - f_c/F_{cEx}} \right) \quad \text{using ASD}$$

$$= \left(\frac{1}{1 - f_{cu}/F_{cExn}} \right) \quad \text{using LRFD}$$

This amplification factor is made up of two terms that measure the P-Δ effect for a bending moment about the strong axis (x-axis). The intent is to have the amplification factor increase as

1. Axial force P increases.

2. Deflection Δ due to bending about the x axis increases.

Obviously, the compressive stress $f_c = P/A$ increases as the load P increases. As f_c becomes larger, the amplification factor will increase.

The increase in the amplification factor due to an increase in Δ may not be quite as clear. The increase for Δ is accomplished by the term F_{cE}. F_{cE} is defined as the value obtained from the Euler buckling stress formula evaluated using the column slenderness ratio for the axis about which the bending moment is applied. Thus, if the transverse loads cause a moment about the x axis, the *slenderness ratio about the x axis* is used to determine F_{cE}. The notation used in this book for this quantity is F_{cEx}.

Figure 7.6a (Sec. 7.4) shows both the Ylinen column equation and the Euler equation. For purposes of beam-column analysis, it should be understood that the adjusted column design value F'_c is defined by the Ylinen formula, but the amplification factor for P-Δ makes use of the Euler formula. For use in the amplification factor, the value given by the expression for F_{cEx} is applied over the entire range of slenderness ratios. In other words, F_{cEx} goes to ∞ as the slenderness ratio becomes small, and F_{cEx} approaches 0 as $(l_e/d)_x$ becomes large.

The logic in using F_{cEx} in the amplification factor for P-Δ is that the deflection will be large for members with a large slenderness ratio. Likewise, Δ will be small as the slenderness ratio decreases. Thus, F_{cEx} produces the desired effect on the amplification factor.

It is necessary for the designer to clearly understand the reasoning behind the amplification factor. In a general problem there are two slenderness ratios: one for the x axis $(l_e/d)_x$ and one for the y axis $(l_e/d)_y$. In order to analyze combined stresses, the following convention should be applied:

1. Column buckling is governed by the larger slenderness ratio, $(l_e/d)_x$ or $(l_e/d)_y$, and the adjusted column design value F'_c is given by the Ylinen formula.

2. When the bending moment is about the x axis, the value of F_{cE} for use in the amplification factor is to be based on $(l_e/d)_x$.

EXAMPLE 7.16   Interaction Equation for Beam-Column with Moment about x Axis

Figure 7.17a   Deflected shape of beam showing P-Δ moment. The computed bending stress f_b is based on the moment M from the moment diagram. The moment diagram considers the effects of the transverse load w, but does not include the secondary moment $P \times \Delta$. The P-Δ effect is taken into account by amplifying the computed bending stress f_b.

By cutting the beam-column in Fig. 7.17a and summing moments at point A, it can be seen that a moment of $P \times \Delta$ is created that *adds* to the moment M caused by the transverse load. In a beam with an axial tension force, this moment *subtracts* from the normal bending moment, and may be conservatively ignored.

The general interaction formula (Eq. 3.9-3 in the NDS) reduces to the following form for the common case of an axial compressive force combined with a bending moment about the x axis:

$$\left(\frac{f_c}{F'_c}\right)^2 + \left(\frac{1}{1 - f_c/F_{cEx}}\right)\frac{f_{bx}}{F'_{bx}} \leq 1.0 \qquad \text{(ASD)}$$

$$\left(\frac{f_{cu}}{F'_{cn}}\right)^2 + \left(\frac{1}{1 - f_{cu}/F_{cExn}}\right)\frac{f_{bxu}}{F'_{bxn}} \leq 1.0 \qquad \text{(LRFD)}$$

where f_c = actual (ASD) compressive stress
    $= P/A$
  f_{cu} = factored (LRFD) compressive stress
    $= P_u/A$
  F'_c = adjusted ASD compressive design value as given by Ylinen formula (Secs. 7.4 and 7.5). Consider critical slenderness ratio $(l_e/d)_x$ or $(l_e/d)_y$. The critical slenderness ratio produces the smaller value of F'_c.
    $= F_c(C_D)(C_M)(C_t)(C_F)(C_P)(C_i)$
  F'_{cn} = adjusted LRFD compressive value as given by Ylinen formula
    $= F_{cn}(\phi_c)(\lambda)(C_M)(C_t)(C_F)(C_P)(C_i)$

F_{cn} = nominal LRFD compressive design value

   = $F_c \cdot K_F$

K_F = $2.16/\phi_c = 2.40$

f_{bx} = actual working (ASD) bending stress

   = M_x/S_x

f_{bxu} = factored (LRFD) bending stress

   = M_{xu}/S_x

F'_{bx} = adjusted ASD bending value about x axis considering effects of lateral torsional buckling (Sec. 6.3)

F'_{bxn} = adjusted LRFD bending value about x axis considering effects of lateral torsional buckling

ϕ_c = resistance factor for compression (LRFD)

   = 0.90

ϕ_b = resistance factor for bending (LRFD)

   = 0.85

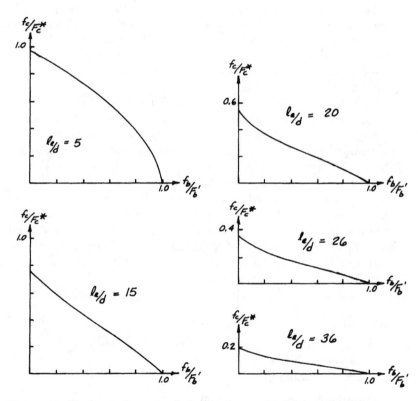

Figure 7.17b   Five interaction curves for beam-columns with different slenderness ratios.

For sawn lumber,

$F'_{bx} = F_b(C_D)(C_M)(C_t)(C_F)(C_L)(C_i)$ for ASD

$F'_{bxn} = F_{bxn}(\phi_b)(\lambda)(C_M)(C_t)(C_F)(C_L)(C_i)$ for LRFD

For glulam the smaller of the following bending stress values should be used,

$$F'_{bx} = F_b(C_D)(C_M)(C_t)(C_L)$$
$$\quad = F_b(C_D)(C_M)(C_t)(C_V) \text{ for ASD}$$
$$F'_{bxn} = F_{bn}(\lambda)(C_M)(C_t)(C_L)$$
$$\quad = F_{bn}(\lambda)(C_M)(C_t)(C_V) \text{ for LRFD}$$

F_{cEx} = Euler-based elastic buckling stress. Because transverse loads cause a bending moment about the x axis, F_{cE} is based on the slenderness ratio for the x axis, that is, $(l_e/d)_x$.

$$F_{cEx} = \frac{0.822E'_{x\,min}}{[(l_e/d)_x]^2} \quad \text{for ASD}$$

$$F_{cExn} = \frac{0.822E'_{x\,min\,n}}{[(l_e/d)_x]^2} \quad \text{for LRFD}$$

The interaction formula for a beam-column takes into account a number of factors, including column buckling, lateral torsional buckling, and the P-Δ effect. It is difficult to show on a graph, or even a series of graphs, all of the different variables in a beam-column problem. However, the interaction plots in Fig. 7.17b are helpful in visualizing some of the patterns. The graphs are representative only, and results vary with specific problems. The ordinate on the vertical axis shows the effect of different slenderness ratios in that the ratio of f_c/F_c^* decreases as the slenderness ratio increases.

The interaction formula in Example 7.16 covers the common problem of axial compression with a bending moment about the strong axis. This can be viewed as a special case of the Zahn general interaction equation. The general formula has a third term which provides for consideration of a bending moment about the y axis. See Example 7.17.

The concept of a general interaction equation can be carried one step further. The problems considered up to this point have involved axial compression plus bending caused by transverse loads. Although this is a comprehensive design expression, Zahn's expanded equation permits the compressive force in the column to be applied with an eccentricity. Thus in the general case, the bending moments about the x and y axes can be the result of transverse bending loads *and* an eccentrically applied column force.

The general loading condition is summarized as follows:

1. Compressive force in member = P
2. Bending moment about x axis
   a. Moment due to transverse loads = M_x
   b. Moment due to eccentricity about x axis = $P \times e_x$
3. Bending moment about y axis
   a. Moment due to transverse loads = M_y
   b. Moment due to eccentricity about y axis = $P \times e_y$

A distinction is made between the moments caused by transverse loads and the moments from an eccentrically applied column force. This distinction is necessary because the bending stresses that develop as a result of the eccentric load are subject to an *additional P-Δ* amplification factor. The general interaction formula may appear overly complicated at first glance, but taken term by term, it is straightforward and logical. The reader should keep in mind that greatly simplified versions of the interaction formula apply to most practical loading conditions (e.g., the version in Example 7.16). The simplified expression is obtained by setting the appropriate stress terms in the general interaction equation equal to zero.

EXAMPLE 7.17   General Interaction Formula for Combined Compression and Bending

Figure 7.18a Axial compression plus bending about x and y axes.

The member in Fig. 7.18a has an *axial* compression load, a transverse load causing a moment about the x axis, and a transverse load causing a moment about the y axis. The following interaction formula from the NDS (NDS Sec. 3.9.2) is used to check this member:

$$\left(\frac{f_c}{F'_c}\right)^2 + \frac{f_{bx}}{F'_{bx}(1 - f_c/F_{cEx})} + \frac{f_{by}}{F'_{by}[1 - f_c/F_{cEy} - (f_{bx}/F_{bE})^2]} \leq 1.0 \quad \text{(ASD)}$$

$$\left(\frac{f_{cu}}{F'_{cn}}\right)^2 + \frac{f_{bxu}}{F'_{bxn}(1 - f_{cu}/F'_{cExn})}$$

$$+ \frac{f_{byu}}{F'_{byn}[1 - f_{cu}/F'_{cEyn} - (f_{bxu}/F'_{bEn})^2]} \leq 1.0 \quad \text{(LRFD)}$$

where f_{by} = actual working (ASD) bending stress about y axis
  $= M_y/S_y$
 f_{byu} = factored (LRFD) bending stress about y axis
  $= M_{yu}/S_y$
 F'_{by} = adjusted ASD bending value about y axis (Sec. 6.4)
 F'_{byn} = adjusted LRFD bending value about y axis
 F_{cEy} = Euler elastic buckling value based on the slenderness ratio for the y axis $(l_e/d)_y$

$$F_{cEy} = \frac{0.822 E'_{y\,\min}}{[(l_e/d)_y]^2} \quad \text{for ASD}$$

$$F_{cEyn} = \frac{0.822 E'_{y\,\min n}}{[(l_e/d)_y]^2} \quad \text{for LRFD}$$

F_{bE} = elastic buckling value considering lateral torsional buckling of beam; F_{bE} based on the beam slenderness factor R_B (Sec. 6.3)

$$F_{bE} = \frac{1.20E'_{y\,\text{min}}}{R_B^2} \qquad \text{for ASD}$$

$$F_{bEn} = \frac{1.20E'_{y\,\text{min}\,n}}{R_B^2} \qquad \text{for LRFD}$$

Other terms are as previously defined.

If one or more of the loads in Fig. 7.18a do not exist, the corresponding terms in the interaction formula are set equal to zero. For example, if there is no load causing a moment about the y axis, f_{by} is zero and the third term in the interaction formula drops out. With $f_{by} = 0$, the interaction formula reduces to the form given in Example 7.16.

On the other hand, if the axial column force does not exist, f_c becomes zero, and the general formula becomes an interaction formula for biaxial bending (i.e., simultaneous bending about the x and y axes).

In the interaction problems considered thus far, the bending stresses have been caused only by transverse applied loads. In some cases, bending stresses may be the result of an eccentric column force. See Fig. 7.18b. The development of a generalized interaction formula for beam-columns with transverse *and* eccentric bending stresses is shown below. In the general formula, the amplification for P-Δ effects introduced in Example 7.16 is applied the same way. However, the bending stresses caused by the *eccentric* column force are subjected to an additional P-Δ amplification.

BEAM-COLUMN   SECTION

Figure 7.18b   Bending moment M due to transverse loads plus eccentric column force $P \times e$. Note that the eccentric moment $P \times e$ is a computed (first-order) bending moment, and it should not be confused with the second-order P-Δ moment. The eccentric moment in Fig. 7.18b causes a bending stress about the x axis.

Cross-Sectional Properties

$$A = bd$$

$$S = \frac{bd^2}{6}$$

Bending Stresses

$$\text{Total moment} = \text{transverse load } M + \text{eccentric load } M$$
$$= M + Pe$$

Bending stress due to transverse bending loads:

$$f_b = \frac{M}{S}$$

Bending stress due to eccentrically applied column force:
Computed stress:

$$\frac{Pe}{S} = \frac{Pe}{bd^2/6} = f_c\left(\frac{6e}{d}\right)$$

Amplified eccentric bending stress:

$$f_c\left(\frac{6e}{d}\right) \times (\text{amplification factor}) = f_c\left(\frac{6e}{d}\right)\left[1 + 0.234\left(\frac{f_c}{F_{cE}}\right)\right]$$

Bending stress due to transverse loads + amplified eccentric bending stress

$$= f_b + f_c\left(\frac{6e}{d}\right)\left[1 + 0.234\left(\frac{f_c}{F_{cE}}\right)\right]$$

General Interaction Formula

The interaction formula is expanded here to include the effects of eccentric bending stresses. Subscripts are added to the eccentric terms to indicate the axis about which the eccentricity occurs. This general form of the interaction formula is given in NDS Sec. 15.4.

$$\left(\frac{f_c}{F_c'}\right)^2 + \frac{f_{bx} + f_c(6e_x/d_x)[1 + 0.234(f_c/F_{cEx})]}{F_{bx}'(1 - f_c/F_{cEx})}$$

$$+ \frac{f_{by} + f_c(6e_y/d_y)\left[1 + 0.234(f_c/F_{cEy}) + 0.234\left[\dfrac{f_{bx} + f_c(6e_x/d_x)}{F_{bE}}\right]^2\right]}{F_{by}'\left[1 - f_c/F_{cEy} - \left[\dfrac{f_{bx} + f_c(6e_x/d_x)}{F_{bE}}\right]^2\right]} \leq 1.0$$

For this general interaction equation to be used in LRFD, the factored stresses and adjusted LRFD values need to be used.

In the general interaction formula in Example 7.17, it is assumed that the eccentric load is applied at the *end* of the column. In some cases, an eccentric compression force may be applied through a *side bracket* at some point between the ends of the column. The reader is referred to NDS Sec. 15.4.2 for an approximate method of handling beam-columns with side brackets.

Several examples of beam-column problems are given in the remaining portions of this chapter using both ASD and LRFD.

7.13   Design Problem: Beam-Column

In the first example, the top chord of the truss analyzed in Example 7.15 is considered using ASD. The top chord is subjected to bending loads caused by (D + S) being applied along the member. The top chord is also subjected to axial compression, which is obtained from a truss analysis using tributary loads to the truss joints.

A 2 × 8 is selected as the trial size. In a beam-column problem, it is often convenient to divide the stress calculations into three subproblems. In this approach, somewhat independent checks on axial, bending, and combined stresses are performed. See Example 7.18.

The first check is on *axial* stresses. Because the top chord of the truss is attached directly to the roof sheathing, lateral buckling about the weak axis of the cross section is prevented. Bracing for the strong axis is provided at the truss joints by the members that frame into the top chord.

The compressive stress is calculated at two different locations along the length of the member. First, the column design value F'_c adjusted for column stability is checked away from the joints, using the gross area in the calculation of f_c. The second calculation involves the stress at the net section at a joint compared with the adjusted value F^*_c without the reduction for stability.

In the *bending* stress calculation, the moment is determined by use of the horizontal span of the top chord and the load on a horizontal plane. It was shown in Example 2.6 (Sec. 2.4) that the moment obtained using the *horizontal plane method* is the same as the moment obtained using the inclined span length and the normal component of the load.

The final step is the analysis of *combined* stresses. The top chord has axial compression plus bending about the strong axis, and the simple interaction formula from Example 7.16 applies. The trial member is found to be acceptable.

EXAMPLE 7.18   Beam-Column Design Using ASD

Design the top chord of the truss shown in Fig. 7.16a in Example 7.15 using ASD. The axial force and bending loads are reproduced in Fig. 7.19a. Use No. 1 Southern Pine. MC ≤ 19 percent, and normal temperatures apply.

Connections will be made with a single row of ¾-in. diameter bolts. The top chord is braced laterally throughout its length by the roof sheathing. Trusses are 4 ft.-0 in. o.c. Design values and section properties are to be obtained from the NDS Supplement.

NOTE: Two load combinations must be considered in this design: D-only and D + S. It has been predetermined that the D + S combination controls the design and only those calculations associated with this combination are included in the example.

Try 2 × 8. Reference design values in NDS Supplement Table 4B for Southern Pine in the Dimension lumber size category are size-specific:

$$F_c = 1650 \text{ psi}$$

$$F_b = 1500 \text{ psi}$$

$$E_{min} = 620,000 \text{ psi}$$

Figure 7.19a Loading diagram for top chord of truss. Section view shows lateral support by roof sheathing.

Because the reference design values are size-specific, most grades of Southern Pine have the appropriate size factors already incorporated into the published values. For these grades, the size factors can be viewed as defaulting to unity:

$$C_F = 1.0 \text{ for compression parallel to grain}$$
$$C_F = 1.0 \text{ for bending}$$

Some grades of Southern Pine, however, require size factors other than unity. Section properties:

$$A = 10.875 \text{ in.}^2$$
$$S = 13.14 \text{ in.}^3$$

Axial

1. *Stability check.* Column buckling occurs away from truss joints. Use gross area.

$$f_c = \frac{P}{A} = \frac{4960}{10.875} = 456 \text{ psi}$$

$(l_e/d)_y = 0$ because of lateral support provided by roof diaphragm

$$\left(\frac{l_e}{d}\right)_x = \frac{8.39 \text{ ft} \times 12 \text{ in./ft}}{7.25 \text{ in.}} = 13.9$$

$$E'_{\min} = E_{\min}(C_M)(C_t) = 620,000(1.0)(1.0)$$

$$= 620,000 \text{ psi}$$

For visually graded sawn lumber,

$$c = 0.8$$

$$F_{cE} = \frac{0.822 E'_{\min}}{[(l_e/d)_{\max}]^2} = \frac{0.822(620{,}000)}{(13.9)^2} = 2638 \text{ psi}$$

$$F_c^* = F_c(C_D)(C_M)(C_t)(C_F)(C_i)$$
$$= 1650(1.15)(1.0)(1.0)(1.0)(1.0) = 1898 \text{ psi}$$

$$\frac{F_{cE}}{F_c'} = \frac{2638}{1898} = 1.390$$

$$\frac{1 + F_{cE}/F_c^*}{2c} = \frac{1 + 1.390}{2(0.8)} = 1.494$$

$$C_P = \frac{1 + F_{cE}/F_c^*}{2c} - \sqrt{\left(\frac{1 + F_{cE}/F_c^*}{2c}\right)^2 - \frac{f_{cE}/F_c^*}{c}}$$

$$= 1.494 - \sqrt{(1.494)^2 - 1.390/0.8} = 0.791$$

$$F_c' = F_c(C_D)(C_M)(C_t)(C_F)(C_P)(C_i)$$
$$= 1650(1.15)(1.0)(1.0)(1.0)(0.791)(1.0)$$
$$= 1501 \text{ psi} > 456 \text{ psi} \qquad OK$$

2. *Net section check.* Assume the hole diameter is $\frac{1}{16}$ in. larger than the bolt (for stress calculations only). See Fig. 7.19*b*.

$$A_n = 1.5(7.25 - 0.8125) = 9.66 \text{ in.}^2$$

$$f_c = \frac{P}{A_n} = \frac{4960}{9.66} = 514 \text{ psi}$$

HOLE DIAMETER

Figure 7.19*b* Net section of top chord at connection.

At braced location there is no reduction for stability.

$$F_c' = F_c^* = F_c(C_D)(C_M)(C_t)(C_F)(C_i)$$

$$= 1650(1.15)(1.0)(1.0)(1.0)(1.0)$$

$$= 1898 \text{ psi} > 514 \text{ psi} \qquad OK$$

Bending

Assume simple span (no end restraint). Take span and load on horizontal plane (refer to Example 2.6 in Sec. 2.4).

$$M = \frac{wL^2}{8} = \frac{0.176(7.5)^2}{8} = 1.24 \text{ ft-k} = 14.85 \text{ in.-k}$$

$$f_b = \frac{M}{S} = \frac{14{,}850}{13.14} = 1130 \text{ psi}$$

The beam has full lateral support. Therefore, l_u and R_B are zero and the lateral stability factor is $C_L = 1.0$. In addition, the spacing of the trusses is 4 ft o.c., and the adjusted bending design value does not qualify for the repetitive member increase, and $C_r = 1.0$.

$$F_c' = F_b(C_D)(C_M)(C_t)(C_L)(C_F)(C_r)(C_i)$$

$$= 1500(1.15)(1.0)(1.0)(1.0)(1.0)(1.0)(1.0)$$

$$= 1725 \text{ psi} > 1130 \text{ psi} \quad OK$$

Combined Stresses

There is no bending stress about the y axis, and $f_{by} = 0$. Furthermore, the column force is concentric, and the general interaction formula reduces to

$$\left(\frac{f_c}{F_c'}\right)^2 + \frac{f_{bx}}{F_{bx}'(1 - f_c/F_{cEx})} \leq 1.0$$

The load duration factor C_D for use in the interaction formula is based on the shortest-duration load in the combination, which in this case is snow load. A C_D of 1.15 for snow was used in the individual checks on the axial stress and bending stress. Therefore, the previously determined values of F_c' and F_b' are appropriate for use in the interaction formula.

In addition to the adjusted design values, the combined stress check requires the elastic buckling stress F_{cE} for use in evaluating the amplification factor. The bending moment is about the strong axis of the cross section, and the P-Δ effect is measured by the slenderness ratio about the x axis [i.e., $(l_e/d)_x = 13.9$].

The value of F_{cE} determined earlier in the example was for the column buckling portion of the problem. Column buckling is based on $(l_e/d)_{max}$. The fact that $(l_e/d)_{max}$ and $(l_e/d)_x$ are equal, is a coincidence in this problem. In other words, F_{cE} for the column portion of the problem is based on $(l_e/d)_{max}$, and F_{cE} for the P-Δ analysis is based on the axis about which the bending moment occurs, $(l_e/d)_x$. In this example, the two values of F_{cE} are equal, but in general they could be different.

$$F_{cEx} = F_{cE} = 2638 \text{ psi}$$

$$\left(\frac{f_c}{F_c'}\right)^2 + \left(\frac{1}{1 - f_c/F_{cEx}}\right)\frac{f_{bx}}{F_{bx}'} = \left(\frac{456}{1501}\right)^2 + \left(\frac{1}{1 - 456/2638}\right)\frac{1130}{1725}$$

$$(0.304)^2 + 1.21(0.655) = 0.884 < 1.0 \quad OK$$

$$\boxed{Use \quad 2 \times 8 \quad No. 1 \quad SP}$$

NOTE 1: The load combination of D + S was predetermined to control the design and for brevity only those calculations associated with this combination were provided. It can be determined that f_c = 179 psi and f_b = 360 psi results from D-*only*, and F'_c = 1257 psi and F'_b = 1350 psi with C_D = 0.9. The D-*only* axial compression plus bending interaction results in a value of 0.306 versus 0.884 for D + S.

NOTE 2: The simplified analysis used in this example applies only to trusses with pinned joints. For a truss connected with metal plate connectors, a design approach should be used which takes the continuity and partial rigidity of the joints into account (Ref. 7.6).

It should be noted that all stress calculations in this example are for loads caused by the design load of (D + S). For this reason, the load duration factor for snow (1.15) is applied to each *individual* stress calculation as well as to the *combined* stress check. The designer is cautioned that the critical load combination (D + S) was predetermined in this example. To be complete, the designer should also check D-*only* with the corresponding C_D for permanent load of 0.9. The same logic would apply to LRFD for 1.2D + 1.6S with λ = 0.8 and for 1.4D with λ = 0.6.

In some cases of combined loading, the *individual* stresses (axial and bending) may not both be caused by the same load. In this situation, the respective C_D or λ values are used in evaluating the individual stresses. However, in the *combined* stress calculation, the same C_D or λ is used for all components. Recall that the C_D for the *shortest*-duration load applies to the entire combination and λ is directly tied to specific load combinations. The appropriate rules for applying C_D or λ should be followed in checking both individual and combined stresses. The application of different C_D's or λ's in the individual stress calculations is illustrated in some of the following examples.

7.14   Design Problem: Beam-Column Action in a Stud Wall Using LRFD

A common occurrence of beam-column action is found in an exterior bearing wall. Axial column stresses are developed in the wall studs by vertical gravity loads. Bending stresses are caused by lateral wind or seismic forces.

For typical wood frame walls, the wall dead load is so small that the design wind force usually exceeds the seismic force. This applies to the normal wall force only, and seismic may be critical for parallel-to-wall (i.e., shearwall) forces. In buildings with large wall dead loads, the seismic force can exceed the wind force. Large wall dead loads usually occur in concrete and masonry (brick and concrete block) buildings.

In the two-story building of Example 7.19, the studs carry a number of axial compressive loads including dead, roof live, and floor live loads. In *load case 1*, the various possible combinations of gravity loads are considered.

Each vertical load combination should theoretically be checked using the appropriate C_D or λ. In this example, LRFD is used and the load case of 1.4D can be eliminated by inspection. The results of the other two vertical loadings are close, and the critical combination can be determined by evaluating the ratio f_c/F'_c for each combination. The loading with the larger stress ratio is critical. Only the calculations for the critical combination of $1.2D + 1.6L + 0.5L_r$ are shown in the example for load case 1.

Load case 2 involves both vertical loads and lateral forces. According to the IBC load combinations, roof live load and floor live load are considered simultaneously with lateral forces (Ref. 7.5 and Sec. 2.17).

EXAMPLE 7.19   Combined Bending and Compression in a Stud Wall Using LRFD

Check the 2 × 6 stud in the first-floor bearing wall in the building shown in Fig. 7.20*a* using LRFD. Consider the given vertical loads and lateral forces. Lumber is No. 2 DF-L. MC ≤ 19 percent and normal temperatures apply. Design values are to be in accordance with the NDS.

The following gravity loads are given:

$$\text{Roof: } D = 10 \text{ psf}$$
$$L_r = 20 \text{ psf}$$
$$\text{Wall: } D = 7 \text{ psf}$$
$$\text{Floor: } D = 8 \text{ psf}$$
$$L = 40 \text{ psf}$$

The following lateral force is also given:

$$W = 27.8 \text{ psf horizontal}$$

Figure 7.20a   Transverse section showing exterior bearing walls.

The applicable LRFD load combinations are

$$1.4D \qquad\qquad\qquad\qquad \lambda = 0.6$$

$$1.2D + 1.6L + 0.5L_r \qquad\qquad \lambda = 0.8$$

$$1.2D + 1.6L_r + (0.5L \text{ or } 0.8W) \quad \lambda = 0.8$$

$$1.2D + 1.6W + 0.5L + 0.5L_r \quad \lambda = 1.0$$

Try 2 × 6 No. 2 DF-L (Dimension lumber size):

Nominal values determined from NDS Supplement Table 4A:

$$F_{bn} = F_b \cdot K_F = F_b\,(2.16/\phi_b) = F_b(2.54)$$

$$= 900(2.54) = 2.29 \text{ ksi}$$

$$F_{cn} = F_c \cdot K_F = F_c\,(2.16/\phi_c) = F_c(2.40)$$

$$= 1350\,(2.40) = 3.24 \text{ ksi}$$

$$F_{c\perp n} = F_{c\perp} \cdot K_F = F_{c\perp}\,(1.875/\phi_c) = F_{c\perp}(2.083)$$

$$= 625\,(2.083) = 1.30 \text{ ksi}$$

$$E_{minn} = E_{min} \cdot K_F = E_{min}(1.5/\phi_s) = E_{min}(1.765)$$

$$= 580{,}000\,(1.765) = 1024 \text{ ksi}$$

Size factors:

$$C_F = 1.3 \qquad \text{for bending}$$

$$C_F = 1.1 \qquad \text{for compression parallel to grain}$$

Section properties:

$$A = 8.25 \text{ in.}^2$$

$$S = 7.56 \text{ in.}^3$$

Load case 1: Gravity Loads Only

Tributary width of roof and floor framing to the exterior bearing wall is 8 ft.

Dead loads:

$$\text{Roof D} = 10 \text{ psf} \times 8 \text{ ft} = 80 \text{ lb/ft}$$

$$\text{Wall D} = 7 \text{ psf} \times 20 \text{ ft} = 140 \text{ lb/lft}$$

$$\underline{\text{Floor D} = 8 \text{ psf} \times 8 \text{ ft} \;\; = 64 \text{ lb/ft}}$$

$$w_D = 284 \text{ lb/ft}$$

Live loads:

$$\text{Roof } L_r = 20 \text{ psf} \times 8 \text{ ft} = 160 \text{ lb/ft}$$
$$\text{Floor } L = 40 \text{ psf} \times 8 \text{ ft} = 320 \text{ lb/ft}$$

LOAD COMBINATIONS:

Calculate the axial load on a typical stud. Studs are spaced 16 in. = 1.33 ft o.c.

$$1.4D = [1.4(284)](1.33)$$
$$= 0.529 \text{ k} \qquad\qquad \lambda = 0.6$$

$$1.2D + 1.6L + 0.5L_r = [1.2(284) + 1.6(320) + 0.5\,(160)](1.33)$$
$$= 1.24 \text{ k} \qquad\qquad \lambda = 0.8$$

$$1.2D + 1.6L_r + 0.5L = [1.2(284) + 1.6(160) + 0.5(320)](1.33)$$
$$= 1.01 \text{ k} \qquad\qquad \lambda = 0.8$$

The combination of D-*only* can be eliminated by inspection. Since the λ's are identical for the second two combinations, the $1.2D + 1.6L + 0.5L_r$ load combination provides the critical vertical loading. Design calculations for this combination only are shown. The designer is responsible for determining the critical load combinations, including the effects of the time effect factor λ. The axial stress in the stud and the bearing stress on the wall plate are equal.

COLUMN CAPACITY:

Sheathing provides lateral support about the weak axis of the stud. Therefore, check column buckling about the x axis only ($L = 10.5$ ft and $d_x = 5.5$ in.):

$$\left(\frac{l_e}{d}\right)_y = 0 \qquad \text{because of sheathing}$$

$$\left(\frac{l_e}{d}\right)_{max} = \left(\frac{l_e}{d}\right)_x = \frac{10.5 \text{ ft} \times 12 \text{ in./ft}}{5.5 \text{ in.}} = 22.9$$

$$E'_{min\,n} = E_{min\,n}(\phi_s)(C_M)(C_t)(C_i) = 1024(0.85)(1.0)(1.0)(1.0)$$
$$= 870 \text{ ksi}$$

For visually graded sawn lumber,

$$c = 0.8$$

$$F_{cEn} = \frac{0.822E'_{min\,n}}{(l_e/d)^2} = \frac{0.822(870)}{(22.9)^2} = 1.36 \text{ ksi}$$

$$F_{cn}^* = F_{cn}(\phi_c)(\lambda)(C_M)(C_t)(C_F)(C_i)$$
$$= 3.24(0.90)(0.8)(1.0)(1.0)(1.1)(1.0) = 2.57 \text{ ksi}$$

$$\frac{F_{cEn}}{F^*_{cn}} = \frac{(1.36)}{(2.57)} = 0.531$$

$$\frac{1 + F_{cEn}/F^*_{cn}}{2c} = \frac{1 + 0.531}{2(0.8)} = 0.957$$

$$C_P = \frac{1 + F_{cEn}/F^*_{cn}}{2c} - \sqrt{\left(\frac{1 + F_{cEn}/F^*_{cn}}{2c}\right)^2 - \frac{F_{cEn}/F^*_{cn}}{c}}$$

$$= 0.957 - \sqrt{(0.957)^2 - 0.531/0.8} = 0.455$$

$$F'_{cn} = F_{cn}(\phi_c)(\lambda)(C_M)(C_t)(C_F)(C_P)(C_i)$$

$$= 3.24(0.90)(0.8)(1.0)(1.0)(1.1)(0.455)(1.0) = 1.17 \text{ ksi}$$

$$P_n = F'_{cn}A$$

$$= (1.17)(8.25)$$

$$= 9.63 \text{ k} > 1.24 \text{ k} \qquad OK$$

BEARING OF STUD ON WALL PLATES:

For a bearing length of $1\frac{1}{2}$ in. on a stud more than 3 in. from the end of the wall plate:

$$C_b = \frac{l_b + 0.375}{l_b} = \frac{1.5 + 0.375}{1.5} = 1.25$$

$$F'_{c\perp n} = F_{c\perp n}(\phi_c)(\lambda)(C_M)(C_t)(C_b) = 1.30(0.90)(0.8)(1.0)(1.0)(1.25) = 1.17 \text{ ksi}$$

$$P_{\perp n} = F'_{c\perp n}A$$

$$= (1.17)(8.25)$$

$$= 9.67 \text{ k} > 1.24 \text{ k} \qquad OK$$

$$\therefore \text{Vertical loads} \qquad OK$$

Load case 2: Gravity Loads + Lateral Forces

BENDING:

Wind governs over seismic. Force to one stud:

$$\text{Wind} = 27.8 \text{ psf}$$

For 1.6W: $w_u = 1.6(27.8 \text{ psf} \times 1.33 \text{ ft}) = 59.2 \text{ lb/ft}$

For 0.8W: $w_u = 0.8(27.8 \text{ psf} \times 1.33 \text{ ft}) = 29.7 \text{ lb/ft}$

For 1.6W: $M_u = \dfrac{w_u L^2}{8} = \dfrac{59.2(10.5)^2}{8} = 816$ ft-lb $= 9.79$ in.-k

$$f_{bu} = \dfrac{M_u}{S} = \dfrac{9.79}{7.56} = 1.29 \text{ ksi}$$

The stud has full lateral support provided by sheathing. Therefore, l_u and R_B are zero, and the lateral stability factor is $C_L = 1.0$. The time effect factor for the load combination involving 1.6W is $\lambda = 1.0$, and the repetitive-member factor is 1.15.

$$F'_{bn} = F_{bn}(\phi_b)(\lambda)(C_M)(C_t)(C_L)(C_F)(C_r)(C_i)$$

$$= 2.29(1.0)(1.0)(1.0)(1.0)(1.3)(1.15)(1.0) = 2.91 \text{ ksi}$$

$$M_n = F'_{bn}S$$

$$= (2.91)(7.56)$$

$$= 22.0 \text{ in.-k} > 9.79 \text{ in.-k} \qquad OK \qquad \text{based on 1.6W}$$

Figure 7.20*b* Loading for beam-column analysis.

AXIAL:

Two load combinations must be considered for combined load: $1.2D + 1.6L_r + 0.8W$ and $1.2D + 1.6W + 0.5L + 0.5L_r$.

For $1.2D + 1.6L_r + 0.8W$: $P_u = [1.2(284) + 1.6(160)](1.33)$

$$= 794 \text{ lb} = 0.794 \text{ k}$$

$$f_{cu} = \dfrac{P_u}{A} = \dfrac{0.794}{8.25} = 0.096 \text{ ksi}$$

For 1.2D + 1.6W + 0.5L + 0.5L$_r$: $P_u = [1.2(284) + (0.5)320 + 0.5(160)](1.33)$

$$= 774 \text{ lb} = 0.774 \text{ k}$$

$$f_{cu} = \frac{P_u}{A} = \frac{0.774}{8.25} = 0.094 \text{ ksi}$$

As noted previously in the example, the time effect factor λ is different for these two load combinations. for the first load combination, 1.2D + 1.6L$_r$ + 0.8W, $\lambda = 0.8$, and for the second 1.2D + 1.6W + L + 0.5L$_r$, $\lambda = 1.0$. Accordingly, the adjusted LRFD values will need to be determined for each situation.

The slenderness ratio about the y axis is zero because of the continuous support provided by the sheathing. The column slenderness ratio and the elastic buckling value that were determined previously apply to the problem at hand:

$$\left(\frac{l_e}{d}\right)_{max} = \left(\frac{l_e}{d}\right)_x = 22.9$$

$$F_{cEn} = 1.36 \text{ ksi}$$

For   1.2D + 1.6L$_r$ + 0.8W

The column stability factor is identical to what was determined previously.

$$C_p = 0.455$$

$$F'_{cn} = 1.17 \text{ ksi}$$

$$P'_n = F'_{cn}A$$

$$= (1.17)(8.25)$$

$$= 9.67 \text{ k} > 0.794 \text{ k} \qquad OK$$

For 1.2D + 1.6W + 0.5L + 0.5Lr: $F_{cEn} = 1.36$ ksi

$$F^*_{cn} = F_{cn}(\phi_c)(\lambda)(C_M)(C_t)(C_F)(C_i)$$

$$= 3.24(0.90)(1.0)(1.0)(1.0)(1.1)(1.0) = 3.20 \text{ ksi}$$

$$\frac{F_{cEn}}{F^*_{cn}} = \frac{(1.36)}{(3.20)} = 0.424$$

$$\frac{1 + F_{cEn}/F^*_{cn}}{2c} = \frac{1 + 0.424}{2(0.8)} = 0.890$$

$$C_p = \frac{1 + F_{cEn}/F^*_{cn}}{2c} - \sqrt{\left(\frac{1 + F_{cEn}/F^*_{cn}}{2c}\right)^2 - \frac{F_{cEn}/F^*_{cn}}{c}}$$

$$= 0.890 - \sqrt{(0.890)^2 - 0.424/0.8} = 0.378$$

$$F'_{cn} = F_{cn}(\phi_c)(\lambda)(C_M)(C_t)(C_F)(C_P)(C_i)$$

$$= 3.24(0.90)(1.0)(1.0)(1.0)(1.1)(0.378)(1.0)$$

$$F'_{cn} = 1.22 \text{ ksi}$$

$$P'_n = F'_{cn}A$$

$$= (1.22)(8.25)$$

$$= 10.0 \text{ k} > 0.774 \text{ k} \qquad OK$$

COMBINED STRESS:

The simplified interaction formula from Example 7.16 (Sec. 7.12) applies:

$$\left(\frac{f_{cu}}{F'_{cn}}\right)^2 + \frac{f_{bxu}}{F'_{bxn}\left(1 - f_{cu}/F_{cExn}\right)} \le 1.0$$

Recall that the adjusted column design value F'_{cn} is determined using the maximum slenderness ratio for the column $(l_e/d)_{\max}$, and the Euler buckling stress F_{cExn} for use in evaluating the P-Δ effect is based on the slenderness ratio for the axis with the bending moment. In this problem, the bending moment is about the strong axis of the cross section, and $(l_e/d)_x$, coincidentally, controls both F'_{cn} and F_{cExn}. In general, one slenderness ratio does not necessarily define both of these quantities.

The value of F_{cEn} determined earlier in the example using $(l_e/d)_x$ is also F_{cExn}.

$$F_{cExn} = F_{cEn} = 1.36 \text{ ksi}$$

For $1.2D + 1.6L_r + 0.8W$

$1.2D + 1.6L_r$ produces the axial stress f_{cu} and $0.8W$ results in the bending stress f_{bxu}:

$$f_{cu} = 0.096 \text{ ksi (as determined previously)}$$

$$F'_{cn} = 1.17 \text{ ksi (as determined previously)}$$

$$\text{Wind} = 27.8 \text{ psf}$$

$$w_u = 0.8(27.8 \text{ psf} \times 1.33 \text{ ft}) = 29.7 \text{ lb/ft}$$

$$M_u = \frac{w_u L^2}{8} = \frac{(29.7)(10.5)^2}{8} = 409 \text{ ft-lb} = 4.90 \text{ in.-k}$$

$$f_{bxu} = \frac{M}{S} = \frac{4.90}{7.56} = 0.649 \text{ ksi}$$

The adjusted LRFD bending design value was previously determined for $\lambda = 1.0$ as $F'_{bxn} = 3.34$ ksi. The time effect factor for this load combination is $\lambda = 0.8$. All other adjustment factors remain the same, including the beam stability factor since the stud is fully braced in the weak axis direction. Therefore,

$$F'_{bxn} = 2.91(0.8) = 2.33 \text{ ksi}$$

$$\left(\frac{f_{cu}}{F'_{cn}}\right)^2 + \left(\frac{1}{1 - f_{cu}/F_{cExn}}\right)\frac{f_{bxu}}{F'_{bxn}} = \left(\frac{0.096}{1.17}\right)^2$$

$$+ \left(\frac{1}{1 - 0.096/1.36}\right)\frac{0.649}{2.33}$$

$$= 0.307 < 1.0 \quad OK$$

For $1.2D + 1.6W + 0.5L + 0.5L_r$

The axial stress f_{cu} results from $1.2D + 0.5L + 0.5L_r$ and the bending stress f_{bxu} is caused by 1.6W:

$$f_{cu} = 0.096 \text{ ksi (as determined previously)}$$

$$F'_{cn} = 1.22 \text{ ksi (as determined previously)}$$

$$f_{bxu} = 1.29 \text{ ksi (as determined previously)}$$

$$F'_{bxn} = 2.91 \text{ ksi (as determined previously)}$$

$$\left(\frac{f_{cu}}{F'_{cn}}\right)^2 + \left(\frac{1}{1 - f_{cu}/F_{cExn}}\right)\frac{f_{bxu}}{F'_{bxn}} = \left(\frac{0.096}{1.22}\right)^2$$

$$+ \left(\frac{1}{1 - 0.096/1.36}\right)\frac{1.29}{2.91}$$

$$= 0.486 < 1.0 \quad OK$$

2×6 No. 2   DF-L   exterior bearing wall   OK

Although several load cases were considered, the primary purpose of Example 7.17 is to illustrate the application of the interaction formula for beam-columns applied to a stud wall. The reader should understand that other load cases, including uplift due to wind, may be required in the analysis of a bearing wall subject to lateral forces.

7.15   Design Problem: Glulam Beam-Column Using ASD

In this example using ASD, a somewhat more complicated bracing condition is considered. The column is a glulam that supports roof dead and live loads as well as lateral wind forces. See Example 7.20.

In load case 1, the vertical loads are considered and $(D + L_r)$ is the critical loading. The interesting aspect of this problem is that there are different unbraced lengths for the x and y axes. Lateral support for the strong axis is provided at the ends only. However, for the weak axis the unbraced length is the height of the window.

In load case 2, the vertical dead load and lateral wind force are considered. Bending takes place about the strong axis of the member. The bending analysis includes a check of lateral stability using the window height as the unbraced length. In checking combined stresses, a C_D of 1.6 for wind is applied to all components of the interaction formula. Note that C_D appears several times in the development of the adjusted column design value, and F'_c must be reevaluated for use in the interaction formula.

The example makes use of a member that is an axial-load glulam combination. Calculations show that the bending stress is more significant than the axial stress, and it would probably be a more efficient design to choose a member from the glulam bending combinations instead of an axial combination. However, with a combined stress ratio of 0.79, the given member is somewhat understressed.

EXAMPLE 7.20   Glulam Beam-Column Using ASD

Check the column in the building shown in Fig. 7.21a using ASD procedures for the given loads. The column is an axial combination 2 DF glulam (combination symbol 2) with tension laminations (F_{bx} = 1700 psi). The member supports the tributary dead load, roof live load, and lateral wind force. The wind force is transferred to the column by the window framing in the wall.

The lateral force is the inward or outward wind pressure:

$$W = 22.2 \text{ psf}$$

The seismic force is not critical. Reference glulam design values are to be taken from the NDS Supplement Table 5B. C_M = 1.0, and C_t = 1.0.

Glulam Column

$5\frac{1}{8} \times 7\frac{1}{2}$ *Axial combination 2 DF glulam:*

A = 38.4 in.^{2} F_c = 1950 psi

S_x = 48 in.^{3} F_{bx} = 1700 psi   (requires tension laminations)

                   $E_{x\,min} = E_{y\,min}$ = 830,000 psi

Figure 7.21a Glulam column between windows subject to bending plus axial compression.

Load Case 1: Gravity Loads

$$D = 5 \text{ k}$$

$$D + L_r = 5 + 4 = 9 \text{ k}$$

The $(D + L_r)$ combination was predetermined to be the critical vertical loading condition, and the load duration factor for the combination is $C_D = 1.25$. (The D-only combination should also be checked.)

$$f_c = \frac{P}{A} = \frac{9000}{38.4} = 234 \text{ psi}$$

Neglect the column end restraint offered by wall sheathing for column buckling about the y axis. Assume an effective length factor (Fig. 7.9) of $K_e = 1.0$ for both the x and y axes.

$$\left(\frac{l_e}{d}\right)_x = \frac{1(16 \text{ ft} \times 12 \text{ in./ft})}{7.5 \text{ in.}} = 25.6$$

$$\left(\frac{l_e}{d}\right)_y = \frac{1(8 \text{ ft} \times 12 \text{ in./ft})}{5.125 \text{ in.}} = 18.7 < 25.6$$

The larger slenderness ratio governs the adjusted column design value. Therefore, the strong axis of the column is critical, and $(l_e/d)_x$ is used to determine F'_c.

$$E_{x\ min} = E_{y\ min} = 830{,}000 \text{ psi}$$

$$E'_{min} = E_{min}(C_M)(C_t) = 830{,}000(1.0)(1.0)$$

$$= 830{,}000 \text{ psi}$$

For glulam:

$$c = 0.9$$

$$F_{cE} = \frac{0.822E'_{min}}{[(l_e/d)_{max}]^2} = \frac{0.822(830{,}000)}{(25.6)^2} = 1041 \text{ psi}$$

$$F^*_c = F_c(C_D)(C_M)(C_t)$$

$$= 1950(1.25)(1.0)(1.0) = 2438 \text{ psi}$$

$$\frac{F_{cE}}{F^*_c} = \frac{1041}{2438} = 0.427$$

$$\frac{1 + F_{cE}/F^*_c}{2c} = \frac{1 + 0.427}{2(0.9)} = 0.793$$

$$C_P = \frac{1 + F_{cE}/F^*_c}{2c} - \sqrt{\left(\frac{1 + F_{cE}/F^*_c}{2c}\right)^2 - \frac{F_{cE}/F^*_c}{c}}$$

$$= 0.793 - \sqrt{(0.793)^2 - 0.427/0.9} = 0.400$$

$$F'_c = F_c(C_D)(C_M)(C_t)(C_P)$$

$$= 1950(1.25)(1.0)(1.0)(0.400)$$

$$= 975 \text{ psi} > 234 \text{ psi} \qquad OK$$

The member is adequate for vertical loads.

Load Case 2: D + W

The two applicable load combinations for this design are (D + W) and D + 0.75 (L$_r$ + W). It has been predetermined that D + W governs the interaction, and only those calculations associated with the combination are provided. The load duration factor for (D + W) is taken as 1.6 throughout the check for combined stresses.

AXIAL (DEAD LOAD):

$$f_c = \frac{P}{A} = \frac{5000}{38.4} = 130 \text{ psi}$$

From load case 1:

$$\left(\frac{l_e}{d}\right)_{\text{max}} = \left(\frac{l_e}{d}\right)_x = 25.6$$

$$F_{cE} = \frac{0.822 E'_{\text{min}}}{[(l_e/d)_{\text{max}}]^2} = 1041 \text{ psi}$$

$$F_c^* = F_c(C_D)(C_M)(C_t)$$
$$= 1950(1.6)(1.0)(1.0) = 3120 \text{ psi}$$

$$\frac{F_{cE}}{F_c^*} = \frac{1041}{3120} = 0.334$$

$$\frac{1 + F_{cE}/F_c^*}{2c} = \frac{1 + 0.334}{2(0.9)} = 0.741$$

$$C_P = \frac{1 + F_{cE}/F_c^*}{2c} - \sqrt{\left(\frac{1 + F_{cE}/F_c^*}{2c}\right)^2 - \frac{F_{cE}/F_c^*}{c}}$$
$$= 0.741 - \sqrt{(0.741)^2 - 0.334/0.9} = 0.319$$

$$F_c' = F_c(C_D)(C_M)(C_t)(C_P)$$
$$= 1950(1.6)(1.0)(1.0)(0.319)$$
$$= 995 \text{ psi}$$

$$\text{Axial stress ratio} = \frac{f_c}{F_c'} = \frac{130}{995} = 0.131$$

BENDING (WIND):

The window headers and sills span horizontally between columns. Uniformly distributed wind forces to a typical header and sill are calculated using a 1-ft section of wall and the tributary heights shown in Fig. 7.21b.

$$\text{Wind on header} = w_1 = (22.2 \text{ psf})(6.5 \text{ ft}) = 144 \text{ lb/ft}$$

Horizontal reaction of two headers on center column (Fig. 7.21c):

$$P_1 = (144 \text{ lb/ft})(12 \text{ ft}) = 1728 \text{ lb}$$

$$\text{Wind on sill} = w_2 = (22.2 \text{ psf})(5.5 \text{ ft}) = 122 \text{ lb/ft}$$

Horizontal reaction of two sills on center column:

$$P_2 = (122 \text{ lb/ft})(12 \text{ ft}) = 1464 \text{ lb}$$

WALL FRAMING & TRIB HEIGHTS

Figure 7.21b Wall framing showing wind pressure heights of 6.5 ft and 5.5 ft to window header and still, respectively

LOAD & SHEAR & MOMENT

Figure 7.21c Load, shear, and moment diagrams for center column subject to lateral wind forces. Concentrated forces are header and sill reactions from window framing.

From the moment diagram in Fig. 7.21c,

$$M_x = 7318 \text{ ft-lb} = 87.8 \text{ in.-k}$$

$$f_b = \frac{M}{S} = \frac{87,816}{48} = 1830 \text{ psi}$$

Bending is about the strong axis of the cross section. The adjusted bending design value for a glulam is governed by the smaller of two criteria: volume effect or lateral stability (see Example 6.11 in Sec. 6.4). The wind pressure can act either inward or outward, and tension laminations are required on both faces of the glulam.

Adjusted design value criteria:

$$F_b' = F_b(C_D)(C_M)(C_t)(C_L)$$

$$F_b' = F_b(C_D)(C_M)(C_t)(C_V)$$

Compare C_L and C_V to determine the critical design criterion.

Beam stability factor C_L

If the beam-column fails in lateral torsional buckling as a beam, the cross section will move in the plane of the wall between the window sill and header. Thus, the unbraced length for beam stability is the height of the window.

$$l_u = 8 \text{ ft} = 96 \text{ in.}$$

The loading condition for this member does not match any of the conditions in NDS Table 3.3.3. However, the effective length given in the footnote to this table may be conservatively used for any loading.

$$\frac{l_u}{d} = \frac{96}{7.5} = 12.8$$

$$7 \le 12.8 \le 14.3$$

$$\therefore \; l_e = 1.63l_u + 3d$$

$$= 1.63(96) + 3(7.5) = 179 \text{ in.}$$

$$R_B = \sqrt{\frac{l_e d}{b^2}} = \sqrt{\frac{179(7.5)}{(5.125)^2}} = 7.15$$

$$F_{bE} = \frac{1.20 E_{y \, \text{min}}'}{R_B^2} = \frac{1.20(830,000)}{(7.15)^2} = 19,483 \text{ psi}$$

$$F_b^* = F(C_D)(C_M)(C_t)$$

$$= 1700(1.6)(1.0)(1.0) = 2720 \text{ psi}$$

$$\frac{F_{bE}}{F_b^*} = \frac{19,483}{2720} = 7.16$$

$$\frac{1 + F_{bE}/F_b^*}{1.9} = \frac{1 + 7.16}{1.9} = 4.29$$

$$C_L = \frac{1 + F_{bE}/F_b^*}{1.9} - \sqrt{\left(\frac{1 + F_{bE}/F_b^*}{1.9}\right)^2 - \frac{F_{bE}/F_b^*}{0.95}}$$

$$= 4.29 - \sqrt{(4.29)^2 - 7.16/0.95} = 0.993$$

Volume factor C_V
For DF glulam, $x = 10$.

$$C_V = \left(\frac{21}{L}\right)^{1/x}\left(\frac{12}{d}\right)^{1/x}\left(\frac{5.125}{b}\right)^{1/x} \leq 1.0$$

$$= \left(\frac{21}{16}\right)^{0.1}\left(\frac{12}{7.5}\right)^{0.1}\left(\frac{5.125}{5.125}\right)^{0.1}$$

$$= 1.077 > 1.0$$

$$\therefore C_V = 1.0$$

Neither beam stability nor volume effect has a significant impact on this problem. However, the smaller value of C_L and C_V indicates that stability governs over volume effect.

$$C_L = 0.993$$

$$F_b' = F_b(C_D)(C_M)(C_t)(C_L)$$

$$= 1700(1.6)(1.0)(1.0)(0.993)$$

$$= 2700 \text{ psi}$$

$$\text{Bending stress ratio} = \frac{f_b}{F_b'} = \frac{1830}{2700} = 0.678$$

COMBINED STRESSES:

The bending moment is about the strong axis of the cross section, and the amplification for P-Δ is measured by the column slenderness ratio about the x axis.

NOTE: It is a coincidence that the adjusted compressive design value F_c' and the amplification factor for the P-Δ effect are both controlled by $(l_e/d)_x$ in this problem. Recall that F_c' is governed by $(l_e/d)_{max}$, and the P-Δ effect is controlled by the slenderness ratio for the axis about which the bending moment is applied.

$$\left(\frac{l_e}{d}\right)_{\substack{\text{bending} \\ \text{moment}}} = \left(\frac{l_e}{d}\right)_x = 25.6$$

$$F_{cEx} = \frac{0.822E_{min}'}{[(l_e/d)_x]^2} = \frac{830{,}000}{(25.6)^2} = 1041 \text{ psi}$$

$$\text{Amplification factor} = \frac{1}{1 - f_c/F_{cEx}} = \frac{1}{1 - 130/1041} = 1.14$$

$$\left(\frac{f_c}{F'_c}\right)^2 + \left(\frac{1}{1 - f_c/F_{cEx}}\right)\frac{f_b}{F'_b} = (0.131)^2 + 1.14(0.678)$$

$$= 0.79 < 1.0 \quad OK$$

$5\frac{1}{8} \times 7\frac{1}{2}$ axial combination 2 DF glulam with tension laminations (F_{bx} = 1700 psi) is OK for combined bending and compression.

NOTE: The critical load combination for the combined load was predetermined to be D + W. If the D + 0.75(L + L_r + W) combination were used, an interaction value of considerably less than 0.79 would result.

7.16 Design for Minimum Eccentricity

The design procedures for a column with an axial load were covered in detail in Secs. 7.4 and 7.5. A large number of interior columns and some exterior columns qualify as axial-load-carrying members. That is, the applied load is *assumed* to pass directly through the centroid of the column cross section, and, in addition, no transverse bending loads are involved.

Although many columns can theoretically be classified as axial-load members, there may be some question about whether the load in practical columns is truly a concentric axial load. In actual construction there may be some misalignment or nonuniform bearing in connections that causes the load to be applied eccentrically.

Some eccentric moment probably develops in columns which are thought to support concentric axial loads only. The magnitude of the eccentric moment, however, is unknown. Many designers simply ignore the possible eccentric moment and design for axial stresses only. This practice may be justified because practical columns typically have square-cut ends. In addition, the ends are attached with connection hardware such that the column end conditions do not exactly resemble the end conditions of an "ideal" pinned-end column. With the restraint provided by practical end conditions, the effective column length is somewhat less than the actual unbraced length. Thus the possible effect of an accidental eccentricity may be compensated by normal field end conditions.

However, Ref. 7.4 states that the possible eccentric moment should not be ignored, and it suggests that columns should be designed for some minimum eccentricity. The minimum eccentricity recommended is similar to the minimum eccentricity formerly required in the design of axially loaded reinforced-concrete columns. In this approach, the moment is taken as the compressive load times an eccentricity of 1 in. or one-tenth the width of the column (0.1d), whichever is larger. The moment is considered independently about both principal axes.

In the design of wood columns, there is no Code requirement to design for a minimum eccentric moment. The suggestion that some designers may provide for an eccentric moment in their column calculations is presented here for information only. Including an eccentric moment in the design of a column is definitely a more conservative design approach. Whether or not eccentricity should be included is left to the judgment of the designer.

7.17   Design Problem: Column with Eccentric Load Using ASD

Example 7.21 demonstrates the use of the interaction formula for eccentric loads. The load is theoretically an axial load, but the calculations are expanded to include a check for the minimum eccentricity discussed in Sec. 7.16. The same interaction formula would be used in the case of a known eccentricity.

The problem uses ASD to illustrate the significant effect of an eccentricity. Without the eccentricity, the member capacity is simply evaluated by the axial stress ratio of 0.69. However, the combined stress ratio is 0.90 for an eccentricity about the x axis and 0.93 for an eccentricity about the y axis. The combined stress ratios are much closer to the full member capacity, which is associated with a value of 1.0.

This example also demonstrates that F'_{by} for a member in the Beams and Stringers size category does not equal F'_{bx}. The reduction factor for F_{by} varies with grade.

EXAMPLE 7.21   Column Design for Minimum Eccentricity Using ASD

The column in Fig. 7.22a is an interior column in a large auditorium. The design roof D + S loads are theoretically axial loads on the column. Because of the importance

STRONG
AXIS

WEAK
AXIS

COLUMN   BRACING

CONDITIONS

Figure   7.22a Sawn   lumber column   with   different   bracing conditions for x and y axes.

of the column, it is desired to provide a conservative design with a minimum eccentricity of $0.1d$ or 1 in. Bracing conditions are shown. Lumber is Select Structural DF-L, and C_M and C_t equal 1.0.

$$D = 20 \text{ k}$$
$$\underline{S = 50 \text{ k}}$$
$$P_{TL} = 70 \text{ k} \qquad \text{(total load governs over D-\textit{only})}$$

A load duration factor of $C_D = 1.15$ applies throughout the problem.

Try 10 × 14 Sel. Str. DF-L.

The trial size is in the B&S size category. Recall that a member in the Beams and Stringers size category has cross-sectional dimensions of 5 in. or greater and a width that exceeds the thickness by more than 2 in. Reference design values are obtained from NDS Supplement Table 4D:

DF-L in this size category may be graded under two different sets of lumber grading rules. If any reference design value conflict occurs in Table 4D, use the smaller value (conservative):

$$F_c = 1100 \text{ psi}$$
$$F_{bx} = 1600 \text{ psi}^*$$
$$E_{x\,min} = 580{,}000 \text{ psi}$$

Section properties:

$$b = 9.5 \text{ in.}$$
$$d = 13.5 \text{ in.}$$
$$A = 128.25 \text{ in.}^2$$
$$S_x = 288.6 \text{ in.}^3$$
$$S_y = 203.1 \text{ in.}^3$$

Axial

$$f_c = \frac{P}{A} = \frac{70{,}000}{128.25} = 546 \text{ psi}$$

$$\left(\frac{l_e}{d}\right)_x = \frac{24 \text{ ft} \times 12 \text{ in./ft}}{13.5 \text{ in.}} = 21.3$$

$$\left(\frac{l_e}{d}\right)_y = \frac{16 \text{ ft} \times 12 \text{ in./ft}}{9.5 \text{ in.}} = 20.2$$

---

*Reference bending design values for members in the B&S size category are for bending about the x_axis. When bending is about the weak axis, a size factor is applied as provided in Table 4D of the NDS Supplement. For the Select Structural grade, $C_F = 0.86$ for bending and 1.0 for all other properties.

$E_{x\,min} = E_{y\,min}$, and the larger slenderness ratio governs the critical buckling design value. Therefore, the strong axis is critical.

$$E'_{min} = E_{min}(C_M)(C_t) = 580{,}000(1.0)(1.0)$$

$$= 580{,}000 \text{ psi}$$

For visually graded sawn lumber,

$$c = 0.8$$

Determine adjusted compressive design value:

$$F_{cE} = \frac{0.822E'_{min}}{[(l_e/d)_{max}]^2} = \frac{0.822(580{,}000)}{(21.3)^2} = 1048 \text{ psi}$$

The size factor for compression parallel to grain applies only to Dimension lumber, and C_F defaults to unity for a B&S.

$$F_c^* = F_c(C_D)(C_M)(C_t)(C_F)$$

$$= 1100(1.15)(1.0)(1.0)(1.0) = 1265 \text{ psi}$$

$$\frac{F_{cE}}{F_c^*} = \frac{1048}{1265} = 0.828$$

$$\frac{1 + F_{cE}/F_c^*}{2c} = \frac{1 + 0.828}{2(0.8)} = 1.143$$

$$C_P = \frac{1 + F_{cE}/F_c^*}{2c} - \sqrt{\left(\frac{1 + F_{cE}/F_c^*}{2c}\right)^2 - \frac{F_{cE}/F_c^*}{c}}$$

$$= 1.143 - \sqrt{(1.143)^2 - 0.828/0.8} = 0.623$$

$$F_c' = F_c(C_D)(C_M)(C_t)(C_P)$$

$$= 1100(1.15)(1.0)(1.0)(0.623)$$

$$F_c' = 788 \text{ psi} > 546 \text{ psi} \quad OK$$

Alternatively, the axial stress ratio is shown to be less than 1.0:

$$\frac{f_c}{F_c'} = \frac{546}{788} = 0.69 < 1.0$$

The member is adequate for the axial load.

Eccentric Load about Strong Axis

Figure 7.22b   Column load applied with eccentricity about x axis.

AXIAL:

The axial stress check is unchanged for this load case.

BENDING:

There are no transverse loads, and the only bending stress is due to the eccentric column force (see Fig.7.22b).

$$e_x = 0.1d = 0.1(13.5) = 1.35 \text{ in.} > 1.0 \text{ in.}$$

$$\text{Eccentric } f_{bx} = \frac{P e_x}{S_x} = f_c\left(\frac{6e_x}{d_x}\right) = 546\left(\frac{6 \times 1.35}{13.5}\right) = 327 \text{ psi}$$

Size factor

$$C_F = \left(\frac{12}{d}\right)^{1/9} = \left(\frac{12}{13.5}\right)^{1/9} = 0.987$$

Lateral stability

The eccentric moment is about the strong axis of the cross section. Lateral torsional buckling may occur in a plane perpendicular to the plane of bending. Therefore, the unbraced length for lateral stability is 16 ft. Determine l_e in accordance with footnote 1 to NDS Table 3.3.3.

$$l_u = 16 \text{ ft} = 192 \text{ in.}$$

$$\frac{l_u}{d} = \frac{192}{13.5} = 14.2$$

$$7.0 \le 14.2 \le 14.3$$

$$\therefore l_e = 1.63 l_u + 3d$$

$$= 1.63(192) + 3(13.5) = 353 \text{ in.}$$

$$R_B = \sqrt{\frac{l_e d}{b^2}} = \sqrt{\frac{353(13.5)}{(9.5)^2}} = 7.27$$

$$F_{bE} = \frac{1.20 E'_{y\,min}}{R_B^2} = \frac{1.20(580,000)}{(7.27)^2} = 13,160 \text{ psi}$$

$$F_b^* = F_b(C_D)(C_M)(C_t)(C_F)$$

$$= 1600(1.15)(1.0)(1.0)(0.987) = 1816 \text{ psi}$$

$$\frac{F_{bE}}{F_b^*} = \frac{13,160}{1816} = 7.249$$

$$\frac{1 + F_{bE}/F_b^*}{1.9} = \frac{1 + 7.249}{1.9} = 4.341$$

$$C_L = \frac{1 + F_{bE}/F_b^*}{1.9} - \sqrt{\left(\frac{1 + F_{bE}/F_b^*}{1.9}\right)^2 - \frac{F_{bE}/F_b^*}{0.95}}$$

$$= 4.341 - \sqrt{(4.341)^2 - 7.249/0.95} = 0.992$$

$$F'_{bx} = F_b(C_D)(C_M)(C_t)(C_L)(C_F)(C_r)$$

$$= 1600(1.15)(1.0)(1.0)(0.992)(1.0)(1.0)$$

$$= 1826 \text{ psi} > 327 \text{ psi} \quad OK$$

COMBINED STRESSES:

There are two amplification factors for combined stresses when all or part of the bending stress is due to an eccentric load.

Amplification factor for eccentric bending stress

The current check on eccentric bending moment is about the x axis, and the amplification factor is a function of the slenderness ratio for the x axis.

$$\left(\frac{l_e}{d}\right)_x = 21.3$$

The Euler elastic buckling stress was evaluated previously using this slenderness ratio in the axial stress portion of the problem.

$$F_{cEx} = F_{cE} = 1048 \text{ psi}$$

$$\frac{f_c}{F_{cEx}} = \frac{546}{1048} = 0.521$$

$$(\text{Amp Fac})_{ecc} = 1 + 0.234\left(\frac{f_c}{F_{cEx}}\right) = 1 + 0.234(0.521) = 1.122$$

General P-Δ amplification factor

$$\text{Amp Fac} = \frac{1}{1 - f_c/F_{cEx}} = \frac{1}{1 - 546/1048} = 2.088$$

$$\left(\frac{f_c}{F'_c}\right)^2 + \left(\frac{1}{1 - f_c/F_{cEx}}\right)\frac{f_b + f_c(6e_x/d_x)[1 + 0.234(f_c/F_{cEx})]}{F'_{bx}}$$

$$= (0.693)^2 + (2.088)\left[\frac{0 + 327(1.122)}{1826}\right]$$

$$= 0.90 \; < \; 1.0 \qquad OK$$

Eccentric Load about Weak Axis

$e_y = 1''$

$P = 70 k$

9.5"

Figure 7.22c   Column load applied with eccentricity about *y* axis.

AXIAL:

The axial stress check remains the same.

BENDING:

The only bending stress is due to the eccentric column force (Fig. 7.22c).

$$e_y = 0.1d = 0.1(9.5) = 0.95 \text{ in.} < 1.0 \text{ in.}$$

$$\therefore \; e_y = 1.0 \text{ in.}$$

$$\text{Ecc. } f_{by} = \frac{Pe_y}{S_y} = f_c\left(\frac{6e_y}{d_y}\right) = 546\left(\frac{6 \times 1.0}{9.5}\right) = 345 \text{ psi}$$

Determine the adjusted bending design value for the y axis. Even with an unbraced length of 24 ft, there is no tendency for lateral buckling when the moment is about the y axis. The depth for bending about the y axis is 9.5 in.

$$F'_{by} = F_b(C_D)(C_M)(C_t)(C_F)$$

$$= 1600(1.15)(1.0)(1.0)(0.86)$$

$$= 1582 \text{ psi} > 345 \text{ psi} \quad OK$$

COMBINED STRESSES:

Amplification factor for eccentric bending stress

The eccentric bending moment being considered is about the y axis, and the amplification factor is a function of the slenderness ratio for the y axis.

$$\left(\frac{l_e}{d}\right)_y = 20.2$$

$$F_{cEy} = \frac{0.822E'_{min}}{[(l_e/d)_y]^2} = \frac{0.822(580,000)}{(20.2)^2} = 1167 \text{ psi}$$

$$\left(\frac{f_c}{F'_c}\right)^2 + \frac{f_{by} + f_c(6e_y/d_y)[1 + 0.234(f_c/F_{cEy})]}{F'_{by}[1 - f_c/F_{cEy} - (f_{bx}/F_{bEx})^2]}$$

$$= (0.693)^2 + \frac{0 + 546[6(1.0)/9.5][1 + 0.234(546/1167)]}{1582[1 - 546/1167 - (0/13,160)^2]}$$

$$= 0.93 > 1.0 \quad OK$$

7.18 References

[7.1] American Forest and Paper Association (AF&PA). 2005. *National Design Specification for Wood Construction and Supplement.* ANSI/AF&PA NDS-2005, AF&PA, Washington DC.

[7.2] American Institute of Steel Construction (AISC). 2005. *Steel Construction Manual,* 13th ed., AISC, Chicago, IL.

[7.3] Bohnhoff, D.R., Moody, R.C., Verill, S.P., and Shirek, L.F. 1991. "Bending Properties of Reinforced and Unreinforced Spliced Nailed-Laminated Posts," Research Paper FPL-RP-503, Forest Products Laboratory, Forest Service, U.S.D.A., Madison, WI.

[7.4] Gurfinkel, G. 1981. *Wood Engineering,* 2nd ed., Kendall/Hunt Publishing (available through Southern Forest Products Association, Kenner, LA).

[7.5] International Codes Council (ICC). 2006. *International Building Code,* 2006 ed., ICC, Falls Church, VA.

[7.6] Truss Plate Institute (TPI). 2002. National Design Standard for Metal Plate Connected Wood Truss Construction, ANSI/TPI 1-2002, TPI, Madison, WI.

[7.7] Zahn, J.J. 1991. "Biaxial Beam-Column Equation for Wood Members," *Proceedings of Structures Congress '91*, American Society of Civil Engineers, Reston, Verginia. pp. 56–59.

[7.8] Zahn, J.J. 1991. "New Column Design Formula." *Wood Design Focus*, Vol. 2, No. 2.

7.19 Problems

Design values and section properties for the following problems are to be in accordance with the 2005 NDS. Dry service conditions, normal temperatures, and bending about the strong axis apply unless otherwise indicated.

Some problems require the use of a personal computer. Problems that are solved using spreadsheet or equation-solving software can be saved and used as templates for other similar problems. Templates can have many degrees of sophistication. Initially, a template may only be a hand (i.e., calculator) solution worked on a computer. In a simple template of this nature, the user will be required to provide many of the lookup functions for such items as

    Reference design values

    Lumber dimensions

    Load duration or time effect factor

    Wet service factor

    Size factor

    Volume factor

As users gain experience with their software, the template can be expanded to perform lookup and decision-making functions that were previously done manually.

Advanced computer programming skills are not required to create effective templates. Valuable templates can be created by designers who normally do only hand solutions. However, some programming techniques are helpful in automating *lookup* and *decision-making* steps.

The first requirement for a template is that it operate correctly (i.e., calculate correct values). Another major requirement is that the input and output be structured in an orderly manner. A sufficient number of intermediate answers should be displayed and labeled so that the solution can be verified by hand.

7.1   A 3×8 member in a diaphragm resists a tension force of 20 k caused by the lateral wind pressure. Lumber is Select Structural DF-L. A single line of $\frac{7}{8}$-in. diameter bolts is used to make the connection of the member to the diaphragm. $C_M = 1.0$, $C_t = 1.0$, and $C_i = 1.0$.

    Find: *a.* The maximum axial tension load using ASD.

                b. The factored axial tension capacity using LRFD.

7.2   A $5\frac{1}{8} \times 15$ DF axial combination 5 glulam is used as the tension member in a large roof truss. A single row of 1-in. diameter bolts occurs at the net section of the member. Loads are a combination of dead and snow (ASD: D + S; LRFD: 1.2D + 1.6S). Joints are assumed to be pin-connected. MC = 10 percent. $C_t = 1.0$.

Find: *a.* The maximum axial tension load using ASD.
        b. Repeat part *a* except that the MC = 15 percent.
        c. Repeat part *a* except that the MC = 18 percent.
        d. Repeat part *a* except that the member is a bending combination 24 F-V8 glulam.
        e. The factored axial tension capacity using LRFD.
        f. Repeat part *e* except that the MC = 15 percent.
        g. Repeat part *e* except that the MC = 18 percent.
        h. Repeat part *e* except that the member is a bending combination 24F-V8 glulam.

7.3 The truss in Fig. 7.A has a 2 × 4 lower chord of Select Structural Spruce-Pine-Fir (South). The loads shown are the result of D = 20 psf and S = 55 psf. There is no reduction of area for fasteners. $C_M = 1.0$, $C_t = 1.0$, and $C_i = 1.0$. Joists are assumed to be pin-connected.* Trusses are spaced 4 ft on center.

Find: *a.* Check the tension in the member using ASD.
        b. Check the tension in the member using LRFD.

Figure 7.A

7.4 Use the hand solution to Prob. 7.1 as a guide to develop a personal computer spreadsheet or equation-solving software template to solve similar problems using ASD or LRFD.
    a. Consider only the specific criteria given in Prob. 7.1.
    b. Expand the template to handle any sawn lumber size. The template is to include a list (i.e. database) of reference design values for all size categories (Dimension lumber, B&S, P&T) of Select Structural DF-L.
    c. Expand the database in part *b* to include all grades of DF-L from No. 2 through Select Structural.

---

*For trusses with joints which are not pinned (such as toothed metal plates and others), the continuity of the joints must be taken into consideration. For the design of metal-plate-connected trusses, see Ref. 7.6.

7.5 The truss in Fig. 7. A has a 2 × 6 lower chord of No. 1 DF-L. In addition to the loads shown on the sketch, the lower chord supports a ceiling dead load of 5 psf (20 lb/ft). There is no reduction of member area for fasteners. Joints are assumed to be pin-connected. $C_M = 1.0$, $C_t = 1.0$, and $C_i = 1.0$.

Find: *a.* Check the combined tension and bending in the member using ASD.
   b. Check the combined tension and bending in the member using LRFD.

7.6 The truss in Fig. 7.B supports the roof dead load of 16 psf shown in the sketch. Trusses are spaced 24 in o.c., and the roof live load is to be in accordance with the IBC. Lumber is No. 2 DF-L. Fasteners do not reduce the area of the members. Truss joints are assumed to be pin-connected. $C_M = 1.0$, $C_t = 1.0$, and $C_i = 1.0$.

Find: *a.* The required member size for the tension (bottom) chord using ASD.
   b. The required member size for the tension (bottom) chord using LRFD.

Figure 7.B

7.7 Repeat Prob. 7.6 except that in addition there is a ceiling dead load applied to the bottom chord of 8 psf (16 lb/ft). Neglect deflection.

7.8 The door header in Fig. 7.C supports a dead load of 120 lb/ft and a roof live load of 120 lb/ft. Lumber is No. 2 Hem-Fir. $C_M = 1.0$, $C_t = 1.0$, and $C_i = 1.0$. There are no bolt holes at the point of maximum moment. Lateral stability is not a problem.

Find: *a.* Using ASD check the given member size under the following loading conditions:
      Vertical loads only
      IBC-required combinations of vertical loads and lateral forces

Figure 7.C

      b. Which loading condition is the most severe?

      c. Repeat parts *a* and *b* using LRFD. The tension force resulting from wind $T = 8$ k is unfactored.

7.9 Repeat Prob. 7.8 except the unbraced length is one-half of the span length (that is, $l_u = 0.5L$).

7.10 Use the hand solution to Problems 7.8 and 7.9 as a guide to develop a personal computer spreadsheet or equation-solving software template to solve similar problems using ASD or LRFD.

      a. Consider only the specific criteria given in Problems 7.8 and 7.9.

      b. Expand the template to handle any span length and any unbraced length. The user should be able to choose any trial size of Dimension lumber and any grade of Hem-Fir from No. 2 through Sel. Str.

7.11 A 4×4 carries an axial compressive force caused by dead, live, and roof live loads. Lumber is No. 1 DF-L. $C_M = 1.0$, $C_t = 1.0$, and $C_i = 1.0$.

    Find: The maximum column load for each ASD load combination if the unbraced length of the member is

      a. 3 ft

      b. 6 ft

      c. 9 ft

      d. 12 ft

7.12 Repeat Prob. 7.11 except use LRFD to find the factored column capacity for each load combination.

7.13   *a.* Repeat Prob. 7.11 except that the member is a 4×6.

      b. Repeat Prob. 7.12 except that the member is a 4×6.

7.14 A 6×8 carries an axial compressive force caused by dead and snow loads. Lumber is No. 1 DF-L. $C_M = 1.0$ and $C_t = 1.0$.

    Find: The maximum column load for each ASD load combination if the unbraced length of the member is

      a. 5 ft

      b. 9 ft

      c. 11 ft

      d. 15 ft

      e. 19 ft

7.15 Repeat Prob. 7.14 except use LRFD to find the factored column capacity for each load combination.

7.16 Use the hand solution to Probs. 7.11 through 7.15 as a guide to develop a personal computer spreadsheet or equation-solving software template to solve similar problems using ASD or LRFD.

      a. The user should be able to specify any sawn lumber member size, column length, and reference design values of F_c and E_{min}. Initially limit the template to

No. 1 DF-L, and assume that the user will look up and provide the appropriate size factor (if required) for compression.

   b. Expand the template to access a database of reference design values for any size and grade of DF-L sawn lumber from No. 2 through Sel. Str. Include provisions for the database to furnish the appropriate size factor.

7.17   *Given:*   The glulam column in Fig. 7.D with the following information:

$$D = 20 \text{ k} \qquad L = 90 \text{ k}$$
$$L_r = 40 \text{ k} \qquad L_2 = 10 \text{ ft}$$
$$L_1 = 22 \text{ ft} \qquad L_3 = 12 \text{ ft}$$

The loads are axial forces and the member is axial combination 2 DF glulam without special tension laminations. The column effective length factor is $K_e = 1.0$. $C_M = 1.0$, and $C_t = 1.0$.

Find: a. Using ASD, is the column adequate to support the design load?
           b. Using LRFD, is the column adequate to support the factored design load?

STRONG AXIS   WEAK AXIS

Figure 7.D

7.18   *Given:*   The glulam column in Fig. 7.D with the following information

$$D = 20 \text{ k} \qquad L = 90 \text{ k}$$
$$L_r = 40 \text{ k} \qquad L_2 = 10 \text{ ft}$$
$$L_1 = 24 \text{ ft} \qquad L_3 = 14 \text{ ft}$$

The loads are axial forces and the member is axial combination 2 DF glulam without special tension laminations. The column effective length factor is $K_e = 1.0$. $C_M = 1.0$, and $C_t = 1.0$.

Find: a. Using ASD, is the column adequate to support the design load?
           b. Using LRFD, is the column adequate to support the factored design load?

7.19 Use the hand solution to Prob. 7.17 or 7.18 as a guide to develop a personal computer spreadsheet or equation-solving software template to solve similar problems using ASD or LRFD.

    a. Initially the template may be limited to axial combination 2 DF-L, and assume that the user will look up and provide the reference design values for the material. The template should handle different loads and unbraced lengths for the x and y axes.

    b. Expand the template to access a database of reference design values for any DF-L glulam combination. Consider either *axial* combinations or *bending* combinations as assigned.

7.20 A sawn lumber column is used to support axial loads of D = 20 k and S = 55 k. Use No. 1 DF-L. The unbraced length is the same for both the x and y axes of the member. The effective length factor is $K_e = 1.0$ for both axes. $C_M = 1.0$, $C_t = 1.0$ and $C_i = 1.0$.

Find: The minimum column size using ASD if the unbraced length is

    a. 8 ft
    b. 10 ft
    c. 14 ft
    d. 18 ft
    e. 22 ft
    A computer-based template may be used provided sufficient output is displayed to allow hand checking.

7.21 Repeat Prob. 7.20 except use LRFD to find the minimum column size.

7.22 *Given:*   The glulam column in Fig. 7.D with the following information:

$$D = 10 \text{ k} \qquad L = 45 \text{ k}$$
$$L_r = 20 \text{ k} \qquad L_2 = 10 \text{ ft}$$
$$L_1 = 26 \text{ ft} \qquad L_3 = 16 \text{ ft}$$

The minimum eccentricity described in Sec. 7.16 is to be considered. The member is a bending combination 24F-1.8E glulam. The column effective length factor is $K_e = 1.0$. $C_M = 1.0$, and $C_t = 1.0$.

Find: a. Using ASD, is the column adequate to support the design loads?
           b. Using LRFD, is the column adequate to support the factored design loads?

7.23 An 8 × 12 column of No. 1 S-P-F(S) has an unbraced length for buckling about the strong x axis of 16 ft and an unbraced length for buckling about the weak y axis of 8 ft. $C_M = 1.0$ and $C_t = 1.0$.

Find: *a.* The maximum axial loads using ASD considering D-only and D + L.
           b. The factored axial capacity using LRFD considering 1.4D and 1.2D + 1.6L.

7.24   A stud wall is to be used as a bearing wall in a wood-frame building. The wall carries axial loads caused by roof dead and live loads. Studs are 2 × 4 Construction-grade Hem-Fir and are located 16 in. o.c. Studs have sheathing attached. $C_M = 1.0$, $C_t = 1.0$, and $C_i = 1.0$.

Find: The maximum load per lineal foot of wall using ASD for each load combination if the wall height is
           a. 8 ft
           b. 9 ft
           c. 10 ft

7.25   Repeat Prob. 7.25 except use LRFD to find factored capacity per lineal foot of wall.

7.26   A stud wall is to be used as a bearing wall in a wood-frame building. The wall carries axial loads caused by roof dead and snow loads. Studs are 2 × 6 No. 2 Southern Pine and are 24 in. o.c. Studs have sheathing attached. $C_M = 1.0$, $C_t = 1.0$, and $C_i = 1.0$.

Find: The maximum load per lineal foot of wall using ASD for each load combination if the wall height is
           a. 10 ft
           b. 14 ft

7.27   Repeat Prob. 7.26 except use LRFD to find factored capacity per lineal foot of wall.

7.28   *Given:*   The exterior column in Fig. 7.E is a 6 × 10 Select Structural DF-L. It supports a vertical load due to a girder reaction and a lateral wind force from the horizontal wall framing. The lateral force causes bending about the strong axis of the member, and wall framing provides continuous lateral support about the weak axis. The following values are to be used:

Figure 7.E

$$D = 5 \text{ k} \qquad L = 16 \text{ ft}$$
$$S = 15 \text{ k} \qquad C_M = 1.0$$
$$W = 200 \text{ lb/ft} \qquad C_t = 1.0$$

Find: *a.* Check the column for combined compression and bending using ASD.
b. Check the column for combined compression and bending using LRFD.

7.29   Repeat Prob. 7.28 except that the following values are to be used:

$$D = 5 \text{ k} \qquad L = 21 \text{ ft}$$
$$S = 15 \text{ k} \qquad C_M = 1.0$$
$$W = 100 \text{ lb/ft} \qquad C_t = 1.0$$

7.30   The truss in Fig. 7.A has a 2×10 top chord of No. 2 Hem-Fir. The top of the truss is fully supported along its length by roof sheathing. There is no reduction of area for fasteners. $C_M = 1.0$, $C_t = 1.0$, and $C_i = 1.0$. Joints are assumed to be pin-connected.

Find: *a.* Check combined compression and bending in the top chord using ASD.
b. Check combined compression and bending in the top chord using LRFD.

7.31   A truss is similar to the one shown in Fig. 7.A except the span is 36 ft, $D = 10$ psf and $S = 20$ psf. The top chord is a 2×10 of No. 2 Hem-Fir, and it is laterally supported along its length by roof sheathing. There is no reduction of member area for fasteners. $C_M = 1.0$, $C_t = 1.0$, and $C_i = 1.0$. Joints are assumed to be pin-connected.

Find: *a.* Check combined compression and bending in the top chord using ASD.
b. Check combined compression and bending in the top chord using LRFD.

7.32   A 2×6 exterior stud wall is 14-ft tall. Studs are 16 in. o.c. The studs support the following vertical loads per foot of wall:

$$D = 800 \text{ lb/ft}$$
$$L = 800 \text{ lb/ft}$$
$$L_r = 400 \text{ lb/ft}$$

In addition, the wall carries a uniform wind force of 15 psf (horizontal). Lumber is No. 1 DF-L. $C_M = 1.0$, $C_t = 1.0$, and $C_i = 1.0$. Sheathing provides lateral support in the weak direction.

Find: *a.* Check the studs using ASD and the IBC-required load combinations. Neglect uplift.
b. Check the studs using LRFD and the IBC-required load combinations. Neglect uplift.

Chapter

8

Wood Structural Panels

8.1 Introduction

Plywood, oriented strand board (OSB), waferboard, composite panels, and structural particleboard, collectively referred to as *wood structural panels*, are widely used building materials with a variety of structural and nonstructural applications. Some of the major structural uses include

1. Roof, floor, and wall sheathing

2. Horizontal and vertical (shearwall) diaphragms

3. Structural components
   a. Lumber-and-plywood beams
   b. Stressed-skin panels
   c. Curved panels
   d. Folded plates
   e. Sandwich panels

4. Gusset plates
   a. Trusses
   b. Rigid frame connections

5. Preservative treated wood foundation systems

6. Concrete formwork

Numerous other uses of wood structural panels can be cited, including a large number of industrial, commercial, and architectural applications.

As far as the types of buildings covered in this text are concerned, the first two items in the above list are of primary interest. The relatively high load capacities and the ease with which panels can be installed have made wood structural panels widely accepted for use in these applications. The other topics listed above are beyond the scope of this text. Information on these and other subjects is available from APA—The Engineered Wood Association.

8.1

This chapter essentially serves as a turning point from the design of the vertical-load-carrying system (beams and columns) to the design of the lateral-force-resisting system (LFRS) (horizontal diaphragms and shearwalls). Wood structural panels provide this transition because they are often used as structural elements in *both* systems.

In the vertical-load-carrying system, structural panels function as the *sheathing* material. As such, it directly supports the roof and floor loads and distributes these loads to the framing system. See Example 8.1. Wall sheathing, in a similar manner, distributes the normal wind force to the studs in the wall. In the LFRS, wood structural panels serve as the shear-resisting element.

EXAMPLE 8.1   Wood Structural Panels Used as Sheathing

Figure 8.1   Plywood used to span between framing.

The most common forms of wood structural panels used for floor or roof sheathing are plywood and OSB. The term *sheathing load*, as used in this book, refers to loads that are normal to the surface of the sheathing. See Fig. 8.1. Sheathing loads for floors and roofs include dead load and live load (or snow load). For walls, the wind force is the sheathing load. Typical sheathing applications use panels continuous over two or more spans. For common joist spacings and typical loads, *design aids* have been developed so that the required sheathing can be chosen without having to perform beam design calculations. A number of these design aids are included in technical publications available from APA—The Engineered Wood Association (Refs. 8.5 and 8.6). When required, cross-sectional properties for a 1-ft-wide section of a structural panel can be obtained from the ASD/LRFD Manual for Engineered Wood Construction (Ref. 8.1).

The required *thickness* of the panel is often determined by *sheathing-type loads* (loads normal to the surface of the panel). On the other hand, the *nailing* requirements for the panel are determined by the *unit shears* in the horizontal or vertical diaphragm. When the shears are high, the required panel thickness may be governed by the diaphragm unit shears instead of by the sheathing loads.

It should be noted that the required thicknesses for roof, floor, and wall sheathing are determined using the provisions presented in Chap. 9 of the NDS (Ref. 8.2), and *design aids* provided in the ASD/LRFD Manual for Engineered Wood Constuction (Ref. 8.1) or APA technical publications (Refs. 8.5 and 8.6). Prior to 2001, provisions for engineering design with wood structural panels were not included in the NDS whatsoever. Requirements for wood structural panels were determined from design aids provided in Code tables (e.g., Ref. 8.16) or APA literature (e.g., Ref. 8.15). The NDS limits its scope to include plywood, OSB, and composite panels. For design with other types of panels, design aids may be available through APA—The Engineered Wood Association.

It is important to realize that wood structural panel design aids are based on beam calculations or concentrated load considerations, whichever are more critical. The need may arise for structural calculations if the design aids do not cover a particular situation. However, calculations for wood structural panels are usually necessary only for the design of a structural-type component (e.g., a lumber-and-plywood beam or stressed-skin panel).

This chapter introduces wood structural panel properties and grades and reviews the procedures used to determine the required thickness and grade of panels for *sheathing applications*. Some of the design aids for determining sheathing requirements are included, but the calculation of stresses in wood panels is beyond the scope of this text. However, the basic behavior of structural panels is explained, and some of the unique design aspects of wood panels are introduced. Understanding these basic principles is necessary for the proper use of wood structural panels.

Plywood and OSB constitute the majority of total wood structural panel production, although other panel products may also be used in structural applications. However, plywood is still the standard by which the other panel products are judged. Because of its traditional use in structural applications, plywood and its use as a sheathing material are covered first (Secs. 8.3 to 8.7), and other panel types, including new-generation panels, are introduced in Sec. 8.8. Chapter 9 continues with an introduction to *diaphragm design*, and Chap. 10 covers *shearwalls*. There the calculations necessary for the design of the LFRS are treated in considerable detail.

8.2 Panel Dimensions and Installation Recommendations

The standard size of wood structural panels is 4 ft × 8 ft. Certain manufacturers are capable of producing longer panels, such as 9 ft, 10 ft, or 12 ft, but the standard 4 ft × 8 ft dimensions should be assumed in design unless the availability of other sizes is known.

Plywood and the other panel products are quite dimensionally stable. However, some change in dimensions can be expected under varying moisture conditions, especially during the early stages of construction when the material is adjusting to local atmospheric conditions. For this reason, installation instructions for many roof, floor, and wall sheathing applications recommend a clearance between panel edges and panel ends. See Example 8.2.

EXAMPLE 8.2 Panel Installation Clearances

TYPICAL PANEL SHEATHING

Figure 8.2 Clearance between wood structural panels.

Many panel sheathing applications for roofs, floors, and walls recommend an *edge* and *end spacing* of $\frac{1}{8}$ in. to permit panel movement with changes in moisture content (MC). Other spacing provisions may apply, depending on the type of panel, application, and moisture content conditions. Refer to the ASD/LRFD Manual for Engineered Wood Construction (Ref. 8.1) or the APA *Engineered Wood Construction Guide*—(Ref. 8.4) for specific recommendations.

The tolerances for panel length and width depend on the panel type. Typical tolerances are $+0, -\frac{1}{16}$ in., and $+0, -\frac{1}{8}$ in. Some panel grade stamps include the term *sized for spacing*, and in this case the larger tolerance ($+0, -\frac{1}{8}$ in.) applies.

The installation clearance recommendations explain the negative tolerance on panel dimensions. By having the panel dimension slightly less than the stated size, the clearances between panels can be provided while maintaining the basic 4-ft module that the use of a wood structural panel naturally implies.

Another installation recommendation is aimed at avoiding *nail popping*. This is a problem that principally affects floors, and it occurs when the sheathing is nailed into green supporting beams. As the lumber supports dry, the members shrink and the nails appear to "pop" upward through the sheathing. This can cause problems with finish flooring (especially vinyl resilient flooring and similar products). Squeaks in floors may also develop.

Popping can be minimized by proper nailing procedures. Nails should be driven flush with the surface of the sheathing if the supporting beams are dry. If the supports are green, the nails should be "set" below the surface of the sheathing and the nail holes should not be filled. Squeaks can also be reduced by field-gluing the panels to the supporting beams. For additional information, contact APA—The Engineered Wood Association.

Wood structural panels are available in a number of standard thicknesses ranging from $\frac{1}{4}$ to $1\frac{1}{8}$ in. The tolerances for thickness vary depending on the thickness and surface condition of the panel. Panels with veneer faces may have several different surface conditions including unsanded, touch-sanded, sanded, overlaid, and others. See the appropriate specification for thickness tolerances.

8.3 Plywood Makeup

A plywood panel is made up of a number of veneers (thin sheets or pieces of wood). Veneer is obtained by rotating *peeler logs* (approximately $8\frac{1}{2}$-ft long) in a lathe. A continuous veneer is obtained as the log is forced into a long knife. The log is simply unwound or "peeled." See Example 8.3. The veneer is then clipped to the proper size, dried to a low moisture content (2 to 5 percent), and graded according to quality.

EXAMPLE 8.3   Fabrication of Veneer

Figure 8.3   Cutting veneer from peeler log.

The log is rotary-cut or peeled into a continuous sheet of *veneer*. See Fig. 8.3. Thicknesses range between $\frac{1}{16}$ and $\frac{5}{16}$ in. As with sawn lumber, the veneer is graded visually by observing the size and number of defects. Most veneers may be repaired or patched to improve their grade. Veneer grades are discussed in Sec. 8.5.

The veneer is spread with glue and cross-laminated (adjacent layers have the wood grain at right angles) into a plywood panel with an *odd* number of *layers*. See Example 8.4. The panel is then cured under pressure in a large hydraulic press. The glue bond obtained in this process is stronger than the wood in the plies. After curing, the panels are trimmed and finished (e.g., sanded) if necessary. Finally, the appropriate grade-trademark is stamped on the panel.

EXAMPLE 8.4   Plywood Cross-Laminated Construction

3 LAYER
3 PLY CONSTRUCTION

Figure 8.4a

3 LAYER
4 PLY CONSTRUCTION

Figure 8.4b

In its simplest form, plywood consists of 3 plies. Each ply has wood grain at right angles to the adjacent veneer (Fig. 8.4a).

An extension of the simple 3-ply construction is the 3-layer 4-ply construction (Fig. 8.4b). The two center plies have the grain running in the same direction. However, the basic concept of cross-laminating is still present because the two center plies are viewed as a single layer. It is the *layer* that is *cross-laminated*.

Three-layer construction is used in the thinner plywood panels. Depending on the thickness and grade of the plywood, 5- and 7-layer constructions are also fabricated. Detailed information on plywood panel makeup is contained in Ref. 8.18.

It is the cross-laminating that gives plywood its unique strength characteristics. It provides increased dimensional stability over wood that is not cross-laminated. Cracking and splitting are reduced, and fasteners, such as nails and staples, can be placed close to the edge without a reduction in load capacity.

In summary, *veneer* is the thin sheet of wood obtained from the peeler log. When *veneer* is used in the construction of plywood, it becomes a *ply*. The cross-laminated pieces of wood in a plywood panel are known as *layers*. A layer is often simply an individual ply, but it can consist of more than one ply.

The direction of the grain in a finished panel must be clearly understood. See Example 8.5. The names assigned to the various layers in the makeup of a plywood panel are

1. *Face*—outside ply. If the outside plies are of different veneer quality, the face is the better veneer grade.

2. *Back*—the other outside ply.

3. *Crossband*—inner layer(s) placed at right angles to the face and back plies.

4. *Center*—inner layer(s) parallel with outer plies.

EXAMPLE 8.5   Direction of Grain

Figure 8.5 Standard plywood layup.

In standard plywood construction, the face and back plies have the grain oriented parallel to the 8-ft dimension of the panel. Crossbands are inner plies that have the grain at right angles to the face and back (i.e., parallel to the 4-ft dimension). If a panel has more than three layers, some inner plies (centers) will have grain that is parallel to the face and back.

When the stress in plywood is parallel to the 8-ft dimension, there are more "effective" plies (i.e., there are more plies with grain parallel to the stress *and* they are placed farther from the neutral axis of the panel). The designer should be aware that different section properties are involved, depending on how the panel is turned. This is important even if stress calculations are not performed.

If structural calculations are required, the cross-laminations in plywood make stress analysis somewhat complex. Wood is stronger parallel to the grain than perpendicular to the grain. This is especially true in tension, where wood has little strength across the grain; it is also true in compression but to a lesser extent. In addition, wood is much stiffer parallel to the grain than perpendicular to the grain. The modulus of elasticity across the grain is approximately $\frac{1}{35}$ of the modulus of elasticity parallel to the grain.

Because of the differences in strength and stiffness, the *plies that have the grain parallel to the stress are much more effective* than those that have the grain perpendicular to the stress. In addition, the odd number of layers used in plywood construction causes further differences in section properties for one direction (say, parallel to the 8-ft dimension) compared with the section properties for the other direction (parallel to the 4-ft dimension). Thus, two sets of cross-sectional properties apply to plywood. One set is used for stresses parallel to the 8-ft dimension, and the other is used for stresses parallel to the 4-ft dimension.

Even if the sheathing thickness and reference uniform load capacity are read from a table (structural calculations not required), the orientation of the panel and its directional properties are important to the proper use of the plywood. To illustrate the effects of panel orientation, two different panel layouts are considered. See Fig. 8.6. With each panel layout, the corresponding 1-ft-wide beam cross section is shown. The bending stresses in these beams are parallel to the span. For simplicity, the plywood in this example is 3-layer construction.

In the first example, the 8-ft dimension of the plywood panel is parallel to the span (sheathing spans between joists). When the plywood is turned this way (face grain perpendicular to the supports), it is said to be used in the *strong direction*. In the second example, the 4-ft dimension is parallel to the span of the plywood (face grain parallel to the supports). Here the panel is used in the *weak direction*.

From these sketches it can be seen that the cross section for the strong direction has more plies with the grain oriented parallel to the span. In addition, these plies are located at a greater distance from the neutral axis. These two facts explain why the effective cross-sectional properties are larger for plywood oriented with the long dimension of the panel perpendicular to the supports.

8.4   Species Groups for Plywood

A large number of species of wood can be used to manufacture plywood. See Fig. 8.7. The various species are assigned, according to strength and stiffness, to one of five different groups. Group 1 species have the highest strength characteristics, and Group 5 species have the lowest strength properties. Reference design

Figure 8.6 Strong and weak directions of plywood. Wood grain that is *parallel* to the span and stress is *more effective* than wood grain that is perpendicular to the span and stress.

values have been determined for species groups 1 to 4, and plywood made up of these species can be used in structural applications. Group 5 has not been assigned reference design values.

The specifications for the fabrication of plywood allow the mixing of various species of wood in a given plywood panel. This practice allows the more complete

Group 1	Group 2	Group 3	Group 4	Group 5
Apitong	Cedar, Port	Alder, Red	Aspen	Basswood
Beech,	Orford	Birch, Paper	Bigtooth	Poplar,
American	Cypress	Cedar, Alaska	Quaking	Balsam
Birch	Douglas	Fir,	Cativo	
Sweet	Fir 2 (a)	Subalpine	Cedar	
Yellow	Fir	Hemlock,	Incense	
Douglas	Balsam	Eastern	Western	
Fir 1 (a)	California	Maple,	Red	
Kapur	Red	Bigleaf	Cottonwood	
Keruing	Grand	Pine	Eastern	
Larch,	Noble	Jack	Black	
Western	Pacific	Lodgepole	(Western	
Maple, Sugar	Silver	Ponderosa	Poplar)	
Pine	White	Spruce	Pine	
Caribbean	Hemlock,	Redwood	Eastern	
Ocote	Western	Spruce	White	
Pine, South.	Lauan	Engelmann	Sugar	
Loblolly	Almon	White		
Longleaf	Bagtikan			
Shortleaf	Mayapis			
Slash	Red			
Tanoak	Tangile			
	White			
	Maple, Black			
	Mengkulang			
	Meranti,			
	Red (b)			
	Mersawa			
	Pine			
	Pond			
	Red			
	Virginia			
	Western			
	White			
	Spruce			
	Black			
	Red			
	Sitka			
	Sweetgum			
	Tamarack			
	Yellow-			
	Poplar			

(a) Douglas Fir from trees grown in the states of Washington, Oregon, California, Idaho, Montana, Wyoming, and the Canadian Provinces of Alberta and British Columbia shall be classed as Douglas Fir No. 1. Douglas Fir from trees grown in the states of Nevada, Utah, Colorado, Arizona and New Mexico shall be classed as Douglas Fir No. 2.
(b) Red Meranti shall be limited to species having a specific gravity of 0.41 or more based on green volume and oven dry weight.

Figure 8.7 Species of wood used in plywood. (*Courtesy of APA—The Engineered Wood Association.*)

usage of raw materials. If it should become necessary to perform stress calculations, the reference design values for plywood calculations have been simplified for use in design. This is accomplished by providing reference design values based on the species group of the face and back plies. The species group of the outer plies is included in the grade stamp of certain grades of plywood (Sec. 8.7). Effective section properties are calculated for Group 4 inner plies (the weakest species group allowed in structural plywood). The assumption of Group 4 inner plies is made regardless of the actual makeup. Reference design values and effective cross-sectional properties are given in the APA *Plywood Design Specification* (PDS, Ref. 8.15).

Although plywood grades have not yet been covered, it should be noted that some grade modifications can be added to the *sheathing grades* that limit the species used in the plywood. For example, the term STRUCTURAL I can be added to certain plywood grades to ensure increased strength properties. The addition of the STRUCTURAL I designation restricts all veneers in the plywood to Group 1 species. Thus larger section properties apply to plywood with this designation, because the inner plies of Group 1 species (rather than Group 4 species) are used in calculations.

Besides limiting the species of wood used in the manufacture of plywood, the STRUCTURAL I designation requires the use of exterior glue and provides further restrictions on layup, knot sizes, and repairs over the same grades without the designation. STRUCTURAL I should be added to the plywood grade specification when the increased strength is required, particularly in shear or cross-panel properties (parallel to the 4-ft dimension).

Before the methods for determining the required plywood grade and thickness for sheathing applications can be reviewed, some additional topics should be covered. These include veneer grades, exposure durability classifications, and plywood grades.

8.5   Veneer Grades

The method of producing the veneers which are used to construct a plywood panel was described in Sec. 8.3. Before a panel is manufactured, the individual veneers are graded according to quality. The *grade of the veneers* is one of the factors that determine the *grade of the panel*.

The six basic veneer grades are designated by a letter name:

N   Special-order "natural finish" veneer. Not used in ordinary structural applications.

A   Smooth, paintable surface. Solid-surface veneer without knots, but may contain a limited number (18 in a 4 ft × 8 ft veneer) of neatly made repairs.

B   Solid-surface veneer. May contain knots up to 1 in. in width across the grain if they are both sound and tight-fitting. May contain an unlimited number of repairs.

C-plugged   An improved grade of C veneer that meets more stringent limitations on defects than the normal C veneer. For example, open defects such as knotholes may not exceed $\frac{1}{4}$ in. by $\frac{1}{2}$ in. Further restrictions apply.

C   May contain open knotholes up to 1 in. in width across the grain and occasional knotholes up to $1\frac{1}{2}$ in. across the grain. This is the minimum-grade veneer allowed in exterior-type plywood.

D   Allows open knotholes up to $2\frac{1}{2}$ in. in width across the grain and occasional knotholes up to 3 in. across the grain. This veneer grade is not allowed in exterior-type plywood.

The veneer grades in this list are given in order of decreasing quality. Detailed descriptions of the growth characteristics, defects, and patching provisions for each veneer grade can be found in Ref. 8.18. Although A and B veneer grades have better surface qualities than C and D veneers, on a structural basis A and C grades are more similar. Likewise, B and D grades are similar, structurally speaking. The reason for these structural similarities is that C veneers can be upgraded through patching and other repairs to qualify as A veneers. See Example 8.6. On the other hand, a B veneer grade can be obtained by upgrading a D veneer. The result is more unbroken wood fiber with A and C veneers.

Except for the special "Marine" exterior grade of plywood, A- and B-grade veneers are used only for face and back veneers. They may be used for the inner plies, but, in general, C and D veneers will be the grades used for the inner plies. It should be noted that D veneers represent a large percentage of the total veneer production, and their use, where appropriate, constitutes an efficient use of materials.

EXAMPLE 8.6   Veneer Grades and Repairs

A-grade veneers are smooth and firm and free from knots, pitch pockets, open splits, and other open defects. A-grade veneers can be obtained by upgrading (repairing) C-grade veneers.

Another upgraded C veneer is C-plugged veneer. Although it has fewer open defects than C, it does not qualify as an A veneer.

B veneers are solid and free from open defects with some minor exceptions. B veneers can be obtained by upgrading (repairing) D-grade veneers. A and B veneers have similar surface qualities, but A and C are structurally similar. Likewise, B and D grades have similar strength properties. See Figs. 8.8a and 8.8b.

Figure 8.8a   A- and C-grade veneers are structurally similar.

TYP. PATCH IN A "B" VENEER

TYP. KNOTHOLE IN A "D" VENEER

Figure 8.8b B- and D-grade veneers are structurally similar.

8.6 Exposure Durability Classifications

Exposure durability classifications apply to plywood fabricated under PS 1 (Ref. 8.18), as well as to panels (both all-veneer plywood and other panels) that are performance-rated. See Sec. 8.7 for additional information on PS 1 and performance-rated panels. For plywood manufactured under PS 1, the primary exposure durability classifications are: Exterior and Interior. *Exterior* plywood is glued with an insoluble "waterproof" glue *and* is constructed with a minimum of C-grade veneers. It will retain its glue bond when repeatedly wet and dried. Exterior plywood is required when it will be permanently exposed to the weather or when the moisture content (MC) in use will exceed 18 percent, either continuously or in repeated cycles. In high MC conditions, the use of preservative treated panels should be considered.

Interior plywood may be used if it is not exposed to the weather and if the MC in service does not continuously or repeatedly exceed 18 percent. Interior plywood can be manufactured with exterior, intermediate, or interior glue, but it is generally available with exterior glue. Thus, *plywood manufactured with exterior glue is not necessarily classified as Exterior plywood.* If a plywood panel contains a D-grade veneer, it cannot qualify as an exterior panel even if it is manufactured with exterior glue. The reason for this veneer grade restriction is that the knotholes allowed in the D veneer grade are so large that the glue bond, even with exterior glue, may not stand up under continuous exposure to the weather. Such exposure may result in localized delamination in the area of the knothole.

Interior plywood with any glue type is intended for use in interior (protected) applications. Interior plywood bonded with exterior glue is known as *Exposure 1* and is intended for use where exposure to moisture due to long construction delays may occur.

Under APA's performance-rated panel system, plywood and other wood structural panel products are either Exterior or Exposure 1, (Ref. 8.19). For additional information on durability application recommendations, see Refs. 8.4 and 8.7.

8.7   Plywood Grades

For many years, the specifications covering the manufacture of plywood have been prescriptive in nature. This means that a method of constructing a plywood panel was fully described by the specification. For a given grade of plywood the species group, veneer grades, and other important factors were specified.

U.S. Product Standard PS 1—Construction and Industrial Plywood (Ref. 8.18) covers the manufacture of traditional *all-veneer* panels known as plywood. For many years PS 1 was a prescriptive-only specification. Although PS 1 still contains prescriptive requirements (a recipe for manufacturing plywood), it now also contains requirements for plywood that can be qualified on the basis of performance tests.

The concept of a *performance standard* was adapted to manufacturing of wood structural panels because a prescriptive type of specification did not lend itself to the development of some of the newer panel products (Sec. 8.8). These panels can be manufactured in a number of different ways using a variety of raw materials. Rather than prescribing how a panel product is to be constructed, a performance standard specifies what the product must do (e.g., load-carrying capability, dimensional stability, and ability to perform satisfactorily in the presence of moisture).

Although performance rating was developed for these newer panel products, it was noted that plywood can also be performance rated. The performance rating of traditional all-veneer plywood has resulted in the development of newer, thinner thicknesses. For example, $^{15}/_{32}$ in. now replaces $^{1}/_{2}$ in., $^{19}/_{32}$ in. replaces $^{5}/_{8}$ in., and $^{23}/_{32}$ in. replaces $^{3}/_{4}$ in. For more information see APA's publication *Performance Rated Panels* (Ref. 8.13).

There are a large number of grades of plywood. Several examples are given here, but for a comprehensive summary of plywood grades and their appropriate uses, the reader is referred to the APA *Engineered Wood Construction Guide* (Ref. 8.4).

Each plywood panel is stamped with a grade-trademark that allows it to be fully identified. A *sanded panel* will have an A- or B-veneer-grade face ply. The back ply may be an A, B, C, or D veneer. The grade-trademark on a sanded plywood panel will include the following:

1. Veneer grade of the face and back

2. Minimum species group (highest species group number from Fig. 8.7) of the outer plies

3. Exposure durability classification

These items essentially identify the plywood. See Example 8.7. Other information included in the stamp indicates the product standard, the manufacturer's mill number (000 shown), and the abbreviation of the "qualified inspection and testing agency." APA—The Engineered Wood Association is the agency that provides this quality assurance for many of the plywood manufacturers in the United States.

EXAMPLE 8.7   Sanded Plywood Panel

A-C GROUP 1

EXTERIOR

000

PS 1-95

Figure 8.9   Sanded plywood panel. (*Courtesy of APA—The Engineered Wood Association.*)

A typical grade-trademark for a *sanded panel* is shown in Fig. 8.9. Outer plies of A- and B-grade veneers will be sanded. C-plugged will be touch-sanded. Others will be unsanded. For the given example only one side of the panel will be fully sanded.

The sanding operation improves the surface condition of the panel, but in doing so it reduces the thickness of the outer veneers by a measurable amount. In fact, the relative thickness of the layers is reduced by such an amount that different cross-sectional properties are used in strength calculations for sanded, touch-sanded, and unsanded panels. See Table 2 in the *Plywood Design Specification* (Ref. 8.15).

Although a sanded plywood grade can be used in a structural application, it is normally not used because of cost. Plywood used in structural applications is often covered with a finish material, and a less expensive plywood grade may be used.

The plywood *sheathing grades* are normally used for roof, floor, and wall sheathing. These are

C-C

C-D

Note that C-C is Exterior-type plywood, and C-D is generally available with exterior glue which qualifies it as Exposure 1. Where added strength is required, STRUCTURAL I can be added to the designation:

C-C STR I

C-D STR I

This grade modification affects both reference design values and effective section properties for the plywood.

A sheathing grade of plywood has several different items in the grade-trademark compared with a sanded panel, including the panel thickness and span rating. See Example 8.8.

The *span rating* on sheathing panels is a set of two numbers. The number on the left in the span rating is the maximum recommended span in inches when the plywood is used as roof sheathing. The second number is the maximum recommended span in inches when the plywood is used as subflooring. For example, a span rating of 48/24 indicates that the panel can be used to span 48 in. in a roof system and 24 in. as floor sheathing. In both roof and floor applications it is assumed the panel will be continuous over two or more spans. The purpose of the span rating is to allow the selection of a proper plywood panel for sheathing applications without the need for structural calculations. Reference roof and floor sheathing loads are covered in later sections of this chapter.

The use of the span rating to *directly* determine the maximum recommended span for a given panel requires that the plywood be oriented in the strong direction (i.e., the long dimension of the panel perpendicular to the supports). In addition, certain plywood *edge support* requirements must be satisfied in order to apply the span rating without a reduction in the recommended span.

EXAMPLE 8.8   Plywood Sheathing Grade

Figure 8.10 A sheathing panel (such as C-C) is unsanded. (*Courtesy of APA—The Engineered Wood Association.*)

A typical grade-trademark for a *sheathing grade* of plywood is shown in Fig. 8.10. The stamp indicates the panel is APA Rated Sheathing, *and* it conforms to the Product Standard PS 1. Because it conforms to PS 1, this is an all-veneer (i.e., plywood) panel. Some APA Rated Sheathing is not traditional all-veneer plywood. These wood structural panels are usually manufactured with some form of reconstituted wood product (Sec. 8.8), and the Product Standard PS 1 will not be referenced in the grade-trademark of these other panels.

The example in Fig. 8.10 also indicates that the plywood is C-C Exterior with the STRUCTURAL I upgrade. The panel is $^{23}\!/_{32}$ in. thick, and it has a *span rating* of 48/24. Other information includes the manufacturer's mill number (000 shown), and the panel conforms to the *performance standard* PRP-108 (Ref. 8.14).

The ASD/LRFD Manual for Engineered Wood Construction (Ref. 8.1) and the International Building Code (IBC) (Ref. 8.16) provide design information for plywood, as well as oriented strand board (OSB) and composite panels. Table M9.1-1 in the ASD/LRFD Manual presents general information regarding grades, thicknesses, and typical uses for various panel types. IBC Table 2304.7(3) provides span ratings for structural panels in both roof and floor applications. The fact to note at this time is that a given span rating may be found on panels of *different* thicknesses. The various thicknesses that comprise a given span rating are listed in Table 5 of the *Panel Design Specification* (Ref. 8.7). This table provides both the predominant thickness and alternative thicknesses for each span rating. The span rating theoretically accounts for the equivalent strength of panels fabricated from different species of wood. Thus, the same span rating may be found on a thin panel that is fabricated from a strong species of wood *and* a thicker panel that is manufactured from a weaker species. Practically speaking, however, plywood for a given span rating is usually constructed so that the thinner (or thinnest) of the panel thicknesses possible will be generally available.

This fact is significant because of the dual function of structural panels in many buildings. The minimum *span rating* should be specified for sheathing loads, *and* the minimum *panel thickness* as governed by lateral diaphragm design (or the minimum thickness compatible with the span rating) should be specified. Thus, both span rating and panel thickness are required in specifying a sheathing grade of plywood or other structural panel.

To summarize, panels can be manufactured with different thicknesses for a given span rating. Generally speaking, the thickness that is *available* is the smaller of those listed for a given span rating. If the minimum span rating and minimum thickness are specified, then a panel with a larger span rating and/or thickness may properly be used in the field.

8.8 Other Wood Structural Panels

Reference to wood structural panels in addition to plywood has been made in previous sections of this chapter. These other panel products include composite panels, waferboard, oriented strand board (OSB), and structural particleboard. APA Rated Sheathing and APA Rated Sturd-I-Floor panels include plywood and all the others mentioned. These are recognized by the IBC under *performance standards* such as U.S. Product Standard PS-2 (Ref. 8.19) and APA PRP-108 (Ref. 8.14). Alternatively, certain wood structural panels may be produced under a prescriptive type of specification such as ANSI A208.1 (Ref. 8.3). These panels can be used for structural roof, floor, and wall sheathing applications. In addition, they can be used to resist lateral forces in horizontal diaphragms and shearwalls.

Wood structural panels that are not all-veneer plywood usually involve some form of reconstituted wood product. A brief description of these panels is given here. See Fig. 8.11.

OSB is a nonveneer panel manufactured from reconstituted wood strands or wafers. The strand-like or wafer-like wood particles are compressed and bonded with phenolic resin. As the name implies, the wood strands or wafers

Composite
Panels of reconstituted wood cores bonded between veneer
face and back plies.

Waferboard
Panels of compressed wafer-like particles or flakes randomly oriented.

Oriented Strand Board (OSB)
Panels of compressed strand-like or wafer-like particles arranged in
layers oriented at right angles to one another.

Structural Particleboard
Panels made of small particles usually arranged in layers by particle
size, but not usually oriented.

Figure 8.11   APA Rated Sheathing. Of the four types of sheathing other than plywood, oriented strand board (OSB) is the most widely used for structural applications. (*Courtesy of APA—The Engineered Wood Association.*)

are directionally oriented. The wood fibers are arranged in perpendicular layers (usually three to five) and are thus cross-laminated in much the same manner as plywood. OSB is manufactured from both softwood species and hardwood species, as well as mixed species.

OSB was first produced in the early 1980s and has grown to become a major part of the structural wood products industry. Today, OSB is considered in many applications to be interchangeable with plywood. As a simple example of this, refer to Table M9.2-1 from the ASD/LRFD Manual where bending stiffness and strength design values are provided for given span ratings and thicknesses, not for panel type. Table 5 in the *Load-Span Tables for APA Structural-Use Panels* (Ref. 8.5) shows that OSB can be used for any span rating, just as plywood can be used for any span rating.

Composite panels (or COMPLY) are recognized by the NDS (Ref. 8.2), along with plywood and OSB, as wood structural panels. Composite panels have a veneer face and back and a reconstituted core. They are typically produced with five layers, such that the center layer is also wood veneer. The reconstituted core is formed from low-quality wood fiber, often from a recycled source.

Waferboard is a nonveneer panel manufactured from reconstituted wood wafers. These wafer-like wood particles or flakes are compressed and bonded with phenolic resin. The wafers may vary in size and thickness, and the direction of the grain in the flakes is usually randomly oriented. The wafers may also be arranged in layers according to size and thickness. Waferboard is the predecessor to OSB, but is not recognized by the NDS as a wood structural panel. APA—The Engineered Wood Association maintains design information regarding use of waferboard.

Structural particleboard is another panel product not recognized by the NDS. It is a nonveneer panel manufactured from small particles (as opposed to larger wafers or strands) bonded with resins under heat and pressure. One primary disadvantage of structural particleboard is its susceptibility to moisture and dimensional instability.

FRP (*fiber reinforced plastic*) *plywood* is the latest addition to the suite of wood structural panel products. FRP sheets are bonded to plywood panels. Currently, the application of FRP to wood panels is limited to plywood, since the bonding of the FRP overlay requires a reasonably smooth surface to adhere and avoid delamination. FRP plywood has been used in situations requiring long-lasting durability performance against wearing and weathering. Additional information regarding FRP plywood may be obtained from APA—The Engineered Wood Association.

As noted, some wood structural panels that are not plywood may involve the use of veneers. For example, composite panels have outer layers that are veneers and an inner layer that is a reconstituted wood core. Other nonplywood panels are completely nonveneer products.

A typical grade-trademark for a nonveneer performance rated panel includes a number of the same items found in a plywood sheathing stamp. See Example 8.9. However, the grade-trademark found on panels that are not all-veneer plywood does not contain reference to PS 1, and nonveneer panels will not have veneer grades (e.g., C-D) shown in the stamp. Panel grades covered in PS 2 include Sheathing, Structural I Sheathing, and Single Floor. These designations also appear in the grade-trademarks when panels conform to this standard.

EXAMPLE 8.9   Nonveneer Sheathing Grade

Figure 8.12   Nonveneer sheathing grade. (*Courtesy of APA—The Engineered Wood Association.*)

A typical grade-trademark for a *nonveneer* panel is shown in Fig. 8.12. The stamp indicates that the panel is APA Rated Sheathing, and it has a span rating of 32/16 (Sec. 8.7). In addition, the thickness ($^{15}\!/_{32}$ in.) and durability classification (Exposure 1) are shown.

Other information includes the manufacturer's mill number (000 shown), and the panel is "sized for spacing" (Sec. 8.2). The panel conforms to APA's performance standard PRP-108 (Ref. 8.14).

8.9 Roof Sheathing

Wood structural panels account for much of the wood roof sheathing used in the United States. These materials are assumed to be continuous over two or more spans. Plywood, OSB, and other panels with directional properties are normally used in the strong direction (long dimension of the panel perpendicular to the supports). However, in panelized-roof systems (Sec. 3.2) the panels are often turned in the weak direction for sheathing loads. In this latter case, a thicker panel may be required, but this type of construction results in a savings in labor. In addition, higher diaphragm shears can be carried with the increased panel thickness.

The *span rating* described in Sec. 8.7 appears in the grade-trademark of both traditional plywood and APA's performance rated sheathing. Recall that plywood, OSB, and other wood structural panels may be performance rated.

Table M9.2-1 in the ASD/LRFD Manual provides bending stiffness and strength capacities for wood structural panels. These capacities can be used to determine the appropriate product (span rating and/or thickness) for a specific application. However, traditionally plywood and other wood structural panels have been viewed more like proprietary products and thus are selected by evaluating adjusted capacities versus actual applied loads. Tables in APA publications (Refs. 8.5 and 8.6) provide reference uniform load capacities on *sheathing, single floor,* and *sanded panels*, respectively. Note that these tables all assume normal load duration, $C_D = 1.0$. For other durations of load, the adjusted ASD load capacities governed by bending and shear are determined by multiplying the tabulated reference values by the appropriate load duration factor. The reference load capacities governed by deflection would not be adjusted for load duration.

Nominal load capacities for LRFD can be determined using the tabulated reference load capacities in APA publications (Refs. 8.5 and 8.6). The reference load capacities governed by deflection are directly applicable to LRFD, since deflection calculations are identical in ASD and LRFD (see Sec. 6.6 and Sec. 6.9). However, the reference load capacities governed by bending and shear must be multiplied by a format conversion factor, $K_F = 2.16/\phi$, to determine the nominal LRFD load capacities for bending and shear. As specified in NDS App. N, the LRFD format conversion factors for bending and shear are:

$$\text{Bending:} \quad K_F = 2.16/\phi_b = 2.16/0.85 = 2.54$$

$$\text{Shear:} \quad K_F = 2.16/\phi_v = 2.16/0.75 = 2.88$$

Finally, the nominal LRFD capacities for bending and shear must be multiplied by the appropriate time effect factor (λ) and resistance factor (ϕ) to determine adjusted LRFD load capacities (see Ref. 8.2). The adjusted LRFD load capacities can then be compared with factored load combinations.

The span rating can be used *directly* to determine the sheathing requirements for panels used in the strong direction under typical roof live loading conditions.

When used directly, the actual span agrees with the roof span in the two-number span rating. For larger roof loads, the span rating can be used *indirectly* to determine panel sheathing requirements. See Example 8.10.

EXAMPLE 8.10   Determination of Panel Sheathing Requirements

The span rating can be used directly to specify sheathing. When used directly the actual span, or support spacing, agrees with the roof span rating (the first number in the two-number span rating).

The span rating can also be used indirectly to determine sheathing requirements. In this example, the roof span in the span rating (the first number in the two-number span rating) will not agree with the actual span or support spacing. To illustrate this, consider a roof system with a sheathing span of 24 in., a dead load of 10 psf, a snow load of 75 psf, and a total load deflection limit of $L/240$. Determine the minimum span-rated sheathing using ASD procedures. Repeat the problem using LRFD procedures.

> Note that reference load capacity tables in APA publications (Refs. 8.5 and 8.6) are for normal load duration, $C_D = 1.0$. For the D + S load combination, $C_D = 1.15$. To use Table 1 in the *Load-Span Tables for APA Structural-Use Panels* (Ref. 8.5) for a load duration other than the normal duration, simply multiply the "governed by bending" and "governed by shear" values in the table by the appropriate load duration factor.
>
> The panel with a span rating of 40/20 is able to meet the demand. The 40/20 placed on 24-in. center-to-center supports allows 146 psf > 85 psf considering the $L/240$ deflection limit, $[(130)(1.15) = 149.5 \text{ psf}] > 85$ psf considering bending, and $[(236)(1.15) = 271.4 \text{ psf}] > 85$ psf considering shear. In this case, the span rating of 40/20 does not match the actual span (24 in.).

> Now repeat the problem using LRFD procedures. The tabulated load capacities governed by deflection in Table 1 of Ref. 8.5 are directly applicable to both ASD and LRFD, based on the unfactored load combination D + S = 85 psf. However, for bending and shear the factored load combination 1.2D + 1.6S = 1.2(10) + 1.6(75) = 132 psf must be used in LRFD. The tabulated reference values for bending and shear must also be converted to nominal LRFD values. For the 40/20 span rating from Table 1 (Ref. 8.5):
>
> $$\text{Bending:} \quad K_F\,(130 \text{ psf}) = 2.54(130) = 330.4 \text{ psf}$$
>
> $$\text{Shear:} \quad K_F\,(236 \text{ psf}) = 2.88(236) = 679.7 \text{ psf}$$
>
> According to NDS Appendix N, the time effect factor $\lambda = 0.8$ applies for the 1.2D + 1.6S load combination. The resistance factors for bending and shear are $\phi_b = 0.85$ and $\phi_v = 0.75$. Therefore, the adjusted LRFD capacities for bending and shear for the 40/20 span rating are:
>
> $$\text{Bending strength:} \quad \lambda\phi_b\,(330.4 \text{ psf}) = 0.8(0.85)(330.4) = 225 \text{ psf} > 132 \text{ psf} \quad OK$$
>
> $$\text{Shear strength:} \quad \lambda\phi_v\,(679.7 \text{ psf}) = 0.8(0.75)(679.7) = 408 \text{ psf} > 132 \text{ psf} \quad OK$$

> As noted above, the deflection calculations are identical in both ASD and LRFD:
>
> $L/_{240}$ deflection limit: 146 psf > 85 psf   *OK*
>
> Therefore, a panel with a 40/20 span rating placed on 24-in. center-to-center supports can support loads of D = 10 psf and S = 75 psf using either ASD or LRFD procedures.

The maximum spans defined by the span rating assume that the edges parallel to the span are supported in some fashion. This is referred to as panels *with edge support*. Typically, panel clips are used to provide this edge support, but lumber blocking or another mechanism may be used. Panels *without edge support* have reduced maximum spans. The maximum spans for panels *with* and *without* edge support are provided in Table M9.4-1 of the ASD/LRFD Manual. See Example 8.11.

Panel edge support is intended to limit differential movement between adjacent panel edges. Consequently if some form of edge support is not provided, a thicker panel or a reduced spacing of supports will be required.

A point about the support of panel edges should be emphasized. The use of tongue-and-groove (T&G) edges or panel clips is accepted as an alternative to lumber blocking for sheathing loads only. If blocking is required for diaphragm action (Chap. 9), panel clips or T&G edges cannot be substituted for lumber blocking (except $1\frac{1}{8}$ -in.-thick 2-4-1 plywood with stapled T&G edges, Ref. 8.17).

EXAMPLE 8.11   Roof Sheathing Edge Support Requirements

Alternative forms of support for panel edge that is perpendicular to roof framing (Fig. 8.13).

a. *Unsupported edge.* In most cases, the NDS requires closer roof joist or rafter spacing than given by the *span rating* when the 8-ft panel edges are not supported. Note that panel thickness may be increased as an alternative to providing support to all edges (or the spacing between roof framing members may be reduced).

b. *Lumber blocking.* Cut and fitted between roof joists or rafters may be reduced.

c. *T&G edges.*

d. *Panel clips.* Metal H-shaped clips placed between plywood edges. The IBC, in footnote f of Table 2304.7(3), requires one panel clip midway between supports when the span is less than 48 in. o.c. Two equally spaced clips are required between supports 48 in. o.c.

Use of lumber blocking, T&G edges, or panel clips constitutes edge support for vertical loads. T&G edges and panel clips cannot be used in place of blocking if blocking is required for diaphragm action. Exception: $1\frac{1}{8}$ -in.-thick plywood with properly stapled T&G edges qualifies as a blocked diaphragm (Ref. 8.17). Diaphragms are covered in Chap. 9.

ROOF JOIST

SPAN OF SHEATHING

①

THIS PLYWOOD EDGE MAY BE SUPPORTED OR UNSUPPORTED DEPENDING ON SPAN

PLYWOOD EDGE SUPPORTED BY ROOF FRAMING

PLYWOOD ORIENTED IN STRONG DIRECTION. ∴ SPAN RATING GIVES ROOF SPAN DIRECTLY

TYPICAL PLYWOOD ROOF

a.   c.   d.

2x4 BLOCKING

BLOCKING MAY BE TURNED FLAT WHEN REDUCED FASTENER PENETRATION IS TAKEN INTO ACCOUNT FOR SHEAR

b.

SECTION ① ALTERNATIVES

Figure 8.13 Support conditions for 8-ft panel edge.

It was noted at the beginning of this section that panels may be oriented in the weak direction in panelized roof systems. The span rating does not apply directly when panels are used in this manner. However, the design tables in APA publications (Refs. 8.5 and 8.6) include values for panels placed such that the strength axis is parallel to the supports.

8.10 Design Problem: Roof Sheathing

A common roof sheathing problem involves supports that are spaced 24 in. o.c. See Example 8.12. This building is located in a non-snow load area. Consequently the sheathing is designed for a roof live load of 20 psf. (Because the sheathing spans only 24 in., there is no tributary area reduction for roof live load.)

Part 1 of the example considers panels oriented in the strong direction, and several alternatives are suggested.

In part 2, plywood sheathing requirements for a panelized roof are considered. In this case, the number of plies used in the construction of the plywood panels is significant. If 5-ply construction is used, the effective section properties in the weak direction are larger. In 4-ply (3-layer) construction, the section properties are smaller, and either stronger wood (STRUCTURAL I) or a thicker panel is required.

EXAMPLE 8.12   Roof Sheathing with a 24-in. Span

The structural sheathing in a wood roof system supports a dead load of $D = 8$ psf and roof live load of $L_r = 20$ psf. The roof framing members are spaced 24 in. o.c., and the roof system must satisfy a live load deflection limit of $L/240$ and a total load deflection limit of $L/180$. Using ASD procedures, determine the minimum structural panel required for (1) structural panels oriented with the long dimension (strength axis) perpendicular to the supports (strong direction) and (2) structural panels oriented with the long dimension (strength axis) parallel to the supports (weak direction). Repeat the problem using LRFD procedures.

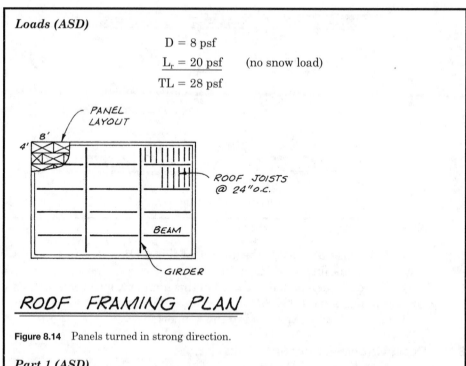

Figure 8.14   Panels turned in strong direction.

Part 1 (ASD)

For the roof layout shown, determine the panel sheathing requirements.

Panel sheathing grades appropriate for this application are Sheathing EXP 1 and STRUCTURAL I Sheathing EXP 1. Both grades are intended for roof applications.

The STRUCTURAL I designation means the panel is made with only exterior glue and Group 1 species. STRUCTURAL I should be used when the added shear capacity and cross-panel strength are necessary for lateral forces. Note that OSB and plywood are both available for these grades. See Table M9.1-1 in the ASD/LRFD Manual.

In Fig. 8.14, the panels are oriented in the strong direction. From Table 1 in Ref. 8.5, the required span rating is 24/0. The adjusted ASD uniform load capacities are as follows:

Load governed by: Adjusted ASD uniform load:
   $L/240$ deflection limit   39 psf > 20 psf (live load deflection)   *OK*
   $L/180$ deflection limit   52 psf > 28 psf (total load deflection)   *OK*
   Bending strength   (52)(1.25) = 65 psf > 28 psf   *OK*
   Shear strength   (147)(1.25) = 184 psf > 28 psf   *OK*

Table 5 in the *Panel Design Specification* (Ref. 8.7) indicates that 24/0 panels may be either $\frac{3}{8}$-, $\frac{7}{16}$-, $\frac{15}{32}$-, or $\frac{1}{2}$-in. thick, with the $\frac{3}{8}$-in. panel being the predominant nominal thickness.

Table M9.4-1 in the ASD/LRFD Manual indicates that with unblocked edges the maximum allowable span for 24/0 panels is 20 in. Therefore, edge support may be provided by

1. Blocking
2. T&G edges (available in plywood thicknesses $\frac{15}{32}$-in. and greater)
3. H-shaped metal clips (panel clips)

See Fig. 8.13.

Although $\frac{3}{8}$-in. panels qualify for a 24/0 span rating, $\frac{7}{16}$-in. (or $\frac{1}{2}$-in.) panels are often used to span 24 in. for the roof in this type of building. Also from Table 5 in Ref. 8.7, $\frac{7}{16}$-in. panels qualify for a span rating of 24/16. If panels with a span rating of 24/16 are used, the maximum unsupported edge length is 24 in., which is equal to the actual span of 24 in. Therefore, edge support is not required for this alternative. When roofing is to be guaranteed by a performance bond, the roofing manufacturer should be consulted for minimum thickness and edge support requirements.

Summary

The *minimum* panel requirement for this building is $\frac{3}{8}$-in. sheathing with a span rating of 24/0, and edges supported.

An alternative structural panel is $\frac{7}{16}$-in. sheathing with a span rating of 24/16 without edge support.

Blocking may be required for either choice for lateral forces (diaphragm action). See Chap. 9.

NOTE: The construction of the panel itself was not specified. That is, any wood structural panel meeting the span rating could be used in this application, including plywood or OSB.

Part 2 (ASD)

In the above building, assume that wood structural panels are is to be used in a panelized roof. Panels will be turned 90 degrees to the orientation shown in Fig. 8.14, so that the long dimension of the panel is parallel to the supports (joists). Here the span rating cannot be used directly because panels are oriented in the weak direction. From Table 1 in Ref. 8.5, the required span rating is 48/24. The adjusted ASD uniform load capacities are as follows:

Load governed by: Adjusted ASD uniform load:

$L/240$ deflection limit	24 psf > 20 psf (live load deflection)	*OK*
$L/180$ deflection limit	33 psf > 28 psf (total load deflection)	*OK*
Bending strength	(38)(1.25) = 47.5 psf > 28 psf	*OK*
Shear strength	(162)(1.25) = 202.5 psf > 28 psf	*OK*

Table 5 in Ref. 8.7 indicates that 48/24 panels may be either $^{23}/_{32}$-, $^{3}/_{4}$-, or $^{7}/_{8}$-in. thick, with the $^{23}/_{32}$-in. panel being the predominant nominal thickness.

Part 1 (LRFD)

Now repeat the problem using LRFD procedures. For bending and shear the LRFD factored load is $1.2D + 1.6L_r = 1.2(8) + 1.6(20) = 42$ psf. The tabulated reference values for bending and shear must be converted to nominal LRFD values. For the 24/0 span rating from Table 1 in Ref. 8.5:

$$\text{Bending:} \quad K_F\,(52 \text{ psf}) = 2.54(52) = 132.1 \text{ psf}$$

$$\text{Shear:} \quad K_F\,(147 \text{ psf}) = 2.88(147) = 423.4 \text{ psf}$$

According to NDS Appendix N, the time effect factor $\lambda = 0.8$ applies for the $1.2D + 1.6L_r$ load combination. Resistance factors for bending and shear are $\phi_b = 0.85$ and $\phi_v = 0.75$. The adjusted LRFD load capacities are:

$L/240$ deflection limit	39 psf > 20 psf (live load deflection)	*OK*
$L/180$ deflection limit	52 psf > 28 psf (total load deflection)	*OK*
Bending strength	$\lambda\phi_b\,(132.1) = 0.8(0.85)(132.1) = 90$ psf > 42 psf	*OK*
Shear strength	$\lambda\phi_v\,(423.4) = 0.8(0.75)(423.4) = 254$ psf > 42 psf	*OK*

Based on the information from Table 5 in Ref. 8.7, the minimum panel requirement is $^{3}/_{8}$-in. sheathing with a span rating of 24/0 and edges supported, using either LRFD or ASD procedures. An alternative structural panel is $^{7}/_{16}$-in. sheathing with a span rating of 24/16 without edge support, using either LRFD or ASD.

Part 2 (LRFD)

If the long dimension of the panel is parallel to the supports, the required span rating is 48/24. The nominal LRFD load capacities for bending and shear are determined from tabulated reference load capacities in Table 1 of Ref. 8.5:

$$\text{Bending:} \quad K_F\,(38 \text{ psf}) = 2.54(38) = 96.6 \text{ psf}$$

$$\text{Shear:} \quad K_F\,(162 \text{ psf}) = 2.88(162) = 466.6 \text{ psf}$$

Using a time effect factor of $\lambda = 0.8$ and resistance factors for bending and shear of $\phi_b = 0.85$ and $\phi_v = 0.75$, the adjusted LRFD load capacities are:

$L/240$ deflection limit	24 psf > 20 psf (live load deflection)	*OK*
$L/180$ deflection limit	33 psf > 28 psf (total load deflection)	*OK*

Bending strength: $\lambda\phi_b\,(96.6) = 0.8(0.85)(96.6) = 66$ psf > 42 psf   *OK*

Shear strength: $\lambda\phi_v\,(466.6) = 0.8(0.75)(466.6) = 280$ psf > 42 psf   *OK*

Based on Table 5 in Ref. 8.7, a $^{23}\!/_{32}$-in. structural panel with a span rating of 48/24 is adequate using either LRFD or ASD procedures.

8.11   Floor Sheathing

Wood structural panels are used in floor construction in two ways. One system involves *two layers* of panels, and the other system involves a *single layer*.

The terms used to refer to these different panel layers are:

1. *Subfloor*—the bottom layer in a two-layer system

2. *Underlayment*—the top layer in a two-layer system

3. *Combined subfloor-underlayment*—a single-layer system

A finish floor covering such as vinyl sheet or tile, ceramic tile, hardwood, or carpeting is normally provided. See Ref. 8.4 for specific installation recommendations.

In the two-layer system, the subfloor is the basic structural sheathing material. See Example 8.13. It may be either a sheathing grade of plywood or a nonveneer panel. Recall the two-number span rating from Sec. 8.7. For panels with this span rating, the second number is the recommended span in inches when the panel is used as a subfloor. Maximum floor live loads depend on the type of panel.

As stated for roof sheathing, Table M9.2-1 in the ASD/LRFD Manual provides the basic bending stiffness and strength design values for sheathing. Table 1 in Ref. 8.5 provides nominal uniform load capacities for sheathing at various spans or center-to-center spacing of the joists supporting the floor sheathing.

It should be noted that typical wood structural panel applications for floor sheathing are not controlled by *uniform* load criteria, but instead are based on deflection under concentrated loads and how the floor *feels* to building occupants. These and other subjective criteria relate to user acceptance of floor sheathing. For additional information, see Refs. 8.5 and 8.6.

EXAMPLE 8.13   Two-Layer Floor Construction

UNDERLAYMENT-grade panels have C-plugged face veneer and special C-grade innerply construction to resist indentations. Typical underlayment thickness is $\frac{1}{4}$ in. for remodeling or use over a panel subfloor and $\frac{3}{8}$ to $\frac{1}{2}$ in. for use over a lumber subfloor or new construction.

When finish flooring has some structural capacity, the underlayment layer is not required. Wood strip flooring and lightweight concrete are examples of finish flooring which do not require the use of underlayment over the subfloor.

Figure 8.15   Floor construction using a separate subfloor and underlayment.

In subfloor construction, panels must be used in the strong direction and must be continuous over two or more spans. Differential movement between adjacent unsupported panel edges is limited by one of the following:

1. T&G edges
2. Blocking
3. $\frac{1}{4}$ -in. underlayment with panel edges offset over the subfloor
4. $1\frac{1}{2}$ in. of lightweight concrete over the subfloor
5. Finish floor of $\frac{3}{4}$ -in. wood strips

As discussed for roof sheathing, the span rating of sheathing can be used *directly* to determine the panel sheathing requirements. For larger loads, the span rating can be used *indirectly* to determine panel sheathing requirements. See Example 8.10.

In two-layer floor construction, the top layer is a grade of panel known as UNDERLAYMENT. The underlayment layer lies under the finish floor covering and on top of the subfloor. It is typically $\frac{1}{4}$ -to $\frac{1}{2}$ -in. thick, and its purpose is to provide a solid surface for the direct application of nonstructural floor finishes. UNDERLAYMENT-grade panels are touch-sanded to provide a reasonably smooth surface to support the non-structural finish floor.

Single-layer floor construction is sometimes known as combination subfloor-UNDERLAYMENT because one layer serves both functions. Single-layer floor systems may use thicker grades of Underlayment and C-C Plugged Exterior plywood, composite panels, or some form of nonveneer panel. APA's performance rated panels for single-layer floors are known as Sturd-I-Floor and include plywood, composite, and nonveneer panels. The *span rating* for panels intended to be used in single-layer floor systems is composed of a single number. Here the span rating is the recommended maximum floor span in inches. See Example 8.14.

The reference bending stiffness and strength design values for single-layer floor panels are given in Table M9.2-1 of the ASD/LRFD Manual. Table 2 in the *Load-Span Tables for APA Structural-Use Panels* (Ref. 8.5) provides reference uniform load capacities for single-layer floor panels at various spans or center-to-center spacing of the framing supporting the panels. Also, as with any span-rated product, the span rating of single-layer floor panels can be used *indirectly* to determine panel sheathing requirements for spans other than the rated spans. See Example 8.10.

EXAMPLE 8.14   Single-Layer Floor Panels

Figure 8.16a Typical grade-trademarks for subfloor-underlayment. (*Courtesy of APA—The Engineered Wood Association.*)

Figure 8.16b Typical grade-trademarks for nonveneer single-layer floor panel. (*Courtesy of APA—The Engineered Wood Association.*)

A typical grade-trademark for a panel combination subfloor-underlayment is shown in Fig. 8.16a. In this example, the panel can be identified as plywood because PS 1 is referenced. It is an UNDERLAYMENT grade of plywood that is an interior type with exterior glue (Exposure 1). The panel is APA Rated Sturd-I-Floor, which is also qualified under the performance standard PRP-108. The span rating is 48 in. o.c., and the panel thickness is $1\frac{1}{8}$ in. Other information in the stamp includes the manufacturer's mill number (000 shown), the panel sized for spacing ($+0, -\frac{1}{8}$ in. tolerance on panel dimensions), and it has T&G edges.

A typical grade-trademark for a nonveneer single-layer floor panel is shown in Fig. 8.16b. Notice that the grade-trademark references PS 2, rather than PS 1. Like the panel grade-trademark in Fig. 8.16a, this panel is performance rated under PRP-108. It has a span rating of 20 in. o.c., and the panel is $\frac{19}{32}$-in. thick. It has a durability rating of Exposure 1. The panel is sized for spacing and has T&G edges. By taking into account the T&G edges and the fact that it is sized for spacing, the panel has a net width of $47\frac{1}{2}$ in.

Single-layer floor systems can be installed with nails. However, the APA glued floor system increases floor stiffness and reduces squeaks due to nail popping (Sec. 8.2). This system uses a combination of field gluing and nailing of floor panels to framing members. Table M9.4-2 of the ASD/LRFD Manual provides recommended minimum nail sizes, types, and spacings for various panel applications. For additional information, see Ref. 8.10.

In addition to sheathing and single-layer floor panels, the NDS references generic sanded Exterior plywood panels that are not span-rated. The reference bending stiffness and strength design values for sanded plywood panels are given for each thickness in Table M9.2-1 of the ASD/LRFD Manual. Table 3 in the *Load-Span Tables for PS-1 Plywood* (Ref. 8.6) provides reference uniform load capacities for sanded panels at various spans or center-to-center spacing of the framing supporting the panels.

8.12   Design Problem: Floor Sheathing

In this example a typical floor sheathing problem for an office building is considered. See Example 8.15. The floor utilizes a two-layer floor system with a separate subfloor and underlayment. The subfloor is chosen from the sheathing grades using the two-number span rating described in Sec. 8.7. A $\frac{1}{4}$-in. plywood UNDERLAYMENT-grade panel is used over the subfloor. If the joints of the underlayment are staggered with respect to the joints in the subfloor, no special edge support is required for the subfloor panels.

A single-layer panel floor could be used as an alternative. Plywood combination subfloor-underlayment (rather than a nonveneer panel) is recommended in Ref. 8.4 when the finish floor is a resilient nontextile flooring or adhered carpet without pad. The span rating for a combination subfloor-underlayment panel consists of a single number in the grade-trademark.

EXAMPLE 8.15   Floor Sheathing with 16-in. Span

Figure 8.17   Floor construction requires panels in strong direction.

The structural sheathing in a wood floor system supports a dead load of 12 psf, a distributed partition load of 15 psf, and a floor live load of 50 psf. The floor framing members are spaced 16-in. o.c. (Fig. 8.17), and the floor system must satisfy a live load deflection limit of $L/360$ and a total load deflection limit of $L/240$. Using ASD procedures, select an appropriate structural panel for the subfloor. Repeat the problem using LRFD procedures.

Loads (ASD):

$$\text{Floor dead load} = 12 \text{ psf}$$

$$\text{Partition live load} = 15 \text{ psf}$$

$$\underline{\text{Floor live load} = 50 \text{ psf}}$$

$$\text{TL} = 77 \text{ psf}$$

For the floor layout, determine the sheathing requirements for vertical loads, assuming a separate plywood subfloor and underlayment construction. A resilient-tile finish floor will be used.

Panel sheathing grades appropriate for this application are Sheathing EXP 1 and STRUCTURAL I Sheathing EXP 1. Both grades are intended for floor applications. The STRUCTURAL I designation means the panel is made with only exterior glue and Group 1 species. STRUCTURAL I should be used when the added shear capacity and cross-panel strength are necessary for lateral forces. Note that OSB and plywood are both available for these grades. See Table M9.1-1 in the ASD/LRFD Manual.

From the floor framing plan, the plywood is oriented in the strong direction over two or more spans. Therefore the span rating applies.

$$\text{Joist spacing} = \text{panel span} = 16 \text{ in.}$$

$$\text{Req'd span rating} = \text{roof span/floor span}$$

$$= 24/16 \text{ or } 32/16$$

Alternatively, the floor sheathing may be selected by comparing the design loads with the reference uniform load capacities from Table 1 in Ref. 8.5. For a live load deflection limit of $L/360$ and a total load deflection limit of $L/240$, it is noted that the 24/0 span-rated panel satisfies the design requirements:

Load governed by:	Adjusted ASD uniform load capacity:	
$L/360$ deflection limit	98 psf > 65 psf (live load deflection)	*OK*
$L/240$ deflection limit	147 psf > 77 psf (total load deflection)	*OK*
Bending strength	117 psf > 77 psf	*OK*
Shear strength	228 psf > 77 psf	*OK*

Although the 24/0 panel is within engineering accuracy for uniform loads, the 32/16 span rating is selected for this application since the 32/16 panel is available as a 3-ply, 4-ply, or 5-ply plywood panel as well as OSB, allowing greater flexibility in product choice. Furthermore, Table 5 in Ref. 8.7 indicates that $15/32$-, $1/2$-, $19/32$-, or $5/8$-in. thick panels are available for this span rating, with the $15/32$-in thickness being the most common. Finally, the selection of a 32/16 panel due to the 16-in. o.c. spacing of floor joists addresses APA span recommendations for concentrated floor loads in addition to uniform loads (see Refs. 8.5 and 8.6). Therefore, the wood structural panel selected for the subflooring in this problem is

$$\tfrac{15}{32}\text{-in. }32/16\text{ span-rated sheathing}$$

Because of the type of finish floor, $\tfrac{1}{4}$-in. or thicker UNDERLAYMENT-grade panels should be installed over the subfloor. Underlayment panel edges should be offset with respect to subfloor edges to minimize differential movement between subfloor panels.

Other possible subfloor choices exist including nonveneer panels. As an alternative, a single-layer floor system can be used.

Now repeat the problem using LRFD procedures. For bending and shear the LRFD factored load is 1.2D + 1.6L = 1.2(12) + 1.6(65) = 118 psf. The tabulated reference values for bending and shear must be converted to nominal LRFD values. For the 24/0 span rating from Table 1 in Ref. 8.5:

Bending: K_F (117 psf) = 2.54(117) = 297.3 psf

Shear: K_F (228 psf) = 2.88(228) = 656.6 psf

According to NDS Appendix N, the time effect factor $\lambda = 0.8$ applies for the 1.2D + 1.6L load combination. Resistance factors for bending and shear are $\phi_b = 0.85$ and $\phi_v = 0.75$. The adjusted LRFD load capacities are:

$L/360$ deflection limit	98 psf > 65 psf (live load deflection)	OK
$L/240$ deflection limit	147 psf > 77 psf (total load deflection)	OK
Bending strength	$\lambda\phi_b$ (297.3) = 0.8(0.85)(297.3) = 202 psf > 118 psf	OK
Shear strength	$\lambda\phi_v$ (656.6) = 0.8(0.75)(656.6) = 394 psf > 118 psf	OK

Although the 24/0 sheathing is within engineering accuracy for uniform loads, a $\tfrac{15}{32}$-in.-thick 32/16 span-rated sheathing is selected to address concerns regarding concentrated floor loads (see Refs. 8.5 and 8.6) and to provide greater flexibility in product choice.

8.13 Wall Sheathing and Siding

Wood structural panels can be used in wall construction in *two basic* ways. In *one method*, the panels serve a structural purpose only. They are attached directly to the framing and serve as sheathing to distribute the normal wind force to the studs, and they may also function as the basic shear-resisting elements if the wall is a shearwall. See Example 8.16. Finish siding of wood or other material is then attached to the outside of the wall.

Typical sheathing grades of plywood and a variety of nonveneer panels are used when finish siding will cover the sheathing. In wall construction, the long dimension of the panel can be either parallel or perpendicular to the studs (supports). For shearwall action all panel edges must be supported. This is provided by studs in one direction and wall plates or blocking between the studs in the other direction.

EXAMPLE 8.16   Wood Structural Panel Sheathing with Separate Siding

Figure 8.18   Separate sheathing and siding.

In this system, the wood structural panels are basically a structural wall element. Wind forces normal to the wall are carried by the sheathing to the wall studs. In some cases, the minimum panel requirements are increased if the face grain is not perpendicular to the studs. Finish siding is applied over the sheathing.

If the panel sheathing also functions as a shearwall (lateral forces parallel to wall), panel edges not supported by wall framing must be blocked and nailed (see Chap. 10). *Minimum panel nailing* is 6d common or galvanized box nails at 6 in. o.c. at supported edges and 12 in. o.c. along intermediate supports (studs). See Table M9.4-2 in the ASD/LRFD Manual. Heavier nailing may be required for shearwall action.

Finish siding material can be attached by nailing through the wood structural panel sheathing into the wall framing, or it may be attached by nailing directly into the sheathing. See Ref. 8.4 for specific recommendations.

In the *second method* of using wood structural panels in wall construction, a single-panel layer is applied as combined sheathing-siding. When panels serve as the siding as well as the structural sheathing, a panel siding grade such as APA Rated Siding may be used. See Example 8.17.

Common types of panel siding used for this application are the APA proprietary products known as APA Rated Siding-303. These are exterior plywood panels available with a variety of textured surface finishes. A special type of 303 panel siding in known as Texture 1-11 (T 1-11) and is manufactured in $\frac{19}{32}$-and $\frac{5}{8}$-in. thicknesses only. It has shiplap edges to aid in weather-tightness and to maintain surface pattern continuity. T 1-11 panels also have $\frac{3}{8}$-in.-wide grooves cut into the finished side for decorative purposes. A net panel thickness of $\frac{3}{8}$ -in.

is maintained at the groove. The required thickness of wood structural panels for use as combined sheathing and siding is the same as when used as structural sheathing only, except as noted in footnote c of Table 25 of the APA *Engineered Wood Construction Guide* (Ref. 8.4). APA Rated Siding includes a single-number span rating in the grade-trademark which indicates the maximum spacing of studs. Additional information on Rated Siding is available in Ref. 8.4.

EXAMPLE 8.17   Plywood Combined Sheathing-Siding

Figure 8.19a   Combined sheathing-siding.

Figure 8.19b Typical grade-trademarks for combined sheathing-siding. (*Courtesy of APA—The Engineered Wood Association.*)

In the combined sheathing-siding system (Fig. 8.19a), the wood structural panel siding usually has a textured surface finish. These medium density overlay (MDO) finishes include rough-sawn, brushed, and smooth finish for painting. In addition to different surface textures, most siding panels are available with grooving for decorative effect.

Typical grade-trademarks for combined sheathing-siding panels are shown in Fig. 8.19b. The examples are for APA Rated Siding. A single-number span rating (e.g., 16 or 24 o.c.) indicates maximum recommended stud spacing when the panels are oriented vertically. For additional information including an explanation of siding face grades, see Ref. 8.10.

Wood structural panel siding is usually installed with the 8-ft panel dimension oriented vertically. However, these panels can also be installed horizontally. Various panel joint details can be used for protection against the weather (see Fig. 8.19c).

Figure 8.19c Panel edge details for combined sheathing-siding.

Nailing requirements for combined sheathing-siding are similar to the nailing for sheathing in the two-layer wall system (Example 8.16). However, hot-dipped galvanized nails are normally used to reduce staining. Casing nails may be used where the presence of a common or box nail head is objectionable. Additional information on plywood nailing for walls is included in Chap. 10. For siding $\frac{1}{2}$-in. thick or less, 6d nails are used. For thicker siding the nail size is increased to 8d nails.

Heavier nailing may be required for shearwall action. For shearwall design, the panel thickness to be used is the *thickness where nailing occurs*. If grooves are nailed (see Fig. 8.19c), the net thickness at the groove is used.

8.14   Stress Calculations for Wood Structural Panels

The design aids that allow the required thickness of wood structural panel sheathing to be determined without detailed design calculations have been described in the previous sections. For most practical sheathing problems, these methods are adequate to determine the required grade and panel thickness of a wood structural panel.

In the design of structural components such as box beams with wood structural panel webs and lumber flanges, foam-core sandwich panels, or stress-skin panels, it will be necessary to use reference design values and effective cross-section panel properties in design calculations.

If it becomes necessary to perform structural calculations for wood structural panels, the designer must become familiar with a number of factors which interrelate to define the panel's structural capacity. Structural design calculations for wood structural panels are similar to those for solid lumber or glulam, as presented in Chaps. 4 to 7. Basic design requirements for wood structural panels are presented in NDS Chap. 9, including the applicability of adjustment factors (see NDS Table 9.3.1).

The ASD/LRFD Design Manual for Engineered Wood Construction (Ref. 8.1) and the *Panel Design Specification* (Ref. 8.7) provide tables of reference design values for structural panels. The grade and construction factors (C_G) are identified as "multipliers for STRUCTURAL I panels" at the bottom of each table of panel design values for various panel construction types (e.g., 3-, 4-, or 5-ply plywood, or OSB). Load duration factors (C_D) for ASD are specified in the NDS, along with format conversion factors (K_F), time effect factors (λ), and resistance factors (ϕ) for LRFD. *The Panel Design Specification* (Ref. 8.7) provides wet service factors (C_M), temperature factors (C_t), and panel size factors (C_s).

The grade and construction adjustment factor compensates for the STRUCTURAL I panel grade within the various panel construction types. Note that the reference design values for wood structural panels have been tabulated with the lowest possible strength value. That is, the grade and construction adjustment factor is defined such that $C_G \geq 1.0$ and can be conservatively assumed to be 1.0. However, for STRUCTURAL I panels, C_G is often greater than 1.0, and may be as high as 1.6.

The reference bending and tension design values tabulated in the ASD/LRFD Manual are for panels 24 in. or greater in width, w. The width here is defined as the dimension of the panel perpendicular to the applied stress. The panel size adjustment factor, C_s, reduces the reference design values when the width is less than 24 in. See Ref. 8.7.

$$w > 24 \text{ in.} \qquad C_s = 1.0$$

$$8 \text{ in.} < w < 24 \text{ in.} \qquad C_s = 0.25 + 0.0313\, w$$

$$w \leq 8 \text{ in.} \qquad C_s = 0.5$$

It will be helpful to review several factors which are unique to structural panel calculations. These will provide a useful background to the designer even if structural calculations are not required. Although some of these points have been introduced previously, they are briefly summarized here.

Panel section properties. The basic cross-section properties for wood structural panels are provided in Table 6 of the *Panel Design Specification* (Ref. 8.7). The section properties in Table 6 are based on a *gross* rectangular cross section with a unit 1-ft width. However, since wood structural panels are composite materials, *effective* design values are more appropriate for use in design. Rather than tabulating effective unit stresses, which would then be used with the cross-section properties of Table 6, reference stiffness and strength design values are provided for wood structural panels. See ASD/LRFD Manual Tables M9.2-1, M9.2-2, M9.2-3, and M9.2-4.

The variables which affect the reference design values are

1. Direction of stress

2. Panel grade and composition

The *direction of stress* relates to the two-directional behavior of wood structural panels because of their cross-laminated construction. Because of this type of construction, two sets of properties are tabulated. One applies when wood structural panels are stressed parallel to the face grain, and the other applies when they are stressed perpendicular to the face grain.

Finally, the composition and grade of a panel influence the *effective* properties. The panel grade and construction adjustment factor, C_G, specifically addresses types and grades of structural panels.

Stress calculations. Prior to 2001, provisions for engineering design with wood structural panels were not included in the NDS. While specifying wood structural panels with design aids (such as those presented in the previous sections) has not changed significantly, stress calculations have subtly changed. In the past when stress calculations were required, *effective* cross-sectional properties were used with tabulated unit stresses. The ASD/LRFD Manual does *not* tabulate effective cross-sectional properties or unit stresses for panels. Rather, reference stiffness and strength design values are tabulated for wood structural panels. For example, rather than tabulating an effective section modulus, S, and a unit bending stress, F_b, the ASD/LRFD Manual tabulates the reference bending strength design value, F_bS, per unit width of panel. See ASD/LRFD Manual Tables M9.2-1 through M9.2-4.

The use of reference design values is similar to ordinary beam design calculations. See Example 8.18.

EXAMPLE 8.18   Plywood Beam Loading and Section Properties

Consider a section of plywood acting as a beam element.

Figure 8.20a Plywood with load normal to surface of panel.

Structural Panels under Normal (Sheathing) Loads

For loads normal to the surface of the panel (Fig. 8.20a), design capacities are obtained from the ASD/LRFD Manual Table M9.2-1. The reference bending strength, $F_b S$, and the reference bending stiffness, EI, are tabulated for a unit 1-ft width of panel. For shear requirements, see Example 8.19.

Figure 8.20b Plywood with load in plane of panel.

Plywood under In-Plane Loads

Procedures for designing structural members with plywood loaded in the plane of the panel (Fig. 8.20b) are not provided in the NDS. However, procedures are given in the *Plywood Design Specification*, Supplements 2 and 5 (Refs. 8.9 and 8.12). Plywood used in this manner is typically found in fabricated box beams using lumber flanges and plywood webs (Fig. 8.20c).

FABRICATED
BEAM

Figure 8.20c   Lumber and plywood box beam.

Another factor that is unique to structural calculations for wood structural panels is that there are two different reference shear design values. The different shear design values are a result of the cross-laminations. The type and direction of the loading determine the type of shear involved, and the appropriate reference shear design value must be used in checking the capacity.

The ASD/LRFD Manual refers to these shear design values as

1. Shear in a plane perpendicular to the plies, or *shear through the thickness* of the panel

2. Shear *in the plane* of the plies, or *rolling shear*

The first type of shear occurs when the load is in the plane of the panel, as in a diaphragm. See Fig. 8.21*a* in Example 8.19. This same type of stress occurs in fabricated box beams using plywood webs. In the latter case, shear through the thickness of the plywood is the result of flexural (horizontal) shear. The design procedures for fabricated lumber and plywood box beams are covered in PDS Supplement 2 (Ref. 8.9).

Rolling shear can also be visualized as the horizontal shear in a beam, but in this case the loads are normal to the surface of the panel (as with sheathing loads). See Fig. 8.21*b*. The shear is seen to be "in the plane of the plies" rather than "through the thickness." With this type of loading the wood fibers that are at right angles to the direction of the stress tend to slide or *roll* past one another, hence the name *rolling shear*. If the stress is parallel to the face plies, the fibers in the inner crossband(s) are subjected to rolling shear.

The ASD/LRFD Manual Tables M9.2-3 and M9.2-4 provide different reference design values for the calculation of shear through the thickness and rolling shear. See Example 8.19.

EXAMPLE 8.19   Types of Shear in Plywood

Consider a section of plywood subjected to direct shear.

Figure 8.21a Shear through the thickness.

Shear through the Thickness

Shear through the thickness can occur from the type of loading shown in Fig. 8.21a (as in diaphragm and shearwall action) or from flexural shear (horizontal shear) caused by the type of loading shown in Fig. 8.20b. The stress calculated for these types of loading conditions is based on the *effective thickness for shear*. See ASD/LRFD Manual Table M9.2-4.

Figure 8.21b Rolling shear is shear in the plane of the plies.

Rolling Shear

Rolling shear occurs in the ply (or plies) that is (are) at right angles to the applied stress (Fig. 8.21b). This type of stress develops when plywood is loaded as shown in Fig. 8.20a. The stress is shear due to bending (horizontal shear) and is calculated from the rolling shear constant Ib/Q. See ASD/LRFD Manual Table M9.2-3.

In Example 8.20, the adjusted design values for a given wood structural panel are determined. The purpose is to illustrate the use of the ASD/LRFD Manual design value tables and adjustment factors unique to wood structural panels.

EXAMPLE 8.20   Adjusted Design Values of Wood Structural Panels

Determine the adjusted ASD and LRFD design values per unit foot of width for a 32/16 span-rated STRUCTURAL I grade 5-ply 5-layer plywood used to support a (D + L) load combination in dry conditions. Assume the panel will have a width of at least 24 in.

Adjusted ASD Bending Stiffness

Two bending stiffness values are provided in Table M9.2-1 of the ASD/LRFD Manual, one for the stress parallel to the strength axis (i.e., strength axis perpendicular to the supports) and one for the stress perpendicular to the strength axis (i.e., strength axis parallel to the supports). The former is the strong orientation for the panel.

Strong orientation:
$$EI' = EI(C_M)(C_t)(C_G)$$
$$= (125{,}000)(1.0)(1.0)(1.0)$$
$$= 125{,}000 \text{ lb-in.}^2/\text{ft width}$$

Weak orientation:
$$EI' = EI(C_M)(C_t)(C_G)$$
$$= (25{,}000)(1.0)(1.0)(1.6)$$
$$= 40{,}000 \text{ lb-in.}^2/\text{ft width}$$

Note that most reference design values for structural panels are based on an assumed 1-ft width of panel.

Adjusted ASD Bending Strength

As with the stiffness, two bending strength design values are provided in Table M9.2-1 of the ASD/LRFD Manual, one for the stress parallel to the strength axis and one for the stress perpendicular to the strength axis. The former is the strong orientation for the panel.

Strong orientation:
$$F_bS' = F_bS(C_D)(C_M)(C_t)(C_G)(C_s)$$
$$= (445)(1.0)(1.0)(1.0)(1.0)(1.0)$$
$$= 445 \text{ lb-in./ft width}$$

Weak orientation:
$$F_bS' = F_bS(C_D)(C_M)(C_t)(C_G)(C_s)$$
$$= (165)(1.0)(1.0)(1.0)(1.5)(1.0)$$
$$= 248 \text{ lb-in./ft width}$$

Adjusted ASD Axial Stiffness

The axial stiffness design values are provided in the ASD/LRFD Manual Table M9.2-2.

Strong orientation:
$$EA' = EA(C_M)(C_t)(C_G)$$
$$= (4{,}150{,}000)(1.0)(1.0)(1.0)$$
$$= 4{,}150{,}000 \text{ lb/ft width}$$

Weak orientation:
$$EA' = EA(C_M)(C_t)(C_G)$$
$$= (3{,}600{,}000)(1.0)(1.0)(1.0)$$
$$= 3{,}600{,}000 \text{ lb/ft width}$$

Adjusted ASD Axial Tension

The axial tension design values are provided in the ASD/LRFD Manual Table M9.2-2.

Strong orientation:

$$F_t A' = F_t A(C_D)(C_M)(C_t)(C_G)(C_s)$$
$$= (3650)(1.0)(1.0)(1.0)(1.0)(1.0)$$
$$= 3650 \text{ lb/ft width}$$

Weak orientation:

$$F_t A' = F_t A(C_D)(C_M)(C_t)(C_G)(C_s)$$
$$= (1650)(1.0)(1.0)(1.0)(1.0)(1.0)$$
$$= 1650 \text{ lb/ft width}$$

Adjusted ASD Axial Compression

In addition to the axial stiffness and tension design values, axial compression design values are also provided in the ASD/LRFD Manual Table M9.2-2.

Strong orientation:

$$F_c A' = F_c A(C_D)(C_M)(C_t)(C_G)$$
$$= (5350)(1.0)(1.0)(1.0)(1.0)$$
$$= 5350 \text{ lb/ft width}$$

Weak orientation:

$$F_c A' = F_c A(C_D)(C_M)(C_t)(C_G)$$
$$= (4650)(1.0)(1.0)(1.0)(1.0)$$
$$= 4650 \text{ lb/ft width}$$

Adjusted ASD In-Plane Shear

In-plane shear occurs when load is applied normal to the surface of the panels as with floors or roofs. Two in-plane shear design values are provided in Table M9.2-3 of the ASD/LRFD Manual, one for the stress parallel to the strength axis and one for the stress perpendicular to the strength axis.

Strong orientation:

$$F_s(Ib/Q)' = F_s(Ib/Q)(C_D)(C_M)(C_t)(C_G)$$
$$= (215)(1.0)(1.0)(1.0)(1.4)$$
$$= 301 \text{ lb/ft width}$$

Weak orientation:

$$F_s(Ib/Q)' = F_s(Ib/Q)(C_D)(C_M)(C_t)(C_G)$$
$$= (165)(1.0)(1.0)(1.0)(1.0)$$
$$= 165 \text{ lb/ft width}$$

Adjusted ASD Shear Rigidity through the Thickness

Shear through the thickness occurs when load is applied in the plane of the panels, as with shearwalls or diaphragms. Shear rigidity design values are provided in Table

M9.2-4 of the ASD/LRFD Manual. While two shear rigidity design values are provided in Table M9.2-4—one for the stress parallel to the strength axis and one for the stress perpendicular to the strength axis—the reference design values and the adjusted design values are identical for both panel orientations.

$$G_v t_v' = G_v t_v (C_M)(C_t)(C_G)$$

$$= (40{,}500)(1.0)(1.0)(1.1)$$

$$= 44{,}550 \text{ lb/in. of panel depth}$$

Adjusted ASD Shear through the Thickness

Shear design values through the thickness are also provided in Table M9.2-4 of the ASD/LRFD Manual. As with shear rigidity, two shear design values are provided—one for the stress parallel to the strength axis and one for the stress perpendicular to the strength axis—but the reference and adjusted design values are identical for both panel orientations.

$$F_v t_v' = F_v t_v (C_D)(C_M)(C_t)(C_G)$$

$$= (93)(1.0)(1.0)(1.0)(1.1)$$

$$= 102 \text{ lb/in. of shear-resisting panel length}$$

Additional discussion of shear through the thickness is provided in Chaps. 9 and 10.

Now repeat the problem using LRFD procedures. The time effect factor for the load combination of 1.2D + 1.6 L is $\lambda = 0.8$. The time effect factor applies to strength design values, but not to stiffness or rigidity design values. There is no difference between the stiffness and rigidity design values for ASD and LRFD.

Adjusted LRFD Bending Stiffness

Strong orientation: $EI' = EI(C_M)(C_t)(C_G) = 125 \text{ k-in}^2/\text{ft width}$

Weak orientation: $EI' = EI(C_M)(C_t)(C_G) = 40 \text{ k-in}^2/\text{ft width}$

Adjusted LRFD Bending Strength

Strong orientation: $(F_b S)_n = F_b S(K_F)$

$$= (445)(2.16/\phi_b)$$

$$= (445)(2.54)$$

$$= 1131 \text{ lb-in/ft width}$$

$$= 1.131 \text{ k-in/ft width}$$

$$(F_bS)'_n = (F_bS)_n(\phi_b)(\lambda)(C_M)(C_t)(C_G)(C_s)$$

$$= (1.131)(0.85)(0.8)(1.0)(1.0)(1.0)(1.0)$$

$$= 0.769 \text{ k-in/ft width}$$

Weak orientation: $(F_bS)_n = F_bS(K_F)$

$$= (165)(2.16/\phi_b)$$

$$= (165)(2.54)$$

$$= 419 \text{ lb-in/ft width}$$

$$= 0.419 \text{ k-in/ft width}$$

$$(F_bS)'_n = (F_bS)_n(\phi_b)(\lambda)(C_M)(C_t)(C_G)(C_s)$$

$$= (0.419)(0.85)(0.8)(1.0)(1.0)(1.5)(1.0)$$

$$= 0.428 \text{ k-in/ft width}$$

Adjusted LRFD Axial Stiffness

Strong orientation: $EA' = EA(C_M)(C_t)(C_G) = 4150 \text{ k/ft width}$

Weak orientation: $EA' = EA(C_M)(C_t)(C_G) = 3600 \text{ k/ft width}$

Adjusted LRFD Axial Tension

Strong orientation: $(F_tA)_n = F_tA(K_F)$

$$= (3650)(2.16/\phi_t)$$

$$= (3650)(2.7)$$

$$= 9855 \text{ lb/ft width}$$

$$= 9.855 \text{ k/ft width}$$

$$(F_tA)'_n = (F_tA)_n(\phi_t)(\lambda)(C_M)(C_t)(C_G)(C_s)$$

$$= (9.855)(0.8)(0.8)(1.0)(1.0)(1.0)(1.0)$$

$$= 6.31 \text{ k/ft width}$$

Weak orientation: $(F_tA)_n = F_tA(K_F)$

$$= (1650)(2.16/\phi_t)$$

$$= (1650)(2.7)$$

$$= 4455 \text{ lb/ft width}$$

$$= 4.455 \text{ k/ft width}$$

$$(F_t A)'_n = (F_t A)_n (\phi_t)(\lambda)(C_M)(C_t)(C_G)(C_s)$$

$$= (4.455)(0.8)(0.8)(1.0)(1.0)(1.0)$$

$$= 2.85 \text{ k/ft width}$$

Adjusted LRFD Axial Compression

Strong orientation: $(F_c A)_n = F_c A(K_F)$

$$= (5350)(2.16/\phi_c)$$

$$= (5350)(2.4)$$

$$= 12{,}840 \text{ lb/ft width}$$

$$= 12.84 \text{ k/ft width}$$

$$(F_c A)'_n = (F_c A)_n (\phi_c)(\lambda)(C_M)(C_t)(C_G)$$

$$= (12.84)(0.9)(0.8)(1.0)(1.0)(1.0)$$

$$= 9.24 \text{ k/ft width}$$

Weak orientation: $(F_c A)_n = F_c A(K_F)$

$$= (4650)(2.16/\phi_c)$$

$$= (4650)(2.4)$$

$$= 11{,}160 \text{ lb/ft width}$$

$$= 11.16 \text{ k/ft width}$$

$$(F_c A)'_n = (F_c A)_n (\phi_c)(\lambda)(C_M)(C_t)(C_G)$$

$$= (11.16)(0.9)(0.8)(1.0)(1.0)(1.0)$$

$$= 8.04 \text{ k/ft width}$$

Adjusted LRFD In-Plane Shear

Strong orientation: $(F_s Ib/Q)_n = F_s(Ib/Q)(K_F)$

$$= (215)(2.16/\phi_v)$$

$$= (215)(2.88)$$

$$= 619 \text{ lb/ft width}$$

$$= 0.619 \text{ k/ft width}$$

$$(F_s Ib/Q)'_n = (F_s Ib/Q)_n (\phi_v)(\lambda)(C_M)(C_t)(C_G)$$

$$= (0.619)(0.75)(0.8)(1.0)(1.0)(1.4)$$

$$= 0.52 \text{ k/ft width}$$

Weak orientation: $(F_s Ib/Q)_n = F_s(Ib/Q)(K_F)$

$$= (165)(2.16/\phi_v)$$

$$= (165)(2.88)$$

$$= 475 \text{ lb/ft width}$$

$$= 0.475 \text{ k/ft width}$$

$$(F_s Ib/Q)'_n = (F_s Ib/Q)_n (\phi_v)(\lambda)(C_M)(C_t)(C_G)$$

$$= (0.475)(0.75)(0.8)(1.0)(1.0)(1.0)$$

$$= 0.285 \text{ k/ft width}$$

Adjusted LRFD Shear Rigidity through the Thickness

$$G_v t_v' = G_v t_v (C_M)(C_t)(C_G) = 44.55 \text{ k/in of panel depth}$$

Adjusted LRFD Shear through the Thickness

$$(F_v t_v)_n = F_v t_v (K_F)$$

$$= (93)(2.16/\phi_v)$$

$$= (93)(2.88)$$

$$= 268 \text{ lb/in of shear-resisting panel length}$$

$$= 0.268 \text{ k/in of shear-resisting panel length}$$

$$(F_v t_v)'_n = (F_v t_v)_n (\phi_v)(\lambda)(C_M)(C_t)(C_G)$$

$$= (0.268)(0.75)(0.8)(1.0)(1.0)(1.1)$$

$$= 0.177 \text{ k/in of shear-resisting panel length}$$

Additional design information on structural calculations for wood structural panels is available from APA—The Engineered Wood Association. The primary design references produced by the APA are the *Panel Design Specification* (Ref. 8.7) and the *Plywood Design Specification* (Ref. 8.15). Supplements to the PDS are available (Refs. 8.8, 8.9, 8.10, 8.11, and 8.12) that provide design information and examples for a variety of applications.

8.15   References

[8.1]   American Forest and Paper Association (AF&PA). 2006. ASD/LRFD *Manual for Engineered Wood Construction* 2006 ed., AF&PA, Washington DC.

[8.2]   American Forest and Paper Association (AF&PA). 2005. *National Design Specification for Wood Construction and Supplement*, ANSI/AF&PA NDS-2005, AF&PA, Washington DC.

[8.3]   American National Standards Institute (ANSI). 1999. Particleboard, ANSI Standard A208.1-99, ANSI, New York, NY.

[8.4]   APA—The Engineered Wood Association. 2005. *Engineered Wood Construction Guide*, Form No. E30T, APA-The Engineered Wood Association, Tacoma, WA.

[8.5]   APA—The Engineered Wood Association. 2001. *Load-Span Tables for APA Structural-Use Panels*, Form No. Q225D, APA-The Engineered Wood Association, Tacoma, WA.

[8.6]   APA—The Engineered Wood Association. 1998. *Load-Span Tables for PS-1 Plywood*, Form No. Z802K, APA-The Engineered Wood Association, Tacoma, WA.

[8.7]   APA—The Engineered Wood Association. 2004. *Panel Design Specification*, Form No. D510A, APA—The Engineered Wood Association, Tacoma, WA.

[8.8]   APA—The Engineered Wood Association. 1995. *PDS Supplement #1: Design and Fabrication of Plywood Curved Panels*, Form No. S811N, APA-The Engineered Wood Association, Tacoma, WA.

[8.9]   APA—The Engineered Wood Association. 1998. *PDS Supplement #2: Design and Fabrication of Glued Plywood-Lumber Beams*, Form No. S812R, APA-The Engineered Wood Association, Tacoma, WA.

[8.10] APA—The Engineered Wood Association. 1996. *PDS Supplement #3: Design and Fabrication of Plywood Stressed-Skin Panels*, Form No. U813L, APA-The Engineered Wood Association, Tacoma, WA.

[8.11] APA—The Engineered Wood Association. 1993. *PDS Supplement #4: Design and Fabrication of Plywood Sandwich Panels*, Form No. U814H, APA-The Engineered Wood Association, Tacoma, WA.

[8.12] APA—The Engineered Wood Association. 1995. *PDS Supplement #5: Design and Fabrication of All-Plywood Beams*, Form No. H815E, APA-The Engineered Wood Association, Tacoma, WA.

[8.13] APA—The Engineered Wood Association. 1999. *Performance Rated Panels*, Form No. F405L, APA—The Engineered Wood Association, Tacoma, WA.

[8.14] APA—The Engineered Wood Association. 2002. *Performance Standards and Qualification Policy for Structural-Use Panels*, APA PRP-108, Form No. E445S, APA-The Engineered Wood Association, Tacoma, WA.

[8.15] APA—The Engineered Wood Association. 1998. *Plywood Design Specification (PDS)*, Form No. Y510T, APA-The Engineered Wood Association, Tacoma, WA.

[8.16] International Codes Council (ICC). 2006. *International Building Code*, 2006 ed., ICC, Falls Church, VA.

[8.17] ICC Evaluation Service, Inc. 2005. "Standards for Structural-Use Panels," Legacy Report No. NER-108, International Codes Council (ICC) (Available from APA–The Engineered Wood Association, Tacoma, WA).

[8.18] National Institute of Standards and Technology (NIST). 1995. *U.S. Product Standard PS 1-95—Construction and Industrial Plywood*, U.S. Dept. of Commerce, Gaithersburg, MD (Available from APA—The Engineered Wood Association, Tacoma, WA).

[8.19] National Institute of Standards and Technology (NIST). 2004. *U.S. Product Standard PS 2-04—Performance Standard for Wood-Based Structural-Use Panels*, U.S. Dept. of Commerce, Gaithersburg, MD (Available from APA—The Engineered Wood Association, Tacoma, WA).

8.16   Problems

8.1   What two major types of loading are considered in designing a wood structural panel roof or floor system that also functions as part of the LFRS?

8.2   Regarding the fabrication of plywood panels, distinguish between (1) veneer, (2) plies, and (3) layers.

8.3   In the cross section of plywood panel shown in Fig. 8A, label the names used to describe the 5 plies.

Figure 8A

8.4   Plywood panels (4 ft × 8 ft) of 5-ply construction (similar to that shown in Fig. 8A) are used to span between roof joists that are spaced 16 in. o.c.

Find: *a.* Sketch a plan view of the framing and the plywood, showing the plywood oriented in the strong direction.

b. Sketch a 1-ft-wide cross section of the sheathing. Shade the plies that are effective in supporting the "sheathing" loads.

8.5   Repeat Prob. 8.4 except that the plywood is oriented in the weak direction.

8.6   How are the species of wood classified in the fabrication of plywood?

8.7   The plywood in Fig. 8A has a grade stamp that indicates *Group* 2.

Find: *a.* What plies in the panel contain Group 2 species?

b. If not all plies are of Group 2 species, what is assumed for the others?

8.8   What is the meaning of the term STRUCTURAL I? How is it used?

8.9   What are the veneer grades? Which veneers are the more similar in appearance and surface qualities? Which veneers are structurally more similar? State the reasons for similarities.

8.10   What is the most common type of glue used in the fabrication of plywood (interior or exterior)?

8.11   List and briefly describe the various types of wood structural panels other than plywood.

8.12   What is the difference between a prescriptive specification and a performance standard for the production of wood structural panels?

8.13   Describe the exposure durability classifications for (*a*) plywood manufactured in accordance with Product Standard PS 1 and (*b*) wood structural panels manufactured in accordance with Product Standard PS 2.

8.14   What is the difference between the construction of interior and exterior plywood?

8.15   Briefly describe the *span rating* found in the grade-trademark of the following panels:
a. Sheathing grades
b. APA Rated Sturd-I-Floor
c. APA Rated 303 Siding

8.16 Explain the significance of the following designations that may be found in the grade-trademarks of wood structural panels:
a. PS 1
b. PS 2

8.17 What APA tables provide load/span information for the following?
a. Sheathing
b. Single floor
c. Group 1 sanded panels

8.18 What are the sheathing grades of plywood? Are these sanded, touch-sanded, or unsanded panels? What veneer grades are used for the face and back plies of a sanded plywood panel?

8.19 The spacing of rafters in a roof is 48 in. o.c. Roof dead load = 5 psf. Snow load = 30 psf. Roof sheathing is to be a sheathing grade of plywood, and panels are oriented in the strong direction.

    Find: The minimum grade, span rating, thickness, and edge support requirements for the roof sheathing using ASD procedures. Repeat using LRFD procedures.

8.20 The spacing of joists in a flat roof is 24 in. o.c. Roof dead load = 8 psf. Snow load = 100 psf. Roof sheathing is to be APA Rated Sheathing (which can be either plywood or a nonveneer panel). Panels are oriented in the strong direction.

    Find: The minimum span rating, thickness, and edge support requirements for the roof sheathing using ASD procedures. Repeat using LRFD procedures.

8.21 The spacing of joists in a floor system is 16 in. o.c. Floor dead load = 15 psf. Floor live load = 185 psf.

    Find: The required panel grade, span rating, thickness, edge support requirements, and panel orientation for the floor sheathing using ASD procedures. Repeat using LRFD procedures.

8.22 Repeat Prob. 8.21 except that the joist spacing in 24 in. o.c.

8.23 Describe the construction of UNDERLAYMENT-grade plywood. How is it used in a floor system?

8.24 A single-layer (combined subfloor-underlayment) floor system is used to support a floor dead load of 10 psf and a floor live load of 75 psf. Panels with a span rating of 16 in. will be used to span between floor joists that are 16 in. o.c.

    Find: *a.* Based on ASD procedures, is an APA nonveneer panel able to support these loads, or is a plywood panel necessary?
              b. What are the advantages of field-gluing floor panels to the framing members before they are nailed in place?
              c. What is meant by nail popping, and how is the problem minimized?

8.25 Describe the following types of plywood used on walls (include typical grades):
a. Plywood sheathing
b. Plywood siding
c. Combined sheathing-siding

8.26   Grooved plywood (such as Texture 1-11) is used for combined sheathing-siding on a shearwall. Nailing to the studs is similar to that in Fig. 8.19c, the detail for a vertical plywood joint.

  Find: What thickness of the plywood is to be used in the shearwall design calculations?

8.27   Regarding the calculation of stresses in wood structural panels,
  a. What is shear through the thickness?
  b. What is rolling shear?

8.28   Determine the adjusted LRFD capacities per unit foot width for $^{15}\!/_{32}$-in. thick OSB that is 32/16 span-rated. Assume the OSB is used in dry conditions, and that the panel will be at least 24-in. wide. The OSB must support a (1.2D + 16L) load combination.

8.29   Determine the adjusted ASD capacities per unit for width for $^{19}\!/_{32}$-in. thick COMPLY that is 16 in. o.c. span-rated. Assume the COMPLY is used in dry conditions, and that the panel will be at least 24-in. wide. The COMPLY must support a (D + S) load combination.

9.1 Introduction

The lateral forces that act on conventional wood-frame buildings (bearing wall system) were described in Chap. 2, and the distribution of these forces was covered in Chap. 3. In a typical case, the lateral forces were seen to be carried by the wall framing to the diaphragms at the top and bottom of the wall sections. A diaphragm acts as a beam in the plane of a roof or floor that spans between shearwalls. See Fig. 9.1.

The examples in Chap. 3 were basically force calculation and force distribution problems. In addition, the calculation of the unit shear in a diaphragm was illustrated. Although the unit shear is a major factor in a diaphragm design, there are a number of additional items that must be addressed.

The basic design considerations for a diaphragm are

1. Sheathing thickness
2. Diaphragm nailing
3. Chord design
4. Collector (strut) design
5. Diaphragm deflection
6. Tie and anchorage requirements

The first item is often governed by loads normal to the surface of the sheathing (i.e., by sheathing loads). This subject is covered in Sec. 6.18 for lumber diaphragms and in Chap. 8 for wood structural panel diaphragms. The nailing requirements, on the other hand, are a function of the unit shear. The sheathing thickness and nailing requirements may, however, both be governed by the unit shears *when the shears are large.*

In this chapter, the general behavior of a diaphragm is described, and the functions of the various items mentioned above are explained. This is followed by

DEFLECTED SHAPE
OF DIAPHRAGM

R

R

REACTION
PROVIDED
BY SHEARWALL

w = TRANSVERSE
LATERAL FORCE

END SHEAR WALL

Figure 9.1   Typical roof diaphragm.

detailed design considerations for the individual elements. Tie and anchorage requirements are touched upon, but these are treated more systematically in Chap. 15. Shearwalls are covered in Chap. 10.

Figures illustrating diaphragm design methodologies are generally applicable to both wind and seismic forces. In some cases a figure will note that it specifically uses wind or seismic forces, while in other cases the force is not stated as being one or the other. In addition, figures and examples will address a combination of allowable stress design (ASD) and load and resistance factor design (LRFD) methods. Use of either method requires use of the ASD or LRFD load combinations, introduced in Sec. 2.17.

9.2   Basic Diaphragm Action

A diaphragm can be defined as a large, thin structural element that is loaded in its plane. It is an assemblage of elements which typically includes

1. Roof or floor sheathing

2. Framing members supporting the sheathing

3. Boundary or perimeter members

When properly designed and connected together, this assemblage will function as a horizontal beam that distributes force to the vertical resisting elements in the lateral-force-resisting system (LFRS). In Sec. 3.3, it was discussed that diaphragm flexibility needs to be considered in order to develop a model for analysis of the structure in order to distribute seismic forces to the shearwalls. In wood light-frame buildings, it is common to designate the diaphragm as either flexible or rigid for purposes of force distribution. This chapter will first introduce flexible diaphragm analysis, which is the simpler and more common of the two methods. In Sec. 9.11, the code criteria for categorizing diaphragms will be discussed, and a rigid diaphragm analysis will be demonstrated.

The diaphragm must be designed for lateral forces in both the transverse and longitudinal directions. See Example 9.1. Like all beams, a diaphragm must be designed to resist both shear and bending. In general, a diaphragm can be thought of as being made up of a shear-resisting element (the roof or floor sheathing) and boundary members. There are two types of boundary members in a diaphragm (chords and collectors or struts), and the direction of the applied force determines the function of the members (Fig. 9.2). The *chords* are designed to *carry the moment* in the diaphragm.

EXAMPLE 9.1   Diaphragm Forces and Boundary Members

Figure 9.2   Diaphragm boundary members: chords and collectors.

Diaphragm *boundary members* change functions depending on the direction of the lateral force.

Chords are boundary members that are perpendicular to the applied force. *Collectors* (*struts*) are boundary members that are parallel to the applied force.

An analogy is often drawn between a diaphragm and a steel wide-flange (W shape) beam. In a steel beam, the flanges resist most of the moment, and the web essentially carries the shear. In a diaphragm, the sheathing corresponds to the web, and the chords correspond to the flanges. The chords are designed to carry axial forces created by the moment. These forces are obtained by resolving the internal moment into a couple (tension and compression forces). See Fig. 9.3. The shear is assumed to be carried entirely by the sheathing

Figure 9.3   Shear and moment resistance in a diaphragm.

material. Proper nailing of the sheathing to the framing members is essential for this resistance to develop.

The *collectors* (*struts*) are designed to transmit the diaphragm *reactions to the shearwalls*. This becomes a design consideration when the supporting shearwalls are shorter in length than the diaphragm. See Example 9.2. Essentially, the unsupported diaphragm unit shear over an opening in a wall must be transmitted to the shearwall elements by the collector.

EXAMPLE 9.2   Function of Collector (Strut)

Figure 9.4 Collector (strut) over wall opening.

In Fig. 9.4 the left wall has no openings, and roof shear is transferred directly to the wall. The right wall has a large opening and the collector transfers the roof diaphragm shear over the opening to the shearwalls. Here the collector member is assumed to function in both tension and compression. When shearwall lengths d and e are equal, the collector forces T and C are equal. Collector forces for other ratios of d and e are covered in Sec. 9.6.

The *collector* is also commonly known as a *strut* or *drag strut*. These names come from the concept that the collector drags or collects the diaphragm shear into the shearwall. For structures in Seismic Design Categories (SDC) C through F, the IBC (Ref. 9.10) and ASCE 7 (Ref. 9.4) require use of special seismic forces (*seismic forces with overstrength* or *maximum seismic load effect* E_m) and special seismic load combinations for design of collectors, their splices, and their connections to shearwalls or other vertical resisting elements. An exception to ASCE 7 Sec. 12.10.2.1 exempts collectors and their splices and connections in structures, or portions thereof, braced entirely by light-frame shearwalls. These members need only be designed for diaphragm F_{px} forces.

Equations for the maximum seismic load effect E_m are found in ASCE 7 Sec. 12.4.3, which also includes ASD and LRFD special seismic load combinations for use with the E_m forces. When conformance with the IBC is required, however, IBC Sec. 1605 requires that the special seismic load combinations of IBC Sec. 1605.4 be used instead of the ASCE 7 load combinations. The IBC equations are simplified versions of the ASCE 7 LRFD load combinations, not requiring the consideration of simultaneous snow load S or horizontal earth pressures H. As long as LRFD load combinations are used, the ASCE 7 equation will result in the same or more conservative requirements. If conformance to IBC is required, use of the ASCE 7 ASD combinations is not recommended.

The equations for maximum seismic load effect determined from ASCE 7 Eqs. 12.4-5 through 12.4-7 are:

$$E_m = \Omega_o Q_E + 0.2 S_{DS} D$$
$$E_m = \Omega_o Q_E - 0.2 S_{DS} D$$

where Ω_o is the overstrength factor from ASCE 7 Table 12.2-1, S_{DS} is the short-period design spectral response acceleration introduced in Chap. 2, and D is the dead load. The term $\Omega_o Q_E$ is the horizontal component of seismic force, while $0.2 S_{DS} D$ acts vertically.

The special seismic force E_m was first introduced as seismic design force Type 6 in Sec. 3.4. This force, like other seismic forces, must be used in load combinations with applicable load factors. ASCE Sec. 12.4.3.2 specifically lists the applicable load combinations, with both the horizontal and vertical components of the overstrength force included. The LRFD load combinations will be used for examples in this book because they will meet the needs of both ASCE 7 and IBC users. The LRFD (strength design) load combinations are:

$$(1.2 + 0.2 S_{DS})D + \Omega_o Q_E + L + 0.2S$$
$$(0.9 + 0.2 S_{DS})D + \Omega_o Q_E + 1.6H$$

The special E_m force level and load combinations are used for a limited number of special elements that are thought to have a major impact on the seismic performance of a building. Besides collectors, the other special elements occurring in wood-frame construction are elements providing vertical support under shearwalls that occur in an upper floor but do not continue down to the foundation (elements supporting discontinuous shearwalls, discussed in Sec. 16.3). The concept of using the higher E_m force and the special load combinations is that elements which are considered critical to the performance of the building should be designed for the maximum seismic force that is expected in the structure, not the typical IBC design force. Design for this force level will ensure that these critical elements are not the weak link (first item to fail) in the structural system. These special requirements for collectors are illustrated in Sec. 9.7.

9.3 Shear Resistance

The shear-resisting element in the diaphragm assemblage is the roof or floor sheathing. This can be either lumber sheathing or wood structural panels. The majority of wood diaphragms use wood structural panel sheathing because of the economy of installation and the relatively high allowable unit shears it provides. Because of its wide acceptance and for simplicity, this chapter deals primarily with wood structural panel diaphragms. Lumber-sheathed diaphragms are covered in Ref. 9.3.

For wood structural panel diaphragms, the starting point is the determination of the required panel thickness for sheathing loads, that is, dead, live, and snow loads perpendicular to the plane of the sheathing (Chap. 8). It was previously noted that these loads often determine the final panel thickness. This is especially true for floors because the sheathing loads (due to vertical dead and live loads) are larger and the deflection criteria are more stringent than those for roofs.

The shear capacity for wood structural panel diaphragms is the shear "through the thickness" (Fig. 8.21a). However, in most cases the shear capacity of the panel is not the determining factor. Unit diaphragm shears are usually limited by the *nail capacity* in the wood, rather than the strength of the panel.

The nail spacing required for a diaphragm may be different at various points in the diaphragm. Because of the importance of the panel nailing requirements, it is necessary for the designer to clearly understand the nailing patterns used in diaphragm construction.

There are two primary sources for information on the capacities and construction of diaphragms and shearwalls that will be used in this book. One source is Chap. 23 of the IBC (Ref. 9.10) and the other is the American Forest & Paper Association (AF&PA) *Special Design Provisions for Wind and Seismic* (SDPWS, Ref. 9.3) a supplement to the AF&PA NDS (Ref. 9.2). Either of these sources contains provisions adequate for design of most buildings, and either may be used to meet the requirements of the IBC.

The simplest nailing pattern is found in *unblocked diaphragms.* See Fig. 9.5a. An unblocked diaphragm is one that has two of the panel edges *not supported* by lumber framing. These edges may be completely unsupported, or there may be some other type of edge support such as tongue and groove (T&G) edges or panel clips (Fig. 8.13). For an *unblocked diaphragm,* the *standard nail spacing* requirements are:

DIAPHRAGM BOUNDARY

THIS PANEL JOINT NOT CONTINUOUS

CONTINUOUS PANEL JOINT

4' X 8' WOOD STRUCTURAL PANEL

NAILING TO INTERMEDIATE FRAMING MEMBERS IS ALSO KNOWN AS "FIELD" NAILING. SPACING IS 12" O.C. FOR ROOFS AND FLOORS (6" O.C. WHEN SUPPORTS ARE SPACED 48" OR GREATER) .

MAX. NAIL SPACING AT DIAPH BOUNDARY AND SUPPORTED EDGES IS 6" O.C.

THIS PANEL EDGE IS NOT SUPPORTED BY BLOCKING AND IS KNOWN AS THE UNBLOCKED EDGE – OTHER TYPE OF EDGE SUPPORT MAY BE REQUIRED FOR SHEATHING LOADS

FLOOR OR ROOF JOISTS

PARTIAL ROOF OR FLOOR PLAN

Figure 9.5a   Unblocked diaphragm.

1. Supported panel edges—6-in. o.c.

2. Along intermediate framing members (known also as field nailing)—12–in. o.c. (except 6-in. o.c. is required when supports are spaced 48 in. or greater—see footnote b to IBC Table 2306.3.1, or SDPWS Sec. 4.2.7.1.1, Item 2).

When a diaphragm has the panel edges supported with *blocking*, the *minimum* nailing requirements are the same as for the unblocked diaphragms. There are however, more supported edges in a blocked diaphragm. See Fig. 9.5*b*.

Although the minimum nailing requirements (i.e., maximum nail spacing) are the same for both blocked and unblocked diaphragms, the allowable unit shears are much higher for blocked diaphragms. These higher unit shears are a result of the more positive, direct transfer of stress provided by nailing all four edges of the plywood panel. See Fig. 9.5*c*. Example 9.3 further discusses diaphragm nailing and layout.

EXAMPLE 9.3   Wood Structural Panel Diaphragm Nailing

Required nail size is a function of the panel thickness and shear requirements. The panel is shown spanning in the strong direction for normal (sheathing) loads. See Fig. 9.5*a* for an unblocked diaphragm and Fig. 9.5*b* for a blocked diaphragm.

The *direction of the continuous panel joint* and the *direction of the unblocked edge* are used to determine the load case for diaphragm design. Different panel layouts and framing

PARTIAL ROOF OR FLOOR PLAN

Figure 9.5b   Blocked diaphragm.

THIS RESISTANCE PROVIDED BY NAILING ALONG
INTERMEDIATE FRAMING MEMBERS –UNBLOCKED DIAPHRAGM

ISOLATED PANEL FROM FIG.9.5.a

ALL PANEL EDGES SUPPORTED AND NAILED
– BLOCKED DIAPHRAGM

ISOLATED PANEL FROM FIG.9.5.b

Figure 9.5c   Shear transfer in a wood structural panel diaphragm.

arrangements are possible. The continuous panel joint and the unblocked edge are not necessarily along the same line. The diaphragm *load case* (described in Example 9.4) is considered in blocked diaphragms as well as unblocked diaphragms.

From a comparison of the isolated panels in Fig. 9.5c, it can be seen that the addition of nailing along the edge of a blocked panel will produce a much stronger diaphragm.

In diaphragms using relatively *thin* panels, there is a tendency for the panel to buckle. This is caused by the edge bearing of panels in adjacent courses as they rotate slightly under high diaphragm forces.

When loaded to failure, the nails in a wood structural panel diaphragm deform, and the *head* of the nail may pull through the face of the panel. Other possible modes of failure include pulling the nail through the *edge* of the panel and *splitting* the framing members to which the panel is attached. A minimum distance of $\frac{3}{8}$ in. from the edge of a panel to the center of a nail is required in order to use the diaphragm shear values tabulated in the IBC and SDPWS tables. Where the framing members are wide enough to allow increased edge distance, the diaphragm will be able to withstand larger deformations prior to failure (the *deformation capacity* is increased).

When loaded to failure, there are several possible modes of failure that can occur in a wood structural panel diaphragm. Perhaps the most common type of failure is by the nail head pulling through the panel. See Fig. 9.5d.

Figure 9.5d Nail deformation at ultimate load. (*APA—The Engineered Wood Association.*)

The 6-in. nail spacing at supported panel edges is used for all *unblocked* diaphragms. For *blocked* diaphragms, however, nail spacing and the resulting tabulated diaphragm shear value can vary. Much higher allowable unit shears can be obtained by using a decreased nail spacing. Diaphragm shear values are tabulated in both IBC Table 2306.3.1 and SDPWS Tables A4.2A and A4.2B (the same SDPWS shear values are repeated in SDPWS Tables A4.2A and A4.2B, along with G_a values that apply to plywood rather than oriented strand board). There are several important differences between the IBC table and the SDPWS tables. First, the IBC tabulated values are ASD allowable shear values intended to be used only with ASD load combinations, while the SDPWS values are nominal unit shear capacities that must be adjusted for use with either ASD or LRFD load combinations. Second, the SDPWS separates the tables into Part A, applicable to seismic design and Part B, applicable to wind design. The values in Part B are 40 percent higher than Part A, based on historic wind design and a better understanding of wind load demand. The IBC only tabulates one value, applicable to seismic loads, and per IBC Sec. 2306.3.2, permits the tabulated values to be increased by 40 percent for wind load design.

The reader is reminded that wind and seismic loads must be used in load combinations with appropriate load factors. Similarly, the tabulated shear values must be adjusted for use with ASD or LRFD load combinations. Other adjustment factors that may apply will be discussed.

Tabulated shear values for diaphragms apply to framing of Douglas Fir-Larch or Southern Pine. If the diaphragm sheathing is nailed into framing members of another species, an adjustment is required to account for reduced nail capacity. Reduction values for other species of wood are given in footnote a to IBC Table 2306.3.1 and in footnote 2 to the SDPWS tables.

Tabulated shear values in the IBC and SDPWS are already adjusted for the short duration of loading applicable to wind and seismic loads, as this duration is most common for diaphragm design. This means that the duration of load factor $C_D = 1.6$ for ASD design is already included, as is the time effect factor $\lambda = 1.0$ for LRFD design. No further increase for wind or seismic loading is permitted. A decrease is required if the diaphragm will be loaded for a longer duration, as addressed by IBC Table 2306.3.1 footnote h. While this is not specifically discussed in the SDPWS, the same concept applies.

When the design unit shears are large, the required *nail spacing* for *blocked diaphragms* must be carefully interpreted. The nail spacing along the following lines must be considered:

1. Diaphragm boundary

2. Continuous panel joints

3. Other panel edges

4. Intermediate framing members (field nailing)

The field-nailing provisions are always the same: 12-in. o.c. for roofs and floors (except 6-in. o.c. spacing is required when framing is spaced at 48 in. or greater).

The nail spacing along the other lines may all be the same (6-in. o.c. maximum), or some may be different. Shear values are tabulated for nail spacing as small as 2-in o.c.

A great deal of information is incorporated into the IBC and SDPWS diaphragm tables. Because of the importance of these tables, the remainder of this section deals with a review of these tables. The tables are divided into several main parts. The top parts give shear values for diaphragms which have the STRUCTURAL I designation. The values in the bottom half of the tables apply to *all other wood structural panel grades*.

The latter represents the large majority of wood structural panels, and it includes both sanded and unsanded grades. STRUCTURAL I is stronger (and more expensive) and should be specified when the added strength is required. The tabulated values for STRUCTURAL I are 10 percent larger than the values for other grades.

The IBC and SDPWS tables include blocked and unblocked diaphragm values. These are combined in IBC Table 2306.3.1. The center columns address blocked diaphragms, while the two right-hand columns address unblocked diaphragms. In the SDPWS, Table 4.2A addresses blocked diaphragms, while Table 4.2B addresses unblocked diaphragms.

Throughout the diaphragm tables, reference is made to the *load cases*. Six load cases are defined. Although somewhat different panel arrangements can be used, all layouts can be classified into one of the six load cases. Because the proper use of the tables depends on the load case, the designer must be able to determine the load case for any layout under consideration.

The *load case* essentially depends on two factors. The *direction of the lateral force* on the diaphragm is compared with the direction of the

1. Continuous panel joint

2. Unblocked edge (if blocking is not provided)

For example, in load case 1 the applied lateral force is perpendicular to the continuous panel joint and it is also perpendicular to the unblocked edge. Load cases 1 through 4 consider the various combinations of these alternative arrangements. Load cases 5 and 6 have continuous panel joints running in both directions. Load cases 1 through 4 are more common than cases 5 and 6.

The panel layout must be shown on roof and floor framing plans along with the nailing and blocking requirements. It should be noted that the load case is defined by the criteria given above, and it does not depend on the direction of the panel. See Example 9.4.

Example 9.4   Diaphragm Load Cases

Determine the *load cases* for two diaphragms (Fig. 9.6a and b) in accordance with the IBC or SDPWS diaphragm tables. Consider both transverse and longitudinal directions.

Figure 9.6a Partial roof framing plan.

Panels in Fig. 9.6a are oriented in the strong direction for sheathing loads. In Fig. 9.6b, the same building is considered except that the panel layout has been revised. The panels in Fig. 9.6b are oriented in the weak direction for sheathing loads.

Transverse Direction

Transverse force in Fig. 9.6a is perpendicular to continuous panel joint and parallel to unblocked edge.

$$\therefore \text{ load case 2}$$

Figure 9.6b Partial roof framing plan.

Longitudinal Direction

Longitudinal force in Fig. 9.6*a* is parallel to continuous panel joint and perpendicular to the unblocked edge.

$$\therefore \text{ load case 4}$$

Practically speaking, for an unblocked diaphragm there is load case 1 and *all* others.

When the criteria defining the load case are considered, it is seen that the load cases are the same in Fig. 9.6*b* as in Fig. 9.6*a*:

Transverse—case 2

Longitudinal—case 4

For a given diaphragm problem, two different load cases apply: one for the transverse load and the other for the longitudinal load. The lateral force in the transverse direction normally produces the larger unit shears in a diaphragm (this may not be the case for shearwalls), but the tabulated unit shear capacities may be different for the two load cases. Thus, the diaphragm shears should be checked in both directions.

Other factors regarding the makeup and use of the diaphragm tables should be noted. For example, the width of the framing members must be considered. Tabulated unit shears for diaphragms using 2-in. nominal framing members are approximately 10 percent less than for diaphragms using wider faming.

The *nail spacing* used for the *diaphragm boundary* on blocked diaphragms is also the nail spacing required along the *continuous panel joint* for load cases 3 and 4. This nail spacing is also to be used at all panel edges for load cases 5 and 6. This information is found in the diaphragm table headers, under the title *blocked diaphragms* in the IBC table and under the headings *seismic* and *wind* in SDPWS Table 4.2A. If load case 1 or 2 is involved, the diaphragm boundary nail spacing is not required at the continuous panel joints, and the somewhat larger nail spacing for "other panel edges" may be used. The selection of the proper nailing specifications is illustrated in several examples later in the chapter.

When the spacing of nails is close, the diaphragm tables require special precautions to avoid the splitting of lumber framing. For example, under certain circumstances, 3-in. (or wider) nominal framing is required when the spacing of nails is 3 in. or less. For the IBC diaphragm table, this requirement is found in footnote c, and in the SDPWS it is found in Sec. 4.2.7.1.1, Item 3.

A point should be made about the combination of nail sizes and panel thicknesses in the diaphragm tables. There are times when the nail size and panel thickness for a given design will not agree with the combinations listed in the table. If the unit shears in the tables are to be used without further justification, the following procedure should be followed: If a thicker panel than that given in the table is used, the unit shear from the table should be based on the nailing used. No increase in shear is permitted because of the increased thickness.

On the other hand, if a larger nail size is used for a panel thickness given in the table, the unit shear from the table should be based on the panel thickness.

No increase in shear is permitted because of the increased nail size. Thus, without further justification, the combination of panel thickness and nail size given in the diaphragm tables is "compatible." Increasing only one parameter does not necessarily provide an increase in tabulated unit shear. Reference 9.5 describes a method of calculating diaphragm unit shear capacities by principles of mechanics using the panel shear values and nail strength values.

Another consideration related to nail size has to do with the penetration of the sheathing nails into the framing members. The diaphragm tables specify a minimum nail penetration into the framing that must be achieved in order to use the table value. The required penetration is $1\frac{1}{4}$ in. for 6d nails, $1\frac{3}{8}$ in. for 8d nails, and $1\frac{1}{2}$ in. for 10d nails. Where gun-driven nails are used, the nail length must be sufficient to provide this penetration. In unusual circumstances when it is not possible to provide this penetration, nail provisions of the NDS provide information on reduced nail capacity and on nail penetrations below which there is no recognized capacity.

Finally, it should be noted that the other "approved" types of panel edge support (such as T&G edges or panel clips—Fig. 8.13) are substitutes for lumber blocking when considering *sheathing loads* only. They do not qualify as substitutes for blocking for *diaphragm design* (except $1\frac{1}{8}$-in.-thick 2-4-1 panels with properly stapled T&G edges).

There are several additional items of information in the IBC diaphragm tables that bear discussion. First, Table 2306.3.1 specifies diaphragm sheathing nail sizes using nail pennyweight designations followed by length and shank diameter minimum dimensions. Nails used must meet or exceed the specified length and diameter. It has always been specified that diaphragms be nailed with common nails or hot-dipped galvanized box nails. These nails will meet the specified shank diameter. There has been a significant history of diaphragms and shearwalls being unintentionally constructed with nails with the same pennyweight designation, but smaller diameter (i.e., 8d box nails with D = 0.113 in. rather than 8d common with D = 0.131 in) resulting in a lesser capacity. The wide range of diameters on available gun nails has made ensuring construction with the correct nail size even more complex. The code has published minimum diameters in order to assist designers, builders and inspectors in complying with the code intent.

Second, Table 2306.3.1 contains tabulated diaphragm values for stapled diaphragms as well as nailed diaphragms. These are staples fabricated from heavy 16 gage wire, and penetrating into the framing not less than 1 in. The capacity per staple is somewhat less than a nail, however with closer spacing stapled diaphragms can provide values similar to nailed diaphragms. Attention to staple orientation is needed during construction; orientation of the staple crown parallel to the long direction of the framing member helps to ensure that the staple legs will be embedded in framing. Footnote f specifies minimum crown width and staple orientation. Some restrictions on use of stapled roof diaphragms in high-wind areas may apply due to lower capacity to resist wind suction loads.

Finally, IBC Table 2306.3.2 provides values for nailed and stapled high-load diaphragms. In order to achieve the higher load, these diaphragms have more complex construction requirements. Multiple rows of sheathing nails and staples are required at each panel edge. Stagger of the fasteners is required from row to row, and wider framing members are required in order to adequately place all of the fasteners. These diaphragms require close attention to design, detailing, and construction.

9.4   Diaphragm Chords

Once the diaphragm web has been designed (sheathing thickness and nailing), the flanges or chord members must be considered. The determination of the axial forces in the chords is described in Fig. 9.3. The axial force at any point in the chords can be determined by resolving the moment in the diaphragm at that point into a couple (equal and opposite forces separated by a lever arm—the lever arm is the distance between chords):

$$T = C = \frac{M}{b}$$

The tension chord is often the critical member. There are several reasons for this. One is that the adjusted design value in compression is often larger than the adjusted design value in tension. This assumes that the chord is laterally supported and that column buckling is not a factor. This is usually the case, but the possible effects of column instability in the compression chord must be evaluated.

Another reason that the tension chord may be critical is that the chords are usually not continuous single members for the full length of the building. In order to develop the chord force, the members must be made effective by splicing separate members together. This is less of a problem in compression because the ends of chord members can transmit loads across a splice in end bearing. Tension splices, on the other hand, must be designed.

Because the magnitude of the chord force is calculated from the diaphragm moment, the *magnitude of the chord force follows the shape of the moment diagram.* The design force for any connection splice can be calculated by dividing the moment in the diaphragm at the location of the splice by the distance separating the chords. A simpler, more conservative approach would be to design all diaphragm chord splices for the maximum chord force.

It should also be noted that each chord member must be capable of functioning in either tension or compression. The applied lateral load can change direction and cause tension or compression in either chord.

Some consideration should be given at this point to what elements in a building can serve as the chords for a diaphragm. In a wood-frame building with stud walls, the doubled top plate is usually designed as the chord. See Example 9.5. This type of construction is accepted by contractors and carpenters as standard practice. Although it may have developed through tradition, the concept behind its use is structurally sound.

The top plate members are not continuous in ordinary buildings unless the plan dimensions of the building are very small. Two plate members are used so

EXAMPLE 9.5   Diaphragm Chord—Double Top Plate

Figure 9.7 Diaphragm chord in building with wood-roof system and wood-frame walls. Code minimum lap splice length is 4 ft to form a continuous chord member. Nail or bolt connection must be designed to transmit chord force T from one plate member to the other.

The double top plate in a wood-frame wall is often used as the chord for the diaphragm (Fig. 9.7). Splices are offset so that one member is effective in tension at a splice. Connections for anchoring the diaphragm and shearwall together are covered in Chap. 15.

that the splice in one plate member can be staggered with respect to the splice in the other member. This creates a continuous chord with at least one member being effective at any given point. When chord forces are large, more than two plate members may be required.

In order for the top plate members to act as a chord, they must be adequately connected together. If the chord forces are small, this connection can be made with nails, but if the forces are large, the connection will require the use of wood screws, bolts, or steel straps. These are connection design problems, and the procedures given in Chaps. 12 and 13 can be used for the design of these splices.

It should also be noted that the chord forces are usually the result of wind or seismic forces. When used in ASD load combinations, a load duration factor C_D of 1.6 will apply, and when used in LRFD combinations, a time effect factor λ of 1.0 will apply. Other adjustment factors specified by the NDS may be applicable.

Wood structural panel diaphragms are often used in buildings that have concrete tiltup walls or masonry (concrete block or brick) walls. In these buildings, the chord is often made up of continuous horizontal reinforcing steel in the masonry or concrete wall. See Example 9.6. If the masonry or concrete is assumed to function in compression only (the usual assumption), the tension chord is critical. The stress in the steel is calculated by dividing the chord force by the cross-sectional area of the horizontal wall steel that is placed at the diaphragm level.

EXAMPLE 9.6   Diaphragm Chord—Wall Steel

PARTIAL MASONRY WALL
CONNECTIONS

Figure 9.8 Typical connections of diaphragm to masonry walls.

The chord for the diaphragm usually consists of horizontal reinforcing steel in the wall at or near the level of the diaphragm. Attempts to design the wood top plate or ledger to function as the chord are usually considered inappropriate because of the larger stiffness of the masonry or concrete walls.

Development of the chord requires that the diaphragm be adequately attached (anchored) to the shearwalls. In this phase of the design, the spacing of anchor bolts, blocking, and nailing necessary to transfer the forces between these elements are considered. The *details in* Fig. 9.8 *are not complete*, and only included here to illustrate the diaphragm chords. Anchorage connections are covered in detail in Chap. 15.

The double plate in wood walls and the horizontal steel in concrete and masonry walls are probably the two most common elements used as diaphragm chords. However, other building elements can be designed to serve as the chord. As an example of another type of chord member, consider a large window in the front longitudinal wall of a building. See Example 9.7.

EXAMPLE 9.7   Header Acting as a Diaphragm Chord

PLAN

HEADER CONNECTION TO WALL

$$\text{MAX. M} = \frac{wL^2}{8}$$

M_1 M_1

M

DIAPHRAGM BOUNDARY NAILING

DIAPHRAGM SHEATHING

BLOCKING

ANCHORAGE NAILING

ROOF FRAMING

WINDOW HEADER

SECTION ①

Figure 9.9 The header over an opening in a wall may be used as a diaphragm chord.

Over the window, the header serves as the chord. It must be capable of resisting the maximum chord force in addition to gravity loads. The maximum chord force is

$$T = C = \frac{\text{max.}\, M}{b}$$

The connection of the header to the wall must be designed for the chord force at that point:

$$T_1 = C_1 = \frac{M_1}{b}$$

NOTE: For simplicity, the examples in this book determine the chord forces using the dimension b as the width of the building. Theoretically, b is the dimension between the centroids of the diaphragm chords, and the designer may choose to use this smaller, more conservative dimension.

The top plate in this longitudinal wall is not continuous. Here the window header supports the roof or floor framing directly, and the header may be designed to function as the chord. The header must be designed for both vertical loads and the appropriate combination of vertical loads and lateral forces. The connection of the header to the shearwall also must be designed. (Note: If a cripple stud wall occurs over the header and below the roof framing, a double top plate will be required over the header. In this case, the plate can be designed as the chord throughout the length of the wall.)

The proper functioning of the chord requires that the diaphragm be effectively anchored to the chords and the supporting shearwalls. *Anchorage* can be provided by systematically designing for the transfer of gravity loads and lateral forces. This approach is introduced in Sec. 10.8, and the design for anchorage is covered in detail in Chap 15.

9.5 Design Problem: Roof Diaphragm

In Example 9.8, a wood structural panel roof diaphragm is designed for a one-story building. The maximum unit shear in the transverse direction is the basis for determining the blocking and *maximum* nailing requirements for the diaphragm. However, the shear in the diaphragm is not constant, and it is possible for the nail spacing to be increased in areas of reduced shear. Likewise, it is possible to omit the blocking where the actual shear in the diaphragm is less than the adjusted unit shear capacity for an *unblocked* diaphragm. The locations where these changes in diaphragm construction can take place are easily determined from the unit shear diagram.

The use of changes in nailing and blocking is fairly common, but these changes can introduce problems in construction and inspection. If changes in diaphragm construction are used, they should be clearly shown on the framing plan. In addition, if blocking is omitted in the center portion of the diaphragm, the requirements for some other type of panel edge support must be considered.

In this building, the chord members are the horizontal reinforcing bars in the masonry walls and a check of the chord stress in these bars is shown. The diaphragm is also checked for the lateral force in the longitudinal direction, and the effect of the change in the diaphragm load case on tabulated unit shears is demonstrated.

EXAMPLE 9.8   Wood Structural Panel Roof Diaphragm

The design forces being considered are seismic, and so are calculated at a strength level in accordance with IBC and ASCE 7 equations. The force notations therefore have a subscript "u" to denote ultimate. Roof framing consists of light-frame wood trusses at 24-in. o.c. The panel layout is given (Fig. 9.10a), and the shear and moment in the diaphragm are summarized below the framing plan. In this example, only the exterior walls are assumed to be shearwalls.

ROOF FRAMING PLAN

Figure 9.10a Roof plan showing transverse lateral force to diaphragm. Shear and moment diagrams for the diaphragm are also shown.

1. *Panel sheathing loads.*

$$\text{Roof Live Load } L_r = 20 \text{ psf}$$

$$\text{Roof Dead Load } D = 8 \text{ psf}$$

Although the roof framing is different, the panel sheathing spans the same distance (24 in.) as the sheathing in Fig. 8.14. The panel choices given in Example 8.12, part 1, apply to the example at hand:

> *Use* $\frac{7}{16}$-in. Sheathing EXP 1

2. *Diaphragm nailing.* Although this panel is available in STRUCTURAL I, the STRUCTURAL I upgrade will not be used in this problem. If a substantially higher unit shear were involved *or* if the nailing could be considerably reduced, then STRUCTURAL I should be considered. The top chord of the roof truss will be a 2 × 4 oriented vertically. This will be more than adequate to provide the minimum nail penetration required for the sheathing nails.

    Panel load case (transverse lateral force):

    The lateral force is *perpendicular to the continuous panel joint and perpendicular to the unblocked edge.*

$$\therefore \text{ load case 1}$$

From this point, the shear capacity of the diaphragm can be evaluated using either ASD provisions or LRFD provisions. This example will first look at the ASD solution, using IBC Table 2306.3.1. Following this, the LRFD design solution using the SDPWS will be illustrated.

If blocking is not used, the maximum allowable ASD unit shear is

$$\text{Allow. } v = 230 \text{ lb/ft}$$

This comes from IBC Table 2306.3.1 for case 1 in the unblocked diaphragm columns, and using $\frac{7}{16}$-in. rated sheathing and 8d common nails. Recall that IF the example used wind loads, IBC would permit the tabulated value to be increased 40 percent. Before comparing this to the calculated unit shear in the diaphragm, the seismic forces in the diaphragm need to be considered in ASD load combinations with applicable load factors. The ASCE 7 equations for seismic force are:

$$E = \rho Q_E + 0.2 S_{DS} D$$

and

$$E = \rho Q_E - 0.2 S_{DS} D$$

As done in the Chap. 3 examples, we will start by looking at the full set of applicable load combinations. The load combinations are simplified by the fact that generally only wind or seismic forces will be causing loading in the plane of the shearwalls and diaphragms. ASD load combinations with E are:

$$D + H + F + (W \text{ or } 0.7E)$$

$$D + H + F + 0.75(W \text{ or } 0.7E) + 0.75L + 0.75(L_r \text{ or } S \text{ or } R)$$

$$0.6D + 0.7E + H$$

With only E acting in the plane of the diaphragm, the forces are simplified to 0.7E for ASD design of the diaphragm shear capacity. The redundancy factor ρ is always taken as 1.0 when determining diaphragm F_{px} forces. Using applicable load factors, the ASD load combinations using story F_{px} forces are simplified to:

$$v = 0.7E = 0.7 \ (550 \ \text{lb/ft}) = 385 \ \text{lb/ft}$$

The actual shear of 385 lb/ft is bigger than the allowable shear of 230 lb/ft, so the case 1 unblocked diaphragm is not adequate for the maximum diaphragm shear.

An increased allowable shear can be obtained by increasing the panel thickness or using STRUCTURAL I panels. Another method of increasing shear capacity used here is to provide blocking and increased nailing (i.e., reduced nail spacing). From the blocked diaphragm portion of IBC Table 2306.3.1, the following design is chosen:

> *Use* $^{15}\!/_{32}$-in. Sheathing EXP 1 with 10d common nails
> at 4 in. o.c. boundary[*]
> 6 in. o.c. all other panel edges
> 12 in. o.c. field
> Blocking required.
> Use 2 × 4 minimum

Allow. $v = 385$ lb/ft   *OK*

The above nailing and blocking requirements can safely be used throughout the entire diaphragm because these were determined for the maximum unit shear. The variation of diaphragm shear can be shown on a sketch. The *unit shear diagram* is the total shear diagram divided by the diaphragm width (Fig. 9.10*b*).

Some designers prefer to take into account the reduced unit shears toward the center of the diaphragm (away from the shearwall reactions). If this is done, the nail

Figure 9.10*b*   Unit shear diagram. Similar triangles may be used to determine locations of possible changes in diaphragm construction. Note that both the unit shear forces and the allowable unit shears in this diagram are at an ASD level.

---

[*]Because this diaphragm is load case 1, the 4-in. nail spacing is required at the diaphragm boundary only. The 6-in. nail spacing is used at continuous panel joints as well as other panel edges.

spacing can be increased to the IBC maximum allowed spacing (i.e., 6-in. o.c. at all edges including the diaphragm boundary) where the shear drops off to the corresponding allowable shear for this nailing. This location can be determined by similar triangles. For example,

$$\text{Allow. } v = 290 \text{ lb/ft}$$

when the nail spacing at the diaphragm boundary is 6-in. o.c. and blocking is provided. Then

$$x_1 = \frac{290}{7.00} = 41.4\,\text{ft}$$

This refinement can be carried one step further, and the location can be determined where blocking can be omitted. Although $\frac{7}{16}$ in. rated sheathing was selected in part 1 of this example for sheathing loads, and 8d common nails were chosen for the initial check of unblocked diaphragm values, the final blocked diaphragm solution changed to $\frac{15}{32}$ in. rated sheathing and 10d common nails. In order to reduce the amount of variation, the center unblocked portion of the diaphragm will also use $\frac{15}{32}$ in. sheathing and 10d nails, slightly increasing the allowable unit shear.

$$\text{Allow. } v = 255 \text{ lb/ft} \qquad \text{for unblocked diaphragm}$$

Then

$$x_2 = \frac{255}{7.00} = 36.4 \text{ ft}$$

If these changes in diaphragm construction are used, the locations calculated above should be rounded off to some convenient dimensions. These locations and the required types of construction in the various segments of the diaphragm must be clearly shown on the roof framing plan. In this regard, the designer must weigh the savings in labor and materials gained by the changes described above against the possibility that the diaphragm may be constructed improperly in the field. The more variations in nailing and blocking, the greater the chance for error in the field. Increased nail spacing and omission of blocking can represent substantial savings in labor and materials, but inspection becomes increasingly important.

Part 2 of this example will now be repeated with a solution using SDPWS Tables 4.2A and 4.2B. If blocking is not used, the maximum tabulated nominal unit shear for seismic load per Table 4.2B is

$$\text{nominal } v = 460 \text{ lb/ft}$$

This comes from case 1 for unblocked diaphragms, using $\frac{7}{16}$-in. rated sheathing and 8d common nails. As noted in footnote 1 of the table, nominal unit shears must be adjusted per SDPWS Sec. 4.2.3 for use with either ASD or LRFD load combinations. This section specifies that the LRFD factored resistance is determined by multiplying the tabulated nominal shear by a resistance factor ϕ_D of 0.80. This gives:

$$\text{factored resistance } v = 0.80\,(460) = 368 \text{ lb/ft}$$

Before comparing this to the calculated unit shear in the diaphragm, the seismic forces in the diaphragm need to be considered in LRFD load combinations with applicable load factors. The ASCE 7 equations for seismic force are:

$$E = \rho Q_E + 0.2 S_{DS} D$$

and

$$E = \rho Q_E - 0.2 S_{DS} D$$

As done in the Chap. 3 examples, we will start by looking at the full set of applicable load combinations. The load combinations are simplified by the fact that generally only wind or seismic forces will be causing loading in the plane of the shearwalls and diaphragms. LRFD load combinations with E are:

$$1.2D + 1.0E + L + 0.2S$$

$$0.9D + 1.0E + 1.6H$$

With only E acting in the plane of the diaphragm, the forces are simplified to 1.0E for LRFD design of the diaphragm shear capacity. The redundancy factor ρ is always taken as 1.0 when determining diaphragm F_{px} forces. Using applicable load factors, the LRFD load combinations using story F_{px} forces are simplified to:

$$v = 1.0E = 1.0 \ (550 \ \text{lb/ft}) = 550 \ \text{lb/ft}$$

The LRFD shear of 550 lb/ft is bigger than the factored resistance of 368 lb/ft, so the case 1 unblocked diaphragm is not adequate for the maximum diaphragm shear. As was done in the ASD example, $^{15}/_{32}$-in. rated sheathing with 10d nails at 4 in. boundary, 6 in. other edges will be checked next. For this value, SDPWS Table 4.2A is used. The tabulated nominal shear is 770 plf. When multiplied by a resistance factor ϕ_D of 0.8, this gives a factored resistance of 616 plf, which is greater than the LRFD load of 550 plf. This solution is adequate using LRFD methods as well as ASD methods. As was done in the ASD solution, it is possible to identify LRFD shear levels at which it is possible to change to 6 in. boundary nailing, and at which it is possible to drop blocking and use an unblocked diaphragm. These steps will not be repeated for LRFD design.

3. *Chord.* The strength level axial force in the chord is obtained by resolving the diaphragm moment into a couple:

$$T_u = C_u = \frac{M_u}{b} = \frac{832 \ \text{k-ft}}{50 \ \text{ft}} = 16.6 \ \text{k}$$

Depending on the type of wall system, various items can be designed to function as the *chord* member.

In this example, the walls are made from 8-in. grouted concrete block units, and the horizontal wall steel (two No. 5 bars) will be checked for chord stresses. (Examples of wood chords are covered elsewhere).

The significant area of masonry available to carry compression forces means that compression will not control design. The reinforcing steel area provided to carry tension chord forces may be critical, however, and will be checked in accordance with applicable masonry design provisions. The masonry provisions applicable under the 2006 IBC are found in a standard developed jointly by the American Concrete Institute (ACI), the American Society of Civil Engineers (ASCE), and The Masonry Society (TMS). The standard is titled *Building Code Requirements for Masonry Structures* (ACI 530-05/ASCE 5-05/ TMS 402-05)(Ref. 9.1). Like the NDS, the standard addresses both ASD and LRFD design provisions. This example will check the tension chord using the ASD provisions. As in part 2 of this example, with the redundancy factor ρ taken as 1.0 for diaphragm design, and only earthquake loads acting on the chord, the ASD load combinations can be reduced to:

$$T = 0.7E = 0.7 \, (16.6 \text{ k}) = 11.6 \text{ k}$$

The stress in the two No. 5 reinforcing bars shown in Fig. 9.10*c* can be calculated as

$$f_s = \frac{T}{A_s} = \frac{11.6}{2(0.31)} = 18.7 \text{ ksi}$$

FOR CONNECTION DESIGN
CONSIDERATIONS SEE CH 15

SECTION
LONGITUDINAL WALL AT ROOF

Figure 9.10c   Horizontal wall steel used as diaphragm chord.

From Sec. 2.3.2 of the masonry standard, the allowable stress is 20 ksi for grade 40 or 50 reinforcing steel, and 24 ksi for grade 60. Regardless of grade, the two No. 5 bars will be adequate. Note that if this had been design of a collector rather than a chord, the maximum seismic load effect E_m would need to be used.

The design of other wall-reinforcing steel is important in masonry design and is beyond the scope of this book. Steps must be taken to ensure that cross-grain bending (Example 6.1) in the ledger does not occur. The design of anchor bolts and other connections for attaching the diaphragm to the shearwall is covered in Chap. 15.

4. *Longitudinal lateral force.* The strength level longitudinal force and the corresponding shear and moment diagrams are shown in Fig. 9.10*d*.

The unit shear in a diaphragm is often critical in the transverse direction. However, the longitudinal shear should be checked, especially if changes in nailing occur. In any

Figure 9.10d Roof plan showing longitudinal lateral force to diaphragm. Shear and moment diagrams are also shown.

event, the longitudinal direction must be analyzed for anchorage forces for connecting the diaphragm and shearwall.

Unit shear:

$$v_u = \frac{V_u}{b} = \frac{16{,}380}{110} = 149 \text{ lb/ft}$$

As was done for the transverse loads, the longitudinal diaphragm shear loads need to be used with the ASCE 7 earthquake load equations, and in ASD or LRFD load combinations. This will be illustrated using ASD load combinations. As in part 2 of this example, the ASD unit shear can be simplified to

$$v = 0.7E = 0.7(149 \text{ lb/ft}) = 104 \text{ lb/ft}$$

The load in the longitudinal direction is parallel to the continuous panel joint and parallel to the unblocked edge,

$$\therefore \text{ load case 3}$$

If blocking is omitted in the center portion of the diaphragm (up to 36 ft on either side of the centerline; see Fig. 9.10b), the allowable unit shear for an unblocked diaphragm must be used. From IBC Table 2306.3.1, the allowable ASD unit shear for $^{15}/_{32}$ in. rated sheathing, unblocked, case 3, with 10d common nails is

$$\text{Allow. } v = 190 \text{ lb/ft} > 104 \text{ lb/ft} \quad OK$$

If the unit shear in the longitudinal direction had exceeded 190 lb/ft, then blocking would have been required for the longitudinal lateral force. When blocking is required for load cases other than cases 1 and 2, the continuous panel joints must be nailed with the same spacing as the diaphragm boundary. The chord strength level force in the transverse walls is less than the chord force in the longitudinal walls:

$$T_u = C_u = \frac{M_u}{b} = \frac{205}{110} = 1.86 \text{ k} < 16.6 \text{ k}$$

$$\therefore \text{ two No. 5 bars} \quad OK \text{ for chord around entire building}$$

5. *Nailing along ridge.* The panel sheathing in this building is attached directly to the top chord of the light-frame wood roof trusses (see Fig. 9.10e). These trusses probably have sufficient strength and stiffness to maintain diaphragm continuity at the ridge.

Figure 9.10e   Diaphragm detail at ridge.

However, it is desirable to provide a positive attachment of the sheathing through doubled 2 × blocking or a shaped 3 × or 4 × block.

See Ref. 9.8 for a discussion of diaphragm design considerations at ridges in steeply pitched roofs.

9.6   Distribution of Lateral Forces in a Shearwall

In order to design diaphragm collectors, it is necessary to know how lateral forces will distribute within a wall line that includes window and door openings. This discussion will introduce the most common of three possible methods used for design of wood structural panel shearwalls and the collector force assumptions that go with this method. The same approach to collector forces may be used with concrete or masonry shearwalls. All three methods of designing wood structural panel shearwalls are addressed in Chap. 10.

The segmented shearwall approach assumes that lateral forces are carried by the full-story-height sections of shearwalls (i.e., Segments 1, 2, and 3 in Fig. 9.11), while other areas of the wall do not carry lateral forces. It is generally assumed that the unit shear (lb/ft) in each of the full-height segments is the same; this is generally reasonable because all of the wall piers are constrained by the collector to deflect the same amount under lateral loading. The collector is then designed to take the uniform unit shear coming from the roof or floor diaphragm and deliver it to the full-story-height shearwall segments. This will be illustrated in Example 9.9.

EXAMPLE 9.9   Distribution of Shear in a Wood Shearwall

In a wood shearwall, the wall is usually assumed to be made up of *separate* segments or shear panels. The shear panels are connected together so that they deflect laterally the same amount at the diaphragm level. Collector (strut) members provide the connection between the shear panels.

The three shear panels in the wall shown in Fig. 9.11 all have the same height h. Double full-height studs or posts are provided at both ends of the shear panels and appropriate shearwall nailing is provided (Chap. 10). Filler panels (wall elements above and below openings which are ignored in the analysis) may be nailed in a manner similar to the shear panels.

ELEVATION
WOOD SHEARWALL

Figure 9.11   Unit shears in a wood-frame shearwall are normally assumed to be uniform.

For the typical wood shearwall, the resistance to the applied lateral force is assumed to be uniform throughout the combined length of the shear panels. Thus,

$$v_1 = v_2 = v_3 = \frac{R}{\Sigma b}$$

This approach is appropriate when the aspect ratio (h/b) of each shearwall panel does not exceed 2. Where shearwall panel aspect ratios exceed 2, IBC Table 2305.3.4 footnote a requires the reduction of the tabulated allowable unit shear values for seismic forces. This topic will be discussed in Chap. 10.

Several types of load resistance develop in a wall subjected to a lateral force. These include bending resistance and shear resistance. For wood-frame walls constructed as described above, the shear resistance is the significant form, and the unit shear in this case approaches a uniform distribution. This is only true, however, when the height-to-width (aspect) ratios conform to the code limitations and the walls are of reasonably similar widths.

In buildings with reinforced concrete or masonry shearwalls, the distribution of forces to the various elements in the wall differs from the uniform distribution assumed for wood shearwalls. Concrete and masonry walls may have significant combined bending and shear resistance. The combined resistance is measured by the *relative rigidity* of the various piers in the wall. Under a lateral force the piers are forced to undergo the same lateral deformation. The force required to deflect a long pier is larger than the force calculated on the basis of its length only.

Thus, in a concrete or masonry wall, the *unit shear* in a wider pier is greater than the unit shear in a narrower wall segment. The relative rigidity is a function of the height-to-width ratio h/b of the pier and not just its width. The relative rigidity and the question of force distribution are related to the *stiffness* of a structural element: The stiffer (more rigid) the element, the greater the force it "attracts." See Example 9.10. The calculation of relative rigidities for these wall elements is a concrete or masonry design problem and is beyond the scope of this text.

EXAMPLE 9.10   Distribution of Shear in a Concrete or Masonry Shearwall

In a concrete or masonry wall, the piers are analyzed to determine their *relative rigidities*. The smaller the height-to-width ratio h/b of a pier, the larger the relative rigidity. The larger the relative rigidity, the greater the percentage of total force R carried by the pier. Pier 2 in the wall shown in Fig. 9.12 has a greater rigidity and a greater unit shear (not just total shear) than pier 1.

Figure 9.12   Unit shears in a concrete or masonry shearwall depend on relative rigidities of wall piers.

In buildings with concrete or masonry walls, the loads over openings in the walls can be supported in two different ways. In the first system, there is a concrete or masonry wall element over the openings in the wall. See Fig. 9.13*a* in Example 9.11. This element can be designed as a beam to carry the gravity loads across the opening. Concrete and masonry beams of this type are known as *lintels*. In the second case, a wood *header* may be used to span the opening (Fig. 9.13*b*).

EXAMPLE 9.11   Lintel and Header between Shearwalls

A concrete or masonry lintel, or a wood (or steel) header, can be used to span over an opening in a wall (see Fig. 9.13). The member over the opening must be designed for:

1. Vertical loads
2. Combined vertical and lateral loads.
   a. For lateral forces parallel to the wall, the force in the header is the collector (strut) force.
   b. For lateral forces perpendicular to the wall, the force in the header is the diaphragm *chord* force.

Figure 9.13*a*   Wall with concrete or masonry lintel.

Figure 9.13*b*   Wall with wood header.

The chord will be designed using the IBC or ASCE 7 *basic load combinations*; however, the collector will need to be designed using the *special (overstrength) seismic load combinations*, as discussed in Sec. 9.2. These will be discussed further in Sec. 9.7.

A distinction is made between these systems because in the first case the load from the diaphragm is transmitted to the shearwalls through the lintel, and the diaphragm design is unaffected. In the case of the wood header, however, the unit shear from the diaphragm must be transmitted through the header/collector (strut) into the shearwalls.

In summary, the distribution of the *unit shear* in a wall subjected to a lateral force will be as follows:

1. Wood shearwalls—uniform distribution to all shear panels with $h/b \geq 2$

2. Concrete and masonry shearwalls—distribution is a function of the relative rigidities of the wall piers

For problems in this book involving concrete or masonry walls, it is assumed that the distribution of wall forces and the unit shears are known by previous analysis.

9.7   Collector (Strut) Forces

The perimeter members of a diaphragm that are parallel to the applied lateral force are the collectors (struts). These members are also known as *drag struts* or *ties*, and their function is illustrated in Fig. 9.4 (Sec. 9.2). The term collector is generally used in this book because it is the term frequently used in the IBC and ASCE 7. Essentially, the collector pulls or drags the unit shear in an unsupported segment of a diaphragm (the portion over an opening in a shearwall, for example) into the supporting elements of the shearwall.

The members in a building that serve as the collectors are typically the same members that are used as the *chords* for the lateral force in the perpendicular direction. Thus, the design of the chord and the collector for a given wall may simply involve the design of the same member for different forces. In fact, for a given perimeter member, the chord force is compared with the collector force, and the design is based on the critical force.

The wood design principles for collectors and chords are covered in Chap. 7. These members are either tension or compression members, and they may or may not have bending. Bending may result from vertical loads, as in the header in Example 9.11. Where vertical dead loads D are acting, the vertical component of the seismic force $0.2S_{DS}D$ must also be considered in accordance with ASCE 7 Sec. 12.4.2, or Sec. 12.4.3 for the maximum earthquake load effect. Connections are also an important part of the collector design. The member itself may be sufficiently strong, but little is gained unless the strut is adequately connected to the supporting shearwall elements. The design of connections is addressed in later chapters, but the calculation of the force in the collector is covered now.

The building in Fig. 9.4 has an end shearwall with an opening located in the *center* of the wall. This presents a simple problem for the determination of the collector forces because the total force is shared equally by both shear panels. This would be true for the equal-length piers in a concrete or masonry wall also. The force in the collector is maximum at each end of the wall opening. This maximum force is easily calculated as the unit shear in the diaphragm times one-half the length of the wall opening.

The force in the collector varies linearly throughout the length of the collector. At one end of the opening the collector is in tension, and at the other end it is in compression. Since the lateral force can come from either direction, the ends may be stressed in either tension or compression. The magnitude of the collector force in the two shear panels is not shown in Fig. 9.4. However, the collector force in the top of the shearwall decreases from a maximum at the wall opening to zero at the outside end of the shearwall.

A somewhat more involved problem occurs when an opening is not symmetrically located in the length of the wall. See Example 9.12. Here a greater portion of the total wall load is carried by the longer shear panel. A corresponding larger percentage of the unsupported diaphragm shear over the opening must be transmitted by the collector to the longer panel.

The magnitude of the force at some point in the collector can be determined in several ways. Two approaches are shown in Example 9.12. The *first method* simply involves summing forces on a free-body diagram (FBD) of the wall. The cut is taken at the point where the collector force is required. The other forces acting on the FBD are the unit shears in the roof diaphragm and the unit shears in the shearwall.

The *second method* provides a plot of the force in the collector along the length of the building. With this diagram the collector force at any point can be readily determined.

The procedure illustrated in Example 9.12 for plotting the collector force can be used for wood, concrete or masonry shearwalls. The only difference is that for masonry or concrete walls, the unit shear in the walls will probably be different.

The two methods presented in Example 9.12 graphically demonstrate the functioning of a collector. These methods are reasonable for the example shown. However, when there are multiple wall openings, some degree of engineering judgment must be exercised in determining the design force for the collector.

Before proceeding with Example 9.12, the discussion of seismic force level for design of collectors needs to be revisited. In Sec. 9.2, it was noted that the IBC considers *collectors* to be critical to the behavior of the lateral force resisting system (LFRS). Because of this, a special seismic force level E_m and special seismic load combinations are required for collectors in Seismic Design Categories (SDC) C and higher (ASCE 7 Sec. 12.10.2.1). However, buildings with a LFRS constructed *entirely with wood or steel light-frame shearwalls* are exempted from the special seismic force requirement.

Because Example 9.12 uses wood shearwalls, the special seismic forces and load combinations do not need to be used. If the building in Example 9.12 had masonry walls and used seismic forces, the strength level force would have to be multiplied by $\Omega_o = 2.5$. The collector design principles demonstrated in Example 9.12 apply regardless of whether the *basic Code forces* or the E_m forces are used.

The calculation of collector forces is applicable to both wind and seismic design. The notation shown in Example 9.12 is allowable stress design. Strength level forces and shears would have "u" subscripts included in the notation.

EXAMPLE 9.12   Calculation of Force in Collector

Determine the magnitude of the collector force throughout the length of the front wall in the building shown in Fig. 19.14a. The walls are wood-frame construction. The forces are wind.

Figure 9.14a   Free body diagram of front longitudinal wall. The forces shown are (1) unit shear in roof diaphragm v_R, (2) unit shear in shearwalls v_W.

Method 1: FBD Approach

The force in the collector at any point can be determined by cutting the member and summing forces. Two examples are shown (Fig. 9.14b and c). One cut is taken through the header, and the other is taken at a point in the wall.

Figure 9.14b Free body diagram of front longitudinal wall taken at point A.

a. Collector force at point A (40 ft from left wall):

$$\Sigma F_x = 0$$

$$T_A + 100(40) - 150(30) = 0$$

$$T_A = +500 \text{ lb} \quad (\text{tension})$$

NOTE: The lateral force can act toward the right or left. If the direction is reversed, the strut force at A will be in compression.

b. Collector force at point B (100 ft from left wall):

Figure 9.14c Free body diagram of front longitudinal wall taken at point B.

$$\Sigma F_x = 0$$

$$C_B + 150(30 + 30) - 100(100) = 0$$

$$C_B = 1000 \text{ lb} \quad (\text{compression})$$

NOTE: This force would be tension if the direction of the lateral force were reversed.

Method 2: Collector Force Diagram

Construction of the collector (drag strut) force diagram (see Fig. 9.14*d*) involves

a. Drawing the unit shear diagrams for the roof diaphragm and shearwalls.

b. Drawing the *net* unit shear diagram by subtracting the diagrams in *a*.

c. Constructing the strut force diagram by using the *areas* of the net unit shear diagram as *changes* in the magnitude of the strut force. The strut force is zero at the outside ends of the building.

PLAN

a. PLOT OF UNIT SHEARS
(LB / FT.)

b. NET UNIT SHEARS
(LB / FT.)

c. COLLECTOR FORCE
(LB)

Figure 9.14d   Collector force diagram. The final plot (*c.* collector force) shows variation in the magnitude of the strut force over the entire length of the front longitudinal wall.

The relationship between the net unit shear diagram and the collector force diagram is similar to the relationship between the shear and moment diagrams for a beam.

NOTE: Signs of forces are not important because forces can reverse direction.

The values obtained by taking FBDs at points A and B can be verified by similar triangles on the collector force diagram.

A third approach to collector force calculation for one-story buildings is to determine the maximum force that might develop in a collector regardless of its location along the length of a wall. This force can be taken as the *unit shear in the diaphragm times the length of the wall opening.* If this is done for the building in Example 9.12, a collector force of $T = 100 \times 40 = 4000$ lb is obtained. Connections at both ends of the collector as well as the member itself would then conservatively be designed for this force.

Although a considerably higher design force is determined in this example (4000 lb versus 2500 lb and 1500 lb), the method has the advantage of being conservative and easy to apply. When there are multiple openings in a wall, the designer can extend this method to determine a conservative value for the collector force by taking the unit shear in the diaphragm times the sum of the lengths of the wall openings. If the force obtained in this manner is overly conservative, a more detailed analysis can be made.

Once the collector forces have been determined, they can be compared with the chord forces at corresponding points along the wall to determine the critical design condition.

9.8 Diaphragm Deflections

The deflection of a diaphragm has been illustrated in a number of sketches in Chap. 9. This topic is the object of some concern because the walls that are attached to (actually supported by) the diaphragm are forced to deflect out-of-plane along with the diaphragm. See Example 9.13. If the walls are constructed so they can undergo or accommodate these deformations without failure, there is little need for concern. However, a potential problem exists when the deflection is large or when the walls are constructed so rigidly that they can tolerate little deflection. In this case, the diaphragm imposes an out-of-plane deflection on the wall that may cause the wall to be overstressed.

As discussed in Sec. 3.3, there is another reason for calculating the diaphragm deflection, which has to do with classifying the diaphragm as being either *flexible* or *rigid*. This will be discussed further in Sec. 9.11.

There are two methods to evaluate out-of-plane wall displacement resulting from the in-plane deflection of the diaphragm. The first method that traditionally was the most widely used is basically a rule of thumb, in that no attempt is made to calculate the magnitude of the actual deflection. This method makes use of the span-to-width ratio of the diaphragm. In this simple approach, the span-to-width ratio L/b is checked to be less than some acceptable limit. For example,

IBC Table 2305.2.3 sets an upper limit on the span-to-width ratio for blocked wood structural panel diaphragms of 4.

Thus, this approach assumes that if the *proportions* of the diaphragm are *reasonable*, deflection will not be excessive. Although no numerical value for the deflection of the diaphragm is calculated, this method has traditionally been applied. As a minimum, conformance to the code required L/b ratio limits should be demonstrated.

EXAMPLE 9.13   Diaphragm Deflections and Span-to-Width Ratios

Figure 9.15   Plan and section views showing the deflections caused by lateral force on diaphragm.

Wood-frame walls are capable of accommodating larger out-of-plane deflections than are more rigid wall systems such as concrete or masonry.

Diaphragm Proportions

The traditional method of controlling deflection was to limit the span-to-width ratio L/b for the diaphragm. For blocked horizontal wood structural panel diaphragms, $L/b \leq 4.0$. For other types of wood diaphragms, see IBC Sec. 2305.2.3. See Fig. 9.15.

The calculation of a numerical diaphragm deflection can be in accordance with either IBC Sec. 2305.2.2 or SDPWS Sec. 4.2.2. While the concept of both of these is the same, there are some differences. The IBC deflection method will be illustrated first, followed by discussion of the differences. The total deflection that occurs in an ordinary beam is made up of two deflection components: bending and shear. However, for most practical beams, the shear deflection is negligible, and only the bending deflection is calculated.

In a diaphragm, the total deflection is the sum of the deflections caused by a number of factors in addition to bending and shear. All these factors may contribute a significant amount to the total deflection. In the IBC formula, the deflection of a diaphragm is made up of four parts:

$$\text{Total } \Delta = \Delta_b + \Delta_v + \Delta_n + \Delta_c$$

where Δ_b = bending deflection
Δ_v = shear deflection of sheathing panel
Δ_n = deflection due to nail slip (deformation)
Δ_c = deflection due to slip in chord connection splices

where:

$$\Delta_b = \frac{5vL^3}{8EAb}$$

$$\Delta_v = \frac{vL}{4Gt}$$

$$\Delta_n = 0.188Le_n$$

$$\Delta_c = \frac{\Sigma(\Delta_c X)}{2b}$$

where A = area of chord cross section, in.^{2}
b = diaphragm width, ft
E = modulus of elasticity of chords, psi
e_n = nail deformation, in. (per IBC Table 2305.2.2(1)
Gt = panel rigidity through the thickness in lb/in [IBC Table 2305.2.2(2)]
L = diaphragm length, ft
v = maximum shear caused by design loads in direction under consideration, lb/ft

$\Sigma(\Delta_c X)$ = sum of the product of individual chord-splice slip values of the diaphragm and its distance to the nearest support, in.-ft.

Example 9.14 illustrates use of this method, which applies to calculation of deflection for a blocked diaphragm at either an ASD or LRFD force level. SDPWS provides methods for calculating deflections for unblocked diaphragms.

EXAMPLE 9.14   Calculation of Diaphragm Deflections

Figure 9.16   Plan of diaphragm.

For the building shown in Fig. 9.16, the total lateral force is determined to be 55 k from a separate analysis. All framing is assumed to be Douglas Fir-Larch and the sheathing to be 32/16 span rating 3-ply plywood panels with 8d common nails. The chords are assumed to be two Douglas Fir-Larch No. 1 grade 2 × 8s spliced together.

The following equations are used to evaluate the four deflection components for a diaphragm:

$$\Delta_b = \frac{5vL^3}{8EAb}$$

$$\Delta_v = \frac{vL}{4Gt}$$

$$\Delta_n = 0.188Le_n$$

$$\Delta_c = \frac{\Sigma(\Delta_c X)}{2b}$$

Considering each deflection component separately:

Bending

$$\Delta_b = \frac{5vL^3}{8EAb}$$

where $v = 55{,}000 \text{ lb}/(2 \times 80 \text{ ft}) = 344 \text{ lb/ft}$
  $L = 180 \text{ ft}$
  $E = 1.7 \times 10^6 \text{ psi for No. 1 DFL } 2 \times 8$
  $A = 2(1.5 \text{ in.} \times 7.25 \text{ in.})$
    $= 21.75 \text{ in.}^2$
  $b = 80 \text{ ft}$

$$\Delta_b = \frac{5vL^3}{8EAb} = \frac{5(344)(180)^3}{8(1.7 \times 10^6)(21.75)(80)} = 0.424 \text{ in.}$$

Shear

$$\Delta_v = \frac{vL}{4Gt}$$

$Gt = 27{,}000 \text{ lb/in. for 32/16 3-ply plywood sheathing}$

$$\Delta_v = \frac{vL}{4Gt} = \frac{(344)(180)}{4(27{,}000)} = 0.573 \text{ in.}$$

Nail Slip

$$\Delta_n = 0.188Le_n$$

It is assumed here that the diaphragm is fabricated and used in the dry condition (framing member MC $\leq$ 19 percent). From IBC Table 2305.2.2(1), the tabulated value of e_n for 8d and 180 lb/nail is 0.05 in. This is increased by 20 percent per footnote a, since STRUCTURAL I sheathing is not being used, and divided by 2 per footnote d, based on framing moisture content (MC):

$$e_n = 0.05(1.20)/2 = 0.030 \text{ in.}$$
$$\Delta_n = 0.188 \, Le_n = 0.188(180)(0.030) = 1.015 \text{ in.}$$

Chord Slip

$$\Delta_c = \frac{\Sigma(\Delta_c X)}{2b}$$

The average slip Δ_c in a chord splice connection is assumed to be $\frac{1}{32}$ in., which is half of the $\frac{1}{16}$-in. allowable oversize for the bolt holes. It is assumed that the splice connections are located every 20 ft along a chord. Since there are two chords (tension and compression), the summation can simply be made for one chord and doubled. Therefore,

$$\Sigma(\Delta_c X) = 2[\tfrac{1}{32}(20 + 40 + 60 + 80 + 80 + 60 + 40 + 20)] = 25 \text{ in.-ft}$$

$$\Delta_c = \frac{\Sigma(\Delta_c X)}{2b} = \frac{25}{2(80)} = 0.156 \text{ in.}$$

Total Deflection

$$\Delta = 0.424 + 0.573 + 1.015 + 0.156$$

$$\Delta = 2.168 \text{ in.}$$

Considered over a 180 ft length of the diaphragm, this deflection is quite small: $L/1000$. Also a point of consideration is the out-of-plane bending of the wall. Since the base of a wall is essentially a pin connection, the diaphragm deflections cause little bending in the walls. The primary issue, however, is potential vertical instability of the wall caused by this deflection, or the P-Δ effect.

If the deflection of the diaphragm is considered to be excessive, increased nailing or the use of STRUCTURAL I panels are two possible remedies to consider.

The calculation of diaphragm deflection in accordance with Sec. 4.2.2 of the SDPWS is similar; however, the deflection terms Δ_v and Δ_n are combined into a single term for in-plane shear deformation, given by:

$$\frac{0.25vL}{1000G_a}$$

where G_a is provided in the SDPWS nominal shear capacity tables. Use of the SDPWS is particularly helpful where unblocked diaphragms are used because G_a values for unblocked diaphragms are tabulated. The commentary to the SDPWS explains the basis of the G_a values.

For the example diaphragm, the 32/16 rated sheathing will be assumed to be $^{15}\!/_{32}$ in. 3-ply plywood. Per SDPWS Table A4.2A, the tabulated G_a value is 9.5 kips/in. The deflection due to the combined shear terms is:

$$\frac{0.25(344)(180)}{1000(9.5)} = 1.629 \text{ in.}$$

This gives a total mid-span diaphragm deflection of:

$$0.424 + 1.629 + 0.156 = 2.209 \text{ in.} \sim 2.168 \text{ in.}$$

Therefore, the result of either formula gives the same answer.

One problem with evaluating out-of-plane wall deformations by applying the diaphragm deflection calculation is that there are no definite criteria against which the calculated deflection must be checked. The simple deflection limits used for ordinary beams (such as $L/180$ and $L/360$) are not applied to diaphragm deflections. These types of limits do not evaluate the out-of-plane effects that the deformations impose on the walls supported by the diaphragm.

Reference 9.8 suggests a possible deflection criterion for diaphragms that support concrete or masonry walls, but generally the evaluation of the calculated deflection is left to the judgment of the designer. It should be noted that the limits on the span-to-width ratio given in IBC Table 2305.2.3 are a requirement of the code, and they must be satisfied.

In this introduction to diaphragm design, the limits on the span-to-width ratio will be relied upon for controlling deflections. As noted, this is fairly common practice in ordinary building design. Diaphragm deflections are examined in greater detail in Sec. 16.10.

9.9 Diaphragms with Interior Shearwalls

All of the diaphragms considered up to this point have been supported by shearwalls that are the exterior walls of the building. Although it is convenient to introduce diaphragms by using this basic system, many buildings make use of interior shearwalls in addition to exterior shearwalls. This section will address analysis of buildings with flexible diaphragms and interior shearwalls. Analysis of buildings with rigid diaphragms is covered in Secs. 9.11 and 16.10.

The roof or floor assembly is assumed to act as a number of separate diaphragms. See Example 9.15. As a result they are treated as simply supported beams that span between the respective shearwall supports. Thus, the *shear in the diaphragm* can be determined using the methods previously covered for buildings with exterior shearwalls. The difference is that the span of the diaphragm is now measured between adjacent parallel walls and not simply between the exterior walls of the building. With these separate diaphragms there will be different unit shears in the diaphragms unless the spans and loads happen to be the same for the various diaphragms.

The *forces to the shearwalls* are calculated as the sum of the reactions from the diaphragms that are supported by the shearwall. Thus, for an exterior wall the shearwall force is the reaction of one diaphragm. For an interior shearwall, however, the force to the wall is the sum of the diaphragm reactions on either side of the wall.

The chord force for the diaphragms is determined in the same manner as for a building with exterior shearwalls. The moments in the respective diaphragms are calculated as simple beam moments. Once the chord forces are determined, the chords are designed in accordance with appropriate procedures for the type of member and materials used. The diaphragm design procedure shown in Example 9.15 is equally applicable with wind or seismic forces. The example uses strength level forces, denoted with a "u" subscript. Resulting seismic unit shears must be used in appropriate ASD or LRFD load combinations.

EXAMPLE 9.15   Flexible Diaphragm with Interior Shearwall

Determine the unit shear in the diaphragm and shearwalls in the building shown in Fig. 9.17*a* for the transverse lateral force of 300 lb/ft. Also calculate the diaphragm chord forces. The applied lateral forces used in this example are seismic, and are therefore at a strength level.

Diaphragm

Flexible diaphragms are assumed to span between the exterior and interior shearwalls. The continuity at the interior wall is disregarded, and two simply supported diaphragms are analyzed for internal unit shears. See Fig. 9.17*b*.

Figure 9.17a Plan view of building with interior and exterior shearwalls. Deflected shapes of assumed separate diaphragms are shown.

Figure 9.17b Free body diagrams of diaphragms.

$$V_{u1} = V_{u2L} = \frac{w_u L_1}{2} = \frac{(0.300)70}{2} = 10.5 \text{ k}$$

$$v_{u1} = v_{u2L} = \frac{V_{u1}}{b} = \frac{10{,}500}{60} = 175 \text{ lb/ft}$$

$$V_{u2R} = V_{u3} = \frac{w_u L_2}{2} = \frac{(0.300)130}{2} = 19.5 \text{ k}$$

$$v_{u2R} = v_{u3} = \frac{V_{u2R}}{b} = \frac{19{,}500}{60} = 325 \text{ lb/ft}$$

The nailing requirements for each diaphragm can be determined separately on the basis of the respective unit shears. Note that both diaphragms 1 and 2 will have their own nailing requirements, including boundary nailing along line 2. These nailing provisions must be clearly shown on the design plans.

As an alternative to double boundary nailing (i.e., two rows of boundary nailing), it is possible to design the nailing along line 2 for the combined shear from diaphragms 1 and 2:

$$v_{u2} = \frac{10,500 + 19,500}{60} = 500 \text{ lb/ft}$$

Shearwalls

The unit shears in the shearwalls can be determined by calculating the reactions carried by the shearwalls. Shearwall reactions must balance the loads (shears) from the diaphragm. See Fig. 9.17c.

SHEAR WALL REACTIONS

Figure 9.17c   Plan view showing free body diagrams of the three transverse shearwalls.

Unit wall shears (no openings in walls):

$$v_{u1} = \frac{R}{b} = \frac{10,500}{60} = 175 \text{ lb/ft}$$

$$v_{u2} = \frac{30,000}{60} = 500 \text{ lb/ft}$$

$$v_{u3} = \frac{19,500}{60} = 325 \text{ lb/ft}$$

Alternate shearwall reaction calculation:

Because the diaphragms are assumed to be simply supported, the shearwall reactions can also be determined using the tributary widths to the shearwalls. See Fig. 9.17d.

$$R_{u1} = (0.300)35 = 10.5 \text{ k}$$

$$R_{u2} = (0.300)100 = 30 \text{ k}$$

$$R_{u3} = (0.300)65 = 19.5 \text{ k}$$

Figure 9.17d   Lateral force diagram showing tributary widths to shearwalls (schematic similar to Fig. 9.17a).

NOTE: The calculated unit shears for the shearwalls are from the diaphragm reactions. If the lateral force is a seismic force, there will be an additional wall shear generated by the dead load of the shearwall itself (see Chap. 3).

Diaphragm Chord Forces

The distribution of force in the chord follows the shape of the simple beam moment diagram for the diaphragms. See Fig. 9.17e.

Figure 9.17e   Moment diagrams for determining the chord forces in the diaphragms.

$$T_{u1} = C_{u1} = \frac{M_1}{b} = \frac{184}{60} = 3.1 \text{ k}$$

$$T_{u2} = C_{u2} = \frac{M_2}{b} = \frac{634}{60} = 10.6 \text{ k}$$

The simple span procedure for handling diaphragms and interior shearwalls illustrated in Example 9.15 is probably the most common approach used to design the general class of buildings covered in this book. However, Ref. 9.8 suggests that the continuity of the diaphragm at an interior shearwall should not be disregarded. Additional diaphragm research is needed to better understand the interaction of the various components in a lateral-force-resisting system.

Additional design considerations are required for seismic forces on buildings that are *structurally irregular*. An example of an irregularity is an interior shearwall on the second floor of a two-story building that is offset horizontally from the interior shearwall on the first floor. This and other types of seismic irregularities are covered in Chap. 16.

9.10   Interior Shearwalls with Collectors

If a diaphragm is supported by an interior shearwall that is not the full width of the building, or if there are openings in the wall, a collector will be required. As with the collectors in exterior walls (Sec. 9.7), the collector will transfer the unit shear from the unsupported portion of the diaphragm into the shearwall. In the case of an interior shearwall, the collector force will be the sum of the forces from the diaphragms on both sides of the shearwall. See Example 9.16.

The distribution of the force in the collector throughout its length can be plotted from the areas of the net unit shear diagram. From this plot it can be seen that the critical axial force in the collector is at the point where it connects to the shearwall. The member and the connection must be designed for this force. In addition, the effects of combined gravity loads and lateral forces must be considered.

As discussed previously, the code has established special seismic forces E_m and special seismic load combinations for design of collectors. The building discussed in Sec. 9.7 was exempt from these special requirements because the primary LFRS was constructed entirely of light-frame wood shearwalls. The building in Example 9.16 has masonry walls, and is not exempt.

From ASCE 7 Sec. 12.4.3, the special seismic forces are:

$$E_m = \Omega_o Q_E + 0.2 S_{DS} D$$

and

$$E_m = \Omega_o Q_E - 0.2 S_{DS} D$$

From IBC Sec. 1605.4, the corresponding *special seismic load combinations* are:

$$1.2D + f_1 L + 1.0\ E_m \tag{16-22}$$

$$0.9D + 1.0\ E_m \tag{16-23}$$

In Example 9.16, the collector force will first be calculated as Q_E and then be multiplied by Ω_o to obtain E_m.

EXAMPLE 9.16   Building with Interior Collector

The building of Example 9.15 is modified in this example so that the interior shearwall is not the full width of the building. Determine the collector force for the interior shearwall. See Fig. 9.18.

Assume:

$$\Omega_o = 2.5 \text{ (ASCE 7 Table 12.2-1)}$$

Also assume that the shears and reactions from Example 9.15 are strength level seismic design forces.

$$\text{Unit shear in shearwall:} \quad v_{u2} = \frac{R_{u2}}{b} = \frac{30{,}000}{40} = 750 \text{ lb/ft}$$

Figure 9.18 Construction of collector force diagram. The steps involved are similar to those introduced in Fig. 9.14d.

The interior shearwall carries the reactions of the diaphragms. Part of this force is transferred to the shearwall by the collector. The maximum collector force is 10 k and is obtained from the collector force diagram in Fig. 9.18.

Collector Force Adjustment

$$E_m = \Omega_o Q_E = 2.5(10{,}000) = 25{,}000 \text{ lb strength level}$$

The Ω_o multiplier adjusts the collector member force to the estimated maximum earthquake force that can develop. IBC requires use of the special seismic force E_m for both allowable stress and strength design methods:

$$1.2D + f_1 L + E_m \tag{16-22}$$

and

$$0.9D + E_m \tag{16-23}$$

where f_1 is 1.0 for floors in places of public assembly or floor live loads greater than 100 psf, and f_1 is 0.5 for other floor live loads.

It is important to note that it is *not intended* for the special seismic forces to be compared to ASD allowable forces. By requiring use of the Sec. 1605.4 equations for special seismic load combinations, the IBC is requiring that design for the maximum seismic load effect E_m be at a strength level exclusively. Past editions of the IBC have provided conversion factors to allow ASD allowable values to be soft-converted to a strength (LRFD) level; however, these are not provided in the 2006 IBC. This means that members requiring design for the special seismic load combinations will have to be designed using the LRFD provisions of the NDS and, if appropriate, the SDPWS.

ASCE 7 provides complete sets of special seismic load combinations for both ASD and LRFD force levels, and allows use of an additional allowable capacity increase when using the ASD equations.

When designing in accordance with the IBC, it is important that IBC load combinations from Equations 16-22 and 16-23 be used. Use of the ASCE 7 load combinations would result in different design forces.

Not all interior walls are necessarily shearwalls. For an interior wood-frame wall to function as a shearwall, it must have sheathing of adequate strength, and it must be nailed to the framing so that the shear resistance is developed. In addition, the height-to-width ratio h/b of the shear panel must be less than the limits given in IBC Table 2305.3.4. (Procedures for designing shearwalls are covered in Chap. 10). The diaphragm must also be connected to the shearwall so that the shears are transferred.

In addition to these strength requirements, some consideration should be given to the relative widths and locations of the various walls. To illustrate this point, consider an interior wall with a width that is very small in comparison with the adjacent parallel wall. See Example 9.17.

Because of the large width b_e of the right exterior wall, it will be a more rigid support for the diaphragm than the small interior wall. The large rigidity of the exterior wall will "attract" and carry a large portion of the lateral force tributary to both walls.

Because of this, it would be a better practice to ignore the small interior wall and design the nearby exterior wall for the entire tributary lateral force. The small interior wall should then be sheathed and nailed lightly to ensure that it does not interfere with the diaphragm span.

Attempts can be made to compare the relative rigidities of the various shearwalls. This is similar to the rigidity analysis of the various piers in a single concrete or masonry wall (Sec. 9.6). However, in wood-frame construction the question of whether to consider a wall as an "effective" shearwall is normally a matter of judgment. The relative widths of the shearwalls and their proximity to other effective walls are used in making this determination.

EXAMPLE 9.17   Effectiveness of Interior Shearwalls

PLAN

Figure 9.19   Comparison of effective shearwalls.

For the building illustrated in Fig. 9.19, the *calculated* load to the interior shearwall using tributary widths is

$$R_i = \left(\frac{a}{2} + \frac{b}{2}\right)w$$

The load to the right exterior wall would be much smaller:

$$R_e = \left(\frac{b}{2}\right)w$$

Under *actual* loading, however, the load carried by the right exterior wall will approach

$$R_e = \left(\frac{L}{2}\right)w$$

The reason for this is that the large exterior wall is more rigid than the interior wall. Thus, in order to distribute the force from the diaphragm to the walls using tributary widths, the shearwalls must be *effective*.

The effectiveness of a wall in providing true shearwall action depends on several factors, including

1. Relative lengths of the walls. If b_i and b_e were more nearly equal, the interior wall would be more effective.
2. Relative location of the walls. If b were larger in comparison to a, the interior shearwall would also begin to function as a separate support.

The evaluation of shearwall effectiveness is, to a large extent, a matter of judgment. Walls which are judged to be ineffective (i.e., ignored as shearwalls) should be constructed so that they do not interfere with the assumed action of the diaphragm.

9.11 Diaphragm Flexibility

The topic of diaphragm flexibility was first introduced in Chap. 3. It was noted that diaphragms are designated as flexible, semirigid, or rigid for the purpose of determining lateral force distribution to shearwalls and for determination of wall to diaphragm anchorage forces. In Chap. 3, ASCE 7 Sec. 12.3 provisions regarding designation of diaphragms were discussed in detail. It was noted that wood diaphragms will often be designated as flexible, and modeled as simple-span beams spanning between shearwalls. There will be times, however, when it will be necessary to model wood diaphragms as rigid. This section will illustrate force distribution with a rigid diaphragm.

With a rigid diaphragm, there is a torsional moment that must be considered. A *torsional moment* is developed if the centroid of the applied lateral force does not coincide with the center of resistance (center of rigidity) of the supporting shearwalls. See Example 9.18. The effect of the torsional moment is to cause wall shears in addition to those that would be developed if no eccentricity existed. This is essentially a combined stress problem in that shears are the result of direct loading and an eccentric moment.

EXAMPLE 9.18   Comparison of Flexible and Rigid Diaphragms

Basic classification of diaphragms:

1. Flexible
2. Rigid

Forces to supporting walls are simple beam (tributary) reactions for a *flexible* diaphragm (Fig. 9.20a).

Forces to walls supporting a *rigid* diaphragm R are the sum of the forces due to direct shear proportional to the wall stiffness (as if P were applied at the center of rigidity) R_d and eccentric forces R_{ecc} due to the torsional moment $P(e)$ (Fig. 9.20b). These forces are discussed in more detail in Sec. 16.10.

Figure 9.20*a*   Distribution of lateral force in a flexible diaphragm.

Figure 9.20*b*   Distribution of lateral force in a rigid diaphragm.

NOTE: The force due to rotation adds to the direct shear force in the open front wall and subtracts from the force in the rear wall.

EXAMPLE 9.19   Rotation in an Open Sided Building

In buildings with no effective shearwall on one side, rotation of the diaphragm can be used to carry the torsional moment into the walls perpendicular to the applied lateral force (not allowed for buildings with concrete or masonry walls). Even in entirely wood-framed buildings, the code limits the depth of the diaphragm normal to the open side to 25 ft or two-thirds the diaphragm width, whichever is smaller. (For one-story buildings, the width is limited to 25 ft or the full diaphragm width, whichever is smaller.)

Rear wall must carry the entire longitudinal load:

$$R = wL$$

Force to transverse walls due to rotation:

$$\text{Torsion} = T = Re = \frac{wL^2}{2}$$

Resolve torsion into a couple separated by the distance between transverse walls:

$$F_1 = F_2 = \frac{T}{b} = \frac{wL^2}{2b}$$

For more information, see the APA publication *Plywood Diaphragms* Ref. 9.5.

Figure 9.21   Plan of building with open side (i.e., no effective shearwall along one side).

Example 9.19 illustrates the use of a rigid diaphragm analysis for a structure that has one wall completely open. The only way that this structure can resist lateral forces is by the rigid diaphragm acting in rotation to transfer forces to the back wall.

It should be noted that this is not necessarily a desirable design technique. In fact, the *code does not allow the consideration of rotation of a wood diaphragm in a building with concrete or masonry walls*. The required approach in these types of buildings, and the desirable approach in all cases, is to provide a shearwall along the open side of the building.

The question of rotation brings up an important point. The designer, insofar as practical, should strive for structural *symmetry*. In Fig. 9.21, it is quite unlikely that the design could be modified to have walls of exactly the same length at the front and back of the building. However, the design of a shearwall with an effective length sufficient to develop unit shears of a reasonable magnitude along the open side is recommended. The addition of this amount of symmetry will aid in reducing damage due to rotation.

Problems involving rotation contributed to a number of the failures in residential buildings in the 1971 San Fernando, 1989 Loma Prieta, and 1994 Northridge earthquakes. Recent research has investigated the behavior of this type of building, referred to as "open front" or "tuck-under parking" buildings, as well as approaches to retrofit of vulnerable existing buildings (Refs. 9.9, 9.11, and 9.12). For additional recommendations on the design of wood diaphragms, see Refs. 9.5, 9.6, and 9.8.

The rigid diaphragm analysis in Example 9.19 was a simple case with a straightforward solution. A flexible diaphragm analysis would currently be more common in practice. It is recommended that the reader analyze several example buildings using both the flexible and rigid diaphragm analyses to develop a feeling for how much the distribution of shear changes between the two approaches.

While the force distribution to shearwalls will be different in each building studied, the reader will be able to develop an understanding of what type of plan layout results in significant redistribution of forces to shearwalls between the two methods. The layouts resulting in significant differences generally have a substantial distance between their center of mass and center of rigidity. Where this occurs, it is recommended that the diaphragms and shearwalls be examined using a rigid diaphragm analysis. Another approach is to perform both flexible and rigid diaphragm analyses and design each element for the larger of the resulting forces. Rigid diaphragm analysis is discussed further in Chap. 16.

9.12   References

[9.1] American Concrete Institute (ACI), American Society of Civil Engineers (ASCE), and The Masonry Society (TMS), 2005. *Building Code Requirements for Masonry Structures* (ACI 530-05/ ASCE 5-05/TMS 402-05), ACI, Farmington Hills, MI; ASCE, Reston, VA, and TMS, Boulder, CO.

[9.2] American Forest and Paper Association (AF&PA), 2005. *National Design Specification for Wood Construction,* ANSI/AF&PA NDS-2005, AF&PA, Washington DC.

[9.3] American Forest and Paper Association (AF&PA), 2005. *Special Design Provisions for Wind and Seismic* (SDPWS) Supplement to the NDS, 2005 ed., AF&PA, Washington DC.

[9.4] American Society of Civil Engineers (ASCE), 2005. *Minimum Design Loads for Buildings and Other Structures (ASCE 7-05),* ASCE, Reston, VA.

[9.5] APA—The Engineered Wood Association, Rev. 2000. Research Report *138—Plywood Diaphragms,* APA, Tacoma, WA.

[9.6] APA—The Engineered Wood Association, 2004. *Design / Construction Guide—Diaphragms and Shear Walls,* APA, Tacoma, WA.

[9.7] Applied Technology Council (ATC), 1976. A Methodology for Seismic Design and Construction of Single Family Dwellings, ATC, Redwood City, CA.

[9.8] Applied Technology Council (ATC), 1981. *Guidelines for Design of Horizontal Wood Diaphragms* (ATC-7), ATC, Redwood City, CA.

[9.9] Cobeen, K., J. Russell, and J.D. Dolan, 2004. *Recommendations for Earthquake Resistance in the Design and Construction of Woodframe Buildings* (CUREE W-30), Consortium of Universities for Research in Earthquake Engineering, Richmond, CA.

[9.10] International Codes Council (ICC), 2006. *International Building Code*, 2006 ed., ICC, Falls Church, VA.

[9.11] International Code Council (ICC), 2006. *International Existing Building Code* (IEBC), 2006 ed., ICC, Falls Church, VA.

[9.12] Mosalam, K., et al., 2002. *Seismic Evaluation of an Asymmetric Three-Story Woodframe Building* (CUREE W-19), Consortium of Universities for Research in Earthquake Engineering, Richmond, CA.

9.13 Problems

Note to instructors: Many of the following problems can be solved using either ASD or LRFD methods. Where this is true, both solutions have generally been provided in the solution set. It is suggested that students be advised whether to use ASD, LRFD, or their choice of the two for homework solutions.

For simplicity, the following diaphragm design problems all use wood structural panel sheathing. Sheathing nailing and tabulated unit shears are to be taken from IBC or SDPWS. Dry-service conditions, normal temperatures, and bending about the strong axis apply unless otherwise noted. The loads given in a problem are to be used with the IBC ASD or LRFD *basic load combinations*.

Some problems require the use of a computer equation solving program. Problems that are solved on a computer can be saved and used as a *template* for other similar problems. Templates can have many degrees of sophistication. Initially, a template may only be a hand (i.e., calculator) solution worked on a spreadsheet. In a simple template of this nature, the user will be required to provide many of the lookup functions for such items as

Tabulated diaphragm unit shears (lb/ft)

Diaphragm nailing specifications

Tabulated design values for wood members

As the user gains experience with computer solutions, a template can be expanded to perform many of the lookup and decision-making functions that were previously done manually.

Advanced computer programming skills are not required to create effective templates. Valuable templates can be created by designers who normally do only hand solutions. However, some programming techniques are helpful in automating *lookup* and *decision-making* steps.

The first requirement is that a computer template operate correctly (i.e., calculate correct values). Another major consideration is that the input and output should be structured in an orderly manner. A sufficient number of intermediate answers are to be displayed and labeled so that the solution can be verified by hand.

9.1 *Given:* The single-story commercial building in Fig. 9.A. The wood structural panel is $^{15}/_{32}$-in. thick 32/16 span-rated sheathing unblocked. Nails are 8d common. Critical lateral forces (wind) are

$$w_T = 320 \text{ lb/ft}$$
$$w_L = 320 \text{ lb/ft}$$

Vertical loads are

$$\text{Roof Dead Load D} = 10 \text{ psf}$$

$$\text{Snow S} = 30 \text{ psf}$$

Find: *a.* For the transverse direction:
1. Roof diaphragm unit shear
2. Panel load case (state criteria)
3. Required nailing and tabulated ASD allowable shear (from IBC) or tabulated nominal shear (from SDPWS) adjusted to ASD capacity per SDPWS.
4. Unit shear due to LRFD load combinations, required nailing and tabulated nominal shear (from SDPWS) adjusted to LRFD capacity per SDPWS.

b. For the longitudinal direction:
1. Roof diaphragm unit shear
2. Panel load case (state criteria)
3. Required nailing and tabulated ASD allowable shear (from IBC) or tabulated nominal shear (from SDPWS) adjusted to ASD capacity per SDPWS.
4. Unit shear due to LRFD load combinations, required nailing and tabulated nominal shear (from SDPWS) adjusted to LRFD capacity per SDPWS.

c. If blocking is not required, do the edges of the panel need some other type of edge support?

9.2 *Given:* The single-story commercial building in Fig. 9.A. The wood structural panel is $^{15}\!/_{32}$-in. thick 32/16 span-rated STRUCTURAL I sheathing. Critical lateral forces (strength level seismic) are

ROOF FRAMING PLAN

Figure 9.A

$$w_u = w_T = 320 \text{ lb/ft}$$
$$w_u = w_L = 570 \text{ lb/ft}$$

Find: a. Roof diaphragm unit shear.
        b. Is the IBC span-to-width ratio satisfied? Show criteria.
        c. Is blocking required for lateral forces? If it is required, calculate at what point it can be omitted. Show all criteria such as diaphragm load case, tabulated unit shears, and nailing requirements.
        d. The maximum adjusted chord force.

9.3 Use the hand solution to Prob. 9.1 or 9.2 as a guide to develop a computer *template* to solve similar problems.
    a. Consider only the specific criteria given in the problems.
    b. Expand the template to handle any diaphragm unit shear covered in the IBC or SDPWS. In other words, input the values from the IBC or SDPWS into a computer template, and have the template return the required nail specification and blocking requirements. Macros may be used as part of the template.

9.4 *Given:* The single-story wood-frame warehouse in Fig. 9.B. Roof Sheathing of $\frac{3}{8}$-in. 24/0 span-rated STRUCTURAL I sheathing is adequate for vertical "sheathing" loads. Strength level seismic lateral forces to roof diaphragm are

$$w_u = w_T = 320 \text{ lb/ft}$$
$$w_u = w_L = 570 \text{ lb/ft}$$

Find: a. Design the roof diaphragm, considering lateral forces in both directions. Show all criteria including design shears, load cases, tabulated shears, and nailing requirements. If blocking is required, determine at what points it can be omitted.

ROOF FRAMING PLAN

Figure 9.B

b. Calculate the maximum chord forces for both lateral forces. Also calculate the chord forces at the ends of all headers in the exterior walls.

c. Plot the distribution of the strut force for each of the walls with openings. Compare the strut forces and chord forces to determine the critical loading.

9.5 Use the hand solution to Prob. 9.4 as a guide to expand the computer template from Prob. 9.3. The expanded template should evaluate the diaphragm *chord forces* and *collector* forces. In addition to maximum chord forces, the template should allow the user to select various arbitrary locations for the evaluation of chord and strut forces.

9.6 Repeat Prob. 9.4 except that the panel is $^{15}\!/_{32}$-in. 32/16 span-rated STRUCTURAL I sheathing and the lateral forces are

$$w_u = w_T = 630 \text{ lb/ft}$$
$$w_u = w_L = 790 \text{ lb/ft}$$

9.7 *Given:* The building in Fig. 9.A and the strength level seismic lateral forces are

$$w_u = w_T = 320 \text{ lb/ft}$$
$$w_u = w_L = 570 \text{ lb/ft}$$

Find: a. The maximum adjusted chord forces. Also determine the adjusted chord forces at each end of the wall openings.

b. Plot the distribution of the strut forces for the two walls that have openings. Compare the magnitudes of the adjusted strut forces with the chord forces.

NOTE: A computer template may be used to solve this problem. Verify results with hand solution.

9.8 *Given:* Assume that the framing plan in Fig. 9.A is to be used for the second-floor framing of a two-story retail sales building. Lateral wind forces are

$$w_T = 450 \text{ lb/ft}$$
$$w_L = 450 \text{ lb/ft}$$

Floor live load L is to be in accordance with the IBC. Panel is STRUCTURAL I.

Find: a. Required span-rating and panel thickness.

b. Required panel nailing. Check unit shears in both transverse and longitudinal directions. Show all criteria in design calculations.

c. Calculate the maximum chord force and drag force in the building. Assume that three 2 × 6s of No. 1 DF-L are used as the top plate in the building. Check the maximum stress in this member if it serves as the chord and strut. At any point in the plate, two 2 × 6s are continuous (i.e., only one member is spliced at a given location). A single row of 1-in.-diameter bolts is used for the connections. $C_M = 1.0$, and $C_t = 1.0$.

        d. Assuming splices in the chords are spaced every 20 ft, determine the deflection of the diaphragm. Assume the structure is constructed and maintained in the dry condition.

NOTE: Design of the bolts is not part of this problem.

9.9   The one-story building in Fig. 9.C has the following loads:

       Roof dead load D = 12 psf
       Wall dead load D = 150 pcf (normal-weight concrete)
       Wind = 20 psf
       Strength level seismic force = 0.200W
       Tributary wall height to roof diaphragm = 10 ft
       Roof live load L_r in accordance with IBC Table 1607.1 or ASCE 7 Table 4-1

Diaphragm chord consists of two No.5 continuous horizontal bars in the wall at the diaphragm level.

Find: *a.* Required panel span-rating and thickness.
        b. Critical lateral forces.
        c. Required plywood nailing based on the diaphragm unit shears. Show all design criteria. Consider forces in both directions.
        d. Determine the deflection of the diaphragm. Assume the structure is constructed and maintained in the dry condition. Assume $^{15}/_{32}$ 3-ply plywood rated sheathing with 32/16 span rating.
        e. Check the stress in the chord reinforcing steel. Allow. f_s = 20 ksi

ROOF FRAMING PLAN

Figure 9.C

9.10 *Given:* The one-story building in Fig. 9.D. Roof sheathing is $^{15}\!/_{32}$-in. 32/26 span-rated panels spanning between roof joists spaced 24 in. o.c. Nails are 10d common. For the transverse strength level seismic force, the panel layout is load case 1.

PLAN

Figure 9.D

Find: a. Design the roof diaphragm, assuming that only the exterior walls are effective in resisting the lateral force. Show all design criteria including span-width ratio, nailing, and blocking requirements. Omit blocking where possible. Do not consider deflection of the diaphragm.

b. Redesign the roof diaphragm, assuming that the interior *and* exterior walls are effective shearwalls. Show all criteria, except deflection.

c. Compare the maximum chord forces for *a* and *b*.

d. Plot the collector force diagram for the interior shearwall.

9.11 *Given:* The header/strut over the opening in the interior shearwall in Fig. 9.D is a 4 × 14 No. 1 DF-L. $C_M = 1.0$ and $C_t = 1.0$. The vertical load is

$$\text{Dead load } w_D = 100 \text{ lb/ft}$$
$$\text{Roof live load } w_{Lr} = 160 \text{ lb/ft}$$

The axial force is to be determined from a plot of the drag strut force.

Find: Analyze the member for combined stresses. Full lateral support is provided for the weak axis by the roof framing.

10

Shearwalls

10.1 Introduction

Shearwalls comprise the vertical elements in the lateral-force-resisting system (LFRS). They support the diaphragm and transfer the lateral forces down into the foundation. The procedures for calculating the forces to the shearwalls are covered in Chaps. 3 and 9.

Wood diaphragms are often used in buildings with masonry or concrete shearwalls as well as buildings with wood-frame shearwalls. However, the design of masonry and concrete shearwalls is beyond the scope of this book, and this chapter deals with the design of wood-frame shearwalls only.

A number of sheathing materials can be used to develop shearwall action in a wood-frame wall. These include

1. Wood structural panels [e.g., plywood and oriented strand board (OSB)]

2. Gypsum wallboard (drywall)

3. Interior and exterior plaster (stucco)

4. Fiberboard (including fiber-cement panels)

5. Lumber sheathing (diagonal or horizontal sheathing)

There may be other acceptable materials, but these are representative. When the design forces (wall shears) are relatively small, the normal wall covering material may be adequate to develop shearwall action. However, when the unit shears become large, it is necessary to design special sheathing and nailing to develop the required capacity. Wood structural panels provide greater shearwall capacity than the other sheathing types listed above.

Many structures with wood-frame walls have a number of interior walls. This is especially true for residential buildings. It is a common design practice to choose some portion of these walls to design as shearwalls resisting lateral forces. Generally, the longer walls and the walls that are stacked from story to story are selected first, and other walls are added if necessary to meet code requirements.

The strength and stiffness provided by other "partition" walls is generally neglected in lateral force design, and shearwall fastening and connectors are not provided. This is viewed as a conservative approach to providing code required strength in new building design. A more detailed look at contributions of non-shearwalls is sometimes appropriate in a more detailed structural evaluation. This is an advanced topic that is beyond the scope of this book.

10.2   Basic Shearwall Action

Essentially a shearwall cantilevers from the foundation and is subjected to one or more lateral forces. See Example 10.1. As the name implies, the basic form of resistance is that of a shear element (Fig. 3.5c). The concept of "shearwall segments" in a wood-frame wall leads to the assumption that the lateral force is generally distributed uniformly throughout the total length of all segments (Sec. 9.6). Exceptions will be discussed in Sec. 10.6.

EXAMPLE 10.1   Cantilever Action of a Shearwall

SHEARWALL – FBD, SHEAR AND MOMENT

Figure 10.1   Deflected shape of one-story shearwall. Also shown are cantilever loading diagram (schematic), and shear and moment diagrams.

Figure 10.1 shows a typical shearwall segment for a one-story building. If there are additional stories, there will be additional forces applied to the shearwall at the diaphragm levels. If the lateral force is a seismic force, there will also be an inertial force generated by the mass of the shearwall (Chap. 3).

A variety of sheathing materials can be used to develop shearwall action for resisting lateral forces. The IBC (Ref. 10.10) sets limits on the height-to-width ratios h/b for shearwall segments in order to develop effective shear (deep-beam) resistance. Limits on h/b vary with different sheathing materials. See IBC Table 2305.3.4.

A number of items should be considered in the design of a shearwall. These include

1. Sheathing thickness
2. Shearwall nailing
3. Chord design (tension and compression)
4. Collector or strut design (tension and compression)
5. Anchorage requirements (hold-downs and shear)
6. Shear panel proportions
7. Deflection

These factors are essentially the same as those required for a diaphragm, but there are some differences in procedures.

The *sheathing thickness* depends on the type of material used in the wall construction. Loads normal to the surface and the spacing of studs in the wall may determine the thickness of the sheathing, but the unit shear often controls the thickness. In other cases, the sheathing thickness may be governed by the required fire rating of a wall. *Shearwall nailing* or stapling requirements are a function of the unit shear in the wall and the materials of construction.

As with a diaphragm, the *chords* are designed to carry the moment, and are required at both ends of a shearwall. The *collector* (or *strut*) for a shearwall is the same member as the collector for the diaphragm. It is the connecting link between the shearwall and the diaphragm for loads parallel to the shearwall. Design forces for the collector are covered in detail in Chap. 9.

The *proportions* of a shearwall are measured by its height-to-width ratio h/b. In buildings with two or more stories, the height of the shearwall is the vertical clear distance between diaphragms. The IBC sets upper limits on the height-to-width ratio for various wall sheathing materials used as shearwall segments. Specific values of these limits will be covered in the sections dealing with the various construction materials. Segments are considered as effective shearwall elements if these limits are not exceeded. Shearwalls which satisfy the height-to-width ratio criteria are considered by many designers to also have proportions that will not allow the segment to deflect excessively under load (this is similar to the span-to-width ratios for diaphragms—Sec. 9.8). Nonetheless, IBC provisions for calculating shearwall deflections and related limitations on story drift are code requirements. Calculation of shearwall deflections is addressed in Sec. 10.11.

10.3 Shearwalls Using Wood Structural Panels

Perhaps the most common wood-frame shearwall is the type that uses wood structural panel sheathing or plywood siding. Today, however, other wood panel products provide additional choices for shearwall construction. Allowable unit shears for various types of wood structural panel shearwalls are given in both IBC Table 2306.4.1 (reproduced in Appendix C of this book) and Special Design Provisions for Wind and Seismic (SDPWS) (Ref. 10.3) Tables 4.3A and A4.3A. Included in these tables are design values for shearwalls constructed with STRUCTURAL I sheathings and plywood panel siding complying with PS 1, PS 2, NER-108 (Refs. 10.13 to 10.15). See Secs. 8.7 and 8.8 for a description of these panel products. The IBC table includes a number of footnotes that are not included in the SDPWS table. The user of the IBC table must adhere to the information in these footnotes.

Tabulated unit shears for shearwalls assume that the framing members are Douglas Fir-Larch or Southern Pine. If the shearwall nailing penetrates into framing members of other species of wood, an adjustment that accounts for a reduced nail capacity in lower density species is required. Reduction factors for other species of wood are given in a footnote to the unit shear tables.

As with diaphragms, wood structural panel shearwalls resisting wind loads are permitted to use unit shear capacities 40 percent higher than for seismic loads. The allowable unit shears in IBC Table 2306.4.1 are permitted to be increased 40 percent for wind design per IBC Sec. 2306.4.1. As was done for diaphragm tables, the SDPWS separately tabulates unit shears for seismic and wind design, with the wind values including the 40 percent increase.

The IBC and SDPWS tables are for short-term (wind and seismic) forces. If a shearwall is used to support loads of longer duration, the tabulated unit shears will have to be reduced (see Sec. 4.15 of this book). For example, if the load under consideration were determined to have a 10-year cumulative duration ($C_D = 1.0$), the tabulated allowable shears would be reduced by a factor of $1/1.6$. In addition to adjustment for load duration, this means that the 40 percent increase for wind loads cannot be included. In order to have similar design values when using the SDPWS, it is important that the tabulated seismic values be adjusted for longer load duration, not the higher wind values.

Wood structural panel sheathing is often provided on only one side of a wall, and then finish materials are applied to both sides of the wall. If the design unit shears are higher than can be carried with a single layer, another layer of sheathing can be installed on the other side of the wall. See Fig. 10.2*a*. If the same nail size and spacing is used, the addition of a second layer of sheathing doubles the shear capacity of the wall. The capacity of the wall covering (gypsum

FINISH WALL COVERING –
SHEAR CAPACITY NEGLECTED   SHEARWALL NAILING

STUD

WOOD STRUCTURAL
PANEL SHEATHING–
ALLOWABLE UNIT SHEARS
FROM IBC TABLE 2306.4.1

DOUBLE SHEATHING WITH
EQUAL NAILING GIVES DOUBLE
THE SHEAR CAPACITY FROM
IBC TABLE 2306.4.1

Figure 10.2*a*   Wood structural panel sheathing.

Figure 10.2b Plywood siding.

wallboard, plaster, stucco, and the like) is generally not additive to the shear capacity of the wood structural panel sheathing. There is one exception, however. For wind design only, both the IBC and SDPWS recognize the addition of gypsum wallboard interior finish and wood structural panel exterior sheathing (i.e., on opposite faces of the wall).

Plywood siding and certain code-recognized proprietary structural wood panel sidings can also be used to resist shearwall forces. This siding can be nailed directly to the studs, or it may be installed over a layer of $\frac{5}{8}$-in. gypsum sheathing. See Fig. 10.2b. A wall construction that gives a 1-hour fire rating uses 2 × 4 studs at 16-in. o.c. with $\frac{5}{8}$-in. Type X gypsum sheathing and minimum $\frac{3}{8}$-in. plywood siding on the outside, and $\frac{5}{8}$ in. Type X gypsum wallboard on the interior. When grooved plywood panels are used, the thickness to be considered in the shearwall design is the net thickness where the plywood is nailed (i.e., the thickness at groove or shiplap edge of the panel).

In a few instances it is theoretically possible to nail certain grooved plywood siding panels through the full thickness of the panel (i.e., away from the grooves or edges). In these cases, thicker wall studs are necessary at the boundaries of the panels to provide sufficient panel and framing edge nailing distance. Sample calculations for unit shear values based upon the plywood strength and fastener capacity are given in Ref. 10.19.

The following discussion in Example 10.2 provides a more detailed discussion of wood structural panel shearwalls.

EXAMPLE 10.2   Wood Structural Panel Shearwalls

Tabulated unit shears for wood structural panel sheathing and plywood siding are given in the IBC and SDPWS. Both address sheathing or siding attached directly to framing. The IBC also addresses sheathing or siding applied over $\frac{1}{2}$- or $\frac{5}{8}$-in. gypsum board.

Tabulated shear values are given for panels nailed with *common* or *galvanized box nails* and *galvanized casing nails*. Finishing nails are not allowed for shearwalls.

It will be helpful at this point to review the organization of the IBC and SDPWS tables. The IBC table starts with sheathing type and thickness, minimum fastener penetration into framing, and minimum fastener sizes on the left hand side. As discussed in Sec. 9.3 for diaphragms, detailed information on minimum fastener size has been added to the IBC table in an effort to reduce the occurrence of construction with smaller than required fasteners. The center portion of the table provides ASD allowable shears for wood structural panel sheathing applied directly over the wall framing, as is most commonly used. The right-hand portion of the table provides revised fastener requirements and tabulated shears where gypsum wallboard is applied to the bare wall framing and the wood structural panel is installed over the wallboard. This construction approach is less commonly used, but sometimes required in order to meet code fire resistance requirements. The SDPWS does not include provisions for installing wood structural panel sheathing over gypsum wallboard.

SDPWS Tables include similar information on sheathing and fastening in the left-hand side of the table. The middle portion provides tabulated nominal unit shear capacities for seismic loads, and the left-hand portion the 40 percent higher values for wind loads. Table 4.3A is indicated as being applicable to wood structural panels (the OSB), excluding plywood, while Table A4.3A is applicable to plywood. For a given sheathing and fastening, the tabulated nominal unit shear capacities are the same; however, G_a values used for shearwall deflection equations differ between Tables 4.3A and A4.3A. Footnote 1 directs the user to SDPWS Sec. 4.3.3 for adjustment of the tabulated nominal unit shears to ASD or LRFD shear capacities.

Although galvanized box nails can be used to install plywood siding, the size of the nail head may be objectionable from an appearance standpoint. Therefore, unit shears are tabulated for plywood siding installed with *galvanized casing nails*. Casing nails have a head diameter that is smaller than that of common or box nails but larger than that of finishing nails. Because of the tendency of the nail head to pull through the siding, the tabulated unit shears are considerably smaller when casing nails are used instead of galvanized box nails. The various types of nails are described more fully in Chap. 12.

Tabulated values apply to shear panels installed either horizontally or vertically. In the tables it is also assumed that all panel edges are supported by and edge nailed to either the wall studs, plates, or blocking. Footnotes give adjustments to the tabulated shears where appropriate. The tables also give special requirements designed to avoid the splitting of lumber framing when closely spaced nails are used. For example, when the nail spacing is 2-in. o.c., 3-in. nominal (or wider) framing is required at adjoining panel edges.

The additional footnotes found in IBC Table 2306.4.1 require staggering of edge nailing to different framing members when shearwalls are sheathed on both faces; specify 3 × studs at abutting panel edges when the unit shear is above 350 plf; reference requirements for sill plate size, and anchorage and steel plate washers; and they give specifics for nail galvanizing and staple crown width.

In Sec. 9.7, it was noted that there are three methods recognized by the IBC and SDPWS for design of wood light frame shearwalls. The first, the segmented shearwall method, is the most common; however, use of the other two methods is increasing. A brief introduction of the three methods will be provided at this point, and further details of the second and third method will be provided in Sec. 10.7.

The *segmented shearwall method* (Fig. 10.3, top) assumes that each shearwall is a cantilevered element, fixed at the base and free to deflect sideways at the top. This is the behavior that was illustrated in Fig. 10.1. When using this method, sections of full height wall between doors and windows are designated as the

Figure 10.3   Three methods for design of wood light frame shearwalls.

shearwall segments. Each segment is modeled as an independent cantilever, and each segment is designed for the moment at the base, resolved into tension and compression reactions of the shearwall chord members.

The second method (Fig. 10.3, center) involves design of the entire wall, including the opening, as a shearwall. The IBC uses the term *design for force transfer around openings* for this method. In this approach, the sheathed wall portions above and below openings act as coupling beams, transferring vertical shear across the opening. A rational analysis of the shearwall and boundary members is required. Special detailing is required at all sides of the opening and extending beyond to develop the opening boundary members. The detailing generally involves increased sheathing nailing, and added blocking and strapping. Tie-down devices anchoring the shearwall chord members are generally only provided at each end of the wall, resulting in fewer anchorage devices (tie-downs) than the segmented shearwall.

The third method (Fig. 10.3, bottom) is called the *perforated shearwall method*. This is a semi-empirical method developed from shearwall testing. Like the second method, the perforated shearwall utilizes the full wall, including wall openings. Unlike the second method, a somewhat lower level of detailing is required. With lower detailing requirements than the first or second method, the wall capacity and wall stiffness are lower. The perforated shearwall method uses a capacity adjustment factor to recognize the lower capacity and stiffness.

The following discussion will focus on the first method—the *segmented shearwall method*. Detailed discussion of the second and third methods is provided in Sec. 10.7.

IBC Table 2305.3.4 sets an upper limit on h/b of 3.5 for wood structural panel shearwalls resisting wind forces and 2.0 for seismic forces. As per the footnote to IBC Table 2305.3.4, the aspect ratio h/b for seismic forces may be increased up to 3.5 provided the tabulated allowable unit shears are reduced by the multiplier $2b/h$. Regardless, these are the largest height-to-width ratios permitted for wood-frame shearwalls. For other sheathing materials, the maximum value for h/b is as low as 1.0 (Sec. 10.4).

These limits can be important in the selection of sheathing material for a shearwall. See Example 10.3. The larger the allowable height-to-width ratio h/b, the shorter the permitted width of a shearwall for a given wall height. Thus, if the width available for a shearwall segment is restricted, the use of a wood structural panel shearwall may be required in order to satisfy the design unit shear within height-to-width limitations (Fig. 10.4).

EXAMPLE 10.3   Height-to-Width Ratios for Wood-Frame Shearwalls

Shearwalls with wood structural panels considering wind have an upper limit on h/b of 3.5. Thus, for a 10-ft-high shearwall, the minimum width is 2.86 ft (Segment 1). Shearwalls with wood structural panels for seismic forces have an upper limit on h/b of 2.0 without reduction in tabulated values. Here a 10-ft high shearwall must have a width of at least 5.0 ft (Segment 2).

Some common wall coverings can be used in shearwall design. For example, gypsum wallboard and plaster walls have a limit of h/b of 2.0 with blocking provided at all panel edges. If blocking is not provided, the limit on h/b reduces to 1.5. Unit shear values for gypsum wallboard and plaster are provided in IBC Table 2306.4.5. See also Sec. 10.4.

$$\left(\frac{h}{b}\right)_1 = \frac{10}{2.86} = 3.50 \qquad \left(\frac{h}{b}\right)_2 = \frac{10}{5.0} = 2.0$$

Figure 10.4 Minimum shearwall widths for different height-to-width ratios. Shearwall segments with smaller widths are ignored in design.

The APA Design/Construction Guide (Ref. 10.4) provides additional discussion on topics related to wood structural panel shear walls.

10.4 Other Sheathing Materials

It was mentioned at the beginning of this chapter that wall materials other than wood structural panels can be used to develop shearwall action. If the design shears are high, wood structural panel sheathing may be required. However, if the design forces are relatively low, other wall coverings may have sufficient strength to carry the shears directly.

The IBC and ASCE 7 have made several significant changes to seismic bracing of light-frame buildings that use sheathing materials other than wood structural panels. First, ASCE 7 Table 12.2-1 assigns a much lower R-factor to light-frame wall systems when sheathed with other than wood structural panel sheathing. The R-factors are 2 for a bearing-wall system and 2.5 for a building-frame system; as compared to 6.5 and 7, respectively, using wood structural panel sheathing. Second, ASCE 7 Table 12.2-1 limits this system to a maximum building height of 35 ft in Seismic Design Category (SDC) D, and prohibits the system in SDCs E and F. Finally, IBC and SDPWS prohibit the use of fiberboard and particleboard sheathing in SDC D and higher.

For wood light-frame systems, common bracing materials are stucco (exterior plaster), gypsum lath and plaster (interior plaster), gypsum wallboard (drywall), particleboard, fiberboard, and diagonal lumber sheathing. Additional proprietary bracing materials are becoming commonly used; however, rather than being incorporated directly into the building code, proprietary bracing materials are evaluated on a case-by-case basis by the evaluation services arm of the International Code Council (ICC).

Allowable unit shears, construction requirements, and other provisions are found in IBC Sec. 2306.4. Lumber sheathed shearwalls, addressed in Sec. 2306.4.2, refer to the IBC diaphragm section for allowable unit shears and construction requirements. No SDC limits are placed on lumber sheathed shearwalls, other than the ASCE 7 Table 12.2-1 system limits, which prohibit their use in SDC E and F. A shearwall aspect ratio maximum limit of 2 can be found in IBC Table 2305.3.4.

In the past, lumber sheathing was used extensively for wood-frame shearwalls. Lumber sheathing can be applied horizontally (known as *straight sheathing*), but is relatively weak and flexible. Diagonally applied sheathing is considerably stronger and stiffer because of its triangulated (truss) action. See Fig. 10.5. Shearwalls using lumber sheathing have largely been replaced with wood structural panel shearwalls. Because of their relatively limited use, the design of lumber-sheathed walls is not covered in this text.

Particleboard and fiberboard shearwalls are addressed in IBC Secs. 2306.4.3 and 2306.4.4, respectively, with allowable unit shears given in Tables 2306.4.3 and 2306.4.4. Use of both materials for shearwalls is prohibited in SDC D, E, and F. For particleboard shearwalls, IBC Table 2305.3.4 indicates that the aspect ratio limits are the same as for wood structural panel shearwalls. For fiberboard shearwalls, an aspect ratio maximum limit of 1.5 is applicable. Fiberboard wall sheathing is often also used for insulation.

IBC Sec. 2306.4.5 addresses the use of stucco, interior plaster over gypsum lath, and gypsum wallboard (drywall). Allowable unit shears, given in Table 2306.4.5, range from 60 to 250 plf. Both blocked and unblocked sheathing panel edges are addressed. Recent cyclic testing of shearwall components has included gypsum wallboard and stucco, alone and in combination with other materials. See Refs. 10.9, 10.12, and 10.16.

Figure 10.5 Lumber-sheathed shearwalls.

The SDPWS provides design values for bracing materials other than wood structural panel in both Tables 4.3A and 4.3B. Table 4.3A, along with wood structural panel sheathing and siding, includes both fiberboard and particleboard. The other materials are covered in Table 4.3B. Aspect ratios are given in Table 4.3.4. Construction requirements and SDC limits for these materials are provided in Sec. 4.3.7.

10.5 Shearwall Chord Members

The vertical members at each end of a segmented shearwall are the chords. See Example 10.4. Like a diaphragm, the chords are designed to carry flexure, while the sheathing web carries shear. The overturning moment at the base of the segmented shearwall is resolved into a couple that creates axial tension and compression forces in the end posts or studs that serve as chord members. See Fig. 10.6 free body diagrams (FBDs). When considering only the horizontal component of wind or seismic forces, the tension and compression forces will be equal.

EXAMPLE 10.4   Shearwall Chord Members

For a wood structural panel segmented shearwall with an aspect ratio (h/b) not greater than 2, the unit shear is generally assumed to be uniform throughout. For a lateral wind force, the unit shear is the same at the top and bottom of the two shearwall segments. The gross overturning moment (OM) is calculated at the base of each shearwall segment:

$$M_1 = v \times b_1 \times h$$

$$M_2 = v \times b_2 \times h$$

Chord forces due to the horizontal component of E or W are calculated by dividing gross OMs by the lengths of the shearwall segments:

$$T_1 = C_1 = \frac{M_1}{b_1} = vh$$

$$T_2 = C_2 = \frac{M_2}{b_2} = vh$$

Thus, neglecting the effects of dead load resistance, the chord forces for the two segments are equal:

$$T_1 = C_1 = T_2 = C_2 = vh$$

A somewhat modified (but similar) procedure is used to calculate the gross chord forces when additional lateral forces are involved, such as a lateral force due to the mass of the shearwall itself (Fig. 10.7d).

Chord forces for the two segments are *not equal* when combined with other loads in load combinations.

Shearwall chord forces can be analyzed by using the expressions developed in this example or by using the basic statics approach described in Example 2.9 (Sec. 2.10). The chord forces should be in agreement regardless of the method used.

SHEARWALL WITH TWO PANELS

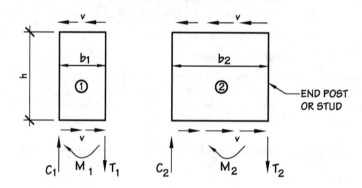

FBDs OF SHEAR PANELS

Figure 10.6 Elevation view of wall with two shearwall segments. Below are FBDs of shearwall segments. The moment at the base of a segment is resolved into a *couple*. The couple forms two concentrated chord forces: one in tension and the other in compression.

Some engineers will include overturning resistance due to dead load in the FBD, while other engineers will neglect the dead load resistance. Neglecting the dead load results in a simpler calculation, and corresponds well with testing indicating that shearwall capacity is influenced little by resisting dead load. Including the dead load results in a reduced uplift reaction.

The force in the tension chord can be conservatively estimated with dead load neglected. From Sec. 2.17 of this book, uplift due to seismic forces will most commonly be controlled by ASD load combination 8: $0.6D + 0.7E$. With the dead load neglected this simplifies to $0.7E$ and includes only tension forces due to shearwall overturning.

The force in the compression chord, however, can be underestimated when dead load is neglected. Care must be taken to provide for the downward compression force for lateral and gravity loads using Sec. 2.17 load combinations.

Where dead load resistance is included in seismic chord forces, it is also necessary to include the vertical component of E: $0.2S_{DS}D$. The vertical component will act upward for tension, increasing tension forces, and act downward for compression, increasing compression forces. Per ASCE 7 Sec. 12.4.2.2, the vertical component can only be neglected when S_{DS} is less than or equal to 0.125g, or when considering the foundation to soil interface.

If the chord members bear on the sill plate, compression perpendicular to the grain of the wall plates may be a design consideration. Lateral forces to a shearwall may come from either direction, and chords must be designed for both tension and compression.

Tests conducted by APA demonstrated that compression perpendicular to grain on wall plates under a shearwall chord can, in fact, be a limiting design consideration. The tests were conducted on heavily loaded shearwalls that had sheathing on both sides of the wall. See Refs. 10.17, 10.18, and 10.19 for details.

The usual assumption is that the shear is uniform throughout the width of a segmented wood-frame shearwall. This common assumption is used throughout this book. However, one reference indicates that a nonuniform distribution of shear may develop depending on how shearwall overturning is handled. See Ref. 10.6 for details about this alternative approach.

In order to ensure that the segments in a shearwall function as assumed (i.e., separate segments of equal height); double full-height studs or posts should be used for the chords at the ends of the shearwall segments (Example 9.9 in Sec. 9.6). These tend to emphasize the vertical continuity of the segments. Recent testing (Ref. 10.8) has indicated that the sheathing and nailing of wall areas, neglected in the structural calculations (above and below windows and above doors), can significantly reduce wall deflection and anchorage forces. Where practical, it is recommended that these areas be sheathed and fastened in a manner similar to the portions designated as shearwalls.

10.6   Design Problem: Shearwall

In Example 10.5, the exterior transverse walls of a rectangular building are designed for a seismic force. The building meets the conditions of many "typical" buildings as outlined in Example 3.5 in Sec. 3.4. The *strength-level* seismic base shear coefficient for a building with wood bearing walls and wood structural panel sheathing is assumed as 0.200.

There is a roof overhang on the left end of the building that develops a smaller, uniformly distributed, lateral seismic force than the main portion of the roof diaphragm. The uniform seismic force for the overhang is smaller because it only includes the weight of the roof assembly, and not the building walls. The reaction of the left shearwall is made up of the cantilevered roof overhang force plus the tributary force of the main roof diaphragm.

The total shear at the midheight of the transverse walls is made up of the tributary roof diaphragm reactions plus the seismic force due to the dead load of the top half of the shearwall.

In this example one wood structural panel shearwall is used for the left wall. In the right wall, two shearwall segments are designed using the same type of sheathing as the left wall, but the nailing is increased to meet the higher shear requirements.

The maximum chord force occurs in the walls with the higher unit shears. Calculations indicate that one 2 × 8 stud is sufficient to resist the tension chord force, but two 2 × 8 studs or a 4 × 8 post are required to carry the compression chord force. In order to function together, the two studs must be effectively connected together and attached to the sheathing.

EXAMPLE 10.5   Shearwall Design

Design the two transverse shearwalls in the building shown in Fig. 10.7a. Studs are 2 × 8s of No. 2 Douglas Fir-Larch (DF-L), and walls are stucco over wood structural panels. The wall dead load D is 20 psf. The strength-level seismic force coefficient is 0.200g. The diaphragm uniform forces are 364 plf and 280 plf for the main diaphragm and overhang, respectively. MC ≤ 19 percent, $C_M = 1.0$, $C_t = 1.0$.

Figure 10.7a   Plan view of building showing *strength-level* lateral seismic force to the diaphragm.

Diaphragm Reactions

Strength-level forces to shearwalls from the diaphragm are calculated on a tributary-width basis.

Wall 1: R_u = reaction from overhang + reaction from main roof

$$= (280 \times 8) + \left(364 \times \frac{100}{2}\right) = 20.44 \text{ k}$$

Wall 2: $R_u = 364 \times \dfrac{100}{2} = 18.2 \text{ k}$

Shearwall 1

ELEVATION WALL 1

Figure 10.7*b* Elevation of left transverse wall (shearwall 1). Seismic force shown is at *strength level*.

Calculate wall seismic force, using the dead load of the *top half* of the wall (Chap. 3). Using the top half of the wall follows dynamic theory (Fig. 2.13*b*). Some designers use the *entire* wall dead load (conservative). See Fig. 10.7*b*.

$$R_{1u} = 0.200\text{W} = 0.200(20 \text{ psf} \times 8 \times 60) = 1.92 \text{ k}$$

The total strength-level design seismic shear force is

$$V_u = R_u + R_{1u} = 20.44 + 1.92 = 22.36 \text{ k}$$

$$v_u = \frac{V_u}{b} = \frac{22,360}{60} = 373 \text{ lb/ft}$$

As was done for diaphragms in Chap. 9, the shearwall shear loads need to be used with the ASCE 7 earthquake load equations, and in ASD or LRFD load combinations. This will be illustrated using ASD load combinations. For shear forces in the plane of the wall, the ASD combination can be simplified to 0.7E. As a result:

$$v = 0.7v_u = 0.7(373 \text{ lb/ft}) = 261 \text{ lb/ft}$$

NOTE: The seismic coefficient of 0.200 used in this problem is for wood structural panel shearwalls. Particleboard and fiberboard are prohibited in SDC D. Although gypsum wallboard, stucco and diagonal lumber sheathing are permitted to be used, the R-factor would have to be reduced from 6.5 to 2, increasing the seismic force coefficient to 0.65. This example will use wood structural panel sheathing.

From IBC Table 2306.4.1, $\frac{3}{8}$ in. rated sheathing with 6d common nails (2 in. × 0.113 in.) at 4-in. o. c. edge nailing:

$$\text{Allow. } v = 300 \text{ lb/ft} > 261 \text{ lb/ft} \quad OK$$

This can be checked against SDPWS Table 4.3A. The tabulated nominal unit shear capacity is 600 lb/ft. Per SDPWS Sec. 4.3.3, the allowable unit shear is determined by dividing the tabulated nominal unit shear by 2. This also results in v allowable equal to 300 lb/ft.

> *Use* $\frac{3}{8}$-in. wood structural panel sheathing with 6d common nails at 4-in. o.c. panel edges
> 12-in. o.c. field
> Blocking required

Using the SDPWS, the shear capacity could also be checked using LRFD load combinations that result in:

$$1.0\,v_u = 373 \ \text{lb/ft}$$

In this case, SDPWS Sec. 4.3.3 specifies that the tabulated nominal shear capacity is to be multiplied by a resistance factor ϕ_D of 0.80, giving a factored resistance of 480 lb/ft. This more than meets the required resistance of 373 lb/ft.

Shearwall 2

Figure 10.7c Elevation of right transverse wall (shearwall 2). Seismic force shown is at *strength level*.

ELEVATION WALL 2

The total *strength-level* design seismic shear force is conservatively calculated using R_1 from shearwall 1 (Fig. 10.7c):

$$V_u = R_u + R_{1u} = 18.20 + 1.92 = 20.12 \ \text{k}$$

$$v_u = \frac{V_u}{b} = \frac{20,120}{30 + 10} = 503 \ \text{lb/ft}$$

Again, the shearwall shear loads need to be used with the ASCE 7 earthquake load equations and in ASD or LRFD load combinations. For shear forces in the plane of the wall, the ASD load combination can be simplified to 0.7E. As a result:

$$v = 0.7(503) = 352 \ \text{lb/ft}$$

> *Use* $\frac{3}{8}$-in. wood structural panel sheathing with 6d common nails at 3 in. o.c. panel edges
> 12-in. o.c. field
> Blocking required

Allow. $v = 390 \ \text{lb/ft} > 352 \ \text{lb/ft}$ \qquad *OK*

NOTE: For SDC D, E, and F, IBC and SDPWS contain unit shear values above which 3× framing members are required at foundation sill plates and abutting panel joints. This value is 350 plf ASD per IBC Table 2306.4.1 Footnote i, or a nominal value of 700 plf per SDPWS Sec. 4.3.7.1, Item 3c, which also corresponds to an ASD value of 350 plf. This requirement originally came from the UBC, where it was triggered based on a chosen tabulated allowable unit shear greater than 350 plf. Based on this intended use, 3× framing would clearly be triggered for the shearwall in this example, since the tabulated ASD capacity is 390 plf. IBC Table 2306.4.1 Footnote i uses the term "shear design values" that does not clearly indicate whether required unit shear or tabulated unit shear is intended to be the trigger. The SDPWS specifically identifies the trigger as required nominal shear capacity. Based on the SDPWS, 3× framing would just be triggered for the shearwall in this example. Use of 3× framing is highly recommended anytime the tabulated ASD shear capacity exceeds 350 plf. Shearwall testing has identified the splitting of 2× studs at abutting panel edges as a weak link in shearwall behavior.

Tension Chord

The chord stresses are critical in wall 2 (Fig. 10.7*d*) because of the higher wall shears. The tension chord can be checked conservatively using the gross OM:

WALL 2 – CHORD FORCES

Figure 10.7*d* Free body diagrams of panels in shearwall 2. Seismic forces and unit shears shown at the *strength level*.

The forces and unit shears shown in Fig. 10.7*d* are the *strength level* magnitudes. To design the tension chord, the unit shears must be converted for use in ASD by multiplying by 0.7. Assume $\rho = 1.0$

$$\text{Load at top of shearwall (roof):} \quad v = 0.7(v_u) = 0.7(455) = 319 \text{ lb/ft}$$

$$\text{Load from top half of shearwall:} \quad v = 0.7(v_u) = 0.7(48) = 33.6 \text{ lb/ft}$$

$$T = C = vh = 319(16) + (33.6)12 = 5.51 \text{ k} \quad \text{(using gross OM)}$$

Check the tension stress in one 2×8 stud (See Fig. 10.7*e*):

$$\text{Net area} = A_n = 1\tfrac{1}{2} \left(7\tfrac{1}{4} - 1\right) = 9.38 \text{ in.}^2$$

$$f_t = \frac{T}{A_n} = \frac{5510}{9.38} = 587 \text{ psi}$$

Figure 10.7*e* Section through shearwall chord member.

Lumber is No. 2 DF-L. A 2 × 8 is in the Dimension lumber size category, and tabulated design values and size factors are from NDS Supplement Table 4A (Ref. 10.2):

$$F_t = 575 \text{ psi} \qquad C_F = 1.2 \qquad \text{for } F_t \text{ in a No. 2 } 2 \times 8$$

$$F_c = 1350 \text{ psi} \qquad C_F = 1.05 \qquad \text{for } F_c \text{ in a No. 2 } 2 \times 8$$

$$F_{c\perp} = 625 \text{ psi}$$

$$E_{min} = 580,000 \text{ psi}$$

Adjusted ASD tension design value:

$$F'_t = F_t(C_D)(C_M)(C_t)(C_F)$$

$$= 575(1.6)(1.0)(1.0)(1.2)$$

$$= 1104 \text{ psi} > 598 \text{ psi} \qquad OK$$

> One 2 × 8 is adequate for the tension chord of the shearwall.

The stress in the tension chord shown above is based on the gross chord force of 5.5 k. If the tension chord had been overstressed using the gross OM, the resisting dead load moment could be considered. However, if consideration is given to the dead load D, for seismic forces the vertical component of the seismic force must also be considered. In order to properly consider the combination of vertical loads and horizontal forces, ASCE 7 Eqs. 12.4-1 and 12.4-2 are required for the seismic force:

$$E = E_h + E_v = \rho Q_E + 0.2S_{DS}D$$

And

$$E = E_h - E_v = \rho Q_E - 0.2S_{DS}D$$

These equations for seismic force need to be combined with IBC basic load combinations 16-12, 16-13, and 16-15 from IBC Sec. 1605.3.1:

$$D + H + F + (W \text{ or } 0.7E)$$

$$D + H + 0.75(W \text{ or } 0.7E) + 0.75L + 0.75(L_r \text{ or } S \text{ or } R)$$

And

$$0.6D + 0.7E + H$$

For the building in this example, the only loads acting are seismic E, dead load D, and roof live load L_r. As a result the above load combinations can be simplified to:

$$D + 0.7E$$

$$D + 0.75(0.7E) + 0.75(L_r)$$

And

$$0.6D + 0.7E$$

When the ASD basic load combinations are used, the force being calculated becomes an ASD force. This will be clarified by use of the variables F, T, and C, rather than F_u, T_u, and C_u. In this example, loads L, S, and R are taken as zero. Because the last equation has the least dead load acting, it will control design of the tension chord for shearwall uplift:

$$T = 0.6D - 0.7(\rho Q_E) - 0.7(0.2S_{DS}D)$$

Assuming $S_{DS} = 1.3g$ and $\rho = 1.0$ gives the following:

$$T = 0.6D - 0.7(1.0)Q_E - 0.7(0.2)(1.3)D$$

$$= 0.6D - 0.7Q_E - 0.18D$$

$$= 0.42D - 0.7Q_E$$

Some thought has to be given to the correct signs for the tension force. The goal is to combine the tension uplift from the shearwall (given a negative sign for tension) with the minimum possible counteracting dead load. For that intent, the shearwall uplift and dead load are in opposing directions. Even with the vertical seismic component, inclusion of the dead load will still be reducing the net tension force due to $0.7Q_E$. As a result, it is still conservative to ignore the effect of dead load. Using a similar procedure, the seismic force combination for compression can be solved as:

$$C = D + 0.7E$$

$$= D + 0.7Q_E + 0.7(0.2)(1.3)D$$

$$= 1.18D + 0.7Q_E$$

Or

$$C = D + 0.75(0.7E) + 0.75(L_r)$$

$$= D + 0.525Q_E + 0.525(0.2)(1.3)D + 0.75L_r$$

$$= 1.14D + 0.525Q_E + 0.75L_r$$

In this case, the compression force is increased when the full load combination is considered. It is not conservative to ignore the effect of vertical loads for compression design.

Although not required in this problem, the reduced tension chord force will be illustrated for comparison. Two approaches can be used to evaluate the effect of the resisting moment:

1. Basic statics using FBDs (refer to Example 2.9 in Sec. 2.10).
2. Subtraction of the dead load tributary to the tension chord from the gross chord force

The second method is illustrated in this example. However, both methods give the same results, and it is suggested that the reader apply both procedures and compare the results. The second method is quick and easy to apply to a one-story shearwall. In the case of a multistory shearwall, the second method requires careful attention to be applied properly, and the first method may be easier to use until experience is gained.

For a given unit shear v, the shorter-width panel in a given shearwall (10 ft in this example) will be critical. A wall dead load of 20 psf was given in the problem statement. In addition, assume that the wall is a bearing wall that supports a portion of the roof. Say 12-ft-long roof joists (not shown) frame into the right end wall, and the roof dead load = 15 psf. For the 10-ft-long shearwall, the tributary dead load to the tension chord is one-half the shear panel length (5 ft).

$$\text{Roof D} = 15 \text{ psf} \times (12/2) \text{ ft} = 90 \text{ lb/ft}$$

$$\underline{\text{Wall D} = 20 \text{ psf} \times 16 \text{ ft} = 320 \text{ lb/ft}}$$

$$\text{Total D} = 410 \text{ lb/ft of wall}$$

Using the just calculated ASD load combination for members critical in tension (uplift), D = 410 lb/ft ×(5 ft tributary length) = 2050 lb, and Q_E = 5510 lb, the ASD level tension chord force is calculated as:

$$T = 0.42D - 0.7Q_E$$

$$T = 0.42(2050) - 0.7(5510)$$

$$T = 861 - 3857 = -2996 \text{ lb, use 3000 lb}$$

Here the dead load and uplift (overturning) force are acting in opposite directions, thus the plus sign for the dead load and the minus sign for the seismic force. In this case, the change in the chord force due to dead load is about 20 percent of the gross uplift force. It is important to note that proper application of the ASD load combinations requires that the seismic forces be converted to an ASD level at this step. This is made clear by using the symbol T, rather than T_u. Converting the forces a second time could result in significant undersizing of members or connections.

NOTE: A single 2× stud may not provide adequate thickness for fastening to the anchor brackets (hold-downs). Conventional construction uses either double 2× studs or a single 4× stud for all shearwall chords to provide greater anchorage thickness. However, compression may control the design and require either double 2× studs or a single 4× stud.

Compression Chord

Using the just calculated ASD load combinations for members critical in compression (downward reaction), D = 410 lb/ft ×(5 ft tributary length) = 2050 lb, Q_E = 5510 lb, and L_r = 20 psf × (12 ft/2)(5 ft tributary length) = 600 lb.

$$C = 1.18D + 0.7Q_E$$

$$= 1.18(2050) + 0.7(5510)$$

$$= 2419 + 3975 = 6276 \text{ lb}$$

Or

$$C = 1.14D + 0.525Q_E + 0.75L_r$$

$$= 1.14(2050) + 0.525(5510) + 0.75(600)$$

$$= 2337 + 3893 + 450 = 5680 \text{ lb}$$

In this case the first load combination controls, giving a compression chord force of 6276 lb. Here all force components are acting in the same direction (downward), so all have a positive sign.

Check the column allowable load of one stud. Sheathing provides lateral support about the weak axis.

$$\left(\frac{l_e}{d}\right)_y = 0$$

$$\left(\frac{l_e}{d}\right)_x = \frac{16 \text{ ft} \times 12 \text{ in./ft}}{7.25} = 26.5$$

From NDS Table 4.3.1, the ASD adjusted allowable compression stress is calculated as:

$$F'_c = F_c(C_D)(C_M)(C_t)(C_F)(C_P)$$

The column stability factor for solid columns is calculated in accordance with NDS Sec. 3.7.1 as follows:

$$C_p = \frac{1 + (F_{cE}/F_c^*)}{2c} - \sqrt{\left[\frac{1 + (F_{cE}/F_c^*)}{2c}\right]^2 - \frac{F_{cE}/F_c^*}{c}}$$

where

$$F_{cE} = \frac{0.822E'_{min}}{(l_e/d)^2}$$

$$E'_{min} = E_{min}(C_M)(C_t)$$

$$= 580,000(1.0)(1.0)$$

$$= 580,000 \text{ psi}$$

$$F_{cE} = \frac{0.822(580,000)}{(26.5)^2}$$

$$= 679 \text{ psi}$$

From NDS Table 4A, $F_c = 1350$ psi, $C_F = 1.05$ for a 2 × 8, C_D is taken as 1.6, and C_M and C_t are set at 1.0.

$$F_c^* = F_c(C_D)(C_M)(C_t)(C_F)$$

$$= 1350(1.6)(1.0)(1.0)(1.05)$$

$$= 2268 \text{ psi}$$

$$F_{cE}/F_c^* = 679/2268 = 0.299$$

Per NDS Sec. 3.7.1, for sawn lumber the value of c is taken as 0.8.

$$C_P = \frac{1 + (F_{cE}/F_c^*)}{2c} - \sqrt{\left[\frac{1 + (F_{cE}/F_c^*)}{2c}\right]^2 - \frac{F_{cE}/F_c^*}{c}}$$

$$= \frac{1 + 0.299}{2(0.8)} - \sqrt{\left[\frac{1 + 0.299}{2(0.8)}\right]^2 - \frac{0.299}{0.8}}$$

$$= 0.812 - \sqrt{0.659 - 0.374} = 0.278$$

$$F'_c = F_c(C_D)(C_M)(C_T)(C_F)(C_P)$$

$$= 1350(1.6)(1.0)(1.0)(1.05)(0.278)$$

$$= 631 \text{ psi}$$

Check bearing perpendicular to the grain on the wall plate: A stud serving as a shear-wall chord occurs at the end of the wall plate. Therefore, $C_b = 1.0$.

$$F'_{c\perp} = F_{c\perp}(C_M)(C_t)(C_b)$$

$$= 625(1.0)(1.0)(1.0)$$

$$= 625 \text{ psi} < 631 \text{ psi}$$

$$\therefore F'_{c\perp} \quad \text{governs}$$

$$\text{Adj. } P = F'_c A = 0.625 \ (1.5 \times 7.25) = 6.8 \text{ k} \cong 6.88 \text{ k} \qquad OK$$

In this case, the use of one 2× stud would result in an overstress of about 1 percent. This would generally be considered within the certainty with which the forces can be calculated, and therefore would be acceptable. For a variety of reasons relating to detailing, it is often desirable to have a double 2× or a 4× post at the ends of shearwalls, even when a 2× stud can be shown to be adequate by calculation.

> *Use* Two 2 ×8 studs of No. 2 DF-L for all shearwall chords

NOTE: The column calculations given above for the compression chord are conservative. The maximum compressive force at the base of the wall was used for the design load. However, the base of the chord is attached at the floor level, and column buckling is prevented at that point. Some reduced axial column capacity occurs at the critical location for buckling.

If the length of the shorter shearwall segment in Wall 2, Fig. 10.6c, had been 4.6 ft instead of 10 ft, the aspect ratio would be 16 ft/4.6 ft = 3.5. Segments with this aspect ratio are still permitted to be designed as shearwalls; however, in accordance with IBC Table 2305.3.4, footnote a, use of a reduced allowable unit shear is required for seismic forces. The reduction factor is $2b/h = 2(4.6)/16 = 0.58$. The unit shear reduction is based on observed larger deflections in high aspect ratio shearwalls.

When unit shears are calculated in walls combining longer and shorter length segments, one approach is to calculate an effective length for wall segments with aspect ratios greater than 2. The effective length of the 4.6 ft segment would be reduced to 4.6(0.58) = 2.7 ft. The unit shear would then be calculated using the full long segment length of 30 ft plus the effective short segment length of 2.7 ft. The sheathing, nailing, and other fastenings should be the same for both segments, and be based on the unit shear calculated using the effective length.

10.7   Alternate Shearwall Design Methods

In Sec. 10.3 and Fig. 10.3, three code-recognized methods for designing wood structural panel shearwalls were introduced. So far this chapter has described design using the first and most common method, the segmented shearwall methods. Design using the other two methods will now be introduced.

Both alternate design methods are intended for walls that include openings for doors or windows. Both consider the behavior of the entire wall including the openings, rather than considering each full-height segment to be an individual shearwall. Both alternate design methods apply only to shearwalls sheathed with wood structural panels, and require that sheathing be provided on all sheathable areas of the wall.

First, *shearwalls designed for force transfer around openings* will be introduced. These shearwalls use sheathing above and below the openings as coupling beams connecting the full-height wall segments. The only places in the IBC that specifically mention this shearwall type are Secs. 2305.3.4 and 2305.3.5 where shearwall aspect ratios are discussed. SDPWS Sec. 4.3.5.2 specifically recognizes this as a permitted shear wall type. The designer is required to develop a rational analysis and design sheathing and boundary members accordingly. No other limits are imposed by the code on this shearwall type. Example 10.6 illustrates this design method.

EXAMPLE 10.6   Shearwall Designed for Force Transfer around Openings

Consider the shearwall illustrated in Fig. 10.8*a* and design the wall using force transfer around openings.

Figure 10.8*a*   Dimensions for shearwall.

The design seismic shear force of 3500 lb is at an ASD level. Design using force transfer around openings means that the entire wall will be considered to overturn as a unit, resulting in the free body diagram shown in Fig. 10.8*b*.

In order for the sheathing areas above and below the opening to be able to provide this behavior, the sheathing areas need to be capable of acting like beams, carrying shear forces, and having flexural continuity at each end (Fig. 10.8*c*).

The shear forces are carried by the sheathing. Flexural forces are resisted by framing members at the opening (header and 2× plates) and continuity into the full height wall at each end is provided by strapping for tension forces and blocking for compression forces.

Figure 10.8b Free body diagram of shearwall designed for force transfer around openings.

Figure 10.8c Beam behavior of sheathed areas above and below window.

There is some variation possible in the rational analysis provided for this shearwall design method; the following is a common approach. The shearwall is divided into eight different zones A to H as follows (Fig. 10.8d):

The shear in each of these zones will be established. The shear in segments B and G can first be determined by statics by making a horizontal cut at wall mid-height and summing horizontal forces (Fig. 10.8e):

Similarly the shear in zones D and E can be determined by making a vertical cut and summing vertical forces (Fig. 10.8f).

Next, an assumption will be made as to where the coupling beam has a zero moment. Because the full-height shearwall segments on each side of the opening are the same length, the point of zero moment will be set at mid-length of the opening. Here the chord tension force T and compression force C can be set to zero (Fig. 10.8g).

Taking Line 2 above the window and starting with T = 0 at the midpoint, the boundary member force at the upper right-hand corner of the window (Lines 2 and 7) can be

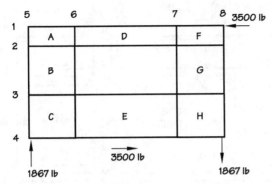

Figure 10.8d   Zones for rational analysis of shearwall.

B & G: HORIZ. v = 3500 / (4' + 4') = 438 plf

Figure 10.8e   Horizontal cut at window mid-height and summing of horizontal forces.

D & E: VERT v = 1867 / (1.5' + 2.5') = 467 plf

Figure 10.8f   Vertical cut at window mid-length and summing of vertical forces.

F & H: HORIZ v = (1635 / 4') = 409 plf = 438 − 29 plf
A & C: HORIZ v = 29 plf by symmetry

Figure 10.8g Complete rational analysis.

calculated as the uniform shear in zone D of 467 plf times the distance, 3.5 ft. The force in Line 2 needs to reduce back down to 0 at Lines 2 and 8. As a result, between 7 and 8 a uniform force of 409 plf needs to be transferred back out of the boundary member. 438 plf is already known to go into zone G. The difference (438 − 409 = 29 plf) will go into zone F. The same pattern can be used to calculated boundary member force and unit shears at the other three corners of the opening. The resulting shears are as follows (Fig. 10.8h):

The resulting analysis can be checked by making sure that the vertical reactions on Lines 5 and 8 add up to 1867 lb, and on Lines 6 and 7 add up to 0. This check is made and the rational analysis is verified.

Once this analysis is complete, design of the shearwall follows the same steps as were used for design of a segmented shearwall; however, careful attention is needed for boundary elements, which are more complex in this wall.

The shearwall aspect ratio will be checked first. For this design method, IBC Sec. 2305.4.5 specifies the aspect ratio to be the clear height of the window opening divided by the width of sheathing at the pier adjacent to the window. In this case, the ratio is 4 ft/4 ft = 1. This is less than the maximum ratio of 2 permitted without allowable shear reduction and much less than the ratio of 3.5 permitted with allowable shear reduction.

Next, required sheathing and nailing will be determined. It was noted at the beginning of this example that the seismic force is already adjusted to an ASD level. The maximum unit shear in segments D and E is 467 plf. From IBC Table 2306.4.1, $^{15}/_{32}$ in. rated sheathing nailed with 8d common nails at 3 in.-o.c. gives an ASD unit shear capacity of 490 plf. It is easiest to use this nailing throughout the shearwall. An alternate approach is to identify zones where less nailing can be used; however, this level of detail results in additional effort during construction and inspection.

LINE 6 & 7 CHECK = (467 – 29) X (1.5' + 2.5') – (438) X (4') = 0 LB OK

LINE 5 & 8 CHECK = (29) X (1.5' + 2.5') – (438) X (4') = 1868 LB OK

Figure 10.8h Check on rational analysis results.

Like the segmented shearwall, boundary members need to be designed for tension and compression forces. Vertical Lines 5 and 8 at each end of the shearwall require design for tension and compression, as was done for the segmented shearwall in Example 10.5. Sheathing edge nailing to these members is required. In addition, Lines 6 and 7 serve as boundary members. Line 6 at Line 1 starts out with a force of 0. Between Lines 1 and 2, the force increases at a rate of $467 - 29 = 438$ plf to 657 lb. Between Lines 2 and 3, the unit shear direction reverses and the forces decreases by 438 plf $\times$ 4 ft = 1752 lb, with a resulting force of 1095 lb. There are commonly two or three studs supporting the header at window and door openings, so the capacity of chord members is generally not an issue. Edge nailing of the sheathing does need to be provided to these chord members over the full height of Lines 6 and 7.

Similar boundary or chord member forces can be calculated along Lines 2 and 3 using the same approach. Because no continuous framing exists along Lines 2 and 3, it is common to provide continuous straps to carry tension loads and continuous blocking to carry compression loads. Edge nailing of the sheathing is required over the full length of these lines also. Nailing through the steel strap into the sheathing and framing is often used to meet the edge nailing requirement. Typical boundary detailing requirements are illustrated in Fig. 10.8i.

It should be noted that dead load resistance to overturning is not considered in this shearwall design method. When designing for force transfer around openings, additional checks will need to be made for adequate load paths for gravity loads using ASD or LRFD load combinations.

Figure 10.8*i*   Boundary members and fastening for shearwall designed with continuity around openings.

The second alternate shearwall design approach, the *perforated shearwall*, will now be introduced. This design approach also considers the entire wall including door and window openings. The primary advantage of this design approach is a reduction in the number of required tie-down anchors. Rather than adding blocking, strapping and additional edge nailing to provide continuity around the openings, this method instead recognizes the somewhat lower capacity and higher deflection of the wall when this detailing is not used. Example 10.7 illustrates design of the Example 10.6 wall using the perforated shearwall method.

EXAMPLE 10.7   Perforated Shearwall Design

Consider the shearwall illustrated in Fig. 10.9*a* and design the wall using the perforated shearwall method. This is the same wall as used in Example 10.6; however, the ASD seismic force has been reduced from 3500 lb to 3000 lb.

The concept of a perforated shearwall is that it relies on the inherent continuity provided by typical construction rather than requiring special detailing. As a result, each

Figure 10.9a Dimensions of perforated shearwall.

Figure 10.9b Perforated shearwall segments fully and partially restrained for overturning.

full-height sheathed segment has a different degree of restraint for overturning forces, and as a result has a varying capacity. See Fig. 10.9b. In the wall used for the example problem, and for the direction of load shown, the right-hand segment has a tie-down that resists wall uplift, and can be considered fully restrained for overturning; therefore, the right-hand segment can reach full design capacity. The left-hand segment develops a lower degree of overturning restraint, and therefore has reduced capacity. The shear resistance adjustment factor C_o is used to adjust the shearwall resistance for the combination of fully restrained and partially restrained segments.

The design of the perforated shearwall will follow the provisions of IBC Sec. 2305.3. Using Secs. 2305.3.4 and 2305.3.5, the aspect ratio of the full-height segments of the perforated shearwall are checked. Like the segmented shearwall, the height is taken as the full clear height of the framed wall—8 ft in this example. The width is taken as the width of solid sheathing—4 ft in this example. The resulting aspect ratio of 8/2 = 2 ft is within acceptable limits for wind and seismic forces, and does not require the reduction of tabulated unit shear capacities.

Next, the perforated shearwall requirements of Sec. 2305.3.8.2 are followed, starting with limitations on the use of the perforated shearwall method. Unlike design for force transfer around openings, a number of limitations are applicable when using perforated shearwall design. From Sec. 2305.3.8.2.1, the following list of limitations is checked:

1. A full-height perforated shearwall segment must be provided at each end of the perforated shearwall—this is met in the example wall, with a wall segment with an aspect ratio of 2 being provided at each end.
2. The allowable shear selected from IBC Table 2306.4.1 cannot exceed 490 plf—this will be checked as the design proceeds.
3. Where out-of-plane offsets occur, the section of wall on each side of the offset needs to be designed as a separate perforated shearwall. This is not an issue with the example problem.
4. Collectors are to be provided for the full length of the perforated shearwall.
5. The perforated shearwall is to have uniform top and bottom elevations
6. The perforated shearwall height is not to exceed 20 ft.

Of this list, all requirements are met except Item 2, which will have to be checked later in the design.

Next, Sec. 2305.3.8.2.2 addresses perforated shearwall capacity:

1. The percentage of full-height sheathing is calculated as the sum of widths of full-height wall segments meeting the aspect ratio requirement ΣL_i, divided by the out-to-out width of the wall. In this case, (4 ft + 4 ft)/15 ft = 0.53 or 53 percent.
2. The ratio of the maximum opening height to the wall clear height is taken as 4 ft/ 8 ft or a maximum opening height of H/2 where H is the wall clear height.
3. The unadjusted shearwall resistance is taken from Table 2306.4.1, and adjustment for aspect ratio is made if needed (not needed in this example). As in Example 10.6, $^{15}\!/_{32}$ in. rated sheathing with 8d common nails at 3 in.-o.c. will be used, giving a tabulated unit shear capacity of 490 plf.
4. The shear resistance adjustment factor C_o is taken from Table 2305.3.8.2. For this example, a maximum opening height of H/2 and percentage of full height sheathing of 50 percent results in C_o = 0.80. This gives an adjusted shear resistance of 490 plf × 0.80 = 392 plf.
5. The perforated shearwall shear resistance is calculated by multiplying the adjusted shear resistance of 392 plf times the sum of full-height wall segments ΣL_i = 8 ft. This gives a shearwall resistance of 3136 lb, which is greater than the ASD seismic force of 3000 lb.

Perforated shearwall anchorage and load path are addressed in IBC Secs. 2305.3.8.2.4 through 2305.3.8.2.8. Tension chords and tie-downs are provided at each end of the perforated shearwall, at the first full-height shearwall segment. The ASD tension chord uplift force due to overturning is calculated as:

$$T = \frac{VH}{C_o \Sigma L_i}$$

For the example problem:

V = 3000 lb ASD seismic load
H = 8 ft wall clear height
C_o = 0.80 from Table 2305.3.8.2
ΣL_i = 8 ft

$$T = 3750 \text{ lb}$$

Unit shear transfer into the top and at the bottom of full-height shearwall segments is calculated as:

$$v = \frac{V}{C_o \Sigma L_i}$$

$$v = 469 \text{ plf}$$

A unit uplift force, t, along the bottom of full-height shearwall segments is calculated as:

$$t = v = 469 \text{ plf}$$

A compression chord force, C, at each end of each full-height shearwall segment is calculated as:

$$C = T = 3750 \text{ lb}$$

The anticipated load path connections for the loading direction shown are illustrated in Fig. 10.9c. Locations of edge nailing at vertical boundary members are shown in Fig. 10.9d.

Figure 10.9c Load path and anchorage reactions for perforated shearwall for loading direction shown.

Figure 10.9d Vertical boundary members for perforated shearwall.

Perforated shearwall design in accordance with the SDPWS will be similar to the Example 10.7 use of the IBC provisions. Although this example used ASD design, LRFD could just as easily be used.

It should be noted that dead load resistance to overturning is not considered in the perforated shearwall design method; the design formulas do not include any consideration of dead load or other gravity loads. When designing using the perforated shearwall method, additional checks will need to be made for adequate load path for gravity loads using ASD or LRFD load combinations. See Refs. 10.7 and 10.11 for additional information on the perforated shearwall method.

The perforated shearwall design method is new and only recently adopted into the code. The main advantage of the perforated shearwall design method is not in design capacity or simplicity of design, but rather that it requires a reduced number of tie-downs as compared to the segmented shearwall design method.

The perforated approach considers the entire wall to act as a single unit. Using the perforated shearwall design method requires the use of only two tie-downs to the foundation—one at each end of the wall (i.e., at the building corners). The segmented shearwall design method requires each end of each full-height segmented wall to be anchored to the foundation with a tie-down bracket. The shearwall in Example 10.7 had a single opening and four tie-downs would be required using the segmented approach versus two tie-downs using the perforated approach. There is a corresponding loss in unit shear capacity using the perforated shearwall design method due to reduced restraint against overturning. This reduced capacity is offset by the economic advantage of reducing the number of tie-downs.

Requirements for distributed anchorage along the bottom of perforated shearwalls are different than for segmented shearwalls. Rather than providing shear anchorage for an average unit shear calculated as V/b, the average unit shear is further multiplied by $1/C_o$. For the example problem configuration with the door opening, this would result in the average unit shear being multiplied by 1.25. This adjustment is made because, in the perforated shearwall method, shearwall segments at the ends of walls that have tie-down brackets develop somewhat higher capacities. Because of this variation, the unit shear at the base of the wall is higher in some location than in others. Locations of highest unit shear will generally be different when the seismic force reverses direction. For this reason, an amplified unit shear is required for base anchorage throughout the perforated shearwall length.

In addition, between the tie-down anchors, the wall base must be fastened for a unit uplift force equal to the amplified unit shear force. This fastening can be provided by the anchor bolts at the foundation level. At an upper framed floor level, this fastening could be sill plate nails in withdrawal, tie-down straps, or a number of other devices. See the commentary to the NEHRP

Provisions (Ref. 10.5) for additional information regarding anchorage of perforated shearwalls.

10.8 Anchorage Considerations

Seven design considerations for shearwalls were listed in Sec. 10.2. The first five of these items have been covered (forces for designing the collector were covered in Chap. 9). The sixth item (anchorage) is considered now. Generally speaking, anchorage refers to the tying together of the main elements so that the building will function as a unit in resisting the design loads. Although gravity loads need to be considered, the term *anchorage* emphasizes the design for lateral forces. This includes uplift caused by overturning moments.

Anchorage can be systematically provided by considering the transfer of the following loads:

1. Vertical (gravity) loads

2. Lateral forces parallel to a wall

3. Lateral forces perpendicular to a wall

If this systematic design approach is followed where the diaphragm connects to the various shearwalls, and where the shearwalls are attached to the foundation, the building will naturally be tied together. See Example 10.8.

EXAMPLE 10.8   Basic Anchorage Criteria

The process of tying the building together can be approached systematically by considering loads and forces in the three principal directions of the building. The critical locations are where the diaphragms connect to the shearwalls and shearwalls tie into the foundation. See Fig. 10.10.

ANCHORAGE FORCES
1. VERTICAL LOAD
2. LATERAL FORCES
   a. PARALLEL TO WALL
   b. PERPENDICULAR TO WALL

Figure 10.10   Anchorage forces in three principal directions.

The remainder of the chapter provides an introduction to anchorage considerations. The transfer of gravity loads is first summarized. For lateral forces, it is convenient to separate the discussion of anchorage into two parts. The transfer of lateral forces at the *base of the shearwall* is fairly direct, and this chapter concludes with a consideration of these anchorage problems. The anchorage of the *diaphragm to the shearwall* is somewhat more complicated, and this problem is covered in detail in Chap. 15.

Anchorage is basically a connection design problem. The scope of the discussion in this chapter is limited to the calculation of the forces involved in the connections. Once the magnitudes of the forces are known, the techniques of Chaps. 12 through 14 can be used to complete the problem.

There is one more type of anchorage, particularly related to wind loads, that is not specifically addressed in this chapter but bears mention. This is anchorage for direct wind uplift. Section 2.12 of this book addressed calculation of wind forces on components and cladding, including forces on roof components. Load combinations including roof dead load and wind uplift load can result in net uplift on roof components. Where this occurs, tension uplift anchorage needs to be provided to roof framing members to resist uplift forces. This uplift anchorage is often provided by using proprietary steel straps. Where resisting dead load is minimal, it may be necessary to continue this anchorage all the way to the building foundation. Roof framing is anchored to the tops of the wall studs. The bottoms of the wall studs are anchored to the tops of studs in the story below, and so on until anchorage is provided to the foundation. This anchorage is important for the performance of wind-resisting systems.

10.9   Vertical (Gravity) Loads

As noted, the term *anchorage* implies connection design for lateral forces. However, this can be viewed as part of an overall connection design process. In connection design, the first loads that come to mind are the gravity loads. These start at the roof level and work their way down through the structure. The "flow" of these forces is fairly easy to visualize. See Example 10.9.

The magnitudes of the loads required for the design of these connections can usually be obtained from the beam and column design calculations. The hardware used for these connections is reviewed in Chap. 14.

EXAMPLE 10.9   Connections for Gravity Loads

Design for gravity loads involves the progressive transfer of loads from their origin at the lightest framing members, through the structure, and eventually into the foundation. Consider the "load path" for the transfer of gravity loads in the roof framing plan (Fig. 10.11). This requires the design of the following connections:

    a. Joist connections to roof beam and bearing wall

    b. Roof beam connections to girder and bearing wall

    c. Girder connections to column and header

TYPICAL ROOF FRAMING PLAN

Figure 10.11 Connections for the transfer of gravity loads.

  d. Header connection to column
  e. Column connection to footing, and stud connection to wall plate (these connections are not shown on the sketch)

10.10 Lateral Forces Parallel to a Wall

A shearwall resists lateral forces that are parallel to the wall. In doing so, it cantilevers from the foundation, and both a shear and a moment are developed at the base of the wall. Anchorage for the shear and moment is provided by separate connections.

The connections for resisting the *moment* will be considered first. It will be recalled that the moment is carried by the chords, and the chord forces are obtained by resolving the moment into a couple. See Fig. 10.12. The chord forces may be calculated from the gross overturning moment, or the resisting moment provided by the dead load may be considered. Both procedures are illustrated in Example 10.5 (Sec. 10.6).

For connection design, the tension chord is generally of main concern. The tension chord force and the connection at the base of the tension chord will become important for

1. Tall shear segments (i.e., when h is large)

2. Narrow segments (i.e., when b is small)

3. Small resisting dead loads

When these conditions exist, a large tension uplift force develops at the base of the wall. If those conditions do not exist, there may be no uplift force at the base of the wall.

SHEAR PANEL

SHEAR PANEL AND
FOUNDATION

TIE-DOWN ANCHOR DETAIL

Figure 10.12   Anchorage for shearwall moment. The FBD on the left is used to design the anchorage connection of the shearwall segment to the foundation. The FBD on the right is used to check overall moment stability. The anchor bracket detail involves a connection between the bracket and the shearwall chord, and between the bracket and the foundation. Fastener spacing requirements (end distance and center-to-center) are important. See Chaps. 12 and 13 for details.

The tension tie for the shearwall anchorage is often made with an engineered or prefabricated metal bracket commonly called a tie-down or hold-down. These brackets are attached to the chord member with nails, bolts, or lag bolts, and to the foundation with an anchor bolt. The number and size of nails, bolts, or lag bolts, and the required embedment length for the anchor bolt, depend on the design force.

A recent move away from using bolts to attach the chord to the anchor bracket has occurred due to observations following the Northridge earthquake. Significant slip appeared to result between the stud and bracket due to oversized holes and wood crushing in the chord around the bolts. This slip in the chord-to-anchor connection led to failure in the nailed connection between the wood panel sheathing and the stud. An "unzipping" of the nailing along the panel edge was caused by a slip in the connection between the chord and the anchor bracket. Additionally, tie-down posts often were observed to split out from the bolt hole to the bottom of the chord, leaving the bolt ineffective. Nails and lag bolts are becoming more popular to attach the chord to the anchor bracket.

Connections of this type can be designed by using the wood design methods of Chaps. 12 and 13. However, prefabricated brackets of this type usually have recognized load values that can be obtained from code evaluation reports.

After the shearwall has been anchored to the foundation for the tension chord force, the overall stability of the wall system must be checked. In this case, the dead load of the foundation is included in the resisting moment. In some cases, the footing may also have to be designed for the compressive chord force. This becomes a concern with a large chord force, especially in areas with low soil-bearing values.

The connection for transferring the *shear* to the foundation is considered now. This connection is normally made with a series of anchor bolts. See Example 10.10 and Fig. 10.13. These bolts are in addition to the anchor bolts used for the chord forces.

EXAMPLE 10.10   Anchorage for Shearwall Shear

The anchor bolt requirement can be determined by first assuming a size of anchor bolt and determining the adjusted load per bolt. If the adjusted load per anchor bolt is Z' (parallel-to-grain design value), and the total lateral force parallel to the shearwall segment is the unit shear times the segment width ($v \times b$), the required number of anchor bolts is

$$N = \frac{v \times b}{Z'}$$

The average spacing is approximately

$$\text{Spacing} = \frac{b}{N} = \frac{bZ'}{vb} = \frac{Z'}{v}$$

(The spacing is approximate because starting and ending anchor bolts must be set in from the segment ends a sufficient distance to clear the chords and tie-down brackets.)

The code minimum anchor bolt requirement for wood-frame walls is given in IBC Sec. 2308.6 as $\frac{1}{2}$-in. diameter anchor bolts at 6 ft-0-in. o.c. Section 2308.12.9 increases the minimum anchor bolt size to $\frac{5}{8}$-inch for SDC E. Additionally, a properly sized nut and washer or steel plate washer must be used on each anchor bolt.

WALL SHEAR = v

SHEATHING
STUD
ANCHOR BOLT
SILL PLATE
CONCRETE
FOUNDATION

CHORD
UPLIFT
ANCHORAGE
(FIG. 10.12)

ANCHOR BOLTS
FOR SHEAR –
RESISTING FORCE
OF ANCHOR BOLT
SHOWN

NAILING –
SHEATHING
TO SILL PLATE

SHEARWALL SHEAR ANCHORAGE

Figure 10.13   A separate set of anchor bolts is provided for shear transfer. These anchor bolts are in addition to those provided for the chord uplift connections.

A minimum of two bolts or anchor straps are required per wall plate and one bolt or strap is required within 12 in. of, and at least 4 in. from, the end of each plate piece. Anchor bolts for shearwalls are usually larger and more closely spaced than the code minimum.

It will be recalled that the shear is essentially carried by the wall sheathing. The attachment of the sheathing to the bottom wall plate will transfer the shear to the base of the wall. The anchor bolts, then, are simply designed to transfer the shear from the bottom plate into the foundation.

Although the design of bolted connections is covered in a later chapter, the basic design procedure is briefly described here. The strength of the anchor bolt is determined as the bolt capacity parallel to the grain in the wood *or* by the strength of the anchor bolt in the concrete footing. The capacity of the bolt in the wood plate is covered in Chap. 13, and the capacity of the anchor bolt in the concrete footing is calculated in accordance with IBC Chap. 19. The smaller of these two values is used to determine the size and number of anchor bolts.

The 2006 IBC requires that capacities for anchor bolt anchorage to concrete be calculated in accordance with ACI 318 (Ref. 10.1) Appendix Chap. D. This is a calculation intensive approach that considers a number of modes of potential failure. ACI 318 Appendix D further requires that capacity of the steel bolt section governs over concrete capacity for seismic forces in SDC C through F. IBC permits concrete capacity to govern provided an additional load factor of 2.5 is used. With these changes in the 2006 IBC, it may be possible for concrete capacity rather than wood sill plate capacity to govern anchor bolt design. The wood sill plate portion of anchor bolt design, along with edge distance and bolt spacing requirements are covered in Chap. 13.

Based on observations of foundation sill splitting in laboratory shearwall tests and in buildings following the Northridge earthquake, requirements for thicker foundation sills and for steel plate washers on anchor bolts have been introduced in SDC D, E, and F. These requirements can be found in IBC Sec. 2305.3.11. The steel plate washers are required in all shearwalls in SDC D, E, and F. The 3× foundation sills are required for ASD design shears greater than 350 plf. Both the thicker sill and the plate washer help to delay or avoid the cross-grain splitting of the sill plate that can occur when the sheathing nailing, at the uplifting end of the wall, pulls up on the foundation sill.

Because the bottom wall plate is supported directly on concrete, there is a potential problem with termites and decay. If a hazard exists, the code requires that protection be provided. This is accomplished by using lumber for the sill plate which has been preservative-treated or wood that has a natural resistance to decay (e.g., foundation redwood). An introduction to preservative-treated lumber is given in Sec. 4.9.

10.11 Shearwall Deflection

The calculated deflection of a shearwall is used to verify that code maximum drift limits are being met, and is also used as part of the determination of a flexible diaphragm for purposes of seismic force distribution. For discussion of diaphragm classification, see Sec. 3.3 and Example 9.18 in Sec. 9.11. Code-imposed drift limits have their basis in serviceability, providing some limit on story deformations in order to restrict damage to nonstructural components and systems. Deflection and drift requirements are given in ASCE 7 Sec. 12.12. ASCE 7 Table 12.12-1 provides drift limits that vary by system type and Occupancy Category. For light-frame shearwall buildings in Occupancy Categories I or II, the permissible drift is 0.020 times the story height. ASCE 7 Sec. 12.8.6 contains provisions for calculating story drift. The deflection of Level x at the center of mass is calculated as:

$$\delta_x = \frac{C_d \delta_{xe}}{I}$$

where C_d is a deflection amplifier given in ASCE 7 Table 12.2-1 (the R-factor table), and δ_{xe} is the deflection at *strength level* determined using an elastic analysis. For light-frame bearing-wall systems with wood structural panel sheathing, C_d is 4.0, and δ_{xe} is the shearwall deflection introduced in Chap. 3 as Δ_s. See Fig. 10.14.

The drift criteria of ASCE 7 Secs. 12.8.6 and 12.12 are generally checked at the building center of mass. Where the building has a torsional irregularity, ASCE 7 Sec. 12.12.1 requires that the building drift be the largest difference in deflection along any building edge. The computation, however, only makes sense where a rigid diaphragm design has been used. For wood light-frame

Figure 10.14 Deflection of a shearwall is termed *story drift* Δ_s.

buildings where the diaphragm has been idealized as flexible in accordance with IBC Sec. 1613.6.1, it will be necessary to check the drift of each shearwall against permitted story drift. This is a new requirement in the 2006 IBC.

The IBC provides a method for calculating the deflection of a shearwall. The method accounts for bending and shear in the wall assembly, as well as nail deformation and anchorage slip (see Fig. 10.15).

$$\Delta_s = \Delta_b + \Delta_v + \Delta_n + \Delta_a$$

where Δ_b = bending deflection of the shearwall
Δ_v = shear deflection of the shearwall
Δ_n = deflection of the shearwall due to nail slip (deformation)
Δ_a = deflection of the shearwall due to anchorage slip and rotation

A shearwall acts similar to a cantilevered I-beam. The chords resist the moment analogous to the flanges of an I-beam, and axial elongation and shortening of the chords cause deflection. Therefore, the bending rigidity of a shearwall is related to the axial rigidity of the chords EA. The *bending deflection* of a shearwall is then determined from

$$\Delta_b = \frac{8vh^3}{EAb}$$

where v = shear force at the top of the wall (lb/ft)
h = height of the wall (ft)
b = width of the wall (ft)
E = modulus of elasticity of the chord (psi)
A = cross-sectional area of chord (in.^{2})

Figure 10.15 Deflection components of a shearwall. (a) *Bending deflection* Δ_b. Shearwall chords carry the moment and act like the flanges of a deep I-beam. The bending deflection of the shearwall is related to the axial shortening and elongation of the compression and tension chords (flanges). (b) *Shear deflection* Δ_v. Wall sheathing carries the shear in a shearwall and act like the web of a deep I-beam. The shear deflection of the shearwall shown is that of a pure shear element. (c) *Nail slip deflection* Δ_n. Slip in the nailed connections between the sheathing and framing allows individual sheathing panels to move relative to the framing and relative to each other. This slip reduces the effectiveness of the sheathing to resist shear and results in additional shearwall deflection. (d) *Anchorage slip deflection* Δ_a. Slip in the anchorage detail will allow rigid body rotation of the shearwall and deflection Δ_a. Slip in the anchorage can occur both between the chord and the anchor bracket, and between the anchor bracket and the foundation.

While the chords resist the bending moment, the sheathing in a shearwall acts like the web of an I-beam and resists the applied shear. The *shear deflection* is then related to the shear rigidity Gt of the sheathing:

$$\Delta_v = \frac{vh}{Gt}$$

where Gt = panel rigidity through the thickness in lb/in [IBC Table 2305.2.2(2)]

If the sheathing were rigidly attached to the studs *and if* the shearwall were rigidly attached to the foundation, the bending and shear terms Δ_b and Δ_v would account for *all* the shearwall deflection. Neither of these conditions exists in typical wood shearwalls, and therefore the two additional deflection terms Δ_n and Δ_a must be accounted for in design. The prediction of these deflections relies on experimental data and empirical relationships.

Slip in the connections between the sheathing and framing members reduces the effectiveness of the sheathing in resisting shear by allowing individual sheathing panels to slip relative to each other. The deflection due to *nail slip* Δ_n is estimated from the relationship

$$\Delta_n = 0.75 h e_n$$

where e_n = nail deformation (in.).

The nail deformation e_n is the slip between the stud and the sheathing at the design load. It is dependent on fastener type, minimum fastener penetration, maximum fastener load V_n, and whether green or dried (seasoned) lumber is used at the time of fabrication. Whether the lumber is initially green or dry, it is assumed that the shearwall will be dry in service and the lumber will dry in place (dry use).

The nail deformation e_n can be determined from IBC Table 2305.2.2(1). The fastener load V_n is determined by dividing the maximum unit shear force v by the number of nails per foot at the interior panel edges. The values listed are for STRUCTURAL I wood structural panels. If the sheathing is not STRUCTURAL I, then the resulting nail deformation e_n should be increased by 20 percent.

Slip or rotation in the anchorage detail permits rigid body rotation of the shearwall. The calculation of the deflection of the shearwall due to slip or rotation in the anchorage detail Δ_a is similar to the calculation of the shearwall deflection due to nail slip. A design slip is estimated for the connection and the deflection of the shearwall is proportioned by the height-to-width ratio of the wall

$$\Delta_a = \frac{h}{b} d_a$$

where d_a = anchorage slip (in.).

The anchorage slip d_a is the total slip between the chord and the anchor bracket, and the anchor bracket and the foundation at design load. For many prefabricated anchor brackets, manufacturers provide information regarding slip. For the most part, the anchor slip values provided by manufacturers include the deflection measured between the anchor bolt and the strap, or bracket, at the location of fastening to the wood post. Not included is the slip due to deformation of fasteners to the post. For purposes of the examples in this text, a vertical anchor slip of $\frac{1}{4}$ in. will be used. This includes approximately $\frac{1}{16}$ in. for anchor bolt and bracket deformation and $\frac{1}{16}$ in. for deformation of the fasteners to the post. Alternatively, data from shearwall testing could be used to determine anchor slip.

In summary, the total story drift of a shearwall is calculated from

$$\Delta_s = \Delta_b + \Delta_v + \Delta_n + \Delta_a = \frac{8vh^3}{EAb} + \frac{vh}{Gt} + 0.75he_n + \frac{h}{b}d_a$$

The use of this equation for the calculation of the shearwall deflection is illustrated in Example 10.11. The shearwall for the building shown in Fig. 10.7 was designed in Example 10.5. In Example 10.11, the deflection or story drift of the left transverse wall (shearwall 1) is calculated. From this, the maximum inelastic response displacement is determined and compared to the IBC limit.

EXAMPLE 10.11   Shearwall Deflection

Determine the deflection Δ_s of the left transverse wall (shearwall 1) of the building shown in Example 10.5. The shearwall is shown in Fig. 10.16 and is 60 ft long and 16 ft high. Two No. 2 DF-L 2×8 studs are used as the shearwall chords and $\frac{3}{8}$-in. wood structural panel sheathing is used with 6d common nails at 3-in. o.c. along panel edges and 12-in. o.c. in the field.

Figure 10.16   Elevation of shearwall. Seismic force shown is at *strength level.*

NOTE: IBC Sec. 1617.3 requires all shearwall deflections to be calculated using earthquake forces without reduction. That is, *strength-level* seismic forces are used to determine shearwall displacements. Considering the load combination D + 0.7E + L + (L_r or S or R), the 0.7 reduction factor shall not be used. No live load, snow load, or rain load is specified, thus the load combination reduces to D + E.

From Example 10.5:

strength-level shear force at top of wall

$$v_u = \frac{V_u}{b} = \frac{20,440}{60} = 341 \text{ lb/ft}$$

chord properties

$$E = 1,600,000 \text{ psi}$$
$$A = 2(1.5)(7.25) = 21.75 \text{ in.}^2$$

To determine the shear rigidity, the span-rating of the sheathing is required. The span-rating for the $\frac{3}{8}$-in.sheathing was not specified in Example 10.5; however, IBC Table 2304.7(3) indicates 24/0 is the typical span-rating available for this thickness. From IBC Table 2305.2.2(2) the rigidity is given as

$$G_v t_v = 25{,}000 \text{ lb/in.}$$

Note that exterior glue is assumed for the sheathing, which is common.

To calculate the nail deformation e_n, the shear force per nail V_n must be determined by dividing the *strength-level* unit shear force v_u by the number of nails per foot at the interior panel edges. The *strength-level* unit shear is determined using load combination D + E. For this calculation, the seismic force including the entire wall dead load is added to the shear at the top of the wall v_u (see Example 10.5). The wall dead load D = 20 psf. The unit shear of the wall is determined using the seismic base shear coefficient 0.200 (see Example 10.5):

$$v_{\text{wall}} = 0.200W = 0.200Dh = 0.200(20 \text{ psf}) \ (16 \text{ ft}) = 64 \text{ lb/ft}$$

The total unit shear is

$$v = 341 + 64 = 405 \text{ lb/ft}$$

With a nail spacing of 3-in. o.c., 4 nails per ft are used along the panel edges.

$$(V_n)_u = \frac{405}{4} = 101.25 \text{ lb/nail}$$

Since the sheathing used is not STRUCTURAL I, the value of e_n must be increased by 20 percent.

$$e_n = 1.2(0.03) = 0.036$$

The story drift Δ_s can now be determined

$$\Delta_s = \Delta_b + \Delta_v + \Delta_n + \Delta_a$$

$$= \frac{8v_u h^3}{EAb} + \frac{v_u h}{Gt} + 0.75he_n + \frac{h}{b}d_a$$

$$= \frac{8(405)(16)^3}{(1{,}600{,}000)(21.75)(60)} + \frac{(405)(16)}{(25{,}000)} + 0.75(16)0.036 + \frac{16}{60}(0.125)$$

$$= 0.0064 + 0.26 + 0.43 + 0.033$$

$$\Delta_s = 0.73 \text{ in.}$$

According to IBC Sec. 1617.3, the calculated shearwall displacement is multiplied by a *deflection amplification factor* C_d from IBC Table 1617.6.2. The deflection amplification factor for wood-frame walls is $C_d = 4$, thus

$$\Delta = C_d \Delta_s = (4)(0.73) = 2.92 \text{ in.}$$

The IBC limit for shearwall displacement is 0.02 times the shearwall height.

$$\Delta_{\text{limit}} = 0.020(16 \times 12) = 3.84 \text{ in.} > 2.92 \text{ in.} \qquad OK$$

10.12   Lateral Forces Perpendicular to a Wall

In addition to being designed for the shearwall forces covered previously in this chapter, wall must be designed for forces that are normal to the wall. The wall must be adequately anchored to the foundation and the roof diaphragm to resist design forces normal to the wall due to tributary wind and seismic forces. See Example 10.12.

EXAMPLE 10.12   Force Perpendicular to a Wall

Figure 10.17   Section through wall showing normal wall forces.

The procedures for determining wind and seismic forces were presented in Chap. 2. See Example 2.17. As illustrated in Fig. 10.17, the distribution of normal wall forces resulting from wind pressures and seismic forces differ significantly. From an analysis of the forces acting normal to the wall, the main wall system may be designed as well as the anchorage to the footing (floor reaction).

Anchorage considerations at the diaphragm level are discussed in Chap. 15.

In the case of the lateral force *parallel* to the wall, the sheathing is assumed to transfer the force into the bottom plate of the wall. However, for the lateral force *normal* to the wall, the sheathing cannot generally be relied upon to transfer these forces. Connections should be designed to transfer the reaction from the stud to the bottom plate and then from the plate to the foundation. See Example 10.13 and Fig. 10.18.

EXAMPLE 10.13   Anchorage for Perpendicular to Wall Force

ANCHORAGE FOR LOADS
NORMAL TO THE WALL

Figure 10.18 Two connections transfer normal wall forces to foundation: (1) nail connection between stud and bottom wall plate, (2) anchor bolt connection of wall plate to foundation. Anchor bolt design is governed by either the perpendicular to wall forces shown above or by the parallel to wall forces in Fig. 10.13.

The lateral force normal to the wall is usually transferred from the *stud to the bottom wall plate* by nails. The nailing schedule in IBC Table 2304.9.1 gives the standard connection for the stud to the bottom (sole) plate as four 8d common or box toenails. An alternate connection is two 16d common end nails.

The capacity of these connections can be evaluated by the techniques of Chap. 12. If the capacity of the standard stud-to-plate connection is less than the applied load, some other type of connection (such as a framing anchor; see Chap. 14) will be required.

The connection of the *plate to the foundation* is accomplished with anchor bolts. The anchor bolt requirement can be determined by first assuming a size of anchor bolt and determining the adjusted load per bolt.

If the adjusted load per anchor bolt is Z' (perpendicular-to-grain design value), and the total foundation reaction (Example 10.12) is the reaction per foot R times the width of the wall b, then the required number of anchor bolts is

$$ N = \frac{R \times b}{Z'} $$

This number must be compared with the required number of bolts determined for anchoring the parallel-to-wall forces (Example 10.10). The larger number of bolts controls.

The standard method of connecting wall studs to the bottom plate is with toenails or nails driven through the bottom plate into the end of the stud. The latter connection is often used when a shearwall segment is preframed and then lifted into place. Methods for evaluating the capacity of nailed connections are given in Chap. 12.

10.13   References

[10.1] American Concrete Institute (ACI). 2005. *Building Code Requirements for Structural Concrete*, 2005 ed. (ACI 318-05) ACI, Farmington Hills, MI.

[10.2] American Forest and Paper Association (AF&PA). 2005. *National Design Specification for Wood Construction* and *Supplement*, ANSI/AF&PA NDS-2005, AF&PA, Washington DC.

[10.3] American Forest and Paper Association (AF&PA). 2005. *Special Design Provisions for Wind and Seismic* (SDPWS) Supplement to the NDS, 2005 ed. AF&PA, Washington, DC.

[10.4] APA-The Engineered Wood Association. 2004. Design/Construction Guide – Diaphragms and Shear Walls, APA, Tacoma, WA.

[10.5] Building Seismic Safety Council (BSSC). 2003. *NEHRP Recommended Provisions for Seismic Regulations for New Buildings and Other Structures*, 2003 ed., BSSC, Washington DC.

[10.6] Diekmann, Edward F. 1998. "Diaphragms and Shearwalls," *Wood Engineering and Construction Handbook*, 3rd ed., K. F. Faherty and T. G. Williamson, eds., McGraw-Hill, New York, NY.

[10.7] Douglas, Bradford K., and Sugiyama, H. 1995. *Perforated Shearwall Design Approach*. American Forest and Paper Association (AF&PA), Washington DC.

[10.8] Fischer, David et al. 2001. *Shake Table Tests of a Two-Story Woodframe House* (CUREE Publication No. W-06), Consortium of Universities for Research in Earthquake Engineering (CUREE), Richmond, CA.

[10.9] Gatto, Kip, and Chia-Ming Uang. 2002. *Cyclic Response of Woodframe Shearwalls: Loading Protocol and Rate of Loading Effects* (CUREE Publication No. W-13), Consortium of Universities for Research in Earthquake Engineering (CUREE), Richmond, CA.

[10.10] International Code Council (ICC). 2006. *International Building Code*, 2006 ed., ICC, Falls Church, VA.

[10.11] Line, Philip, and Douglas, Bradford K. 1996. *Perforated Shearwall Design Method*, International Wood Engineering Conference, New Orleans, LA, pp. 2-345-2-349.

[10.12] McMullin, Kurt M., and Merrick, Dan. 2002. *Seismic Performance of Gypsum Walls: Experimental Test Program* (CUREE Publication No. W-15), Consortium of Universities for Research in Earthquake Engineering (CUREE), Richmond, CA.

[10.13] National Evaluation Service Committee. 1997. "Standards for Structural-Use Panels," Report No. NER-108, Council of American Building Officials (Available from APA—The Engineered Wood Association, Tacoma, WA).

[10.14] National Institute of Standards and Technology (NIST). 2004. *U.S. Product Standard PS 2-04—Performance Standard for Wood-Based Structural-Use Panels*, U.S. Dept. of Commerce, Gaithersburg, MD (reproduced by APA—The Engineered Wood Association, Engineered Wood Systems, Tacoma, WA).

[10.15] National Institute of Standards and Technology (NIST). 1995. *U.S. Product Standard PS 1-95—Construction and Industrial Plywood*, U.S. Dept. of Commerce, Gaithersburg, MD (reproduced by APA—The Engineered Wood Association, Engineered Wood Systems, Tacoma, WA).

[10.16] Pardoen, G., Waltman, A., Kazanjy, R., Freund, E., and Hamilton, C. 2003. *Testing and Analysis of One-Story and Two-Story Shearwalls Under Cyclic Loading* (CUREE Publication No. W-25), Consortium of Universities for Research in Earthquake Engineering (CUREE), Richmond, CA.

[10.17] Rose, John D. 1998. *Laboratory Report 158—Preliminary Testing of Wood Structural Panel Shearwalls Under Cyclic (Reversed) Loading*, APA—The Engineered Wood Association, Engineered Wood Systems, Tacoma, WA.

[10.18] Rose, J. D., and Keith, E. D. 1996. *Laboratory Report 157—Wood Structural Panel Shearwalls with Gypsum Wallboard and Window/Door Openings*, APA—The Engineered Wood Association, Engineered Wood Systems, Tacoma, WA.

[10.19] Tissel, John R. 1996. *Laboratory Report 154—Structural Panel Shearwalls*, APA—The Engineered Wood Association, Engineered Wood Systems, Tacoma, WA.

10.14   Problems

The following shearwall problems all involve some type of wood structural panel. Nailing and tabulated unit shears are taken from the IBC or SDPWS. Design of wood members in the following problems is to be in accordance with the NDS. Dry-service conditions, normal temperatures, and bending about the strong axis apply unless otherwise noted. A load duration factor of 1.6 is to be used for problems involving wind or seismic forces.

Some problems require the use of spreadsheet or equation-solving software. Problems that are solved using such software can be saved and used as a *template* for other similar problems. Templates can have many degrees of sophistication. Initially, a template may only be a hand (i.e., calculator) solution worked using spreadsheet or equation-solving software. In a simple template of this nature, the user will be required to provide many of the lookup functions for such items as

Tabulated shearwall unit shear (lb/ft)

Shearwall nailing specifications

Tabulated design values for wood members

As the users gain experience with their software, a template can be expanded to perform many of the lookup and decision-making functions that were previously done manually.

Advanced computer programming skills are not required to create effective templates. Valuable templates can be created by designers who normally do only hand solutions. However, some programming techniques are helpful in automating *lookup* and *decision-making* steps.

The first requirement is that a template operate correctly (i.e., calculate correct values). Another major consideration is that the input and output be structured in an orderly manner. A sufficient number of intermediate answers are to be displayed and labeled in a template so that the solution can be verified by hand.

10.1 *Given:* The single-story wood-frame building in Fig. 10.A. The critical lateral wind force in the longitudinal direction is given. Sheathing is $^{15}\!/_{32}$-in. STRUCTURAL I with 8d common nails.

     Find: a. The unit shear in the shear segment along Line 1.
          b. The required nailing for the shearwall along Line 1.
          c. The net maximum uplift force at the base of the wall along Line 1. Consider the resisting moment provided by the weight of the wall (assume no roof load frames into longitudinal walls). Wall dead load D = 20 psf.
          d. Are the proportions of the shear segment in the above wall within the limits given in the IBC?

10.2 *Given:* The single-story wood-frame building in Fig. 10.A. The critical lateral wind force in the longitudinal direction is given. Wood structural panel sheathing is designed to function as a shearwall. Studs are spaced 16-in. o.c.

     Find: a. The minimum required thickness of sheathing, assuming that the face grain may run either horizontally or vertically. (Sheathing is used under a nonstructural wall covering.)
          b. The unit shear in the shear segment along Line 1.
          c. Using the segment sheathing thickness determined in *a*, select the required spacing of 6d common nails for the shearwall along Line 1. Indicate blocking requirements (if any). Check the proportions of the shear segment.

w_L= 250 LB/FT

PLAN

ROOF DIAPHRAGM

BLOCKING
BETWEEN
ROOF JOISTS

TOP PLATES

STUDS

CRIPPLE STUDS

CHORD

HEADER

ELEVATION WALL 1

Figure 10.A

d. The maximum chord force in the shear segment along Line 1, neglecting any resisting moment. Is a doubled 2 ×6 stud adequate to resist this chord force? Lumber is No. 2 DF-L. A single row of $\frac{3}{4}$-in.-diameter bolts is used for the anchorage connection at the base of the wall. $C_M = 1.0$, and $C_t = 1.0$.

e. The uplift force for the design of the connection at the base of the tension chord. Consider the resisting moment provided by the wall and the roof dead load supported by the wall. The wall supports a 5-ft tributary width of roof dead load. Roof D = 10 psf. Wall D = 15 psf.

10.3 *Given:* The single-story wood-frame building in Fig. 10.A. The *transverse* lateral wind force (not shown in the sketch) is 375 lb/ft. The dead load of the roof and the wall is 250 lb/ft along the wall. Texture 1-11 (Sec. 8.13) plywood siding is applied directly to studs spaced 16-in o.c.

Find: a. Unit shear in the roof diaphragm and in the shearwalls along lines A and B.

b. The required nailing for the plywood siding to function as a shear segment. Use galvanized casing nails. Show all criteria. Check the proportions of the shear segment.

c. The net design uplift force at the base of the shear segment chords.

10.4   *Given:*   The wood-frame building in Fig. 10.A. The critical lateral wind force in the longitudinal direction is given. The exterior wall finish is stucco ($\frac{7}{8}$-in. portland cement plaster over woven wire mesh).

      Find: Determine if the longitudinal wall along Line 2 is capable of functioning as a shearwall for the given lateral force. Specify nailing. Check diaphragm proportions. Show all criteria.

10.5   Use the hand solution to Prob. 10.1, 10.2, or 10.3 as a guide to develop a computer *template* to solve similar problems.

a. Consider only the specific criteria given in the problems.

b. Expand the template to handle any unit shear covered in IBC Table 2306.4.1. In other words, input the values from the tables into a spreadsheet table *or* database and have the template return the required nail specification and blocking requirements.

10.6   *Given:*   The one-story wood-frame building in Fig. 10.B. The critical lateral force in the longitudinal direction is the given *strength-level* (unfactored or not reduced) seismic force. The roof overhang has a higher lateral force because of the larger dead load of the plastered eave. Neglect the seismic force generated by the dead load of the wall itself. Wall sheathing is $\frac{15}{32}$ in. STRUCTURAL I with 10d common or galvanized box nails. $C_M = 1.0$ and $C_t = 1.0$.

PLAN

Figure 10.B

Find: *a.* Design the shear segments in the wall along Line 1. Show all criteria including nailing, blocking, and shear segment proportions.
    b. Check shear segment chords using two 2 × 6s of No. 2 DF-L. Consider the resisting moment provided by a wall dead load of 20 psf. Assume no roof dead load frames into wall. A single row of $\frac{3}{4}$-in.-diameter bolts is used to anchor the chord to the foundation.
    c. Plot the distribution of the strut force along Line 1. Check the stresses caused by the maximum strut force if a double 2 × 6 wall plate serves as the collector (strut). A single row of $\frac{7}{8}$-in.-diameter bolts is used for the splices. Lumber is No. 2 DF-L.

10.7 *Given:* The one-story wood-frame building in Fig. 10.B. The longitudinal *strength-level* (unfactored or not reduced) seismic force to the roof diaphragm is given. Texture 1-11 (Sec. 8.13) plywood siding is applied over $\frac{5}{8}$-in. gypsum sheathing for the wall along Line 2. In addition to the load shown, consider the seismic force generated by the dead load of the top half of the wall. Assume wall D = 15 psf throughout and a seismic coefficient of 0.200.

Find: *a.* Design the shear segments in the wall along Line 2, using galvanized box nails. Show all criteria.
    b. Check the shear segments chords, using two 2 × 4s of No. 2 DF-L. Neglect any resisting moment. A single row of $\frac{5}{8}$-in.-diameter bolts is used to anchor the chord to the foundation. $C_M = 1.0$ and $C_t = 1.0$.
    c. Plot the distribution of the strut force along Line 2. Check the stresses caused by the maximum strut force if a double 2 ×4 wall plate serves as the strut. A single row of $\frac{5}{8}$-in.-diameter bolts is used for the splices. Lumber is No. 2 DF-L.

10.8 *Given:* The one-story wood-frame building in Fig. 10.B. Roof dead load D = 10 psf. Wall dead load D = 18 psf (assume constant throughout). Wind = 20 psf. Seismic coefficient = 0.650

Find: *a.* Calculate the critical lateral force in the transverse direction.
    b. The exterior wall covering for the transverse walls is stucco ($\frac{7}{8}$-in. Portland cement plaster on expanded metal lath). Determine if these walls can function adequately as shearwalls. Consider the shear at the midheight of the wall. Show all criteria and give construction requirements.

9.10 *Given:* The one-story wood-frame building in Chap. 9, Fig. 9.D. The transverse lateral wind force is w_r = 355 lb/ft. The height of the building is 10 ft.

Find: *a.* The unit shears in the three transverse shearwalls.
    b. Design the three transverse shearwalls, using $\frac{3}{8}$-in. wood structural panel sheathing and 8d common or galvanized box nails. Studs are 16-in. o.c. Show all criteria including shearwall proportions.
    c. Calculate the chord forces for each of the three transverse shearwalls. Because the layout of the roof framing is not completely given, calculate the chord forces using the gross overturning moments.

10.10 *Given:* The elevation of the two-story wood-frame shearwall in Fig. 10.C. The wind force reactions from the roof diaphragm and second-floor diaphragm are

ELEVATION WALL

Figure 10.C

shown in the sketch. Vertical loads carried by the wall are also shown in the sketch (dead load values include the weight of the wall). Wall sheathing is $^{15}/_{32}$-in. STRUCTURAL I with 8d common or galvanized box nails.

Find: *a.* The required nailing for the shear segment. Give full specifications.
  b. The required tie (collector strut) forces for the connections at *A*, *B*, *C*, and *D*.
  c. The net design uplift force for the tension chord at the first- and second-floor levels.
  d. The maximum force in the compression chord at the first- and second-floor levels. Consider the effects of both gravity loads and lateral forces. Assume that the header is supported by the shear segment chord.
  e. If the allowable shear load on one anchor bolt is 1500 lb, determine the number and approximate spacing of anchor bolts necessary to transfer the shear from bottom wall plate to the foundation. These bolts are in addition to those required for uplift.

10.11 *Given:* The wood-frame shearwall in Fig. 10.D. The lateral wind force from the roof diaphragm to the end shear segment is given. Gravity loads are also shown on the sketch. The dead load includes both roof and wall dead loads. The shear segment has $^{15}/_{32}$-in. sheathing.

Find: *a.* Determine the required nailing for the sheathing.
  b. Calculate the net uplift force for the shear segment chord. Does this occur at column *A* or *B*?
  c. Determine the maximum compressive force in the shear segment chord. Consider the effects of gravity loads. Assume that the header is supported by the shear segment chord. What is the critical load combination? Specify which chord is critical.

ELEVATION

Figure 10.D

    d. Assume that the allowable shear load per anchor bolt is 1000 lb. Determine the required number and approximate spacing of anchor bolts necessary to transfer the wall shear from the bottom plate to the foundation. These bolts are in addition to those required for uplift.

10.12 *Given:* The elevation of the two-story wood-frame shearwall in Fig. 10.E. Vertical loads and lateral wind forces are shown in the sketch. Wall framing is 2 in. nominal. Studs are 16-in. o.c. Wall sheathing is $^{15}\!/_{32}$-in. STRUCTURAL I with 8d common or galvanized box nails.

    Find: *a.* The required nailing for the first- and second-story shear segments.
            b. The net uplift force at the base of shear segments 1 and 2.
            c. The maximum compressive force at the end of the first-floor shear segment. State the critical load combination.
            d. Assume that the allowable shear load of an anchor bolt parallel to the wall is 1500 lb. Determine the required number and approximate spacing of anchor bolts necessary to transfer the shear from the wall into the foundation. These bolts are in addition to those required for uplift.

10.13 Repeat Prob. 10.12 except that shear segment 1 is 12 ft long and lies directly below shear segment 2.

10.14 *Given:* The two-story shearwall system shown in Fig. 10.E. Wall framing is 2 in. nominal. Studs are 16-in. o.c. Wall sheathing is $^{15}\!/_{32}$-in. STRUCTURAL I with 8d common or galvanized box nails with nailing 6-in. o.c. along

ELEVATION

Figure 10.E

edges and 12-in. o.c. in the field. The second-story windows are 4 ft high, and the first-floor door is 8 ft high. The loads shown in Fig. 10.E are not applicable to this problem. Assume that seismic forces govern.

Find: The maximum lateral loads R_R and R_2 using the perforated shearwall design method.

10.15 *Given:* The one-story wood-frame building shown in Fig. 10.B. The *strength-level* (unfactored or not reduced) seismic force in the longitudinal direction is shown. Wall framing is No. 2 grade DF-L 2 × 8s, 16-in. o.c. The chords are two 2 × 8 nailed together. Wall sheathing is $^{15}/_{32}$-in. STRUCTURAL I with 8d galvanized box nails 3-in. o.c. at the edges and 12-in. o.c. in the field. Wall dead load D = 20 psf. A low-slip anchorage detail is used with $d_a = 0.125$ in.

Find: *a.* The story drift Δ_s of wall Line 1.
       b. The story drift Δ_s of wall Line 2.
       c. Compare the results of parts *a* and *b* to the IBC limit.

Wood Connections—Background

11.1 Introduction

The proper design of connections plays a critical role in the overall performance of wood structures. Post-disaster surveys following earthquakes and hurricanes have repeatedly shown that structural failures often occur due to inadequately designed or improperly fabricated connections. The importance of connection design is evident in the NDS (Ref. 11.2), with over 45 percent of the document devoted to fasteners and connections.

The NDS presents an engineering mechanics approach for the design of laterally loaded connections known as the *yield limit model for dowel-type fasteners.* The yield limit model is equally applicable for connection design in either allowable stress design (ASD) or load and resistance factor design (LRFD). The primary differences between ASD and LRFD from the standpoint of the designer involve the use of a format conversion factor K_F and resistance factor ϕ for LRFD, and the use of a time effect factor λ for LRFD instead of the load duration factor C_D for ASD. The designer has the option of applying the yield limit equations directly (which is best done on a computer or programmable calculator) or using a set of fastener capacity tables based on the yield limit equations.

Chapter 11 starts by introducing the common types of fasteners used in typical wood connections, and the notation system for connection design is described. Greater detail about the connections and the required adjustment factors is included in the chapters that deal with specific fasteners. The chapter concludes with a review of the general concepts of the yield limit theory. Important strength considerations are included along with a summary of the various possible yield modes that are considered in this approach to connection design.

11.2 Types of Fasteners and Connections

There are a wide variety of fasteners and many different types of joint details that can be used in wood connections. When structural members are attached with fasteners or some other type of hardware, the joint is said to be a *mechanical connection.*

Mechanical connections are distinguished from connections made with *adhesives.* Adhesives are normally used in a controlled environment, such as a glulam, plywood, structural composite lumber, or wood I-joist manufacturing plant under the control of a formal quality assurance program. In some cases adhesives are used in the field in combination with nails for attaching sheathing to framing members. However, since adhesives and mechanical fasteners exhibit different levels of connection rigidity, increases in connection capacity are not typically associated with their combined usage. The connections covered in this discussion are limited to mechanical connections.

The design of connections using the following types of fasteners is covered in this book:

1. Nails

2. Bolts

3. Lag bolts or lag screws

4. Split ring and shear plate connectors

Some background material is introduced here, but much more detailed descriptions of the fasteners and design requirements for connections are given in Chaps. 12 and 13.

Connections are generally classified according to the direction of loading. *Shear connections* have the load applied *perpendicular* to the length of the fastener. See Example 11.1. These connections are further classified as to the number of shear planes. The most common applications are *single shear* and *double shear,* but additional shear planes are possible. According to the NDS (Sec. 11.3.8), connections with more than two shear planes are to be analyzed by evaluating each shear plane as a single-shear connection. The design value for the weakest shear plane is then multiplied by the number of shear planes to obtain the total connection capacity.

EXAMPLE 11.1   Typical Wood Connections Subjected to Shear

The most common fasteners used in wood connections are nails, bolts, and lag bolts. In shear connections the load acts perpendicular to the length of the fasteners. One or more shear planes occur. Split ring and shear plate connectors are also used in shear-type connections to provide additional bearing area on the wood members for added load capacity. See Figs. 11.1*a* to 11.1*d.*

The symbol Z represents the reference fastener load capacity for these types of connections. *Shear* connections are also known as *laterally loaded* connections. The other basic type of connections in wood design is withdrawal connections (Example 11.2).

NAIL CONNECTION

Figure 11.1a Single-shear (two member) wood-to-wood *nail connection*. A similar connection can be made with a steel side plate (wood-to-metal).

BOLT CONNECTIONS

Figure 11.1b Single-shear (two-member) and double-shear (three-member) *bolt connections*. Wood-to-wood connections are shown, but connections with steel side plates are common (wood-to-metal).

LAG BOLT

Figure 11.1c Single-shear (two-member) wood-to-wood *lag bolt connection*. Another common application is with a steel side plate (wood-to-metal)

SPLIT RING CONNECTOR

Figure 11.1d *Split ring and shear plate connectors* fit into precut grooves. A split ring connector (shown in sketch) is used only for a wood-to-wood connection. Shear plate connectors can be used for either wood-to-wood or wood-to-metal connections. See Fig. 13.27 for photographs of split ring and shear plate connectors in partially assembled joints.

The second major type of loading in a wood connection has the applied load *parallel* to the length of the fastener, and the fastener is loaded in tension. When nails and lag bolts are subjected to this type of loading, the concern is that the fastener may pull out of the wood member. *Withdrawal loading* is the term applied to this situation (see Example 11.2). Reference withdrawal values are tabulated in the NDS (Ref. 11.2) for nails, wood screws, and lag bolts.

Bolts may be loaded in tension, but such connections are not subject to withdrawal. The strength of the fastener in tension is usually not critical, but technically it should be verified. Compression perpendicular to grain under the head of a bolt or lag bolt is a normal design consideration. A washer or plate of sufficient size must be provided between the wood member and the fastener head so that the adjusted bearing design value $F'_{c\perp}$ is not exceeded. Refer to Sec. 6.8 for a review of adjusted compressive strength perpendicular to grain. Finally, it should be noted that split ring and shear plate connectors are basically used in shear connections and are not subjected to tension.

EXAMPLE 11.2 Typical Wood Connections Loaded in Withdrawal and Tension

Withdrawal loading attempts to pull a nail or lag bolt (lag screw) out of the main (holding) wood member (Fig. 11.2*a*). The symbol W represents the reference fastener load capacity for these types of connections. Note that the main member is the member that receives the pointed end of the fastener. In a withdrawal connection the load is parallel to the length of the fastener, and the fastener is stressed in tension. Bolts can be loaded in tension, but they are not subject to withdrawal (Fig. 11.2*b*).

A washer of adequate size must be provided under the head and nut of a bolt loaded in tension, and under the head of a lag bolt loaded in withdrawal, so that the adjusted compression design value perpendicular to the grain is not exceeded.

Figure 11.2*a*   Nail and lag bolt connections subject to withdrawal loading.

TENSION CONNECTION

Figure 11.2*b*   Bolts used in tension.

The notation system for fastener design is based on the two types of loading just described. The following symbols are used:

Z = reference design value for single fastener subjected to lateral *shear* load

W = reference design value for single fastener subjected to *withdrawal* load

The term *reference design value* refers to the basic load capacity for a fastener as defined by a table or equation in the NDS. As with all tabulated properties for wood design, the reference fastener design value applies to a specific set of conditions. These idealized conditions include such things as loads of *normal* (10-year) *duration*, lumber that is *initially dry* and *remains dry* in service, no sustained exposure to high *temperature*, and so on. The idealized conditions for connections are similar to the conditions that apply to the tabulated design values for a wood member such as a beam. Thus, the reference fastener design value is simply the starting point for determining the adjusted load capacity for a particular structural application.

Adjustment factors for load duration (ASD) or time effect (LRFD), moisture content conditions, temperature, and a number of other design considerations may be required to convert the reference design value to an *adjusted design value* for a single fastener. A prime is added to the symbol to indicate that the appropriate adjustments have been applied:

Z' = Z × (product of applicable adjustment factors for shear connection)

W' = W × (product of applicable adjustment factors for withdrawal connection)

It should be noted that the NDS provides design criteria for a number of fasteners in addition to those listed above. The appropriate adjustment factors for all fasteners covered in the NDS are summarized in NDS Table 10.3.1, *Applicability of Adjustment Factors for Connections.* The specific adjustment factors for nails, bolts, lag bolts, and split ring and shear plate connectors are reviewed in detail in Chaps. 12 and 13.

The fasteners covered in this book are those that are most likely to be found in a typical wood frame building. Other fasteners included in the NDS are

1. Wood screws
2. Metal plate connectors
3. Timber rivets
4. Drift bolts and drift pins
5. Spike grids

Wood screws are similar to lag bolts, but wood screws are smaller in diameter, have a slightly tapered shank, and do not have a hex (or square) bolt head. Wood screws are inserted with a screwdriver or screw gun, but lag bolts are installed with a wrench. Wood screws have adjusted design values for lateral load comparable to nails. Due to the gripping action of threads, wood screws have significantly larger adjusted design values for withdrawal load than nails. They have some application in manufactured housing, but are not often used in conventional wood frame construction. For larger design loads, lag bolts are the more likely choice.

Metal plate connectors are frequently found in wood buildings that use prefabricated light-frame wood trusses. A photograph of metal plate connected trusses was given in Fig. 6.33b. Metal plate connectors are light-gage metal plates that connect the truss members together. Metal plate connectors are also known as *nail plates* or *truss plates.* Truss plates are typically produced by pressing or stamping sheet metal to form teeth or barbs which protrude from the side of the plate. See Fig. 11.3. The protrusions serve as the "nails" for the truss plate and become embedded in the wood members during fabrication of a wood truss. Alternatively, conventional nails can be driven through holes punched (without protrusions) in nail plates.

The load capacity of a particular type of truss plate is determined by testing, and the load rating is usually given in pounds per square inch of contact area between the metal plate and the wood member. When a truss is analyzed, the forces to be transferred between the wood members are determined. The required size of metal plate connector is then obtained by dividing the force by the adjusted design value (load per square inch) for the truss plate. The overall size and configuration of a metal plate connector must provide the necessary contact areas for each member in the joint. Truss plates can be found in many wood buildings, but they can be viewed as a specialty item. The design of metal plate connectors is usually handled by a truss manufacturer, and fabrication is typically restricted to a manufacturing plant operating under a formal quality control program. See Ref. 11.11 for information.

Figure 11.3   Metal plate connectors. *(Courtesy of the FPL.)*

Timber rivets (also called *glulam rivets*) are a type of fastener introduced in the 1997 edition of the NDS. The reader should not confuse a timber rivet with a rivet in steel construction. They do not look anything alike. Timber rivets are hardened steel fasteners shaped somewhat like a nail, but with a countersunk head and a rectangular-shaped cross section with rounded corners. The cross-sectional dimensions are approximately $\frac{1}{8}$ in. by $\frac{1}{4}$ in., and rivet lengths range from $1\frac{1}{2}$ in. to $3\frac{1}{2}$ in. Timber rivets are driven with a hammer through predrilled holes in steel side plates. They are installed with the longer ($\frac{1}{4}$ in.) cross-sectional dimension parallel-to-grain to prevent splitting in wood members. Their primary application is for resisting large loads in multiple rivet, heavy glulam timber connections.

Drift bolts and *drift pins* are unthreaded steel rods that are driven into a hole bored through one wood member and into the adjacent member. *Spike grids* are intended for use in wood pile and pole construction. Drift pins and spike grids are also considered specialty fasteners. A number of fasteners are used in wood construction in addition to those included in the NDS. Information on a variety of fasteners may be obtained from the *Wood Handbook* (Ref. 11.8), the *Timber Construction Manual* (TCM) (Ref. 11.4), the International Staple, Nail, and Tool Association (see Ref. 11.9), Ref. 11.6, and catalogs of connection hardware suppliers. Some of the hardware used in conjunction with wood fasteners is covered in Chap. 14.

11.3   Yield Model for Laterally Loaded Fasteners

In the past, the capacity of laterally loaded nails, bolts, and lag bolts was obtained from fastener capacity tables given in earlier editions of the NDS.

These tables were also included in building codes, typically in abbreviated form. The load values in these tables were obtained from empirical formulas which are summarized in the NDS Commentary (Ref. 11.1) and various editions of the *Wood Handbook* (Ref. 11.8).

The empirical formulas were based on the *proportional limit* strength of connections that were tested at the Forest Products Laboratory (FPL) during the 1930s. The empirical formulas from the *Wood Handbook* contain parameters that are important to connection performance such as fastener diameter and length. However, the empirical expressions simply reproduce load values that were obtained from laboratory testing programs. These formulas cannot be derived using principles of engineering mechanics and may not be appropriate for all connections.

In the 1940s an approach to connection analysis was developed in Europe that was based on the yielding of the various elements in the connection. In the 1980s a number of papers were published by FPL and other researchers that confirmed the yield limit theory. These studies concluded that an engineering mechanics approach based on yield limit theory is appropriate for analyzing *dowel-type fasteners* in wood connections (Refs. 11.7 and 11.10).

A dowel is traditionally thought to be a cylindrical rod of wood or metal that is placed in a hole which has been drilled in a wood member or members. However, in the context of the NDS, the term *dowel-type fastener* simply refers to fasteners with a generally cylindrical shape such as bolts, screws, and nails.

The idea is that these fasteners act similar to a dowel as they bear against the grain in a wood member under a lateral load. The term *lateral load* here refers to a load that is perpendicular to the length of the fastener, that is, a *shear connection*. The load may or may not include the effects of lateral wind or earthquake forces. The yield limit model applies only to shear connections and not to fasteners loaded in withdrawal. Withdrawal connections are still covered by empirical formulas provided in the NDS.

In utilizing the yield limit theory, the various yield modes that can occur in a given type of connection are analyzed. A load capacity is computed for each of the various modes. The yield limit is then taken as the smallest of these load capacities. This critical load defines the *yield mechanism* for the connection. It should be noted that the yield limit equations for Z in the NDS include appropriate factors of safety for allowable stress design (ASD).

An engineering mechanics approach to connection design, such as the yield limit model, has both advantages and disadvantages. The principal advantage is that the designer can mathematically analyze a connection of practically any configuration. Thus, the designer is not limited to joint configurations and details that are given in the NDS or Code tables. In the past, if a connection did not match the details in the table, the designer was forced to either revise the connection or extrapolate values from a table that may not have been appropriate.

In addition, the yield limit model is able to take into account the strengths of different fastener materials. In the older empirical method, a certain fastener strength was simply presumed. However, the yield theory directly incorporates the yield strength of a fastener into the analysis. A final advantage is that the yield limit model can be used to evaluate the strength of a connection involving wood members of different species, or a connection between a wood member and another material.

On the negative side, the application of the yield model to connection design can involve the use of some rather cumbersome equations. The number of equations varies depending on whether the fastener is in single shear or double shear. However, regardless of the particular conditions, application of the yield limit model in equation form requires the consideration of a number of possible yield modes. The equations can be lengthy and awkward to solve by hand on a calculator. Fortunately, the yield limit equations are easy to program and solve on a computer using any of the popular spreadsheet or equation-solving software.

The NDS contains a complete set of yield limit equations for the analysis of dowel-type fasteners. However, the NDS takes a balanced approach to this problem. It recognizes that some designers may not need or want the sophistication provided by the formal yield limit equations. Thus, for the designer who does only an occasional wood connection design, or for the designer who simply does not want to explicitly evaluate the yield limit equations, the NDS also provides fastener capacity tables for various connection configurations.

The NDS fastener capacity tables are based on the yield limit equations, but are simplified from the general theory by reducing the number of parameters used to obtain load capacity. For example, the tables apply to connections with all members from the same species of wood, and they cover only a limited number of species (specific gravities). The tables for bolts and lag bolts also handle only parallel- and perpendicular-to-grain loadings. Therefore, the tables are simple, conservative, and relatively easy to use. They apply to a certain set of common, but limited, conditions.

The fastener capacity tables are limited in scope, and the real advantage of the yield limit model comes in the automated evaluation of the yield equations. In this way all of the details affecting a particular connection may be taken into account. In addition, there are some fastener problems that *require* use of the yield limit equations. For example, the NDS bolt tables cover only parallel- and perpendicular-to-grain loading, and the equations will be necessary to evaluate loading at some other angle to grain.

A programmed solution to the yield limit equations using a spreadsheet or equation-solving software can be easier than looking up a value in the NDS fastener tables. Solving the yield limit equations is, in fact, an elementary task on a spreadsheet. After the equations have been solved once, a *template* is created which can be used for other similar problems. The template can also be expanded to include some of the other design provisions such as the adjustment factors for wood connections.

The remaining portions of this chapter summarize several of the important properties that affect the behavior of dowel-type connections. The purpose is to develop a general understanding of the rationale behind the theory, without attempting to derive the lengthy equations. The possible yield modes are also reviewed for typical two-member and three-member connections.

11.4   Factors Affecting Strength in Yield Model

It has been stated that the yield limit model is appropriate for the design of nails, bolts, lag bolts, and wood screws used in shear connections. For reasons given in Sec. 11.2, the design of wood screw connections is not covered in this book. Split ring and shear plate connectors may not behave as dowel-type fasteners, and because of their rather limited use, they are addressed only briefly (Chap. 13).

The yield limit model for dowel-type fasteners is based upon engineering mechanics, and it uses *connection geometry* and *material properties* to evaluate strength. The primary factors used to compute the reference design value Z include

1. Fastener diameter D
2. Dowel bearing length l
3. Fastener bending yield strength F_{yb}
4. Dowel bearing strength of wood member F_e
   a. Specific gravity (dry) of wood member G
   b. Angle of load to grain for wood member θ
   c. Relative size of fastener (large or small diameter)
5. In the case of metal side plates, the dowel bearing strength F_e of the metal

The notation in the above list is usually expanded by adding subscripts m and s to the appropriate terms to indicate whether the symbols apply to the main member or the side member. For example, l_m, F_{em}, G_m, and θ_m refer to the dowel bearing length, dowel bearing strength, specific gravity, and angle of load to grain for the *main member*. Likewise, l_s, F_{es}, G_s, and θ_s represent the corresponding properties for the *side member*. *Member thicknesses* (t_m and t_s) were labeled in Fig. 11.1. For bolted connections, the dowel bearing length l and member thickness t are identical. However, for nail, wood screw, and lag bolt connections the dowel bearing length in the main member is typically less than the main member thickness ($l_m < t_m$). In a two-member bolted connection, the main member is the thicker member; and in a three-member connection, the main member is the center member. In a two-member nail or lag bolt connection, the member holding the pointed end of the fastener is the main member. The other members in the connections are the side members.

The *bending yield strength of the fastener* F_{yb} is a property that is used to predict the load capacity of a mechanism that involves the formation of a plastic hinge (Sec. 11.6) in the fastener. The NDS provides typical values of bending yield strength for various types of fasteners. These values may be used in the event that material properties for a specific fastener are not available. Values of F_{yb} for the fasteners given in the NDS are representative of those found in the field and are generally thought to be conservative. See NDS Appendix I for additional information.

The *dowel bearing strength* F_e is the strength property of the members in a connection that resists *embedding* of a dowel. Designers familiar with structural steel can best relate this property to the available bearing strength (R_n/Ω for ASD, or ϕR_n for LRFD) of a steel plate in a bolted connection.

In wood design, the value of F_e depends on a number of different factors. The formulas for dowel bearing strength are given in Sec. 11.5. It is important to understand that the members in a connection can have different dowel bearing strengths. Because F_e depends on the specific gravity of the wood, one way to have different dowel bearing strength in the same connection is to have main and side members from different species.

The angle of load to grain θ may also cause F_e to be different for the main and side members. Several typical uses of bolts and lag bolts in shear-type connections were illustrated in Example 11.1 (Sec. 11.2). The connections in this earlier example were made up of only wood members, and in each case the load was applied parallel to the length of the member. The load, therefore, was parallel to the grain, and the angle of load to grain was zero ($\theta = 0$ degrees).

The wood members in a connection may have different angles of load to grain. For example, the load from a fastener can cause bearing parallel to the grain in one member and bearing perpendicular to the grain in another member. Other connections may involve loading at an angle θ other than 0 or 90 degrees to the grain. See Fig. 11.4a in Example 11.3.

EXAMPLE 11.3 Examples of Different Dowel Bearing Strengths in the Same Connection

Different dowel bearing strengths can result from different angles of load to grain or the use of side members of different materials.

Figure 11.4a Depending on fastener size, the angle of load to grain may have an effect on the dowel bearing strength of a wood member. In the sketch on the left, the load transmitted by the fastener is parallel to the grain in the horizontal wood member and perpendicular to the grain in the vertical member. On the right, the load is at an angle θ to the direction of the grain in the wood member.

Figure 11.4b Dowel-type fasteners are often used in wood connections with steel side plates. Single-shear and double-shear applications are common in wood-to-metal connections. The dowel bearing strength F_e for a steel plate is obviously much larger than F_e for a wood member.

The dowel bearing strength for a wood member is greatest when the load is parallel to the grain and weakest for a load perpendicular to the grain. For intermediate angles, the bearing strength lies between the parallel- and perpendicular-to-grain values. (Exception: small-diameter fasteners have the same dowel bearing strength regardless of angle of load to grain—see Sec. 11.5.)

In other common connections, metal side members are used. These can be full-size structural steel members, but often the side plates are some form of metal connection hardware. See Fig. 11.4b. The dowel bearing strength of a steel plate is much greater than the dowel bearing strength of a wood member.

If a metal side member is used in a connection, the plate is to be designed in accordance with recognized engineering practice. For example, a steel side plate should meet the requirements of the American Institute of Steel Construction (Ref. 11.3) or the American Iron and Steel Institute (Ref. 11.5). In addition, a multiple of the *ultimate tensile strength* of the steel F_u is used to define the dowel bearing strength of metal plates for use in the yield limit equations for a wood connection.

The major factors that affect the reference design value Z of an individual dowel-type fastener according to the yield limit model have been briefly introduced. Some additional information regarding dowel bearing strength and the bending of fasteners in a connection is given in the following sections.

Recall that there are a number of other factors that may influence the adjusted design value Z' for a connection. Some of these factors apply to all types of connections (e.g., wet service factor C_M). Some factors apply only to certain types of fasteners (e.g., penetration depth factor C_d for split ring and shear plate connectors). Other factors apply only for specific design procedures (e.g., the load duration factor C_D for ASD, versus the time effect factor λ for LRFD). In particular, the format conversion factor K_F and the resistance factor ϕ_z apply only for LRFD and have the same magnitude for all types of fasteners in the NDS:

$$\phi_z = 0.65$$

$$K_F = 2.16/\phi_z = 3.32$$

As explained in Chap. 4, the format conversion factor K_F adjusts from an ASD reference design value to a nominal LRFD design value. Other adjustment factors for connections are covered in subsequent chapters dealing with specific fastener types.

11.5 Dowel Bearing Strength

The *property of wood* that affects the reference design value Z of a nail, bolt, or lag bolt is known as *dowel bearing strength*. In much of the research literature, this property is termed *embedding strength*, hence the notation F_e. It is related to the crushing strength of the wood member under loading from a dowel subjected to a shear load. As noted in the previous section, the dowel bearing strength varies with the *specific gravity* of the wood and the relative size of the dowel. For large-diameter fasteners, the *angle of load to grain* also affects the dowel bearing strength.

Values of specific gravity G for the various combinations of lumber species are given in NDS Table 11.3.2A. Dowel bearing strengths F_e are given in NDS Table 11.3.2 for various values of G, and in NDS Table 11.3.2B for plywood and oriented strand board (OSB) panels. Dowel bearing strengths for structural composite lumber products must be obtained from the manufacturer's technical literature or Code evaluation reports. In practice the dowel bearing strengths for lumber, glulam, or panel products in a connection problem will typically be obtained directly from NDS Tables 11.3.2 and 11.3.2B. This will undoubtedly be the case for a hand solution of the yield limit equations. In contrast, if the equations for Z are programmed on the computer or solved on a spreadsheet, the *formulas for F_e* given in Example 11.4 may be useful.

Tests have shown that the dowel bearing strengths for *large-diameter* fasteners depend on the angle of load to grain. Consequently, there is one formula for dowel bearing strength for parallel-to-grain loading ($F_{e\parallel}$) and another for perpendicular-to-grain loading ($F_{e\perp}$). For intermediate angles of load to grain, the Hankinson formula is used to obtain a dowel bearing strength ($F_{e\theta}$) that lies between $F_{e\parallel}$ and $F_{e\perp}$. The Hankinson formula was first introduced in Sec. 6.8 in a form for evaluating the adjusted bearing design value in a wood member. The same expression, with a simple change in notation, is appropriate for determining the dowel bearing strength for fastener design.

Small-diameter dowels do not exhibit the directional strength characteristics of large fasteners, and a single expression defines dowel bearing strength regardless of the direction of loading. The shank diameter separating the categories of large dowels and small dowels is $\frac{1}{4}$ in. Large-diameter fasteners (such as bolts and lag bolts) are installed in drilled holes. In contrast, small dowel-type fasteners (such as nails) are usually driven into the wood with a hammer or pneumatic gun.

EXAMPLE 11.4   Formulas for Dowel Bearing Strength

Numerical values of dowel bearing strength are listed in the NDS:

Sawn lumber and glulam	NDS Table 11.3.2
Plywood and OSB	NDS Table 11.3.2B

$F_{e\parallel}$ - PARALLEL TO GRAIN

$F_{e\perp}$ - PERPENDICULAR TO GRAIN

$F_{e\theta}$ - ANGLE TO GRAIN θ

F_e - ANY ANGLE TO GRAIN

LARGE   DOWELS   **SMALL DOWELS**

Figure 11.5   Definition of dowel bearing strength depends on relative size of fastener. The dowel bearing strength F_e for *large*-diameter fasteners is a function of angle of load to grain. Specific gravity and diameter are also important. The F_e for *small*-diameter fasteners depends only on specific gravity.

The values in the NDS tables are based on the following expressions:

LARGE DOWELS ($D \geq 0.25$ in.):

Parallel-to-grain loading: $F_{e\parallel} = 11{,}200(G)$

Perpendicular-to-grain loading: $F_{e\perp} = 6100(G^{1.45})\,D^{-0.5}$

The Hankinson formula gives F_e at an angle of load to grain θ. See Fig. 11.5.

$$F_{e\theta} = \frac{F_{e\parallel}\,F_{e\perp}}{F_{e\parallel}\sin^2\theta + F_{e\perp}\cos^2\theta}$$

SMALL DOWELS ($D \geq 0.25$ in.):

All angles of load to grain: $F_e = 16{,}600\,G^{1.84}$

where F_e = dowel bearing strength; subscripts $\parallel$, $\perp$, and θ indicate strengths parallel to grain, perpendicular to grain, and at an angle to grain θ
  G = specific gravity (dry) of wood member
  D = diameter of dowel-type fastener

The yield limit model is capable of handling a wide variety of wood connections including those in which the main and side members have different dowel bearing strengths. Several situations are described in Sec. 11.4 where F_{em} and F_{es} are not equal. On the other hand, in a number of design conditions the dowel bearing strength will be the same for the main and side members.

The formulas for dowel bearing strength in Example 11.4 are provided in the footnotes to NDS Table 11.3.2. Again, these formulas may be more convenient than the tables for F_e if the yield limit equations for Z are programmed on a computer. The dowel bearing strength formulas are evaluated for several fasteners in Example 11.5 to simply confirm the NDS tables.

EXAMPLE 11.5   NDS Dowel Bearing Values

Use the appropriate formulas from Example 11.4 to evaluate the dowel bearing strength values in NDS Table 11.3.2 for the following fasteners. Note that the values for F_e in the NDS tables are rounded to the nearest 50 psi.

LARGE DOWELS

Bolts:

a. $\frac{5}{8}$-in.-diameter bolt in Douglas Fir-Larch (DF-L) with load applied parallel to grain ($\theta = 0$ degrees) Values of specific gravity are available in NDS Table 11.3.2A. For DF-L the specific gravity is 0.50.

$$F_{e\parallel} = 11{,}200(G) = 11{,}200 \times 0.50 = 5600 \text{ psi}$$

This value agrees with $F_{e\parallel}$ in NDS Table 11.3.2.

b. $\frac{3}{4}$-in.-diameter bolt in DF-L with load applied perpendicular to grain ($\theta = 90$ degrees)

$$F_{e\perp} = 6100(G)^{1.45}D^{-0.5}$$

$$= 6100(0.50)^{1.45}(0.75)^{-0.5}$$

$$= 2578 \text{ psi} \approx 2600 \text{ psi} \quad \text{from NDS Table 11.3.2}$$

SMALL DOWELS

Nails: 16d common nail at any angle of load to grain in DF-L

$$F_e = 16{,}600(G)^{1.84} = 16{,}600(0.50)^{1.84}$$

$$= 4637 \text{ psi} \approx 4650 \text{ psi} \quad \text{from NDS Table 11.3.2}$$

In the case of a connection between a wood member and a metal side plate, it is necessary to have the equivalent of an embedding strength value for the metal member. Steel design standards specify bearing strength of 2.4 times the ultimate tensile strength ($2.4F_u$) for hot-rolled steel plates (Ref. 11.3) and 2.2 times the ultimate tensile strength ($2.2F_u$) for cold-formed steel plates (Ref. 11.5). As noted in NDS Appendix I, steel bearing strengths are reduced by a factor of 1.6 to obtain dowel bearing strengths F_e that can be used directly in NDS *yield limit equations*. The 1.6 reduction factor permits F_e values for steel to be used in conjunction with applicable load duration or time effect factors for wood members in a connection. Thus, the dowel bearing strengths for steel members in connections are $F_e = 1.5F_u$ for hot-rolled steel plates and $F_e = 1.375F_u$ for cold-formed steel plates. In general, steel plates with thickness less than $\frac{1}{4}$ in. are assumed to be cold-formed steel, and plates with thickness greater than or equal to $\frac{1}{4}$ in. are assumed to be hot-rolled steel. For wood members in connections, F_e is defined as in Example 11.4.

This section concludes with a brief discussion of the dowel bearing capacity of a fastener in a *glulam* beam and in *wood structural panel* sheathing. It will be recalled from Chap. 5 that higher-quality laminating stock is located at the outer tension and compression zones in a glulam bending member, and lower-quality material is placed in the inner core. In fact, the manufacturing specifications permit the mixing of different species of wood in the same member (i.e., the inner core may be from a weaker species than the outer laminations). Although mixing different species groups is permitted, most current practice is to manufacture a glulam member with material from a single species group.

If a glulam beam is produced from a single species group, a single specific gravity applies to the member. Values of G can be obtained from the NDS tables that were previously described. Dowel bearing strength for the glulam may be computed using the expressions given in Example 11.4, or the value of F_e may be read from NDS Table 11.3.2.

In the event that a glulam member is manufactured with material from more than one species group, there will be a different dowel bearing strength for each species group. See Example 11.6. A conservative approach to connection design is to use F_e based on the specific gravity for the inner core laminations. To take advantage of a larger F_e for the outer laminations, the designer will have to investigate the layup of a given member. In other words, to use the dowel bearing strength for the outer laminations, the designer must be assured that the fastener, or group of fasteners, is located entirely within the laminations with the higher F_e.

EXAMPLE 11.6   Dowel Bearing Strength in Glulam Beams

Figure 11.6   Distribution of laminations in a glulam beam follows bending stress distribution.

Most glulam beams produced today are manufactured with laminating stock from a single species group, and F_e is a function of a single specific gravity. In a member with outer laminations from one species group and inner laminations from another, two different specific gravities apply. See Fig. 11.6. To use the larger F_e, the fasteners must be located entirely within the laminations from the denser species group.

Wood structural panels such as plywood and OSB may include plies from a single species group or from multiple species groups. Preliminary research (Ref. 11.7) indicates that when all plies are from the same species group, the dowel bearing strength for a small-diameter fastener ($D < 0.25$ in.) in plywood can be obtained from the formula given in Example 11.4. However, often the species make up of plywood or OSB panels is not known by designers. According to NDS Table 11.3.2B, $F_e = 4650$ psi and $G = 0.5$ can be conservatively used for all grades of OSB and for STRUCTURAL I or Marine grades of plywood. For all other grades of plywood, $F_e = 3350$ psi and $G = 0.42$ should be used. These F_e values are based on dowel bearing tests for nails in panel products conducted by APA—The Engineered Wood Association.

11.6 Plastic Hinge in Fastener

Dowel bearing strength is the primary measure of strength of the wood members in a connection. In addition to this, the strength of the fastener can affect the yield limit mechanism of a connection. The mode of yielding in a fastener is related to the formation of one or more *plastic hinges*. Designers who are familiar with the behavior of structural steel members already understand the concept of a plastic hinge, but it is reviewed here briefly.

At normal temperatures, structural steel behaves in a ductile manner. In other words, the material can undergo large deformations before rupture occurs. This behavior is best seen on a stress-strain diagram for steel. See Fig. 11.7. The horizontal yield plateau is characteristic of the material. The ability of steel to undergo large deformations without rupture is essential to the formation of a plastic hinge.

In structural steel design, a plastic hinge is usually introduced by examining the bending stress diagram in a beam as the member is loaded through several stages. See Example 11.7. In order for a plastic hinge to form, the beam must be braced so that buckling does not occur.

Figure 11.7 Typical stress-strain diagram for a ductile steel.

EXAMPLE 11.7   Formation of a Plastic Hinge in a Steel Beam

Figure 11.8   Formation of a plastic hinge in a structural steel beam. In a simply supported beam, the formation of a single plastic hinge causes a failure or collapse
mechanism to develop.

The loading diagram for the W-shape beam has a concentrated load at the midspan.
Several bending stress diagrams are given to show various stages as the load is increased
(Fig. 11.8). The steps in the formation of a plastic hinge are as follows:

1. Small load P causes a small bending stress f_b. The member is elastic, and the stress
    diagram is triangular with zero stress at the neutral axis and maximum stress at
    the outside fibers. The maximum bending stress is below the yield point.

2. Larger bending load causes the maximum bending stress to reach yield at the out-
    side fibers ($f_b = F_y$). The moment in the beam that corresponds to this stress dia-
    gram is the yield moment M_y, and the load is known as the yield load P_y.

3. As the concentrated load is increased, the beam does not collapse (assuming that
    buckling is prevented). Because of the ductile nature of the material, the outer
    fibers simply continue to deform with no increase in stress. However, the inner
    fibers increase in stress until yielding occurs. This pattern of increasing the stress
    of the inner fibers continues until all of the fibers eventually yield.

4. When all fibers of the cross section have reached F_y, a *plastic hinge* has formed. The
    moment at the cross section is known as the *plastic moment M_p*, and the cross sec-
    tion at the point of maximum moment is incapable of carrying any additional
    moment. Thus, the plastic hinge functions as a real hinge as far as additional load-
    ing is concerned.

When a single plastic hinge forms in a simply supported beam, a failure mechanism develops. The plastic hinge at the mid-span causes the formation of an unstable linkage (three
hinges on a straight line) or failure mechanism. Note that in the sketch of the linkage, a

real hinge is shown as an open circle, and the plastic hinge is a solid circle. The load that causes a collapse mechanism to form is known as the *ultimate* or *collapse load*.

Consider what happens to the member as the load is increased to the point where the bending stress reaches yield F_y at the outside fibers. The beam does not collapse simply because the outer fibers in the member reach F_y. In fact, the member can carry additional load beyond this point as inner portions of the cross section reach yield. Because of the ductile nature of the material, this behavior continues until the entire section yields.

The deformation of the outside fibers of the beam takes place over a relatively small portion of the beam length. On either side of this localized yielding, the beam remains relatively straight. This behavior leads to the concept of a plastic hinge.

A plastic hinge corresponds to a rectangular bending stress block in which all of the fibers have reached the yield stress F_y. By definition, a real hinge is a structural element that does not transmit moment. In contrast, a plastic hinge develops the plastic moment capacity M_p of the cross section. However, once M_p is reached at the point of local maximum moment, additional bending moment cannot be developed by the member. Thus, a plastic hinge functions similar to a real hinge as far as *additional* loading is concerned.

The cross-sectional property that measures the plastic moment capacity of a member is known as the plastic section modulus Z. The plastic moment capacity is the product of the plastic section modulus and the yield stress ($M_p = ZF_y$). This relationship is analogous to the one for maximum elastic moment, which is the elastic section modulus times the yield stress ($M_y = SF_y$). Additional information on plastic section modulus can be found in books on steel structures.

Now that the concept of a plastic hinge has been introduced, the idea of a failure mechanism can be addressed in more detail. In a steel beam, a failure mechanism is a structural form that cannot support a load. It is characterized by large, unbounded displacements. In a beam or similar structure, the failure mechanism is associated with the collapse load.

The method of structural analysis that deals with the determination of collapse mechanisms is known as *plastic analysis*. Plastic analysis involves adding a sufficient number of plastic hinges to an initially stable structure to form a failure mechanism. The most elementary form of a collapse mechanism for a steel beam is one with three hinges on a straight line. Once this form develops, the structure is unstable. One of the objectives in plastic analysis is to determine the collapse load, which is the smallest load that will cause a failure mechanism to develop.

The linkage mechanism in Fig. 11.8 was created by the formation of a single plastic hinge in the beam. Whether or not a single plastic hinge causes the development of a collapse mechanism depends on the support conditions for the member. In a simply supported beam, only one plastic hinge is needed to produce a failure mechanism. However, for other support conditions, additional plastic hinges may be required. See Example 11.8. The beam in this example is fixed at one end and pinned at the other, and the development of a collapse mechanism occurs in several stages.

The first plastic hinge logically occurs at the point of maximum moment. However, the first plastic hinge does not produce a failure mechanism. For additional loading (beyond the loading required to cause the first plastic hinge), the beam appears to be a member that is pinned at both ends. There is a real hinge at one end of the member and a plastic hinge at the other end. A collapse mechanism with three hinges in line requires the formation of a second plastic hinge. The second plastic hinge develops under the concentrated load.

EXAMPLE 11.8   Formation of Two Plastic Hinges in a Steel Beam

A plastic hinge first develops at the fixed end, which is the point of maximum moment. However, a failure mechanism does not occur until an increased load causes a second plastic hinge under the concentrated load. The linkage created by the second plastic hinge has three hinges on a straight line: one real hinge and two plastic hinges. See Fig. 11.9.

Figure 11.9   Steel beam fixed at one end and pinned at the other. Two plastic hinges are required to develop a failure mechanism.

The formation of a plastic hinge in a structural steel member and the development of a failure mechanism have been discussed here for orientation and review. There are some major differences between the behavior of a dowel-type steel fastener used in a wood connection and the structural steel beams just reviewed. One difference is that the cross section of a fastener is circular. However, the configuration of the cross section does not change the concept of a plastic hinge.

Another difference is that the mechanism in a wood connection is not that of a structural steel beam. The behavior of a dowel-type fastener is more complex. In addition, there is a reserve capacity in a wood connection; hence, the term *yield limit* is applied to wood connections rather than *collapse load*.

However, the elementary beam examples are useful to define the concept of a plastic hinge and a mechanism. At this point it should simply be understood that dowel-type fasteners (bolts, lag bolts, screws and nails) of a ductile metal can develop one or more plastic hinges as a *yield mechanism* is formed in a wood connection

The NDS contains a complete set of yield limit equations to evaluate the load capacity of the various yield mechanisms. Some of the mechanisms involve crushing of the wood fibers (exceeding the dowel bearing strength), and others are the result of a plastic hinge forming in the fastener combined with crushing of the wood fibers. The designer need not be fully conversant with the derivation of the yield limit equations, but an understanding of the various yield modes is important. In particular, ductile yield modes that involve the formation of plastic hinges in fasteners (e.g., Mode III and Mode IV) may be preferred for resisting earthquake and other dynamic loads. A general summary of these modes is given in Sec. 11.7 for single-shear and double-shear wood connections.

11.7 Yield Limit Mechanisms

Several properties critical to the performance of dowel-type fasteners have been introduced in this chapter. These include connection geometry, dowel bearing strength of the members, and a plastic hinge in a fastener. These factors are used in the *yield limit equations* for laterally loaded nails, bolts, and lag bolts to evaluate the reference design value Z for a single fastener.

To summarize the general procedure, there are several possible yield modes for dowel-type fasteners. An equation is provided in the NDS to evaluate the load capacity for each mode. The reference design value Z for a particular fastener is then taken as the smallest value considering all of the appropriate modes.

The remaining portion of this chapter describes the various yield modes that can occur in shear-type connections. The yield modes are common to all dowel-type connections, and the yield limit mechanisms described here will be referenced in subsequent chapters. Some modes that apply to single-shear (two-member) connections do not occur in double-shear (three-member) connections because of the symmetric loading that takes place in double-shear connections. See Example 11.9. The yield limit equations are given in Example 11.10 for single-shear and double-shear connections.

Mode I: Dowel bearing failure under uniform bearing. The bearing stress for Mode I yielding is uniform because in this yield mechanism the fastener does not rotate or bend. The dowel bearing strength of a wood member is simply exceeded under uniform bearing. The possibility of Mode I yielding must be considered in each wood member in the connection. Consequently, a Mode I mechanism is further classified as Mode I$_m$ if the dowel bearing strength is exceeded in the main member. If the overstress occurs in the side member, the mechanism is Mode I$_s$.

The need to check for a bearing yield mechanism in each member stems from the fact that member thickness and dowel bearing strength can be different for the main and side members (i.e., the products of $l_m F_{em}$ and $l_s F_{es}$ may not be equal). Different strengths can be the result of mixing different species of wood in a connection or of having a different angle of load to grain for the main and side members.

It should be recalled that there is also a definition of dowel bearing strength for steel members as well as wood members. The F_e for a steel member is a necessary parameter for the general yield limit equations.

Mode II: Dowel bearing failure under nonuniform bearing. The second principal yield mechanism also causes the dowel bearing strength to be exceeded. In this case, the fastener remains straight, but it undergoes rigid body rotation. Rotation causes a nonuniform bearing stress, and the wood fibers are crushed at the outside face of each member. Because of symmetry, Mode II yielding does not occur in a double-shear connection.

Mode III: Plastic hinge located near each shear plane. In a Mode III mechanism, the dowel bends and a plastic hinge forms in the fastener. In a single-shear connection, the plastic hinge can occur within the side member or the main member, near the shear plane. There are two yield limit equations associated with Mode III, and the mechanisms are further classified as Mode III$_m$ and Mode III$_s$.

If the plastic hinge occurs inside one member, there is a corresponding crushing of the wood fibers in the other member. This is one reason why the yield mechanisms in a wood connection are more complex than those for the steel beams described in Sec. 11.6. The mechanism is classified as Mode III$_m$ when the dowel bearing strength is exceeded in the main member. Likewise, Mode III$_s$ indicate a crushing of the wood fibers in the side member. The notation emphasizes the member in which the dowel bearing strength is exceeded, rather than the member which contains the plastic hinge.

In a double-shear connection, two plastic hinges may form (one near each shear plane) with crushing of the wood fibers in the side members. This is known as a Mode III$_s$ mechanism. A Mode III$_m$ mechanism does not occur in double-shear connections.

Mode IV: Two plastic hinges near each shear plane. The final mechanism covered by the yield model involves the formation of two plastic hinges at each shear plane. Mode IV yielding can occur in both single-shear and double-shear connections.

EXAMPLE 11.9   Yield Modes

Dowel-type fasteners can yield in several modes. See Fig. 11.10. These modes are also sketched in *NDS Appendix I.*

Figure 11.10   Four basic yield modes in dowel-type connections.

To fully cover the four basic yield modes, there are a total of six yield limit equations for single-shear connections. The two additional equations are required because Mode I and Mode III mechanisms can occur in either the main or side members. In double-shear connections, Mode II and Mode III_m mechanisms do not occur, and three-member connections are fully covered by four yield limit equations.

EXAMPLE 11.10   Yield Limit Equations

Single-Shear Connections

Mode I_m (NDS equation 11.3-1):

$$Z = \frac{Dl_m F_{em}}{R_d}$$

Mode I_s (NDS equation 11.3-2):

$$Z = \frac{Dl_s F_{es}}{R_d}$$

Mode II (NDS equation 11.3-3):

$$Z = \frac{k_1 Dl_s F_{es}}{R_d}$$

Mode III_m (NDS equation 11.3-4):

$$Z = \frac{k_2 Dl_m F_{em}}{(1 + 2R_e)R_d}$$

Mode III_s (NDS equation 11.3-5):

$$Z = \frac{k_3 Dl_s F_{em}}{(2 + R_e)R_d}$$

Mode IV (NDS equation 11.3-6):

$$Z = \frac{D^2}{R_d}\sqrt{\frac{2F_{em}F_{yb}}{3(1 + R_e)}}$$

Double-Shear Connections

Mode I_m (NDS equation 11.3-7):

$$Z = \frac{Dl_m F_{em}}{R_d}$$

Mode I_s (NDS equation 11.3-8):

$$Z = \frac{2Dl_s F_{es}}{R_d}$$

Mode III_s (NDS equation 11.3-9):

$$Z = \frac{2k_3 Dl_s F_{em}}{(2 + R_e)R_d}$$

Mode IV (NDS equation 11.3-10):

$$Z = \frac{2D^2}{R_d}\sqrt{\frac{2F_{em}F_{yb}}{3(1 + R_e)}}$$

where Z = reference design value for a single-fastener connection (Z is taken as the smallest value from all applicable yield limit equations), lb

$$k_1 = \frac{\sqrt{R_e + 2R_e^2(1 + R_t + R_t^2) + R_t^2 R_e^3} - R_e(1 + R_t)}{(1 + R_e)}$$

$$k_2 = -1 + \sqrt{2(1 + R_e) + \frac{2F_{yb}(1 + 2R_e)D^2}{3F_{em}l_m^2}}$$

$$k_3 = -1 + \sqrt{\frac{2(1 + R_e)}{R_e} + \frac{2F_{yb}(2 + R_e)D^2}{3F_{em}l_s^2}}$$

$$R_e = \frac{F_{em}}{F_{es}}$$

$$R_t = \frac{l_m}{l_s}$$

D = fastener diameter, in.
l_m = dowel bearing length in main member, in.
l_s = dowel bearing length in side member, in.
F_{em} = dowel bearing strength of main member, psi
F_{es} = dowel bearing strength of side member, psi
F_{yb} = bending yield strength of fastener, psi
R_d = coefficient specified in NDS Table 11.3.1B for reducing connection yield capacity to a reference design value

As discussed in Chaps. 12 and 13, the reduction coefficient R_d ranges from 2.2 to 5.0 depending on the fastener diameter, yield mode, and angle between direction of load and direction of grain orientation in a wood member.

The yield limit model for fasteners thus covers four general modes of yielding. In a single-shear connection, Mode I and Mode III can occur in either the main member or side member, and these bring the total possible number of yield mechanisms to six. As noted, there is one yield equation for each yield mechanism. Because of symmetry there are only four yield mechanisms for a double-shear connection.

The NDS Commentary (Ref 11.1) provides additional information on the historical development of the yield limit equations and a comprehensive summary of the supporting research.

In practice, the designer will establish a trial configuration for a particular wood connection. This defines the geometry and strength parameters for use in determining the reference design value Z for a single fastener. This reference strength is the smallest load capacity obtained from all of the appropriate yield limit equations for the connection. The reference strength is then subject to the other adjustments that account for the number of fasteners, duration of load or time effect, moisture content, and so on. The adjusted fastener load with all adjustments taken into account is given the symbol Z'. Once the adjusted load has been determined, a decision can be made about the trial joint configuration.

The basic concept of the yield model approach to wood connection design has been introduced in Chap. 11. With this background, the reader can explore the detailed design provisions for the common fasteners used in wood connections. In spite of the rather technical nature of the general yield theory, it should be recalled that the designer is not forced to obtain load capacities for all connections by evaluating the yield limit equations. The NDS contains a number of fastener capacity tables that allow the designer to read directly a reference fastener design value for many common applications. The tables are based on the appropriate yield limit equations and a simplified set of criteria. The intent of the NDS is to provide the occasional user with an alternative to programming the equations or solving the equations by hand. Although the tables provide appropriate design values for many connections, the tables do not cover all loading conditions, and use of the yield limit equations cannot be avoided entirely. It has been noted previously that the yield limit equations are easy to solve on a modern spreadsheet program, and a spreadsheet template may involve less effort than reading the NDS tables.

11.8   References

[11.1] American Forest and Paper Association (AF&PA). 1999. *Commentary on the National Design Specification for Wood Construction*, 1997 ed., AF&PA, Washington DC.

[11.2] American Forest and Paper Association (AF&PA). 2005. *National Design Specification for Wood Construction and Supplement*, ANSI/AF&PA NDS-2005, AF&PA, Washington DC.

[11.3] American Institute of Steel Construction (AISC). 2005. *Steel Construction Manual,* 13th ed., AISC, Chicago, IL.

[11.4] American Institute of Timber Construction (AITC). 2005. *Timber Construction Manual*, 5th ed., John Wiley & Sons Inc., Hoboken, NJ.

[11.5] American Iron and Steel Institute (AISI). 2002. *Cold-Formed Steel Design Manual*. 2002 ed., AISI, Washington, DC.

[11.6] American Society of Civil Engineers (ASCE). 1996. *Mechanical Connections in Wood Structures,* Manual on Engineering Practice No. 84, ASCE, Reston, VA.

[11.7] Aune, P., and Patton-Mallory, M. 1986. "Lateral Load-Bearing Capacity of Nailed Joints Based on Yield Theory," Research Papers FPL 469 and 470, FPL, Forest Service, USDA., Madison, WI.

[11.8] Forest Products Laboratory (FPL). 1999. *Wood Handbook: Wood as an Engineering Material*, General Technical Report 113, FPL, Forest Service, USDA., Madison, WI.

[11.9] ICC Evaluation Service Inc. 2004. "Power-Driven Staples and Nails for Use in All Types of Building Construction," Report No. NER-272, International Codes Council (ICC) (Available from the International Staple, Nail, and Tool Association, ISANTA, Chicago, IL).

[11.10] Patton-Mallory, M. 1989. "Yield Theory of Bolted Connections Compared with Current U.S. Design Criteria," *Proceedings of the 2nd Pacific Timber Engineering Conference*, University of Auckland, New Zealand.

[11.11] Truss Plate Institute (TPI). 2002. *National Design Standard for Metal Plate Connected Wood Truss Construction*, ANSI/TPI 1-2002, TPI, Madison, WI.

11.9   Problems

11.1   Define the following terms and briefly describe how the terms relate to wood connections:

a. Dowel-type fastener

b. Plastic hinge

    c. Dowel bearing strength
    d. Yield mechanism
    e. Angle of load to grain

11.2   Sketch the yield mechanisms that may occur in a single-shear (two-member) connection that uses a dowel-type fastener. Along with the sketch, give a brief description of each mode.

11.3   Sketch the yield mechanisms that may occur in a double-shear (three-member) dowel-type connection. Give a brief description of each mode.

11.4   Describe the general concept behind the yield model theory of connection analysis.

11.5   Define the following symbols as used in the yield model for dowel-type fasteners: D, l_s, F_{em}, F_u, G_s, θ_s, F_{yb}.

11.6   Use the appropriate expressions from Example 11.4 to compute the dowel bearing strengths for the following bolt connections. Compare the results with the values in NDS Table 11.3.2.
    a. $\frac{1}{2}$-in.-diameter bolt in Hem-Fir with load applied parallel to grain
    b. $\frac{1}{2}$-in.-diameter bolt in Hem-Fir with load applied perpendicular to grain
    c. $\frac{1}{2}$-in.-diameter bolt in Hem-Fir with angle of load to grain of 40 degrees
    d. 1-in.-diameter bolt in DF-L with load applied parallel to grain
    e. 1-in.-diameter bolt in DF-L with load applied perpendicular to grain
    f. 1-in.-diameter bolt in DF-L with angle of load to grain of 60 degrees

11.7   Use the appropriate expressions from Example 11.4 to compute the dowel bearing strengths for the following lag bolt (lag screw) connections. Compare the results with values in NDS Table 11.3.2.
    a. $\frac{5}{8}$-in.-diameter lag bolt in Southern Pine with load applied parallel to grain
    b. $\frac{5}{8}$-in.-diameter lag bolt in Southern Pine with load applied perpendicular to grain
    c. $\frac{5}{8}$-in.-diameter lag bolt in Southern Pine with angle of load to grain of 55 degrees
    d. $\frac{5}{8}$-in.-diameter lag bolt in DF-L with load applied parallel to grain
    e. $\frac{5}{8}$-in.-diameter lag bolt in DF-L with load applied perpendicular to grain
    f. $\frac{5}{8}$-in.-diameter lag bolt in DF-L with angle of load to grain of 45 degrees

11.8   Use the appropriate expressions from Example 11.4 to compute the dowel bearing strengths for the following nail connections. Compare the results with the values in NDS Table 11.3.2.
    a. 8d nail in Southern Pine with load applied parallel to grain
    b. 8d nail in Southern Pine with load applied perpendicular to grain
    c. 8d nail in Southern Pine with angle of load to grain of 30 degrees
    d. 10d nail in DF-L with load applied parallel to grain
    e. 10d nail in DF-L with load applied perpendicular to grain
    f. 10d nail in DF-L with angle of load to grain of 45 degrees

11.9   Some manufacturers supply threaded dowel-type steel fasteners with diameters of 0.263 in., 0.268 in., 0.294 in., 0.32 in., and 0.372 in. Use the appropriate expressions from Example 11.4 to compute the dowel bearing strengths for these five fasteners in the following wood members:

a. Hem-Fir with load applied parallel to grain
b. Hem-Fir with load applied perpendicular to grain
c. DF-L with load applied parallel to grain
d. DF-L with load applied perpendicular to grain
e. Southern Pine with load applied parallel to grain
f. Southern Pine with load applied perpendicular to grain

Nailed and Stapled Connections

12.1 Introduction

The general use of nails in the construction of horizontal diaphragms and shear-walls was addressed in previous chapters. In addition, the two basic forms of nailed connections in structural applications (*shear* and *withdrawal*) were described in the introductory chapter on wood connections. It is recommended that the reader review the joint configurations, nomenclature, and yield modes for these connections in Chap. 11 before proceeding with the detailed design provisions given here.

The usual practice in the United States is to design nailed connections when the loads to be transmitted are relatively small, and to use other types of fasteners (e.g., bolts) for joints with larger loads. This is the general approach to connection design covered in this book. However, this contrasts with recent practice of some designers in Europe and New Zealand to also use nails in connections subjected to large forces. Obviously connections with heavy loads require the use of a large number of nails, and special fabrication practices are necessary. The preference for these types of connections is based on the expected ductile behavior of nailed joints and a corresponding concern about the possible lack of ductility in certain bolted connections, especially those involving large-diameter bolts.

The basic concept of a nail attaching one member to another member has undergone little change, but many developments have occurred in the configurations of nails and in the methods of installation. For example, Ref. 12.14 lists over 235 definitions and descriptions relating to the use of nails. The majority of these nails are fasteners for special applications, and the structural designer deals with a relatively limited number of nail types. Power-driving equipment has contributed substantially to the advances in this basic type of wood connection.

The different types of nails are distinguished by the following characteristics:

1. Nail head

2. Shank

Figure 12.1 Definitions of nail terms.

3. Nail point

4. Material type

5. Surface condition

The first three items have to do with the configuration of the nail. See Fig. 12.1, in which four typical nail heads are shown along with the widely used diamond-shaped nail point. Many other types of nails are described in Refs. 12.6, 12.8, and 12.14.

  This chapter introduces the main types of nails used in structural applications. The yield limit equations for laterally loaded nails are summarized, and the modification factors for determining adjusted fastener design values are reviewed. Design procedures for the various types of nailed connections are outlined.

12.2 Types of Nails

The NDS (Ref. 12.2) provides structural design guidelines for the following basic types of nails:

1. Common nails and spikes

2. Box nails

3. Sinker nails

4. Threaded hardened-steel nails

The sizes of common, box, and sinker nails are specified by the pennyweight of the nail, abbreviated d. The pennyweight for different types of nails specifies the length, shank diameter, and head size of the nail. The sizes of the nails listed above are given in NDS Appendix L.

Common, box, and sinker nails are all fabricated from low-carbon steel wire with a plain shank and a diamond point. Common nails and box nails have a flat head, while sinker nails have a countersunk head as shown in Fig. 12.1. For a given pennyweight, common nails and box nails have the same length, while sinker nails are $\frac{1}{8}$ in. or $\frac{1}{4}$ in. shorter. All three types of nails have different diameters for a given pennyweight. Threaded nails are typically fabricated from high-carbon steel wire and have an annularly or helically threaded shank, a flat head, and a diamond point (some low-carbon threaded steel nails have sinker heads). See Fig. 12.2.

Common nails represent the basic structural nail. Design values for common wire nails are for a nail surface condition that is described as *bright*. This term is used to describe nails with a natural bare-metal finish, and design values for this surface condition can be considered conservative for other surface conditions. For a given pennyweight, the diameter of common wire nails is larger than the diameter of box nails, and design values are correspondingly higher. Because of their larger diameter, common nails have less tendency to bend when driven manually.

Box nails and *sinker nails* are also widely used. They are available in several different surface conditions including bright, galvanized (zinc-coated), and cement-coated. Design values for nails in the bright condition may conservatively be used for other surface conditions. Coated nails are discussed below. The wire diameter of box nails and sinker nails is smaller than that for common nails, and the design values are correspondingly smaller. Box nails and sinker nails also have a greater tendency to bend during driving.

Common spikes are similar in form to common wire nails except that they have a larger diameter. Because of their larger diameter, spikes have higher design

HELICAL THREAD
ALSO SPIRAL AND SCREW SHANK

ANNULAR THREAD
ALSO RING SHANK

Figure 12.2 Threaded hardened-steel nails.

values, but the spacing requirements between fasteners are greater in order to avoid splitting of the wood. Predrilling of holes may also be necessary to avoid splitting.

Threaded hardened-steel nails are often made from high-carbon steel and are heat-treated and tempered to provide greater strength. These nails are especially useful when the lumber moisture content (MC) will vary during the service life of the structure. The threaded shank provides greater withdrawal capacity in these circumstances. In addition, the higher yield strength values associated with high-carbon steel provide improved lateral (shear) resistance.

Design values for threaded hardened-steel nails are based on the wire diameter, except that for annularly threaded nails with threads on the shear plane, lateral design values should be based on the root diameter of the nail.

The structural designer may encounter some types of *nails not covered in the NDS*. These may include

1. Finishing nails

2. Casing nails

3. Cooler nails

4. Zinc-coated nails

5. Cement-coated nails

Finishing nails are slender, bright, low-carbon steel nails with a finishing head and a diamond point. These nails are used for finish or trim work, and they may be used for the installation of nonstructural paneling. However, because of the small nail head, finishing nails are not allowed for the attachment of structural sheathing or siding (Sec. 10.3).

Casing nails are slender, low-carbon steel nails with a casing head and a diamond point. These are available with a zinc coating (galvanizing) to reduce staining of wood exposed to the weather. Galvanized casing nails can be used for the structural attachment of plywood siding when the appearance of a flat nail head is considered objectionable. Because of the smaller diameter of the nail head, nominal shearwall values for casing nails in plywood siding are smaller than for common or galvanized box nails (Ref. 12.3) (Sec. 10.3).

Cooler nails are slender, low-carbon steel nails with a flat head and diamond point, and are usually cement-coated. The head diameter is the same as or smaller than that of a common wire nail of the same length. When gypsum wallboard (drywall) of the required thickness is properly installed with the Code-required size and number of cooler nails, a wall may be designed as a shear panel using the nominal unit shear capacities from Table 4.3B of the *Special Design Provisions for Wind and Seismic Supplement* (Ref. 12.3) (Sec. 10.4).

It was noted earlier that nails can be obtained with a number of different surface conditions. The bright condition was previously described, and several others are introduced in the remainder of this section.

Zinc-coated (galvanized) nails are intended primarily for use where corrosion and staining resistance are important factors in performance and appearance. In addition, resistance to withdrawal may be increased if the zinc coating is evenly distributed. However, extreme irregularities of the coating may actually reduce withdrawal resistance. It should also be pointed out that the use of galvanized nails to avoid staining is not fully effective. Where staining should be completely avoided for architectural purposes, stainless steel nails can be used. The expense of stainless steel nails is much greater than that of galvanized nails.

Cement-coated nails are coated by tumbling or submerging in a resin or shellac (cement is not used). Resistance to withdrawal is increased because of the larger friction between the nail and the wood. However, there are substantial variations in the uniformity of cement coatings, and much of the coating may be removed in the driving of the nail. These variables cause large differences in the relative resistance to withdrawal for cement-coated nails. The increased resistance to withdrawal for these types of nails is, in most cases, only temporary.

12.3   Power-Driven Nails and Staples

Major advances in the installation of driven fasteners such as nails and staples have been brought about by the development of pneumatic, electric, and mechanical guns. The use of such power equipment greatly increases the speed with which fasteners can be installed. The installation of a nail or other fastener has been reduced to the pull of a trigger and can be accomplished in a fraction of a second. See Fig. 12.3.

Figure 12.3   Typical power fastening equipment. On the left, nailing through bottom wall plate with Senco SN325+ framing nailer. On the right, stapling plywood sheathing with Senco M series stapler. (*Courtesy of Senco Products, Inc.*)

Conventional round flathead nails can be installed with power equipment. To accomplish this, the nails are assembled in clips or coils that are fed automatically into the driving gun. The round head on these types of fasteners prevents the shanks of the nails from coming in close contact with adjacent nails, and therefore, a relatively large clip is required for a given number of nails.

The size of the clip can be reduced by modifying the shape of the conventional nail. A number of different configurations have been used to make more compact fastener clips. *Wire staples, T nails,* and *modified roundhead nails* are all in current use. See Fig. 12.4. Fasteners such as staples are often made from relatively thin wire. The tendency of this type of fastener to split the wood is reduced. In addition, standard power-driven nails are often thinner and shorter than corresponding pennyweight sizes of hand-driven nails.

Testing programs have resulted in Code recognition of a number of power-driven fasteners for use in structural design. For example, the International Staple, Nail, and Tool Association (ISANTA) has structural design values published in NER 272 (Ref. 12.12) and ESR 1539 (Ref 12.13). ICC evaluation service reports (ESR) are published by the International Codes Council (ICC) under the name of the ICC Evaluation Service. Recognition for other power-driven fasteners has also been obtained.

The literature giving design information is extensive and is not reproduced in this text. The remainder of this chapter deals specifically with the basic types of nails covered in the NDS. The designer, however, should be able to locate *equivalent* power-driven fasteners in the references.

Figure 12.4 Typical power-driven fasteners. (*a*) Wire staples, (*b*) T nails, (*c*) modified roundhead nails. (*Photograph by Mike Hausmann.*)

It should be pointed out that the setting of power-driving guns is an important factor in the performance of a fastener. Care should be taken that the driving gun does not cause the head of a fastener to penetrate below the surface of the wood. Fasteners improperly installed will have a substantially lower load-carrying capacity. Precautions should also be taken in the manual driving of nails, and the head of a nail should not be "set" deeply below the surface of the wood. Because of the rapidity and consistency obtained with power-driving equipment, the installation setting is of particular importance with power-driven fasteners. The type of member will affect the setting of the gun. For example, when plywood roof sheathing is installed on a panelized roof, the gun should be adjusted for nailing into the different supporting members (2×'s, 4×'s, and glulams).

12.4   Yield Limit Equations for Nails

A laterally loaded nail connection is usually a *single-shear* connection between a wood or steel *side member* and a wood *main member*. See Example 12.1. Recall from Chap. 11 that the dowel bearing strength F_e for a given species of wood is constant for a small-diameter fastener regardless of the angle of load to grain. Therefore, the angle of load to grain θ is not a concern in the design of nailed connections. Furthermore, if the wood members in a connection are *both from the same species*, the dowel bearing strengths are the same ($F_{em} = F_{es}$). Most of the design value tables for wood-to-wood connections in the NDS are limited to this simplified condition.

The dowel bearing strength for a nail connection is a function of the specific gravity of the wood (Example 11.4 in Sec. 11.5). Different values of F_e occur when the main member and side member are from different species. Connections involving mixed species are easily accommodated in the yield limit equations by using the corresponding dowel bearing strengths for the main and side members.

Mixing different species of wood in a nailed connection is a common occurrence. For example, plywood manufactured from Douglas Fir (specific gravity of 0.50) may be nailed to Hem-Fir sawn-lumber framing (specific gravity of 0.43). The respective dowel bearing strengths in this connection are $F_{es} = 4650$ psi for the plywood side member and $F_{em} = 3500$ psi for the framing or main member (see Example 12.5 later in this chapter).

If a nailed connection uses a metal side plate, the dowel bearing strengths for the main member and side member are also different. The bearing strength of a cold-formed steel member may conservatively be taken as 1.375 times the ultimate tensile strength of the steel side plate $F_e = 1.375F_u$.

EXAMPLE 12.1   Single-Shear Nail Connections

The most common nailed connections are single-shear (two-member) joints with a wood main member and either a wood or a cold-formed steel side member. The nail passes through the *side member* and into the main member. The *main member* or *holding member* receives the pointed end of the fastener. The *penetration* is the length of the nail in the main member. The penetration is computed as the length of the nail minus the thickness of the side member. If the nail extends beyond the main member, the penetration is taken as the thickness of the main member (see Fig. 12.5a):

$$\text{Penetration } p = l - t_s \le t_m$$

The *dowel bearing length* of the nail in the main member is used in yield limit equations to determine the reference design value Z for a nail connection. According to NDS 11.3.4 the dowel bearing length in the main member l_m is equivalent to the nail penetration when the nail penetration is at least 10 times the shank diameter of the nail, $p \ge 10D$. Alternatively, the dowel bearing length in the main member is taken as the penetration minus the length of the tapered tip of the nail when $6D \le p < 10D$. According to the American Forest and Paper Association (AF&PA) *Technical Report 12—General Dowel Equations for Calculating Lateral Connection Values*, the length of tapered tip for

Figure 12.5a   Wood-to-wood and wood-to-metal connections.

Figure 12.5b   Angle of load to grain does not affect reference strength of laterally loaded nail.

nails is typically in the range of $1.3D$ to $2D$ (see Ref. 12.4). Thus the dowel bearing length in the main member can be determined by:

$$l_m = \begin{cases} p \le t_m & \text{when } p \ge 10D \\ p - 2D \le t_m & \text{when } 6D \le p < 10D \end{cases}$$

$$\therefore l_m = \begin{cases} l - t_s \le t_m & \text{when } p \ge 10D \\ l - t_s - 2D \le t_m & \text{when } 6D \le p < 10D \end{cases}$$

The *penetration* of a nail is defined as the distance that the nail extends into the main member. In order to use the reference design value Z of a nail according to the yield limit equations, the penetration must be a minimum of 6 times the shank diameter of the nail, $p \ge 6D$. The *dowel bearing length* of a nail in the main member is defined as the portion of the nail penetration that is used in yield limit equations to determine the reference design value Z for a nail connection. The dowel bearing length l_m is defined in Example 12.1 for different ranges of nail penetration in the main member.

There are a number of possible nail connections in addition to those shown in Fig. 12.5a. For example, nails can be used in three-member connections, and the NDS provides the appropriate method for obtaining the reference design value Z for joints in double shear. Another type of connection occurs between a plywood panel and light-gage metal studs. In this case the wood panel becomes the side member, and the sheet-metal stud is the main member because it receives the point of the fastener. Fasteners for this application may be nails, pins, or screws. These and other types of connections can be analyzed using yield limit theory. The reader should contact APA—The Engineered Wood Association for information on this latter type of connection.

The purpose of this book is to provide a general introduction to nailed connections and the yield limit model, and it is not feasible to address all possible types of nailed connections. Only the basic *single-shear* connections in Fig. 12.5a are covered in this book. Six possible yield modes are considered, and the yield limit equations from the NDS for these modes are given in Example 12.2.

As an alternative to using the yield limit equations, reference design values for nails for a number of common applications are available in the NDS tables. These tables cover single-shear connections for three basic types of nails: common, box, and sinker nails. The fastener capacity tables are based on the yield limit equations with nail penetration of $p = 10D$ in the main member, and are thus directly applicable for connections with $p \ge 10D$. For nail connections with $6D \le p < 10D$, a footnote to the NDS tables provides a conservative reduction factor of $p/10D$ to account for the reduced nail penetration in the main member. The reduction factor may be used to adjust tabulated design values, in lieu of calculating a reduced connection capacity directly from yield limit equations. The NDS tables cover *wood-to-wood* connections (for the case that both members are of the same species), *wood-to-metal* connections, and *wood-to-panel* connections. See NDS Tables 11N to 11R.

The reference design values for many nailed connection problems are covered by the NDS tables. However, once a software program is developed or a spreadsheet template is created, it will often be more convenient to obtain nail design values from the computer. In addition, some problems will not be covered by the NDS tables, and application of the yield limit equations will be *required*. The examples in Sec. 12.5 illustrate the step-by-step solution of the yield limit equations.

EXAMPLE 12.2   Yield Limit Equations for Nail in Single Shear

The basic strength of a single dowel-type fastener subjected to a lateral load is known as the *reference design value* and is given the symbol Z. The reference design value is taken as the smallest load capacity obtained by evaluating all of the yield limit equations for a given type of connection. The *adjusted design value* Z' for a single fastener may then be obtained by multiplying the reference design value by the appropriate adjustment factors described in Sec. 12.6.

Six yield modes apply to a single-shear nail connection, as given in NDS Tables 11.3.1A and 11.3.1B. The yield modes are sketched in Fig. 11.10 (Sec. 11.7).

Mode I_m (NDS Eq. 11.3-1):

$$Z = \frac{Dl_m F_{em}}{K_D}$$

Mode I_s (NDS Eq. 11.3-2):

$$Z = \frac{Dl_s F_{es}}{K_D}$$

Mode II (NDS Eq. 11.3-3):

$$Z = \frac{k_1 Dl_s F_{es}}{K_D}$$

Mode III_m (NDS Eq. 11.3-4):

$$Z = \frac{k_2 Dl_m F_{em}}{(1 + 2R_e)K_D}$$

Mode III_s (NDS Eq. 11.3-5):

$$Z = \frac{k_3 Dl_s F_{em}}{(2 + R_e)K_D}$$

Mode IV (NDS Eq. 11.3-6):

$$Z = \frac{D^2}{K_D}\sqrt{\frac{2F_{em}F_{yb}}{3(1 + R_e)}}$$

where Z = reference design value for nail (Z is to be taken as smallest value from six yield limit equations), lb

$$k_1 = \frac{\sqrt{R_e + 2R_e^2(1 + R_t + R_t^2) + R_t^2 R_e^3} - R_e(1 + R_t)}{(1 + R_e)}$$

$$k_2 = -1 + \sqrt{2(1 + R_e) + \frac{2F_{yb}(1 + 2R_e)D^2}{3F_{em}l_m^2}}$$

$$k_3 = -1 + \sqrt{\frac{2(1 + R_e)}{R_e} + \frac{2F_{yb}(2 + R_e)D^2}{3F_{em}l_s^2}}$$

$$R_e = \frac{F_{em}}{F_{es}}$$

$$R_t = \frac{l_m}{l_s}$$

$$p = l - t_s \leq t_m$$

  = penetration of nail in main member

l_m = dowel bearing length of nail in main member (member holding point), in.

t_s = thickness of side member, in.

t_m = thickness of main member, in.

l_s = dowel bearing length of nail in side member, in.

l = length of nail, in.

F_{es} = dowel bearing strength of side member, psi

F_{em} = dowel bearing strength of main (holding) member, psi

F_{yb} = bending yield strength of fastener, psi

D = diameter of nail or spike, in.

K_D = reduction coefficient for fasteners with $D < 0.25$ in.

$$= \begin{cases} 2.2 & \text{for } D \leq 0.17 \text{ in.} \\ 10D + 0.5 & \text{for } 0.17 \text{ in.} < D < 0.25 \text{ in.} \end{cases}$$

Values of F_e may be obtained from NDS Tables 11.3.2 or 11.3.2B, or they may be computed using the equation given in Sec. 11.5 of this book.

The yield limit equations in the NDS can be used to obtain reference design values for any fastener with a known bending yield strength F_{yb}. It will be recalled that the bending yield strength of a dowel-type fastener is one of the major factors relating to the formation of a plastic hinge in a connection (Sec. 11.6). The term *bending yield strength* indicates that the yield stress is obtained from a bending test of the nail rather than a tension test. See Fig. 12.6. The bending yield strength is obtained from an *offset* taken parallel to the initial slope of the load deflection curve. The nail is loaded as a simply supported beam with a concentrated load at the midspan (Ref. 12.7).

To provide guidance to the designer regarding the strength of nails generally available in the marketplace, a limited survey of box and common nails was conducted (NDS Appendix I). The survey found that bending yield strength is higher for smaller-diameter nails. The apparent reason for the higher increase

Figure 12.6 Test used to determine bending yield strength of a nail. An offset to the load deflection curve is used to obtain P_y and Δ_y from which the bending yield strength F_{yb} is computed.

in F_{yb} is the additional work-hardening of the material that takes place in the manufacturing of small-diameter nails.

Based on the results of the nail survey, the following formula for the bending yield strength of low-carbon steel nails and spikes (Ref. 12.11) was developed:

$$F_{yb} = 130.4 - 213.9D$$

According to NDS Appendix I, the bending yield strength is assumed to be 30 percent higher for threaded hardened-steel nails:

$$F_{yb} = 1.3(130.4 - 213.9D)$$

In these equations, D is the diameter of the fastener (in.), and F_{yb} is the bending yield strength (ksi). Note that F_{yb} in other equations is in psi. The 30 percent increase for threaded hardened-steel nails is a soft conversion based on previous NDS design values. The diameters and lengths of common, box, and sinker nails are summarized for comparison in Fig. 12.7a. These dimensions are also available in NDS Appendix L. The NDS fastener capacity tables for nails are based on the above formulas, but values of F_{yb} in NDS Tables 11N to 11R were conservatively *truncated*. Formula values and truncated values of F_{yb} are compared in Fig. 12.7b. The truncated values of F_{yb} are listed in the footnotes to the NDS tables.

When connection design values are obtained from yield limit equations or from NDS tables, the design process is referred to as *engineered construction*. It has not been common practice for manufacturers to identify the bending yield strength of fasteners used in wood construction. However, the primary specification for nails and spikes in the United States (Ref. 12.8) now includes supplementary provisions that enable purchasers to specify *engineered construction nails* with minimum bending yield strengths. Since specification of bending yield strength for fasteners is a relatively new practice, it is recommended that

Sizes of Common, Box, and Sinker Nails

Pennyweight	Length in.		Wire diameter, in.		
	Common and Box nails	Sinker nails	Common nails	Box nails	Sinker nails
6d	2	1.875	0.113	0.099	0.092
7d	2.25	2.125	0.113	0.099	0.099
8d	2.5	2.375	0.131	0.113	0.113
10d	3	2.875	0.148	0.128	0.120
12d	3.25	3.125	0.148	0.128	0.135
16d	3.5	3.25	0.162	0.135	0.148
20d	4	3.75	0.192	0.148	0.177
30d	4.5	4.25	0.207	0.148	0.192
40d	5	4.75	0.225	0.162	0.207
50d	5.5	—	0.244	—	—
60d	6	5.75	0.263	—	0.244

Figure 12.7a   Table of nail sizes.

Common, Box, and Sinker Nails

Nail diameter, in.	F_{yb} from formula, ksi	Truncated F_{yb} used in NDS tables, ksi
0.099	109	100
0.113	106	100
0.120	105	100
0.128	103	100
0.131	102	100
0.135	102	100
0.148	99	90
0.162	96	90
0.177	93	90
0.192	89	80
0.207	86	80
0.225	82	80
0.244	78	70
0.263	74	70

Figure 12.7b   Bending yield strengths for nails used to develop reference design values Z in NDS Tables 11N to 11R. Values of F_{yb} may be verified in the footnotes to NDS tables.

designers either specify *engineered construction nails* or identify the minimum F_{yb} for each type of nail used in the design.

Several numerical examples of nail connections using the yield limit equations are given in the next section.

12.5   Applications of Yield Limit Equations

Example 12.3 evaluates the reference design value of a 16d common nail in a connection between two sawn lumber members. The smallest value from the six yield limit equations is obtained from NDS Eq. 11.3-6, which indicates that a Mode IV yield governs. This type of yield mode has two plastic hinges in the fastener. See Fig. 11.10 (Sec. 11.7) for a sketch of the yield mode.

Because both members are the same species, the design value of the connection can be verified using NDS Table 11N.

EXAMPLE 12.3   Yield Limit Equations for Single-Shear Wood-to-Wood Nail Connection

Determine the reference design value for the nail in Fig. 12.8. Both pieces of lumber are Douglas- Fir-Larch (DF-L), and the nail is a 16d common nail. The angle of load to grain does not affect the problem and is not specified.

Figure 12.8   Single-shear wood-to-wood connection.

Nail dimensions are obtained from the table in Fig. 12.7*a* or from NDS Appendix L:

$$D = 0.162 \text{ in.}$$

$$l = 3.5 \text{ in.}$$

Penetration

$$l_s = t_s = 1.5 \text{ in.}$$

$$p = l - t_s \le t_m$$

$$= 3.5 - 1.5 = 2.0 \text{ in.} < 5.5 \text{ in.}$$

$$10D = 10(0.162) = 1.62 \text{ in.} < 2.0 \text{ in.}$$

$$\therefore l_m = p = l - t_s = 2.0 \text{ in.}$$

The bending yield strength of a 16d common nail is assumed to be

$$F_{yb} = 90 \text{ ksi} = 90,000 \text{ psi}$$

This value can be obtained from the table in Fig. 12.7*b* or from NDS Table 11N.

Specific gravity is the same for both main and side members and is obtained from NDS Table 11.3.2A:

$$G_m = G_s = 0.50$$

Dowel bearing strength can be computed or read from NDS Table 11.3.2:

$$F_{em} = F_{es} = 4650 \text{ psi}$$

Coefficients for use in yield limit equations:

$$K_D = 2.2$$

$$R_t = \frac{l_m}{l_s} = \frac{2}{1.5} = 1.33$$

$$R_e = \frac{F_{em}}{F_{es}} = \frac{4650}{4650} = 1.0$$

$$1 + R_e = 1 + 1.0 = 2.0$$

$$1 + 2R_e = 1 + 2(1.0) = 3.0$$

$$2 + R_e = 2 + 1.0 = 3.0$$

$$k_1 = \frac{\sqrt{R_e + 2R_e^2(1 + R_t + R_t^2) + R_t^2 R_e^3} - R_e(1 + R_t)}{(1 + R_e)}$$

$$= \frac{\sqrt{1.0 + 2(1.0^2)(1 + 1.33 + 1.33^2) + (1.33^2)(1.0^3)} - 1.0(1 + 1.33)}{2.0}$$

$$= 0.4916$$

$$k_2 = -1 + \sqrt{2(1 + R_e) + \frac{2F_{yb}(1 + 2R_e)D^2}{3F_{em}l_m^2}}$$

$$= -1 + \sqrt{2(2.0) + \frac{2(90,000)(3.0)(0.162^2)}{3(4650)(2.0^2)}}$$

$$= 1.0625$$

$$k_3 = -1 + \sqrt{\frac{2(1 + R_e)}{R_e} + \frac{2F_{yb}(2 + R_e)D^2}{3F_{em}l_s^2}}$$

$$= -1 + \sqrt{\frac{2(2.0)}{1.0} + \frac{2(90,000)(3.0)(0.162^2)}{3(4650)(1.5^2)}}$$

$$= 1.1099$$

Yield limit equations:

Mode I_m (NDS Eq. 11.3-1):

$$Z = \frac{Dl_m F_{em}}{K_D} = \frac{0.162(2.0)(4650)}{2.2} = 685 \text{ lb}$$

Mode I_s (NDS Eq. 11.3-2):

$$Z = \frac{Dl_s F_{es}}{K_D} = \frac{0.162(1.5)(4650)}{2.2} = 514 \text{ lb}$$

Mode II (NDS Eq. 11.3-3):

$$Z = \frac{k_1 Dl_s F_{es}}{K_D} = \frac{0.4916(0.162)(1.5)(4650)}{2.2} = 253 \text{ lb}$$

Mode III_m (NDS Eq. 11.3-4):

$$Z = \frac{k_2 Dl_m F_{em}}{(1 + 2R_e)K_D} = \frac{1.0625(0.162)(2.0)(4650)}{3.0(2.2)} = 243 \text{ lb}$$

Mode III_s (NDS Eq. 11.3-5):

$$Z = \frac{k_3 Dl_s F_{em}}{(2 + R_e)K_D} = \frac{1.1099(0.162)(1.5)(4650)}{3.0(2.2)} = 190 \text{ lb}$$

Mode IV (NDS Eq. 11.3-6):

$$Z = \frac{D^2}{K_D} \sqrt{\frac{2F_{em}F_{yb}}{3(1 + R_e)}}$$

$$= \frac{(0.162)^2}{2.2} \sqrt{\frac{2(4650)(90{,}000)}{3(2.0)}} = 141 \text{ lb}$$

Reference design value is selected as the smallest value from the yield limit equations:

$$\boxed{Z = 141 \text{ lb}}$$

This agrees with the value listed in NDS Table 11N.

The second example of the yield limit equations uses an 8d box nail to connect a metal side plate to a sawn lumber main member of Southern Pine. See Example 12.4. The dowel bearing strength for a cold-formed steel plate may conservatively be taken as 1.375 times the ultimate tensile strength of the steel. In this problem a Mode III_s yield mechanism controls, which indicates that a single plastic hinge forms in the nail.

EXAMPLE 12.4   Yield Limit Equations for Single-Shear Wood-to-Metal Nail Connection

Determine the reference design value for the nail in Fig. 12.9. The side member is a 12-gage plate of ASTM A653 Grade 33 steel, and the main member is a 4 × 4 of Southern

Pine. The fastener is an 8d box nail. The angle of load to grain does not affect nailed connections and is not given.

Figure 12.9 Single-shear wood-to-metal connection.

Nail dimensions are obtained from the table in Fig. 12.7a or NDS Appendix L:

$$D = 0.113 \text{ in.}$$

$$l = 2.5 \text{ in.}$$

Penetration:

$$l_s = t_s = 0.105 \text{ in. (12-gage plate thickness from Ref. 12.5 or NDS Table 11N)}$$

$$p = l - t_s \leq t_m$$

$$= 2.5 - 0.105 = 2.40 \text{ in.} < 3.5 \text{ in.}$$

$$10D = 10(0.113) = 1.13 \text{ in.} < 2.40 \text{ in.}$$

$$\therefore l_m = p = l - t_s = 2.40 \text{ in.}$$

Bending yield strength of an 8d box nail is assumed to be

$$F_{yb} = 100{,}000 \text{ psi}$$

Specific gravity for Southern Pine is obtained from NDS Table 11.3.2A:

$$G_m = 0.55$$

Dowel bearing strength for Southern Pine can be computed or read from NDS Table 11.3.2:

$$F_{em} = 5550 \text{ psi}$$

Dowel bearing strength of steel side member:

$$F_{es} = 1.375F_u = 61{,}850 \text{ psi}$$

Coefficients for yield limit equations:

$$K_D = 2.2$$

$$R_t = \frac{l_m}{l_s} = \frac{2.4}{0.105} = 22.86$$

$$R_e = \frac{F_{em}}{F_{es}} = \frac{5550}{61,850} = 0.0897$$

$$1 + R_e = 1 + 0.0897 = 1.0897$$

$$1 + 2R_e = 1 + 2(0.0897) = 1.1795$$

$$2 + R_e = 2 + 0.0897 = 2.0897$$

$$k_1 = \frac{\sqrt{R_e + 2R_e^2(1 + R_t + R_t^2) + R_t^2 R_e^3} - R_e(1 + R_t)}{(1 + R_e)}$$

$$= \frac{\sqrt{0.0897 + 2(0.0897^2)(1 + 22.86 + 22.86^2) + (22.86^2)(0.0897^3)} - 0.0897(1 + 22.86)}{1.0897}$$

$$= 0.8287$$

$$k_2 = -1 + \sqrt{2(1 + R_e) + \frac{2F_{yb}(1 + 2R_e)D^2}{3F_{em}l_m^2}}$$

$$= -1 + \sqrt{2(1.0897) + \frac{2(100,000)(1.1795)(0.113^2)}{3(5550)(2.40^2)}}$$

$$= 0.4869$$

$$k_3 = -1 + \sqrt{\frac{2(1 + R_e)}{R_e} + \frac{2F_{yb}(2 + R_e)D^2}{3F_{em}l_s^2}}$$

$$= -1 + \sqrt{\frac{2(1.0897)}{0.0897} + \frac{2(100,000)(2.0897)(0.113^2)}{3(5550)(0.105^2)}}$$

$$= 6.3049$$

Yield limit equations:

Mode I_m (NDS Eq. 11.3-1):

$$Z = \frac{Dl_m F_{em}}{K_D} = \frac{0.113(2.4)(5550)}{2.2} = 684 \text{ lb}$$

Mode I_s (NDS Eq. 11.3-2):

$$Z = \frac{Dl_s F_{es}}{K_D} = \frac{0.113(0.105)(61,850)}{2.2} = 334 \text{ lb}$$

Mode II (NDS Eq. 11.3-3):

$$Z = \frac{k_1 Dl_s F_{es}}{K_D} = \frac{0.8287(0.113)(0.105)(61,850)}{2.2} = 276 \text{ lb}$$

Mode III_m (NDS Eq. 11.3-4):

$$Z = \frac{k_2 D l_m F_{em}}{(1 + 2R_e)K_D} = \frac{0.4869(0.113)(2.40)(5550)}{1.1795(2.2)} = 282 \text{ lb}$$

Mode III_s (NDS Eq. 11.3-5):

$$Z = \frac{k_3 D l_s F_{em}}{(2 + R_e)K_D} = \frac{6.3049(0.113)(0.105)(5550)}{2.0897(2.2)} = 90 \text{ lb}$$

Mode IV (NDS Eq. 11.3-6):

$$Z = \frac{D^2}{K_D} \sqrt{\frac{2F_{em}F_{yb}}{3(1 + R_e)}}$$

$$= \frac{(0.113)^2}{2.2} \sqrt{\frac{2(5550)(100,000)}{3(1.0897)}} = 107 \text{ lb}$$

Reference design value is selected as the smallest value from the yield limit equations:

$$\boxed{Z = 90 \text{ lb}}$$

This agrees with the value listed in NDS Table 11P.

The first two nail examples were given to clarify the use of the yield limit formulas and to verify the NDS tables. The concept behind the tables is to cover many of the common nail applications. However, the yield limit equations allow the designer to evaluate a much wider variety of connections. The final nail problem in this section requires use of the yield equations because the specific conditions are not addressed by the tables.

The 10d common nail in Example 12.5 is driven through a $\frac{1}{2}$-in. thickness of Douglas Fir (DF) plywood into a sawn lumber framing member of Hem-Fir. Because different species are involved, there are different dowel bearing strengths for the main and side members. The yield mechanism is Mode III_s because NDS Eq. 11.3-5 produces the smallest reference design value.

EXAMPLE 12.5   Yield Limit Equations for Single-Shear Plywood-to-Lumber Nail Connection

Determine the reference design value for the nail in Fig. 12.10. The side member is $\frac{1}{2}$-in. STRUCTURAL I (STR I) DF plywood panel, and the main member is a 2 × 6 piece of Hem-Fir. The fastener is a 10d common nail. The angle of load to grain does not affect nailed connections and is not given.

Nail dimensions are obtained from the table in Fig. 12.7a or NDS Appendix L:

$$D = 0.148 \text{ in.}$$

$$l = 3.0 \text{ in.}$$

Figure 12.10 Single-shear plywood-to-lumber connection.

Penetration: The nail is driven into the edge of the 2×6, and $t_m = 5.5$ in.

$$l_s = t_s = 0.5 \text{ in.}$$

$$p = l - t_s \leq t_m$$

$$= 3.0 - 0.5 = 2.5 \text{ in.} < 5.5 \text{ in.}$$

$$10D = 10(0.148) = 1.48 \text{ in.} < 2.5 \text{ in.}$$

$$\therefore l_m = p = l - t_s = 2.5 \text{ in.}$$

Bending yield strength of a 10d common nail is assumed to be

$$F_{yb} = 90{,}000 \text{ psi}$$

Specific gravity for Hem-Fir is obtained from NDS Table 11.3.2A:

$$G_m = 0.43$$

Dowel bearing strength for Hem-Fir can be computed or read from NDS Table 11.3.2:

$$F_{em} = 3500 \text{ psi}$$

Because plywood is STR I, the panel contains veneer from all Group I species, and the specific gravity for DF can be used to determine dowel bearing strength. Values are obtained from NDS Table 11.3.2B:

$$G_s = 0.50$$

$$F_{es} = 4650 \text{ psi}$$

Coefficients for yield limit equations:

$$K_D = 2.2$$

$$R_t = \frac{l_m}{l_s} = \frac{2.5}{0.5} = 5.0$$

$$R_e = \frac{F_{em}}{F_{es}} = \frac{3500}{4650} = 0.7527$$

$$1 + R_e = 1 + 0.7527 = 1.7527$$

$$1 + 2R_e = 1 + 2(0.7527) = 2.5054$$

$$2 + R_e = 2 + 0.7527 = 2.7527$$

$$k_1 = \frac{\sqrt{R_e + 2R_e^2(1 + R_t + R_t^2) + R_t^2 R_e^3} - R_e(1 + R_t)}{(1 + R_e)}$$

$$= \frac{\sqrt{0.7527 + 2(0.7527^2)(1 + 5.0 + 5.0^2) + (5.0^2)(0.7527^3)} - 0.7527(1 + 5.0)}{1.7527}$$

$$= 1.3156$$

$$k_2 = -1 + \sqrt{2(1 + R_e) + \frac{2F_{yb}(1 + 2R_e)D^2}{3F_{em}l_m^2}}$$

$$= -1 + \sqrt{2(1.7527) + \frac{2(90,000)(2.5054)(0.148^2)}{3(3500)(2.5^2)}}$$

$$= 0.9120$$

$$k_3 = -1 + \sqrt{\frac{2(1 + R_e)}{R_e} + \frac{2F_{yb}(2 + R_e)D^2}{3F_{em}l_s^2}}$$

$$= -1 + \sqrt{\frac{2(1.7527)}{0.7527} + \frac{2(90,000)(2.7527)(0.148^2)}{3(3500)(0.5^2)}}$$

$$= 1.9651$$

Yield limit equations:

Mode I_m (NDS Eq. 11.3-1):

$$Z = \frac{Dl_m F_{em}}{K_D} = \frac{0.148(2.5)(3500)}{2.2} = 589 \text{ lb}$$

Mode I_s (NDS Eq. 11.3-2):

$$Z = \frac{Dl_s F_{es}}{K_D} = \frac{0.148(0.5)(4650)}{2.2} = 156 \text{ lb}$$

Mode II (NDS Eq. 11.3-3):

$$Z = \frac{k_1 Dl_s F_{es}}{K_D} = \frac{1.3156(0.148)(0.5)(4650)}{2.2} = 206 \text{ lb}$$

Mode III_m (NDS Eq. 11.3-4):

$$Z = \frac{k_2 Dl_m F_{em}}{(1 + 2R_e)K_D} = \frac{0.9120(0.148)(2.5)(3500)}{2.5054(2.2)} = 214 \text{ lb}$$

Mode III_s (NDS Eq. 11.3-5):

$$Z = \frac{k_3 Dl_s F_{em}}{(2 + R_e)K_D} = \frac{1.9651(0.148)(0.5)(3500)}{2.7527(2.2)} = 84 \text{ lb}$$

Mode IV (NDS Eq. 11.3-6):

$$Z = \frac{D^2}{K_D} \sqrt{\frac{2F_{em}F_{yb}}{3(1 + R_e)}}$$

$$= \frac{(0.148)^2}{2.2} \sqrt{\frac{2(3500)(90,000)}{3(1.7527)}} = 109 \text{ lb}$$

Reference design value is selected as the smallest value from the yield limit equations:

$$\boxed{Z = 84 \text{ lb}}$$

Whether the NDS tables or the yield limit equations are used, the determination of the reference design value Z is the first step in developing the adjusted design value Z' for a single nail. Other important strength properties and the required adjustment factors that may be necessary to obtain the adjusted design value for a single fastener are covered in the next section.

12.6   Adjustment Factors for Laterally Loaded Nails

The yield limit equations for laterally loaded nailed connections were reviewed in Sec. 12.5. As a reminder, *reference design value Z* is the notation assigned to the allowable stress design (ASD) load capacity of a single fastener as obtained from the *yield equations*. In order to be used in the design of a particular connection, the reference design value must be converted to an *adjusted design value Z'*.

This is done by multiplying the reference design value by a series of adjustment factors. Depending on the circumstances, the adjustment factors could have a major or minor effect on the adjusted design value. The adjustment factors required in the design of connections are summarized in NDS Table 10.3.1. If the particular design conditions agree with the conditions for the reference design value, the adjustment factors for ASD simply default to unity.

The base conditions associated with the reference design value of a nail connection are as follows:

1. Load is normal (10-year) duration.

2. Wood is initially dry at the time of fabrication of a connection and remains dry in service.

3. Temperature range is normal.

4. Penetration of nail is at least 6 fastener diameters.

5. Nail is driven into the side grain of the main (holding) member.

6. Nail is not part of a diaphragm or shearwall nailing (Chaps. 9 and 10).

7. Nail is not driven as a toenail.

If a nailed connection is used under exactly these conditions, the adjustment factors are all unity and $Z' = Z$ for ASD.

For a laterally loaded nail connection that does not satisfy the base conditions described above, the following adjustment factors may be required for ASD.

$$Z' = Z(C_D C_M C_t C_{eg} C_{di} C_{tn})$$

where C_D = load duration factor (Sec. 4.15)
C_M = wet service factor (Sec. 4.14)
C_t = temperature factor (Sec. 4.20)
C_{eg} = end grain factor
C_{di} = diaphragm factor
C_{tn} = toenail factor

In load and resistance factor design (LRFD), the reference design value must be multiplied by a format conversion factor K_F and a resistance factor ϕ. The K_F factor adjusts the reference design value Z to a nominal design value Z_n appropriate for LRFD. As part of the conversion to an LRFD format, the K_F factor adjusts from a normal (10-year) load duration basis to a short-term (5 to 10 minutes) load duration basis.

The load duration factor C_D is not applicable for LRFD calculations. Instead a time effect factor λ is selected based on the governing LRFD load combination, as specified in NDS Appendix N (Ref. 12.2) (Sec. 4.16). All other adjustment factors are identical for both LRFD and ASD. Thus, for a laterally loaded nail connection the following adjustment factors may be required for LRFD:

$$Z_n = Z\,(K_F)$$

$$Z'_n = Z_n(\phi_z \lambda C_M C_t C_{eg} C_{di} C_{tn})$$

The total adjusted design value for a connection using two or more nails is the sum of the individual adjusted design values Z'. This summation is appropriate if the nails are all of the same type and size because each fastener will have the same yield mode. The adjusted design capacity for the connection is, then, the number of nails N times the adjusted design value for one fastener:

$$\text{Adj. } P = N(Z')$$

A number of adjustment factors have been described in previous sections, and these are reviewed only briefly. Other adjustments apply only to nailed connections and require greater explanation.

Resistance factor ϕ_z (LRFD only).   The resistance factor for all fasteners in the NDS is $\phi_z = 0.65$. Resistance factors are specified in NDS Appendix N (Ref. 12.2) for connections and various wood properties.

Format conversion factor K_F (LRFD only).   Format conversion factors for adjusting reference design values to nominal LRFD design values are specified in NDS

Appendix N (Ref. 12.2). The format conversion factor for all fasteners in the NDS is:

$$K_F = \frac{2.16}{\phi_z} = \frac{2.16}{0.65} = 3.32$$

Load duration factor C_D (ASD only).   For most loading conditions, the load duration factors for connections are the same as those applied to reference design values in the design of structural members such as beams, tension members, and columns. However, there is one exception to this general rule which relates to *impact loads*. Although impact loads are not part of the usual design loads for many wood structures, they are required in certain applications. Impact loads are applied suddenly and last for a very short period of time (2.0 seconds or less load duration). See Fig. 4.13 in Sec. 4.15. An example of impact loading is the increased force caused by the starting or stopping of a crane as it lifts or lowers a load. In the design of a wood *member* under impact loading, the designer may increase the reference design value by a factor of $C_D = 2.0$.

Most laboratory tests of connections are for load durations of 5–10 minutes, providing ample support for using $C_D = 1.6$ under short-term loads. However, there is little laboratory test data for connections under impact load conditions. Therefore, in the design of a wood *connection* for impact loading the NDS conservatively limits the load duration factor to $C_D = 1.6$.

Time effect factor λ (LRFD only).   For most LRFD load combinations, the time effect factor is the same for structural members and connections. See Sec. 4.16. The only exception is for the load combination with an *impact* live load (L) multiplied by a load factor of 1.6. See NDS Appendix N (Ref. 12.2). When designing a wood *member* for this LRFD load combination, a time effect factor of $\lambda = 1.25$ is used. However, when a wood *connection* is designed for this load combination the NDS limits the time effect factor to $\lambda = 1.0$. The reason for this limitation is identical to the reason for limiting C_D to 1.6 for impact loads in ASD.

Wet service factor C_M.   The concept of a design value adjustment based on moisture content (MC) was introduced in Sec. 4.14. It will be recalled that the wet service factors for member design vary with the type of design values. Different adjustments also apply in connection design depending on the type of fastener and the MC of the wood at several stages. The MC is considered (1) at the time of fabrication of the joint and (2) in service. The reference design value applies to nailed connections in which the wood is initially dry and remains dry in service. For connection design the term *dry* refers to wood that has an MC of 19 percent or less.

If the MC exceeds the dry limits, either at the time of fabrication *or* in service, a reduction in load capacity is required for common nails, box nails, sinker nails, and common wire spikes. For laterally loaded connections $C_M = 0.7$ for these conditions if the fastener diameter is less than $\frac{1}{4}$ in. Values of C_M are listed in NDS Table 10.3.3, *Wet Service Factors, C_M, for Connections.*

Temperature factor C_t.   If wood is used in an application with sustained exposure to higher-than-normal temperature conditions, the reference design values for member design and the reference design values for connections are to be reduced by the appropriate temperature factor C_t. Refer to Sec. 4.20 for a brief review of the temperature factor. Numerical values of C_t for connections are obtained from NDS Table 10.3.4. The temperature range for most wood-frame buildings does not require adjustment of design values, and C_t can usually be set equal to one.

Nail penetration.   The penetration of a nail was defined in Fig. 12.5a as the distance that the nail extends into the main or holding member. The dowel bearing length of a nail in the main member l_m was defined as a function of nail penetration p in Example 12.1. To obtain any design value Z for a nail, the nail penetration in the main member must be at least 6 times the diameter of the nail ($p \geq 6D$). In previous editions of the NDS, only four yield limit equations (rather than six) were specified for nailed connections, and a penetration depth factor was provided for nail penetration in the range $6D \leq p \leq 12D$. However, all six yield limit equations are specified for nailed connections in the 2005 NDS, and the penetration depth factor is no longer required since dowel bearing length in the main member l_m is explicitly addressed in the yield limit equations.

End grain factor C_{eg}.   The reference design value applies to a laterally loaded connection with the nail driven into the *side grain* of the main member. Recall that the main member receives the pointed end of the fastener. In this basic shear connection, the load is perpendicular to the length of the fastener, and the axis of the nail is perpendicular to the wood fibers in the holding member. This is the strongest and most desirable type of nailed connection. See Fig. 12.11a in Example 12.6. When a nail is driven into the side grain, the end grain factor does not apply or may be set equal to one.

In another type of single-shear connection, the nail penetrates into the *end grain* of the holding member (Fig. 12.11b). The load is still perpendicular to the length of the nail, but the fastener axis is now parallel to the grain. This is a much weaker type of connection. To determine the adjusted design value, the reference design value is multiplied by the end grain factor $C_{eg} = 0.67$.

EXAMPLE 12.6   Comparison of Side Grain and End Grain Nailing

Lateral Resistance in Side Grain

Two examples of nails used in the *basic* shear connection are shown in Fig. 12.11a. The basic connection has the nail driven into the *side grain* (i.e., perpendicular to the grain) of the holding member, and the load is applied perpendicular to the length of the nail. For nails driven into the side grain, $C_{eg} = 1.0$.

Figure 12.11a   Nails in side grain.

Lateral Resistance in End Grain

The second main type of lateral load connection has the nail driven into the end grain (i.e., parallel to the grain) of the holding member, and the load is applied perpendicular to the length of the nail. The front and side views of an end grain connection are shown in Fig. 12.11b with two different sets of lateral loads. For nails driven into the end grain, $C_{eg} = 0.67$.

Figure 12.11b   Nails in end grain.

Diaphragm factor C_{di}.   Design of nailing for horizontal diaphragms and shearwalls using standard tables was covered in detail in Chaps. 9 and 10. The nominal unit shear capacities in Tables 4.2A, 4.2B, 4.3A, A.4.2A, A.4.2B, and A.4.3A of the *Special Design Provisions for Wind and Seismic Supplement* (Ref. 12.3) are the result of shearwall and diaphragm tests conducted by APA. As an alternative to designing diaphragms and shearwalls with these tables, the designer may compute the adjusted unit shear. The adjusted unit shear will be controlled by the shear capacity of the structural panel *or* by the lateral design value of the fasteners.

Tests on diaphragms and shearwalls have demonstrated that the capacity of nails used in diaphragms is greater than the capacity of nails used in other structural connections. The diaphragm factor C_{di} is used to adjust the reference design value Z to an adjusted design value for use in diaphragm and shearwall design. Note that the diaphragm factor is only used for nails that attach structural panel sheathing to framing members; C_{di} does not apply to other connections that may simply be a part of the diaphragm. For example, C_{di} is not used in the design of diaphragm chord splices or drag strut connections made with nails. See Ref. 12.15 for an example of how to compute the adjusted ASD unit shear for a diaphragm using C_{di} and other appropriate factors. See also the note in Example 15.4 (Sec. 15.3).

Toenail factor C_{tn}.   In many cases it is not possible to nail directly through the side member into the holding member. In these circumstances toenails may be used. See Example 12.7. Toenails are nails that are driven at an angle of 30 degrees to the side member and are started approximately one-third of the nail length from the intersection of the two members.

Anyone who has ever driven a toenail realizes that the installation details shown in Fig. 12.12a are only approximated in the field. However, the geometry given allows the designer to evaluate the dowel bearing length in the side member l_s and the penetration $p_L = l_m$ of the toenail into the holding member for use in the yield limit equations. Penetration requirements for a laterally loaded toenail are the same as those for a conventional shear connection. Common applications of toenails are for stud to wall plate, beam to wall plate, and blocking to plate connections.

The capacity of a toenail is obtained by multiplying the reference design value Z by the toenail factor C_{tn}. The toenail factor is $C_{tn} = 0.83$ for a laterally loaded toenail, and a different toenail factor applies to withdrawal connections (Sec. 12.12). The toenail factor does not apply to conventional laterally loaded connections. In other words, C_{tn} defaults to unity ($C_{tn} = 1.0$) for the ordinary single-shear connections in Fig. 12.11a.

Finally, it will be noted that some authors draw a distinction between the types of connections shown in Fig. 12.12a and b. The difference has to do with the direction of the grain in the side member. For example, assume that the vertical member in Fig. 12.12a is a wall stud. The toenail, therefore, goes through the *end grain* of the side member. In contrast, the toenails in Fig. 12.12b run through the *side grain* of the beams and the blocking. When this difference is recognized, the connection in Fig. 12.12a is referred to as a *toenail*, and the connections in Fig. 12.12b are termed *slant nails*.

EXAMPLE 12.7   Toenail Connection

Dimensions for toenails for use in yield limit equations:

l = length of nail
l_s = dowel bearing length in side member for toenail design
  = projected length of toenail in side member
  = one-third nail length
  = $l/3$
p_L = penetration of toenail for *laterally loaded* connection
  = projected length of toenail in main (holding) member
  = $l \cos 30 - l_s \leq t_m$

Figure 12.12a   Toenail installation details and geometry.

l_m = dowel bearing length of toenail in main member

$$= \begin{cases} p_L & \text{when } p_L \geq 10D \\ p_L - \text{projected length of tapered tip} & \text{when } 6D \leq p_L < 10D \end{cases}$$

$$= \begin{cases} l\cos 30 - l_s \leq t_m & \text{when } p_L \geq 10D \\ l\cos 30 - l_s - 2D\cos 30 \leq t_m & \text{when } 6D \leq p_L < 10D \end{cases}$$

There are two different definitions of penetration for toenails. The penetration p_L of a toenail in a laterally loaded connection is the *projected* length of the toenail in the main member. The definition of penetration p_W for a toenail in a *withdrawal* connection changes to the actual length of the nail in the main member. Although the yield limit

Figure 12.12b   Typical applications of toenails. On the left, toenails connect beams to top plate of stud wall. On the right, toenails attach blocking to wall plate.

equations do not apply to withdrawal connections, the two definitions of nail penetration are shown in Fig. 12.12a for comparison.

$$p_W = \text{penetration of toenail for a connection } \textit{loaded in withdrawal}$$

$$= \text{total length of nail in main member.}$$

$$= l - \frac{l/3}{\cos 30} \le \frac{t_m}{\cos 30}$$

The capacity of a toenail in a withdrawal connection is related to the surface area of the nail in the holding member. Nail connections loaded in withdrawal are addressed in Sec. 12.12.

According to Ref. 12.9, toenails require a reduction in design value ($C_{tn} = 0.83$), and slant nails do not ($C_{tn} = 1.0$). There is a problem with semantics here, because field personnel refer to both types of installations as toenails. Perhaps to avoid confusion, the NDS does not recognize the difference between the two applications, and $C_{tn} = 0.83$ is currently used for both.

Several numerical examples will now be given to demonstrate the practical design of nails in single-shear connections. In many cases the reference design values for the nails can be determined from either the yield limit equations or the NDS nail tables.

12.7   Design Problem: Nail Connection for Knee Brace

The type of brace used in this example is typical for a number of unenclosed structures including carports, sheds, and patios. See Example 12.8. The design load is the lateral seismic force, and it is conservatively assumed that there are no offsetting dead loads acting on the brace. The lateral seismic force is assumed to be carried equally by all of the columns, and the reaction at the base of the column is determined first.

The horizontal component of force in the brace is then calculated using the free body diagram (FBD) of a column. After the horizontal component is known, the axial force in the brace can be evaluated. The force in the brace is used to determine the required number of nails to attach the brace. The same connection will be used at the top and bottom of the brace.

The reference design value Z for the nail is obtained from the applicable NDS table, and the problem essentially deals with the determination of the adjusted design value Z'.

EXAMPLE 12.8   Knee Brace Connection

The carport shown in Fig. 12.13a uses 2 × 6 knee braces to resist a longitudinal seismic force of $w_E = 108$ lb/ft distributed along the 22 ft width of the structure. Determine the number of 16d common nails required for the connection of the brace to the 4 × 4 post using

Figure 12.13a Lateral force in longitudinal direction for ASD.

ASD procedures. Material is Southern Pine lumber that is dry at the time of construction. Normal temperatures apply. Repeat the design using LRFD procedures.

The magnitude of the distributed seismic force for ASD is determined from IBC basic load combinations:

$$w = 0.7w_E = 0.7\ (108) = 76\ \text{lb/ft}$$

Force to one row of braces:

$$R = \frac{wL}{2} = 76\left(\frac{22}{2}\right) = 836\ \text{lb}$$

Assume that the force is shared equally by all braces (Fig. 12.13b).

$$\Sigma M_0 = 0$$

$$3H - 209(10) = 0$$

$$H = 697\ \text{lb}$$

$$B = \sqrt{2}\,H = \sqrt{2}\ (697)$$

$$= 985\ \text{lb axial force in knee brace}$$

$$= \text{force on nailed connection}$$

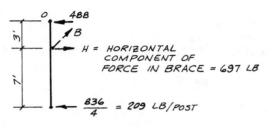

Figure 12.13b FBD to determine axial force B in brace for ASD.

FBD OF COLUMN

The reference design value for a 16d common nail in Southern Pine can be evaluated using the yield limit equations (Sec. 12.4), or it can be obtained from NDS Table 11N. Reference design value from NDS Table 11N

$$Z = 154 \text{ lb/nail}$$

Figure 12.13c Nail connection of brace to column.

Adjustment Factors

Penetration

According to the footnotes for NDS Table 11N, the tabulated reference design values Z are applicable for connections with nail penetration of $p \geq 10D$

$$10D = 10(0.162) = 1.62 \text{ in.} < 2.0 \text{ in.} \quad OK$$

Moisture content

Because the building is "unenclosed," the brace connection may be exposed to the weather, and the severity of this exposure must be judged by the designer. Assume that a reduction for high moisture content is deemed appropriate, and the wet service factor C_M is obtained from NDS Table 10.3.3.

$$C_M = 0.7$$

Load duration

The load duration factor recommended in the NDS for seismic forces is $C_D = 1.6$.

Other adjustment factors

All other adjustment factors for adjusted ASD nail capacity do not apply to the given problem, and each can be set equal to unity:

$C_t = 1.0$ because normal temperature range is assumed

$C_{eg} = 1.0$ because nails are driven into side grain of holding member

$C_{di} = 1.0$ because connection is not part of nailing for diaphragm or shearwall

$C_{tn} = 1.0$ because nails are not toenailed

Adjusted ASD design value for 16d common nail in Southern Pine:

$$Z' = Z(C_D C_M C_t C_{eg} C_{di} C_{tn})$$
$$= 154(1.6)(0.7)(1.0)(1.0)(1.0)$$
$$= 172 \text{ lb/nail}$$

Required number of nails:

$$N = \frac{B}{Z'} = \frac{985}{172} = 5.7$$

> *Use*   six 16d common nails each end of knee brace for high-moisture conditions.[*]

If the reduction for wet service is not required, $C_M = 1.0$. The revised connection is

$$Z' = 154(1.6)(1.0) = 246 \text{ lb/nail}$$

$$N = \frac{985}{246} = 4.0$$

> *Use*   four 16d common nails each end of knee brace if moisture is not a concern.

---

[*]The use of preservative treated wood should be considered (Sec. 4.9).

Now repeat the problem using LRFD procedures. Refer to Fig. 12.13, noting that force magnitudes must be based on *factored* loads (w_u, R_u, H_u, B_u) for LRFD. The magnitude of the distributed *factored* seismic force is $w_u = 1.0w_E = 108$ lb/ft.

Force to one row of braces for LRFD:

$$R_u = \frac{w_E L}{2} = 108\left(\frac{22}{2}\right) = 1188 \text{ lb}$$

or $\dfrac{1188}{4} = 297$ lb per brace

$$\Sigma M_0 = 0$$

$$3H_u - 297\,(10) = 0$$

$$H_u = 990 \text{ lb}$$

$$B_u = \sqrt{2}\, H_u = 1400 \text{ lb } \textit{factored} \text{ axial force in knee brace}$$

$$= \textit{factored} \text{ load on nailed connection}$$

Reference design value from NDS Table 11N

$$Z = 154 \text{ lb/nail}$$

The nominal design value for LRFD is:

$$Z_n = Z\,(K_F) = 154\,(3.32) = 511 \text{ lb/nail}$$

Adjustment Factors

The time effect factor for LRFD load combinations in which seismic load is dominant is $\lambda = 1.0$. The load duration factor C_D is not applicable for LRFD. As noted in Sec. 12.6, the resistance factor is $\phi_z = 0.65$ for all fasteners in the NDS. The other adjustment factors are identical for both LRFD and ASD.

Adjusted LRFD design value for a 16d common nail in Southern Pine:

$$Z'_n = Z_n(\phi_z \lambda C_M C_t C_{eg} C_{di} C_{tn})$$

$$= 511(0.65)(1.0)(0.7)(1.0)(1.0)(1.0)(1.0)$$

$$= 233 \text{ lb/nail}$$

Required number of nails:

$$N = \frac{B_u}{Z'_n} = \frac{1400}{233} = 6.0$$

> *Use* six 16d common nails each end of knee brace for high-moisture conditions (*same as for ASD*).

If the reduction factor for wet service is not required, $C_M = 1.0$ The revised connection is

$$Z'_n = 511(0.65) = 332 \text{ lb/nail}$$

$$N = \frac{B_u}{Z'_n} = \frac{1400}{332} = 4.2$$

> *Use*   five 16d common nails each end of knee brace if moisture is not a concern (*one more nail than required for ASD*).

For an unenclosed building of this nature, the designer must evaluate the exposure of the connections and the MC of the members. This example illustrates the effects of different assumed moisture conditions. The load duration factor of $C_D = 1.6$ for seismic forces in ASD was also illustrated, as well as the time effect factor of $\lambda = 1.0$ for seismic forces in LRFD.

Other adjustment factors for converting Z to Z' default to 1.0 in this example for ASD. The reasons for the defaults to unity are listed in this first nail example, but in subsequent problems the adjustments are set equal to 1.0 without detailed explanation. Similarly, the format conversion factor $K_F = 3.32$ and resistance factor $\phi_z = 0.65$ for LRFD will simply be given without further explanation in subsequent connection design problems that use LRFD procedures.

The number of nails that can be accommodated in a connection may be limited by the spacing requirements (joint details). The layout of this connection is considered in Sec. 12.14.

12.8 Design Problem: Top Plate Splice

In Example 12.9, a splice in the double top plate of a wood frame wall is designed. The forces in the plate are caused by the horizontal diaphragm action of the roof. The longitudinal wall plate is designed for the *chord force* (produced by the transverse lateral force) or the *collector* (*strut*) *force* (produced by the longitudinal lateral force). The larger of the two forces is used for the connection design. In this building, the chord force controls, but the strut force could have been the critical force if the opening in the wall had been somewhat longer than 14 ft.

In buildings with larger lateral forces, the magnitude of the chord or strut force may be too great for a nailed connection. In this case, bolts or some form of connection hardware could be used for the plate splice.

At any splice, either the top or bottom member of the chord will be continuous. If a check of the stress in the chord were shown, the tension force would be divided by the cross-sectional area of *one* of the plate members. This example concentrates on the design of the connection only, and the stress in the chord or strut member is not included.

An interesting point about the yield limit equations is highlighted in this example. The results of three of the six required equations are the same as those in a previous nail problem (Example 12.3). However, the values given by the Mode I_m, Mode II, and Mode III_m equations are considerably less in the current example. The lower values of Z are the result of a reduced nail penetration caused by a main member that is only 1.5 in. thick. In Example 12.9, the values of Mode I_m and Mode I_s are the same because the dowel bearing lengths in the main and side members are equal. The values of Mode III_m and Mode III_s are also identical for the same reason.

The designer needs to be aware of the subtle way this problem changed from the earlier example. The outcome was the same (that is, $Z = 141$ lb in both examples), but the results could have been different.

It should be noted that the tables in the NDS *do not* take into account the thickness of the main member. In other words, the tables assume that $l_m = p = 10D$ and the nail does not protrude out of the holding member. This does not affect the result of Example 12.9 because the Mode I_m, Mode II, and Mode III_m yield mechanisms do not control. Consequently, the reference design value agrees with the value listed in NDS Table 11N. However, this may not always be the case.

EXAMPLE 12.9   Top Plate Splice

The building shown in Fig. 12.14a has stud walls with a double 2× top plate around the entire building. Lateral seismic forces to the horizontal diaphragm include a transverse load of $w_{ET} = 200$ lb/ft distributed along the 40 ft length of the building and a longitudinal load of $w_{EL} = 229$ lb/ft distributed along the 32 ft width of the building. Splices in the double top plate for the *longitudinal wall* are to be designed for the horizontal diaphragm chord force or drag force (whichever is larger) using ASD procedures. Lumber

Figure 12.14a Lateral forces to roof diaphragm for ASD. Consider force in each direction separately.

is dry DF-L that will remain dry in service. Normal temperatures apply. Connection is to be made with 16d common nails. Repeat the design using LRFD procedures.

The magnitudes of the distributed seismic forces for ASD are:

$$w_T = 0.7w_{ET} = 0.7(200) = 140 \text{ lb/ft}$$

$$w_L = 0.7w_{EL} = 0.7(229) = 160 \text{ lb/ft}$$

Transverse Lateral Force (ASD)

Top plate of wall serves as diaphragm chord.
   Moment in diaphragm:

$$M = \frac{w_T L^2}{8} = \frac{140(40)^2}{8} = 28{,}000 \text{ ft-lb}$$

Chord force:

$$T = C = \frac{M}{b} = \frac{28{,}000}{32} = 875 \text{ lb}$$

For a discussion of horizontal diaphragm chords and collectors (struts), see Chap. 9.

Longitudinal Lateral Force (ASD)

Wall plate serves as drag strut (Fig. 12.14b).
   Shearwall reaction:

$$R = \frac{w_L L}{2} = \frac{160(32)}{2} = 2560 \text{ lb}$$

Wall with opening has the critical collector (strut) force.
   Roof diaphragm unit shear:

$$v_r = \frac{R}{b} = \frac{2560}{40} = 64 \text{ lb/ft}$$

Unit shear in shearwall:

$$v_r = \frac{R}{b} = \frac{2560}{6 + 20} = 98.5 \text{ lb/ft}$$

Splice Connection

$$\text{Max. chord force} = 875 \text{ lb} \quad \text{governs}$$

$$\text{Max. strut force} = 689 \text{ lb}$$

If a minimum lap of 4 ft is used for the top plate splice, the connection within the lap must transmit the full chord force.
   The yield limit equations for a 16d common nail in Fig. 12.14c are the same as the yield limit equations in Example 12.3 with one exception: The Modes I_m, II, and III_m yield mechanisms depend on the penetration of the nail. In this problem the nail protrudes

Figure 12.14b   Force in collector (strut) along wall with opening for ASD.

Figure 12.14c   Splice in top wall plate.

beyond the thickness of the main member, and the furnished penetration is less. NDS Eqs. 11.3-1, 11.3-3, and 11.3-4 are reevaluated here, and the results of the other three yield limit equations are obtained from the previous example.

Information from Example 12.3 that applies to the problem at hand:

$$D = 0.162 \text{ in.}$$

$$l = 3.5 \text{ in.}$$

$$F_{yb} = 90{,}000 \text{ psi}$$

$$G_m = G_s = 0.50$$

$$F_{em} = F_{es} = 4650 \text{ psi}$$

$$l_s = t_s = 1.5 \text{ in.}$$

$$K_d = 2.2$$

$$R_e = \frac{F_{em}}{F_{es}} = \frac{4650}{4650} = 1.0$$

$$1 + R_e = 2.0$$

$$1 + 2R_e = 3.0$$

$$2 + R_e = 3.0$$

Reduced nail penetration:

$$p = l - t_s \le t_m$$

$$= 3.5 - 1.5 = 2.0 \text{ in.} > 1.5 \text{ in.}$$

$$\therefore p = 1.5 \text{ in.}$$

$$10D = 10(0.162) = 1.62 \text{ in.} > 1.5 \text{ in.}$$

$$l_m = l - t_s - 2D \le t_m$$

$$= 3.5 - 1.5 - 2(0.162) = 1.68 \text{ in.} > 1.5 \text{ in.}$$

$$\therefore l_m = 1.5 \text{ in.}$$

Revised values of R_t, k_1, and k_2 with dowel bearing length of 1.5 in.:

$$R_t = \frac{l_m}{l_s} = \frac{1.5}{1.5} = 1.0$$

$$k_1 = \frac{\sqrt{R_e + 2R_e^2(1 + R_t + R_t^2) + R_t^2 R_e^3} - R_e(1 + R_t)}{(1 + R_e)}$$

$$= \frac{\sqrt{1.0 + 2(1.0^2)(1 + 1.0 + 1.0^2) + (1.0^2)(1.0^3)} - 1.0(1 + 1.0)}{2.0}$$

$$= 0.4142$$

$$k_2 = -1 + \sqrt{2(1 + R_e) + \frac{2F_{yb}(1 + 2R_e)D^2}{3F_{em}l_m^2}}$$

$$= -1 + \sqrt{2(2.0) + \frac{2(90{,}000)(3.0)(0.162^2)}{3(4650)(1.5^2)}}$$

$$= 1.1099$$

Mode I_m (NDS Eq. 11.3-1):

$$Z = \frac{Dl_mF_{em}}{K_D} = \frac{0.162(1.5)(4650)}{2.2} = 514 \text{ lb}$$

Mode II (NDS Eq. 11.3-3):

$$Z = \frac{k_1 Dl_s F_{es}}{K_D} = \frac{0.4142(0.162)(1.5)(4650)}{2.2} = 213 \text{ lb}$$

Mode III_m (NDS Eq. 11.3-4):

$$Z = \frac{k_2 Dl_m F_{em}}{(1 + 2R_e)K_D} = \frac{1.1099(0.162)(1.5)(4650)}{3.0(2.2)} = 190 \text{ lb}$$

These are less than the values for Mode I_m, Mode II, and Mode III_m for the connection in Example 12.3.

Results of other yield limit equations from Example 12.3:

Mode I_s (NDS Eq. 11.3-2)

$$Z = 514 \text{ lb}$$

Mode III_s (NDS Eq. 11.3-5)

$$Z = 190 \text{ lb}$$

Mode IV (NDS Eq. 11.3-6)

$$Z = 141 \text{ lb}$$

Refer to Fig. 11.10 in Sec. 11.7 for sketches of the yield modes. Notice that the same value of Z is obtained for Mode I_m and Mode I_s for the top plate splice. Similarly, the same value of Z is obtained for Mode III_m and Mode III_s. This is to be expected because of the equal thicknesses of the main and side members. The length of the nail is the same in both members.

The reference design value is taken as the smallest value from the yield limit formulas:

$$Z = 141 \text{ lb}$$

Adjustment factors for ASD in this problem are all unity except the load duration factor. For seismic forces in ASD the NDS recommends

$$C_D = 1.6$$

Required penetration to use value of Z:

$$6D = 6(0.162) = 0.97 \text{ in.} < 1.5 \text{ in.} \quad OK$$

Adjusted ASD design value for a 16d common nail in DF-L:

$$Z' = Z(C_D C_M C_t C_{eg} C_{di} C_{tn})$$

$$= 141(1.6)(1.0)(1.0)(1.0)(1.0)(1.0)$$

$$= 225 \text{ lb/nail}$$

Required number of nails:

$$N = \frac{T}{Z'} = \frac{875}{225} = 3.9$$

> *Use*   four 16d common nails between splice points.

Now repeat the problem using LRFD procedures. Refer to Figs. 12.14 and 12.15, noting that force magnitudes must be based on *factored* loads for LRFD. The magnitudes of the distributed *factored* seismic forces are $w_{Tu} = 1.0 w_{ET} = 200$ lb/ft and $w_{Lu} = 1.0 w_{EL} = 229$ lb/ft.

Factored Transverse Lateral Force (LRFD)

$$M_u = \frac{w_{Tu} L^2}{8} = \frac{200(40)^2}{8} = 40,000 \text{ ft-lb}$$

$$T_u = C_u = \frac{M_u}{b} = \frac{40,000}{32} = 1250 \text{ lb}$$

Factored Longitudinal Lateral Force (LRFD)

Factored shearwall reaction:

$$R_u = \frac{w_{Lu} L}{2} = 229 \left(\frac{32}{2} \right) = 3664 \text{ lb}$$

Factored roof diaphragm unit shear:

$$v_{ru} = \frac{R_u}{b} = \frac{3664}{40} = 91.6 \text{ lb/ft}$$

Factored unit shear in shearwall:

$$v_{ru} = \frac{R_u}{b} = \frac{3664}{20 + 6} = 140.9 \text{ lb/ft}$$

Splice Connection

Max. factored chord force = 1250 lb   governs

Max. factored strut force = 986 lb

Figure 12.15 Factored force in collector (strut) along wall with opening for LRFD.

The reference design value from the yield limit equations for a 16d common nail in DF-L is:

$$Z = 141 \text{ lb}$$

The nominal design value for LRFD is:

$$Z_n = Z\, K_F = 141(3.32) = 468 \text{ lb}$$

The time effect factor for LRFD load combinations in which seismic load is dominant is $\lambda = 1.0$. Adjusted LRFD design value for a 16d common nail in DF-L:

$$Z'_n = Z_n(\phi_z \lambda C_M C_t C_{eg} C_{di} C_{tn})$$

$$= 468\,(0.65)(1.0)(1.0)(1.0)(1.0)(1.0)(1.0)$$

$$= 304 \text{ lb/nail}$$

Required number of nails:

$$N = \frac{T_u}{Z'_n} = \frac{1250}{304} = 4.1$$

> *Use*   five 16d common nails between splice points *(one more nail than required for ASD).*

Although it was not done in this example, the designer can specify the location of the splices in the top plate. If these locations are clearly shown on the plans, the design forces (chord and collector loads) that occur at these points may be used for connection design. In this example, the *maximum* forces were used, and consequently the top plate splices may occur at any point along the length of the wall.

12.9 Design Problem: Shearwall Chord Tie

The requirements for anchoring a shearwall panel to the foundation were discussed in Chap. 10. A similar problem occurs in the attachment of a second-story shearwall to the supporting first-story shearwall. The same basic forces must be considered.

In Example 12.10, the tension chord of the second-story shearwall is spliced to the chord of the first-story shearwall. This splice is required because in typical *platform* construction the studs are not continuous from the roof level to the foundation. The first-story studs stop at the double wall plate that supports the second-floor framing. Separate second-story studs are then constructed on top of the second-floor platform.

The purpose of the splice, then, is to develop a continuous shearwall chord from the roof level to the foundation. Splices of this type are referred to as *continuity ties* or splices. Various types of connections can be used for these ties. In this example, the connection consists of a steel strap connected to the first- and second-story chords with 8d box nails. For larger forces, bolts or lag screws can be used in place of nails. Calculations to verify the tension capacity of the strap are not shown in this example, but they would normally be required in design. In addition, the layout and spacing requirements of the nails in the connection should be considered.

Chord forces can be calculated using either the gross overturning moment or the design overturning moment. Both are illustrated, and the choice of which method to use in a particular situation is left to the judgment of the designer.

EXAMPLE 12.10   Shearwall Chord Tie

Design the connection tie between the first- and second-story shearwall chords for the right-end wall using ASD procedures (Fig. 12.16a). A 12-gage ASTM A653 Grade 33 steel strap with 8d box nails is to be used. Lumber is dry Southern Pine that will remain dry in service. Normal temperatures apply. Wind force reactions from the horizontal roof and second-floor diaphragms are given. Repeat the problem using LRFD procedures.

> *Chord Force (ASD)*
>
> The chord forces at the second-story level are caused by the lateral forces at the roof level. See Fig. 12.16b.
>
> From the gross overturning moment OM (refer to Example 10.4),
>
> $$\text{Gross } T = C = vh = (150)(12) = 1800 \text{ lb} \quad \text{(conservative)}$$
>
> From the design OM (refer to Example 16.4),
>
> $$\text{Design OM} = \text{gross OM} - 0.6 \text{ (dead load RM)}$$
> $$= (150 \times 10)(12) - 0.6(140 \times 10)(5)$$
> $$= 13.8 \text{ ft-k}$$
>
> $$\text{Net } T = \frac{M}{b} = \frac{13{,}800}{10} = 1380 \text{ lb}$$

Figure 12.16a Shearwall chord connection.

Alternate calculation

$$\text{Design } T = \text{gross } T - 0.6(\text{trib. DL})$$

$$= 1800 - 0.6(140 \times 5) = 1380 \text{ lb}$$

FBD OF SHEAR PANEL

Figure 12.16b Shearwall overturning for ASD.

Nail Connection

The reference design value for an 8d box nail installed through a metal side plate into a Southern Pine main member was computed using the yield limit formulas in Example 12.4. The same value applies to the problem at hand.

$$Z = 90 \text{ lb}$$

Load duration factor for wind is

$$C_D = 1.6$$

All other adjustment factors are unity for this problem, and the adjusted design value is

$$Z' = Z(C_D) = 90(1.6) = 144 \text{ lb/nail}$$

For the gross overturning moment:

$$N = \frac{1800}{144} = 12.5$$

> *Use* thirteen 8d box nails each end of steel strap into multiple studs.

For the design overturning moment:

$$N = \frac{1380}{144} = 9.6$$

> *Use* ten 8d box nails each end of steel strap into multiple studs.

Now repeat the problem using LRFD procedures. Refer to Fig. 12.16. Note that force magnitudes must be based on *factored* loads for LRFD. The magnitude of the *factored* wall shear due to lateral wind loads is $v_u = 1.6v = 1.6 (150) = 240$ lb/ft, and the magnitude of the *factored* distributed dead load at the top of the wall is $w_{Du} = 0.9w_D = 0.9(140) = 126$ lb/ft.

Chord Force (LRFD)

From the *factored* gross overturning moment OM,

$$\text{Gross } T_u = C_u = v_u h = (240)(12) = 2880 \text{ lb} \quad \text{(conservative)}$$

From the *factored* design OM

$$\text{Design OM} = \text{gross OM} - 0.9 \,(\text{dead load RM})$$

$$= (240 \times 10)(12) - 0.9\,(140 \times 10)(15)$$

$$= 22.5 \text{ ft-k}$$

$$\text{Net } T_u = \frac{M_u}{b} = \frac{22{,}500}{10} = 2250 \text{ lb}$$

Nail Connection (LRFD)

The reference design value is:

$$Z = 90 \text{ lb}$$

The nominal design value for LRFD is:

$$Z_n = Z\,(K_F) = 90(3.32) = 299 \text{ lb}$$

The time effect for LRFD load combinations in which wind load is dominant is $\lambda = 1.0$. Adjusted LRFD design value:

$$Z'_n = Z_n(\phi_z \lambda) = 299(0.65)(1.0) = 194 \text{ lb/nail}$$

For the *factored* gross overturning moment:

$$N = \frac{T_u}{Z'_n} = \frac{2880}{194} = 14.8$$

> *Use*   fifteen 8d box nails each end of steel strap into multiple studs (*two more nails than required for ASD*).

For the *factored* design overturning moment:

$$N = \frac{T_u}{Z'_n} = \frac{2250}{194} = 11.6$$

> *Use*   twelve 8d box nails each end of steel strap into multiple studs (*two more nails than required for ASD*).

12.10   Design Problem: Laterally Loaded Toenail

The example to illustrate the design capacity of a toenail connection is the Code attachment of a stud to the bottom plate of a wall. See Example 12.11. The nailing schedule in IBC Tale 2304.9.1 lists four 8d common toenails as one of the acceptable methods of attaching a stud to the sole plate. Design forces and special considerations for this attachment in areas of high seismic risk were discussed in Sec. 10.12.

The force on the given connection is wind perpendicular to the wall. The penetration of the nail and the thickness of the side member for use in the yield limit equations are determined using the geometry from Fig. 12.12*a* (Sec. 12.6). Recall that l_s and l_m are to be taken as the *projected* lengths of the nail in the side and main members.

The geometry thus established is not directly covered by the NDS fastener capacity tables, and the yield limit expressions are evaluated in the example. This would normally be done using a computer or programmable calculator, and the full numerical example is given for illustration.

EXAMPLE 12.11   Capacity of Toenail Connection

Use ASD procedures to determine the maximum wind force reaction at the base of a wood frame shearwall that uses the standard Code toenailing of four 8d common nails (Fig. 12.17). Lumber is dry DF-L that will remain dry in service. Normal temperatures apply. Repeat the problem using LRFD procedures to determine the maximum *factored* wind force reaction at the base of the shearwall.

Geometry:

Dimensions of an 8d common nail are obtained from the table in Fig. 12.7*a* or NDS Appendix L:

$$D = 0.131 \text{ in.}$$

$$l = 2.5 \text{ in.}$$

Figure 12.17   Toenail connection at bottom of stud subject to lateral wind force.

For a toenail, the thickness of the side member and penetration are determined using the geometry given in Fig. 12.12a. The bottom wall plate is a nominal 2×, and $t_m = 1.5$ in.

$$l_s = \frac{l}{3} = \frac{2.5}{3} = 0.83 \text{ in.}$$

$$p_L = l \cos 30 - l_s \leq t_m$$

$$= 2.5 \cos 30 - 0.83 \leq 1.5 \text{ in.}$$

$$\therefore p_L = 1.33 \text{ in.}$$

$$10D = 10(0.131) = 1.31 \text{ in.} < 1.33 \text{ in.}$$

$$\therefore l_m = p_L = 1.33 \text{ in.}$$

Strength properties:

Bending yield strength of an 8d common nail is obtained from the table in Fig. 12.7b or NDS Table I1:

$$F_{yb} = 100,000 \text{ psi}$$

Specific gravity and dowel bearing strength are the same for both the main member and side member and are obtained from NDS Tables 11.3.2A and 11.3.2:

$$G_m = G_s = 0.50$$

$$F_{em} = F_{es} = 4650 \text{ psi}$$

Coefficients for use in yield limit equations:

$$K_D = 2.2$$

$$R_e = \frac{F_{em}}{F_{es}} = \frac{4650}{4650} = 1.0$$

$$1 + R_e = 2.0$$

$$1 + 2R_e = 3.0$$

$$2 + R_e = 3.0$$

$$R_t = \frac{l_m}{l_s} = \frac{1.33}{0.83} = 1.5981$$

$$k_1 = \frac{\sqrt{R_e + 2R_e^2(1 + R_t + R_t^2) + R_t^2 R_e^3} - R_e(1 + R_t)}{(1 + R_e)}$$

$$= \frac{\sqrt{1.0 + 2(1.0^2)(1 + 1.5981 + 1.5981^2) + (1.5981^2)(1.0^3)} - 1.0(1 + 1.5981)}{2.0}$$

$$= 0.5623$$

$$k_2 = -1 + \sqrt{2(1 + R_e) + \frac{2F_{yb}(1 + 2R_e)D^2}{3F_{em}l_m^2}}$$

$$= -1 + \sqrt{2(2.0) + \frac{2(100,000)(3.0)(0.131^2)}{3(4650)(1.33^2)}}$$

$$= 1.1015$$

$$k_3 = -1 + \sqrt{\frac{2(1 + R_e)}{R_e} + \frac{2F_{yb}(2 + R_e)D^2}{3F_{em}l_s^2}}$$

$$= -1 + \sqrt{\frac{2(2.0)}{1.0} + \frac{2(100,000)(3.0)(0.131^2)}{3(4650)(0.83^2)}}$$

$$= 1.2501$$

Yield limit equations:

Mode I_m (NDS Eq. 11.3-1):

$$Z = \frac{Dl_m F_{em}}{K_D} = \frac{0.131(1.33)(4650)}{2.2} = 369 \text{ lb}$$

Mode I_s (NDS Eq. 11.3-2):

$$Z = \frac{Dl_s F_{es}}{K_D} = \frac{0.131(0.83)(4650)}{2.2} = 231 \text{ lb}$$

Mode II (NDS Eq. 11.3-3):

$$Z = \frac{k_1 Dl_s F_{es}}{K_D} = \frac{0.5623(0.131)(0.83)(4650)}{2.2} = 130 \text{ lb}$$

Mode III_m (NDS Eq. 11.3-4):

$$Z = \frac{k_2 Dl_m F_{em}}{(1 + 2R_e)K_D} = \frac{1.1015(0.131)(1.33)(4650)}{3.0(2.2)} = 135 \text{ lb}$$

Mode III_s (NDS Eq. 11.3-5):

$$Z = \frac{k_3 Dl_s F_{em}}{(2 + R_e)K_D} = \frac{1.2501(0.131)(0.83)(4650)}{3.0(2.2)} = 96 \text{ lb}$$

Mode IV (NDS Eq. 11.3-6):

$$Z = \frac{D^2}{K_D}\sqrt{\frac{2F_{em}F_{yb}}{3(1 + R_e)}}$$

$$= \frac{(0.131)^2}{2.2}\sqrt{\frac{2(4650)(100,000)}{3(2.0)}} = 97 \text{ lb}$$

Reference design value is taken as the smallest value from the yield limit equations:

$$Z = 96 \text{ lb}$$

Adjusted design value (ASD):

Load duration factor for wind is

$$C_D = 1.6$$

Limit for penetration:

$$6D = 6(0.113) = 0.68 \text{ in.}$$

$$0.68 \text{ in.} < 1.33 \text{ in.} \quad OK$$

Toenail factor for laterally loaded toenails:

$$C_{tn} = 0.83$$

All other adjustments do not apply to this problem and are set equal to unity:

$$Z' = Z(C_D C_{tn})$$

$$= 96(1.6)(0.83) = 128 \text{ lb/nail}$$

Maximum wind reaction at base of stud (ASD):

$$R = N \times Z' = 4(128) = \boxed{511 \text{ lb/stud}}$$

Now repeat the problem using LRFD procedures.
The reference design value from the yield limit equations is:

$$Z = 96 \text{ lb}$$

The nominal design value for LRFD is:

$$Z_n = Z\,(K_F) = 96(3.32) = 319 \text{ lb}$$

Adjusted design value (LRFD):

The time effect factor for wind loads is $\lambda = 1.0$.

$$Z'_n = Z_n(\phi_z \lambda C_{tn}) = 319(0.65)(1.0)(0.83)$$

$$= 172 \text{ lb/nail}$$

Maximum wind reaction at base of stud (LRFD):

$$R_{u\,max} = N \times Z'_n = 4(172) = 688 \text{ lb/stud}$$

12.11   Design Problem: Laterally Loaded
Connection in End Grain

The final example of a shear connection involves nailing a member into the end grain of a roof beam. See Example 12.12. This is not a desirable type of connection, but it is frequently seen in the field and serves as an illustration of when the end grain factor C_{eg} is required. There are a number of alternative connections, with nails driven into the side grain, that can be used in place of the given connection. These alternative connections are generally preferred and become necessary for larger loads. The alternative connections may involve some type of metal hardware (Chap. 14).

EXAMPLE 12.12   Nail into End Grain of Roof Beam

Use ASD procedures to determine the maximum load that can be transmitted by the nail connection in Fig. 12.18. Lumber is DF-L that is initially dry and remains dry in service. Normal temperatures apply. The connection is made with two 16d common nails into the end grain of the main member. Loads are (D + S). Repeat the problem using LRFD procedures to determine the maximum *factored* load that can be transmitted by the nail connection assuming the governing load combination is 1.2D + 1.6S.

Figure 12.18   Nails driven into end grain of holding member.

Reference design value:

The reference design value in this example is the same as in Example 12.3. The member sizes are different, but the dimensions and other factors affecting the reference design value are the same in the two problems. Thus, from Example 12.3

$$Z = 141 \text{ lb}$$

Adjusted design value (ASD):

Load duration factor for (D + S):

$$C_D = 1.15$$

End grain factor:

$$C_{eg} = 0.67$$

Other adjustment factors default to unity.

$$Z' = Z(C_D C_{eg}) = 141(1.15)(0.67)$$

$$= 109 \text{ lb/nail}$$

Maximum load on connection (ASD):

$$P_{(D+S)} = N \times Z' = 2 \times 109$$

$$= \boxed{218 \text{ lb} \quad \text{(dry service)}}$$

If the connection is exposed to the weather or if high-humidity conditions exist, the wet service factor will apply, and the maximum load for ASD becomes

$$Z' = Z(C_D C_M C_{eg}) = 141(1.15)(0.7)(0.67)$$

$$= 76 \text{ lb/nail}$$

Max. $P_{(D+S)} = 2 \times 76 = \boxed{152 \text{ lb} \quad \text{(wet service)}}$

Now repeat the problem using LRFD procedures.
The reference design value is:

$$Z = 141 \text{ lb}$$

The nominal design value for LRFD is:

$$Z_n = Z\,(K_F) = 141(3.32) = 468 \text{ lb}$$

Adjusted LRFD design value:

The time effect factor for 1.2D + 1.6S is $\lambda = 0.8$.

$$Z'_n = Z_n(\phi_z \lambda C_{eg}) = 468(0.65)(0.8)(0.67)$$

$$= 163 \text{ lb/nail}$$

Maximum *factored* load on connection (LRFD):

Max. $P_u = N \times Z'_n = 2(163) = \boxed{326 \text{ lb} \quad \text{(dry service)}}$

If the connection is exposed to weather or high humidity conditions, the wet service factor will apply and the maximum *factored* load becomes

$$Z'_n = Z_n(\phi_z \lambda C_M C_{eg}) = 468(0.65)(0.8)(0.7)(0.67)$$

$$= 114 \text{ lb/nail}$$

Max. $P_u = N \times Z'_n = 2(114) = \boxed{228 \text{ lb} \quad \text{(wet service)}}$

12.12   Nail Withdrawal Connections

The concept of a fastener loaded in withdrawal was introduced in Chap. 11 (Fig. 11.2). In the *basic withdrawal connections*, the nail passes through the side member and into the *side grain* of the main member. In this connection the load is parallel to the length of the nail, and the load tends to pull the nail out of the holding member. Generally speaking, connections loaded in withdrawal are weaker and less desirable than the same connections subjected to shear loading. The NDS provides reference design values for fasteners subjected to "pullout," but the usual recommendation is that withdrawal loading of unthreaded fasteners should be avoided (in favor of other more positive types of connections) where practical.

The yield limit model for dowel-type fasteners applies only to shear connections, and the design values for nails in withdrawal are based on empirical test results. The *reference design value* for the basic withdrawal connection is given the symbol W. In addition to nail size and type, the withdrawal capacity depends on the specific gravity of the wood and the penetration p of the nail into the holding member.

NDS Table 11.2C gives the reference withdrawal design value *per inch of penetration* for common nails, box nails, common wire spikes, and threaded hardened-steel nails. If the value from the table is given the symbol w, the reference design value for a single nail is

$$W = w \times p$$

As explained previously, the reference design value W represents the load capacity for a nail used under a set of base conditions, similar to the base conditions for laterally loaded connections:

1. Load is normal (10-year) duration.

2. Wood is initially dry at the time of fabrication of a connection, and it remains dry in service.

3. Temperature range is normal.

4. Nail is driven into the side grain of the main (holding) member.

5. Nail is not driven as a toenail.

To obtain the *adjusted ASD design value* W' for another set of conditions, the reference design value is multiplied by the appropriate adjustment factors:

$$W' = W(C_D C_M C_t C_{tn})$$

where C_D = load duration factor (Secs. 4.15 and 12.6)
$\quad\quad C_M$ = wet service factor (Secs. 4.14 and 12.6)
$\quad\quad C_t$ = temperature factor (Secs. 4.20 and 12.6)
$\quad\quad C_{tn}$ = toenail factor

Similarly, the adjusted LRFD design value W'_n is obtained by first determining the nominal design value W_n, then multiplying the nominal design value by the appropriate adjustment factors:

$$W_n = W\,(K_F)$$

$$W'_n = W_n(\phi_z \lambda C_M C_t C_{tn})$$

where K_F = format conversion factor for connections (Sec. 12.6)
ϕ_z = resistance factor for connections (Sec. 12.6)
λ = time effect factor (Secs. 4.16 and 12.6)

If more than one nail of the same size and type is used, the adjusted withdrawal capacity for the connection is the sum of the individual design values W' for each of the nails. If all nails in the connection are of the same type and size with the same penetration length p, then the adjusted withdrawal design value for one fastener W' times the number of nails N is:

$$P = N(W')$$

Perhaps the most basic point to understand about nails loaded in withdrawal has to do with the direction of the grain of the holding member. See Example 12.13. The reference design values apply only to connections with the nail penetrating into the *side grain* of the holding member (Fig. 12.19a). This has been referred to as the *basic* withdrawal connection. The other possible arrangement is to have the nail penetrate into the *end grain* of the holding member (Fig. 12.19b).

Nails that are driven into the end grain have very low capacities and exhibit considerable variation in withdrawal values. Consequently, the NDS does not provide withdrawal design values for nails driven into the end grain. This point could be emphasized by introducing an end grain factor of $C_{eg} = 0$ into the formula for adjusted withdrawal design values.

EXAMPLE 12.13   Comparison of Withdrawal Connections

Withdrawal from Side Grain

The *basic* withdrawal connection has the nail driven into the *side grain* (i.e., perpendicular to the grain) of the holding member, and the load is parallel to the length of the nail (Fig. 12.19a).

Withdrawal from End Grain—Not Permitted

The second example is a withdrawal connection with the nail driven into the *end grain* (i.e., parallel to the grain) of the holding member. The load is again parallel to the length of the nail (Fig. 12.19b). This is an inherently weak connection, and it is not permitted in design (i.e., it has no design value).

Figure 12.19a Basic withdrawal connection.

Figure 12.19b Withdrawal from end grain—not allowed.

Toenail Loaded in Withdrawal

A *toenail* connection may be loaded in withdrawal as long as the nails penetrate the *side grain* of the holding member. For withdrawal loading, the toenail factor is $C_{tn} = 0.67$ (compared with 0.83 for toenails subjected to shear loading). For a toenail in withdrawal, NDS 11.5.4 indicates that the wet service factor need not be applied. In other words, C_M may be set equal to 1.0 for toenails loaded in withdrawal (see Fig. 12.19c).

Figure 12.19c Toenail connection withdrawal from side grain.

A brief summary of the adjustment factors for withdrawal connections is given. This is followed by two numerical examples.

Resistance factor ϕ_z (LRFD only).   The resistance factor for all fasteners in the NDS is $\phi_z = 0.65$. See Sec. 12.6.

Format conversion factor K_F (LRFD only).   The format conversion factor for all fasteners in the NDS is $K_F = 3.32$. See Sec. 12.6.

Load duration factor C_D (ASD only).   The load duration factor C_D for nail connections subject to withdrawal is the same as that applied in the design of laterally loaded nail connections. Refer to Secs. 4.15 and 12.6 for a review of load duration and the numerical values for C_D.

Time effect factor λ (LRFD only).   The time effect factor λ for nail connections subject to withdrawal is the same as that applied in the design of laterally loaded nail connections. Refer to Secs. 4.16 and 12.6 for a review of time effect factors for LRFD.

Wet service factor C_M.   The concept of an adjustment factor based on moisture content was introduced in Sec. 4.14. In most cases, nails loaded in withdrawal are more negatively affected by variations in moisture content than are other connections. As with shear connections, the moisture content of the wood is considered (1) at the time of fabrication of the joint and (2) in service. The reference design value applies to withdrawal connections in which the wood is initially dry and remains dry in service. For other moisture content conditions, values of C_M are obtained from NDS Table 10.3.3.

The performance of threaded hardened-steel nails in withdrawal is better than other types of nails, and $C_M = 1.0$ for all moisture content conditions. The NDS also specifies $C_M = 1.0$ for all toenail connections loaded in withdrawal, regardless of nail type.

Temperature factor C_t.   If wood is used at higher-than-normal temperatures for an extended period of time, the reference design values for connections are to be reduced by the temperature factor C_t. Numerical values of C_t are obtained from NDS Table 10.3.4. It should be noted that the temperature range in most wood-frame buildings does not require a reduction in fastener load capacity, and C_t is usually taken as 1.0.

Toenail factor C_{tn}.   The installation of toenails was covered earlier in this chapter. The geometry in Fig. 12.12a is used to evaluate penetration in determining the reference withdrawal value W. The reference design value is then reduced by the toenail factor for withdrawal $C_{tn} = 0.67$. Note again that the wet service factor does not apply to toenails loaded in withdrawal ($C_M = 1.0$).

Because withdrawal connections are not the preferred connection, only two brief examples are given. The first is a withdrawal connection with the nail driven perpendicular to the member and into the side grain. See Example 12.14. The second problem demonstrates the procedure for a toenail connection loaded in withdrawal. See Example 12.15.

EXAMPLE 12.14   Basic Withdrawal Connection

Assume that the hanger in Fig. 12.19a is used to suspend a dead load P. The side member is a 2×8, and the holding members are 6×6s. Lumber is Western Hemlock that is initially green, and seasoning occurs in service. Normal temperatures apply. Six 16d box nails are used at each end of the side member. Determine the maximum dead load P using ASD procedures. Repeat the problem using LRFD procedures to determine the maximum *factored* dead load P_u.

Nail dimensions are obtained from the table in Fig. 12.7a or from NDS Appendix L:

$$D = 0.135 \text{ in.}$$

$$l = 3.5 \text{ in.}$$

Specific gravity is obtained from NDS Table 11.3.2A:

$$G = 0.47$$

Reference withdrawal value from NDS Table 11.2C:

$$w = 28 \text{ lb/in.}$$

Penetration:

$$p = l - t_s \le t_m$$

$$= 3.5 - 1.5 \le 5.5 \text{ in.}$$

$$\therefore p = 2.0 \text{ in.}$$

Reference design value:

$$W = p \times w = 2.0(28) = 56 \text{ lb/nail}$$

Adjusted ASD design value:

$$W' = W(C_D C_M C_t C_{tn})$$

$$= 56(0.9)(0.25)(1.0)(1.0) = 12.6 \text{ lb/nail}$$

The load is applied symmetrically, and the maximum ASD load capacity is

$$\text{Max. } P = N(W') = 12(12.6) = \boxed{151 \text{ lb}}$$

Now repeat the problem using LRFD procedures.

Reference design value:

$$W = 56 \text{ lb/nail}$$

Nominal LRFD design value:
$$W_n = W K_F = 56(3.32) = 186 \text{ lb/nail}$$

Adjusted LRFD design value:

The time effect factor for dead load alone is $\lambda = 0.6$.

$$W'_n = W_n(\phi_z \lambda C_M C_t C_{tn})$$

$$= 186(0.65)(0.6)(0.25)(1.0)(1.0) = 18.1 \text{ lb/nail}$$

$$\text{Max. } P_u = N(W'_n) = 12(18.1) = 218 \text{ lb}$$

EXAMPLE 12.15   Toenail Withdrawal Connection

Assume that the connection in Fig. 12.19c has three 8d common nails. Both members are 2-in. nominal DF-L that are initially dry and remain dry in service. Normal temperatures apply. Determine the maximum wind force P using ASD procedures. Repeat the problem using LRFD procedures to determine the maximum *factored* wind force P_u.

Nail dimensions are obtained from the table in Fig. 12.7a or NDS Appendix L:

$$D = 0.131 \text{ in.}$$
$$l = 2.5 \text{ in.}$$

Penetration for a toenail in a withdrawal connection (Fig. 12.12a):

$$p_W = l - \frac{l/3}{\cos 30} \leq \frac{t_m}{\cos 30}$$

$$= 2.5 - \frac{2.5}{3\cos 30} \leq \frac{1.5}{\cos 30}$$

$$= 1.54 \text{ in.} < 1.73 \text{ in.}$$

$$\therefore p_W = 1.54 \text{ in.}$$

Specific gravity form NDS Table 11.3.2A:

$$G = 0.50$$

Reference withdrawal value from NDS Table 11.2C:

$$w = 32 \text{ lb/in.}$$

Reference design value:

$$W = p_W \times w = 1.54(32) = 49.3 \text{ lb/nail}$$

Adjusted ASD design value:

$$W' = W(C_D C_M C_t C_{tn})$$

$$= 49.3(1.6)(1.0)(1.0)(0.67) = 53 \text{ lb/nail}$$

Maximum load capacity:

$$\text{Max. } P = N(W') = 3(53) = \boxed{159 \text{ lb}}$$

Now repeat the problem using LRFD procedures.

Reference design value:

$$W = 49.3 \text{ lb/nail}$$

Nominal LRFD design value:

$$W_n = W K_F = 49.3(3.32) = 164 \text{ lb/nail}$$

Adjusted LRFD design value:

The time effect factor for wind load alone is $\lambda = 1.0$.

$$W'_n = W_n(\phi_z \lambda C_M C_t C_{tn})$$

$$= 164(0.65)(1.0)(1.0)(1.0)(0.67) = 71 \text{ lb/nail}$$

$$\text{Max. } P_u = N(W'_n) = 3(71) = 214 \text{ lb}$$

12.13   Combined Lateral and Withdrawal Loads

In addition to the direct withdrawal problem, there are occasions when nails and spikes are subjected to combined lateral and withdrawal loading. See Fig. 12.20. The resultant load acts at an angle α to the surface of the main member. This situation can occur for nails that attach roof sheathing to framing members on

Figure 12.20   Nail subjected to combined lateral and withdrawal loading.

low-slope roofs. The nail withdrawal loading is due to wind uplift on the roof sheathing, while the simultaneous lateral loading is caused by transfer of wind loads from the walls into the roof diaphragm (Sec. 9.2). For this type of loading, the NDS includes a provision for calculating an adjusted resultant design value Z'_α that must equal or exceed the total design load.

The first step in computing the adjusted resultant design value is to evaluate the reference design value Z of the nail or spike for *shear* loading using the appropriate yield limit equations. Then the reference *withdrawal* design value W is computed using the tabulated value from NDS Table 11.2C and the penetration depth of the nail in the main member ($W = w \times p$). The necessary adjustment factors are applied to the reference values to obtain adjusted values Z' and W'. The interaction between these values is taken into account by the following formula to give the adjusted resultant design value:

$$Z'_\alpha = \frac{Z'W'}{Z'\sin\alpha + W'\cos\alpha}$$

This interaction equation is similar to the Hankinson formula (Secs. 11.5 and 13.16), differing only in the magnitude of the exponents for the trigonometric terms in the denominator.

12.14 Spacing Requirements

The spacing requirements for nail connections given in NDS 11.1.5.6 are rather general. Nail spacing "shall be sufficient to prevent splitting of the wood."

In the past, the layout of nails in a connection has typically been left to the discretion of the field personnel fabricating the connection. However, because an increasingly large number of nails are used in some structural connections, the designer may need to address the layout of fasteners in certain cases. Judgment must be used in the detailing of nail connections because there is not a recognized standard for the placement of nails. The NDS Commentary (Ref. 12.1) provides partial guidance in Table C11.1.5.6.

In order to discuss the spacing of fasteners, it is necessary to define some terms that are used in detailing wood connections. See Fig. 12.21. These definitions apply to nails as well as other types of fasteners, but there are additional clarifications required for bolts and other connectors. See Sec. 13.5.

End distance is the distance, measured parallel to the grain, from the center of the nearest fastener to the end grain of a square-cut piece of lumber. *Edge distance* is the distance, measured perpendicular to the grain, from the center of the nearest fastener to the edge of the member. *Center-to-center distance* (also known as *nail spacing*) is the distance from the center of one fastener to the center of an adjacent fastener in a row. A *row* of fasteners is defined as two or more fasteners in a line parallel to the direction of load, and *row spacing* is the perpendicular distance between rows of fasteners.

In most cases, nails are simply driven into wood members without drilling. However, preboring nail holes may be used to avoid splitting the wood. The NDS

Figure 12.21 Definition of terms for nail spacing. Drawing shows *three rows* of nails.

limits the diameter of a pilot hole to a maximum of 90 percent of the nail diameter for wood with a specific gravity greater than 0.6, and 75 percent of the nail diameter for less dense wood ($G \leq 0.6$). The adjusted lateral design value Z' and withdrawal design value W' apply to connections with or without prebored holes. Preboring is an effective method of avoiding the splitting of wood members.

There are many connection detailing questions that are left unanswered under the NDS Commentary criteria. Although the NDS Commentary addresses a wide range of connection configurations, there are no guidelines given for center-to-center nail spacing when loading is at an angle to grain, nor for end distance requirements when loads are applied perpendicular to grain.

Another problem occurs in members that do not have square-cut ends. See Example 12.16. As the angle θ in Fig. 12.22 becomes small, the perpendicular distance from the end cut on the diagonal member to the center of the nail becomes small if the end distance (parallel to the grain) is kept constant. A similar problem occurs with the edge distance for the vertical member. For an end cut that is not square, the NDS (Ref. 12.2) recommends that the *perpendicular distance from the end cut* should not be less than the *edge distance* for connections utilizing split ring or shear plate connectors. Similar logic may reasonably be applied to the perpendicular distance from the nail to the edge of the vertical member in nailed connections with sloping end cuts.

EXAMPLE 12.16   Nail Spacing

Determine the nail spacing for the knee brace connection designed in Example 12.8 (Sec. 12.7). See Fig. 12.22 for a scale drawing. Apply the NDS Commentary criteria (see Fig. 12.23) to the extent possible. Can size 16d common nails be accommodated?

NOTE: As the angle θ becomes small, the perpendicular distance from the nail to the end cut may become too small. Therefore, use the general recommendation from NDS Fig. 12K that the perpendicular distance from the end cut be not less than the edge distance.

Figure 12.22 Detailing of nail connection for brace.

	Connection configuration		
	Wood side members *without* prebored holes	Wood side members *with* prebored holes **OR** steel side plates *without* prebored holes	Steel side plates *with* prebored holes
center-to-center spacing			
parallel to grain	15D	10D	5D
perpendicular to grain	10D	5D	2.5D
end distance parallel to grain			
tension loading	15D	10D	5D
compression loading	10D	5D	3D
edge distance	2.5D	2.5D	2.5D
row spacing			
nails staggered in adjacent rows	2.5D	2.5D	2.5D
nails *in-line* in adjacent rows	5D	3D	2.5D

Figure 12.23 Recommended minimum nail spacings based on NDS Commentary provisions.

The edge distance guidelines will also be applied to the edge distance for the vertical member, even though the load is not parallel to the axis of the vertical member.

For connections without prebored holes, the NDS Commentary specifies

for the knee brace:

$$\text{Min. end distance} = 15D = 15 \times 0.162 = 2.43 \text{ in.}$$

$$\text{Min. spacing (c. to c. distance)} = 15D = 2.43 \text{ in.}$$

$$\text{Min. row spacing} = 5D = 5 \times 0.162 = 0.81 \text{ in.}$$

for the 4 × 4 post:

$$\text{Edge distance} = 2.5D = 2.5 \times 0.162 = 0.41 \text{ in.}$$

In order to accommodate these spacings, the post width must be:

$$(2 \times 2.43 \text{ in.} \times \cos \theta) + (2 \times 0.41 \text{ in.}) = (2 \times 2.43 \times 0.707) + (2 \times 0.41) = 4.26 \text{ in.}$$

Since the post is only 3.5 in. wide, a six nail pattern cannot be accommodated without prebored holes. However, if nail holes are prebored the spacing and end distance requirements for the knee brace reduce to:

$$\text{Min. end distance} = 10D = 10 \times 0.162 = 1.62 \text{ in.}$$

$$\text{Min. spacing (c. to c. distance)} = 10D = 1.62 \text{ in.}$$

$$\text{Min. row spacing} = 3D = 3 \times 0.162 = 0.49 \text{ in.}$$

The minimum required post width is reduced to:

$$(2 \times 1.62 \text{ in.} \times \cos \theta) + (2 \times 0.41 \text{ in.}) = (2 \times 1.62 \times 0.707) + (2 \times 0.41) = 3.11 \text{ in.}$$

Therefore a six nail pattern with prebored holes meets the NDS Commentary requirements for wood-to-wood connections.

A scale drawing is helpful in determining the spacing requirements for a connection. Because the spacing provisions perpendicular to the load are not specified, the total number of nails that can be used in a limited area is not fully defined.

The spacings given in Example 12.16 represent limited criteria for nail connections from the NDS Commentary combined with engineering judgement.

1. Distance perpendicular to the end cut equal to the edge distance comes from the general spacing recommendations in the NDS for split ring and shear plate connectors.

2. Recommended spacings, end distances, and edge distances for nailed connections with either wood or steel side members and with or without prebored holes are provided in the NDS Commentary.

Clearly there are a number of possible interpretations of nail spacing criteria. Furthermore, other sources offer much different spacing recommendations. For example, without prebored holes, Ref. 12.9 recommends an end distance of $20D$

and other spacings from this reference are generally larger than those cited in the NDS Commentary.

At this point the reader should understand that certain connections will require careful consideration of nail spacing, and that considerable judgment is required in detailing connections. Nail patterns should be chosen by the designer with a basic understanding of the material and the tendency of wood to split if adequate spacing is not provided.

Definitive criteria for spacing in a nail connection do not exist at this time. The criteria included in the NDS Commentary represent committee consensus, but additional research is necessary to develop comprehensive spacing criteria.

12.15 Nailing Schedule

IBC Table 2304.9.1 is a *nailing schedule* that gives the minimum nailing to be used for a number of common connections in wood frame construction. The table covers such items as the nailing of beams (joists and rafters) to the top plate of a wall. For this connection, three 8d toenails are specified (Fig. 12.12b). The required nailing for numerous other connections is covered by this schedule. Copies of a number of IBC tables are included in Appendix C of this book.

The requirements given by the IBC should be regarded as a *minimum*. They apply to conventional wood frame construction as well as construction that has been structurally designed. A larger number of nails (or some other type of connection hardware) may be required when structural calculations are performed. Special consideration should be given to the attachment of the various elements in the lateral-force-resisting system.

12.16 References

[12.1] American Forest and Paper Association (AF&PA). 2006. *Commentary on the National Design Specification for Wood Construction*, 2005 ed., AF&PA, Washington DC.

[12.2] American Forest and Paper Association (AF&PA). 2005. *National Design Specification for Wood Construction and Supplement*, ANSI/AF&PA NDS-2005, AF&PA, Washington DC.

[12.3] American Forest and Paper Association (AF&PA). 2005. *Special Design Provisions for Wind and Seismic Supplement*, ANSI/AF&PA SDPWS-2005, AF&PA, Washington DC.

[12.4] American Forest and Paper Association (AF&PA). 1999. *Technical Report 12—General Dowel Equations for Calculating Lateral Connection Values*, AF&PA, Washington, DC.

[12.5] American Institute of Steel Construction (AISC). 2005. *Steel Construction Manual*, 13th ed., AISC, Chicago, IL.

[12.6] American Society for Testing and Materials (ASTM). 2005. "Standard Terminology of Nails for Use with Wood and Wood-Base Materials," ASTM F547-01, *Annual Book of Standards*, Vol. 01.08, ASTM, Philadelphia, PA.

[12.7] American Society for Testing and Materials (ASTM). 2005. "Standard Test Method for Determining Bending Yield Moment of Nails," ASTM F1575-03, *Annual Book of Standards*, Vol. 01.08, ASTM, Philadelphia, Pa.

[12.8] American Society for Testing and Materials (ASTM). 2005. "Standard Specification for Driven Fasteners: Nails, Spikes, and Staples," ASTM F1667-03, *Annual Book of Standards*, Vol. 01.08, ASTM, Philadelphia, PA.

[12.9] Hoyle, R.J., and Woeste, F.E. 1989. *Wood Technology in the Design of Structures*, 5th ed., Iowa State University Press, Ames, IA.

[12.10] International Codes Council (ICC). 2006. *International Building Code*, 2006 ed., ICC, Falls Church, VA.

[12.11] McLain, T.E. 1991. "Engineered Wood Connections and the 1991 National Design Specification," *Wood Design Focus*, Vol. 2, No. 2.

[12.12] ICC Evaluation Service Inc. 2004. "Power-Driven Staples and Nails for Use in All Types of Building Construction," Report No. NER-272, International Codes Council (ICC) (Available from the International Staple, Nail, and Tool Association, ISANTA, Chicago, IL).

[12.13] ICC Evaluation Service Inc. 2005. Evaluation Service Report No. ESR-1539, ICC (Available from ISANTA, Chicago, IL).

[12.14] Stern, E.G. 1967. *Nails—Definitions and Sizes, a Handbook for Nail Users*, Report No. 61, Virginia Tech, Blacksburg, VA.

[12.15] Tissell, J.R., and Elliot, J.R. 1997. *Plywood Diaphragms,* Research Report 138, APA—The Engineered Wood Association, Engineered Wood Systems, Tacoma, WA.

12.17   Problems

Many of the nail problems in this section require the reference *design value Z* and the adjusted *design value Z'* as either final or intermediate answers. These values should be identified as part of the general solution along with the information specifically requested.

The reference design values for a given nail may be obtained by solving the yield limit equations or by using the NDS nail tables. Tables are to be used only if they apply directly to the design conditions stated in the problem and are to be referenced in the solution. Yield limit equations may be evaluated by either calculator or computer. If the computer is used, output must be sufficient so that all steps in the solution can be verified.

In all problems normal temperatures are assumed, and $C_t = 1.0$. All adjustment factors are to be referenced to the appropriate NDS table or section so that the source of the value may be verified.

12.1 Use spreadsheet or programming software to set up a template for the solution of the yield limit equations for a laterally loaded nail. Consider a single-shear wood-to-wood connection, using the following variables for input:

Nail diameter D
Nail length l
Thickness of side member t_s
Specific gravity of the side member G_s
Thickness of main member t_m
Specific gravity of main member G_m

The template is to do the following:
a. Compute the dowel bearing strength for the side and main members F_{es} and F_{em}.
b. Determine the dowel bearing length of the nail in the main member l_m.
c. Evaluate the coefficients for use in the yield limit equations.
d. Solve the six yield limit equations and select the smallest value as the *reference design value Z*.

12.2 Expand or modify the template from Prob. 12.1 to handle a single-shear connection with a steel side plate. Input remains the same except the ultimate tensile strength of the metal side plate F_u replaces the specific gravity of the side member G_s.

12.3 Expand the template from Prob. 12.1 to handle laterally loaded *toenail* connections *in addition to* conventional shear connections. Input is to essentially remain the same as Prob. 12.1, but some method of distinguishing between a toenail and a nail not driven as a toenail is required.

One approach is to leave the thickness of the side member blank (i.e., $t_s = 0$) to indicate a toenail connection. The spreadsheet would then evaluate l_s for the toenail using the geometry in Fig. 12.12a.

12.4 Expand the template in Prob. 12.1, 12.2, or 12.3 (as assigned) to convert the *reference design value Z* to an *adjusted* ASD *design value Z′*. The input must be expanded to include the appropriate adjustment factors. Include, as a minimum, the adjustments for:
   a. Load duration factor C_D
   b. Wet-service factor C_M
   c. End grain factor C_{eg}
   d. Toenail factor C_{tn} (applies only if expanding Prob. 12.3)

12.5 The shed in Fig. 12.A has double 2×6 knee braces on each column to resist the lateral force. Lumber is Southern Pine. The braces are to be connected to the columns with 16d common nails. Lateral seismic force to roof diaphragm w_{EL} is 98 lb/ft. Heights: $h_1 = 9$ ft, $h_2 = 2$ ft. Use ASD procedures and remember to use the appropriate ASD load factor for seismic forces.

   Find: a. The required number of nails for the brace connection, assuming that $C_M = 1.0$.
           b. The required number of nails for the brace connection, assuming that the connection is exposed to the weather.

Figure 12.A

   c. How would a and b change if helically threaded hardened-steel nails
      with $D = 0.148$ in. and $l = 3.5$ in. were used?

12.6 The shed in Fig. 12.A has double 2×6 knee braces on each column to resist the
lateral force. Lumber is Redwood (open grain). The braces are to be connected
to the columns with 16d common nails. Lateral seismic force to roof diaphragm
is $w_{EL} = 66$ lb/ft. Heights: $h_1 = 12$ ft, $h_2 = 4$ ft. Use LRFD procedures and remember to use the appropriate LRFD load factor for seismic forces.

   Find: a. The required number of nails for the brace connection, assuming that
             $C_M = 1.0$.
           b. The required number of nails for the brace connection, assuming that
             the connection is exposed to the weather.
           c. How would a and b change if helically threaded hardened-steel nails
             with $D = 0.148$ in. and $l = 3.5$ in. were used?

12.7 Repeat Prob. 12.6 except that 16d box nails are to be used. Use ASD procedures,
then repeat using LRFD procedures.

12.8 The single-story wood frame building in Fig. 12.B has a double 2×4 top wall plate
of Southern Pine. This top plate serves as the chord and drag strut along line 1.
Seismic loads are $w_{ET} = 191$ lb/ft and $w_{EL} = 319$ lb/ft. $C_M = 1.0$. Use LRFD procedures and remember to use the appropriate LRFD load factor for seismic forces.

   Find: a. The maximum chord force along line 1.
           b. The maximum drag strut force along line 1.
           c. Determine the number of 16d common nails necessary to splice the
             top plate for the maximum force determined in a and b.

Figure 12.B

12.9 Repeat Prob. 12.8 except that 16d box nails are to be used. Use LRFD procedures
and remember to use the appropriate LRFD load factor for seismic forces.

12.10 The single-story wood frame building in Fig. 12.B has a double 2 × 4 top wall plate of Douglas Fir-Larch. The top plate serves as the chord and drag strut along line 1. Seismic loads are $w_{ET} = 143$ lb/ft and $w_{EL} = 357$ lb/ft. $C_M = 1.0$. Use ASD procedures and remember to use the appropriate ASD load factor for seismic forces.

Find: *a.* The maximum chord force along line 1.
           b. The maximum drag strut force along line 1.
           c. Determine the number of 16d common nails necessary to splice the top plate for the maximum force determined in *a* and *b*.

12.11 Repeat Prob. 12.10 except that 16d box nails are to be used. Use ASD procedures and remember to use the appropriate ASD load factor for seismic forces.

12.12 *Given:* The elevation of the shearwall in Fig. 12.C. The lateral wind force and the resisting dead load of the shearwall are shown on the sketch. The uplift force at the bottom of the wall is anchored to the foundation by a metal strap embedded in the concrete foundation. The 10-gage ASTM A653 Grade 33 steel strap is attached to the shearwall chord (two 2 × 4s) with 10d common nails. Lumber is Hem-Fir. $C_M = 1.0$. Assume that the steel strap is adequate.

      Find: Determine the number of nails required to attach the metal anchor strap to the chord using ASD procedures. Consider the resisting dead load when calculating the anchorage force.

Figure 12.C

12.13 Repeat Prob. 12.12 except that nails are 16d common nails. Use LRFD procedures.

12.14 The two-story wood-frame building in Fig. 12.D has plywood shear panels along line 1. The lateral wind forces are shown in the sketch. Diaphragm span length $L = 30$ ft. Resisting dead loads to line 1 are roof D = 100 lb/ft, floor D = 120 lb/ft, and wall D = 10 psf. Tie-down anchorage is to be with 12-gage ASTM A653 Grade 33 steel straps and 16d common nails. Studs are 2 × 6 of Spruce-Pine-Fir (South). Assume that the metal strap is adequate. $C_M = 1.0$.

Find: The number of nails required for the anchorage connections at *A* and *B* using LRFD procedures.

DIAPHRAGM FORCES
ROOF w_R = 200 LB/FT
2ND FL w_2 = 260 LB/FT

WINDOW

20'   20'   20'

PLAN

ROOF

13'

2ND FLOOR

14'

SHEARWALLS

A

B

ELEV - WALL   1

Figure 12.D

12.15   Repeat Prob. 12.14 except that the diaphragm span length is L = 40 ft. Use ASD procedures.

12.16   The attachment of a roof diaphragm to the supporting shearwall is shown in Fig. 12.E. The shear transferred to the wall is v = 230 lb/ft due to wind load. Toenails in the blocking are used to transfer the shear from the roof diaphragm into the double top plate. Lumber is DF-L. C_M = 1.0.

Find: The required number of 8d common toenails per block if roof beams are 16 in o.c. Use LRFD procedures.

12.17   Repeat Prob. 12.16 except v = 180 lb/ft due to wind, and nails are 8d box toenails. Use ASD procedures.

12.18   The studs in the wood frame wall in Fig. 12.F span between the foundation and the horizontal diaphragm. Roof beams are anchored to the wall top plate with 8d box toenails. Lumber is Hem-Fir. Roof beams are spaced 24 in o.c. Wind = 20 psf. Wall height = 10ft. C_M = 1.0.

Find: The required number of toenails per roof beam using ASD procedures. Repeat using LRFD procedures.

ELEVATION   SECTION

Figure 12.E

Figure 12.F

12.19   The support shown in Fig. 12.G carries a dead load of 500 lb. Lumber is DF-L. Nails are 20d common nails driven into the side grain of the holding member. Use ASD procedures, then repeat using LRFD procedures.

Figure 12.G

Find: The required number of nails, assuming that
  a. $C_M = 1.0$
  b. The connection is exposed to the weather.
  c. The connection is exposed to the weather and threaded hardened-steel nails with $D = 0.177$ in. and $l = 4$ in. are used in place of common nails.

12.20   Determine the IBC minimum required nailing for the following connections. Give nail size, type, and required number.
  a. Joist to sill or girder, toenail
  b. Bridging to joist, toenail each end
  c. Sole plate to joist or blocking, face nail
  d. Top plate to stud, end nail
  e. Stud to sole plate, toenail
  f. Double studs, face nail
  g. Doubled top plates, face nail
  h. Top plates, laps, and intersections; face nail
  i. Ceiling joists to plate, toenail
  j. Continuous header to stud, toenail
  k. Ceiling joists to parallel rafters, face nail
  l. Rafter to plate, toenail

Bolts, Lag Bolts, and Other Connectors

13.1 Introduction

Once the vertical-load-supporting system and lateral-force-resisting system (LFRS) of a building have been designed, attention is turned to the design of the connections. The importance of connection design has been emphasized in previous chapters, and the methods used to calculate the forces on several types of connections have been illustrated.

When the design forces are relatively small, the connections may often be made with nails (Chap. 12). However, bolts, lag bolts, and other connectors are normally used for larger loads. There are several additional reasons why nailed connections were covered in a separate chapter. Recall from Chap. 11 that the design value for a small-diameter fastener, such as a nail, is independent of the angle of load to grain. The angle of load to grain, however, is a major consideration in the design of large-diameter fasteners. The other principal reason is that there are a number of adjustment factors that apply to bolts, lag bolts, and other connectors that do *not* apply to nails.

Connections made with larger-diameter fasteners often involve the use of some form of structural steel hardware. For many common connections this hardware may be available prefabricated from manufacturers or suppliers. Other connections, however, will require the fabrication of special hardware. Prefabricated hardware, when available, is often more economical than made-to-order hardware. Chapter 14 continues the subject of connection design by giving examples of common types of hardware. These examples are accompanied by comments about both good and poor connection layout practices.

The types of fasteners covered in this chapter include

1. Bolts

2. Lag bolts or lag screws

3. Other connectors
   a. Split rings
   b. Shear plates

Examples of several connections using these fasteners were given in the introductory chapter on connections. It is recommended that the reader review the joint configurations, nomenclature, and yield modes that are summarized in Chap. 11 before proceeding with the detailed design provisions given here. Because of their greater use, the majority of the coverage in Chap. 13 is devoted to bolted connections. This is followed by lag bolts. Split ring and shear plate connectors are often viewed as specialty connections, and for this reason their use is covered only briefly.

13.2 Bolt Connections

Most bolts are used in laterally loaded dowel-type connections. These are further described by giving the number of shear planes and the types of structural members connected. For example, the most common connections are either *single-shear* (two-member) (Fig. 13.1) or *double-shear* (three-member) (Fig. 13.2) joints. See Example 13.1. Connections can be *wood-to-wood* or *wood-to-metal*. A single-shear connection also occurs when a wood member is attached by anchor bolts to a concrete foundation or a concrete or masonry wall. These can be referred to as *wood-to-concrete* or *wood-to-masonry* connections, but they are often simply termed *anchor bolt* connections.

EXAMPLE 13.1   Bolts in Shear Connections

Figure 13.1 Single-shear connections. (*a*) Wood-to-wood, (*b*) Wood-to-metal, (*c*) Wood-to-concrete anchor bolt.

Figure 13.2 Double-shear connections. (*a*) Wood-to-wood, (*b*) Wood-to-metal.

In shear connections, the load is perpendicular to the axis of the bolt. The angle of load to grain in the wood can be zero (load parallel to grain), 90 degrees (load perpendicular to grain), or at some intermediate angle θ. The angle of load to grain affects the design value of bolted connections.

Bolted connections can have more than two shear planes (i.e., more than three members). For multiple shear planes, the NDS (Ref. 13.1) states that the design value is to be evaluated by first determining the single-shear capacity for each shear plane. The design value for the connection is determined by multiplying the value for the weakest shear plane by the number of shear planes. In this book, only the common cases of single- and double-shear bolt connections are covered.

Washers of adequate size are to be provided between a wood member and the bolt head, and between the wood member and the nut. Washers can be circular or square. The use of washers applies to bolts in shear as well as to those loaded in tension. For bolts in tension (Fig. 11.2*b* in Sec. 11.2), the size of washer needs to be such that the bearing stress in the wood member does not exceed the adjusted compression perpendicular to grain design value $F'_{c\perp}$.

The size of the washer is not particularly critical for bolts loaded in shear, and the washer simply protects the wood as the nut is tightened. The nut should be snugged tight, but care should be taken so that the nut is not installed too tightly. In other words, the washer should not be embedded in the wood member by overtightening the nut. On the other hand, if a wood member seasons in service, the width of the wood member may shrink so that the nut may need to be retightened after a period of time.

The quality of bolts used in wood connections has not been a major concern in the past. However, with the move to a design procedure based on the yield model for dowel-type fasteners, there is a need to have recognized mechanical properties for bolts. Recall that some yield modes are based on the formation of one or more plastic hinges in the fastener, and this requires bolts with predictable bending yield strength F_{yb} and ductility characteristics.

Designers familiar with steel know that the trend in steel structures is to move away from the use of low-strength A307 bolts and toward the use of high-strength A325 bolts. However, this is not the case in wood structures, and A307 bolts are often used in wood connections. The current problem with specifying A307 bolts is that the ASTM standard for these bolts provides only an ultimate tensile strength F_u, and there is no minimum F_y covered by the specification (Ref. 13.8).

Without a published value for yield strength, there is a conceptual problem in using the yield limit model with ASTM A307 bolts. As a practical matter, A307 bolts have performed satisfactorily in the past, and these fasteners must have a reasonable F_y. The tables in the NDS are based on $F_{yb} = 45$ ksi, which is generally thought to be a conservative value of bending yield strength for A307 bolts.

As an alternative to specifying A307 bolts, the designer may choose to use another bolt specification such as SAE J429, *Mechanical and Material Requirements for Externally Threaded Fasteners* (Ref. 13.11). This specification provides a more complete set of strength criteria for bolts including F_y. Note that F_y is the yield strength determined from a tension test. See NDS Appendix I for an approximate relationship for bending yield strength F_{yb} as a function of tension yield F_y and tension ultimate F_u.

Bolts should be installed in properly drilled holes, and members in a connection should be aligned so that only light tapping is required to insert the bolts. If the bolt holes are too small, an excessive amount of driving will be required to install the bolts, and splitting of the wood may be induced. Obviously, splitting will greatly reduce the shear capacity of a bolted connection. On the other hand, if bolt holes are too large, nonuniform bearing stresses may occur. The NDS specifies that bolt holes are to be a minimum of $\frac{1}{32}$ in. to a maximum of $\frac{1}{16}$ in. larger than the bolt diameter.

The manner in which the holes are bored in the wood also has an effect on load capacity. A bolt hole with a smooth surface develops a higher load with less deformation than one that has a rough surface. See Fig. 13.3. Smooth surfaces are obtained with sharp drill bits, a proper drill speed, and a slow rate of feed. See the *Wood Handbook* (Ref. 13.9) for more information.

A related problem concerning fabrication tolerances can occur for A307 bolts in wood construction. Manufacturers of A307 bolts are permitted to provide *reduced diameter body* bolts having an unthreaded shank diameter approximately 9 percent smaller than the *outside-to-outside* thread diameter (major diameter) of the bolts. This effectively doubles the gap between the bolt shank and the inner surface of the bolt hole when reduced diameter body bolts are installed in connections with properly drilled holes. The use of reduced diameter body bolts may lead to excessive displacement at connections, and can also preclude effective load distribution among individual fasteners in multiple-bolt connections. In order to avoid these problems, it is recommended that designers specify *full diameter body* A307 bolts for connections where overall displacement or load distribution to all fasteners in a group is critical.

Bolt sizes used in wood connections typically range from $\frac{1}{2}$ in. through 1 in. in diameter, in increments of $\frac{1}{8}$ in. In the past, large bolts ($1\frac{1}{4}$-in. and $1\frac{1}{2}$-in. diameter) were also used. However, for a number of years AITC expressed concern about the use of large-diameter bolts in wood members (Ref. 13.4). The reasons

Figure 13.3 Fabrication technique affects the performance of bolt connections. The bolt hole on the left has a rough surface as the result of torn wood fibers due to improper drilling. The hole on the right is smooth and will make a better connection. Both holes were made with the same twist drill bit. The smooth hole was produced using a higher drill speed (rpm) and a lower rate of feed (in./min) through the piece. (*Courtesy of the FPL.*)

cited for this concern are the difficulty in accurately drilling larger holes and the difficulty in aligning large-diameter bolts in a multiple-bolt connection. There is some opinion that the greater stiffness of large-diameter bolts may not allow uniform bearing on all fasteners in a connection.

If there is nonuniform bearing in a connection as the result of inaccuracies in fabrication (Sec. 7.2), small-diameter bolts may be flexible enough to allow a redistribution of stresses in the connection so that the bolts become more uniformly loaded. With large-diameter bolts there is concern that splitting of the wood member may occur before a redistribution of stress to less highly loaded bolts can occur. For these reasons, the NDS effectively restricts the use of large-diameter bolts in wood connections by only providing dowel bearing strengths (F_e) and tabulated design values (Z) for bolts 1 in. and less in diameter. Additional research should clarify the effect of fabrication tolerances on the strength of large- and small-diameter bolts.

13.3   Bolt Yield Limit Equations for Single Shear

The yield limit equations for bolts in single shear are covered in this section. The value obtained from the yield limit equations is the *reference design value* Z for a single bolt. The adjustment factors necessary to convert the reference value to an *adjusted design value* Z' are covered in Sec. 13.5

Many of the coefficients needed to evaluate the yield limit equations have been previously defined. Recall that for a large-diameter fastener, such as a bolt, the dowel bearing strength may be different for the main and side members even if the members are from the same species. This occurs when the angle of load to grain is different for the main and side members ($\theta_m \neq \theta_s$).

Recall also from Sec. 11.5 that the dowel bearing strength parallel to grain $F_{e\parallel}$ depends only on the specific gravity of the wood. However, dowel bearing strength perpendicular to grain $F_{e\perp}$ is a function of both specific gravity and bolt diameter. $F_{e\parallel}$ and $F_{e\perp}$ can be computed given the specific gravity of the wood and the fastener diameter, or values can be obtained from NDS Table 11.3.2, *Dowel Bearing Strengths*. If needed, the dowel bearing strength at an angle of load to grain $F_{e\theta}$ can be determined using the Hankinson formula. The dowel bearing strength of a metal member may be taken equal to 1.5 times the ultimate tensile strength for hot-rolled steel plates ($F_e = 1.5F_u$) or 1.375 times the ultimate tensile strength for cold-formed steel plates ($F_e = 1.375F_u$) (see Sec. 11.5). Finally, in wood-to-concrete connections the NDS recommends a dowel bearing strength (F_e) for concrete of 7500 psi. This value is generally considered appropriate for concrete and masonry with compression strength of $f'_c \geq 2500$ psi (see Sec. 15.4).

Yield limit equations from the NDS for a *bolt* in *single shear* are given in Example 13.2. This type of connection requires six yield limit equations because any of the six possible yield modes (Fig. 11.10 in Sec. 11.7) may control. The equations require the use of several coefficients that are, in turn, defined by some rather lengthy expressions. Again, the use of computer spreadsheet or equation-solving software greatly aids in the solution of the yield limit equations.

EXAMPLE 13.2   Bolt Yield Limit Equations for Single Shear

The following yield limit equations apply to the three types of single-shear bolt connections that were illustrated in Fig. 13.1. Yield modes are sketched in Fig. 11.10 (Sec. 11.7).

The reference design value for one bolt in a single-shear connection between two members is the smallest value obtained from the following six equations.

Mode I_m (NDS Eq. 11.3-1):

$$Z = \frac{Dl_m F_{em}}{4K_\theta}$$

Mode I_s (NDS Eq. 11.3-2):

$$Z = \frac{Dl_s F_{es}}{4K_\theta}$$

Mode II (NDS Eq. 11.3-3):

$$Z = \frac{k_1 Dl_s F_{es}}{3.6K_\theta}$$

Mode III_m (NDS Eq. 11.3-4):

$$Z = \frac{k_2 Dl_m F_{em}}{3.2(1 + 2R_e)K_\theta}$$

Mode III_s (NDS Eq. 11.3-5):

$$Z = \frac{k_3 Dl_s F_{em}}{3.2(2 + R_e)K_\theta}$$

Mode IV (NDS Eq. 11.3-6):

$$Z = \frac{D^2}{3.2K_\theta} \sqrt{\frac{2F_{em}F_{yb}}{3(1 + R_e)}}$$

where Z = reference design value for bolt in single shear (Z is taken as the smallest value from six yield limit equations), lb

$$k_1 = \frac{\sqrt{R_e + 2R_e^2(1 + R_t + R_t^2) + R_t^2R_e^3} - R_e(1 + R_t)}{(1 + R_e)}$$

$$k_2 = -1 + \sqrt{2(1 + R_e) + \frac{2F_{yb}(1 + 2R_e)D^2}{3F_{em}l_m^2}}$$

$$k_3 = -1 + \sqrt{\frac{2(1 + R_e)}{R_e} + \frac{2F_{yb}(2 + R_e)D^2}{3F_{em}l_s^2}}$$

$$R_e = \frac{F_{em}}{F_{es}}$$

$$R_t = \frac{l_m}{l_s}$$

$$K_\theta = 1 + \frac{\theta}{360}$$

D = bolt diameter, in.

l_m = dowel bearing length in main (thicker) member, in.

l_s = dowel bearing length in side (thinner) member, in.

F_{em} = dowel bearing strength of main member, psi:

$$= \begin{cases} F_{e\parallel} & \text{for load parallel to grain} \\ F_{e\perp} & \text{for load perpendicular to grain} \\ F_{e\theta} & \text{for load at angle to grain } \theta \text{ (see Hankinson formula below)} \end{cases}$$

For a concrete or masonry main member, F_e = 7500 psi

F_{es} = dowel bearing strength of side member, psi:

$$= \begin{cases} F_{e\parallel} & \text{for load parallel to grain} \\ F_{e\perp} & \text{for load perpendicular to grain} \\ F_{e\theta} & \text{for load at angle to grain } \theta \text{ (see Hankinson formula below)} \end{cases}$$

For a hot-rolled steel side member, F_{es} = 1.5F_u

For a cold-formed steel side member, F_{es} = 1.375F_u

F_{yb} = bending yield strength of bolt, psi

θ = maximum angle of load to grain ($0 \le \theta \le 90$ degrees) for any member in connection

Dowel bearing strength at an angle of load to grain θ is given by the Hankinson formula:

$$F_{e\theta} = \frac{F_{e\parallel}F_{e\perp}}{F_{e\parallel}\sin^2\theta + F_{e\perp}\cos^2\theta}$$

In wood-to-metal connections, the designer must ensure that the bearing capacity of the steel side member is not exceeded in accordance with recognized steel design practice. In wood-to-concrete or wood-to-masonry connections, the designer must ensure that the bearing capacity of the concrete or masonry is not exceeded in accordance with recognized concrete and masonry design practice. This can be done for concrete in IBC Table 1911.2 (Ref. 13.10). See Example 15.6 (Sec. 15.4) for two sample anchor bolt problems.

As an alternative to solving the yield limit equations, the reference design values for a number of common *single-shear* connections are available in NDS Tables 11A to 11E.

For *wood-to-wood* connections, four reference design values are given in the NDS tables. Four values are listed because the equations used to generate the tables produce a different bolt capacity depending on the *angle of load to grain* for the members in the connection. Values of Z are given for the following conditions:

1. Load *parallel* to grain in *both* main member and side member. The notation in the NDS tables for this condition is $Z_{\parallel}$.
2. Load *perpendicular* to grain in *side* member and parallel to grain in main member, $Z_{s\perp}$.
3. Load *perpendicular* to grain in *main* member and parallel to grain in side member, $Z_{m\perp}$.
4. Load *perpendicular* to grain in *both* main member and side member, $Z_{\perp}$.

For *wood-to-steel* and *wood-to-concrete* connections, the NDS tables provide two values of Z based on the angle of load to grain for the wood member:

1. Load parallel to grain in wood member, $Z_{\parallel}$.
2. Load perpendicular to grain in wood member, $Z_{\perp}$.

Metal side plates are assumed to be $\frac{1}{4}$-in.-thick ASTM A36 steel in NDS Tables 11B and 11D.

The NDS tables may be helpful in providing reference bolt design values for a number of loading conditions without the need to evaluate the yield limit formulas. However, the NDS tables are restricted to parallel- and perpendicular-to-grain loadings, and the yield limit equations must be evaluated for connections with an intermediate angle of load to grain ($0 < \theta < 90$ degrees).

Two examples are given to illustrate the application of the yield limit equations. The first problem verifies a bolt value in one of the NDS tables. See Example 13.3. The load is perpendicular to the grain in the main member and parallel to the grain in the side member. Both wood pieces are of the same species, but there are different dowel bearing strengths for the two members because of the angle of load to grain. The problem is not a complete design because the conditions necessary to convert the reference design value to an adjusted design value are not given. The example is simply intended to illustrate the yield limit equations and the NDS tables.

EXAMPLE 13.3   Reference Design Value: Single-Shear Bolt in Wood-to-Wood Connection

The horizontal load in the connection shown in Fig. 13.4 is parallel to the grain in the horizontal member and perpendicular to the grain in the vertical member. Lumber is Douglas Fir-Larch (DF-L). Determine the reference design value of the single $\frac{3}{4}$-in. full-diameter body A307 bolt.

Figure 13.4 Single-shear wood-to-wood connection.

Summary of Known Values

Summarize the information necessary to evaluate the yield limit equations. Bolt diameter, member thicknesses, and angles of load to grain are obtained from the sketch. Dowel bearing strengths can be computed using the formulas from Example 11.4 (Sec. 11.5), or they can be read from NDS Table 11.3.2. Specific gravity is given in NDS Table 11.3.2A.

$$D = 0.75 \text{ in.}$$

$$F_{yb} = 45 \text{ ksi} \qquad \text{(assumed for A307 bolt)}$$

$$l_m = t_m = 3.5 \text{ in.}$$

$$\theta_m = 90 \text{ degrees}$$

$$F_{em} = 2600 \text{ psi} \qquad (F_{e\perp} \text{ for } \tfrac{3}{4}\text{-in. bolt in DF-L})$$

$$l_s = t_s = 1.5 \text{ in.}$$

$$\theta_s = 0 \text{ degrees}$$

$$F_{es} = 5600 \text{ psi} \qquad (F_{e\parallel} \text{ for bolt in DF-L})$$

$$G_m = G_s = 0.50$$

Coefficients for Yield Limit Equations

$$R_e = \frac{F_{em}}{F_{es}} = \frac{2600}{5600} = 0.464$$

$$1 + R_e = 1 + 0.464 = 1.464$$

$$1 + 2R_e = 1 + 2(0.464) = 1.929$$

$$2 + R_e = 2 + 0.464 = 2.464$$

$$R_t = \frac{l_m}{l_s} = \frac{3.5}{1.5} = 2.333$$

$$1 + R_t = 1 + 2.333 = 3.333$$

$$1 + R_t + R_t^2 = 3.333 + (2.333)^2 = 8.778$$

$\theta = 90$ degrees (Note: θ is the larger of θ_m and θ_s.)

$$K_\theta = 1 + \frac{\theta}{360} = 1 + \frac{90}{360} = 1.25$$

$$k_1 = \frac{\sqrt{R_e + 2R_e^2(1 + R_t + R_t^2) + R_t^2 R_e^3} - R_e(1 + R_t)}{(1 + R_e)}$$

$$= \frac{\sqrt{0.464 + 2(0.464^2)(8.778) + (2.333)^2(0.464^3)} - 0.464(3.333)}{1.464}$$

$$= 0.4383$$

$$k_2 = -1 + \sqrt{2(1 + R_e) + \frac{2F_{yb}(1 + 2R_e)D}{3F_{em}l_m^2}}$$

$$= -1 + \sqrt{2(1.464) + \frac{2(45,000)(1.929)(0.75^2)}{3(2600)(3.5^2)}}$$

$$= 0.9876$$

$$k_3 = -1 + \sqrt{\frac{2(1 + R_e)}{R_e} + \frac{2F_{yb}(2 + R_e)D^2}{3F_{em}l_s^2}}$$

$$= -1 + \sqrt{\frac{2(1.464)}{0.464} + \frac{2(45,000)(2.464)(0.75^2)}{3(2600)(1.5^2)}}$$

$$= 2.663$$

Yield Limit Equations

Mode I_m (NDS Eq. 11.3-1):

$$Z = \frac{Dl_m F_{em}}{4K_\theta} = \frac{0.75(3.5)(2600)}{4(1.25)} = 1365 \text{ lb}$$

Mode I_s (NDS Eq. 11.3-2):

$$Z = \frac{Dl_s F_{es}}{4K_\theta} = \frac{0.75(1.5)(5600)}{4(1.25)} = 1260 \text{ lb}$$

Mode II (NDS Eq. 11.3-3):

$$Z = \frac{k_1 Dl_s F_{es}}{3.6K_\theta} = \frac{0.4383(0.75)(1.5)(5600)}{3.6(1.25)} = 614 \text{ lb}$$

Mode III_m (NDS Eq. 11.3-4):

$$Z = \frac{k_2 Dl_m F_{em}}{3.2(1 + 2R_e)K_\theta} = \frac{0.9876(0.75)(3.5)(2600)}{3.2(1.929)(1.25)} = 874 \text{ lb}$$

Mode III_s (NDS Eq. 11.3-5):

$$Z = \frac{k_3 Dl_s F_{em}}{3.2(2 + R_e)K_\theta} = \frac{2.663(0.75)(1.5)(2600)}{3.2(2.464)(1.25)} = 790 \text{ lb}$$

Mode IV (NDS Eq. 11.3-6):

$$Z = \frac{D^2}{3.2K_\theta}\sqrt{\frac{2F_{em}F_{yb}}{3(1 + R_e)}}$$

$$= \frac{0.75^2}{3.2(1.25)}\sqrt{\frac{2(2600)(45,000)}{3(1.464)}} = 1026 \text{ lb}$$

The reference design value is selected as the smallest value from the yield limit equations:

$$\boxed{Z = 614 \text{ lb}}$$

Compare with value from NDS Table 11A, *BOLTS: Reference Lateral Design Values (Z) for Single Shear (two member) Connections.* Enter table with $t_m = 3.5$, $t_s = 1.5$, $D = 0.75$, and $G = 0.50$:

$$Z_{m\perp} = 610 \text{ lb} \approx 614 \text{ lb} \quad OK$$

The second example is a bolt in single shear that is not covered by the NDS tables, and the yield limit equations must be used. The connection is between a $4\times$ wood member and a $\frac{1}{2}$-in.-thick hot-rolled angle. The angle is A36 steel, and the minimum tensile strength is $F_u = 58$ ksi. Example 13.4 has an angle of load to grain of 45 degrees, and $F_{e\theta}$ is computed with the Hankinson formula.

EXAMPLE 13.4   Reference Design Value: Single-Shear Bolt in Wood-to-Metal Connection

The bolt through steel angle in Fig. 13.5 provides a load to the horizontal wood member at an angle of 45 degrees to the grain. The angle is A36 steel, and the wood member is Hem-Fir. Determine the reference design value of a single $\frac{5}{8}$-in. full-diameter body A307 bolt.

Figure 13.5   Single-shear wood-to-steel connection.

Summary of Known Values

Dowel bearing strengths parallel and perpendicular to grain in Hem-Fir can be computed using the formulas in Example 11.4 (Sec. 11.5), or they can be obtained from NDS Table 11.3.2.

$$D = 0.625 \text{ in.}$$

$$F_{yb} = 45 \text{ ksi} \quad \text{(assumed for A307 bolt)}$$

$$l_m = t_m = 3.5 \text{ in.}$$

$$\theta_m = 45 \text{ degrees}$$

$$F_{em\|} = 4800 \text{ psi}$$

$$F_{em\perp} = 2250 \text{ psi} \quad (F_{e\perp} \text{ for } \tfrac{5}{8}\text{-in. bolt in Hem-Fir})$$

$$l_s = t_s = 0.5 \text{ in.}$$

$$F_{es} = 1.5F_u = 87{,}000 \text{ psi} \quad \text{(for A36 steel)}$$

Coefficients for Yield Limit Equations

$$F_{e\theta} = \frac{F_{e\|}F_{e\perp}}{F_{e\|}\sin^2\theta + F_{e\perp}\cos^2\theta}$$

$$= \frac{4800(2250)}{4800\sin^2 45 + 2250\cos^2 45} = 3064 \text{ psi}$$

$$R_e = \frac{F_{em}}{F_{es}} = \frac{3064}{87{,}000} = 0.03522$$

$$1 + R_e = 1 + 0.03522 = 1.035$$

$$1 + 2R_e = 1 + 2(0.03522) = 1.070$$

$$2 + R_e = 2 + 0.03522 = 2.035$$

$$R_t = \frac{l_m}{l_s} = \frac{3.5}{0.5} = 7.0$$

$$1 + R_t = 1 + 7.0 = 8.0$$

$$1 + R_t + R_t^2 = 8.0 + 7.0^2 = 57.0$$

$$\theta = 45 \text{ degrees}$$

$$K_\theta = 1 + \frac{\theta}{360} = 1 + \frac{45}{360} = 1.125$$

$$k_1 = \frac{\sqrt{R_e + 2R_e^2(1 + R_t + R_t^2) + R_t^2 R_e^3} - R_e(1 + R_t)}{(1 + R_e)}$$

$$= \frac{\sqrt{0.03522 + 2(0.03522^2)(57.0) + 7.0^2(0.03522^3)} - 0.03522(8.0)}{1.035}$$

$$= 0.1367$$

$$k_2 = -1 + \sqrt{2(1 + R_e) + \frac{2F_{yb}(1 + 2R_e)D^2}{3F_{em}l_m^2}}$$

$$= -1 + \sqrt{2(1.035) + \frac{2(45{,}000)(1.070)(0.625^2)}{3(3064)(3.5^2)}}$$

$$= 0.5507$$

$$k_3 = -1 + \sqrt{\frac{2(1 + R_e)}{R_e} + \frac{2F_{yb}(2 + R_e)D^2}{3F_{em}l_s^2}}$$

$$= -1 + \sqrt{\frac{2(1.035)}{0.03522} + \frac{2(45{,}000)(2.035)(0.625^2)}{3(3064)(0.5^2)}}$$

$$= 8.483$$

Yield Limit Equations

Mode I_m (NDS Eq. 11.3-1):

$$Z = \frac{Dl_m F_{em}}{4K_\theta} = \frac{0.625(3.5)(3064)}{4(1.125)} = 1489 \text{ lb}$$

Mode I_s (NDS Eq. 11.3-2):

$$Z = \frac{Dl_s F_{es}}{4K_\theta} = \frac{0.625(0.5)(87{,}000)}{4(1.125)} = 6042 \text{ lb}$$

Mode II (NDS Eq. 11.3-3):

$$Z = \frac{k_1 D l_s F_{es}}{3.6 K_\theta} = \frac{0.1367(0.625)(0.5)(87{,}000)}{3.6(1.125)} = 917 \text{ lb}$$

Mode III$_m$ (NDS Eq. 11.3-4):

$$Z = \frac{k_2 D l_m F_{em}}{3.2(1 + 2R_e)K_\theta} = \frac{0.5507(0.625)(3.5)(3064)}{3.2(1.070)(1.125)} = 958 \text{ lb}$$

Mode III$_s$ (NDS Eq. 11.3-5):

$$Z = \frac{k_3 D l_s F_{em}}{3.2(2 + R_e)K_\theta} = \frac{8.483(0.625)(0.5)(3064)}{3.2(2.035)(1.125)} = 1109 \text{ lb}$$

Mode IV (NDS Eq. 11.3-6):

$$Z = \frac{D^2}{3.2 K_\theta} \sqrt{\frac{2 F_{em} F_{yb}}{3(1 + R_e)}}$$

$$= \frac{0.625^2}{3.2(1.125)} \sqrt{\frac{2(3064)(45{,}000)}{3(1.035)}} = 1022 \text{ lb}$$

The reference design value is selected as the smallest value from the yield limit equations:

$$\boxed{Z = 917 \text{ lb}}$$

13.4 Bolt Yield Limit Equations for Double Shear

The yield limit equations and several numerical examples for bolts in single-shear connections were given in the previous section. In addition to this application, bolts are frequently used in double shear. From a review of the various yield modes in Fig. 11.10 (Sec. 11.7), it should be clear that there are different sets of yield limit equations for double- and single-shear connections.

The yield limit equations for bolt connections in double shear are given in Example 13.5. Because of symmetric behavior, only four yield limit equations are required for a double-shear connection. In addition, steel side members must be designed in accordance with recognized steel design practice (Refs. 13.3 and 13.7).

EXAMPLE 13.5 Yield Limit Equations for Bolts in Double Shear

The following yield limit equations from the NDS apply to the two common types of double-shear connections illustrated in Fig. 13.2. The yield modes are sketched in Fig. 11.10.

Wood-to-Wood Connection

The reference design value for one bolt in a double-shear connection between three wood members is the smallest value obtained from the following four equations. The thicknesses of the *side* members are assumed to be equal. See the NDS for the case of unequal thickness.

Mode I_m (NDS Eq. 11.3-7):

$$Z = \frac{Dl_m F_{em}}{4K_\theta}$$

Mode I_s (NDS Eq. 11.3-8):

$$Z = \frac{2Dl_s F_{es}}{4K_\theta}$$

Mode III_s (NDS Eq. 11.3-9):

$$Z = \frac{2k_3 Dl_s F_{em}}{3.2(2 + R_e)K_\theta}$$

Mode IV (NDS Eq. 11.3-10):

$$Z = \frac{2D^2}{3.2K_\theta}\sqrt{\frac{2F_{em}F_{yb}}{3(1 + R_e)}}$$

where Z = reference design value for bolt in double shear (Z is taken as the smallest value from four yield limit equations), lb

$$k_3 = -1 + \sqrt{\frac{2(1 + R_e)}{R_e} + \frac{2F_{yb}(2 + R_e)D^2}{3F_{em}l_s^2}}$$

$$R_e = \frac{F_{em}}{F_{es}}$$

$$K_\theta = 1 + \frac{\theta}{360}$$

D = bolt diameter, in.

l_m = dowel bearing length in main (center) member, in.

l_s = dowel bearing length in one of the side members, in.

F_{em} = dowel bearing strength of main (center) member, psi:

$$= \begin{cases} F_{e\parallel} & \text{for load parallel to grain} \\ F_{e\perp} & \text{for load perpendicular to grain} \\ F_{e\theta} & \text{for load at angle to grain } \theta \text{ (see Hankinson formula below)} \end{cases}$$

F_{es} = dowel bearing strength of side member, psi:

$$= \begin{cases} F_{e\parallel} & \text{for load parallel to grain} \\ F_{e\perp} & \text{for load perpendicular to grain} \\ F_{e\theta} & \text{for load at angle to grain } \theta \text{ (see Hankinson formula below)} \end{cases}$$

For hot-rolled steel side members, $F_{es} = 1.5F_u$

For cold-formed steel side members, $F_{es} = 1.375F_u$

F_{yb} = bending yield strength of bolt, psi

θ = maximum angle of load to grain ($0 \le \theta \le 90$ degrees) for any member in connection

Dowel bearing strength at an angle of load to grain θ is given by the Hankinson formula:

$$F_{e\theta} = \frac{F_{e\parallel}F_{e\perp}}{F_{e\parallel}\sin^2\theta + F_{e\perp}\cos^2\theta}$$

Wood-to-Metal Connections

For one bolt in a double-shear connection between a *wood main member* and *two steel side plates*, the reference design value is taken as the smallest value obtained from the four yield limit equations given above for a double-shear connection. In addition, the designer must ensure that the bearing capacity of the steel side members is not exceeded in accordance with accepted steel design practice (Refs. 13.3 and 13.7).

A less frequently encountered double-shear connection involves a *steel main (center) member* and *two wood side members*. In this situation a Mode I_m mechanism represents a bearing failure in the steel main member. The yield limit equations are evaluated to determine the reference design value for the bolt. In this case also, the designer must ensure that the bearing capacity of the steel member is adequate in accordance with recognized steel design practice.

As with single shear, the NDS bolt tables may be helpful in providing reference design values for a number of loading conditions without the need to evaluate the yield limit formulas. Values of Z for double-shear connections are given in NDS Tables 11F to 11I. Again, the NDS tables are limited to parallel- and perpendicular-to-grain loading only, and the yield limit equations are required for connections with an intermediate angle of load to grain ($0 < \theta < 90$ degrees).

One example is given to illustrate the yield limit equations for double shear and to gain familiarity with the NDS bolt tables. See Example 13.6. Although the example deals with a connection that can be read from an NDS table, a computer solution on a spreadsheet or other program will normally be used for the cases not covered by a table. The example illustrates the yield limit formulas, and the adjustment factors necessary to convert the reference design value to an adjusted bolt capacity are covered in the next section.

EXAMPLE 13.6   Reference Design Value: Double-Shear Bolt in Wood-to-Metal Connection

Determine the reference design value of the single $\frac{7}{8}$ -in. full-diameter body A307 bolt in Fig. 13.6. The load in the connection is parallel to the grain in the 6 × wood member, and $\frac{1}{4}$-in.-thick steel side plates are used. The wood member is DF-L, and the side plates are A36 steel with $F_u = 58$ ksi.

Summary of Known Values

Dowel bearing strength of the wood member can be computed using the formulas in Example 11.4 (Sec. 11.5), or it can be read from NDS Table 11.3.2. Specific gravity is also available in NDS Table 11.3.2A.

$$D = 0.875 \text{ in.}$$

$$F_{yb} = 45 \text{ ksi} \quad \text{(assumed for A307 bolt)}$$

$$l_m = t_m = 5.5 \text{ in.}$$

$$\theta_m = 0 \text{ degrees}$$

$$F_{em} = 5600 \text{ psi} \quad (F_{e\parallel} \text{ for DF-L})$$

$$G_m = 0.50$$

$$l_s = t_s = 0.25 \text{ in.}$$

$$F_{es} = 1.5 F_u = 87,000 \text{ psi} \qquad \text{(for A36 steel)}$$

Figure 13.6 Double-shear wood-to-steel connection.

Coefficients for Yield Limit Equations

$$R_e = \frac{F_{em}}{F_{es}} = \frac{5600}{87,000} = 0.06437$$

$$1 + R_e = 1 + 0.06437 = 1.064$$

$$2 + R_e = 2 + 0.06437 = 2.064$$

$$\theta = 0 \text{ degrees}$$

$$K_\theta = 1 + \frac{\theta}{360} = 1 + \frac{0}{360} = 1.0$$

$$k_3 = -1 + \sqrt{\frac{2(1 + R_e)}{R_e} + \frac{2F_{yb}(2 + R_e)D^2}{3F_{em}l_s^2}}$$

$$= -1 + \sqrt{\frac{2(1.064)}{0.06437} + \frac{2(45,000)(2.064)(0.875^2)}{3(5600)(0.25^2)}}$$

$$= 11.98$$

Yield Limit Equations

Mode I_m (NDS Eq. 11.3-7):

$$Z = \frac{Dl_m F_{em}}{4K_\theta} = \frac{0.875(5.5)(5600)}{4(1.0)} = 6783 \text{ lb}$$

Mode I_s (NDS Eq. 11.3-8):

$$Z = \frac{2Dl_s F_{es}}{4K_\theta} = \frac{2(0.875)(0.25)(87,000)}{4(1.0)} = 9516 \text{ lb}$$

Mode III_s (NDS Eq. 11.3-9):

$$Z = \frac{2k_3 Dl_s F_{em}}{3.2(2 + R_e)K_\theta} = \frac{2(11.98)(0.875)(0.25)(5600)}{3.2(2.064)(1.0)} = 4444 \text{ lb}$$

Mode IV (NDS Eq. 11.3-10):

$$Z = \frac{2D^2}{3.2K_\theta}\sqrt{\frac{2F_{em}F_{yb}}{3(1 + R_e)}} = \frac{2(0.875^2)}{3.2(1.0)}\sqrt{\frac{2(5600)(45,000)}{3(1.064)}} = 6012 \text{ lb}$$

The reference design value is selected as the smallest value from the yield limit equations:

$$\boxed{Z = 4444 \text{ lb}}$$

Compare with the value from NDS Table 11G:

$$Z_\| = 4440 \approx 4444 \text{ lb} \qquad OK$$

13.5 Adjustment Factors for Bolts

The yield limit equations for single- and double-shear bolt connections were reviewed in previous sections. The term *reference design value* is assigned to the allowable stress design (ASD) capacity of a single bolt as obtained from the yield limit equations. For use in the design of a particular connection, the reference design value Z must be converted to an *adjusted design value Z'*. As with nails, this is done by multiplying the reference value by a series of adjustment factors.

The adjustments convert the base conditions for a bolt to the design conditions for a given problem. Depending on the circumstances, the adjustment factors could have a major or minor effect on the adjusted design value. The adjustment factors required in the design of connections are summarized in NDS Table 10.3.1. If the particular design conditions agree with the conditions for the reference design value, the adjustment factors for ASD simply default to unity.

The base conditions associated with the reference design value are as follows:

1. Load is normal (10-year) duration.
2. Wood is initially dry at the time of fabrication of a connection and remains dry in service.

3. Temperature range is normal.

4. There is one bolt in a line parallel to the applied load (i.e., only one bolt per row).

5. Spacing provisions for bolts (geometry requirements for the connection) are satisfied.

If a bolt connection is used under exactly these conditions, the adjustment factors are all unity, and $Z' = Z$ for ASD.

For a laterally loaded bolt connection that does not satisfy the base conditions described above, the following adjustment factors may be required for ASD:

$$Z' = Z(C_D C_M C_t C_g C_\Delta)$$

where C_D = load duration factor (Sec. 4.15)
   C_M = wet service factor (Sec. 4.14)
   C_t = temperature factor (Sec. 4.20)
   C_g = group action factor
   C_Δ = geometry factor

In load and resistance factor design (LRFD), the reference design value must be multiplied by a format conversion factor K_F and a resistance factor ϕ_z. The K_F factor adjusts the reference design value Z to a nominal design value Z_n appropriate for LRFD. As part of the conversion to an LRFD format, the K_F factor adjusts from a normal (10-year) load duration basis to a short term (5 to 10 minutes) load duration basis. The load duration factor C_D is not applicable for LRFD calculations. Instead a time effect factor λ is selected based on the governing LRFD load combination, as specified in NDS Appendix N (Ref. 13.1) (Sec. 4.16). All other adjustment factors are identical for both LRFD and ASD. Thus, for a laterally loaded bolt connection the following adjustment factors may be required for LRFD:

$$Z'_n = Z_n(\phi_z \lambda C_M C_t C_g C_\Delta)$$

The total adjusted capacity for a connection using two or more bolts is the sum of the individual adjusted design values Z'. This summation is appropriate if the bolts are all the same type and size, because each fastener will have the same yield mode. The capacity for the connection is the number of bolts N times the adjusted design value for one fastener.

$$\text{Adj. } P = N(Z')$$

A number of adjustment factors were described previously and are mentioned only in summary. Other adjustments require greater explanation. Some of these new factors will also be required for other connectors such as lag bolts, split rings, and shear plates.

Resistance factor ϕ_z (LRFD only).   The resistance factor for all fasteners in the NDS is $\phi_z = 0.65$. Resistance factors are specified in NDS Appendix N (Ref. 13.1) for connections and various wood properties.

Format conversion factor K_F (LRFD only). Format conversion factors for converting reference design values to nominal LRFD design values are specified in NDS Appendix N (Ref. 13.1). The format conversion factor for all fasteners in the NDS is:

$$K_F = \frac{2.16}{\phi_z} = \frac{2.16}{0.65} = 3.32$$

Load duration factor C_D (ASD only). The load duration factor C_D for bolted connections is the same as that applied to nailed connections. Refer to Secs. 4.15 and 12.6 for a review of load duration and the numerical values for C_D.

Time effect factor λ (LRFD only). The time effect factor λ for bolted connections is the same as that applied to nailed connections. Refer to Secs. 4.16 and 12.6 for a review of the time effect factor and the numerical values for λ.

Wet service factor C_M. The concept of a design value adjustment based on moisture content (MC) was introduced in Sec. 4.14. However, the wet service factors for member design vary with the type of stress. Different adjustments also apply in connection design depending on the type of fastener and the MC of the wood at several stages. The MC is considered (1) at the time of fabrication of the joint and (2) in service. The reference design value applies to a bolt connection in which the wood is initially dry and remains dry in service. For connection design, the term *dry* refers to wood that has an MC of 19 percent or less. For these conditions, there is no reduction in the design value of a bolt for moisture content, and $C_M = 1.0$. Other MC conditions may require significant reductions of the reference design value (i.e., $C_M < 1.0$).

The proper design of bolted connections requires that the effects of changes in MC be considered in the design of a wood structure. In the past, problems were created in certain bolted connections by designers who were not familiar with the unique characteristics of wood.

It is necessary for the designer to understand the material if it is to be used properly. Wood undergoes dimensional changes as the MC varies. Shrinkage (or swelling) is caused by changes in moisture content below the fiber saturation point (Sec. 4.7). Most of the dimensional change takes place across the grain, and very little occurs parallel to the grain. These volumetric changes require that the *arrangement of bolts* in a connection be taken into account, in addition to the moisture content conditions mentioned above.

Problems involving volume changes develop when bolts are held rigidly in position. Perhaps the best illustration is in a connection that uses a metal side plate and parallel rows of bolts. See Example 13.7. The bolts are essentially fixed in position by the metal side plates (the only movement permitted is that allowed by the clearance between the bolts and the holes). If the lumber is at an initially high moisture content and is allowed to season in place after fabrication of the connection, serious splits and cracks can develop.

The shrinkage that occurs is primarily across the grain. Wood is very weak in tension perpendicular to grain, and the result will be the development of cracking parallel to the grain. Obviously this type of cracking greatly reduces the strength of a connection.

EXAMPLE 13.7   Cross-Grain Cracking

Figure 13.7*a* shows a critical connection for cross-grain cracking. As the wood loses moisture, cross-grain shrinkage occurs, and the bolts are held rigidly in position by the steel plates. The lumber becomes stressed in tension across the grain (a weak state of stress for wood).

Figure 13.7a   Restrained shrinkage causes tension stress perpendicular to grain.

Wet-Service Factors (C_M)

In the NDS, the following definitions apply to connections:
   "Wet" condition—MC > 19 percent
   "Dry" condition—MC ≤ 19 percent

For the type of connection in Fig. 13.7*a*, the following values of C_M are specified in NDS Table 10.3.3:

1. When the lumber is initially dry and remains dry in service, $C_M = 1.0$.
2. When the lumber used to fabricate a connection is initially wet and seasons in place to a dry condition, the reference design value is reduced by 60 percent (i.e., $C_M = 0.40$).

The cross-grain cracking problem can be minimized by proper joint detailing. The use of separate side plates for each line of bolts allows the free movement of the bolts across the grain. Slotted holes in steel side plates can also be used. With proper detailing C_M can be taken as 1.0 for bolt connections in dry-service conditions, or 0.7 for bolt connections in wet service conditions. See Fig. 13.7*b*.

   NDS spacing criteria limit the spacing between outer rows of bolts to a maximum of 5 in. unless separate side plates are used. If separate plates are not used, the 5-in. limit on the spacing across the grain provides a limited width across the grain in which shrinkage can occur.

WOOD
MEMBER

CLEARANCE
FOR SHRINKAGE

SEPARATE SIDE ℞'s ALLOW
FREE MOVEMENT OF BOLTS
ACROSS THE GRAIN

STEEL
SIDE ℞'s

Figure 13.7b   Proper connection detailing eliminates restraint across the grain.

The problem of cross-grain shrinkage can often be eliminated by proper detailing of a connection. This can be accomplished by providing for unrestricted movement of the bolts across the grain. For example, in the case of the steel side plate, separate side members for the two rows of bolts can be used. These separate plates will simply move with the bolts as the MC changes, and cross-grain tension and the associated cracking will be avoided. Another approach is to use slotted holes in the steel side plates so that movement is not restricted. Other examples of proper connection details to avoid splitting are given in Chap. 14.

It was noted that very little dimensional change occurs parallel to the grain. This fact can cause a cross-grain cracking problem when wood members frame together at right angles. See Fig. 13.8. The problem develops when there are

WOOD
MEMBERS

BOLTS

SHRINKAGE OCCURS
⊥ TO GRAIN IN HORIZ
MEMBER BUT NOT
∥ TO GRAIN IN THE
VERTICAL MEMBER

CROSS-GRAIN
CRACKING

SHRINKAGE OCCURS ⊥ TO THE
GRAIN IN THE VERTICAL MEMBER
BUT NOT ∥ TO THE GRAIN IN
THE HORIZ MEMBER

Figure 13.8   Cracking in wood-to-wood connection.

multiple rows of bolts passing through the two members and significant changes in moisture content occur. Here the problem is caused by cross-grain shrinkage in the mutually perpendicular members with little or no shrinkage parallel to the grain. The use of a single fastener for this connection is desirable. The problem is not critical in connections that are initially dry and remain dry in service.

The importance of proper connection detailing can be seen by the magnitude of the wet service factor required when severe cross-grain cracking can develop. Example 13.7 indicates that a C_M of 0.40 is required when these connections are fabricated with wood members that are initially wet and season to a dry condition in service. A summary of the C_M values for wood connections is given in NDS Table 10.3.3, *Wet Service Factors, C_M, for Connections.*

It should be noted that $C_M = 1.0$ for connections where cross-grain cracking is prevented (e.g., connections that are initially dry and remain dry). The NDS also indicates that $C_M = 1.0$ for connections with proper detailing when the lumber is wet at the time of fabrication and will season in place to a dry service condition. Examples of proper connection detailing include:

1. One fastener only

2. Two or more fasteners in a single line parallel to the grain (no shrinkage parallel to grain)

3. Fasteners in two or more lines parallel to the grain with separate side plates for each line (movement not restricted)

Note that there can be a fairly large difference between the initial and final MC in the *dry* range, and shrinkage can occur (Sec. 4.7). This is especially true in dry climate zones. In view of the concern about the effects of cross-grain shrinkage in critical connections, the designer may want to reduce design values when large differences in MC are expected even in the dry range.

As noted earlier, however, the most desirable solution is to detail the connection so that shrinkage cannot induce cross-grain cracking in the first place. Providing for unrestricted movement is a key factor. Recommended connection detailing practices with this intention are given in Chap. 14.

Temperature factor C_t.   If wood is used in an application with sustained exposure to higher-than-normal temperature conditions, the reference design values for connections should be reduced by an appropriate temperature factor C_t. Refer to Sec. 4.20 for a brief review of the temperature factor. Numerical values of C_t for connections are given in NDS Table 10.3.4. The temperature range for most wood frame buildings does not require adjustment of bolt values, and C_t is often unity.

Group action factor C_g.   Research has shown that each bolt in a row does not carry an equal portion of the total load on a connection. This is also true for other large-diameter fasteners such as lag bolts, split ring connectors, and shear plate connectors. To account for this behavior, the reference design value is reduced based on the number of fasteners in a row. A *row* is considered to be a *group of fasteners in a line parallel* to the direction of *loading.* The term *group action* *factor* C_g is the reduction factor that accounts for the nonuniform loading of fasteners. The value of C_g decreases as the number of fasteners in a row increases. The group action factor does not apply to fasteners with $D < \frac{1}{4}$ in.

The NDS defines the group action factor with an equation and provides tables of C_g values for a limited set of conditions. The formula for the group action factor is rather cumbersome to solve by hand on a calculator, and it is preferable to solve the expression on a computer. The definition of group action factor in equation form allows the designer to handle a much wider variety of loading conditions and joint details. See Example 13.8.

The primary consideration in determining C_g is the number of fasteners in a row n. The number of fasteners in a row n should not be confused with the number of fasteners in the connection N. If there is only one row of fasteners, then n and N are equal.

In addition to the number of fastener in a row, the group action factor depends on

Ratio of the stiffnesses of members being connected

Diameter of fastener

Spacing of fasteners in a row

Load/slip modulus for the fastener

Although net area is used to evaluate member stresses at a connection, the distribution of forces to the fasteners is a function of relative stiffness. For an axial force member, the stiffness of a member is the product of *gross* cross-sectional area and modulus of elasticity.

EXAMPLE 13.8   Formula for Group Action Factor C_g

The reference design value for a fastener Z is reduced by the group action factor C_g. The value of C_g decreases as the number of fasteners in a row increases.

Figure 13.9 Definition of terms for C_g. Group action factor primarily depends on the number of fasteners in a row. For sample connection: Number of fasteners in row $= n = 4$; number of fasteners in connection $= N = 8$.

The number of fasteners in a row n is used to determine the group action *factor* C_g. The total number of fasteners N is used to determine the total adjusted capacity for the connection.

$$\text{Adj. } P = N(Z')$$

The formula for C_g is the same for bolts, lag bolts, split rings, and shear plates. The differences in these fasteners are reflected in the load/slip modulus γ for the connection. The group action factor is given by the following formula:

$$C_g = \left[\frac{m(1 - m^{2n})}{n[(1 + R_{EA}m^n)(1 + m) - 1 + m^{2n}]}\right]\left[\frac{1 + R_{EA}}{1 - m}\right]$$

where n = number of fasteners in a row
$\quad R_{EA}$ = smaller ratio of member stiffnesses

$\quad\quad$ = lesser of $\dfrac{E_s A_s}{E_m A_m}$ or $\dfrac{E_m A_m}{E_s A_s}$

$\quad E_m$ = modulus of elasticity of main member, psi
$\quad A_m$ = gross cross-sectional area of main member, in.^{2}
$\quad E_s$ = modulus of elasticity of side member, psi
$\quad A_s$ = gross cross-sectional area of side member, in.^{2} (Note: If there is more than one side member, A_s is the sum of the gross cross-sectional areas of the side members.)

$\quad m = u - \sqrt{u^2 - 1}$

$\quad u = 1 + \gamma\left(\dfrac{s}{2}\right)\left[\dfrac{1}{E_m A_m} + \dfrac{1}{E_s A_s}\right]$

$\quad s$ = center-to-center spacing between adjacent fasteners in a row, in.
$\quad \gamma$ = load/slip modulus for a connection, lb/in.
$\quad\quad$ = $180,000(D^{1.5})$ for dowel-type fasteners such as bolts or lag bolts in a wood-to-wood connection
$\quad\quad$ = $270,000(D^{1.5})$ for dowel-type fasteners such as bolts or lag bolts in a wood-to-metal connection
$\quad\quad$ = $500,000$ for 4-in.-diameter split ring or shear plate connectors
$\quad\quad$ = $400,000$ for $2\frac{1}{2}$-in.-diameter split ring or $2\frac{5}{8}$-in.-diameter shear plate connectors
$\quad D$ = diameter of fastener, in.

As an alternative to the equation for group action factor, the NDS provides expanded tables for C_g that may be used for certain applications. Again, the concept in the NDS is to define terms, to the extent possible, with general formulas that can be readily programmed, *and* to provide tables for use by the designer who does only an occasional wood design problem. Numerical values of C_g are given in NDS Tables 10.3.6A to 10.3.6D.

Geometry factor C_Δ.   The NDS provides spacing requirements for bolts. These are typically given as a number of bolt diameters D (e.g., $4D$ is the required end distance for a wood member in compression). In many cases the NDS provides two sets of spacings. The larger ones can be viewed as *base dimensions* that are required in order to use the full *reference design value* for the fastener. The *geometry factor* C_Δ is the multiplier that is used to reduce the reference design value Z if the spacings furnished in a given connection are less than the base dimensions. Thus, if the base spacing requirements for a connection are all satisfied, the geometry factor defaults to unity ($C_\Delta = 1.0$). Note that C_Δ does not exceed 1.0 if additional space is provided.

The second set of spacings given in the NDS is the *minimum permitted spacings*. For example, the minimum permitted end distance for a member in compression is $2D$. An end distance less than this minimum is unacceptable. The geometry factor could be viewed as zero ($C_\Delta = 0$) if the furnished spacings are less than the minimums.

A practical application of the geometry factor occurs when the furnished spacing lies between the base dimension and the minimum spacings. The geometry factor is a simple straight-line reduction. See Example 13.9. If there is more than one spacing that does not satisfy the basic spacing requirements, C_Δ is governed by the worst case. In other words, the smallest C_Δ for a given connection is to be used for all fasteners in a connection. However, the reductions are not cumulative, and several geometry factors need not be multiplied together.

EXAMPLE 13.9   Geometry factor C_Δ

The geometry factor C_Δ is used to reduce the reference design value for a bolt if the base dimensions are not satisfied. As an illustration, determine the geometry factor for the connection shown in Fig. 13.10 considering end distance only. Assume that all other base dimensions are provided.

Figure 13.10   End distance for compression member.

From NDS Table 11.5.1B, the end distances for a member loaded in axial compression are as follows:

Base end distance = $4D$ = 4(1) = 4 in.   (end distance required to obtain *full* reference design value)

Minimum end distance = $2D$ = 2(1) = 2 in.   (end distance required to obtain *any* design value)

The end distance provided in Fig. 13.10 is between the base dimension of 4 in. and the minimum of 2 in. The geometry factor is the ratio of the furnished end distance to the end distance required to obtain the full design value

$$C_\Delta = \frac{\text{furnished end distance}}{\text{base end distance}} \leq 1.0$$

$$= \frac{3 \text{ in.}}{4 \text{ in.}} = 0.75$$

A geometry factor is to be calculated for each spacing that does not satisfy the base dimensions. However, geometry factors are not cumulative, and if more than one geometry factor is calculated, the *smallest* C_Δ is used for *all* fasteners in a connection. The other geometry factors are disregarded.

Although the NDS allows the bolt spacings to be reduced below the base dimensions (provided the reference design value is reduced using the appropriate C_Δ), it is considered good detailing practice to provide the full base spacings whenever possible.

Now that the concept of the geometry factor has been introduced, the spacing requirements for bolted connections can be considered in detail. The dimensions for locating bolts in a connection are given to the centerline of the bolts. The detailing of a bolted connection requires consideration of

1. End distance

2. Edge distance

3. Center-to-center (c. to c.) spacing of bolts in a row

4. Row spacing

Bolted connections are somewhat more complicated than nailed connections because the direction of loading in relation to the direction of the grain must be considered. Other factors such as bolt slenderness l/D and the sense of force in the member (tension or compression) may also affect the spacing requirements. The *bolt slenderness ratio* is defined as the *smaller* value of

$$\frac{l_m}{D} = \frac{\text{bearing length of bolt in main member}}{\text{bolt diameter}}$$

$$\frac{l_s}{D} = \frac{\text{total bearing length of bolt in side member(s)}}{\text{bolt diameter}}$$

If there is more than one side member, l_s is taken as the combined thickness of the side members.

Because of the complexity, the spacing requirements are best summarized on a sketch. The NDS spacing criteria for *parallel-to-grain loading* are given in Example 13.10. The spacing requirements to obtain the full reference design value ($C_\Delta = 1.0$) are shown first, followed by the minimum permitted spacings in parentheses.

Note that there are different end distances for members stressed in tension and compression. For tension members it further depends on whether the member is a softwood or hardwood species. See Sec. 4.5. The 5-in. limit on the distance between outer rows of bolts was addressed previously (see Wet Service Factor above).

EXAMPLE 13.10   Bolt Spacing Requirements: Parallel-to-Grain Loading

Parallel-to-grain loading spacing requirements (Fig. 13.11) are as follows:

1. *End distance*
   a. Members in tension
      Softwood species = 7D   (3.5D minimum)
      Hardwood species = 5D   (2.5D minimum)
   b. Members in compression = 4D   (2D minimum)
2. *Edge distance:* bolt slenderness = l/D
   a. For $l/D \leq 6.0$: edge distance = $1.5D$
   b. For $l/D > 6.0$; edge distance = $1.5D$ or $0.5 \times$ row spacing, whichever is larger
3. *Center-to-center* spacing = $s = 4D$ ($3D$ minimum)
4. *Row spacing* = $1.5D$. Row spacing between outer rows of bolts shall not exceed 5 in. unless separate side plates are used for each row.

Figure 13.11   Bolt spacing requirements: parallel to grain loading.

The NDS spacing criteria for *perpendicular-to-grain loading* are given in Example 13.11. Again, the spacing requirements to obtain the full reference design value ($C_\Delta = 1.0$) are shown first, followed by the minimum permitted spacings in parentheses. Note that there are different edge distances for the *loaded edge* and the *unloaded edge*. The bolt can be viewed as delivering a load to the wood member. The loaded edge is the edge of the piece toward which the load is acting (i.e., the edge in the direction of bolt displacement for each member). The unloaded edge is the other side of the member.

EXAMPLE 13.11   Bolt Spacing Requirements: Perpendicular-to-Grain Loading

Perpendicular-to-grain loading spacing requirement (Fig. 13.12) are as follows:

1. *End distance* = 4D (2D minimum)
2. *Edge distance*
   a. Loaded edge = 4D
   b. Unloaded edge = 1.5D
3. *Center-to-center* spacing to obtain full design value is governed by the spacing requirements of the member(s) to which the bolts are attached (e.g., steel side plates* or wood member loaded parallel to grain). (3D minimum)
4. *Row spacing*
   a. For $l/D \leq 2.0$: row spacing = 2.5D.
   b. For $l/D \geq 6.0$: row spacing = 5D.
   c. For $2.0 < l/D < 6.0$: interpolate between 2.5D and 5D.

Figure 13.12   Bolt spacing requirements: perpendicular to grain loading.

NOTE: Shear in the member must be checked in accordance with NDS Sec. 3.4.3.

Specific requirements for the spacing of bolts in a connection where the load is at some angle other than 0 or 90 degrees to the direction of the grain are not given in the NDS. It is stated, however, that "the gravity axis of each member shall pass through the center of resistance of the group of fasteners to ensure uniform stress in the main member and a uniform distribution of load to all fasteners." In

---

*The center-to-center and edge spacing requirements for steel plates are given in Ref. 13.3

addition to providing this symmetry, the spacing provisions for both parallel and perpendicular loading can be used as a guide in detailing the connection.

This concludes the basic summary of adjustment factors for bolted connections. Several numerical examples are provided in the following sections to illustrate the design procedures for bolts.

13.6 Tension and Shear Stresses at a Multiple Fastener Connection

In 2001, the NDS introduced specific procedures for evaluating stresses at multiple-fastener connections in wood tension members loaded parallel to grain. The new provisions in NDS Appendix E (Ref. 13.1) are based on recent tests of multiple-fastener connections with bolts spaced closely together. These tests indicate that connections with rows of closely spaced bolts may fail before reaching the connection yield capacity predicted by the yield limit equations, even when the predicted yield capacity is reduced by the appropriate group action factor C_g and geometry factor C_Δ. The failures observed at multiple-bolt connections in wood tension members involve combinations of shear failure and tension failure in the wood, and are analogous to *shear rupture* failure at bolted connections in steel tension members (see Ref. 13.3). Thus, the procedures in NDS Appendix E constitute a final check of connection capacity in wood tension members after the required number of fasteners has been determined based on the yield limit equations, and a fastener pattern has been established based on NDS spacing requirements.

The provisions in NDS Appendix E for multiple-fastener connections include three design checks of connection capacity in wood tension members. The first is a check of *net section tension* as specified in NDS Secs. 3.1.2 and 3.8.1. This provision addresses tensile stress on the net cross-sectional area of a tension member. Net section tension capacity Z'_{NT} is determined based on the adjusted tension design value F'_t for the wood member, the number of fasteners that are placed at the critical cross section of the tension member, and the projected area of each fastener in the cross section. This design check was described in Chap. 7.

The second design check addresses failure of the wood member in shear along the perimeters of each row of bolts. The term for this failure mode is *row tearout*, and it involves the formation of *shear plugs* along each line of bolts oriented parallel to the longitudinal axis of the tension member (see Fig. 13.13). Row tearout capacity Z'_{RT} is based on the adjusted shear design value F'_v for the wood member, the number of fasteners in each row n_i, and the critical distance between fasteners in a row s_{crit}. Row tear-out failure is assumed to occur whenever shear failure initiates between any two fasteners in a row, or between the last fastener in a row and the end of the tension member. Thus, the critical distance between fasteners in a row s_{crit} is defined as the lesser of the minimum spacing between fasteners in a row or the distance from the last fastener in a row to the end of the tension member. For glued laminated timber members, the reference shear design value F_v must be multiplied by a reduction factor of 0.72

Figure 13.13 Row tear-out failure in the main member for a multiple-bolt connection. Shear plug failures occur between the last bolt in a row and the end of the glulam member.

when evaluating shear stresses at a connection (see Ref. 13.6 and footnotes to Tables 5A and 5B in the NDS Supplement, Ref. 13.1).

The third design check specified in NDS Appendix E addresses *group tear-out*, which is a combination of net section tension and row tear-out. Group tear-out involves failure of the wood member around the perimeter of the entire group of fasteners. Thus, group tear-out capacity Z'_{GT} is based on the net section tension capacity of the cross section between outer rows of fasteners, plus the shear capacity along the perimeter of the outer rows of fasteners.

When the connection capacity is governed by row tear-out or group tear-out, the connection capacity can be increased by increasing the spacing between fasteners. However, designers should avoid spacing outer rows of fasteners more than 5 in. apart unless separate side plates are provided, as illustrated in Example 13.7 (Sec. 13.5).

When using LRFD procedures to determine the adjusted capacity Z'_n at a multiple fastener connection, different format conversion factors K_F and resistance factors ϕ must be used for various material strength properties. As stated in Sec. 11.5, adjusted LRFD connection capacities associated with fastener performance include a format conversion factor of $K_F = 3.32$ and a resistance factor of $\phi_z = 0.65$. Since net section tension and group tear-out capacities at a connection are based on the tension strength of a wood member, a format conversion factor of $K_F = 2.7$ must be used to determine the nominal LRFD tension design value F_{tn} and a resistance factor of $\phi_t = 0.8$ must be used to determine the adjusted LRFD tension design value F'_{tn}. Similarly, for row tear-out and group tear-out capacities at a connection, a format conversion factor of $K_F = 2.88$ must be used to determine the nominal LRFD shear design value F_{vn} and a resistance factor of $\phi_v = 0.75$ must be used to determine the adjusted LRFD shear design value F'_{vn} for a wood member. See Sec. 4.24. Since appropriate K_F factors, ϕ factors, and other adjustment factors are used to determine F'_{tn} and F'_{vn}, no additional adjustment factors are used to calculate Z'_{nNT}, Z'_{nRT}, and Z'_{nGT} at connections.

The NDS provisions for row tear-out and group tear-out are based on tests of a limited number of multiple-bolt connection configurations. It is assumed that these provisions are equally applicable to all large diameter fasteners including split ring and shear plate connectors, bolts, and lag bolts that penetrate the entire thickness of a wood member. See Example 13.12.

EXAMPLE 13.12   Tension and Shear Stresses at a Multiple-Fastener Connection

Tension and shear stresses at a multiple-fastener connection can lead to three possible failure modes in wood tension members loaded parallel to grain. In some cases these failure modes may occur prior to achieving the connection yield mode predicted from yield limit equations. The symbol Z' is used to designate the adjusted capacity at a connection, and subscripts are employed to indicate whether the adjusted capacity is based on consideration of net section tension Z'_{NT}, row tear-out Z'_{RT}, or group tear-out Z'_{GT}. When evaluating tension and shear stresses at connections, the adjusted capacity is not multiplied by the number of fasteners in the connection N since the fasteners in critical regions of the connection are directly considered when calculating Z'_{NT}, Z'_{RT}, and Z'_{GT}. Net section tension failure was discussed in Chap. 7. Row tear-out and group tear-out failure modes are illustrated in Fig. 13.14.

Figure 13.14a   Multiple fastener connection.

Figure 13.14b   Row tear-out failure along each row of fasteners; dashed lines indicate locations of shear failure in wood member.

MAIN MEMBER

SHEAR FAILURE LINES

T

TENSION FAILURE LINE

FASTENER LOADS CAUSING BLOCK SHEAR

Figure 13.14c Group tear-out failure around the perimeter of fastener group; dashed lines indicate locations of tension failure and shear failure in wood member.

Net section tension (NDS Eq. E.2-1):

$$Z'_{\text{NT}} = F'_t A_n$$

Row tear-out (NDS Eqs. E.3-2 and E.3-3):

$$Z'_{\text{RT}} = \sum_{i=1}^{n_{\text{row}}} Z'_{\text{RT}_i} = \sum_{i=1}^{n_{\text{row}}} n_i F'_v t s_{\text{crit}}$$

Group tear-out (NDS Eq. E.4-1):

$$Z'_{\text{GT}} = \frac{Z'_{\text{RT}-1}}{2} + \frac{Z'_{\text{RT}-n}}{2} + F'_t A_{\text{group-net}}$$

where F'_t = adjusted tension design value parallel to grain, psi (see Chap. 7)

F'_v = adjusted shear design value, psi (see Chap. 6)

A_n = net cross-sectional area of tension member, in.^{2} (see Chap. 7)

$A_{\text{group-net}}$ = net cross-sectional area of tension member between outer rows of fasteners, in.^{2}

t = thickness of tension member, in.

s_{crit} = critical spacing between fasteners in row i, in.

= the lesser of the minimum spacing between fasteners in row i, or the distance from the last fastener in row i to the end of the tension member

n_i = number of fasteners in row i

n_{row} = number of rows of fasteners in a fastener group

Z'_{RT_i} = adjusted row tear-out capacity for any (ith) row of fasteners in a fastener group, lb

$Z'_{\text{RT}-1}$ = adjusted row tear-out capacity for the first row of fasteners in a fastener group, lb

$Z'_{\text{RT}-n}$ = adjusted row tear-out capacity for the last (nth) row of fasteners in a fastener group, lb

As discussed in Sec. 13.5, a row of fasteners is defined as a group of fasteners in a line parallel to the direction of loading. The formula for calculating the adjusted row tear-out-capacity Z'_{RT} is based on the assumption that shear failure for each row of fasteners occurs along two lines, one failure line on each side of the row and extending from the first fastener in the row to the end of the wood member. Since failure of the entire row may occur when shear failure initiates between any two fasteners in the row, the critical distance between fasteners in a row s_{crit} is used to calculate row tear-out capacity. If the shear stress due to applied load were uniformly distributed between fasteners in a row, the adjusted tear-out capacity for each row of fasteners would be estimated by:

$$Z'_{RT_i} = 2(n_i F'_v ts_{crit})$$

However, experimental test data indicates that the actual row tear-out capacity is approximately half the magnitude calculated using this formula. Therefore, a triangular stress distribution is assumed to occur between adjacent fasteners in a row. This means that the shear stress due to applied load in a multiple-fastener connection is maximum immediately adjacent to a fastener and decreases linearly to zero near the next fastener in the row. Since the actual row tear-out capacity is reached when F'_v is exceeded at any location along a row of fasteners, the adjusted tear-out capacity for each row of fasteners is thus:

$$Z'_{RT_i} = \frac{1}{2}[2(n_i F'_v ts_{crit})] = n_i F'_v ts_{crit}$$

The formula for group tear-out is based on net section tension failure of the area between outer rows of fasteners $A_{group-net}$ combined with shear failure along the perimeter of the outer rows of bolts. Thus, the adjusted row tear-out capacities for each of the outer rows of fasteners (Z'_{RT-1} and Z'_{RT-n}) are divided by 2 since only one shear failure line occurs at each row. According to NDS Sec. 3.1.2, when fasteners are arranged in a staggered pattern the net section tension area $A_{group-net}$ is typically determined assuming that staggered fasteners occur at the same cross section.

13.7   Design Problem: Multiple-Bolt Tension Connection

The multiple-fastener connection at the end of a glulam beam illustrates the importance of comparing connection capacity based on yield limit equations with capacity based on tension and shear stresses in a wood tension member at a connection. See Example 13.13. Even though the NDS base dimensions are met or exceeded in this connection ($C_\Delta = 1.0$), the capacity at the connection is governed by group tear-out and row tear-out capacities. Both group tear-out and row tear-out are associated with shear failure in the wood member along rows of bolts. The group tear-out and row tear-out capacities can be increased by increasing bolt spacing in the connection.

EXAMPLE 13.13   Multiple-Bolt Tension Connection

Determine the adjusted ASD capacity of the multiple-bolt double shear tension connection at the end of the 24F-V4 Douglas- Fir glulam member (from the 24F-1.8E Stress Class) shown in Fig. 13.15a. Two $\frac{1}{4}$-in.-thick A36 steel side plates are attached to the $5\frac{1}{8} \times 12$ glulam member using six 1-in.-diameter A307 bolts in a grid pattern. Each steel side plate is 6-in. wide. The spacing between rows of bolts is 3 in., the spacing between bolts in a row is 4 in., and the distance from the last bolt in each row to the end of the glulam member is 7 in. The axial tension force in the glulam member is a seismic force. The glulam member is initially dry and will remain dry in service. The temperature range is normal. Repeat the problem using LRFD procedures to determine the adjusted LRFD capacity of the multiple-bolt double shear tension connection.

Figure 13.15a Six-bolt tension connection; double shear wood-to-steel.

The reference design value Z for a single-bolt connection can be determined from the yield limit equations for double shear wood-to-metal connections. Alternatively, $Z_{\parallel}$ can be obtained from NDS Table 11I. The following parameters apply for this connection:

$$D = 1.0 \text{ in.}$$
$$F_{yb} = 45 \text{ ksi}$$
$$t_m = 5.125 \text{ in.}$$
$$\theta_m = 0 \text{ degrees}$$
$$G_m = 0.50$$
$$F_{em} = 5600 \text{ psi} \quad (F_{e\parallel} \text{ for bolt in DF-L})$$
$$t_s = 0.25 \text{ in.}$$
$$F_{es} = 1.5F_u = 1.5(58,000) = 87,000 \text{ psi} \quad \text{(for A36 steel plate)}$$

Reference design value from NDS Table 11I:

$$Z_{\parallel} = 5720 \text{ lb}$$

The reader may wish to verify this reference design value by applying the yield limit equations and observing that Mode III_s governs.

Adjusted ASD connection capacity based on bolt yield limit equations:

$$n = 3 \text{ bolts in each now}$$

$$C_D = 1.6 \quad \text{for loading due to wind or seismic}$$

$$C_M = 1.0 \quad \text{given}$$

$$C_t = 1.0 \quad \text{given}$$

The NDS base dimensions and formulas for the geometry factor C_Δ are summarized in Examples 13.9 and 13.10 (Sec. 13.5). In this problem:

End distance = 7 in. = $7D$ (tension member)
c.-to-c. spacing between bolts in a row = s = 4 in. = $4D$
Spacing between rows = 3 in. > $1.5D$
Edge distance = 4.5 in. > $1.5D$ (for $l_m/D = 5.125$)

$$C_\Delta = 1.0 \quad \text{(since all NDS base dimensions are met or exceeded)}$$

The formula and coefficients for the group action factor C_g are summarized in Example 13.8 (Sec. 13.5):

$$\gamma = \text{load/slip modulus for wood-to-metal bolt connection}$$

$$= 270,000(D^{1.5}) = 270,000(1.0)^{1.5} = 270,000 \text{ lb/in.}$$

$$E_m = 1,700,000 \text{ psi} \quad (E_{\text{axial}} \text{ from NDS Supplement Table 5A})$$

$$A_m = (5.125)(12) = 61.5 \text{ in.}^2$$

$$E_m A_m = 1,700,000(61.5) = 104,550,000$$

$$E_s = 29,000,000 \text{ psi}$$

$$A_s = (2)(0.25)(6.0) = 3.0 \text{ in.}^2 \quad \text{(for two steel side plates)}$$

$$E_s A_s = 29,000,000(3.0) = 87,000,000$$

$$R_{EA} = \frac{E_s A_s}{E_m A_m} = \frac{87,000,000}{104,550,000} = 0.8321 \quad \text{(since } E_s A_s < E_m A_m)$$

$$1 + R_{EA} = 1 + 0.8321 = 1.832$$

$$u = 1 + \gamma\left(\frac{s}{2}\right)\left[\frac{1}{E_m A_m} + \frac{1}{E_s A_s}\right]$$

$$= 1 + 270,000\left(\frac{4.0}{2}\right)\left[\frac{1}{104,550,000} + \frac{1}{87,000,000}\right]$$

$$= 1.011$$

$$m = u - \sqrt{u^2 - 1}$$

$$= 1.011 - \sqrt{1.011^2 - 1} = 0.8601$$

$$1 + m = 1 + 0.8601 = 1.860$$

$$1 - m = 1 - 0.8601 = 0.1399$$

$$m^{2n} = 0.8601^{2(3)} = 0.4049$$

$$1 + R_{EA}m^n = 1 + 0.8321(0.8601)^3 = 1.530$$

$$C_g = \left[\frac{m(1 - m^{2n})}{n[(1 + R_{EA}m^n)(1 + m) - 1 + m^{2n}]} \right]\left[\frac{1 + R_{EA}}{1 - m} \right]$$

$$= \left[\frac{0.8601(1 - 0.4049)}{3[(1.530)(1.860) - 1 + 0.4049]} \right]\left[\frac{1.832}{0.1399} \right]$$

$$= 0.993 \qquad \text{(compare with } C_g \approx 0.99 \text{ from NDS Table 10.3.6C)}$$

$$\text{Adj. } P = N(Z') = N(Z)(C_D C_M C_t C_g C_\Delta)$$

$$= (6)(5720)(1.6)(1.0)(1.0)(0.993)(1.0)$$

$$= 54{,}500 \text{ lb}$$

Adjusted ASD capacity based on tension and shear stresses in the glulam member:

The formulas for adjusted net section tension Z'_{NT}, row tear-out Z'_{RT}, and group tear-out Z'_{GT} are summarized in Example 13.12 (Sec. 13.6).

Since the bolts penetrate the wide face of the 24F-V4 glulam member, select the reference shear design value for bending about the strong x-axis and multiply by a reduction factor of 0.72 in accordance with NDS Supplement Table 5A:

$$F_v = (0.72)265 \text{ psi} = 191 \text{ psi}$$

$$F'_v = F_v(C_D C_M C_t)$$

$$= 191(1.6)(1.0)(1.0)$$

$$= 305 \text{ psi}$$

$$F'_t = F_t(C_D C_M C_t)$$

$$= 1100(1.6)(1.0)(1.0)$$

$$= 1760 \text{ psi}$$

Net section tension for ASD (NDS Eq. E.2-1):

$$A_n = 5.125[12 - 2(1.0 + \tfrac{1}{16})] = 50.6 \text{ in.}^2 \qquad \text{(see Fig. 13.15}b\text{)}$$

The $\tfrac{1}{16}$ in. was added to the bolt diameter to account for drilling oversize holes in accordance with NDS Sec. 11.1.2.

$$Z'_{NT} = F'_t(A_n) = 1760(50.6) = 89{,}100 \text{ lb} > 54{,}500 \text{ lb}$$

Figure 13.15b Net section of glulam member (left); portion of net area between rows of bolts ($A_{\text{group-net}}$) for group tear-out (right).

Row tear-out for ASD (NDS Eqs. E.3-2 and E.3-3):

$$s_{\text{crit}} = s = 4.0 \text{ in.}$$

$$Z'_{\text{RT-1}} = Z'_{\text{RT-2}} = nF'_v ts_{\text{crit}} = 3(305)(5.125)(4.0) = 18{,}775 \text{ lb}$$

$$Z'_{\text{RT}} = \sum_{i=1}^{n_{\text{row}}} Z'_{\text{RT}_i} = 18{,}775 + 18{,}775 = 37{,}500 \text{ lb} < 54{,}500 \text{ lb}$$

Group tear-out for ASD (NDS Eq. E.4-1):

$$A_{\text{group-net}} = 5.125[3 - 2(\tfrac{1}{2})(1.0 + \tfrac{1}{16})] = 9.93 \text{ in.}^2 \quad (\text{see Fig. 13.15b})$$

$$Z'_{\text{GT}} = \frac{Z'_{\text{RT-1}}}{2} + \frac{Z'_{\text{RT-2}}}{2} + F'_t A_{\text{group-net}}$$

$$Z'_{\text{GT}} = \frac{18{,}775}{2} + \frac{18{,}775}{2} + 1760(9.93) = 36{,}300 \text{ lb} < 54{,}500 \text{ lb}$$

> The adjusted ASD capacity is 36,300 lb due to group tear-out at the connection.

The net section tension capacity Z'_{NT} at the connection is much larger than the capacity based on yield limit equations for the bolts. However, the group tear-out capacity Z'_{GT} at the connection is approximately 67 percent of the capacity based on yield limit equations for the bolts. The row tear-out capacity Z'_{RT} is only marginally higher than Z'_{GT}. Z'_{GT} could be increased by increasing the spacing between rows of bolts (thus increasing $A_{\text{group-net}}$), as well as the spacing between bolts in a row (thus increasing s_{crit}). Similarly, Z'_{RT} could be increased by increasing the spacing between bolts in a row.

Repeat the design using LRFD procedures.
The reference design value from NDS Table 11I based on yield limit equations is:

$$Z_{\parallel} = 5720 \text{ lb}$$

The nominal design value for LRFD is:

$$Z_n = Z(K_F) = 5720(3.32) = 18{,}990 \text{ lb}$$

The time effect factor for LRFD load combinations in which seismic load is dominant is $\lambda = 1.0$. The resistance factor is $\phi_z = 0.65$ for all fasteners in the NDS. The other adjustment factors are identical for both LRFD and ASD.

Adjusted LRFD capacity for the six bolt connection based on yield limit equations:

$$\text{Adj. } P_n = N(Z'_n) = N(Z_n) \; (\phi_z \lambda C_M C_t C_g C_\Delta)$$

$$= 6(18,990)(0.65)(1.0)(1.0)(1.0)(0.993)(1.0)$$

$$= 73,500 \text{ lb}$$

Adjusted LRFD capacity based on tension and shear stresses in the glulam member:

Since net section tension, row tear-out, and group tear-out are based on failure in the wood member, the format conversion factor K_F and the resistance factor ϕ are based on member design. See Sec. 4.24. For member failure in tension, the format conversion factor is $K_F = 2.7$ and the resistance factor is $\phi_t = 0.8$. For member failure in shear, the format conversion factor is $K_F = 2.88$ and the resistance factor is $\phi_v = 0.75$. The nominal and adjusted LRFD design values for tension and shear are:

$$F_{tn} = F_t \, (K_F) = 1100(2.7) = 2970 \text{ psi}$$

$$F'_{tn} = F_{tn}(\phi_t \lambda C_M C_t)$$

$$= 2970(0.8)(1.0)(1.0)(1.0)$$

$$= 2376 \text{ psi}$$

$$F_{vn} = F_v \, (K_F) = 191(2.88) = 550 \text{ psi}$$

$$F'_{vn} = F_{vn}(\phi_v \lambda C_M C_t)$$

$$= 550(0.75)(1.0)(1.0)(1.0)$$

$$= 412 \text{ psi}$$

Net section tension for LRFD (NDS Eq. E.2-1):

$$Z'_{n\,\mathrm{NT}} = F'_{tn}(A_n) = 2376(50.6) = 120,200 \text{ lb} > 73,500 \text{ lb}$$

Row tear-out for LRFD (NDS Eqs. E.3-2 and E.3-3):

$$Z'_{n\mathrm{RT}-1} = Z'_{n\mathrm{RT}-2} = nF'_{vn}ts_{\mathrm{crit}} = 3(412)(5.125)(4.0) = 25,350 \text{ lb}$$

$$Z'_{n\mathrm{RT}} = \sum_{i=1}^{n_{\mathrm{row}}} Z'_{n\mathrm{RT}_i} = 25,350 + 25,350 = 50,700 \text{ lb} < 73,500 \text{ lb}$$

Group tear-out for LRFD (NDS Eq. E.4-1):

$$Z'_{n\mathrm{GT}} = \frac{Z'_{n\mathrm{RT}-1}}{2} + \frac{Z'_{n\mathrm{RT}-2}}{2} + F'_{tn} A_{\mathrm{group-net}}$$

$$Z'_{n\mathrm{GT}} = \frac{25,350}{2} + \frac{25,350}{2} + 2376(9.93) = 48,900 \text{ lb} < 73,500 \text{ lb}$$

The adjusted LRFD capacity is 48,900 lb due to group tear-out at the connection.

The group tear-out capacity $Z'_{n\mathrm{GT}}$ at the connection is approximately 67 percent of the capacity based on yield limit equations for the bolts (*same as for ASD*).

13.8   Design Problem: Bolted Chord Splice for Diaphragm

For the building in Example 13.14, the top plate of a 2 × 6 stud wall serves as the chord for the roof diaphragm. The connections for the splice in the top plate are to be designed. The magnitude of the chord force is large, and the splice will be made with bolts instead of nails.

The location of the splices in the chord should be considered. In this building, the chord is made up of fairly long members (say 20 ft), and the splices are offset by half the length of the members (10 ft in this example). At some point between the splices, the chord force is shared equally by the two members. Therefore, the connections at each end of a chord member may be designed for one-half of the total chord force.

This is in contrast to the nailed chord splice designed in Example 12.9 (Sec. 12.8). If the lap of the members in the chord is kept to a minimum, the entire chord force must be transmitted by the single connection within the lap. Either method of splicing the chord may be used in practice.

The connection is designed by first assuming a bolt size and determining the reference design value for the fastener. In this problem, Z can be found using the yield limit equations or the NDS bolt tables. Except for t_m and θ_m, the current example is similar to Example 13.3. Therefore, the determination of the reference design value Z is not shown and is left to the reader as an exercise.

With Z known, the approximate number of bolts is estimated, and the adjusted design value for the connection is determined and compared with the design load. As with many practical problems, a number of the possible adjustment factors default to unity. However, the load duration factor and the group action factor are required to obtain Z' for ASD. The format conversion factor, resistance factor, time effect factor, and group action factor are required to determine Z'_n for LRFD.

Part 1 of the example follows the recommended practice of providing the full base dimensions for the end distance and center-to-center spacing. Part 2 is given for illustrative purposes only, to show how the geometry factor is applied should the base dimensions not be provided.

EXAMPLE 13.14   Bolt Splice for Tension Chord

A large wood frame warehouse uses 2 × 6 stud walls. A double 2 × 6 top plate functions as the chord of the horizontal roof diaphragm. See Fig. 13.16a. Determine the number and spacing of $\frac{3}{4}$-in. full-diameter body bolts for the splice connection in the top plate. The bolts are in a single row. Lumber is Select Structural DF-L. The chord force at the connection is the result of seismic loading and has been determined to be E = 13.3 k. The splice is to be designed for one-half of the chord force. The stress in the member is also to be evaluated. $C_M = 1.0$ and $C_t = 1.0$. Repeat the problem using LRFD procedures.

The chord force for ASD procedures is $T = 0.7E = 0.7(13.3) = 9.3$ k

Bolt Design—Part 1

In part 1, spacings are to be determined in accordance with the base dimensions to obtain the full design value for the bolt. The following parameters are known about

Figure 13.16a   Bolt splice connections in top plate chord.

the connection. The reference design value for the bolt can be determined from the yield limit equations for a single shear wood-to-wood connection. As an alternative to the equations, Z can be obtained from NDS Table 11A. Evaluation of Z is left as an exercise for the reader.

$$D = 0.75 \text{ in.}$$
$$F_{yb} = 45 \text{ ksi} \qquad \text{(assumed for A307 bolts)}$$
$$t_m = t_s = 1.5 \text{ in.}$$
$$\theta_m = \theta_s = 0 \text{ degrees}$$
$$F_{em} = F_{es} = 5600 \text{ psi} \qquad (F_{e\parallel} \text{ for bolt in DF-Ł)}$$
$$G_m = G_s = 0.50$$

Reference design value:

$Z = 725 \text{ lb}$ (Mode II yield mechanism given by NDS Eq. 11.3-3. This compares with $Z = 720$ lb in NDS Table 11A.)

Adjustment factors for ASD:

$C_D = 1.6$ for lateral loading due to seismic force

$C_M = 1.0$ given

$C_t = 1.0$ given

$C_g \approx 0.95$ (assumed) before the group action factor can be determined, it is necessary to know the number of bolts. C_g will be verified later.

$C_\Delta = 1.0$ for bolt spacings given by base dimensions

Base dimensions:

$$\text{End distance} = 7D = 7(0.75) = 5.25 \text{ in.} \qquad \text{(tension member)}$$
$$\text{c.-to-c. spacing} = s = 4D = 4(0.75) = 3.0 \text{ in.}$$
$$\text{Bolts will be located in the center of the 2} \times \text{6s.}$$
$$\text{Edge distance} = 0.5(5.5) = 2.75 \text{ in.}$$
$$l/D = 1.5/0.75 = 2.0 < 6.0$$
$$\text{Minimum edge distance} = 1.5D = 1.5(0.75)$$
$$= 1.125 \text{ in.} < 2.75 \text{ in.} \qquad OK$$

Estimate number of bolts for ASD:

$$\text{Load on one bolt group} = \frac{T}{2} = \frac{9300}{2} = 4650 \text{ lb}$$

$$\text{Approx. } Z' = Z(C_D C_M C_t C_g C_\Delta)$$

$$= 725(1.6)(1.0)(1.0)(0.95)(1.0)$$

$$= 1102 \text{ lb/bolt}$$

$$N \approx \frac{4650}{1102} = 4.2 \qquad \text{say 5 bolts}$$

Adjusted design value:
   Verify group action factor by evaluating the equation for C_g for five bolts in a row.

$$\gamma = \text{load/slip modulus for wood-to-wood bolt connection}$$

$$= 180,000(D^{1.5}) = 180,000(0.75^{1.5}) = 116,900 \text{ lb/in}$$

$$E_m = E_s = 1,900,000 \text{ psi}$$

$$A_m = A_s = 8.25 \text{ in}^2$$

$$R_{EA} = 1.0$$

NOTE: R_{EA} is the minimum ratio of member stiffnesses [i.e., the smaller of $(E_m A_m)/(E_s A_s)$ or $(E_s A_s)/(E_m A_m)$]. In this example, the member stiffnesses are equal, hence $R_{EA} = 1.0$.

$$u = 1 + \gamma\left(\frac{s}{2}\right)\left[\frac{1}{E_m A_m} + \frac{1}{E_s A_s}\right]$$

$$= 1 + 116,900\left(\frac{3.0}{2}\right)\left[\frac{1}{1,900,000(8.25)} + \frac{1}{1,900,000(8.25)}\right]$$

$$= 1.022$$

$$m = u - \sqrt{u^2 - 1}$$

$$= 1.022 - \sqrt{1.022^2 - 1} = 0.8097$$

$$1 + m = 1 + 0.8097 = 1.810$$

$$1 - m = 1 - 0.8097 = 0.1903$$

$$m^{2n} = 0.8097^{2(5)} = 0.1211$$

$$1 + R_{EA} = 1 + 1.0 = 2.0$$

$$1 + R_{EA} m^n = 1 + 1.0(0.8097)^5 = 1.348$$

$$C_g = \left[\frac{m(1 - m^{2n})}{n[(1 + R_{EA}m^n)(1 + m) - 1 + m^{2n}]}\right]\left[\frac{1 + R_{EA}}{1 - m}\right]$$

$$= \left[\frac{0.8097(1 - 0.1211)}{5[(1.348)(1.810) - 1 + 0.1211]}\right]\left[\frac{2.0}{0.190}\right]$$

$$= 0.958$$

Adjusted ASD capacity for one splice connection:

$$\text{Adj. } P = N(Z') = N(Z)(C_D C_M C_t C_g C_\Delta)$$

$$= 5(725)(1.6)(1.0)(1.0)(0.958)(1.0)$$

$$= 5560 \text{ lb} > 4650 \text{ lb} \quad OK$$

Use   Five $\frac{3}{4}$-in.-diameter bolts each side of splice.

NOTE: The spacing of bolts must be shown on the design plans.

Bolt Design—Part 2

The bolt spacings given in part 1 are the preferred spacings, even though the adjusted ASD capacity is greater than the applied load. For illustration, determine the effect on the ASD capacity of the connection if reduced spacings are used.

Reduced spacings and geometry factor:

Reduced end distance = 5.0 in. < 7D = 5.25 in.   (base)

> 3.5D = 2.625 in.   (minimum)

$$\therefore \text{ End } C_\Delta = \frac{5.0}{5.25} = 0.952$$

Reduced c.-to-c. spacing = s = 2.5 in. < 4D = 3.0 in.   (base)

> 2D = 1.5 in.   (minimum)

$$\therefore \text{ Center-to-center } C_\Delta = \frac{2.5}{3.00} = 0.833$$

A single geometry factor is used for the entire connection, and the smaller value applies:

$$C_\Delta = 0.833$$

Revised group action factor:

$$u = 1 - \gamma\left(\frac{s}{2}\right)\left[\frac{1}{E_m A_m} + \frac{1}{E_s A_s}\right]$$

$$= 1 - 116,900\left(\frac{2.5}{2}\right)\left[\frac{1}{1,900,000(8.25)} + \frac{1}{1,900,000(8.25)}\right]$$

$$= 1.019$$

$$m = u - \sqrt{u^2 - 1}$$

$$= 1.019 - \sqrt{1.019^2 - 1} = 0.8246$$

$$1 + m = 1 + 0.8246 = 1.825$$

$$1 - m = 1 - 0.8246 = 0.1754$$

$$m^{2n} = 0.8246^{2(5)} = 0.1454$$

$$1 + R_{EA} = 1 + 1.0 = 2.0$$

$$1 + R_{EA}m^n = 1 + 1.0(0.8246)^5 = 1.381$$

$$C_g = \left[\frac{m(1 - m^{2n})}{n[(1 + R_{EA}m^n)(1 + m) - 1 + m^{2n}]} \right]\left[\frac{1 + R_{EA}}{1 - m} \right]$$

$$= \left[\frac{0.8246(1 - 0.1454)}{5[(1.381)(1.825) - 1 + 0.1454]} \right]\left[\frac{2.0}{0.1754} \right]$$

$$= 0.965$$

Adjusted ASD capacity for one splice connection:

$$\text{Adj. } P = N(Z') = N(Z)(C_D C_M C_t C_g C_\Delta)$$

$$= 5(725)(1.6)(1.0)(1.0)(0.965)(0.833)$$

$$= 4660 \text{ lb} > 4650 \text{ lb} \qquad OK$$

Although the adjusted capacity is greater than the applied load, the connection capacity has been substantially reduced (4660 lb versus 5560 lb) as a result of bolt spacing.

Figure 13.16b Net section at splice.

Adjusted ASD capacity based on tension and shear stresses in one 2 × 6 member:

The formulas for net section tension Z'_{NT} and row tear out Z'_{RT} are summarized in Example 13.12 (Sec. 13.6).

Net section tension for ASD (NDS Eq. E.2-1):

$$A_n = 1.5[5.5 - (\tfrac{3}{4} + \tfrac{1}{16})] = 7.03 \text{ in.}^2$$

The $\tfrac{1}{16}$ in. was added to the bolt diameter to account for drilling oversize holes in accordance with NDS Sec. 11.1.2.

$$F'_t = F_t(C_D)(C_M)(C_t)(C_F)$$

$$= 1000(1.6)(1.0)(1.0)(1.3)$$

$$= 2080 \text{ psi}$$

$$Z'_{\text{NT}} = F'_t A_n = 2080(7.03) = 14{,}600 \text{ lb} > 4650 \text{ lb} \qquad OK$$

Row tear-out for ASD (NDS Eqs. E.3-2 and E.3-3):

$$n = \text{five bolts in a single row}$$

$$s_{\text{crit}} = s = 2.5 \text{ in.}$$

$$F'_v = F_v(C_D)(C_M)(C_t)$$

$$= 180(1.6)(1.0)(1.0)$$

$$= 288 \text{ psi}$$

$$Z'_{\text{RT}} = nF'_v t s_{\text{crit}}$$

$$= 5(288)(1.5)(2.5) = 5400 \text{ lb} > 4650 \text{ lb} \quad OK$$

Use double 2×6 Select Structural DF-L for top plate.

Now repeat the problem using LRFD procedures. The *factored* seismic chord force is $T_u = E = 13.3$ k

Bolt Design—Part 1

The reference design value for one bolt is:

$$Z = 725 \text{ lb}$$

The nominal design value for LRFD is:

$$Z_n = Z \, (K_F) = 725(3.32) = 2407 \text{ lb}$$

Estimate number of bolts for LRFD with full base dimensions, assuming $C_g = 0.95$:

$$\text{Load on one bolt group} = \frac{T_u}{2} = \frac{13{,}300}{2} = 6650 \text{ lb}$$

$$\text{Approx. } Z'_n = Z_n(\phi_z \lambda C_M C_t C_G C_\Delta)$$

$$= 2407(0.65)(1.0)(1.0)(1.0)(0.95)(1.0)$$

$$= 1486 \text{ lb}$$

$$N \approx \frac{6650}{1486} = 4.5 \quad say \, 5 \text{ bolts}$$

$$C_g = 0.958 \text{ for full base dimensions (}same \ as \ for \ ASD\text{)}$$

Adjusted LRFD capacity for one splice connection:

$$\text{Adj. } P_n = N(Z_n') = N(Z_n)(\phi_z \lambda C_M C_t C_g C_\Delta)$$

$$= 5(2407)(0.65)(1.0)(1.0)(1.0)(0.958)(1.0)$$

$$= 7490 \text{ lb} > 6650 \text{ lb} \qquad OK$$

> *Use* five $\frac{3}{4}$ in. diameter bolts each side of splice
> for full base dimensions (*same as for ASD*)

Bolt Design—Part 2

Determine the effect of reduced end distance and spacing on the adjusted LRFD capacity.

$$\text{Reduced end distance} = 5.0 \text{ in.}$$

$$\text{Reduced c.-to-c. spacing} = s = 2.5 \text{ in.}$$

$$C_\Delta = 0.833 \text{ (}Same \text{ as for ASD)}$$

$$C_g = 0.965 \text{ for five bolts and reduced spacing (}same \text{ as for ASD)}$$

Adjusted LRFD capacity for one splice connection with five bolts:

$$\text{Adj. } P_n = N(Z_n') = N(Z_n)(\phi_z \lambda C_M C_t C_g C_\Delta)$$

$$= 5(2407)(0.65)(1.0)(1.0)(1.0)(0.965)(0.833)$$

$$= 6290 \text{ lb} < 6650 \text{ lb} \qquad NG$$

$\therefore$ Require an additional bolt for reduced spacing of $s = 2.5$ in. using LRFD procedures.

Revised group action factor for six bolts:

$$n = 6$$

$$u = 1.019$$

$$m = 0.8246$$

$$1 + m = 1 + 0.8246 = 1.825$$

$$1 - m = 1 - 0.8246 = 0.1754$$

$$m^{2n} = 0.8246^{2(6)} = 0.09884$$

$$1 + R_{EA}m^n = 1 + (1.0)(0.8246)^6 = 1.314$$

$$C_g = \left[\frac{m(1 - m^{2n})}{n[(1 + R_{EA}m^n)(1 + m) - 1 + m^{2n}]} \right]\left[\frac{1 + R_{EA}}{1 - m} \right]$$

$$= \left[\frac{0.8246(1 - 0.09884)}{6[(1.314)(1.825) - 1 + 0.09884]} \right]\left[\frac{2.0}{0.1754} \right]$$

$$= 0.943$$

Adjusted LRFD capacity for one splice connection with six bolts:

$$\text{Adj. } P_n = N(Z'_n) = N(Z_n)(\phi_z \lambda C_M C_t C_g C_\Delta)$$

$$= 6(2407)(0.65)(1.0)(1.0)(1.0)(0.943)(0.833)$$

$$= 7370 \text{ lb} > 6650 \text{ lb} \quad OK$$

> *Use*   six $\tfrac{3}{4}$ in. diameter bolts each side of splice for reduced
> end distance and spacing (*one more bolt than for ASD*)

Adjusted LRFD capacity based on tension and shear stresses in one 2x6 member:

Net section tension for LRFD (NDS Eq. E.2-1):

$$F_{tn} = F_t\,(K_F) = 1000(2.7) = 2700 \text{ psi}$$

$$F'_{tn} = F_{tn}(\phi_t \lambda C_M C_t C_F)$$

$$= 2700(0.8)(1.0)(1.0)(1.0)(1.3)$$

$$= 2810 \text{ psi}$$

$$Z'_{n\,\text{NT}} = F'_{tn}(A_n) = 2810(7.03) = 19{,}700 \text{ lb} > 6650 \text{ lb} \quad OK$$

Row tear-out for LRFD (NDS Eqs. E.3-2 and E.3-3) with six bolts and reduced spacing:

$$F_{vn} = F_v\,(K_F) = 180(2.88) = 518 \text{ psi}$$

$$F'_{vn} = F_{vn}(\phi_v \lambda C_M C_t)$$

$$= 518(0.75)(1.0)(1.0)(1.0)$$

$$= 389 \text{ psi}$$

$$Z'_{nRT} = nF'_{vn} t s_{crit} = 6(389)(1.5)(2.5) = 8750 \text{ lb} > 6650 \text{ lb}$$

> *Use*   double 2×6 Select Structural DF-L for top plate.

13.9   Shear Stresses in a Beam at a Connection

The evaluation of shear stresses in a beam is covered in Chap. 6. When a beam is supported by a connection with bolts, lag bolts, split rings, or shear plates, there are additional considerations for horizontal shear (NDS Sec. 3.4.3). The adjusted design shear V'_r is determined using the *effective depth* d_e, which is taken as the depth of the member minus the distance from the unloaded edge of the member to the *center* of the nearest bolt or lag bolt. For timber connectors, d_e is taken as the depth of the beam minus the distance from the unloaded edge to the *edge* of the nearest connector. See Example 13.15.

EXAMPLE 13.15   Shear Stresses at a Connection

The shear in a beam supported by fasteners is checked using the effective depth d_e (Fig. 13.17). For bolts and lag bolts, d_e is the depth of the member minus the distance from the unloaded edge to the center of the nearest fastener. For split ring and shear plate connectors, d_e is the depth of the member minus the distance from the unloaded edge to the nearest edge of a connector. The loaded and unloaded edges are determined by examining the *force on the beam* at the fastener. The loaded edge is the edge toward which the force acts. The unloaded edge is the opposite edge.

Figure 13.17 Definition of effective depth d_e for evaluation of shear in a beam supported by fasteners.

The total shear V in the beam at a connection is computed in the usual manner (Chap. 6). The NDS provides an equation for the adjusted design shear V_r' which must be greater than or equal to the calculated shear force on the member V. When the connection is located less than $5d$ from the end of the member:

$$V_r' = \frac{2}{3}F_v'bd_e\left(\frac{d_e}{d}\right)^2 \geq V$$

When the connection is more than $5d$ from the end of the member:

$$V_r' = \frac{2}{3}F_v'bd_e \geq V$$

For a discussion of the adjusted shear design value F_v', see Chap. 6. For LRFD, the adjusted design shear must be greater than or equal to the *factored* shear force in the member, $V_{rn}' \geq V_u$.

It should be clear that the tendency of a beam to develop a split parallel to the grain will be affected by the magnitude of the shear stress in the member. In addition, the tendency of the member to split will be affected by the proximity of the connection to the end of the beam. Consequently, when the connection is less than 5 times the depth of the beam d from the end of the member, the adjusted design shear is decreased by the square of the ratio of the effective depth to the member depth $(d_e/d)^2$. When the connection is greater than $5d$ from the end of the member, the splitting tendency is reduced and the adjusted design shear is computed without the $(d_e/d)^2$ decrease. The adjusted shear design value is as defined in Chap. 6, except that for glued laminated timber beams the reference shear design value F_v must be multiplied by a reduction factor of 0.72 when evaluating shear stresses at a connection (see footnotes to Tables 5A and 5B in the NDS Supplement, Ref. 13.1).

13.10   Design Problem: Bolt Connection for Diagonal Brace

This final bolt example involves a connection that must be evaluated with the yield limit equations because the NDS bolt tables cover only parallel- and per-pendicular-to-grain loading, and they handle only wood members with the same specific gravity.

The load in the side member in Example 13.16 is parallel to the grain ($\theta = 0$ degrees), and the angle of load to grain for the main member is 45 degrees. In addition, the beam and brace are from different species. Many of the adjustment factors default to unity in this connection problem.

The center of resistance of the fastener coincides with the centroids of the members, and there is not eccentricity in the connection. However, the horizontal shear in the beam should be evaluated using the effective depth to the bolt (Sec. 13.9). Because the connection is more than 5 times the beam depth from the end of the member, the adjusted design shear need not be multiplied by the ratio $(d_e/d)^2$.

EXAMPLE 13.16   Bolt Connection for Diagonal Brace

The beam in Fig. 13.18a supports a dead load of 575 lb at its free end and is connected to the diagonal brace with a single $\frac{3}{4}$-in. full-diameter body bolt. The horizontal beam is a 24F-V4 DF glulam (from the 24F-1.8E Stress Class), and the diagonal brace is No. 1 Hem-Fir sawn lumber. Wood is initially dry and remains dry in service. Normal temperature conditions apply.

Determine if the bolt is adequate, and check the design shear in the beam at the connection using ASD procedures. Assume that other stresses in the members are OK. Repeat the problem using LRFD procedures.

Solve for force in brace.

$$\Sigma\, M_C = 0$$
$$6\, T_V = 575(10)$$
$$T_V = 958\ \text{lb}$$
$$T = \sqrt{2}\, T_V = \sqrt{2}\,(958) = 1355\ \text{lb}$$

Figure 13.18a   Member AC supported by diagonal brace at B.

CONNECTION DETAIL

Figure 13.18b   Bolt connection at B: double shear wood-to-wood

Shear at the connection in the beam:

$$V = P_{DL} = 575 \text{ lb}$$

NOTE: Segment BC of the beam is subjected to combined axial and bending stresses (Chap. 7).

Bolt Capacity

Summary of values for yield limit equations:

$D = 0.75$ in.

$F_{yb} = 45$ ksi   (assumed for A307 bolts)

$t_m = 3.125$ in.

$\theta_m = 45$ degrees

$G_m = 0.50$ (dowel bearing strength may be computed using G or read from NDS Table 11.3.2)

$F_{e\parallel} = 5600$ psi   for bolt in DF-L

$F_{e\perp} = 2600$ psi   for $\frac{3}{4}$-in. bolt in DF-L

$t_s = 1.5$ in.

$\theta_s = 0$ degrees

$G_s = 0.43$

$F_{es} = F_{e\parallel} = 4800$ psi   for bolt in Hem-Fir

Coefficients for yield equations:

$$F_{em} = F_{e\theta} = \frac{F_{e\parallel}F_{e\perp}}{F_{e\parallel}\sin^2\theta + F_{e\perp}\cos^2\theta}$$

$$= \frac{5600(2600)}{5600\sin^2 45 + 2600\cos^2 45} = 3551 \text{ psi}$$

$$R_e = \frac{F_{em}}{F_{es}} = \frac{3551}{4800} = 0.7398$$

$$1 + R_e = 1 + 0.7398 = 1.740$$

$$2 + R_e = 2 + 0.7398 = 2.740$$

$$\theta = \text{larger value of } \theta_m \text{ and } \theta_s$$

$$= 45 \text{ degrees}$$

$$K_\theta = 1 + \frac{\theta}{360} = 1 + \frac{45}{360} = 1.125$$

$$k_3 = -1 + \sqrt{\frac{2(1 + R_e)}{R_e} + \frac{2F_{yb}(2 + R_e)D^2}{3F_{em}l_s^2}}$$

$$= -1 + \sqrt{\frac{2(1.740)}{0.7398} + \frac{2(45,000)(2.740)(0.75^2)}{3(3551)(1.5^2)}}$$

$$= 2.239$$

Yield limit equations:

Mode I$_m$ (NDS Eq. 11.3-7):

$$Z = \frac{Dl_mF_{em}}{4K_\theta} = \frac{0.75(3.125)(3551)}{4(1.125)} = 1850 \text{ lb}$$

Mode I$_s$ (NDS Eq. 11.3-8):

$$Z = \frac{2Dl_sF_{es}}{4K_\theta} = \frac{2(0.75)(1.5)(4800)}{4(1.125)} = 2400 \text{ lb}$$

Mode III$_s$ (NDS Eq. 11.3-9):

$$Z = \frac{2k_3Dl_sF_{em}}{3.2(2 + R_e)K_\theta}$$

$$= \frac{2(2.239)(0.75)(1.5)(3551)}{3.2(2.740)(1.125)} = 1814 \text{ lb}$$

Mode IV (NDS Eq. 11.3-10):

$$Z = \frac{2D^2}{3.2K_\theta}\sqrt{\frac{2F_{em}F_{yb}}{3(1 + R_e)}}$$

$$= \frac{2(0.75^2)}{3.2(1.125)}\sqrt{\frac{2(3551)(45,000)}{3(1.740)}} = 2445 \text{ lb}$$

Reference design value:

$$Z = 1814 \text{ lb}$$

Adjusted ASD design value:

The bolt is located on the center lines of the two members.

The required edge distance to the loaded edge of the horizontal member is 4D.

$$4D = 4(0.75) = 3.0 \text{ in.}$$

$$\text{Edge distance} = \frac{9}{2} = 4.5 \text{ in.} > 3.0 \text{ in.} \quad OK$$

Assume that all other dimensions in the connection (including end distance for the tension member) will be chosen to satisfy the base dimension spacings for the connection in accordance with Sec. 13.5. Thus, $C_\Delta = 1.0$.

The load duration factor for dead load acting alone is $C_D = 0.9$.

$$Z' = Z(C_D C_M C_t C_g C_\Delta)$$

$$= 1814(0.9)(1.0)(1.0)(1.0)(1.0)$$

$$= 1633 \text{ lb} > 1355 \text{ lb} \quad OK$$

$$\boxed{\text{$\frac{3}{4}$-in.-diameter bolt} \quad OK}$$

Adjusted Design Shear for ASD

The design shear stress in the $3\frac{1}{8} \times 9$ glulam beam is not critical, but calculations will be shown to demonstrate the criteria given in Sec. 13.9. Compare *5 times the beam depth* with the distance of the connection from the end of the beam.

$$5d = 5 \times 9 = 45 \text{ in.} < 48 \text{ in.}$$

The joint is more than 5 times the depth from the end. The design shear need not be multiplied by $(d_e/d)^2$. Since the bolts penetrate the wide face of the 24F-V4 glulam beam, select the reference shear design value for bending about the strong x-axis and multiply by a reduction factor of 0.72 in accordance with NDS Supplement Table 5A:

$$F_v = 0.72(265) = 191 \text{ psi}$$

$$F_v' = F_v(C_D C_M C_t)$$

$$= 191(0.9)(1.0)(1.0)$$

$$= 172 \text{ psi}$$

$$d_e = \frac{9.0}{2} = 4.5 \text{ in.}$$

$$V_r' = \frac{2}{3} F_v' b d_e$$

$$= \frac{2}{3}(172)(3.125)(4.5)$$

$$= 1610 \text{ lb} > 575 \text{ lb} \quad OK$$

$$\boxed{\text{Design shear in beam is OK for ASD.}}$$

It should be noted that shear at the connection in the glulam beam was OK since the location of the connection was greater than $5d$ from the end of the beam. If the connection had been located closer than $5d$ to the end of the beam, then the adjusted design shear at the connection would not have been OK since:

$$V'_r = \frac{2}{3} F'_v b d_e \left(\frac{d_e}{d}\right)^2$$

$$= \frac{2}{3}(172)(3.125)(4.5)\left(\frac{4.5}{9}\right)^2$$

$$= 402 \text{ lb} < 575 \text{ lb} \qquad NG$$

Possible solutions, when $V'_r < V$ at a connection, include increasing d_e by locating the bolt below (or above) the neutral axis or increasing $b d_e$ by selecting a larger glulam beam.

Now solve the problem using LRFD procedures. Refer to Figs. 13.18 and 13.19. Note that force magnitudes must be based on *factored* loads for LRFD. The magnitude of the *factored* dead load is:

$$P_u = 1.4 \, P_{DL} = 1.4(575) = 805 \text{ lb}$$

Solve for *factored* force in one brace (LRFD):

$$\Sigma M_C = 0$$

$$6 T_{Vu} = 805 \,(10)$$

$$T_{Vu} = 1342 \text{ lb}$$

$$T_u = \sqrt{2} \, T_{Vu} = \sqrt{2} \,(1342) = 1897 \text{ lb}$$

Factored shear at the connection in the beam:

$$V_u = P_u = 805 \text{ lb}$$

Figure 13.19 Factored shear diagram for Member AC.

Reference design value based on yield limit equations:

$$Z = 1814 \text{ lb}$$

The nominal design value for LRFD is:

$$Z_n = Z\,(K_F) = 1814(3.32) = 6022\text{ lb}$$

The time effect factor for dead load acting alone is $\lambda = 0.6$.

Adjusted LRFD design value:

$$Z'_n = Z_n(\phi_z\lambda C_M C_t C_g C_\Delta)$$

$$= 6022(0.65)(0.6)(1.0)(1.0)(1.0)(1.0)$$

$$= 2349\text{ lb} > 1897\text{ lb}\quad OK$$

$$\boxed{\text{$\tfrac{3}{4}$-in. diameter bolt}\quad OK}$$

Adjusted Design Shear for LRFD

The connection is located more than $5d$ from the end of the beam.

$$F_{vn} = F_v\,(K_F) = 191(2.88) = 550\text{ psi}$$

$$F'_{vn} = F_{vn}(\phi_v\lambda C_M C_t) = 550(0.75)(0.6)(1.0)(1.0) = 247\text{ psi}$$

$$V'_{rn} = \frac{2}{3}F'_{vn}bd_e$$

$$= \frac{2}{3}(247)(3.125)(4.5)$$

$$= 2318\text{ lb} > 805\text{ lb}\quad OK$$

$$\boxed{\text{Design shear in beam is OK for LRFD}}$$

If the connection had been located closer than $5d$ to the end of the beam, then the adjusted design shear at the connection would not have been OK since:

$$V'_{rn} = \frac{2}{3}F'_{vn}bd_e\left(\frac{d_e}{d}\right)^2$$

$$= \frac{2}{3}(247)(3.125)(4.5)\left(\frac{4.5}{9}\right)^2$$

$$= 580\text{ lb} < 805\text{ lb}\quad NG$$

13.11 Lag Bolt Connections

Lag bolts are relatively large-diameter fasteners that have a *wood screw* thread and a square or hexagonal bolt head. See Fig. 13.20*a*. The NDS and the TCM (Refs. 13.1 and 13.6) refer to these fasteners as lag screws. The term *lag bolt* implies a large-diameter fastener with a bolt head, and the term *lag screw* indicates a fastener with a wood screw thread. Both refer to the same type of fastener, and the term *lag bolt* is generally used in this book.

A distinction should be made between large-diameter lag bolts (lag screws) and similar small-diameter fasteners known as *wood screws*. Recall from Chap. 11 that the dowel bearing strength is different for large- and small-diameter fasteners. For lag bolts, $F_{e\perp}$ is a function of diameter. The methods of installation for lag bolts and wood screws are also different. Because of their large size, lag bolts are installed with a wrench, and wood screws are usually installed with a screwdriver or screw gun. The NDS covers the design of both lag bolts and wood screws, but only lag bolts are addressed in this text.

Lag bolts are used when an excessively long bolt would be required to make a connection or when one side of a through-bolted connection may be inaccessible. Lag bolts can be used in shear connections (Fig. 13.20*b*) or withdrawal connections (Fig. 13.20*c*).See Example 13.17. Both applications are used in practice, and the designer should be familiar with the design procedures for both types of problems.

EXAMPLE 13.17   Lag Bolt Connections

Figure 13.20*a* Lag bolt. Large-diameter fastener with wood screw thread and bolt head. Also known as lag screw.

Figure 13.20*b* Lag bolts in single shear connections. Wood-to-wood and wood-to-metal connections are the common applications.

Figure 13.20c Lag bolts in withdrawal connection.

Lag bolts are installed in special prebored holes that accommodate the shank diameter and provide holding material for the thread. Washers or steel plates are required under the fastener head (similar to bolts).

As with the majority of other fasteners used in wood connections, there is not a currently recognized national standard that defines mechanical properties for lag bolts. However, lag bolts are typically manufactured from the same wire stock as bolts, and $F_{yb} = 45$ ksi appears to be conservative for commonly available lag bolts $\frac{3}{8}$ in. or greater in diameter. For smaller diameters, the bending yield strength is assumed to increase because of work-hardening (similar to nails): $F_{yb} = 70$ ksi for $\frac{1}{4}$-in. diameter, and $F_{yb} = 60$ ksi for $\frac{5}{16}$-in. diameter. It is felt that these values can reasonably be used in the yield limit equations until a standard is developed. In the meantime, the designer should list the bending yield strength as part of the design specifications.

The configuration of lag bolts is given in NDS Appendix L *Typical Dimensions for Dowel-Type Fasteners*. This appendix provides a number of important dimensions that are used in the installation and design calculations for a lag bolt. These dimensions include the nominal length, shank diameter and length, root diameter and thread length, and length of tapered tip.

Lag bolts are installed in prebored holes that accommodate the shank and the threaded portion. See Example 13.18. The hole involves drilling with two different-diameter bits. The larger-diameter hole has the same diameter and length as the unthreaded shank of the lag bolt. The diameter of the lead hole for the threaded portion is given as a percentage of the shank diameter, and depends on the specific gravity of the wood.

It is important that lag bolts be installed properly with a wrench. Driving with a hammer is an unacceptable method of installation. Soap or another type of lubricant should be used on the lag bolt or in the pilot hole to facilitate installation and to prevent damage to the fastener.

EXAMPLE 13.18   Installation of Lag Bolt

Figure 13.21a   Hole sizes.

A lag bolt is installed in a pilot hole. The dimensions of the pilot hole in the main or holding member are as follows:
For the shank:

$$L_1 = \text{shank length} - \text{side plate thickness} - \text{washer thickness}$$

$$d_1 = \text{shank diameter}$$

Figure 13.21b   Photographs of lag bolt threads. (*Courtesy of the FPL.*)

For the threaded portion:

$$L_2 \geq \text{thread length}$$

$$d_2 = \begin{cases} (0.65 \text{ to } 0.85) \times d_1 & \text{for wood with } G > 0.6 \\ (0.60 \text{ to } 0.75) \times d_1 & \text{for wood with } 0.5 < G \leq 0.6 \\ (0.40 \text{ to } 0.70) \times d_1 & \text{for wood with } G \leq 0.5 \end{cases}$$

For a given specific gravity range, the larger percentages of d_1 should be used for larger-diameter lag bolts. It is important that lag bolts be installed by turning with a wrench into lead holes of proper size.

The photograph on the left in Fig. 13.21b shows deep, clean-cut threads of a lag bolt that was properly installed in the correct size lead hole. On the right, the threads have shallow penetration into the wood because of an oversize pilot hole.

The shear capacity of a lag bolt is covered by the yield limit equations (Sec. 13.12). The withdrawal capacity is empirically based on test results (Sec. 13.15).

13.12   Yield Limit Equations for Lag Bolts

The yield limit equations for lag bolts in single shear (two-member) connections are identical to the yield limit equations for bolts in single shear, as summarized in Example 13.2. Lag bolts are not generally used in double shear applications. The fastener diameter D in the yield limit equations is typically taken as the root diameter of the threads D_r for most lag bolt connections since lag bolt threads are frequently located near the shear plane where plastic hinges form in the fastener (see Sec. 11.7.). Alternatively, the unthreaded shank diameter may be used for D in the yield limit equations if the moment and bearing behavior at the threaded region of the lag bolt are explicitly addressed in a more detailed analysis (see NDS Sec. 11.3.6 and Ref. 13.2).

According to NDS Sec. 11.3.4, the dowel bearing length in the main member l_m is defined in two ways based on the total length of the lag bolt in the main member (the member that receives the threaded end of the lag bolt). If the lag bolt penetrates a distance of $10D$ or greater into the main (holding) member, then the dowel bearing length l_m is taken as the total length of the lag bolt in the main member. However, if the lag bolt penetrates a distance less than $10D$ into the main member, the dowel bearing length l_m is taken as the total length of the lag bolt in the main member minus the length of the tapered tip E of the lag bolt (see NDS Appendix L). This is similar to the definition of dowel bearing length in the main member for nails. See Example 12.1. The dowel bearing strength F_e can be computed given the specific gravity of the wood, or values can be obtained from NDS Table 11.3.2, *Dowel Bearing Strengths*.

As an alternative to solving the yield limit equations, the reference design value for a lag bolt used in certain common applications may be read from NDS Tables 11J and 11K. Tabulated design values are based on the root diameter of the lag bolt threads D_r and lag bolt penetration in the main member of $p = 8D$. Thus, the NDS table values can be used for connections with p ≥ 8D. For lag bolt connections with $4D \leq p < 8D$, a footnote to the NDS tables provides a conservative reduction factor of $p/8D$ to account for the reduced lag bolt penetration in the main member. The reduction factor may be used to adjust tabulated design values in lieu of calculating a reduced connection capacity directly from yield limit equations. Design values are given for parallel- and perpendicular-to-grain loading. The notation is similar to that used in the bolt tables:

1. Load *parallel* to grain in *both* the main member and the side member, $Z_\parallel$.

2. Load *perpendicular* to grain in the *side* member and parallel to grain in the main member, $Z_{s\perp}$.

3. Load *perpendicular* to grain in the *main* member and parallel to grain in the side member, $Z_{m\perp}$.

4. Load *perpendicular* to grain in *both* the main member and the side member, $Z_\perp$.

When the load is at an intermediate angle of load to grain, the yield limit equations must be evaluated.

One example is given to illustrate the application of the yield limit equations for lag bolts. See Example 13.19. The connection is between a steel side plate and a glulam beam.

EXAMPLE 13.19   Lag Bolt Reference Design Value: Single Shear Wood-to-Metal Connection

Determine the reference design value of a $\frac{5}{8}$-in.-diameter by 6-in.-long lag bolt used to connect a $\frac{1}{4}$-in.-thick metal side plate of A36 steel and a DF-L glulam beam. Load is parallel to grain in the glulam.

Summary of Known Values

Dowel bearing strength can be computed (Sec. 11.5) given the specific gravity of the glulam beam, or it can be read as $F_{e\parallel}$ from NDS Table 11.3.2.

$$L = 6 \text{ in.}$$

$$D_r = 0.471 \text{ in.} \quad [\text{for a } \tfrac{5}{8}\text{-in.-diameter lag bolt (see NDS Appendix L)}]$$

$$E = \tfrac{13}{32} \text{ in.} = 0.406 \text{ in.} \quad (\text{see NDS Appendix L})$$

$$F_{yb} = 45 \text{ ksi} \quad (\text{assumed for } \tfrac{5}{8}\text{-in.-diameter lag bolt})$$

$$\theta_m = 0 \text{ degrees}$$

$$G_m = 0.50$$

$$F_{em} = 5600 \text{ psi} \quad (F_{e\parallel} \text{ for lag bolt in DF-L})$$

$$l_s = t_s = 0.25 \text{ in.}$$

$$10D = 10(0.625) = 6.25 \text{ in.}$$

Total lag bolt length in main member $= L - t_s = 6 - 0.25 = 5.75$ in. < 6.25 in.

$$\therefore l_m = L - t_s - E = 6 - 0.25 - 0.406 = 5.34 \text{ in.}$$

$$F_{es} = 1.5\, F_u = 87{,}000 \text{ psi} \qquad \text{(for A36 steel plate)}$$

Coefficients for Yield Limit Equations

$$R_e = \frac{F_{em}}{F_{es}} = \frac{5600}{87{,}000} = 0.06437$$

$$1 + R_e = 1 + 0.06437 = 1.064$$

$$2 + R_e = 2 + 0.06437 = 2.064$$

$$1 + 2R_e = 1.129$$

$$R_t = \frac{l_m}{l_s} = \frac{5.34}{0.25} = 21.4$$

$$1 + R_t = 1 + 21.4 = 22.4$$

$$1 + R_t + R_t^2 = 22.4 + 21.4^2 = 479.3$$

$$\theta = 0 \text{ degree}$$

$$K_\theta = 1 + \frac{\theta}{360} = 1 + \frac{0}{360} = 1.0$$

$$k_1 = \frac{\sqrt{R_e + 2R_e^2(1 + R_t + R_t^2) + R_t^2 R_e^3} - R_e(1 + R_t)}{(1 + R_e)}$$

$$= \frac{\sqrt{0.06437 + 2(0.06437^2)(479.3) + (21.4^2)(0.06437^3)} - 0.06437(22.4)}{1.064}$$

$$= 0.5626$$

$$k_2 = -1 + \sqrt{2(1 + R_e) + \frac{2F_{yb}(1 + 2R_e)D^2}{3F_{em}l_m^2}}$$

$$= -1 + \sqrt{2(1.064) + \frac{2(45{,}000)(1.129)(0.471^2)}{3(5600)(5.34^2)}}$$

$$= 0.4750$$

$$k_3 = -1 + \sqrt{\frac{2(1 + R_e)}{R_e} + \frac{2F_{yb}(2 + R_e)D^2}{3F_{em}l_s^2}}$$

$$= -1 + \sqrt{\frac{2(1.064)}{0.06437} + \frac{2(45{,}000)(2.064)(0.471^2)}{3(5600)(0.25^2)}}$$

$$= 7.504$$

Yield Limit Equations

Mode I_m (NDS Eq. 11.3-1):

$$Z = \frac{Dl_m F_{em}}{4K_\theta} = \frac{0.471(5.34)(5600)}{4(1.0)} = 3524 \text{ lb}$$

Mode I_s (NDS Eq. 11.3-2):

$$Z = \frac{Dl_s F_{es}}{4K_\theta} = \frac{0.471(0.25)(87,000)}{4(1.0)} = 2561 \text{ lb}$$

Mode II (NDS Eq. 11.3-3):

$$Z = \frac{k_1 Dl_s F_{es}}{3.6K_\theta} = \frac{0.5626(0.471)(0.25)(87,000)}{3.6(1.0)} = 1602 \text{ lb}$$

Mode III_m (NDS Eq. 11.3-4):

$$Z = \frac{k_2 Dl_m F_{em}}{3.2(1 + 2R_e)K_\theta} = \frac{0.4750(0.471)(5.34)(5600)}{3.2(1.129)(1.0)} = 1854 \text{ lb}$$

Mode III_s (NDS Eq. 11.3-5):

$$Z = \frac{k_3 Dl_s F_{em}}{3.2(2 + R_e)K_\theta} = \frac{7.504(0.471)(0.25)(5600)}{3.2(2.064)(1.0)} = 749 \text{ lb}$$

Mode IV (NDS Eq. 11.3-6):

$$Z = \frac{D^2}{3.2K_\theta}\sqrt{\frac{2F_{em}F_{yb}}{3(1 + R_e)}} = \frac{0.471^2}{3.2(1.0)}\sqrt{\frac{2(5600)(45,000)}{3(1.064)}} = 871 \text{ lb}$$

The reference design value is selected as the smaller value from the yield limit formulas:

$$\boxed{Z = 749 \text{ lb}}$$

Compare with value from NDS Table 11K:

$$Z = 750 \text{ lb} \approx 749 \text{ lb}$$

13.13   Adjustment Factors for Lag Bolts in Shear Connections

The yield limit equations for lag bolts in single shear were discussed in Sec. 13.12. The *reference design value Z* represents the ASD capacity of a single lag bolt or lag screw as obtained from the yield limit equations. In the design of a given connection, the reference design value is converted to an *adjusted design value Z'* by a series of adjustment factors.

The adjustment factors convert the base conditions to the conditions for a specific design situation. The adjustment factors required in the design of connections are summarized in NDS Table 10.3.1.

The base conditions associated with the reference design value are as follows:

1. Load is normal (10-year) duration.

2. Wood is initially dry at the time of fabrication of a connection and remains dry in service.

3. Temperature range is normal.

4. There is one lag bolt in a line parallel to the applied load (i.e., only one lag bolt per row).

5. Spacing provisions for lag bolts (geometry requirements for the connection) are satisfied.

6. Penetration of lag bolt in the holding member (not including the tapered tip) is a minimum of 4 times the shank diameter (i.e., $p \geq 4D$).

7. Lag bolt penetrates the side grain (not the end grain) of the main (holding) member.

If a lag bolt is used in a connection under exactly these conditions, the adjustment factors all default to unity and $Z' = Z$ for ASD.

For a shear connection that does not satisfy the base conditions described above, the following adjustment factors may be required for ASD:

$$Z' = Z(C_D C_M C_t C_g C_\Delta C_{eg})$$

where C_D = load duration factor (Sec. 4.15)
C_M = wet service factor (Sec. 4.14)
C_t = temperature factor (Sec. 4.20)
C_g = group action factor (Sec. 13.5)
C_Δ = geometry factor (Sec. 13.5)
C_{eg} = end grain factor

In LRFD, the reference design value must be multiplied by a format conversion factor K_F and a resistance factor ϕ_z. The K_F factor adjusts the reference design value Z to a nominal design value Z_n appropriate for LRFD. As part of the conversion to an LRFD format, the K_F factor adjusts from a normal (10-year) load duration basis to a short term (5 to 10 minutes) load duration basis.

The load duration factor C_D is not applicable for LRFD calculations. Instead a time effect factor λ is selected based on the governing LRFD load combination, as specified in NDS Appendix N (Ref. 13.1) (Sec. 4.16). All other adjustment factors are identical for both LRFD and ASD. Thus, for a laterally loaded lag bolt connection the following adjustment factors may be required for LRFD.

$$Z'_n = Z_n(\phi_z \lambda C_M C_t C_g C_\Delta C_{eg})$$

The total adjusted capacity for a connection using two or more lag bolts is the sum of the individual adjusted design values Z'. This summation is appropriate if the lag bolts are all the same type and size because each fastener will have the same yield mode. The adjusted capacity for the connection is the number of lag bolts N times the adjusted design value for one fastener:

$$\text{Adj. } P = N(Z')$$

A number of adjustment factors are described elsewhere in this book and are mentioned only in summary. Some of these were covered in Chap. 4, and others, which apply to both lag bolts and bolts, were summarized in Sec. 13.5. Only the adjustment factors that are unique to lag bolts are covered in detail.

Resistance factor ϕ_z (LRFD only). The resistance factor for all fasteners in the NDS is $\phi_z = 0.65$. See NDS Appendix N (Ref. 13.1).

Format conversion factor K_F (LRFD only). The format conversion factor for all fasteners in the NDS is $K_F = 3.32$. See NDS Appendix N (Ref. 13.1).

Load duration factor C_D (ASD only). The load duration factor for a lag bolt connection is the same as that applied to the design of other fasteners. Refer to Secs. 4.15 and 12.6 for a review of load duration and the numerical values for C_D.

Time effect factor λ (LRFD only). The time effect factor λ for lag bolt connections is the same as that applied to bolted and nailed connections. Refer to Secs. 4.16 and 12.6 for a review of the time effect factor and the numerical values for λ.

Wet service factor C_M. The concept of an adjustment factor based on moisture content (MC) was introduced in Sec. 4.14. The discussion regarding possible damage in a wood connection due to cross-grain volume changes applies to both lag bolts and bolts. See Sec. 13.5 for a discussion of this potentially serious problem. The numerical values of C_M are also the same for lag bolts and bolts and are given in NDS Table 10.3.3, *Wet Service Factors, C_M, for Connections.*

The first objective in good design practice is to keep the connection dry where possible. Second, the connection should be detailed so that cross-grain cracking does not occur. Wood is very weak in tension perpendicular to grain, and the conditions that cause tension stress across the grain can be eliminated, or minimized, by proper detailing. For example, separate steel side plates for rows of lag bolts parallel to the grain will allow volume changes to occur without locking in stresses.

Temperature factor C_t. If wood is used in an application with sustained exposure to high temperatures, the reference design value for a lag bolt is to be reduced by the temperature factor C_t. Refer to Sec. 4.20 for a brief review of the temperature factor. Numerical values of C_t for connections are given in NDS Table 10.3.4. The temperature range for most wood-frame buildings does not require a reduction of design values, and the temperature factor is often taken as unity.

Group action factor C_g. In a connection that has more than one lag bolt in a row, it is necessary to reduce the reference design value by a multiplier known as the group action factor C_g. The group action factor primarily depends on the number of fasteners in a row. The group action factor for a lag bolt connection is the same as for a bolt connection. Refer to Sec. 13.5 for a discussion and definition of C_g. Note that C_g does not exceed unity ($C_g \leq 1.0$).

Geometry factor C_Δ.   End distance, edge distance, center-to-center spacing, and row spacing requirements for lag bolts are the same as for bolts. Refer to Sec. 13.5 for a discussion of spacing requirements. Specific criteria are given in Example 13.10 for parallel-to-grain loading and in Example 13.11 for perpendicular-to-grain loading.

Examples 13.10 and 13.11 summarize the base dimensions for fastener spacing. It is recommended that the base dimensions be used in designing and detailing a connection whenever practical. However, if conditions do not permit the use of the full base dimensions, it is possible to reduce certain spacings with a corresponding reduction in adjusted design value. The minimum spacing limits are also given in Examples 13.10 and 13.11.

The geometry factor C_Δ is the multiplier that is used to reduce the reference design value Z if the furnished spacings are less than the base dimensions. If the base requirements for a connection are all satisfied, the geometry factor defaults to unity ($C_\Delta = 1.0$). For other conditions, the geometry factor for lag bolts is the same as for bolts. Refer to Sec. 13.5 for the definition of C_Δ for reduced spacings. Note that C_Δ does not exceed unity ($C_\Delta \leq 1.0$).

Lag bolt penetration.   The penetration of a lag bolt is defined as the distance that the fastener extends into the main (holding) member minus the length of the tapered tip. Dimensions of typical lag bolts are given in NDS Appendix L, and the length of tapered tip E is included in the Appendix. Thus, the penetration of a lag bolt can be computed as the length of the lag bolt minus the thickness of the side member minus the washer thickness (if applicable) minus the length of the tapered end:

$$p = L - t_s - t_{\text{washer}} - E$$

To obtain any design value Z for a lag bolt connection, the penetration must be at least 4 times the diameter of the lag bolt ($p \geq 4D$). In previous editions of the NDS, only three yield limit equations (rather than six) were specified for lag bolt connections, and a penetration depth factor was provided for lag bolt penetration in the range $4D \leq p \leq 8D$. However, all six yield limit equations are specified for lag bolt connections in the 2005 NDS. Thus, the penetration depth factor is no longer required since the dowel bearing length of the lag bolt in the main member l_m is explicitly addressed in the yield limit equations.

End grain factor C_{eg}.   The reference design value Z applies to a laterally loaded shear connection with the lag bolt installed into the *side grain* of the main member. Again, the main member is the member that receives the threaded end of the fastener. See Example 13.20 (Fig. 13.22a). This is the strongest and most desirable type of lag bolt connection. When a lag bolt is installed in the side grain, the end grain factor does not apply, or it can be viewed as having a default value of unity ($C_{eg} = 1.0$).

EXAMPLE 13.20   Comparison of Lag Bolts in Side Grain and End Grain

Lateral Resistance in Side Grain

The basic shear connection has the lag bolt installed in the side grain of the main member (Fig. 13.22a). The full reference design value applies to this type of loading, and the end grain factor is $C_{eg} = 1.0$

Figure 13.22a   Lag bolt in side grain.

Lateral Resistance in End Grain

A second type of shear connection has the lag bolt installed in the end grain of the holding member (Fig. 13.22b). In other words, the lag bolt is installed parallel to the grain of the main member. The connection still qualifies as a laterally loaded connection because the load is perpendicular to the axis of the fastener. For lag bolts installed in the end grain, $C_{eg} = 0.67$. For a lag bolt in the end grain, $F_{em} = F_{e\perp}$.

Figure 13.22b   Lag bolt in end grain.

In another type of shear connection, the lag bolt is installed in the end grain of the holding member (Fig. 13.22b). This is a much weaker and less desirable connection. Laterally loaded connections of this type are permitted, but the reference design value is reduced by an end grain factor of $C_{eg} = 0.67$. A similar factor is used in connections with nails driven into the end grain (Sec. 12.6).

This concludes the review of adjustment factors for lag bolts in shear connections. One design example is provided in the next section. Many of the possible adjustment factors in this example are shown to default to unity. Although this occurs frequently in practice, the designer needs to be aware of the base conditions associated with the reference design value. When other circumstances occur, adjustment factors are required.

13.14   Design Problem: Collector (Strut) Splice with Lag Bolts

This example makes use of lag bolts to attach a glulam beam to a concrete shearwall. The load is the result of a seismic force on the horizontal diaphragm. See Example 13.21. The force on the connection is obtained from the collector (strut) force diagram, which is given in the problem. For a review of collector (strut) force diagrams, refer to Secs. 9.7 and 9.10.

The force is transferred from the glulam beam by lag bolts in a shear connection to a steel plate. A second connection transfers the load from the steel plate to the shearwall with anchor bolts. The example is limited to the lag bolt connection, and calculations for the anchor bolts used in the second connection are not illustrated. The reference design value for the lag bolt is obtained from a previous example, and the yield limit equations are not repeated.

The glulam beam in this example is manufactured with laminations that are all from one species. The designer is reminded of special considerations for connections in glulam members that are fabricated from more than one species. These were discussed in Example 11.6 (Sec. 11.5), but do not affect the problem at hand.

EXAMPLE 13.21   Splice with Lag Bolts

The roof beam in the building in Fig. 13.23 serves as the collector (strut) for two horizontal diaphragms. The maximum collector force E = 9000 lb occurs at the connection of the roof beam and the concrete shearwall, and the magnitude of the load is obtained from the collector (strut) force diagram. For ASD, the magnitude of the collector force is $P = 0.7E = 0.7(9000) = 6300$ lb.

Determine the number of $\frac{5}{8}$-in.-diameter, × 6-in. lag bolts necessary to connect the steel splice plate to the Douglas Fir (24F-V4) glulam beam (from the 24F-1.8E Stress Class) using ASD procedures. The reference design value for the lag bolt is obtained from Example 13.19. Assume that the splice plate and the connection to the wall are adequate. A single row of lag bolts is to be used, and the base dimensions for lag bolt spacings are to be summarized. $C_M = 1.0$ and $C_t = 1.0$. Repeat the problem using LRFD procedures.

Figure 13.23   Collector (strut) connection made with lag bolts.

Trial Design (ASD)

The reference design value for a $\frac{5}{8}$-in.-diameter, 6-in. lag bolt from Example 13.19 is governed by Mode III$_s$ yield behavior.

$$Z = 749 \text{ lb}$$

For an initial trial, assume all adjustment factors default to unity except the load duration factor for ASD:

$$\text{Approx. } Z' \approx Z(C_D) = 749(1.6) = 1198 \text{ lb/lag bolt}$$

Approximate number of lag bolts

$$N \approx \frac{P}{Z'} = \frac{6300}{1198} = 5.3 \quad say \quad 6 \text{ lag bolts}$$

Final Design

The possible adjustment factors that can affect the capacity of a lag bolt connection include C_D, C_M, C_t, C_g, C_Δ, and C_{eg}. At this point the following factors are known or assumed:

$C_D = 1.6$ for seismic load

$C_M = 1.0$ given (assumes connection is initially dry and remains dry in service)

$C_t = 1.0$ assumes normal temperature range

$C_{eg} = 1.0$ lag bolts installed in side grain (top face) of glulam

The remaining adjustments will now be considered.

Penetration depth:

Thickness of the metal side plate $t_s = \frac{1}{4}$ in.

A washer is not required since the steel side plate separates the lag bolt head from the wood member.

From NDS Appendix L, the length of the tapered tip is

$$E = {}^{13}\!\!/_{32} \text{ in.}$$

$$p = L - t_s - E$$

$$= 6 - 0.25 - 0.406$$

$$= 5.34 \text{ in.}$$

Minimum penetration to develop the full reference design value is

$$p = 4D = 4(0.625) = 2.5 \text{ in.} < 5.34 \text{ in.} \qquad OK$$

Geometry factor:

Connection is a single row of lag bolts with load parallel to grain. Spacing requirements for lag bolts are the same as for bolts. For parallel-to-grain loading, spacing requirements are summarized in Example 13.10. If there is adequate room, it is recommended that base dimensions be used in detailing a connection.

Base dimensions for a single row of lag bolts in a tension member (softwood species):

End distance $= 7D = 7(0.625) = 4.375$ in.   *use*   5 in.

c.-to-c. spacing $= s = 4D = 4(0.625) = 2.5$ in.   *use*   3 in.

Edge distance $= 1.5D = 1.5(0.625) = 0.94$ in.

The row of lag bolts will be installed along the centerline of the beam.

$$\text{Edge distance} = \frac{5.125}{2} = 2.56 \text{ in.} > 0.94 \text{ in.} \qquad OK$$

The furnished spacings all exceed the base dimensions.

$$\therefore C_\Delta = 1.0$$

Group action factor:

The formula and coefficients for the group action factor are summarized in Example 13.8 (Sec. 13.5).

$$\text{Load/slip modulus } \gamma = 270,000(D^{1.5}) \qquad \text{for lag bolts in a wood-to-metal connection}$$

$$= 270,000(0.625^{1.5}) = 133,400 \text{ lb/in.}$$

Member stiffnesses:

$$E_m = E_{\text{axial}} = 1,700,00 \text{ psi} \qquad \text{(NDS Supplement Table 5A)}$$

$$A_m = 5.125 \times 21 = 107.6 \text{ in.}^2$$

$$E_m A_m = 1,700,000(107.6) = 183,000,000$$

$$E_s = E_{\text{steel}} = 29,000,000 \text{ psi}$$

$$A_s = 0.25 \times 4 = 1.0 \text{ in.}^2$$

$$E_s A_s = 29,000,000(1.0) = 29,000,000$$

Spacing of fasteners in a row

$$s = \text{c.-to-c. spacing} = 3 \text{ in.}$$

Coefficients for C_g:

$$n = \text{number of lag bolts in row} = 6$$

$$u = 1 + \gamma\left(\frac{s}{2}\right)\left[\frac{1}{E_m A_m} + \frac{1}{E_s A_s}\right]$$

$$= 1 + 133,400\left(\frac{3.0}{2}\right)\left[\frac{1}{183,000,000} + \frac{1}{29,000,000}\right]$$

$$= 1.008$$

$$m = u - \sqrt{u^2 - 1}$$

$$= 1.008 - \sqrt{1.008^2 - 1} = 0.8813$$

$$1 + m = 1 + 0.8813 = 1.881$$

$$1 - m = 1 - 0.8813 = 0.1187$$

$$m^{2n} = 0.8813^{2(6)} = 0.2195$$

$$\frac{E_m A_m}{E_s A_s} = \frac{183,000,000}{29,000,000} = 6.309$$

$$\frac{E_s A_s}{E_m A_m} = \frac{1}{6.309} = 0.1585$$

$$R_{EA} = \text{smaller of ratios of member stiffnesses}$$

$$= 0.1585$$

$$1 + R_{EA} = 1 + 0.1585 = 1.159$$

$$1 + R_{EA} m^n = 1 + 0.1585(0.8813)^6 = 1.074$$

$$C_g = \left[\frac{m(1 - m^{2n})}{n[(1 + R_{EA} m^n)(1 + m) - 1 + m^{2n}]} \right] \left[\frac{1 + R_{EA}}{1 - m} \right]$$

$$= \left[\frac{0.8813(1 - 0.2195)}{6[(1.074)(1.881) - 1 + 0.2195]} \right] \left[\frac{1.159}{0.1187} \right]$$

$$= 0.902$$

Adjusted ASD capacity for the lag bolt connection:

$$\text{Adj. } P = N(Z') = N(Z)(C_D C_M C_t C_g C_\Delta C_{eg})$$

$$= 6(749)(1.6)(1.0)(1.0)(0.902)(1.0)(1.0)$$

$$= 6485 \text{ lb} > 6300 \text{ lb} \qquad OK$$

> *Use*   6 lag bolts in a single row
> $\frac{5}{8}$-in. diameter $\times$ 6 in. long
> end distance = 5 in.
> c.-to-c. spacing = 3 in.

Now repeat the problem using LRFD procedures.
The *factored* load at the connection is $P_u = 1.0E = 9000$ lb.
The time effect factor for a load combination dominated by seismic force is $\lambda = 1.0$

Trial Design (LRFD)

The reference design value for one lag bolt is:

$$Z = 749 \text{ lb}$$

The nominal design value for LRFD is:

$$Z_n = Z (K_F) = 749(3.32) = 2487 \text{ lb}$$

For an initial trial, assume all adjustment factors except ϕ_z default to unity (LRFD):

$$\text{Approx. } Z'_n = Z_n(\phi_z\lambda) = 2487(0.65)(1.0) = 1616 \text{ lb}$$

$$N \approx \frac{P_u}{Z'_n} = \frac{9000}{1616} = 5.6 \quad say \quad 6 \text{ bolts}$$

$$C_g = 0.902 \ (same \ as \ for \ ASD)$$

Adjusted LRFD capacity for the lag bolt connection:

$$\text{Adj. } P_n = N(Z'_n) = N(Z_n)(\phi_z\lambda C_M C_t C_g C_\Delta C_{eg})$$

$$= 6(2487)(0.65)(1.0)(1.0)(1.0)(0.902)(1.0)(1.0)$$

$$= 8750 \text{ lb} < 9000 \text{ lb} \quad NG$$

$\therefore$ Require an additional lag bolt for LRFD procedures.

Revised group action factor for seven lag bolts:

$$n = 7$$

$$u = 1.008$$

$$m = 0.8813$$

$$1 + m = 1 + 0.8813 = 1.881$$

$$1 - m = 1 - 0.8813 = 0.1187$$

$$m^{2n} = 0.8813^{2(7)} = 0.1705$$

$$1 + R_{EA}m^n = 1 + (0.1585)(0.8813)^7 = 1.065$$

$$C_g = \left[\frac{m(1 - m^{2n})}{n[(1 + R_{EA}m^n)(1 + m) - 1 + m^{2n}]}\right]\left[\frac{1 + R_{EA}}{1 - m}\right]$$

$$= \left[\frac{0.8813(1 - 0.1705)}{7[(1.065)(1.881) - 1 + 0.1705]}\right]\left[\frac{1.159}{0.1187}\right]$$

$$= 0.868$$

Adjusted LRFD capacity for connection with seven lag bolts:

$$\text{Adj. } P_n = N(Z'_n) = N(Z_n)(\phi_z\lambda C_M C_t C_g C_\Delta C_{eg})$$

$$= 7(2487)(0.65)(1.0)(1.0)(1.0)(0.868)(1.0)(1.0)$$

$$= 9820 \text{ lb} > 9000 \text{ lb} \quad OK$$

> *Use* 7 lag bolts in a row (*one more than for ASD*)
> $\frac{5}{8}$-in. diameter × 6 in. long
> end distance = 5 in.
> c.-to-c. spacing = 3 in.

13.15   Lag Bolts in Withdrawal

The concept of a lag bolt being subject to a withdrawal load was introduced in Chap. 11 (Fig. 11.2a). Withdrawal loading attempts to pull the fastener out of the holding member.

The yield limit equations for dowel-type fasteners apply only to shear-type connections, and the yield limit theory does not apply to lag bolts in withdrawal. The withdrawal values for lag bolts are based on an empirical formula summarized in the NDS.

The *reference design value* for the basic withdrawal connection is given the symbol W. The basic withdrawal connection is with the lag bolt installed in the side grain of the holding member. NDS Table 11.2A gives the reference withdrawal design value *per inch of thread penetration* into the main member.

If the value from the table is given the symbol w and the *thread* penetration for withdrawal is given the symbol p_w, the reference design value for a single lag bolt is

$$W = w \times p_w$$

The penetration p for a shear connection was explained in Sec. 13.13. For shear, the penetration is typically taken as the length of the lag bolt in the main (holding) member minus the length of the tapered end. In other words, in a shear connection, a portion of the unthreaded shank may be included in the penetration length. However, the effective penetration for withdrawal loading p_w depends on the effective thread length in the holding member. See Fig. 13.24. NDS Appendix L gives the length of the threaded portion of the lag bolt T, the length of the tapered tip E, and the length of the full-diameter threaded region $T - E$. Assuming that the threads are all within the main member, the effective penetration for withdrawal is

$$p_w = T - E$$

Figure 13.24   Effective thread length for withdrawal.

The reference design value W represents the ASD capacity for a lag bolt used under a set of base conditions. The base conditions associated with the reference design value are as follows:

1. Load is normal (10-year) duration.
2. Wood is initially dry at the time of fabrication of connection and remains dry in service.
3. Temperature range is normal.
4. Lag bolt is installed in the side grain of the holding member.

To obtain the adjusted ASD design value W' for a different set of conditions, the reference design value is multiplied by adjustment factors:

$$W' = W(C_D)(C_M)(C_t)(C_{eg})$$

where C_D = load duration factor (Sec. 4.15)
C_M = wet service factor (Sec. 4.14)
C_t = temperature factor (Sec. 4.20)
C_{eg} = end grain factor

As with other fasteners, the nominal LRFD design value W_n is obtained by multiplying the reference design value by the format conversion factor for LRFD. Then the adjusted LRFD design value W_n' is obtained by multiplying the nominal design value by the LRFD resistance factor and appropriate adjustment factors:

$$W_n = W(K_F)$$

$$W_n' = W_n(\phi_z \lambda C_M C_t C_{eg})$$

where K_F = format conversion factor for connections (Sec. 13.5)
ϕ_z = resistance factor for connections (Sec. 13.5)
λ = time effect factor (Secs. 4.16 and 12.6)

A discussion of adjustment factors for lag bolts in *shear* connections was given in Sec. 13.13. The majority of this material also applies to lag bolts in *withdrawal* connections.

However, the end grain factor is different for a withdrawal connection. The default condition for $C_{eg} = 1.0$ still applies to the base condition of a lag bolt installed into the *side grain* of the main member. But for a withdrawal connection into the *end grain* of the holding member, the end grain factor is $C_{eg} = 0.75$ instead of the 0.67 value used for shear connections. Connections into the end grain should be avoided where possible. Although the loading shown is not withdrawal, Fig. 13.22a and b illustrates lag bolts installed in the side grain and end grain.

The total adjusted withdrawal capacity for a connection is the sum of the individual adjusted design values for each lag bolt. If all lag bolts in a connection are the same size and have the same penetration for withdrawal p_w, then the adjusted withdrawal capacity for the connection is the number of lag bolts N times the adjusted design value for one fastener:

$$\text{Adj. } P = N(W')$$

An example of a simple lag bolt withdrawal connection is given in Example 13.22.

EXAMPLE 13.22   Lag Bolt Withdrawal Connection

The load P_D is an equipment dead load which is suspended from two Southern Pine supporting beams (Fig. 13.25). The load is centered between the two beams. Determine the adjusted ASD design value based on the strength of the $\frac{1}{2}$ in.-diameter $\times$ 6 in. lag bolts. Lumber is initially dry and is exposed to the weather in service. $C_t = 1.0$. Repeat the problem using LRFD procedures.

Figure 13.25 Withdrawal loading on lag bolts.

Reference design value:

Length of lag bolt in main member $= L - t_s - t^*_{\text{washer}} = 6 - 1.5 - 0.125 = 4.375$ in.

Thread length $= T = 3\frac{1}{2}$ in.   (NDS Appendix L)

All of the threads are inside the main member.

Effective penetration for withdrawal $p_w = T - E = 3\frac{3}{16}$ in.

Specific gravity for Southern Pine:

$$G = 0.55$$

---

*Washer must be sufficiently large to keep the bearing stress below the adjusted compression perpendicular to grain design value of the wood member ($f_{c\perp} \leq F'_{c\perp}$).

From NDS Table 11.2A

$$w = 437 \text{ lb/in.}$$

$$W = w(p_w) = 437(3\,\tfrac{3}{16}) = 1393 \text{ lb/lag bolt}$$

Adjusted ASD design value:

$$W' = W(C_D)(C_M)(C_t)(C_{eg})$$

$$= 1393(0.9)(0.7)(1.0)(1.0)$$

$$= 878 \text{ lb}$$

$$\text{Adj. } P = N(W') = 2(878) = 1756 \text{ lb}$$

Now repeat the problem using LRFD procedures.

Reference design value:

$$W = 1393 \text{ lb/lag bolt}$$

Nominal design value for LRFD:

$$W_n = W(K_F) = 1393(3.32) = 4625 \text{ lb/lag bolt}$$

Adjusted LRFD design value:

The time effect factor for dead load acting alone is $\lambda = 0.6$.

$$W'_n = W_n(\phi_z \lambda C_M C_t C_{eg})$$

$$= 4625(0.65)(0.6)(0.7)(1.0)(1.0)$$

$$= 1262 \text{ lb}$$

$$\text{Adj. } P_n = N(W'_n) = 2(1262) = 2525 \text{ lb}$$

13.16   Combined Lateral and Withdrawal Loads

In addition to the direct withdrawal problem, there are occasions when lag bolts are subjected to combined lateral and withdrawal loading. See Fig. 13.26. The resultant load acts at an angle α to the surface of the main member. The practice for this type of loading is to determine an adjusted resultant design value Z'_α which must equal or exceed the total design load.

The first step in computing the adjusted design value is to evaluate the reference design value Z of the lag bolt for *shear* loading using the appropriate yield limit equations. For the next step, the reference *withdrawal* design value W is

Figure 13.26   Lag bolt subjected to combined lateral and withdrawal loading.

computed using the tabulated value from NDS Table 11.2A and the effective thread length for withdrawal ($W = w \times p_w$). The necessary adjustment factors are then applied to the reference values to obtain adjusted values Z' and W'. The interaction between these values is finally taken into account by the Hankinson formula, to give the adjusted resultant:

$$Z'_\alpha = \frac{Z'W'}{Z'\sin^2\alpha \, + \, W'\cos^2\alpha}$$

13.17   Split Ring and Shear Plate Connectors

Split rings and shear plates are connectors that are installed in precut grooves in wood members. Split ring and shear plate connectors provide a large bearing surface to resist shearing-type forces in a wood connection. A bolt or lag bolt is required through the center of the split ring or shear plate to hold the assembly together. A sketch of the load transfer was shown in the introductory chapter on connections in Fig. 11.1d.

Split rings are only for wood-to-wood connections because the steel ring fits into a groove cut into the mating surfaces of the members being connected. See Fig. 13.27a. The steel ring is "split" to provide simultaneous bearing on the inner core and the outer surface of the groove.

Shear plates can be used for wood-to-metal connections because the shear plate is flush with the surface of the wood. Shear plates may also be used for wood-to-wood connections, but a shear plate is required in each wood member. Both wood-to-wood and wood-to-steel connections are shown in Fig. 13.27b. Shear plate wood-to-wood connections are easier to assemble or disassemble than split-ring connections and may be desirable where ease of erection is important.

EXAMPLE 13.23   Split Ring and Shear Plate Connectors

Split rings are for wood-to-wood connections only. Figure 13.27a is an unassembled joint showing the split ring installed in the center member and the precut groove in one of the side members ready for installation. Also shown are the bolt, washer, and nut required to hold the assembly together. In a three-member joint, a second split ring is required at the mating surface between the center member and the other side member.

Figure 13.27a   Split ring connection. (*Courtesy of the FPL.*)

Joint A at the top of Fig. 13.27b shows a wood-to-wood connection ready for assembly. A shear plate is required in each mating face of the wood members. Therefore, in the three-member joint illustrated, a total of four shear plates is required. Shear plates have the advantage of creating a surface that is flush with the surface of the wood member. Although more connectors are required, joints made with shear plates are easier to assemble than similar split ring connections. Joint B is a three-member wood-to-steel connection. One shear plate is required at each interface between the wood and steel members.

Figure 13.27*b* Shear plate connections. (*Courtesy of the FPL.*)

The adjusted design values for split rings and shear plates are higher than those for bolts or lag bolts. However, split ring and shear plate connectors require special fabrication equipment, and their use is limited by fabrication costs. These types of connectors may be used where relatively large loads must be transferred in a fairly limited amount of space, but they are much less common than simple bolted connections. Split ring and shear plate connectors are more widely used in glulam arches and heavy timber trusses than in shearwall-type buildings. See Example 13.24.

EXAMPLE 13.24   Typical Arch Connection Using Timber Connectors

CROWN CONNECTION

TYPICAL GLULAM ARCH

Figure 13.28a Typical glulam arch connection.

Arch connections are shown in Fig. 13.28*a* and *b* to illustrate how shear plates can be used in the end grain of a member. Reference design values for timber connectors are given in NDS Tables 12.2A and 12.2B. Similar tables are available in the TCM. Information on the design of arches can also be found in the TCM (Ref. 13.6).

THREADED ROD AND SHEAR R's

DOWEL AND SHEAR R's

TIE R WITH BOLTS

TYPICAL   CROWN   CONNECTIONS

Figure 13.28b   Typical crown connections.

For arches with steep slopes, two shear plates can be used with a threaded rod. Washers are counterbored into the arch. For flat arches, two shear plates can be used on a dowel for vertical reactions. The tie plate keeps the arch from separating.

Split rings are available in $2\frac{1}{2}$- and 4-in. diameters, and shear plates are available in $2\frac{5}{8}$- and 4-in. diameters. Other pertinent dimensions are given in NDS Appendix K. For the design of split ring and shear plate connections commercial lumber species are divided into four groups (A to D) in NDS Table 12A. Information on the species groupings for timber connectors in glulam members is available in the TCM (Ref. 13.6) and Ref. 13.5.

Design values for split rings and shear plates are not covered by the yield limit theory that is applied to nails, bolts, and lag bolts. Reference design values for split rings are listed in NDS Table 12.2A, and values for shear plates are found in NDS Table 12.2B. The basic connection for both split rings and shear plates is with the connector installed in the side grain of the wood member. Tabulated reference design values include *parallel-to-grain design value P* and *perpendicular-to-grain design value Q*. Adjustment factors are provided to convert tabulated reference values to adjusted design values P' and Q'.

The Hankinson formula is used to obtain the adjusted design value N' at an angle of load to grain other than 0 or 90 degrees. Further adjustments are used for obtaining design values for connectors installed in the end grain of a member (as in the arch crown connection in Fig. 13.28b). Spacing requirements are important for split rings and shear plates, and design criteria are given in the NDS. A detailed review of this material is beyond the scope of this book. Additional information for obtaining adjusted design values and connector spacings is available in Ref. 13.6.

A brief example will illustrate the use of the NDS tables for timber connectors. See Example 13.25. The connection involves the design of shear plates for the splice of a chord in a horizontal diaphragm.

EXAMPLE 13.25   Shear Plate Connection

Determine the number of 4-in.-diameter shear plates necessary to develop the chord splice in Fig. 13.29 using ASD procedures. Use $\frac{7}{8}$-in.-diameter bolts with the shear plates. The tension force is the result of a seismic load E = 14.3 k. The member is a 24F-V4 DF glulam (from the 24F-1.8E Stress Class) that is dry and remains dry in service. Temperature range is normal. Assume that the metal splice plates and the glulams* are adequate. Repeat the problem using LRFD procedures.

Reference 13.5 indicates that the compression and inner core laminations of a 24F-V4 DF glulam are in timber connector group B (this agrees with NDS Table 12A for DF-L). The seismic force for ASD is $T = 0.7E = 0.7(14.3) = 10.0$ k.

> From NDS Table 12.2B, the reference design value for parallel-to-grain loading is
>
> $$P = 4320 \text{ lb}$$

---

*For tension calculations, the net area of a wood member with shear plates must allow for the projected area of the connectors in addition to the projected area of the bolt hole. See NDS Sec. 3.1.2 and Appendix K.

STEEL SIDE ℞

6¾" WIDE
GLULAM

CHORD
FORCE
$T = 10\,K$

T

SHEAR ℞'s
IN BOTH FACES
OF GLULAM

GLULAM

PIPE
COLUMN

ELEVATION   SECTION

Figure 13.29 Diaphragm chord splice made with shear plates.

The following adjustment factors apply to this problem for ASD:

$C_D = 1.6$ wind or seismic force

$C_M = 1.0$ moisture content factor

$C_t = 1.0$ normal temperature range

$C_g = 1.0$ one shear plate in a row

$C_\Delta = 1.0$ spacings will exceed base dimensions

$C_{st} = 1.11$ steel side plate factor for parallel-to-grain loading in species group B
(NDS Table 12.2.4)

Adjusted ASD design value for one shear plate:

$$P' = P(C_D)(C_M)(C_t)(C_g)(C_\Delta)(C_{st})$$

$$= 4320(1.6)(1.0)(1.0)(1.0)(1.0)(1.11)$$

$$= 7672 \text{ lb}$$

There is a shear plate in each face of the glulam.

$$\text{Adj. } T = 2(7672) = 15{,}300 \text{ lb} > 10{,}000 \text{ lb}\quad OK$$

> *Use*   total of four 4-in.-diameter shear
> plates (two each side of splice).

Now repeat the problem using LRFD procedures.
The *factored* seismic force for LRFD is $T_u = 1.0E = 14.3$ k.

Reference design value for one shear plate:

$$P = 4320 \text{ lb}$$

Nominal design value for LRFD:

$$P_n = P\,(K_F) = 4320(3.32) = 14{,}340 \text{ lb}$$

Adjusted LRFD design value for one shear plate:

The time effect factor for load combinations dominated by seismic load is $\lambda = 1.0$.

$$P'_n = P_n(\phi_z \lambda C_M C_t C_g C_\Delta C_{st})$$

$$= 14{,}340(0.65)(1.0)(1.0)(1.0)(1.0)(1.0)(1.11)$$

$$= 10{,}350 \text{ lb}$$

There is a shear plate in each face of the glulam

Adj. $T_n = 2(10{,}350) = 20{,}700 \text{ lb} > 14{,}300 \text{ lb}$ *OK*

> *Use* total of four 4-in. diameter shear plates (*same as for ASD*)

13.18 References

[13.1] American Forest and Paper Association (AF&PA). 2005. *National Design Specification for Wood Construction and Supplement*, ANSI/AF&PA NDS-2005, AF&PA, Washington DC.
[13.2] American Forest and Paper Association (AF&PA). 1999. *Technical Report 12—General Dowel Equations for Calculating Lateral Connection Values*, AF&PA, Washington DC.
[13.3] American Institute of Steel Construction (AISC). 2005. *Steel Construction Manual*, 13th ed., AISC, Chicago, IL.
[13.4] American Institute of Timber Construction (AITC). 2002. *Bolts in Glued Laminated Timber*, AITC Technical Note 8, AITC, Englewood, CO.
[13.5] American Institute of Timber Construction (AITC). 2004. *DESIGN Standard Specifications for Structural Glued Laminated Timber of Softwood Species*, AITC 117-2004, AITC, Englewood, CO.
[13.6] American Institute of Timber Construction (AITC). 2005. *Timber Construction Manual* (TCM), 5th ed., John Wiley & Sons Inc., Hoboken, NJ.
[13.7] American Iron and Steel Institute (AISI). 2002. *Cold-Formed Steel Design Manual*. 2002 ed., AISI, Washington DC.
[13.8] American Society for Testing and Materials (ASTM). 2005. "Standard Specification for Carbon Steel Bolts and Studs, 60,000 psi Tensile Strength," ASTM A307-04, *Annual Book of Standards*, Vol. 01.08, ASTM, Philadelphia, PA.
[13.9] Forest Products Laboratory (FPL). 1999. *Wood Handbook: Wood as an Engineering Material*, General Technical Report 113, FPL, Forest Service, USDA., Madison, WI.
[13.10] International Codes Council (ICC). 2006. *International Building Code*, 2006 ed., ICC. Falls Church, VA.
[13.11] Society of Automotive Engineers (SAE). 1999. "Mechanical and Material Requirements for Externally Threaded Fasteners," SAE J 429, SAE, Warrendale, PA.

13.19 Problems

Many of the problems require the reference design value Z and the adjusted design value Z' as either final or intermediate answers. These values are to be identified as part of the general solution along with the information specifically requested.

Reference design values may be obtained by solving the yield limit equations or by using the NDS tables. Tables are to be used only if they apply directly to the design conditions stated in the problem and are to be referenced in the solution. Yield limit equations may be evaluated by either calculator or computer.

In all problems, normal temperatures are assumed, and $C_t = 1.0$. All adjustment factors are to be referenced to the appropriate NDS table or section so that the source of the values may be verified.

13.1 Use a computer spreadsheet or programming software to solve the yield limit equations for a bolt in a single shear wood-to-wood connection. Input the following:

Diameter of bolt D (in.)

Bending yield strength of bolt F_{yb} (ksi)

Thickness of main member t_m (in.)

Specific gravity of main member G_m

Angle of load to grain for main member θ_m (degrees)

Thickness of side member t_s (in.)

Specific gravity of side member G_s

Angle of load to grain for side member θ_s (degrees)

The software template is to do the following:

a. Compute the dowel bearing strengths parallel ($F_{e\parallel}$), perpendicular ($F_{e\perp}$), and at angle θ ($F_{e\theta}$) to the grain for both the main and side members.

b. Evaluate the coefficients for use in the yield limit equations: R_e, R_t, θ, K_θ, k_1, k_2, and k_3. Other intermediate answers may be provided to aid in checking.

c. Solve the six yield limit equations, and select the smallest value as the *reference design value Z*.

13.2 Expand or modify the template from Prob. 13.1 to handle a bolt in a single shear connection with a steel side plate (wood-to-metal connection). Input remains the same, except G_s and θ_s are not required, and the ultimate tensile strength of the metal side plate F_u is an input.

13.3 Expand or modify the template from Prob. 13.1 to handle a single shear connection with an anchor bolt (wood-to-concrete connection). Input remains the same, except G_m and θ_m are not required, and F_e for the concrete main member is an input.

13.4 Expand the template in Prob. 13.1, 13.2, or 13.3 (as assigned) to convert the *reference design value Z* to an *adjusted ASD design value Z'*. Input must be expanded to include the appropriate adjustment factors. Include, as a minimum, adjustments for

a. Load duration factor C_D

b. Wet service factor C_M

c. Group action factor C_g

d. Geometry factor C_Δ

NOTE: C_D and C_M may be input by the user, but C_g and C_Δ should be generated by the template with a minimum of input. Necessary input is to be determined as part of the problem.

13.5 Use a computer spreadsheet or programming software to solve the yield limit equations for a double shear wood-to-wood connection, using the following variables for input:

Bolt diameter D (in.)

Bending yield strength of bolt F_{yb} (ksi)

Thickness of main member t_m (in.)

Specific gravity of main member G_m

Angle of load to grain for main member θ_m (degrees)

Thickness of side member t_s (in.)

Specific gravity of side member G_s

Angle of load to grain for side member θ_s (degrees)

The software template is to do the following:

a. Compute the dowel bearing strengths parallel ($F_{e\parallel}$), perpendicular ($F_{e\perp}$), and at angle θ ($F_{e\theta}$) to the grain for both the main and side members.

b. Evaluate the coefficients for use in the yield limit equations: R_e, R_t, θ, K_θ, and k_3. Other intermediate answers may be provided to aid in checking.

c. Solve the four yield limit equations, and select the smallest value as the *reference design value Z*.

13.6 Expand or modify the template from Prob. 13.5 to handle a bolt in a double shear connection with a steel side plate (wood-to-metal connection). Input remains the same, except G_s and θ_s are not required, and the ultimate tensile strength of the metal side plate F_u is an input.

13.7 Expand the template in Prob. 13.5 or 13.6 (as assigned) to convert the *reference design value Z* to an *adjusted* ASD *design value Z'*. Input is to be expanded to include the appropriate adjustment factors. Include as a minimum:

Load duration factor C_D

Wet service factor C_M

Group action factor C_g

Geometry factor C_Δ

NOTE: C_D and C_M may be input by the user, but C_g and C_Δ should be generated by the template with a minimum of input. Necessary input is to be determined as part of the problem.

13.8 Use a computer spreadsheet or programming software to set up the solution of the yield limit equations for a single shear lag bolt connection. Consider both *wood-to-wood* and *wood-to-steel* connections, using the following variables for input:

Diameter of lag bolt D (in.)

Length of lag bolt L (in.)

Bending yield strength of lag bolt F_{yb} (ksi)

Thickness of main member t_m (in.)

Specific gravity of main member G_m

Angle of load to grain for main member θ_m (degrees)

Thickness of side member t_s (in.)

Specific gravity of side member G_s

Angle of load to grain for side member θ_s (degrees)

The software template is to do the following:

a. Compute the dowel bearing strengths parallel ($F_{e\parallel}$), perpendicular ($F_{e\perp}$), and at angle θ ($F_{e\theta}$) to the grain for both the main and side members.

b. Evaluate the coefficients for use in the yield limit equations: R_e, R_t, θ, K_θ, k_1, k_2, and k_3. Other intermediate answers may be provided to aid in checking.

c. Solve the six yield limit equations for a *wood-to-wood* connection or a *wood-to-steel* connection. Select the smallest value as the *reference design value Z*.

13.9   Expand the template in Prob. 13.8 to convert the *reference design value Z* to an *adjusted ASD design value Z'*. Input must be expanded to include the appropriate adjustment factors. Include as a minimum:

Load duration factor C_D

Wet service factor C_M

Group action factor C_g

Geometry factor C_Δ

End grain factor C_{eg}

NOTE: C_D, C_M and C_{eg} may be input by the user, but C_g and C_Δ should be generated by the template with a minimum of input. Necessary input is to be determined as part of the problem.

13.10   The connection in Fig. 13.A uses a single row of five $\frac{3}{4}$-in.-diameter bolts. Lumber is No. 1 Hem-Fir. The load is the result of wind. $C_M = 1.0$

Figure 13.A

Find: a. Adjusted LRFD design value for the bolt connection.

              b. Adjusted LRFD tension capacity of the wood members.

      c. Indicate the base spacing requirements for the bolts.

      d. Adjusted LRFD row tear-out capacity for the connection.

13.11 A connection similar to the connection in Fig. 13.A has a single row of *six* (instead of five) ³⁄₄-in.-diameter bolts. Lumber is No. 1 DF-L. Load is the result of (D + S). $C_M = 1.0$.

    Find: *a.* Adjusted ASD design value for the bolt connection.

               b. Adjusted ASD tension capacity of the wood members.

               c. Indicate the base spacing requirements for the bolts.

               d. Adjusted ASD row tear-out capacity for the connection.

13.12 Repeat Prob. 13.10 using ASD procedures.

13.13 Repeat Prob. 13.11 using LRFD procedures. The load is due to (1.2D + 1.6S).

13.14 Repeat Prob. 13.10 except that the connection is exposed to the weather.

13.15 Repeat Prob. 13.11 except that the connection is exposed to the weather.

13.16 The connection in Fig. 13.B uses side plates of A36 steel and four 1-in. diameter bolts on each side of the splice. Bolt spacing is $4D$ and end distance is $7D$. Lumber is No. 1 DF-L. The force is seismic.

    Find: Adjusted LRFD design value for the bolts if:

              a. $C_M = 1.0$.

              b. Green lumber is used that later seasons in service.

              c. Connection is exposed to the weather.

              d. Lumber is above the FSP in service.

Figure 13.B

13.17 The beam in Fig. 13.C is suspended from a pair of angles (two L 4 × 3 × $\frac{5}{16}$ of A36 steel). The load is (D + L). Lumber is Select Structural Spruce-Pine-Fir (S) that is initially dry and remains dry in service. Assume that the angles are adequate.

Find: *a.* Determine if the bolts are adequate using ASD procedures.
   b. Check the shear stress in the beam at the connection.
   c. Are the base spacing requirements satisfied?

Figure 13.C

13.18 Repeat Prob. 13.17 except the hanger is not vertical. It forms an angle of 60 degrees to the horizontal, upward to the right.

13.19 Repeat Prob. 13.17 except that the lumber is initially at the FSP and it seasons in service.

13.20 The connection in Fig. 13.D uses two rows of $\frac{7}{8}$-in.-diameter bolts with three fasteners in each row. The load carried by the connection is (1.2D + 1.6L). The wood member is a $6\frac{3}{4}$ × 16.5 DF glulam with $G = 0.50$. It is initially dry (at or near the EMC) and remains dry in service.

Find: *a.* Adjusted LRFD connection design value based on the capacity of the bolts. Show the base dimension spacing requirements on a sketch of the connection.

b. Check the stresses in the main member and side plates for the load in (*a*). Assume A36 steel plate which has a tensile yield strength of $F_y = 36$ ksi and an ultimate tensile strength of $F_u = 58$ ksi.

13.21 The connection in Fig. 13.D uses two rows of $\frac{7}{8}$-in.-diameter bolts with *five* (instead of three) fasteners in each row. Load carried by the connection is (D + S). The wood member is a 6 × 14 Select Structural DF-L that is initially wet, and it later seasons in place.

Find: *a.* Adjusted ASD design value for the connection, based on the strength of the bolts.
   b. Adjusted ASD design value for the connection if separate side plates are used for the two rows of bolts. Show the base dimension spacing requirements for the bolts in the connection.
   c. Check the stresses in the main member and side plates for the load in (*b*). Assume A36 steel plate which has a tensile yield strength of $F_y = 36$ ksi and an ultimate tensile strength of $F_u = 58$ ksi.

Figure 13.D

13.22 The connection in Fig. 13.E uses $\frac{3}{4}$-in.-diameter bolts in single shear. Lumber is DF-L that is initially dry (near the EMC) and remains dry in service. The load is caused by (D + L).

Find: *a.* The adjusted ASD design value for the connection, based on the strength of the bolts.
   b. Check tension and shear stresses at the connection in each member and select bolt spacings and end distances to provide the adjusted ASD design value determined in part *a.* Show the spacing requirements for the bolts on a sketch of the connection.

13.23 The connection in Fig. 13.F uses 1-in.-diameter bolts and serves as a tie for lateral wind forces. Lumber is DF-L that is initially dry (near the EMC) and it remains dry in service.

Figure 13.E

Figure 13.F

Find: a. Adjusted LRFD design value for the connection based on the strength
          of the bolts.
          b. Check tension and shear stresses at the connection in the side mem-
          bers. Check shear stresses at the connection in the main member
          assuming the connection is located at least 5d from the end of the
          member. Select bolt spacing, end distance, and edge distances to pro-
          vide the adjusted LRFD design value determined in part *a*. Show the
          spacing requirements for the bolts on a sketch of the connection.
          c. If initially green lumber is used, what will be the effect on the
          adjusted LRFD design value?

13.24 The connection in Fig. 13.G uses one 1-in.-diameter bolt that carries a lateral wind force. Lumber is Southern Pine that is dry ($C_M = 1.0$). Assume that the members are adequate.

Find: Adjusted ASD and LRFD design values for the connection.

Figure 13.G

13.25 The building in Fig. 13.H uses 2×6 wall framing of No. 1 Hem-Fir. The double top wall plate serves as the horizontal diaphragm chord and drag strut. The top plate is fabricated from 20-ft-long lumber, and splices occur at 10-ft intervals. Design lateral forces due to wind are given. $C_M = 1.0$.

Find: Design and detail the connection splice in the top plate of wall 1, using $\frac{3}{4}$-in.-diameter bolts and ASD procedures.

Figure 13.H

13.26 Repeat Prob. 13.25 using LRFD procedures, except that the lumber is No. 1 DF-L and the bolts are $\frac{7}{8}$ in. diameter.

13.27 *Given:* The shearwall and anchorage detail in Fig. 13.I. The gravity loads and the lateral wind force are shown in the sketch. The chords in the shearwall are 4 × 4 members of No. 2 DF-South. Assume the bracket is $\frac{3}{8}$ in. thick and has adequate capacity. $C_M = 1.0$.

   Find: a. Number of $\frac{3}{4}$-in.-diameter bolts necessary to anchor the chord to the hold-down bracket using ASD procedures. Assume an area of 2.5 in.^{2} for the bracket. Sketch the base dimension spacing requirements for the bolts.

   b. Check the tension and shear stresses of the chord at the connection (bolts are in a single row).

   c. Determine the required size of anchor bolt to tie the bracket to the foundation. Use IBC Table 1911.2.

Figure 13.I

13.28 *Given:* The wood shearwall in Fig. 13.I. Lumber is DF-L, and $C_M = 1.0$.

   Find: a. The required number of $\frac{3}{4}$-in.-diameter anchor bolts necessary to transfer the 4-k lateral wind force from the 2 × bottom plate to the

$f'_c = 2500$ psi concrete foundation using ASD procedures. These bolts are in addition to the anchor bolt shown in the sketch for the uplift anchor bracket. See Sec. 10.10 for design considerations for the anchor bolts for this problem.

    b. For the anchor bolts found in part *a*, determine the total ASD foundation reaction (Figs. 10.17 and 10.18) perpendicular to the wall for short-term wind forces.

13.29 The connection of the ledger to the stud wall in Fig. 13.J uses ½-in.-diameter × 6-in. lag bolts. Patio D = 6 psf and L_r = 10 psf. Lumber is DF-L.

    Find: *a.* Determine if the connection is adequate to carry the design loads if the lumber is initially dry and remains dry in service. Use LRFD procedures.

              b. Is the connection adequate if the lumber is exposed to the weather?

SHEATHING

BLOCKING

2× LEDGER

FRAMING

$\frac{1}{2}"\phi × 6"$ LAG
BOLTS @ 32" o.c.

$\frac{7}{8}"$ STUCCO

2×6 STUDS
@ 16" o.c.

16' - 0"

SECTION - PATIO ROOF

Figure 13.J

13.30 Repeat Prob. 13.29 except that the L_r is 20 psf. Use ASD procedures.

13.31 The building in Fig. 13.K uses the connection shown in Section *A* to transfer the lateral wind force (chord force and strut force) from the wall plate to the header. Lumber is No. 1 DF-L, and $C_M = 1.0$.

    Find: Determine the required number of ⅝-in.-diameter × 6-in. lag bolts for the connection using ASD procedures. Repeat using LRFD procedures.

Figure 13.K

13.32 Repeat Prob. 13.31 except that the lateral force transfer is made with the connection shown in Section $A1$. In addition to the number of lag bolts, determine the required number of $\frac{5}{8}$-in.-diameter bolts (A307) necessary to connect the steel splice plate to the double $2 \times$ top plate.

Connection Details and Hardware

14.1 Introduction

Chapters 12 and 13 covered the basic fasteners that are used in the construction of wood buildings. In addition to these fasteners, many wood connections involve some type of metal connection hardware. This hardware may be a simple tension strap, but considerably more complicated devices are often used.

This chapter introduces some typical connection hardware details. A critique of the connection accompanies a number of the details, and particular attention is given to the effects of cross-grain shrinkage. These principles should be given special attention if it is necessary to design, detail, and fabricate special connection hardware. An example of several connection hardware calculations is given after the general review.

In a number of cases, prefabricated connection hardware may be available for common connections. In any event, the basic principles of connection design should be understood by the designer. The chapter concludes with a review of some typical prefabricated connection hardware.

14.2 Connection Details

The hardware used in a connection must itself be capable of supporting the design loads, and the structural steel design principles of Refs. 14.2 and 14.5 should be used to determine the required plate thickness and weld sizes. Some of these principles are demonstrated in examples, but a comprehensive review of steel design is beyond the scope of this book.

Some suggested details for common wood connections are included in Refs. 14.3 and 14.6, with an emphasis on avoiding cross-grain cracking caused by changes in moisture content. These details are aimed at minimizing restraint in connections that will cause splitting. Where restraint does occur, the distance across the grain between the restraining elements should be kept small so that the total shrinkage between these points is minimized. Where possible, the connections should be designed to accommodate the shrinkage (and swelling) of the wood without initiating built-in stresses.

The majority of the details in this section are taken or adapted from Ref. 14.3 by permission of the American Institute of Timber Construction (AITC). The sketches serve as an introduction to the types of connections that are commonly used in wood construction. See Example 14.1. Both good and poor connection design practices are illustrated. The details were developed specifically for glulam construction, where members and loads are often large. The principles are valid for both glulam and sawn lumber of any size, but the importance of these factors increases with increasing member size.

The connection details in Example 14.1 are not to scale and are to be used as a guide only. Actual designs will require complete drawings with dimensions of the connection and hardware. The designs must take into consideration the spacing requirements for bolts (Examples 13.10 and 13.11 in Sec. 13.5) or the other types of fasteners. Examples of the following connections are included in Example 14.1:

1. Typical beam-to-column connections
2. Beam-to-girder saddle connection
3. Beam face hanger connection
4. Beam face clip connection
5. Cantilever beam hinge connection
6. Beam connection for uplift
7. Beam connection to continuous column
8. Beam with notch in tension side
9. Beam with end notch
10. Inclined beam—lower support detail
11. Inclined beam—upper support detail
12. Suspending multiple loads from beam
13. Suspending isolated loads from beam
14. Truss heel connection
15. Additional truss joint considerations
16. Moisture protection at column base

References 14.3 and 14.6 cover a number of connections in addition to these. However, the details given here illustrate the basic principles involved in proper connection design. A numerical example is provided in the following section for the design of a beam-to-column connection. This is followed by some additional considerations for beam hinge connections.

EXAMPLE 14.1   Connections Details[*]

Beam-to-Column Connections

If compression perpendicular to grain $F_{c\perp}$ at the beam reaction is satisfactory, the beam may be supported directly on a column. A T-bracket can be used to tie the various members together (Fig. 14.1a).

The bottom face of the beam bears on the column. If cross-grain shrinkage occurs in length a, cracking or damage in the region of the T-bracket will result. The problem can be minimized by keeping a small.

Figure 14.1a   Beam-to-column connection with T-bracket.

For larger loads, the T-bracket will be replaced with a U-bracket (Fig. 14.1b). The U-bracket can provide a bearing length l_b that is greater than the width of the column. The spacing requirements for the fasteners must also be considered in determining the length of the bracket.

If a separate splice plate is used near the top of the beam (Fig. 14.1a), the effects of the end rotation and beam separation should be considered. Bolts in tightly fitting holes can restrict the movement of the end of the beam, and splitting may be induced. Slotted holes in metal plates and brackets can reduce this problem. If the beam carries an axial drag or chord force, the T-bracket or U-bracket can be designed to transmit the lateral force.

U-brackets (Fig. 14.1b) may be fabricated from bent plates when the bracket does not cantilever a long distance beyond the width of the column. For longer U-brackets, the vertical plates will probably be welded to a thicker base plate. Calculations for a U-bracket connection are given in Sec. 14.3.

[*]Adapted from Ref. 14.3, courtesy of AITC.

Figure 14.1b   Beam-to-column U-bracket hardware.

In U-brackets and similar connection hardware, it is important that the wood member be fully seated on the bearing surface *before* the bolts are installed. This may require that the bottom edges of the beam be shaped to conform to the inside of the bracket. Without such precautions the squared corners of the beam could initially rest on the inside radius of a bent plate (or on a weld), and the member may not fully seat until it is loaded heavily in service. In this case the bolt capacity perpendicular to the grain may become over-stressed as the member moves downward, and a split in the beam may develop.

Beam Saddle Connection

Beam saddle connections are the preferred means of transferring beam reactions to girders (Fig. 14.2a). The beam reaction is transferred by bearing perpendicular to the grain on the bottom of the hanger. The load on the hanger is then transferred by bearing perpendicular to the grain through the top of the saddle to the girder. This type of connection is recommended for larger loads.

Figure 14.2a   Beam hanger with saddle feature providing bearing on the *bottom* of beam and on the *top* of girder.

The tops of beams are shown higher than the girder to allow for shrinkage in beams without affecting roof sheathing or flooring. The idea is to have the tops of beams and girders at the same level after shrinkage occurs. For shrinkage calculations, see Example 4.3 (Sec. 4.7).

Note that in Fig. 14.2a the bolts are located *near* the bearing points to minimize cross-grain effects. Figure 14.2b illustrates the possible effects of cross-grain shrinkage if the bolts are located *improperly*. The bolt in the beam is located some distance away from the initial point of bearing (section 1). After the beam is in service for some time, it loses moisture and shrinks across the grain. The bolt is held rigidly in position by the hanger bracket. Under a relatively small applied load, the bolt capacity may be large enough to support the beam reaction, and the bottom face of the member may shrink *up* from its intended bearing point (section 2).

However, when the member finally becomes heavily loaded, the bearing capacity of the bolt perpendicular to the grain may be exceeded, and a split at the end of the member will develop as the beam seats again on the hanger (section 3). The end split not only reduces the capacity of the member, but it also renders the bolt ineffective in resisting lateral (wind or seismic) forces that may create an axial force in the beam. Improperly located bolts as shown in Fig. 14.2b should be *avoided*.

Although the bolts should be located near the bearing face of the member, care should be taken to avoid boring holes in the high-quality outer lamination in bending glulams. This is especially important when bolts occur in the tension zone of the member.

Figure 14.2b *Avoid* improper bolt location in beam hanger. Three stages are identified that lead to the formation of a split. Proper detailing will avoid splits by placing bolts closer to the beam bearing surface.

Beam Face Hanger Connection

When the reaction of the beam or purlin is relatively small, the hanger can be bolted to the face of the girder (Fig. 14.3). The bolts in the main supporting beam or girder should be placed in the upper half of the member, but not too close to the top of the beam where extreme fiber bending stresses are maximized. This is especially important in the tension zone that occurs on the top side of cantilever girders.

Figure 14.3 Beam hanger with saddle under supported beam. Beam reaction is transferred to girder through fasteners instead of bearing as in Fig. 14.2a.

Beam Face Clip Connection

Face connections without the saddle feature shown in Fig. 14.3 may be used for light loads (Fig. 14.4a). However, construction is more difficult than with the saddle because the beams must be held in place while the bolts are installed. The connection at the end of the supported beam must be checked for shear in accordance with the procedures outlined in Sec. 13.9.

Figure 14.4a Face clip connection is limited to light loads. Beam reaction is transferred through fasteners in both members.

Face clips should be limited to connections with light loads. Larger loads will require a long row of bolts perpendicular to the grain through steel side plates. Glulam timbers, although relatively dry at the time of manufacture, may shrink as the EMC is reached in service. The problems associated with cross-grain tension should be avoided by not using the type of connection in Fig. 14.4b.

Figure 14.4b *Avoid* long rows of bolts perpendicular to grain through steel side plates. Changes in moisture content will result in cross-grain volume changes, and cross-grain tension and splitting may occur. The saddle connection in Fig. 14.2a is the preferred method of transferring heavy loads.

Cantilever Beam Hinge Connection

PROBLEM

For vertical loads, the *suspended* beam in Fig. 14.5a bears on the *bottom* of the hinge connector, and the hinge connector in turn bears on the *top* of the *supporting* (cantilever) member. Cantilever systems may be subjected to lateral forces as well as gravity bending loads. Some designers have used a tension connector with bolts lined up in both members near the bottom of the beam. The problem with this arrangement is that as the wood shrinks, the supporting member will permit the saddle to move downward. The bolts in the cantilever, however, will restrain this movement, and a split may occur. The revised detail (Fig. 14.5b) avoids the problem.

SUGGESTED REVISION

The connection in Fig. 14.5b does not interfere with possible shrinkage in the members and cracking is avoided. It also allows a slight rotation in the joint. A positive tension tie for lateral forces can be provided with a separate tension strap at the middepth of the joint. This tension strap may also be an integral part of the hanger if the holes in the tension tie are slotted vertically when used with deep members. Also see Sec. 14.4.

Beam Connection for Uplift

In this example, gravity loads are transferred by bearing perpendicular to grain. Adequate fastenings must also be provided to carry horizontal and uplift forces. See Fig. 14.6. These loads are usually of a transient nature and are of short duration. The fastener is usually placed toward the lower edge of the member. The distance d_e must be no less than the required perpendicular-to-grain edge distance for the type of fastener used. It must also be large enough so that the shear stresses caused by uplift (calculated in accordance

TYP CANTILEVER BEAM SYSTEM

CANTILEVER CONNECTION

Figure 14.5a *Avoid* cantilever beam hinge connections with improperly located tension tie which causes cracking as cross-grain shrinkage occurs.

REVISED HINGE CONNECTION

Figure 14.5b *Revised* cantilever beam hinge connection with tension ties properly located near top and bottom bearing surfaces.

Figure 14.6 Gravity loads are transferred by bearing. Connection hardware may be necessary for uplift and lateral forces. Proper detailing considers fastener spacing requirements.

with the methods of Sec. 13.9) are not excessive. The end distance e must also be adequate for the type of fastener used.

Beam Connection to Continuous Column

PROBLEM

Connections of the type shown in Fig. 14.7a should not be used on large beams or girders because tension perpendicular to the grain and horizontal shear at the lower fastener tend to cause splitting of the member.

Figure 14.7a Avoid transferring large reactions at the end of a beam with fasteners.

SUGGESTED REVISION

When the design is such that the bending member does not rest on the top of a column, wall, or pilaster, but frames into another member, several methods can be used to support the ends. The preferred method is to transfer the end reaction by bearing perpendicular to grain. See Fig. 14.7*b*.

1. Vertical beam reaction is transferred by bearing perpendicular to grain on the beam seat angle.
2. Positive connection is made between the beam and the column with fasteners in the tie clip.
3. Clip angle at top provides lateral stability to beam, but it is not attached to the top of the beam with fasteners. In other words, the clip angle prevents rotation perpendicular to the plane of the beam, and it thereby braces the member against lateral torsional buckling. However, because there is no fastener in the beam at the top, rotation in the plane of the beam is accommodated as the member is loaded in bending. This provides a stable member that will not split the end.

Figure 14.7b *Revised* connection transfers beam reaction by bearing on the bottom of the member.

Beam with Notch in Tension Side

PROBLEM

The detail in Fig. 14.8 is critical in cantilever framing (negative moment at support). The top *tension* fibers have been cut to provide space for recessed connection hardware or for the passage of conduit or other elements over the top of the beam. This is particularly serious in glulam construction because the tension laminations are critical in the performance of the structure.

SUGGESTED REVISION

Notches in the tension zone are serious stress raisers and should be avoided. Revise the detail in Fig. 14.8 so that the notch in the beam is eliminated. Also see Sec. 6.2.

NOTCH IN
TENSION ZONE

BEAM

COLUMN

Figure 14.8 *Avoid* notching the tension side of a wood member.

Beam with End Notch

PROBLEM

An abrupt notch in the end of a wood member creates two problems (Fig. 14.9a). One is that the effective shear strength of the member is reduced because of the smaller depth d_n and stress concentration. In addition, the exposure of end grain in the notch will permit a more rapid migration of moisture in the lower portion of the member, and a split may develop.

For horizontal shear calculations at the notched end of a beam, see Sec. 6.5.

NOTCHED GIRDER
SHOWING POTENTIAL
SPLIT

CONNECTION
BRACKET

CONCRETE OR
MASONRY WALL
OR PILASTER

Figure 14.9a *Avoid* square cut notches.

SUGGESTED REVISION

Where the height of the top of a beam is limited, the beam seat should be lowered to a pilaster or specially designed set in the wall (Fig. 14.9b). A lateral support connection at the top of the girder is required for stability. There is no fastener between the wood member and lateral support hardware. Therefore, the lateral support connection prevents rotation perpendicular to the plane of the girder, but it allows rotation of the end of the member for in-plane loading.

Figure 14.9b   *Revised* beam support without notch.

Inclined Beam-Lower Support Details

PROBLEM

The condition in Fig. 14.10*a* is similar to the problem in Fig. 14.9*a*, but it may not be as evident. The shear strength of the end of the member is reduced, and the exposed end grain may result in splitting.

Figure 14.10a   *Avoid* end cuts that expose the end grain on the *tension* side of a beam.

SUGGESTED REVISION

Where the end of the beam must be flush with the outer wall, the beam seat should be lowered so that the tapered cut is loaded in bearing (Fig. 14.10*b*).

Figure 14.10b *Revised* taper cut is detailed so that the end grain is stressed in compression.

Inclined Beam-Upper Support Details

PROBLEM

The bird's mouth cut in Fig. 14.11a is commonly used in rafters to obtain a horizontal bearing surface. In major, heavily loaded beams, large cuts of this nature substantially reduce the effective depth of the member at the notch and should be avoided. Stress concentrations also cause high shear stress at the notch and may cause splitting.

Figure 14.11a *Avoid* deep square cut notches in heavily loaded members.

SUGGESTED REVISION

A welded bracket with a sloping seat can be used to avoid notching the member (Fig. 14.11b). The bracket must be properly designed to support the beam reaction. Eccentricity in the connection should be avoided.

Figure 14.11b *Revised* beam support without end notch.

Suspending Multiple Loads from Beam

PROBLEM

Suspending multiple loads below the neutral axis (NA) of a wood member is not recommended (Fig. 14.12a). Tension stresses perpendicular to the grain tend to concentrate at the fastenings. Wood is weak in cross-grain tension, and tension perpendicular to grain can interact with horizontal shear to cause splitting. Hanger loads may be mechanical or electric equipment suspended below the beam, or they may be reactions from joists, purlins, or other members framing into the side of the beam.

Figure 14.12a *Avoid* suspending multiple loads below the neutral axis.

SUGGESTED REVISION 1

It is better design practice to locate the fasteners above the neutral axis, as shown in Fig. 14.12b. Where possible, the loads should be transferred by bearing (Fig. 14.12c).

Figure 14.12b   *Revised* detail with loads suspended above the neutral axis.

SUGGESTED REVISION 2

The preferred method is to suspend loads from the top of a beam or girder, as in Fig. 14.12*c*. This approach transfers the loads by bearing on the top of the beam. It avoids tension perpendicular to grain and stress concentrations from large loads being transferred by fasteners. Transferring loads by compression perpendicular to grain is generally the preferred method of load transfer in all cases, but it becomes increasingly important as the magnitude of the load increases.

Note that this method was recommended previously for the transfer of beam reactions to a girder in Fig. 14.2*a*. The saddle transfers the load by bearing rather than by fasteners.

Figure 14.12c   *Revised* detail with loads transferred by bearing (compression perpendicular to grain).

Suspending Isolated Loads from Beam

The connections shown in Fig. 14.13 are for isolated loads. Recommendations for suspending isolated loads are similar to those for multiple loads.

Figure 14.13 Isolated loads suspended from a beam.

An isolated light load may be suspended below the NA of a beam. However, the load should be limited to one that can be transferred with small fasteners (e.g., small conduit loads). The distance a must equal or exceed the required edge distance, and Ref. 14.3 recommends a minimum of 6 in. Shear stresses in the member must be checked according to Sec. 13.9. It is good practice to locate fasteners above the NA. Even better practice is to detail the connection as shown in Fig. 14.12c.

Truss Heel Connection

PROBLEM

The truss heel connection in Fig. 14.14a has structural steel gusset plates with bolts or shear plates for fasteners. There are two problems with this detail. First, the connection involves an eccentricity because the forces at the joint do not intersect at the center of the group of fasteners. Second, the steel gusset plates hold the fasteners rigidly in position, and this can cause the wood truss members to split as the joint attempts to rotate when the truss is loaded.

Figure 14.14a *Avoid* eccentric forces and rigid truss connections.

SUGGESTED REVISION

A truss connection is shown in Fig. 14.14b in which the force in the top chord is taken by bearing. This connection has the advantage of concentric forces. The single fastener in the top chord member and the clearance provided between the chord members allow free joint rotation as the truss is loaded. Cross-grain shrinkage in the bottom chord of the truss can be eliminated by the proper placement of bolts in vertically slotted holes.

See the TCM (Ref. 14.4) for additional discussion of timber truss design.

NOTE: For the design of light-frame wood trusses using light-gage metal plates, see Ref. 14.8.

Figure 14.14b *Revised* connection detail with concentric forces. The detail also accommodates rotation between truss members.

Additional Truss Joint Considerations

PROBLEM

When the centerlines of members (and correspondingly the axial forces in the truss) do not intersect at a common point, considerable shear and moment may result in the bottom chord (Fig. 14.15a). When these stresses are combined with a high tension stress in the member, failure may occur.

Figure 14.15a *Avoid* truss connections with eccentric forces.

PROBLEM

The welded steel gusset plate in Fig. 14.15b has a rigidity problem similar to the solid gusset plate in Fig. 14.14a. The welded plates are held in position, and splitting of the wood members may occur as the truss attempts to deform under load.

Figure 14.15b *Avoid* rigid truss connections which do not allow rotation between truss members. Prying and splitting may occur.

SUGGESTED REVISION

For trusses with single-piece members, the chords and web members should be in the same plane, and separate straps or gusset plates should be used for the connection. See Fig. 14.15c. The separate gusset plates should accommodate truss deflection and joint rotation without splitting of the wood members.

Steel gusset plates must be properly designed for the applied tension or compression forces. Shims should be provided between the steel plates and the wood members where necessary to maintain gusset plate alignment. For example, shims should be used between the vertical gusset plate and the vertical wood member to account for the thickness of the diagonal gusset plate. Without these shims the vertical gusset plate may become damaged as the bolts are tightened.

See the TCM (Ref. 14.4) for additional discussion of timber truss design.

Figure 14.15c *Revised* connection detail with separate gusset plates. Fastener arrangement allows joint rotation without prying or splitting.

Moisture Protection at Column Base

PROBLEM

Some designers try to conceal the base of a column or an arch by placing concrete around the connection (Fig. 14.16a). Moisture may migrate into the lower portion of the wood and cause decay.

Figure 14.16a *Avoid* joint details that entrap moisture and promote decay.

SUGGESTED REVISION

Detail the column base at the top of the floor level (Fig. 14.16b). Note that the base of the column is separated from concrete by a bottom bearing plate. Where subjected to splashing water, untreated columns should be supported by piers projecting up at least 1 in. (IBC 2304.11.2.7; Ref. 14.7) above the finished floor.

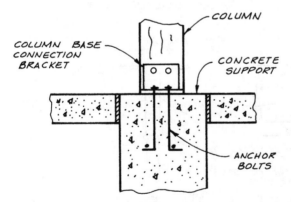

Figure 14.16b *Revised* connection detail does not entrap moisture.

14.3 Design Problem: Beam-to-Column Connection

Example 14.2 illustrates a typical beam-to-column connection. The beam is a $5\frac{1}{8} \times 27$ glulam, and the column is a 4-in. steel tube. Steel tubes and steel pipe columns are commonly used to support wood beams.

   The design of this connection starts with a consideration of the required bearing length for the wood beam. Calculations of this nature were introduced in Chap. 6. The required overall length of the U-bracket for bearing stresses is 8 in. In later calculations this length is shown to be less than the length required to satisfy the bolt spacing criteria.

   The example concludes with a check of the bending stresses in the U-bracket. Because of the small cantilever length, a bent plate is used for the bracket. For longer cantilever lengths, a welded U-bracket will be required. For special connection hardware, complete details of the bracket should be included on the structural plans. However, prefabricated U-brackets of the type covered in this example are generally available from commercial sources.

EXAMPLE 14.2   U-Bracket Column Cap

Design a U-bracket to connect two $5\frac{1}{8} \times 27$ glulam beams to the 4-in. steel tube column. Gravity roof loads are given in Fig. 14.17a. There is no uplift force on the beam-to-column connection.

Figure 14.17a   Beam-to-column connection.

   The lateral seismic force in the glulam beams is E = 8.5 k. Glulam is 24F-V4 Douglas Fir (from the 24F-1.8E Stress Class) which is initially dry and remains dry in service. Temperature range is normal. Use allowable stress design (ASD) procedures and then repeat the problem using load and resistance factor design (LRFD) procedures.

Vertical Load (ASD)

Reaction from one beam:

$$R = \frac{wL}{2} = \frac{600(40)}{2} = 12{,}000 \text{ lb}$$

Adjusted ASD bearing design value:

$$F'_{c\perp} = F_{c\perp}(C_M C_t C_b) = 650(1.0)(1.0)(1.0)$$

$$= 650 \text{ psi}$$

$$\text{Req'd bearing length} = \frac{R}{b \times F'_{c\perp}} = \frac{12{,}000}{5.125 \times 650}$$

$$= 3.6 \text{ in./beam}$$

$$\text{Minimum length of bracket} = 2(3.6) = 7.2 \text{ in.} \qquad say \text{ 8 in. minimum}$$

Lateral Force (ASD)

The lateral force is $0.7E = 5.95$ k for ASD. Assume that the lateral force will be transferred from one glulam to the other through the U-bracket made from $\frac{1}{4}$-in.-thick A36 steel. Use $\frac{3}{4}$-in. diameter bolts in double shear with steel side plates.

$$D = 0.75$$

$$F_{yb} = 45 \text{ ksi} \qquad \text{(assumed for A307 bolt)}$$

$$l_m = 5.125 \text{ in.}$$

$$\theta_m = 0 \text{ degrees}$$

$$F_{em} = 5600 \text{ psi} \qquad (F_e \text{ for DF-L})$$

$$l_s = 0.25 \text{ in.}$$

$$F_{es} = 87{,}000 \text{ psi} = 1.5 \, F_u \qquad \text{for A36 steel}$$

The reference design value for the bolt may be computed as the smallest value from the four yield limit equations for a double shear wood-to-metal connection (Examples 13.5 and 13.6 in Sec. 13.4). However, the given conditions are covered directly by an NDS table, and Z is obtained from NDS Table 11I as

$$Z = 3340 \text{ lb}$$

The reader may wish to verify this value by applying the yield limit equations and observing that Mode III$_s$ governs.

Adjusted ASD design value:
   Load duration factor C_D from the NDS (Ref. 14.1) is 1.6 for seismic forces. Both C_m and C_t are 1.0. A row is defined as a line of fasteners *parallel* to the direction of

loading. Assume $C_g = 1.0$. The geometry factor C_Δ will also be unity because the base dimensions for bolt spacing will be provided as a minimum. Therefore,

$$Z' = Z(C_D C_M C_t C_g C_\Delta)$$

$$= 3340(1.6)(1.0)(1.0)(1.0)(1.0)$$

$$= 5344 \text{ lb/bolt}$$

$$\text{Req'd number of bolts} = \frac{5950}{5344} \approx 1.1$$

> *Use* two $\frac{3}{4}$ -in. diameter bolts each side of splice.

Although there are two bolts on each side of the splice, two rows will be provided with one bolt in each row.

$$\therefore\ C_g = 1.0 \qquad OK$$

Connection Details (ASD and LRFD)

A number of possible connection details can be used. One possible design is included in Fig. 14.17*b*.

Figure 14.17b   Connection details.

Lateral force is carried by the U-bracket in tension, and vertical load is carried by bearing. Spacing requirements are governed by the lateral force (parallel to grain loading criteria—Example 13.10 in Sec. 13.5).

HORIZONTAL SPACING

$$\text{End distance} = 7D = 7(\tfrac{3}{4}) = 5\tfrac{1}{4} \text{ in.}$$

Steel plate edge distance = $1\frac{1}{2}$ in. (typical for bolts in steel)

$$\text{Total bracket length} = (5\tfrac{1}{4} + 1\tfrac{1}{2})\,2 = 13\tfrac{1}{2} \text{ in.}$$

$$13\tfrac{1}{2} \text{ in.} > 8 \text{ in.}$$

Base dimension spacing provisions govern over bearing $(f_{c\perp})$.

VERTICAL SPACING

$$\text{Steel plate edge distance} = 1\tfrac{1}{2} \text{ in.}$$

$$\text{Row spacing} = 1\tfrac{1}{2} \text{ D} = 1\tfrac{1}{2}\,(\tfrac{3}{4}) = 1\tfrac{1}{8} \text{ in.} \qquad say \ 1\tfrac{1}{2} \text{ in.}$$

$$\text{Bolt slenderness} = \frac{l}{D} = \frac{5.125}{0.75} = 6.83 > 6.0$$

$$\therefore \text{ Edge distance*} = 1\tfrac{1}{2} \text{ } D = 1\tfrac{1}{8} \text{ in.}$$

$$or \quad \tfrac{1}{2} \text{ row spacing} = \tfrac{3}{4} \text{ in.} \qquad \text{(the larger governs)}$$

$$Say \quad 1\tfrac{1}{2} \text{ in.}$$

$$\text{Minimum bracket height} = 3(1\tfrac{1}{2}) = 4\tfrac{1}{2} \text{ in.}$$

PROPORTIONS OF U-BRACKET

As a rule of thumb, a bracket height of one-third the length is reasonable:

$$\tfrac{1}{3} \times 13.5 = 4.5 \text{ in.} \qquad \text{(agrees with vertical bolt spacing requirements)}$$

U-Bracket Strength (ASD)

The plan and elevation of the connection detail are shown in Fig. 14.17c. The bracket essentially cantilevers from the face of the column. [If a round column (e.g., a pipe column) is used, the cantilever length is figured from the face of an equivalent square column of equal area.]

The width of the U-bracket is determined by adding a clearance to the width of the beam to allow installation of the member. Reference 14.3 recommends a maximum clearance of $\tfrac{1}{4}$ in. In this example, a clearance of $\tfrac{1}{8}$ in. is used (inside bracket width $= 5\tfrac{1}{8} + \tfrac{1}{8} = 5\tfrac{1}{4}$ in.). See uniform bearing considerations in Fig. 14.1b.

At this point it can be determined whether a bent plate bracket or welded plate bracket should be used (Fig. 14.17d). If the cantilever length of the bracket is small in comparison to the distance between the vertical legs, a bracket of uniform thickness throughout can be used. This can be obtained by folding a plate into the desired shape. The bending stresses in the bracket caused by the cantilever action must be evaluated.

On the other hand, if the cantilever length is long in comparison to the distance between the vertical legs, the bottom plate may have to be thicker than the vertical legs. In this case, the bottom plate of the bracket should first be designed for the bending stresses developed as it spans horizontally between the vertical plates. After the thickness of the bottom plate is determined, the composite section of the U-bracket can be checked for cantilever stresses (as in the analysis of the stresses in the bent plate bracket). The weld between the vertical and horizontal plates must also be

---

*Uplift force was given as zero. If uplift occurs, the edge distance is $4D$ as governed by perpendicular-to-grain criteria (Example 13.11 in Sec. 13.5) or possibly greater if governed by horizontal shear criteria (Sec. 13.9).

Figure 14.17c   U-bracket plan and elevation views.

Figure 14.17d   Alternative ways of forming U-bracket. For lighter loads a folded (bent) plate may be used. Larger loads may require a welded bracket with a thicker bottom plate capable of spanning between the vertical legs.

Figure 14.17e   Dimensions for computing section properties of U-bracket.

designed. The minimum weld size based on the plate thickness (Ref. 14.2) often governs the weld size.

In the problem at hand, a bent plate bracket will be used because the cantilever length of 4.75 in. (Fig. 14.17c) is less than the distance between vertical plates ($5\frac{1}{4}$ in.). Check cantilever bending stresses using $\frac{1}{4}$-in. plate (Fig. 14.17e).

Elastic section properties are conveniently calculated in a table from the formulas

$$I_x = \Sigma I_g + \Sigma A d^2$$

Element	A	y	Ay	$I_g = \dfrac{bh^3}{12}$	d	Ad^2
1	$4.75 \times \frac{1}{4} = 1.19$	2.38	2.82	2.23	0.80	0.76
2	1.19	2.38	2.82	2.23	0.80	0.76
3	$5.25 \times \frac{1}{4} = 1.31$	0.125	0.16	0.01	1.45	2.76
	3.69		5.80	4.47		4.28

$$\bar{y} = \frac{Ay}{A} = \frac{5.80}{3.69} = 1.57 \text{ in.}$$

$$I_x = \Sigma I_g + \Sigma A d^2 = 4.47 + 4.28 = 8.75 \text{ in.}^4$$

$$S_x = \frac{I_x}{(4.75 - \bar{y})} = \frac{8.75}{(4.75 - 1.57)} = 2.76 \text{ in.}^3$$

Since the bracket is stressed in bending about its minor (weak) principal axis, design is based on the plastic moment capacity M_P^* as long as $Z_x \leq 1.6 S_x$ (Ref. 14.2). The plastic neutral axis (PNA) divides the cross section into two equal areas, and is located at a distance y_p above the base of the bracket. Since the bracket is loaded as a cantilever, the cross-sectional area above the PNA is stressed in tension and the area below the PNA is stressed in compression. For a plastic stress distribution in the cross section, the resultant compressive force acts at the centroid of the compression area and the resultant tensile force acts at the centroid of the tension area. The centroids are separated by a distance d_p.

$$y_p(0.25)(2) + (5.25)(0.25) = (4.75 - y_p)\,(0.25)(2)$$

$$y_p = 1.06 \text{ in. above the base of the bracket}$$

$$d_p = \frac{(4.75 - 1.06)}{2} + \frac{(0.125)(5.25)(0.25) + (0.53)(1.06)(0.25)}{(5.25)(0.25) + (1.06)(0.25)} = 2.09 \text{ in.}$$

The plastic section modulus is

$$Z_x = \left(\frac{A}{2}\right) d_p = \left(\frac{3.69}{2}\right) 2.09 = 3.85 \text{ in}^3$$

$$1.6\, S_x = 1.6(2.76) = 4.41 \text{ in}^3 > 3.85 \text{ in}^3 \qquad OK \text{ to use plastic moment capacity}$$

---

* The width-to-thickness ratios of plate elements in the compression zone are less than the limits given in Sec. B4 of Ref. 14.2.

Uniform load on bracket:

$$p = \frac{R}{A} = \frac{12,000 \times 2}{5.125 \times 13.5} = 347 \text{ psi}$$

$$M = \frac{wL^2}{2} = \frac{1780(4.75)^2}{2} = 20 \text{ in.-k} \quad (\text{See Fig. 14.17}f.)$$

LOAD ON CANTILEVER
W = 347 × 5.125 = 1780 LB/IN.

4.75"

Figure 14.17f

For A36 steel plate,

$$\frac{M_P}{\Omega_b} = \frac{F_y Z_x}{1.67} = \frac{(36)(3.85)}{1.67} = 82.9 \text{ in.-k} > 20 \text{ in.-k} \quad OK$$

Use ¼-in.-thick plate U-bracket.

The U-bracket must be welded to the top of the column. Without design loads, the weld size is governed by plate thickness requirements (Ref. 14.2).

Now repeat the design problem using LRFD procedures.

Vertical Load (LRFD)

Reaction from one beam:

$$w_u = 1.2w_D + 1.6w_L = 1.2(300) + 1.6(300) = 840 \text{ lb/ft}$$

$$R_u = \frac{w_u L}{2} = \frac{840(40)}{2} = 16,800 \text{ lb}$$

Nominal LRFD bearing design value:

$$F_{c\perp n} = F_{c\perp} K_F = 650(2.08) = 1352 \text{ psi}$$

Adjusted LRFD bearing design value:

$$F'_{c\perp n} = F_{c\perp n} (\phi_c)(\lambda)(C_M)(C_t)(C_b) = 1352(0.9)(0.8)(1.0)(1.0)(1.0) = 975 \text{ psi}$$

$$\text{Req'd bearing length} = \frac{R_u}{b \times F'_{c \perp n}} = \frac{16,800}{5.125 \times 975} = 3.4 \text{ in./beam}$$

Minimum length of bracket $= 2(3.4) = 6.8$ in.   *say* 7 in. minimum

Note, however, that a bracket length of 13.5 in. was selected for either ASD or LRFD based on connection details.

Lateral Force (LRFD)

The lateral force to be transferred from one glulam to the other is E = 8.5 k for LRFD. The reference design value for $\frac{3}{4}$-in. diameter bolts in double shear with steel side plates is obtained from NDS Table 11I.

$$Z = 3340 \text{ lb}$$

The nominal LRFD design value is:

$$Z_n = Z \, K_F = 3340(3.32) = 11,088 \text{ lb/bolt}$$

The adjusted LRFD design value is:

$$Z'_n = Z_n \, (\phi_z)(\lambda)(C_M)(C_t)(C_g)(C_\Delta)$$

$$= 11,088(0.65)(1.0)(1.0)(1.0)(1.0)(1.0)$$

$$= 7210 \text{ lb/bolt}$$

$$\text{Req'd number of bolts} = \frac{8500}{7210} \approx 1.2$$

Use   two $\frac{3}{4}$ in. diameter bolts each side of splice (*same as for ASD*).

U-Bracket Strength (LRFD)

The dimensions and section properties calculated for ASD are also applicable for LRFD. Uniform load on bracket:

$$p_u = \frac{R_u}{A} = \frac{(16,800)}{(5.125)(13.5)} = 486 \text{ psi}$$

$$w_u = 486(5.125) = 2490 \text{ lb/in}$$

$$M_u = \frac{w_u L^2}{2} = \frac{2490(4.75)^2}{2} = 28 \text{ in.-k}$$

For A36 steel plate,

$$\phi_b M_P = 0.9 F_y Z_x = 0.9(36)(3.85) = 125 \text{ in.-k} > 28 \text{ in.-k}   OK$$

> *Use* $\frac{1}{4}$-in.-thick plate U-bracket (*same as for ASD*).

14.4   Cantilever Beam Hinge Connection

The type of connection used for an internal hinge in a wood cantilever beam system was introduced in Fig. 14.5. There the detailing of the connection was considered to minimize the effects of shrinkage. In the design of this type of

Figure 14.18 Saddle hinge with tension ties for gravity loads.

connection, the equilibrium of the hinge must also be considered. See Fig. 14.18. The saddle hinge with tension ties located near the bearing surfaces is recommended for gravity loads. The tension ties are required to balance the eccentric moment generated by the vertical reactions.

Once the dimensions of the hinge connection have been established, the force on the tension ties can be calculated. This force is used to determine the number and size of bolts in the tension ties, as well as the size and length of the weld between the tension tie and the main portion of the hinge connector.

Other considerations include spacing requirements for the bolts in the connection. The top and bottom bearing plates for the hinge must be capable of spanning horizontally between the vertical plates, and the welds between the vertical and horizontal plates must be designed.

In many cases there will be a tension force at the hinge due to an axial force in the beam. This is in addition to the tension load caused by gravity loads. This typically occurs in buildings with concrete and masonry walls under seismic

SUSPENDED BEAM

HINGE CONNECTOR

WELD

SUPPORTING MEMBER

SEISMIC TIE OR OTHER *FORCE* FROM HORIZONTAL DIAPHRAGM

WELDED SEISMIC TIE STRAPS
BOLTS CAREFULLY LOCATED IN SLOTS
AWAY FROM BEARING SEAT TO
ALLOW FOR SHRINKAGE

HINGE WITH SEISMIC TIE

Figure 14.19   Saddle hinge with seismic tie.

forces. Here the beams are designed as continuous cross ties between horizontal diaphragm chords (Chap. 15). A tension force can also result from diaphragm chord or drag strut action.

The addition of a lateral force to the hinge connection problem can be handled in several ways. One solution is to provide a *welded* splice across the connection. See Fig. 14.19.

In the example shown, the seismic ties are located at the midheight of the hinge connector. Vertically slotted holes are required in the welded seismic ties to allow for cross-grain shrinkage. Note that the bolts must be located properly in the vertical slots to accommodate the movement between the bolts and the points of bearing in the connection. In addition, the nuts on these bolts should be hand-tightened only, to permit movement.

In a different approach, *separate* tension ties for the seismic force are provided across the hinge in a typical double shear bolt connection. This connection is similar to Fig. 14.19, except that the tie plates are continuous across the hinge and are not welded to the hinge connector. In this case, the use of vertically slotted holes is not required because the separate splice plates will not restrain movement across the grain. Shims may be used to isolate the seismic tension tie from the hinge connector.

Other solutions to the seismic force problem are possible, but the two described are representative.

14.5   Prefabricated Connection Hardware

A number of manufacturers produce prefabricated connection hardware for use in wood construction. Catalogs are available from manufacturers or suppliers which show the configurations and dimensions of the hardware. In addition, the catalogs generally indicate allowable loads (adjusted ASD capacities) for connections using the hardware. Hardware manufacturers can also provide equivalent adjusted LRFD connection capacities if requested by the designer. In all cases, the designer should verify that the hardware and the allowable loads are recognized by the local building code authority.

In many cases, prefabricated hardware will be the most economical solution to a connection problem. This is especially true if the hardware is a "shelf item." For example, light-gage steel joist or purlin hangers (similar in configuration to the beam face hanger in Fig. 14.3) are usually available as a stock item. Allowable loads for some hardware have been established through testing, and design values may be higher than capacities determined solely by stress calculations. Obviously, this type of hardware represents an efficient use of materials.

For connections involving larger loads, the appropriate hardware may be available from a manufacturer. However, the designer should have the ability to apply the fastener design principles of the previous chapters *and* basic structural steel design principles (Refs. 14.2 and 14.5) to develop appropriate connection details for a particular job. In some cases it may be more economical to fabricate special hardware, *or* there may not be suitable prefabricated hardware for a certain application.

Many manufacturers offer a number of variations of the hardware illustrated in Example 14.1. The fittings may range from light-gage steel to $\frac{1}{4}$-in. and thicker steel plate. Adjusted ASD and LRFD capacities vary correspondingly. Several other types of connection hardware are generally available from manufacturers. See Example 14.3. These include

1. Framing anchors
2. Diaphragm-to-wall anchors
3. Tie straps

Some hardware that has been illustrated in previous chapters is also available commercially. For example, shearwall tie-down brackets (Fig. 10.12) are produced by a number of manufacturers. Metal knee braces (Fig. 6.27b) and metal bridging (mentioned in Example 6.7) are also available.

With some knowledge of structural steel design and the background provided in Chaps. 11 to 14, the designer should be prepared to develop appropriate connections for the transfer of both vertical loads and lateral forces. Chapters 15 and 16 conclude the discussion of the lateral force design problem with final anchorage considerations and an introduction to some advanced problems in diaphragm design, structural irregularity, and overturning.

EXAMPLE 14.3   Connection Hardware

Framing Anchors

Figure 14.20*a*   Typical framing anchor

Framing anchors (Fig. 14.20*a*) are light-gage steel connection brackets that are usually installed with nails. They have a variety of uses, especially where conventional nail connections alone are inadequate. Framing anchors generally can be bent and nailed in a number of different ways (Fig. 14.20*b*). Adjusted ASD and LRFD capacities depend on the final configuration of the anchor and the direction of loading (different adjusted capacities apply to loadings *A* through *E*).

Figure **14.20***b*   Framing anchor applications.

Diaphragm-to-Wall Anchors

Figure 14.21 Wall anchors are used to tie a concrete or masonry wall to a horizontal diaphragm.

Diaphragm-to-wall anchors (Fig. 14.21) are known commercially by various names (purlin anchors, joist anchors, strap anchors, and so on). These anchors are used to tie concrete or masonry walls to horizontal diaphragms (Chap. 15). Their purpose is to prevent the ledger from being stressed in cross-grain bending (Example 6.1).

Note that the connection in Fig. 14.21 is only part of the total wall-to-diaphragm connection. It carries the lateral force that acts perpendicular to the wall. The anchor-bolt-to-ledger connection in Fig. 6.1*b* is used for vertical loads and lateral forces parallel to the wall. See Example 15.6 (Sec. 15.4) for additional information.

Tie Straps

Figure 14.22 Tie straps.

Tie straps (Fig. 14.22) are typically used to provide a connection across an intervening member. Ties of this type may be required if the beams are part of a horizontal diaphragm (Chap. 15).

14.6 References

[14.1] American Forest and Paper Association (AF&PA). 2005. *National Design Specification for Wood Construction and Supplement*, ANSI/AF&PA NDS-2005, AF&PA, Washington DC.

[14.2] American Institute of Steel Construction (AISC). 2005. *Steel Construction Manual*, 13th ed., AISC, Chicago, IL.

[14.3] American Institute of Timber Construction (AITC). 2003. *Typical Construction Details*, AITC 104-2003, AITC, Englewood, CO.

[14.4] American Institute of Timber Construction (AITC). 2005. *Timber Construction Manual*, 5th ed., John Wiley & Sons Inc., Hoboken, NJ.

[14.5] American Iron and Steel Institute (AISI). 2002. *Cold-Formed Steel Design Manual*, 2002 ed., AISI, Washington DC.

[14.6] APA—The Engineered Wood Association. 2005. *Glulam Connection Details, EWS Form T300*, APA—The Engineered Wood Association, Engineered Wood Systems, Tacoma, WA.

[14.7] International Codes Council (ICC). 2006. *International Building Code*, 2006 ed., ICC, Falls Church, VA.

[14.8] Truss Plate Institute (TPI). 2002. *National Design Standard for Metal Plate Connected Wood Truss Construction*, ANSI/TPI 1-2002, TPI, Madison, WI.

Diaphragm-to-Shearwall Anchorage

15.1 Introduction

Anchorage was defined previously as the tying together of the major elements of a building with an emphasis on the transfer of lateral (wind and earthquake) forces. A systematic approach to anchorage involves a consideration of load transfer in the three principal directions of the building.

Chapter 10 illustrated the anchorage requirements at the base of a wood-frame shearwall. This chapter continues the anchorage problem with a detailed analysis of the connection between the diaphragm and the shearwall. Additional considerations for shearwall anchorage and overturning are addressed in Chap. 16. Several typical anchorage details are analyzed including connections to wood-frame shearwalls and concrete or masonry shearwalls.

Following the subject of diaphragm-to-shearwall anchorage, the subdiaphragm concept is introduced. This design technique was developed to ensure the integrity of a wood structural panel diaphragm that supports seismic forces generated by *concrete* or *masonry walls*. Its purpose is to satisfy the code requirement that "continuous ties" be provided to distribute these larger seismic forces into the diaphragm. ASCE 7 Sec. 12.11.2.2.1 includes subdiaphragm requirements for Seismic Design Categories (SDCs) C and higher. This is also an anchorage problem, but it is unique to buildings with concrete or masonry walls.

15.2 Anchorage Summary

Diaphragm anchorage refers to the design of connections between the diaphragm and the vertical elements of the building. These elements may support the diaphragm, or they may transfer a force to the diaphragm.

A systematic approach to anchorage is described in Example 10.8 (Sec. 10.8). This simply involves designing the connections between the various resisting elements for the following loads and forces:

1. Vertical loads

2. Lateral forces parallel to a wall

3. Lateral forces perpendicular to a wall

In designing for these gravity loads and lateral forces, it is important that a *continuous path* be developed. The path must take the load or force from its source, through the connections, and into the supporting element.

The "path" in the progressive transfer of *vertical loads* is described in Example 10.9 (Sec. 10.9) and is not repeated here. Many of the connections for vertical loads can be made with the hardware described in Chap. 14.

There are several types of *lateral forces parallel* to a wall. See Example 15.1. These include

1. Basic unit shear transfer

2. Collector (strut) force

3. Diaphragm chord force

The methods used to calculate collector (strut) forces and diaphragm chord forces were given in Chap. 9, and several connection designs were demonstrated in Chaps. 12 and 13. These forces are listed here simply to complete the anchorage design summary. However, the connection for the transfer of the unit shear has not been discussed previously, and it requires some additional consideration. The connections used to transfer this shear are different for different types of walls. Typical details of these connections and the calculations involved are illustrated in the following sections.

EXAMPLE 15.1   Anchorage Forces Parallel to Wall

Consider the anchorage forces *parallel* to the right end wall in the building shown in Fig. 15.1a. Three different forces are involved.

Unit Shear Transfer

The unit shear in the roof diaphragm v_R (Fig. 15.1a) must be transferred to the supporting elements (collector and shearwall).

Collector Connection to Shearwall

The force in the collector must be transferred to the shearwall (Fig. 15.1b). The magnitude of the *strength level* collector force may be read from the collector force diagram. In addition to this horizontal force, vertical forces due to dead and live loads need to be considered in design of the collector connection to the shearwall. Further, in SDC C and higher, where the structure is not entirely of light-frame construction, ASCE 7 Sec. 12.10.2.1 requires that the collector and its connection to the wall be capable of resisting the special (amplified) seismic forces of ASCE 7 Sec. 12.4.3.2.

Figure 15.1*a*   Plan view showing lateral force to diaphragm, and free body diagram cut a small distance away from right shearwall and collector.

Figure 15.1*b*   *Collector force* caused by lateral force to diaphragm in the transverse direction.

Diaphragm Chord Connection

The diaphragm chord force in Fig. 15.1*c* is an alternative design criterion to the collector force in Fig. 15.1*b*. The connection between the member over the door and the wall should be designed for the *chord force* or the *collector force,* whichever is larger. As was true with the collector, it is necessary to combine the horizontal chord force with vertical forces due to dead and live loads. Unlike the collector, there is no requirement for the chord member or connection to be designed for special (amplified) seismic forces.

Figure 15.1c   *Diaphragm chord force* caused by lateral force to diaphragm in the longitudinal direction.

There are several types of lateral forces perpendicular to a wall that must be considered when designing the wall anchorage to the diaphragm. For all wall materials and SDCs, design anchorage forces include:

1. Wind forces on the wall component.
2. A force not less than the greater of $F_P = 0.4S_{DS}IW_P$ or $F_P = 0.1\ W_P$, where W_P is the weight of the wall tributary to the anchor. ASCE 7 Sec. 12.11.1.

In addition, ASCE 7 Sec. 12.11.2 specifies that anchorage of concrete or masonry walls must also be designed for:

3. A force not less than $F_P = 400S_{DS}I$ times the number of linear feet tributary to the anchor (i.e., horizontal distance between anchors).
4. A force not less than 280 lb/ft times the number of linear feet tributary to the anchor.

Finally, in SDCs C through F, when concrete or masonry walls are anchored to a flexible diaphragm, the anchorage must also be designed for:

5. A force not less than $F_P = 0.8S_{DS}IW_P$

Wind forces will be at an ASD level. All other wall anchorage forces noted above are at a strength level. Wall anchorage forces are to be used in ASD or LRFD load combinations.

The design considerations listed above are alternate design forces, and anchorage must be designed for the maximum value. These forces act normal (perpendicular) to the wall (either inward or outward) and attempt to separate the wall from the diaphragm. See Example 15.2.

Similar forces were introduced in Examples 2.16 (Sec. 2.16) and 10.12 (Sec. 10.12). In the sketches in these examples, the force labeled *roof reaction* corresponds to the anchorage force being discussed for wind forces and is similar to the anchorage force being discussed for seismic forces.

In wood-frame buildings, wind usually controls the design of anchorage for normal to wall forces. For masonry and concrete walls, however, the seismic force F_p often governs because of the large dead load of these walls. The ASCE 7 minimum force (e.g., 280 lb/ft) for these types of walls ensures a minimum connection regardless of the calculated seismic forces.

EXAMPLE 15.2   Anchorage Forces Perpendicular to Wall

Figure 15.2   Schematic of one-story building. In most buildings, wall framing is assumed to span vertically between the foundation and the roof diaphragm.

Assuming that the wall in this example is of light-frame construction, the required wall to diaphragm anchorage forces are:

1. Wind forces on the wall component.

2a. $F_P = 0.4 S_{DS} I W_P$

2b. $F_P = 0.1\ W_P$

If the walls were of concrete or masonry, additional anchorage design forces would be:

3. $F_P = 400 S_{DS} I$ times the number of linear feet tributary to the anchor.

4. 280 lb/ft times the number of linear feet tributary to the anchor.

In SDCs C through F, when concrete or masonry walls are anchored to a flexible diaphragm, the anchorage must also be designed for:

5. $F_P = 0.8 S_{DS} I W_P$

A large number of connection details can be used to anchor the diaphragm to the walls for both parallel and perpendicular forces. The type of wall framing (wood frame, concrete, or masonry) and size and direction of the framing members for the diaphragm affect the choice of the anchorage connection. Examples of two typical anchorage details are given in Secs. 15.3 and 15.4.

15.3 Connection Details—Diaphragm to Wood-Frame Wall

A typical anchorage connection of a wood structural panel diaphragm to a wood-frame wall is considered here. Two anchorage details are shown in Example 15.3: one for the transverse wall and one for the longitudinal wall.

In practice, additional details will be required. For example, the connection where the girder ties into the transverse wall must be detailed in the structural plans. The vertical reaction of the girder, and the diaphragm chord force caused by lateral forces in the longitudinal direction need to be considered. However, the two connection details given in the sketches are intended to define the problem and be representative of the methods used for anchorage.

As noted previously, the key to the anchorage problem is to provide a continuous *path* for the "flow" of forces in the connection. Each step in this flow is labeled on the sketch, and a corresponding explanation is given. Sample calculations are provided for the transverse wall anchorage detail. See Example 15.4.

EXAMPLE 15.3   Anchorage to Wood-Frame Walls

FRAMING PLAN

Figure 15.3a One-story building with wood roof system and wood-frame stud walls.

Different anchorage details are required for the diaphragm attachment to the transverse and longitudinal walls. More complicated roof or floor framing will require additional anchorage details.

The following details shown are representative only, and other properly designed connections are possible. The path for the transfer of anchorage forces is described for each connection.

Transverse Wall Anchorage

Parallel-to-wall force (diaphragm shear)

a. Diaphragm boundary nailing transfers shear into end joist (Fig. 15.3b)

b. The shear must be transferred from the end joist into the wall plate. The top plate here serves as the diaphragm chord and collector. Various types of connections can be used for this transfer. Framing anchor is shown. Alternate connections not shown are toenails or additional blocking, as in Example 9.5. The designer should exercise caution when specifying toenails since the toenails may split the blocking. For this reason, IBC Sec. 2305.1.4 limits the allowable shear transferred by toenails to 150 lb/ft in SDCs D, E, and F (Ref. 15.7).

c. Shearwall edge nailing transfers shear into wall sheathing.

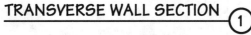

Figure 15.3b   Anchorage connections between roof and transverse shearwall.

Perpendicular-to-wall force (wind critical for most wood-frame walls)

d. Connect stud to double plate for reaction at top of stud. Standard connection from IBC Table 2304.9.1 is two 16d common nails into end grain of stud. Alternate connection for larger forces is a framing anchor.

  e. Full-depth blocking normal to wall is required to prevent the rotation (rolling over) of the joists under this type of force. This blocking is necessary whether or not the diaphragm is a "blocked" diaphragm. Spacing of blocks depends on lateral force (2 to 4 ft is typical). Connect top plate to block for tributary lateral force. Toenails are shown (not allowed for concrete or masonry walls in SDC C and higher per ASCE 7 Sec. 12.11.2.2.3). Alternate connection (not shown) is framing anchor.

  f. Connect block to diaphragm sheathing for same force considered in *e.* For consistency, use the same size nails as for diaphragm nailing.

Longitudinal Wall Anchorage

Parallel-to-wall force (diaphragm shear)

  a. Diaphragm boundary nailing transfers shear into blocking (Fig. 15.3*c*).

  b. Connection of blocking to double plate transfers shear into top plate. Framing anchor is shown. Alternate connections not shown are toenails (limited to 150 lb/ft seismic force in SDCs D, E, and F, per IBC Sec. 2305.1.4) or additional blocking, as in Example 9.5.

  c. Shearwall edge nailing transfers shear from top plate to wall sheathing.

Perpendicular-to-wall force (wind critical for most wood-frame walls)

  d. Same as transverse wall connection *d* (Fig. 15.3*b*).

  e. Connect top plate to joist for the tributary lateral force. Framing anchor is shown.

LONGITUDINAL WALL SECTION ②

Figure 15.3c   Anchorage connection between roof and longitudinal shearwall.

Alternate connection not shown is with toenails (not allowed for concrete or masonry walls in SDC C or higher, per ASCE 7 Sec. 12.11.2.2.3). The connection of the joist to the diaphragm sheathing is provided directly by the diaphragm nailing. Specific nail design similar to the transverse wall connection f is usually not required.

EXAMPLE 15.4   Anchorage of Transverse Wood-Frame Wall

Design the connection for the anchorage of the transverse wall for the building in Example 15.3 (section 1), using the following loads. The letter designations in this example correspond to the designations of the connection details in Fig. 15.3b. Lumber is Douglas Fir-Larch (DF-L), and sheathing is $^{15}/_{32}$-in. STRUCTURAL I with 10d common nails at 6-in. o.c. edge, 12-in. field, with all edges blocked. $C_M = 1.0$ and $C_t = 1.0$.

Known Information

Roof unit shear parallel to wall = 205 lb/ft   (ASD wind or seismic)

Wind force perpendicular to wall = 26 psf   (ASD)

Seismic $S_{DS} = 1.0g$

Wall unit weight = 15 psf

Wall height = 10 ft

Assume allowable load on framing anchor for parallel to wall forces (item b in Fig. 15.3b).

Allow. V = 410 lb/anchor*   (normal duration)

= 410(1.6) = 656 lb   (wind or seismic)

Seismic Force Perpendicular to Wall (Strength)

Minimum connection (ASCE 7 Sec. 12.11.1) = $0.4S_{DS}IWp$

= 0.4(1.0)(1.0)15 = 6 psf

But not less than = $0.1Wp$

= 1.5 psf

Wind force will control design forces for anchorage perpendicular to the wall.

Wall Anchorage Force Perpendicular-to-Wall

Wall anchorage T and C = 26 psf (10 ft wall height/2) = 130 lb/ft

---

*In practice, the capacity of the framing anchor is obtained from the evaluation report giving code-recognized design values for a proprietary product.

Parallel-to-Wall Forces

a. The nailing for item a in Fig. 15.3b is obtained from Table 2306.3.1 of the IBC.

Minimum allowable roof shear = 320(1.4) = 448 lb/ft >205 lb/ft   *OK*

Minimum edge nailing (10d common nails at 6-in. o.c.) will transfer roof shear into end joist.

b. Item b in Fig. 15.3b connects end joist to plate with framing anchors (assume diaphragm width = b = 40 ft):

$$V = vb = 205 \times 40 = 8200 \text{ lb}$$

$$\text{Number of anchors} = \frac{8200}{656} \approx 13$$

$$\text{Spacing of anchors} \approx \frac{40}{13} = 3.08 \text{ ft}$$

Connection b: *Use*   13 framing anchors total. Space anchors at approximately 3 ft-0 in. o.c.

Alternate toenail connection:

Assume 10d common toenails. The yield limit equations from Chap. 12 for the strength of the toenails will be shown for review. In practice, the yield limit equations will be solved on a spreadsheet template or other computer program. For the remaining nail problems in this example, the nominal design value will be obtained from NDS tables (Ref. 15.4).

Dimensions and bending yield strength of 10d common nails can be obtained from the tables in Fig. 12.7a and b (Sec. 12.4) or from NDS Appendix I and Appendix L.

$$D = 0.148 \text{ in.}$$

$$l = 3.0 \text{ in.}$$

$$F_{yb} = 90 \text{ ksi}$$

For a toenail, the thickness of the side member (for use in the yield limit equations) and the penetration are determined using the geometry given in Fig. 12.12a (Sec. 12.6). The thickness of the main member t_m is assumed to be the combined thicknesses of the double 2 $\times$ members forming the top wall plate.

$$t_m = 2(1.5) = 3.0 \text{ in.}$$

$$l_s = \frac{l}{3} = \frac{3.0}{3} = 1.0 \text{ in.}$$

$$p_L = l\cos(30) - l_s \le t_m$$

$$= 3.0\cos(30) - 1.0 \le 3.0$$

$$\therefore p_L = 1.60 \text{ in.}$$

$$10D = 10(0.148) = 1.48 \text{ in.} < 1.60 \text{ in.}$$

$$\therefore l_m = p_L = 1.60 \text{ in.}$$

Coefficients for yield limit equations:

The specific gravity and dowel bearing strength are the same for both the main member and the side member and are obtained from NDS Tables 11.3.2A and 11.3.2:

$$G_m = G_s = 0.50$$

$$F_{em} = F_{es} = 4650 \text{ psi}$$

$$K_D = 2.2 \quad \text{for } D \leq 0.17 \text{ in.}$$

$$R_e = \frac{F_{em}}{F_{es}} = \frac{4650}{4650} = 1.0$$

$$1 + R_e = 1 + 1.0 = 2.0$$

$$1 + 2R_e = 1 + 2(1.0) = 3.0$$

$$2 + R_e = 2 + 1.0 = 3.0$$

$$R_t = \frac{l_m}{l_s} = \frac{1.60}{1.0} = 1.6$$

$$k_1 = \frac{\sqrt{R_e + 2R_e^2(1 + R_t + R_t^2) + R_t^2 R_e^3} - R_e(1 + R_t)}{(1 + R_e)}$$

$$= \frac{\sqrt{1.0 + 2(1.0^2)(1 + 1.6 + 1.6^2) + (1.6^2)(1.0^3)} - 1.0(1 + 1.6)}{2.0}$$

$$= 0.5628$$

$$k_2 = -1 + \sqrt{2(1 + R_e) + \frac{2F_{yb}(1 + 2R_e)D^2}{3F_{em}l_m^2}}$$

$$= -1 + \sqrt{2(2.0) + \frac{2(90{,}000)(3.0)(0.148^2)}{3(4650)(1.60^2)}}$$

$$= 1.081$$

$$k_3 = -1 + \sqrt{\frac{2(1 + R_e)}{R_e} + \frac{2F_{yb}(2 + R_e)D^2}{3F_{em}l_s^2}}$$

$$= -1 + \sqrt{\frac{2(2.0)}{1.0} + \frac{2(90{,}000)(3.0)(0.148^2)}{3(4650)(1.0^2)}}$$

$$= 1.202$$

Yield limit equations:

Mode I_m (NDS Eq. 11.3-1):

$$Z = \frac{Dl_m F_{em}}{K_D} = \frac{0.148(1.6)(4650)}{2.2} = 501 \text{ lb}$$

Mode I_s (NDS Eq. 11.3-2):

$$Z = \frac{Dl_s F_{es}}{K_D} = \frac{0.148(1.0)(4650)}{2.2} = 313 \text{ lb}$$

Mode II (NDS Eq. 11.3-3):

$$Z = \frac{k_1 Dl_s F_{es}}{K_D} = \frac{0.5628(0.148)(1.0)(4650)}{2.2} = 176 \text{ lb}$$

Mode III_m (NDS Eq. 11.3-4):

$$Z = \frac{k_2 Dl_m F_{em}}{K_D(1 + 2R_e)} = \frac{1.081(0.148)(1.60)(4650)}{2.2(3.0)} = 180 \text{ lb}$$

Mode III_s (NDS Eq. 11.3-5):

$$Z = \frac{k_3 Dl_s F_{em}}{K_D(2 + R_e)} = \frac{1.202(0.148)(1.0)(4650)}{2.2(3.0)} = 125 \text{ lb}$$

Mode IV (NDS Eq. 11.3-6):

$$Z = \frac{D^2}{K_D} \sqrt{\frac{2F_{em}F_{yb}}{3(1 + R_e)}}$$

$$= \frac{(0.148)^2}{2.2} \sqrt{\frac{2(4650)(90,000)}{3(2.0)}} = 118 \text{ lb}$$

Reference design value:

The reference design value is taken as the smallest value from the yield limit equations:

$$Z = 118 \text{ lb}$$

Adjustment factors:

Load duration factor from the NDS for wind:

$$C_D = 1.6$$

Limit for penetration:

$$6D = 6(0.148) = 0.89 \text{ in.} \qquad \text{(minimum penetration)}$$

$$0.89 \text{ in.} < 1.6 \text{ in.} \qquad OK$$

Toenail factor for laterally loaded toenails:

$$C_{tn} = 0.83$$

All other adjustment factors default to unity in this problem.

Adjusted ASD design value:

$$Z' = Z(C_D C_M C_t C_{eg} C_{di} C_{tn})$$

$$= 118(1.6)(1.0)(1.0)(1.0)(1.0)(0.83)$$

$$= 157 \text{ lb/nail}$$

$$\text{Req'd spacing} = \frac{157 \text{ lb/nail}}{205 \text{ lb/ft}} = 0.764 \text{ ft} = 9.2 \text{ in.} \quad say \quad 9 \text{ in. o.c.}$$

| Alternate connection *b*:   *Use*   10d common toenails at 9 in. o.c. |

From IBC Table 2304.9.1, the alternate connection *b* using toenails requires a minimum 8d common or box nail at 6-in. o.c. The above connection exceeds this minimum.

c. Plywood shearwall edge nailing transfers shear from wall plate into shearwall (Table 2306.4.1 of the IBC).

Perpendicular-to-Wall Forces

d. Check connection of stud to wall plate for separation force. Assume two 16d common nails through wall plate into end grain of stud (IBC Table 2304.9.1).

  Lumber is DF-L, and the thickness of the side member is the thickness of one of the wall plates ($t_s = 1\frac{1}{2}$ in.). The nominal design value for a 16d nail could be obtained from the yield limit equations. However, the plate-to-stud connection is covered directly in NDS Table 11N. The reader may wish to verify the results of the yield limit equations as an exercise.

$$Z = 141 \text{ lb}$$

The following adjustments apply:

$$C_D = 1.6 \quad \text{for wind}$$

$$C_{eg} = 0.67 \quad \text{nail is driven into end grain of stud.}$$

Check nail penetration:

$$p = l - t_s = 3.5 - 1.5 = 2.0 \text{ in.}$$

$$10D = 10(0.162) = 1.62 \text{ in.} < 2.0 \text{ in.}$$

$$\therefore l_m = p = 2.0 \text{ in.}$$

All other adjustment factors are unity.

$$Z' = Z(C_D C_M C_t C_{eg} C_{di} C_{tn})$$

$$= 141(1.6)(1.0)(1.0)(0.67)(1.0)(1.0)$$

$$= 151 \text{ lb/nail}$$

Adjusted ASD load for stud connection:

$$\text{Adj. } P = N(Z') = 2(151) = 302 \text{ lb/stud}$$

Load on one stud connection is the force per foot of wall times the spacing of the studs.

$$P = 130 \text{ lb/ft} \times 1.33 \text{ ft}$$

$$= 173 \text{ lb/stud} < 302 \text{ lb/stud} \qquad OK$$

e. Assume that blocking is spaced 24-in. o.c.

$$\text{Load per block} = 130 \text{ lb/ft} \times 2 \text{ ft} = 260 \text{ lb}$$

Capacity of 10d common toenails:

$$Z' = 157 \text{ lb/nail} \qquad \text{(from } b)$$

$$\text{Number of toenails} = \frac{260 \text{ lb/block}}{157 \text{ lb/nail}} \approx 2$$

> *Use*   two 10d common toenails per block.
> Blocks at 24 in. o.c.

f. Attach blocking to roof sheathing.

$$\text{Load per block} = 260 \text{ lb} \qquad \text{(from } e)$$

For the general diaphragm nailing 10d common nails in $^{15}/_{32}$-in. STRUCTURAL I sheathing are used. Determine the number of 10d nails required to transmit the load from the block into the roof diaphragm.

The yield limit equations could be evaluated for a 10d common nail through $^{15}/_{32}$-in. sheathing into DF-L blocking. However, a reference design value will be obtained from NDS Table 11Q for $^{15}/_{32}$ in. sheathing:

$$Z = 88 \text{ lb/nail}$$

The following adjustments apply:

$$C_D = 1.6 \quad \text{for wind}$$

$$C_{di} = 1.1 \quad \text{nail is into wood structural panel diaphragm.}$$

Check nail penetration:

$$p = l - t_s = 3.0 - 0.5 = 2.5 \text{ in.}$$

$$10D = 10(0.148) = 1.48 \text{ in.} < 2.5 \text{ in.}$$

$$\therefore l_m = p = 2.5 \text{ in.}$$

Other adjustment factors default to unity.

$$Z' = Z(C_D C_M C_t C_{eg} C_{di} C_{tn})$$

$$= 88(1.6)(1.0)(1.0)(1.0)(1.1)(1.0)$$

$$= 155 \text{ lb/nail}$$

NOTE: Reference 15.11 recommends an 11 percent reduction in nail value calculated using yield limit equations when diaphragm nailing is driven into narrow (2-in. nominal) framing. The 0.9 multiplying factor in the next step is to account for this reduction. Depending on materials, other adjustments may apply. Contact APA for the latest information regarding nail design values for diaphragms using the yield limit equations.

$$\text{Modified } Z' = 155(0.9) = 138 \text{ lb/nail}$$

$$\text{Number of nails} = \frac{260 \text{ lb/block}}{138 \text{ lb/nail}} \approx 2$$

Length of block will depend on spacing of roof joists (assume joists are 24-in. o.c.). If nails are spaced 6-in. o.c.,

$$N = \frac{24}{6} \approx 4 \text{ nails/block} > 2 \qquad OK$$

Use 10d common nails at 6 in. o.c. through wood structural panel into blocking.

15.4 Connection Details—Diaphragm to Concrete or Masonry Walls

The anchorage of concrete tilt-up and masonry (concrete block and brick walls) to wood structural panel diaphragms has been the subject of considerable discussion in the design profession. The use of wood structural panel diaphragms in buildings with concrete or masonry walls is common, especially in one-story commercial and industrial buildings. Some failures of these connections occurred in recent earthquakes (Ref. 15.6), and revised design criteria have been included in ASCE 7 to strengthen these attachments.

ASCE 7 has the following requirements for the anchorage of concrete and masonry walls for seismic forces:

In all SDCs:

1. Interconnection of the wall elements and connection to supporting framing systems shall have sufficient ductility, rotational capacity, or sufficient strength to resist shrinkage, thermal changes, and differential foundation settlement when combined with seismic forces (ASCE 7 Sec. 12.11.1).

2. Where the anchor spacing exceeds 4 ft, walls must be designed to resist bending between anchors (ASCE 7 Sec. 12.11.2).

In addition, in SDCs C and higher (ASCE 7 Sec. 12.11.2.2):

3. Diaphragms shall be provided with continuous ties or struts between diaphragm chords to distribute these anchorage forces into the diaphragms. Diaphragm connection shall be positive, mechanical, or welded. Added chords are permitted to be used to form subdiaphragms to transmit anchorage forces to the main continuous cross-ties. The maximum length-to-width ratio of the structural subdiaphragm shall be 2.5 to 1. Connections and anchorages capable of resisting the prescribed forces shall be provided between the diaphragm and the attached components. Connections shall extend into the diaphragm a sufficient distance to develop the force transfer into the diaphragm (ASCE 7 Sec. 12.11.2.2.3).

4. In wood diaphragms, the continuous ties shall be in addition to the diaphragm sheathing. Anchorage shall not be accomplished by use of toenails or nails subject to withdrawal, nor shall wood ledgers or framing be used in cross-grain tension. The diaphragm sheathing shall not be considered effective as providing the ties or struts required by this section. (ASCE 7 Sec. 12.11.2.2.3).

5. The strength design forces for steel elements of the wall anchorage system, other than anchor bolts and reinforcing steel, shall be 1.4 times the forces otherwise required by this section (ASCE 7 Sec. 12.11.2.2.2).

6. Diaphragm to wall anchorage using embedded straps shall be attached to or hooked around, the reinforcing steel or otherwise terminated so as to effectively transfer forces to the reinforcing steel (ASCE 7 Sec. 12.11.2.2.5).

7. When elements of the wall anchorage system are loaded eccentrically, or are not perpendicular to the wall, the system shall be designed to resist all components of the forces induced by the eccentricity (ASCE 7 Sec. 12.11.2.2.6).

8. When pilasters are present in the wall, the anchorage force at the pilasters shall be calculated considering the additional load transferred to the pilasters. However, the minimum unit anchorage force at a floor or roof shall not be reduced (ASCE 7 Sec. 12.11.2.2.7).

Two different anchorage problems are considered in this section, and the path used to transfer the forces is described. The first anchorage detail involves a *ledger connection* to a wall with a parapet. See Example 15.5. The vertical load must be transferred in addition to the lateral forces. The anchor bolts are designed to carry both the vertical load and the lateral force parallel to the wall. This requires that two different load combinations be checked, and the capacity of the bolts in the wall and in the ledger must be determined.

A special anchorage device is provided to transmit the lateral force normal to the wall into the framing members. This is required to prevent cross-grain bending in the ledger. A numerical example is provided for the anchorage of a masonry wall with a ledger. See Example 15.6.

EXAMPLE 15.5   Typical Ledger Anchorage to Masonry Wall

Anchorage connections to a masonry wall with a ledger, as depicted in Figure 15.4a are:

TYPICAL LEDGER CONNECTION

Figure 15.4a   Anchorage connections to masonry wall with ledger.

a. Vertical loads [D + (L_r or S)] are transferred from roof framing members to ledger with prefabricated metal hangers.

b. Diaphragm unit shear parallel to the wall is transferred to ledger by diaphragm boundary nailing.

c. The vertical load and unit shear from a and b are transferred from the ledger to the wall by the anchor bolts. The strength of the anchor bolts is governed by the capacity of the bolts in the

1. *Wall*

   Concrete walls—ACI 318 Appendix D (Ref. 15.1).

   Masonry walls—ACI 530 (Ref 15.2).

2. *Wood ledger* with bolt in single shear (Chap. 13)

The load cases in IBC Sec. 1605.3.1 can be simplified to the following to determine the bolt ASD loading (fluid load F, retaining load H, floor live load L, self-straining load T, and rain load R can all be taken as zero):

(a) $D + (L_r \text{ or } S)$

(b) $D + (W \text{ or } 0.7E)$

(c) $D + 0.75 (W \text{ or } 0.7E) + 0.75(L_r \text{ or } S)$

(d) $0.6D + W$

(e) $0.6D + 0.7E$

Except for equation (a), both horizontal and vertical forces are included in the load combination equations. Vertical forces will act perpendicular to the ledger grain and horizontal forces will act parallel to the ledger grain. It must be remembered that E has both horizontal and vertical components:

$$E = \rho\, Q_E \pm 0.2S_{DS}D$$

This will be applicable to load combinations *b*, *c*, and *e*.

In addition, for load combinations with two or more variable loads, IBC Sec. 1605.3.1.1 permits the variable loads to be multiplied by 0.75. This is seen in load combination *c*.

In this case, equations *b*, *c*, and *e* could be modified to the following:

(c) $D + 0.75[0.7(\rho\, Q_E + 0.2S_{DS}D)] + 0.75(L_r \text{ or } S)$

$= D + 0.525\rho\, Q_E + 0.11S_{DS}D + 0.75(L_r \text{ or } S)$

(b) $D + 0.7(\rho\, Q_E + 0.2S_{DS}D)$

$= D + 0.7\rho\, Q_E + 0.14S_{DS}D$

(e) $0.6D + 0.7\rho\, Q_E - 0.7(0.2S_{DS}D)$

d. Lateral forces perpendicular to the wall are carried by diaphragm-to-wall anchors (Fig. 14.21 in Sec. 14.5) or similar-type hardware. If the strap is bent down to allow bolting through the framing member, the fold-down distance should be kept small to minimize eccentricity. However, the required edge distance for the fasteners must be

Figure 15.4b Anchor bolt connection between ledger and masonry wall. Vectors shown are resisting forces provided by anchor bolt under gravity load and lateral forces parallel to wall.

provided. The anchors are embedded in the wall, hooked around horizontal wall steel, and nailed, bolted, or lag bolted to the roof framing. The purpose of this separate anchorage connection for normal wall forces is to avoid cross-grain bending in the wood ledger (Example 6.1 in Sec. 6.2).

Subdiaphragm Problem

Once the perpendicular-to-wall force has been transferred into the framing members, it is carried into the sheathing. Because these normal wall forces may be large for concrete and masonry walls, it is important that the framing members extend back into the diaphragm (i.e., away from the wall) in order to make this transfer. The concept of subdiaphragms (Sec. 15.5) was developed to ensure that the framing is sufficiently anchored into the diaphragm.

EXAMPLE 15.6   Anchorage of Concrete Wall with Ledger

Design the connections for the ledger anchorage in Example 15.5 using ASD procedures. Assume SDC D, with $S_{DS} = 1.0g$, and assume seismic controls over wind forces. Assume snow load S is not applicable at this building location. The letter designations in this example correspond to the designations of the connection details in Fig. 15.4a. The following information is given.

Roof dead load D = 60 lb/ft   (along wall)
Roof live load L_r = 100 lb/ft   (along wall)
Roof diaphragm shear = 420 lb/ft   (strength level, parallel to wall)

Construction

$^{15}\!/_{32}$-in. STRUCTURAL I plywood roof sheathing is used.

Roof framing members are 2 × 6 at 24-in. o.c.

Ledger is 4 × 12 DF-L that is dry and remains dry in service. Normal temperature conditions apply.

Walls are 6-in. concrete = 75 psf

Wall height is 12 ft to roof with a one ft parapet above

Wall tributary height to diaphragm is $(^{12}\!/_2) + 1 = 7$ ft

Wall Anchorage Force (Perpendicular to wall)

First the wall weight tributary to the diaphragm needs to be calculated

$$W = (75 \text{ psf}) (7 \text{ ft tributary}) = 525 \text{ lb/ft}$$

Three anchorage forces will be checked to determine the controlling anchorage force for this building. The three forces were introduced in Sec. 15.2 but will be considered in greater detail here.

1. For all wall construction types (ASCE 7 Sec. 12.11.1)

$$F_P = 0.4\,S_{DS}IW_P \geq 0.1\,W_P$$

$$= 0.4\,(1.0)(1.0)(525\ \text{lb/ft}) = 210\ \text{lb/ft}$$

2. For concrete or masonry walls (ASCE 7 Sec. 12.11.2)

$$F_P = 400\,S_{DS}I$$

$$= (400)(1.0)(1.0) = 400\ \text{lb/ft}$$

3. For concrete or masonry walls (ASCE 7 Sec. 12.11.2)

$$F_P = 280\ \text{lb/ft}$$

4. For concrete or masonry walls in SDCs C to F (ASCE 7 Sec. 12.11.2.1)

$$F_P = 0.8\,S_{DS}IW_P$$

$$= 0.8(1.0)(1.0)(525\ \text{lb/ft}) = 420\ \text{lb/ft}$$

From all of these force requirements, the controlling *strength level* design force for the wall anchorage is 420 lb/ft.

Load Combinations

From the load combinations introduced in Example 15.5, and because only dead load D, roof live load L_r, and seismic forces E are being considered, the following load combinations need to be considered:

(a) $D + L_r$

(b) $D + 0.7\rho Q_E + 0.14S_{DS}D$

(c) $D + 0.525\rho Q_E + 0.11S_{DS}D + 0.75L_r$

(e) $0.6D + 0.7\rho Q_E - 0.14S_{DS}D$

With the added information that $S_{DS} = 1.0$ and $\rho = 1.0$, the above combinations can be further simplified to

(a) $D + L_r$

(b) $1.14D + 0.7Q_E$

(c) $1.11D + 0.525Q_E + 0.75L_r$

(e) $0.46D + 0.7Q_E$

Of these load combinations, *a* involves only vertical forces and will use a different duration of load factor C_D of 1.25. Load combinations *b*, *c*, and *e* involve both horizontal and vertical forces and will use a duration of load factor C_D of 1.6.

Anchorage

a. Roof beams are 2-ft o.c. Reaction of roof beam on ledger $D + L_r$:

$$R = (60 + 100)2 = 320\ \text{lb}$$

Choose prefabricated metal hanger from manufacturer with code-recognized capacity greater than 320 lb.

b. From Table 2306.3.1 of the IBC for $^{15}\!/_{32}$-in. STRUCTURAL I plywood, allow. $v = 320$ lb/ft for a *blocked* diaphragm with 2-in. nominal framing and 10d common nails at:

6-in. o.c. edges

12-in. o.c. field

This ASD capacity will be compared to a seismic demand of:

$$0.7Q_E = 0.7 \ (420 \ \text{lb/ft}) = 294 \ \text{plf}$$

The capacity of 320 plf is adequate.

c. Assume $^3\!/_4$-in. anchor bolts at 32-in. o.c. For this anchor bolt spacing, compare the *actual* load and *adjusted ASD capacity* for two cases:

1. Vertical load $D + L_r$
2. Vertical load and lateral force parallel to wall

(1) *Vertical load combination:*

$$D + L_r = (60 + 100)\left(\frac{32}{12}\right) = 427 \ \text{lb/bolt}$$

Based on calculations in accordance with ACI 318 Appendix D, the strength level capacity (ϕ times the nominal strength) of the $^3\!/_4$-in. bolt in concrete is calculated to be 2800 lb.

Multiplying by 0.7 gives an ASD capacity of 1960 lb/bolt for gravity loads (note that other reductions apply for seismic forces). This is greater than the calculated $D + L_r$ demand of 427 lb/bolt.

Check bolt capacity in ledger.

The capacity of a wood-to-concrete connection is defined in NDS Table 11E. The reference design value for a wood-to-concrete connection is taken as the single-shear capacity for a connection with the following definitions:

l_m = dowel bearing length in the main member for use in yield limit equations is taken equal to the embedment length of the anchor bolt in the concrete.

F_{em} = dowel bearing strength of main member for use in yield limit equations is taken equal to the dowel bearing strength of the concrete. The NDS recommends 7500 psi for $f_c' \geq 2500$ psi.

Application of the yield limit equations from Chap. 13 can be used to determine the anchor bolt capacity. In practice, however, NDS Table 11E will be used which provides bolt design values for single shear connections with sawn lumber anchored to concrete. If the particular case is not addressed in the NDS Table, the yield limit equations must be solved. Such is the case when $f_c' < 2500$ psi and the dowel bearing strength of the concrete or masonry is not defined in the NDS. The dowel bearing strength of 7500 psi used in the NDS was derived from several research studies involving the dowel bearing strength of concrete and lateral connections. The research indicated dowel bearing strengths on the order of 2 to 5 times the compressive strength of concrete, and the NDS initially considered adopting a dowel bearing strength for concrete and masonry equal to 3 times the compressive strength ($3 \times f_c'$ or f_m'). However, in consideration of past practice that allowed 6150 psi for dowel bearing strength of concrete and masonry along with the research studies, the NDS adopted a dowel bearing strength of 7500 psi for the purpose of tabulating wood-to-concrete connection values. Although data to support the use of a specific dowel bearing strength when $f'_c < 2500$ psi is currently not available, past practice of using dowel bearing strengths of 6150 psi resulted in satisfactory connection performance regardless of concrete strength (see Ref. 15.5).

For vertical (gravity) loads, the load is perpendicular to grain:

$$D = 0.75 \text{ in.}$$

$$F_{yb} = 45 \text{ ksi} \quad \text{(assumed for A307 bolt)}$$

$$l_s = t_s = 3.5 \text{ in.}$$

$$\theta_s = 90 \text{ degrees}$$

$$F_{es} = 2600 \text{ psi } (F_{e\perp} \text{ for } \tfrac{3}{4}\text{-in. bolt in DF-L from NDS Table 11.3.2)}$$

For an anchor bolt problem:

$$l_m = t_m = 4.0 \text{ in.} \quad \text{(assumed embedment)}$$

$$F_{em} = 7500 \text{ psi}$$

Coefficients for yield limit equations:

$$R_e = \frac{F_{em}}{F_{es}} = \frac{7500}{2600} = 2.885$$

$$1 + R_e = 1 + 2.885 = 3.885$$

$$1 + 2R_e = 1 + 2(2.885) = 6.769$$

$$2 + R_e = 2 + 2.885 = 4.885$$

$$R_t = \frac{l_m}{l_s} = \frac{4.0}{3.5} = 1.143$$

$$1 + R_t = 1 + 1.143 = 2.143$$

$$1 + R_t + R_t^2 = 2.143 + 1.143^2 = 3.449$$

$$\theta_{max} = 90 \text{ degrees}$$

$$K_\theta = 1 + \frac{\theta_{max}}{360} = 1 + \frac{90}{360} = 1.25$$

$$k_1 = \frac{\sqrt{R_e + 2R_e^2(1 + R_t + R_t^2) + R_t^2 R_e^3} - R_e(1 + R_t)}{(1 + R_e)}$$

$$= \frac{\sqrt{2.885 + 2(2.885^2)(3.449) + (1.143^2)(2.885^3)} - (2.885)(2.143)}{3.885}$$

$$= 0.873$$

$$k_2 = -1 + \sqrt{2(1 + R_e) + \frac{2F_{yb}(1 + 2R_e)D^2}{3F_{em}l_m^2}}$$

$$= -1 + \sqrt{2(3.885) + \frac{2(45,000)(6.769)(0.75^2)}{3(7500)(4.0^2)}}$$

$$= 1.953$$

$$k_3 = -1 + \sqrt{\frac{2(1 + R_e)}{R_e} + \frac{2F_{yb}(2 + R_e)D^2}{3F_{em}l_s^2}}$$

$$= -1 + \sqrt{\frac{2(3.885)}{2.885} + \frac{2(45,000)(4.885)(0.75^2)}{3(7500)(3.5^2)}}$$

$$= 0.895$$

Yield limit equations:

Mode I_m (NDS Eq. 11.3-1):

$$Z = \frac{Dl_mF_{em}}{4K_\theta} = \frac{0.75(4.0)(7500)}{4(1.25)} = 4500 \text{ lb}$$

Mode I_s (NDS Eq. 11.3-2):

$$Z = \frac{Dl_sF_{es}}{4K_\theta} = \frac{0.75(3.5)(2600)}{4(1.25)} = 1365 \text{ lb}$$

Mode II (NDS Eq. 11.3-3):

$$Z = \frac{k_1Dl_sF_{es}}{3.6K_\theta} = \frac{0.873(0.75)(3.5)(2600)}{3.6(1.25)} = 1324 \text{ lb}$$

Mode III_m (NDS Eq. 11.3-4):

$$Z = \frac{k_2Dl_mF_{em}}{3.2(1 + 2R_e)K_\theta} = \frac{1.953(0.75)(4.0)(7500)}{3.2(6.769)(1.25)} = 1623 \text{ lb}$$

Mode III_s (NDS Eq. 11.3-5):

$$Z = \frac{k_3Dl_sF_{em}}{3.2(2 + R_e)K_\theta} = \frac{0.895(0.75)(3.5)(7500)}{3.2(4.885)(1.25)} = 902 \text{ lb}$$

Mode IV (NDS Eq. 11.3-6):

$$Z = \frac{D^2}{3.2K_\theta}\sqrt{\frac{2F_{em}F_{yb}}{3(1 + R_e)}} = \frac{0.75^2}{3.2(1.25)}\sqrt{\frac{2(7500)(45,000)}{3(3.885)}} = 1070 \text{ lb}$$

Reference design value is selected as the smallest value from the yield limit formulas:

$$Z = 902 \text{ lb}$$

Load duration factor for the combination $(D + L_r)$ is $C_D = 1.25$. With a minimum edge distance for the loaded edge of $4D = 4(0.75) = 3.0$ in., the full bolt capacity can be used (i.e., $C_\Delta = 1.0$). Other adjustment factors are unity for this problem.

$$Z' = Z \, (C_D C_M C_t C_g C_\Delta)$$

$$= 902 \, (1.25)(1.0)(1.0)(1.0)(1.0)$$

$$= 1127 \text{ lb/bolt} > 427 \text{ lb/bolt} \quad OK$$

(2) *Vertical loads plus lateral force parallel to wall (see Fig. 15.5):*

Recall the loads considered in this example include dead, roof live, and seismic. The applicable ASD load combinations are

(b) $1.14D + 0.7Q_E$

(c) $1.11D + 0.525Q_E + 0.75L_r$

(e) $0.46D + 0.7Q_E$

Consider first load combination e:
Perpendicular-to-grain component q:

$$q = 0.46D = 0.46(60)\left(\frac{32}{12}\right) = 74 \text{ lb/bolt}$$

Parallel-to-grain component p:

$$p = \text{lateral force} = 0.7(420)\left(\frac{32}{12}\right) = 784 \text{ lb/bolt}$$

Resultant load on one bolt:

$$z = \sqrt{784^2 + 74^2} = 787 \text{ lb/bolt}$$

Now considering $1.11D + 0.525Q_E + 0.75L_r$:

Figure 15.5 Anchor bolt in ledger carrying gravity load and lateral force parallel to wall.

Perpendicular-to-grain component q:

$$q = 1.11D + 0.75L_r = [1.11(60 \text{ plf}) + 0.75(100 \text{ plf})]\left(\frac{32}{12}\right)$$

$$= (67 + 75)\left(\frac{32}{12}\right) = 378 \text{ lb/bolt}$$

Parallel-to-grain component p:

$$p = 0.525Q_E = 0.525(420 \text{ plf})\left(\frac{32}{12}\right) = 220.5\left(\frac{32}{12}\right) = 588 \text{ lb/bolt}$$

NOTE: These forces have already been adjusted to an ASD level in order to properly apply load combinations.

Resultant load on one bolt:

$$z = \sqrt{378^2 + 588^2} = 699 \text{ lb/bolt}$$

And finally considering $1.14D + 0.7Q_E$

Perpendicular-to-grain component q:

$$q = 1.14D = 1.14(60 \text{ plf})\left(\frac{32}{12}\right) = (68)\left(\frac{32}{12}\right) = 182 \text{ lb/bolt}$$

Parallel-to-grain component p:

$$p = 0.7Q_E = 0.7(420 \text{ plf})\left(\frac{32}{12}\right) = 294\left(\frac{32}{12}\right) = 784 \text{ lb/bolt}$$

Resultant load on one bolt:

$$z = \sqrt{182^2 + 784^2} = 805 \text{ lb/bolt}$$

The equation b load combination will govern for design of the anchor bolt attachment to concrete. In this case, the equation c load combination with the larger perpendicular-to-grain force will likely control for the wood ledger, and will be used for the balance of this example. The reader is encouraged to check the ledger for the other combinations.

Based on calculations in accordance with ACI 318 Appendix D, the concrete anchorage strength level capacity (ϕ times the nominal strength) of the $\frac{3}{4}$-in. bolt is calculated to be 2800 lb. ACI 318 requires that this also be multiplied by a factor of 0.75 for seismic forces, and IBC requires that the resulting value be divided by 2.5 because steel failure does not control capacity. This results in a seismic strength level capacity of 840 lb/bolt. Multiplying by 0.7 gives an ASD capacity of 588 lb/bolt. Note that this is significantly less than the bolt capacity for gravity loads.

This concrete capacity is less than the calculated bolt force of 805 lb. It appears that the bolt spacing will need to be reduced to 16-in. o.c.

Bolt capacity in ledger:

Evaluation of the anchor bolt capacity in the ledger requires the use of the yield limit equations since the load acts at an angle to the grain. Computer solutions handle this problem by simply changing θ.

Angle of load to grain of θ:

$$\theta = \tan^{-1}\frac{378}{588} = 32.7 \text{ degrees}$$

$D = 0.75$ in.

$F_{yb} = 45$ ksi   (assumed for A307 bolt)

$l_s = t_s = 3.5$ in.

$\theta_s = 32.7$ degrees

$F_{e\|} = 5600$ psi   ($F_{e\|}$ for bolt in DF-L from NDS Table 11.3.2)

$F_{e\perp} = 2600$ psi   ($F_{e\perp}$ for $\frac{3}{4}$-in. bolt in DF-L from NDS Table 11.3.2)

$$F_{es} = F_{e\theta} = \frac{F_{e\|}F_{e\perp}}{F_{e\|}\sin^2\theta + F_{e\perp}\cos^2\theta}$$

$$= \frac{5600(2600)}{5600\sin^2 32.7 + 2600\cos^2 32.7} = 4190 \text{ psi}$$

For an anchor bolt problem:

$$l_m = 4.0 \text{ in.}\quad \text{(embedment)}$$

$$F_{em} = 7500 \text{ psi}$$

Coefficients for yield limit equations:

$$R_e = \frac{F_{em}}{F_{es}} = \frac{7500}{4190} = 1.79$$

$$1 + R_e = 1 + 1.79 = 2.79$$

$$1 + 2R_e = 1 + 2(1.79) = 4.58$$

$$2 + R_e = 2 + 1.79 = 3.79$$

$$R_t = \frac{l_m}{l_s} = \frac{4.0}{3.5} = 1.143$$

$$1 + R_t = 1 + 1.143 = 2.143$$

$$1 + R_t + R_t^2 = 2.143 + 1.143^2 = 3.449$$

$$\theta_{max} = 32.7 \text{ degrees}$$

$$K_\theta = 1 + \frac{\theta_{max}}{360} = 1 + \frac{32.7}{360} = 1.09$$

$$k_1 = \frac{\sqrt{R_e + 2R_e^2(1 + R_t + R_t^2) + R_t^2 R_e^3} - R_e(1 + R_t)}{(1 + R_e)}$$

$$= \frac{\sqrt{1.79 + 2(1.79^2)(3.449) + (1.143^2)(1.79^3)} - (1.79)(2.143)}{2.79}$$

$$= 0.633$$

$$k_2 = -1 + \sqrt{2(1 + R_e) + \frac{2F_{yb}(1 + 2R_e)D^2}{3F_{em}l_m^2}}$$

$$= -1 + \sqrt{2(2.79) + \frac{2(45,000)(4.58)(0.75^2)}{3(7500)(4.0^2)}}$$

$$= 1.495$$

$$k_3 = -1 + \sqrt{\frac{2(1 + R_e)}{R_e} + \frac{2F_{yb}(2 + R_e)D^2}{3F_{em}l_s^2}}$$

$$= -1 + \sqrt{\frac{2(2.79)}{1.79} + \frac{2(45,000)(3.79)(0.75^2)}{3(7500)(3.5^2)}}$$

$$= 0.953$$

Yield limit equations:

Mode I_m (NDS Eq. 11.3-1):

$$Z = \frac{Dl_mF_{em}}{4K_\theta} = \frac{0.75(4.0)(7500)}{4(1.09)} = 5161 \text{ lb}$$

Mode I_s (NDS Eq. 11.3-2):

$$Z = \frac{Dl_sF_{es}}{4K_\theta} = \frac{0.75(3.5)(4190)}{4(1.09)} = 2523 \text{ lb}$$

Mode II (NDS Eq. 11.3-3):

$$Z = \frac{k_1Dl_sF_{es}}{3.6K_\theta} = \frac{0.633(0.75)(3.5)(4190)}{3.6(1.09)} = 1775 \text{ lb}$$

Mode III_m (NDS Eq. 11.3-4):

$$Z = \frac{k_2Dl_mF_{em}}{3.2(1 + 2R_e)K_\theta} = \frac{1.495(0.75)(4.0)(7500)}{3.2(4.58)(1.09)} = 2105 \text{ lb}$$

Mode III_s (NDS Eq. 11.3-5):

$$Z = \frac{k_3Dl_sF_{em}}{3.2(2 + R_e)K_\theta} = \frac{0.953(0.75)(3.5)(7500)}{3.2(3.79)(1.09)} = 1419 \text{ lb}$$

Mode IV (NDS Eq. 11.3-6):

$$Z = \frac{D^2}{3.2K_\theta}\sqrt{\frac{2F_{em}F_{yb}}{3(1 + R_e)}}$$

$$= \frac{0.75^2}{3.2(1.09)}\sqrt{\frac{2(7500)(45,000)}{3(2.79)}} = 1448 \text{ lb}$$

Reference design value is selected as the smallest value from the yield limit formulas:

$$Z = 1419 \text{ lb}$$

Load duration factor from NDS for load combinations involving seismic loads is $C_D = 1.6$. Other adjustment factors are unity for this problem.

$$Z' = Z\,(C_D C_M C_t C_g C_\Delta)$$

$$= 1419\,(1.6)(1.0)(1.0)(1.0)(1.0)$$

$$= 2270 \text{ lb/bolt} > 805 \text{ lb/bolt} \quad \text{at 32-in. o.c.}$$

$$> 403 \text{ lb/bolt} \quad \text{at 16-in.o.c.}$$

> *Use* $\frac{3}{4}$ -in.-diameter anchor bolts at 16 in. o.c.
> controlled by concrete anchorage capacity

d. Provide wall anchors at every other roof framing member (i.e., 4 ft-0 in. o.c.). Perpendicular-to-wall force:

$$P = 0.7(420 \text{ lb/ft}) \times 4 \text{ ft} = 1176 \text{ lb}$$

Choose a wall anchor from manufacturer with code-recognized load capacity that is greater than 1176 lb.

A series of additional design and detailing requirements for anchorage of concrete and masonry walls was introduced in Sec. 15.4. Of interest for selection of an anchorage device is the requirement that the strength design forces for steel elements of the wall anchorage system, other than the anchor bolts and reinforcing steel, shall be 1.4 times the forces otherwise required (ASCE 7 Sec. 12.11.2.2.2). Also of interest when selecting a device is the requirement that the anchorage must address forces toward and away from the diaphragm. An anchorage device capable of resisting both tension and compression forces most directly addresses this issue. Alternatively a compression load path that makes use of bearing can be used. Care should be taken, however, to ensure that wood shrinkage and construction tolerances will not leave gaps in the load path. Retrofit of wall anchorage for existing systems of this type is addressed in Ref. 15.10.

The second anchorage detail example is for a diaphragm that connects to the *top* of a concrete or masonry wall. See Example 15.7. The anchor bolts in the wall are used to transfer the lateral forces both parallel and perpendicular to the wall. These forces can act concurrently if the wall reinforcing steel is also intended to act as the diaphragm chord. The strength of the bolt in the wall must also be checked.

The problem in this connection centers on the need to avoid cross-grain tension in the wood plate when forces normal to the wall are considered. Two possible connections for preventing cross-grain tension are presented.

Calculations for the anchorage of this type of connection are similar in many respects to those for the ledger connection (Example 15.6), and a numerical example is not provided.

EXAMPLE 15.7   Anchorage at Top of Masonry Wall

Parallel-to-Wall Force (Diaphragm Shear)

a. Boundary nailing transfers diaphragm shear from sheathing into blocking.

b. Connection for transferring shear from blocking to plate. Framing anchor is shown. Alternate connection is additional blocking as in Example 9.5 (Sec. 9.4). Toenails

SHEATHING

BLOCKING BETWEEN
FRAMING

ROOF OR
FLOOR
FRAMING

WOOD PLATE

MASONRY OR
CONCRETE WALL

TYPICAL CONNECTION
TO TOP OF WALL

Figure 15.6a   Anchorage connections to top of masonry wall.

are limited to seismic forces not greater than 150 plf in SDCs D, E, and F, per IBC Sec. 2305.1.4.

c1. Shear is transferred from plate to wall by anchor bolts. Bolt strength is governed by wall capacity or by the adjusted design value parallel to grain of the bolt in the wood plate.

Perpendicular-to-Wall Force (Wind, Seismic, or Code Minimum)

c2. Lateral force normal to wall is transferred from wall to plate by anchor bolts. Bolt strength is governed by wall capacity or by the adjusted design value perpendicular to the grain of the bolt in the wood plate.

d. The attachment of the wall plate to the framing presents a problem. The code requires that anchorage of concrete and masonry walls be accomplished without cross-grain bending or *cross-grain tension.*

The connection in Fig. 15.6b is shown with framing anchors on both the right and left sides of the wall plate. In this way, cross-grain tension in the plate is not being relied upon for load transfer. The framing anchors shown with solid lines are those that stress the plate in compression perpendicular to grain.

For a given direction of lateral force, the effective framing anchors cause the wood plate to bear on the anchor bolt in compression. Without fasteners on both sides of the anchor bolt, cross-grain tension would be developed in the plate for one of the directions of loading. The shear in the wood plate should be checked in accordance with the procedures given in Sec. 13.9.

Other connection details can be used to avoid cross-grain tension. One alternative, shown in Fig. 15.6c, avoids the problem by tying the framing member directly to the anchor bolt. Use of this type of anchorage requires close field coordination. The anchor bolts must be set accurately when the wall is constructed to match the future locations of the roof or floor framing.

Figure 15.6b   For each case, the framing anchor shown with solid lines is capable of transferring the lateral force to the anchor bolt by compression perpendicular to grain.

PARTIAL CONNECTION DETAIL
SHOWING ALT. ANCHORAGE
FOR NORMAL TO WALL FORCES

Figure 15.6c   Angle bracket is connected directly to framing and anchor bolt. Therefore, normal wall force is transferred directly to the framing without placing the wood plate on top of wall in cross-grain tension.

Subdiaphragm Anchorage

For a discussion of the subdiaphragm anchorage of concrete and masonry walls, see Example 15.5 and Sec. 15.5.

15.5   Subdiaphragm Anchorage of Concrete and Masonry Walls

The anchorage requirements for concrete and masonry walls were outlined in Sec. 15.4. There it was noted that the code requires *continuous ties* between the diaphragm chords in order to distribute the anchorage forces perpendicular to a wall into the diaphragm (ASCE 7 Sec. 12.11.2.2.1). The concept of subdiaphragms (also known as minidiaphragms) was developed to satisfy this code requirement. The concept is fully developed in Ref. 15.8.

If the framing members in a diaphragm are continuous members from one side of the building to the other, the diaphragm would naturally have continuous ties between the chords. See Example 15.8. This is possible, however, only in small buildings or in buildings with closely spaced trusses. Even if continuous framing is present in one direction, it would normally not be present in the other direction.

EXAMPLE 15.8   Diaphragm with Continuous Framing

Continuous framing in the transverse direction would provide continuous ties between diaphragm chords for the transverse lateral force. For this force, the chords are parallel to the longitudinal walls. Continuous ties distribute the anchorage force back into the diaphragm. The action of a continuous tie is similar in concept to a collector or drag strut. The tie drags the wall force into the diaphragm.

If the spacing of wall anchors exceeds 4 ft, the wall must be designed to resist bending between the anchors. For the lateral force in the longitudinal direction, no intermediate ties are shown perpendicular to the continuous framing (Fig. 15.7). Ties must be designed and spliced across the transverse framing, or the transverse wall must be designed for bending (at the diaphragm level) between the longitudinal shearwalls (not practical in most buildings).

PLAN

Figure 15.7   Building with continuous cross ties between diaphragm chords in one direction provided by continuous framing.

The majority of buildings do not have continuous framing. In a typical panelized roof system, for example, the subpurlins span between purlins, purlins span between girders, and girders span between columns. See Example 15.9. In addition, the tops of the beams are usually kept at the same elevation, and the diaphragm sheathing is nailed directly to the framing members. In order for this type of construction to be used, the lighter beams are not continuous across supporting members. Lighter beams are typically suspended from heavier beams with metal hangers (Chap. 14).

If continuous cross ties are to be provided between diaphragm chords in both directions of the building, a larger number of additional connections will be required. In order to meet the continuous cross tie requirement literally, each beam would have to be spliced for an axial drag-type force across the members to which it frames.

EXAMPLE 15.9   Diaphragm without Continuous Framing

In a building without continuous framing between the diaphragm chords, a large number of cross-tie splices would be required (Fig. 15.8).

a. Subpurlins spliced across purlins.

b Purlins spliced across girders.

c. Girders spliced across hinges.

Splices must be capable of transmitting the wall anchorage force tributary to a given member (Fig. 15.7). With the use of subdiaphragms, the continuous-tie requirement may be satisfied *without* connection splices at every beam crossing.

PLAN

Figure 15.8 Continuous cross ties between diaphragm chords by splicing each framing member would require an excessive number of connections.

The idea of a *subdiaphragm* was developed to ensure the proper anchorage of the wall seismic force without requiring a cross tie splice at every beam crossing. With this method the designer selects a portion of the total diaphragm (known as a subdiaphragm) which is designed as a separate diaphragm. See Example 15.10. The unit *shear* capacity (wood structural panel thickness and nailing) must be sufficient to carry the tributary wall lateral forces, and the subdiaphragm must have its own *chords* and *continuous cross ties* between the chords. Subdiaphragm chords and cross ties are also referred to as *subchords* and *subcross ties* (or *subties*).

The subdiaphragm reactions are in turn carried by the main cross ties, which must be continuous across the full diaphragm. Use of the subdiaphragm avoids the requirement for continuous ties across the full width of the diaphragm except at the subdiaphragm boundaries.

ASCE 7 Sec. 12.11.2.2.1 specifically requires continuous ties or struts between diaphragm chords for SDC C and higher, and specifically permits use of subdiaphragms to meet this requirement. This section also places an upper limit of 2.5 on the subdiaphragm length-to-width ratio.

EXAMPLE 15.10   Subdiaphragm Analysis

The building from the previous example is redrawn in Fig. 15.9a showing the subdiaphragms which will be used for the wall anchorage. The assumed action of the subdiaphragms for resisting the lateral forces is shown in Figs. 15.9b and 15.9c. The subdiaphragm concept is only a computational device that is used to develop adequate anchorage of concrete and masonry walls. It is recognized that the subdiaphragms are actually part of the total diaphragm, and they do not deflect independently as shown.

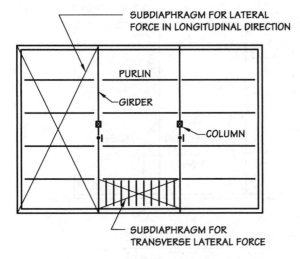

SUBDIAPHRAGM FOR LATERAL
FORCE IN LONGITUDINAL DIRECTION

PURLIN

GIRDER

COLUMN

SUBDIAPHRAGM FOR
TRANSVERSE LATERAL FORCE

PLAN

Figure 15.9a   Subdiaphragms may be used to provide wall anchorage without requiring the large number of connections described in Fig. 15.8.

SUBDIAPHRAGMS FOR
ANCHORING FRONT WALL

Figure 15.9b Subdiaphragms spanning between the continuous cross ties for the complete diaphragm.

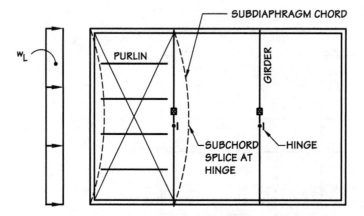

SUBDIAPHRAGMS FOR
ANCHORING END WALL

Figure 15.9c A subdiaphragm can have the same span as the complete diaphragm. However, the subdiaphragm is designed for the wall anchorage force, and the complete diaphragm is designed for the total lateral force to the diaphragm.

Three subdiaphragms for the anchorage of the front longitudinal wall are shown in Fig. 15.9b. Three similar subdiaphragms will also be used for the anchorage of the rear wall. Note that the transverse anchorage force can act in either direction.

The purlin and wall serve as the chords (subchords) for the subdiaphragms. The purlin must be designed for combined bending and axial forces. Because the subdiaphragms are treated as separate diaphragms, the subchord forces are zero at the girder. The purlins, therefore, need not be spliced across the girders.

The subpurlins serve as continuous cross ties (subties) between the subdiaphragm chords and do not have to be spliced cross the purlin. The girders act as continuous cross ties for the complete diaphragm, and a cross-tie splice at the hinge is required. The splice must carry the reactions from two subdiaphragms.

The subdiaphragm for anchorage of the left wall is shown in Fig. 15.9c. A similar subdiaphragm is required for anchoring the right wall. The girder and wall serve as chords for the subdiaphragm. The girder must be designed for combined bending from gravity loads plus axial tension or compression caused by lateral forces.

The splice at the girder hinge must be designed for the subdiaphragm chord force. This is an alternate force to the continuous cross-tie force caused by the transverse lateral force in Fig. 15.9b. The purlins serve as continuous ties between subdiaphragm chords and do not have to be spliced across the girder.

The designer has some flexibility in the choice of what portions of the complete diaphragm are used as subdiaphragms. See Example 15.11. Figure 15.10a in this example shows that a number of small subdiaphragms are used to anchor the transverse wall. In Fig. 15.10b, one large subdiaphragm is used to anchor the same wall force between the longitudinal walls only. For additional examples illustrating subdiaphragm choices, see Refs. 15.8 and 15.9.

The flexibility permitted in the choice of subdiaphragm arrangements results from the fact that the subdiaphragm concept is a computational device only. Independent deflection of the subdiaphragms is not possible, and the analysis is used arbitrarily to ensure adequate anchorage connections.

Various types of connections can be used to provide the splices in the cross ties and chords of the diaphragm and subdiaphragms. Because most beams frame into the sides of the supporting members, compression forces are normally assumed to be taken by bearing. Thus the connections are usually designed for tension. This tension tie can be incorporated as part of the basic connection hardware (e.g., the seismic tie splice on the hinge connection in Fig. 14.19), or a separate tension strap may be provided (Fig. 14.22). The connections may be specially designed using the principles of Chaps. 12, 13, and 14, or some form of prefabricated connection hardware may be used.

Where compression forces are to be transferred through bearing, design and detailing attention is needed to make sure that wood shrinkage and construction tolerances do not disrupt the bearing, and that tension and compression cycling of the joint will not lead to premature failure (such as low-cycle fatigue from buckling of cold-formed steel straps). Alternately, connection devices rated for both tension and compression forces may be used.

EXAMPLE 15.11   Alternate Subdiaphragm Configurations

The subdiaphragm arrangement in Fig. 15.10a uses three small subdiaphragms. The two outside subdiaphragms span between the longitudinal shearwalls and the continuous cross ties for the complete diaphragm. The center subdiaphragm spans between the two continuous cross ties. The line of purlins forming the two main cross ties must be spliced at each girder crossing (in this example a column happens to occur at each of these locations) for the tributary wall anchorage force.

The other purlins in the subdiaphragms serve as continuous ties (subties) between the subdiaphragm chords, and they need not be spliced for lateral forces across the girder line.

The transverse wall and first row of girders serve as the subchords. The subdiaphragm chord force is zero at the ends of the girders, and subchord splices are not required at the columns.

Figure 15.10a   Three small subdiaphragms used to anchor transverse end wall.

In the second subdiaphragm layout (Fig. 15.10b), one large subdiaphragm spans between the exterior walls. Continuous subtie splices are required for the purlins across the first girder only. The subdiaphragm chord formed by the second line of girders must be tied across the columns for the appropriate subchord force.

The choice of the subdiaphragm configuration in practice will depend on the required number of connections when both the transverse and longitudinal forces are considered.

Figure 15.10b   One large subdiaphragm used to anchor transverse end wall.

The reader is reminded that the lateral forces shown in the examples in this section are *wall anchorage forces*. In other words, the forces represent one of the four perpendicular-to-wall forces described in Example 15.2, and they are *not* the lateral forces for designing the main diaphragm.

Subdiaphragms are a significant design concern in structures with concrete or masonry walls. The primary purpose of Examples 15.8 to 15.11 was to define the concept of a subdiaphragm and to illustrate where additional seismic connections may be required.

15.6 Design Problem: Subdiaphragm

In Example 15.12, the wall anchorage requirements for a large one-story building are analyzed. Exterior walls are of reinforced brick masonry construction.

In order to check the unit shear in the subdiaphragms, the construction of the entire roof diaphragm must be known. For this reason, the unit shear diagrams for the complete diaphragm are shown along with the nailing and the tabulated unit shears. Two different nailing patterns are used for the roof diaphragm. The heavier nailing occurs near the ends of the building in areas of high diaphragm shear. This different nailing may be important when the unit shears in the subdiaphragms are checked.

Along the longitudinal wall a series of small subdiaphragms is used to anchor the wall (Fig. 15.11b). Subdiaphragm 1 spans between girders which serve as continuous cross ties for the entire roof diaphragm in the transverse direction. Because subdiaphragm 1 occurs in both areas of diaphragm nailing, the unit shear in the subdiaphragm is first checked against the lower tabulated shear for the complete diaphragm.

The lower shear capacity is found to be adequate for subdiaphragm 1. If the lower shear value had not been adequate, an 8-ft-wide strip of heavier nailing could be added next to the longitudinal wall. An alternative solution would be to increase the size of subdiaphragm 1.

The transverse wall is anchored by a large subdiaphragm that spans between the longitudinal walls (Fig. 15.11c). Continuous cross ties are required only between the subdiaphragm chords. Because a portion of subdiaphragm 2 coincides with the area of lighter nailing for the complete diaphragm, the unit shear is checked against the lower shear capacity of the diaphragm.

The use of the larger-diaphragm shear capacity (425 lb/ft) requires that the heavier boundary nailing be provided at the perimeter of each subdiaphragm. Care must be taken to avoid potential splitting of lumber that might result from the use of closely spaced nails at adjacent subdiaphragm boundaries. See SDPWS Section 4.2.7.

EXAMPLE 15.12   Subdiaphragm Anchorage for Masonry Walls

The complete roof diaphragm for the building in Example 15.11 has been designed for seismic forces in both the transverse and longitudinal directions using ASD procedures. The unit shears and wood structural panel nailing are summarized below. A subdiaphragm analysis is to be performed using subdiaphragm 1 (Fig. 15.11b) for the longitudinal walls and subdiaphragm 2 (Fig. 15.11c) for the transverse walls. The calculation of subdiaphragm unit shears and tie forces is demonstrated.

Given Information

Total wall height is 16 ft. Tributary wall height to roof diaphragm:

$$h = \frac{16}{2} = 8 \text{ ft}$$

Dead loads:

$$\text{Roof dead load D} = 11 \text{ psf}$$

$$\text{Wall dead load D} = 90 \text{ psf}$$

Seismic coefficient: Assume that the strength level seismic coefficient for design of the roof diaphragm is $0.225g$, and S_{DS} is 1.0. This is for the main seismic force system. Calculation of the seismic forces for wall anchorage and subdiaphragms will follow.

Figure 15.11a Lateral forces on entire diaphragm. Unit shear diagrams caused by transverse and longitudinal lateral forces are also shown.

Entire Roof Diaphragm

Seismic force to roof diaphragm (*strength level*):

$$\text{Transverse } w_u = 0.225[11 \text{ psf}(120 \text{ ft}) + 90 \text{ psf } (8 \text{ ft}) (2 \text{ walls})]$$

$$= 621 \text{ lb/ft}$$

$$\text{Longitudinal } w_u = 0.225 [11 \text{ psf}(200 \text{ ft}) + 90 \text{ psf } (8 \text{ ft}) (2 \text{ walls})]$$

$$= 819 \text{ lb/ft}$$

Roof unit shears (*strength level*):

$$\text{Transverse } V_u = \frac{w_u L}{2} = \frac{621(200)}{2} = 62{,}100 \text{ lb}$$

$$v_u = \frac{V_u}{b} = \frac{62{,}100}{120} = 518 \text{ lb/ft}$$

$$\text{Longitudinal } V_u = \frac{w_u L}{2} = \frac{819(120)}{2} = 49{,}140 \text{ lb}$$

$$v_u = \frac{V_u}{b} = \frac{49{,}140}{200} = 246 \text{ lb/ft}$$

Roof unit shears (*allowable stress design level*): Since the shear forces in the roof diaphragm result entirely from seismic forces, the applicable load combinations reduce to 0.7E.

$$\text{Transverse } v = 0.7(518) = 362 \text{ lb/ft}$$

$$\text{Longitudinal } v = 0.7(246) = 172 \text{ lb/ft}$$

See unit shear diagrams (Fig. 15.11a).

Wood structural panel sheathing and nailing:

$\frac{15}{32}$-in. STRUCTURAL I plywood.

All edges supported and nailed into 2-in. minimum nominal framing (i.e., blocking is used).

Assume wood structural panel load cases 2 and 4 (Table 2306.3.1 of the IBC). Note that wood structural panel layout is not shown, but will eventually be included on the plans.

> *Diaphragm nailing a.*   10 d common nails at:
>   4-in. o.c. boundary and continuous panel edges
>   6-in. o.c. other panel edges (blocking required)
>   12-in. o.c. field

$$\text{Allow. } v = 425 \text{ lb/ft} > 362 \text{ lb/ft} \quad OK$$

> *Diaphragm nailing b.*   10d common nails at:
>   6-in. o.c. all edges (blocking required)
>   12-in. o.c. field

$$\text{Allow. } v = 320 \text{ lb/ft}$$

From similar triangles on the unit shear diagram, a unit shear of 320 lb/ft occurs 11.6 ft from the transverse shearwalls.

∴ Use diaphragm *nailing a* for distance of 20 ft from transverse shearwalls (11.6 ft is minimum), and use diaphragm *nailing b* in central 160 ft of diaphragm.

NOTE: In the longitudinal direction two allowable shears apply: 425 lb/ft for *nailing a* and 320 lb/ft for *nailing b*. Both of these allowable shears exceed the actual shear of 172 lb/ft. Therefore, unit shear in the longitudinal direction is not critical.

Subdiaphragm Anchorage of Longitudinal Masonry Wall

Subdiaphragm 1(8 ft × 20 ft):

The seismic forces for wall anchorage were introduced in Sec. 15.4 and illustrated in Example 15.6, which also used S_{DS} = 1.0. From Example 15.6 the critical wall anchorage force was found to be $0.8W_P$, based on ASCE 7 Sec. 12.11.2.1. This seismic force is at the *strength level* and is used to determine the force normal to the wall.

$$w_u = 0.80W_P = 0.80(90 \text{ psf})$$

$$= 72 \text{ psf}$$

Anchorage force is the "roof reaction" which is obtained as the uniform seismic force times the tributary height to the roof level.

$$w_{Tu} = 72 \text{ psf} \times 8 \text{ ft} = 576 \text{ lb/ft} > \text{code minimum of } 280 \text{ lb/ft}$$

Since this force is the result from seismic forces only, the applicable load combinations reduce to 0.7E.

$$w_T = 0.7(576) = 403 \text{ lb/ft} > \text{required wind anchorage (wind analysis not shown)}$$

Subdiaphragm span-to-width ratio:

$$\frac{L}{b} = \frac{20}{8} = 2.5 \quad OK$$

Subdiaphragm unit shear:

A typical 8 ft × 20 ft subdiaphragm (subdiaphragm 1) in the central 100 ft of the building will have the critical unit shear.

$$\text{Max. } V = \frac{wL}{2} = \frac{403(20)}{2} = 4030 \text{ lb}$$

$$v = \frac{V}{b} = \frac{4030}{8} = 504 \text{ lb/ft} > 320 \text{ lb/ft} \quad NG$$

The nailing for the entire roof diaphragm in the center portion of the building (i.e., diaphragm nailing *b*) is not adequate for subdiaphragm 1. There are several possible options. For example, the 8-ft width of subdiaphragm 1 could be increased to 16 ft to reduce the unit shear. In other words, subdiaphragm 1 could be changed from an 8 ft × 20 ft subdiaphragm to a 16 ft × 20 ft subdiaphragm. Although this would reduce the unit shear in the subdiaphragm, it would require the addition of subtie splice connections across the first line of purlins next to the longitudinal wall. An alternative solution could be to provide increased nailing for the original 8 ft × 20 ft subdiaphragm 1. This could be done by extending nailing *a* for the entire roof diaphragm, to include an 8-ft-wide strip along the longitudinal walls. The second solution with increased nailing will be assumed for the following demonstration of subdiaphragm design.

For the following anchorage force calculations, refer to Fig. 15.11*b*.

PLAN

Figure 15.11b Subdiaphragm 1 (8 ft × 20 ft) anchors longitudinal masonry wall for seismic force w_T in transverse direction.

Girder—continuous cross-tie force at columns:

$$T = (403 \text{ lb/ft})(20 \text{ ft}) = 8060 \text{ lb}$$

The continuous-cross-tie connections along the girders at the columns must be designed for a force of 8060 lb

NOTE: The purpose of this example is to define the subdiaphragm problem and to determine the associated design forces. Detailed connection designs for these forces are beyond the scope of the example. See Chaps. 12 to 14 for design principles.

Subpurlin—anchor wall to subpurlin at 4 ft-0 in. o.c.

Every other subpurlin will serve as a continuous cross tie (subtie) for subdiaphragm 1. The masonry wall need not be designed for bending between wall anchors that are spaced 4 ft-0 in. o.c. or less. As with the girder splices above, subpurlin anchors must be designed for a force:

$$T = (403 \text{ lb/ft})(4 \text{ ft}) = 1612 \text{ lb}$$

An anchor bracket with a code-recognized load capacity in excess of this value should be provided between the masonry wall and the 2× subpurlins (subties).

The anchorage brackets may be attached to subpurlins that occur at the edge of a wood structural panel. However, unless the designer is assured of this, it should be assumed that the anchors occur at subpurlins located at the center of a wood structural panel. In this case, the capacity of the diaphragm *field nailing* should be checked against the wall anchorage force. The capacity of a 10d common nail in a wood structural panel diaphragm nailed to DF-L framing is

$Z = 88$ lb/nail (obtained from NDS Table 11Q for $^{15}\!/_{32}$ -in. wood structural panel. As an alternative to the NDS table, yield limit equations may be evaluated.)

The following adjustments apply:

$C_D = 1.6$ for seismic force

$C_{di} = 1.1$ wood structural panel diaphragm nailing (Sec. 12.6)

Nail penetration is adequate by inspection, and all other adjustment factors are unity.

$$Z' = Z\,(C_D C_M C_t C_{eg} C_{di} C_{tn})$$
$$= 88\,(1.6)(1.0)(1.0)(1.0)(1.1)(1.0)$$
$$= 155 \text{ lb/nail}$$

NOTE: Reference 15.11 recommends an 11 percent reduction in nail value when diaphragm nailing is driven into narrow (2-in. nominal) framing. The 0.9 multiplying factor in the next step is to account for this reduction. Depending on materials, other adjustments may apply. Contact APA for the latest information regarding nail design values for diaphragms based on yield limit theory.

Modified $Z' = 0.9(155) = 139$ lb/nail

The usual spacing of nails in the field (i.e., along intermediate framing) of a wood structural panel is 12-in. o.c. Subpurlins are 8 ft long, and at least 8 nails are available to develop the anchorage force.

$$P = 8(139) = 1115 \text{ lb}$$

This sheathing nailing capacity is less than the subpurlin anchorage force of 1612 lb, so additional fastening is required. It is not unusual to specify a row of diaphragm edge nailing (i.e., nailing at the sheathing edge spacing) at each subpurlin anchor. Vigilance is required, however, to ensure that this requirement is understood and installed by the contractor.

Subdiaphragm chord force purlins (refer again to Fig. 15.11*b*):

The first line of purlins parallel to the masonry wall serves as the subchord for sub-diaphragm 1.

$$T_{\max} = C_{\max} = \frac{M}{b} = \frac{wL^2}{8b} = \frac{403(20)^2}{8(8)} = 2520 \text{ lb}$$

A subchord splice across the girders is not required. However, the subchord must be checked for combined axial stress (from the subchord force) plus bending stress (from gravity loads).

Subdiaphragm Anchorage of Transverse Masonry Wall

PLAN

Figure 15.11c   Subdiaphragm 2 (60 ft × 120 ft) anchors transverse masonry wall for seismic force w_L in longitudinal direction.

Subdiaphragm 2 (60 ft × 120 ft)(Fig. 15.11c):

Wall anchorage force (from subdiaphragm 1):

$$w_L = 403 \text{ lb/ft}$$

Span-to-width ratio:

$$\frac{L}{b} = \frac{120}{60} = 2 < 2.5 \qquad OK$$

Subdiaphragm shear:

$$V = \frac{(403)(120)}{2} = 24{,}180 \text{ lb}$$

$$v = \frac{V}{b} = \frac{24{,}180}{60} = 403 \text{ lb/ft} > 320 \text{ lb/ft} \qquad NG$$

Diaphragm nailing b is not adequate for subdiaphragm 2, so to qualify for an allowable shear of 425 lb/ft (nailing a), the boundary nailing of 10d common nails at 4-in. o.c. will need to be provided along the second girder line in addition to the usual boundary nailing locations for nailing a.

Purlin—subtie splice force across the first girder line:

$$T = (403 \text{ lb/ft}) (8 \text{ ft}) = 3224 \text{ lb}$$

All the subtie splices across the first and second girder lines must be designed for this force.

Purlin—anchor wall to purlin at 8 ft-0 in. o.c.:

The connection hardware between the masonry wall and the purlins has the same design forces as the subtie splices just determined: $T = 3224$ lb.

The nailing between the wood structural panel sheathing and the cross tie for subdiaphragm 2 is not as critical as subdiaphragm 1. The maximum subtie force has doubled ($2 \times 1612 = 3224$ lb), but the cross-tie length has far more than doubled (the length is 60 ft instead of 8 ft). In addition, wood structural panel edge nailing (not just the field nailing) will occur along the subties (purlins) due to the wood structural panel layout. Therefore, a less critical subtie nailing situation exists for subdiaphragm 2.

Note that the masonry wall must be designed for bending because the 8-ft spacing of anchors exceeds the 4-ft code limit. This is a masonry design problem and is not included here.

Subdiaphragm chord force in second girder line:

The maximum chord force in subdiaphragm 2 (Fig. 15.11c) is

$$T = C = \frac{M}{b} = \frac{wL^2}{8b} = \frac{403(120)^2}{8(60)} = 12{,}090 \text{ lb}$$

The subdiaphragm chord along the second girder should be spliced at the columns for the subchord force of 12,090 lb, instead of the continuous-cross-tie force of 8060 lb (see calculations for subdiaphragm 1).

An alternate for subdiaphragm 2 is to use several smaller subdiaphragms similar to the first arrangement in Example 15.11.

The design of anchorage for concrete and masonry walls to wood roof systems using the concept of subdiaphragms has been utilized since the early 1970s.

15.7 References

[15.1] American Concrete Institute (ACI), 2005. *Building Code Requirements for Structural Concrete* (ACI 318-05), ACI, Farmington Hills, MI.

[15.2] American Concrete Institute (ACI), American Society of Civil Engineers (ASCE), and The Masonry Society (TMS), 2005. *Building Code Requirements for Masonry Structures* (ACI 530-05/ ASCE 5-05/TMS 402-05), ACI, Farmington Hills, MI; ASCE, Reston, VA; and TMS, Boulder, CO.

[15.3] American Forest and Paper Association (AF&PA). 2005. *Special Design Provisions for Wind and Seismic (SDPWS) Supplement to the NDS,* 2005 ed., AF&PA, Washington DC.

[15.4] American Forest and Paper Association (AF&PA). 2005. *National Design Specification for Wood Construction and Supplement,* ANSI/AF&PA NDS-2005 AF&PA, Washington DC.

[15.5] American Forest and Paper Association (AF&PA). 1999. *Commentary on the National Design Specification for Wood Construction,* 1997 ed., AF&PA, Washington DC.

[15.6] Hamburger, Ronald O., and McCormick, David L. 1994. "Implications of the January 17, 1994 Northridge Earthquake on Tiltup and Masonry Buildings with Wood Roofs," Proceedings, 1994 Spring Seminar, Structural Engineers Association of Northern California, San Francisco, CA.

[15.7] International Codes Council (ICC). 2006. *International Building Code,* 2006 ed., ICC, Falls Church, VA.

[15.8] Sheedy, Paul. 1983. "Anchorage of Concrete and Masonry Walls," *Building Standards,* September/October, 1983, Whittier, CA (reprint available with comments from APA, Tacoma, WA).

[15.9] Sheedy, Paul, and Sheedy, Christine. 1992. "Concrete and Masonry Wall Anchorage," *Buildings Standards,* July/August, 1992, ICBO, Whittier, CA.

[15.10] Structural Engineers Association of Northern California (SEAONC). 2001. *Guidelines for Seismic Evaluation and Rehabilitation of Tilt-up Buildings and Other Rigid Wall/Flexible Diaphragm Structures*. Structural Engineers Association of Northern California, San Francisco, CA.

[15.11] Tissel, J.R., and Elliot, J.R. 1997. Plywood Diaphragms, Research Report 138, APA—The Engineered Wood Association, Engineered Wood Systems, Tacoma, WA.

Chapter

16

Advanced Topics in Lateral Force Design

16.1 Introduction

A number of important issues have been incorporated into recent editions of the code. One significant issue is consideration of whether a structure is *regular* or *irregular*. In addition, specific overturning requirements for earthquake forces have been modified. It is important for the designer to understand the similarities and differences in the overturning requirements for wind and seismic forces.

This chapter provides an introduction to seismic irregularity considerations, gives a more detailed review of the overturning requirements for both wind and seismic forces, and takes a detailed look at the diaphragm flexibility issues introduced in Sec. 9.11.

16.2 Seismic Forces—Regular Structures

Experience has shown that regular structures perform much better in earthquakes than irregular structures. The subject of irregularities has added to the length and complexity of the seismic code. To demonstrate how involved this subject has become, the seismic code identifies 10 types of structural irregularities (ASCE 7 Sec. 12.3.2) (Ref. 16.3). In order to organize these into a workable system, the code summarizes these irregularities in tables, and the tables then reference appropriate sections of the code which give the detailed requirements for each type of irregularity. The commentary to the NEHRP Provisions (Ref. 16.9) expands on each item. Obviously, a comprehensive treatment of structural irregularity is beyond the scope of this introductory text. However, it is essential that the designer of a wood-frame building have some knowledge of this subject.

Several examples of *regular* and *irregular structures* are given in this chapter to illustrate the general concepts. The wood-frame buildings that were considered in previous chapters meet the basic definition of a regular structure. Dealing with this basic type of structure allowed a number of important subjects to be covered without becoming overly complicated. These subjects included diaphragms, shearwalls, chords, collectors (drag struts), and anchorage forces.

However, regular structures are rather limited in plan and elevation, and many buildings do not qualify as regular structures. It is important for the designer to be able to recognize what constitutes a structural irregularity and to know what, if any, additional requirements are imposed because of the irregularity. A number of features that can cause structures to be classified as irregular are addressed in Sec. 16.3.

Regular structures are those that have no significant physical or structural irregularities in either plan view or elevation. For example, a simple rectangular wood-frame building with a uniform distribution of mass can fit this description. However, in order to be classified as structurally regular an additional requirement for multistory buildings is that the shearwalls in each upper story must be located directly over the shearwalls in each lower story. See Example 16.1.

EXAMPLE 16.1   Regular Structure

Regular structures have a history of good performance in resisting earthquake forces. These structures have no *horizontal irregularities* or *vertical irregularities* (see Fig. 16.1).

A building will be classified as structurally regular if

1. The plan is essentially *rectangular*.
2. The *mass* is reasonably uniform over the height.
3. The *shearwalls* in a lower story are directly below the shearwalls in each upper story.

Figure 16.1   Regular structure. Recall from Chaps. 2 and 3 that two different distributions of seismic forces are provided by the code: F_x for shearwall design, and F_{px} for diaphragm design. Although the sketch shows lateral forces in both directions, each principal direction of the structure is considered separately.

Both transverse and longitudinal lateral seismic forces are shown in the sketch. Normal practice is to consider the force in each direction separately.

For regular structures, the seismic forces are computed and distributed as illustrated in Chaps. 2 and 3. The design of the lateral-force-resisting systems (LFRSs) for these buildings follows the procedures given in Chaps. 9 and 10, and the anchorage details are handled as described in Chap. 15.

The IBC load combinations, load combination factors, and load duration factors were covered in Chap. 2. Before proceeding with a discussion of seismic irregularity, several topics from Chap. 2 will be revisited. Recall that the IBC has two sets of load combinations for allowable stress design (ASD). The newer and preferred combination of loads is known as the ASD *basic load combinations*. The older ASD load combinations from the 1994 (and earlier) codes may still be used and are now known as the ASD *alternate basic load combinations*. All of the examples in this book make use of the ASD and LRFD basic load combinations.

16.3   Seismic Forces—Irregular Structures

An *irregular structure* has one or more significant discontinuities in its configuration or in its LFRS. Several examples of irregular structures are given in this section to illustrate the general concept of structural irregularities.

The 10 general types of irregularity identified in the code are classified as either *vertical irregularities* (ASCE 7 Table 12.3-2) or horizontal (plan) *irregularities* (ASCE 7 Table 12.3-1). Detailed descriptions of the irregular features are given in these tables. As mentioned previously, the commentary to the NEHRP Provisions expands these descriptions and helps to clarify the intent of the code.

The discussion in this section will first introduce the irregularities and then discuss the design penalties applied to structures with these irregularities. The first form of irregularity considered in this introduction has an effect on the dynamic properties of a building that is easy to visualize. It is a type of vertical irregularity known as *weight (mass) irregularity* (ASCE 7 Table 12.3-2, Type 2). This is best described by considering the lumped-mass model of the three-story building in Fig. 2.13b (Sec. 2.13). In this figure, the F_x story forces are shown to follow a triangular distribution with zero force at the bottom and maximum force at the roof level. This triangular distribution is obtained if the masses at the story levels are all equal. Obviously, some other distribution will result (using the code formula for F_x) if the masses are not all the same magnitude.

Now, if one of the masses is significantly larger or smaller than the mass at an adjacent level, the dynamic behavior of the system will be altered. Thus, *mass irregularity* is said to exist if the mass of any story is more than 150 percent of the mass of an adjacent story. However, a roof which is lighter than the floor below need not be considered as being mass-irregular. The fundamental mode (first mode) of vibration may not be the controlling mode for a structure that has mass irregularity.

Recall that the code recognizes two basic types of seismic analysis, *static procedures* and *dynamic procedures*. Static procedures (e.g., *equivalent lateral force analysis*) are generally applied to smaller buildings, while dynamic procedures are applied to larger buildings. The code now requires that dynamic procedures be used for more structures than in the past. In Seismic Design Categories (SDCs) D, E, and F, ASCE 7 Table 12.6-1 requires a dynamic analysis procedure for structures with horizontal irregularity Type 1 and vertical irregularity Types 1, 2, and 3. Buildings of light-frame construction not exceeding three stories in height are exempt as long as they are used for standard occupancies (Occupancy Categories I and II). Some irregular wood light-frame buildings greater than three stories, and those used for essential or hazardous facilities, will now require a dynamic analysis procedure. This will be problematic because dynamic analysis tools for wood-frame structures are just now being developed at a research level and still require much further development and calibration before they are appropriate for design application. Because of the exemption, however, the majority of wood light-frame buildings can still use equivalent lateral force analysis.

The example of mass irregularity is given here as a simple illustration of a vertical irregularity. One can easily visualize how the mode shape in Fig. 2.13*b* could be significantly altered by having one of the lumped masses in the system much larger or smaller than the others. The designer of any structure should generally be aware of mass irregularity and should pay particular attention to tying the structure together for lateral force transfer and continuity.

The question of diaphragm flexibility was also discussed in Chap. 9, and it was noted that wood diaphragms have usually been classified as *flexible* diaphragms. If a building uses a flexible diaphragm in its lateral-force-resisting system, the code does not require that *torsional irregularity* (Horizontal Irregularity Type 1) be considered. However, a rigid diaphragm analysis should be considered for buildings with a notable separation between the center of mass and center of rigidity. The separation of the center of mass and center of rigidity is the same characteristic that causes a building to become torsionally irregular, resulting in one side of the building drifting significantly more than the average story drift. It should be noted that few wood buildings will have a torsional irregularity significant enough to qualify as torsionally irregular under the code criteria discussed in Sec. 16.9. Recent research (Refs. 16.7 and 16.11) has suggested that current code design procedures are not adequately identifying the most torsionally irregular wood-frame structures. Until design procedures can be revised, the designer is encouraged to consider irregularity requirements in any building that can be visually identified as having significant torsional eccentricity and low torsional redundancy.

Reentrant corners (ASCE 7 Table 12.3-1) occur frequently in wood-frame structures. Typical L-, T-, and U-plan shaped buildings are common examples of structures with reentrant corners. When *both projections* beyond a reentrant corner are greater than 15 percent of the plan dimension of the structure in a given direction, the building is classified as irregular. See Fig. 16.2. A method

PROJECTING WING 2

P_1

P_2

L_2

REENTRANT CORNER

PROJECTING WING 1

L_1

HORIZONTAL IRREGULARITY WHEN:

$$\frac{P_1}{L_1} > 0.15$$

AND

$$\frac{P_2}{L_2} > 0.15$$

PLAN OF L-SHAPED BUILDING

Figure 16.2   A building with a *reentrant corner* is classified as horizontal irregularity Type 2 when both projections exceed 15 percent of the corresponding side lengths.

of calculating the design forces for collectors and related connections for a typical L-shaped building is demonstrated in Sec. 16.8.

Another type of horizontal irregularity is described as a *diaphragm discontinuity* (ASCE 7 Table 12.3-1, Type 3). An example of a diaphragm discontinuity is a large opening in a floor or roof diaphragm. See Fig. 16.3.

Diaphragms with openings have additional diaphragm chords and collectors (drag struts). These are framing members at the boundary of the diaphragm opening. If a free body diagram is cut through the diaphragm at the opening, the chords at the opening plus the chords at the boundary of the diaphragm carry the moment in the diaphragm. A method of analyzing the diaphragm chord and drag strut forces in a diaphragm with an opening is given in Ref. 16.4 and additional information is available in Refs. 16.5 and 16.6. The evaluation of forces described in these references applies whether or not the opening in the diaphragm is large enough to be classified as a Type 3 horizontal irregularity.

Tests of diaphragms with openings are also reported in Ref. 16.4. These indicate that large forces can develop at the corners of the openings in the diaphragm and that the magnitude of the forces increases with increased opening size.

HORIZONTAL IRREGULARITY WHEN:

$$\frac{A_1}{A} > 0.5$$

PLAN – DIAPHRAGM WITH OPENING

Figure 16.3 A diaphragm with a large opening is a *diaphragm discontinuity* and is known as horizontal irregularity Type 3. The building is classified as having a horizontal irregularity if the size of the opening A_1 is greater than 50 percent of the gross area A.

Design of reinforcing around the opening is required regardless of whether or not design penalties are involved. Limiting the size of the opening will reduce the required reinforcing and improve the performance of the structure.

On the other hand, if an opening is small, Ref. 16.4 offers the following general guide to designers: ". . . When openings are relatively small, chord forces do not increase significantly and it is usually sufficient to simply reinforce perimeter framing and assure that it is continuous. Continuous framing should extend from each corner of the opening both directions into the diaphragm a distance equal to the largest dimension of the opening."

The fourth type of horizontal irregularity is known as an *out-of-plane offset* (ASCE 7 Table 12.3-1, Type 4). This is a particularly serious type of discontinuity because it interrupts the flow in the normal path of lateral forces down through the structure and into the foundation. Because of this, the use of the *special seismic load combinations* and the *special seismic (overstrength) force* E_m is required for supporting members.

To describe this type of horizontal irregularity, consider a two-story building with diaphragms at the second floor and roof levels. See Example 16.2. The supporting walls include an interior shearwall as well as the usual exterior shearwalls. This example focuses on the interior shearwall, and the exterior walls are not shown in detail.

Initially, the building is considered *without* a horizontal irregularity (i.e., an out-of-plane offset does not exist) so that the *normal flow* of forces down through the structures can first be understood. See Fig. 16.4a. Note that the interior

ROOF DIAPHRAGM

SEISMIC FORCES TO
ROOF DIAPHRAGM

INTERIOR SHEARWALLS
ALIGN VERTICALLY

COLLECTOR
AT ROOF LEVEL

ROOF

2ND FLOOR

COLLECTOR
AT 2ND FLOOR

1ST FLOOR

2 – STORY SHEARWALL WITHOUT OFFSET

Figure 16.4a Structurally *regular* two-story building. The alignment of the shear-walls provides a *direct path* for the continuous flow of the lateral force from the roof level down to the foundation. Drag struts function as described in Sec. 9.10.

shearwall on the second story is vertically in line with the shearwall on the first-story level.

Because one shearwall is immediately above the other, the lateral force from the upper wall can be transferred *directly* to the lower wall. Note that the shearwall on the first story carries a lateral force from the second-floor diaphragm in addition to the force from the second-story shearwall. The collectors (drag struts) transfer the shears in the diaphragms to the interior shearwall. See Sec. 9.10 for a review of collectors.

In Fig. 16.4b, the interior shearwalls have been revised so that there is a horizontal offset between the shearwalls on the first- and second-story levels. Two problems are created by the *out-of-plane offset* (see Fig. 16.4b):

1. The force from the second-story wall must travel horizontally through the diaphragm at the second-story level before it can be transferred into the shearwalls on the first story.

2. The resistance to the overturning moment at the base of the second-story interior shearwall requires special consideration.

These are described in greater detail in Figs. 16.4c and d.

OUT-OF-PLANE (HORIZONTAL) OFFSET

Figure 16.4b Interior shearwall with *out-of-plane offset.* This is classified in ASCE 7 Table 12.3-1 as horizontal irregularity Type 4.

PLAN – 2ND FLOOR DIAPHRAGM

Figure 16.4c *Shear* transfer as a result of out-of-plane offset. The shear from the second-story shearwall is transferred as a concentrated force E to the second-floor diaphragm.

ROOF

INCREASED SEISMIC FORCE FROM
ROOF DIAPHRAGM FOR
OVERTURNING MOMENT (OM)

$E_m = \Omega_0 Q_E$

2ND STORY
SHEARWALL

2ND FLOOR

1ST STORY
COLUMNS
CARRY OM

OPEN

1ST FLOOR

MOMENT COUPLE :
$C = T = \dfrac{OM}{b} = \dfrac{Emh}{b}$

C

T

b

Figure 16.4d *Overturning moment* at base of second-story shearwall. Note that the OM is at the *base* of the shearwall and is computed using height h. The overturning at the first-floor level is taken into account by an overturning analysis of the first-story shearwalls. The special seismic forces and load combinations apply only to the structure supporting the shearwall (i.e., first-story columns) and not to the shearwall itself.

EXAMPLE 16.2   Out-of-Plane Offset Added to a Regular Structure

The interior shearwalls in the two-story building in Fig. 16.4a do not involve an out-of-plane offset. The shearwalls on the first and second floors are aligned vertically, one over the other. The revised layout of this building in Fig. 16.4b has the interior shearwall at the first-story level relocated to the left of its original position. A much more critical situation has been created because the shear from the second-story shearwall must now be transferred *horizontally* through the second-floor diaphragm before it can be picked up by the shearwalls on the first floor. The second problem involves the overturning moment at the base of the second-story shearwall. Neither of these problems occurs in the original configuration (Fig. 16.4a), which was structurally regular.

The horizontal seismic force from the roof is carried by the second-story shearwall. At the second-floor diaphragm, this force must be distributed horizontally to the shearwalls at the first-story level. See Fig. 16.4c. Although it is possible to develop a continuous path for the transfer of lateral forces down through the structure, the forces do not follow a direct route. The discontinuity introduced by the out-of-plane offset causes the shear to travel horizontally in its path to the foundation. This places a concentrated force on the second-floor diaphragm that is in addition to the usual distributed force.

The concentrated force in Fig. 16.4c accounts for the transfer of the *shear* at the base of the second-story shearwall. The force to the second-floor diaphragm includes the usual uniformly distributed force plus a concentrated earthquake force E from the second-story shearwall. The second-floor diaphragm is supported by the first-story shearwall reactions R_1, R_2, and R_3.

The *moment* at the base of the second-story shearwall must also be transferred. This overturning moment is resolved into a couple, and the moment is replaced by a pair of equal and opposite forces (one tension and one compression) separated by a lever arm at the ends of the second-story interior shearwall.

One approach to handling this moment is to provide columns in the story directly below these forces for the purpose of carrying the overturning moment to the foundation. See Fig. 16.4*d*. These columns may carry gravity loads in addition to the forces due to overturning. Note that these are isolated *columns* in the first story, and there is *no shearwall* between them. Other solutions to the overturning moment problem are possible.

The overturning moment below the second-story interior shearwall may be carried by columns in the first story. The forces *T* and *C* are determined by dividing the overturning moment OM by the lever arm *b*. An out-of-plane offset requires that the columns carrying the overturning moment be designed for the special seismic load combinations using E_m. The increased seismic force to be considered for overturning is significantly larger than the usual earthquake force E. Detailed discussion of design penalties for irregular structures will follow.

In summary, the discontinuity resulting from an out-of-plane offset causes the lateral *shear force* and the *overturning moment* from the second-story shearwall to follow two different paths. The shear is transferred horizontally through the second-floor diaphragm to the first-story shearwalls, and the overturning moment is transferred vertically to the columns directly below the second-story shearwall. As another example, if a second story shearwall were to be supported on a beam, the beam would require design for load combinations using the magnified E_m force. The design penalty assigned by the code indicates the level of concern about Type 4 irregularities.

The final type of irregularity to be covered in this section is a form of *vertical irregularity* (Type 4 in ASCE 7 Table 12.3-2). Although this type of irregularity is classified as a vertical irregularity, it is similar in concept to the out-of-plane offset (horizontal irregularity) that was just reviewed. An in-plane-offset of a lateral-force-resisting element also interrupts the normal lateral force path of the system.

To illustrate an in-plane discontinuity, again consider the regular structure in Fig. 16.4*a*. The building has shearwalls on the second story that are located directly over the shearwalls on the first story. In Fig. 16.5, this building is reconfigured to have a Type 4 vertical irregularity. The interior shearwalls on the first and second stories are offset by a distance greater than the length of the shearwall elements.

Essentially, the same overturning problem occurs with the in-plane offset in Fig. 16.5 and the out-of-plane offset in Fig. 16.4*d*. For this reason, the code requires that columns or other members supporting the second-story shearwall also be designed for the overturning effects caused by the special seismic (overstrength) force. (More detailed discussion of design penalties for irregular structures follows.)

The discussion of design penalties for irregular structures has been separated from the introduction to the irregularities because multiple design penalties will often apply for a given irregularity.

ROOF DIAPHRAGM

SEISMIC FORCE TO
ROOF DIAPHRAGM

2ND STORY
INTERIOR SHEARWALL

ROOF

COLLECTOR
AT ROOF LEVEL

COLLECTORS
AT 2ND FLOOR
LEVEL

2ND FLOOR

1ST FLOOR

1ST STORY INTERIOR
SHEARWALL

OFFSET

IN-PLANE (VERTICAL) OFFSET

Figure 16.5 An *in-plane-discontinuity in vertical lateral-force-resisting element is* classified as vertical irregularity Type 4 in ASCE 7 Table 12.3-2.

The nine design penalties triggered by horizontal (plan) structural irregularities will be presented first. The first penalty takes the form of a load multiplier of 1.25 on connections of diaphragms to vertical elements, and on collectors, collector splices, and connections of collectors to vertical elements. The increase for collectors and their splices and connections need only be used where the special seismic load combinations using E_m are not already being used. The 1.25 multiplier need not be used in combination with E_m forces. This penalty is found in ASCE 7 Sec. 12.3.3.4, and is applicable to horizontal structural irregularities 1a and 1b through 4 in ASCE 7 Table 12.3-1 (torsional irregularities, reentrant corners, diaphragm discontinuity, and out of plane offset) and applies in SDCs D, E, and F.

The second penalty requires use of a torsional amplification factor for structures having horizontal torsional irregularities, in accordance with ASCE 7 Sec. 12.8.4.3. As previously mentioned, investigation of torsional irregularities is not required for structures using flexible diaphragm analysis models. Where other than flexible diaphragms are used, the amplification factor is required for SDCs C through F.

The third penalty requires the use of a 3-D model in determining forces and displacements, in accordance with ASCE 7 Sec. 12.7.3. This is triggered for horizontal structural irregularities 1a, 1b, 4, and 5 for SDCs B through F.

The fourth penalty requires that story drift limits be checked along all edges of the structure, rather than just at the center of mass, in accordance with ASCE 7 Sec. 12.12.1. This is triggered for horizontal irregularities 1a and 1b in SDCs C through F.

The fifth penalty triggers requirements for dynamic analysis in accordance with ASCE 7 Table 12.6-1. This is triggered in SDCs D through F for all horizontal irregularities; however, equivalent lateral force analysis is permitted for occupancy categories I and II buildings of light-frame construction not exceeding three stories in height.

The sixth penalty imposes requirements for response history modeling in accordance with ASCE 7 Sec. 16.2.2. This modeling approach, however, will seldom be used for wood light-frame construction.

The seventh penalty prohibits extreme torsional irregularities in SDCs E and F. This limit is found in ASCE 7 Sec. 12.3.3.1.

The eighth penalty, found in ASCE 7 Sec. 12.3.3.3, addresses horizontal irregularity 4 (support of discontinuous walls or frames of the LFRS). This is applicable when an element of the LFRS (such as a wall or moment frame) is included at an upper floor but not continued to the foundation, and applies in SDCs B and higher. It is required that columns beams, trusses, or slabs supporting the discontinuous elements be designed using the axial (vertical) loads that can develop in accordance with the special seismic load combinations using E_m. This is required for all materials, including wood light-frame structures. The overstrength force is intended to apply to elements below the discontinuous element but not to the discontinuous element itself. The intent is to protect the system that is providing vertical support against failure, in the event that the discontinuous element experiences E_m level forces. Note that the E_m force is often not used for design of the shearwall tie-downs because this could potentially further strengthen the shearwall system, putting additional force into the beam or other support that is being protected. This penalty is not intended to address the design of the diaphragm supporting the discontinuous element.

The ninth and final penalty applicable to horizontal irregularities is triggered for nonparallel systems. A system is nonparallel when the shearwall or frames are not at right angles to each other. This is of concern because forces acting along one primary axis will induce forces in all of the resisting elements, increasing interaction of forces. This applies in SDCs C and higher in accordance with ASCE 7 Sec. 12.5.3.

The five design penalties triggered by vertical structural irregularities will now be presented. The first penalty, applicable for vertical irregularities 1a and 1b through 3 (soft story and extreme soft story, weight [mass] irregularity, and vertical geometric irregularity), places limitations on acceptable analyses for SDCs D and higher. This penalty was discussed earlier in this section, where it was noted that there is an exemption for light-frame structures of three stories or less. For light-frame structures of four stories or more, this penalty requires use of analysis that is currently at or beyond the state of the art.

The second penalty is the prohibition of an extreme soft story in SDCs E and F. An extreme soft story stiffness is less than 60 percent of the story above. From a theoretical standpoint, it is generally possible to provide adequate shearwall sheathing and nailing so that this is not triggered in wood shearwall buildings. From a practical standpoint, recent research suggests that soft lower story behavior is common in wood shearwall buildings, and may not be possible to avoid.

The third and fourth penalties require special seismic loads at collectors and discontinuous elements, as previously discussed. The fifth and final penalty addresses extreme weak stories, where the strength of a story is less than 65 percent of the strength of the story above. This condition is prohibited in SDCs D through F, and structures with this irregularity are limited to the lesser of two stories or 30 ft in SDCs B and C. This irregularity has the same issues as a soft story. While sheathing nailing can be provided to theoretically eliminate a weak story, in practice weak stories often occur due to the strength contribution of finish materials. Practically occurring weak and soft stories have not been identified as a significant concern in otherwise regular and adequately designed wood light-frame buildings.

When design using the special seismic load combinations is required, the increase in seismic force is Ω_o *times* the usual seismic force, where Ω_o is determined from ASCE 7 Table 12.2-1. The large order of magnitude of this force indicates the severe nature of the triggering of discontinuities in the lateral-force path. The code specifies the increased seismic force for overturning as

$$E_m = \Omega_o Q_E \pm 0.2 S_{DS} D$$

In the case of columns supporting discontinuous shearwalls, the added force requirement is to guard against the collapse of the LFRS that would be triggered by a failure of the columns carrying the overturning moment. This force level was selected because the columns at such a discontinuity must be capable of supporting the LFRS as it is forced well into the inelastic range. Thus, the performance of these columns is essentially a *strength* design requirement, representing an extreme loading condition.

The designer should recall that Sec. 9.7 discussed design of collectors and their connections to shearwalls. Use of the special seismic load combinations and magnified seismic forces E_m is required irrespective of the existence of any structural irregularity. Buildings braced entirely of light-frame wood walls are exempted from the magnified force level; buildings with concrete or masonry walls or mixed systems are not exempt. Where applicable, the design requirements resulting from the E_m force are likely to be more critical than those resulting from the typical design force.

A full understanding of the design requirements for irregular structures requires the reader to first understand the simple regular buildings that are covered in Chaps. 1 to 15. The intent of Sec. 16.3 is to summarize some of the common discontinuities that may be encountered in a typical wood-frame building. The next section continues the discussion of overturning under lateral forces and provides a comparison of the practices for wind and seismic forces.

16.4   Overturning—Background

The general combinations of loads and forces that are required by the IBC were summarized briefly in Sec. 2.17. These combinations essentially define which loads and forces must be applied simultaneously in the design of a building. To a certain extent, a combination reflects the probability that the various loads and forces will occur concurrently. Some of these probabilities were discussed previously and are not repeated here.

As the code requirements for wind and seismic forces have become increasingly complex, the considerations for overturning (moment) stability have also become more involved. The overturning analysis for a shearwall under wind forces was introduced in Example 2.9 (Sec. 2.11), and additional overturning moment problems were covered in Chap. 10. The reader may find it useful to review these sections in preparation for the detailed summary given here.

A comprehensive study of overturning is introduced by first reviewing a simple statics problem. The principles of statics are then expanded to cover moment stability at the base of a wood-frame shearwall.

It is important that the designer understand how to evaluate overturning effects. Initially, this problem is treated purely from a statics approach, without concern for the code-required factor of safety (or load and force combinations). This essentially defines the moment stability problem, and it introduces the analysis necessary to evaluate soil bearing pressures under combined gravity loads and lateral forces. While shearwalls are often modeled as stand alone segments for the purpose of analysis, foundations supporting shearwalls are seldom designed this way. This is because the foundation generally provides considerable strength and stiffness, and it serves to interconnect the many segments of shearwall and other structural and nonstructural portions of the system. As a result, the foundation can be viewed as a continuous beam with many applied loads and forces. For this reason, foundation design for overturning is beyond the scope of this text. The principles of overturning design described in this section, however, can be used in design of foundations.

After the statics of this problem is understood, the code-required factor of safety (or required load and force combinations) is finally introduced. The need for the detailed review here is the result of the seismic code requirements.

16.5   Overturning—Review

Most designers are first exposed to the subject of *moment stability* in a course in statics or engineering mechanics. See Example 16.3. In this type of problem, the dimensions and weight of an object are given. In addition, the coefficient of static friction μ of the surface upon which the block rests is known.

A typical requirement in such a problem is to determine whether the block *moves* or *remains in static equilibrium* when subjected to a lateral force. If the block moves, the problem is to indicate whether it *slides* along the supporting surface or whether it *tips* (overturns). If the block remains at rest, the *factors of safety* against *sliding* and *overturning* are to be calculated.

EXAMPLE 16.3   Typical Moment Stability Problem from a Course in Statics

GIVEN

The dimensions and weight of a block are known. See Fig. 16.6. The coefficient of static friction μ of the surface upon which the block rests is given.

Figure 16.6   Typical statics problem involving moment stability of a block.

FIND

If a lateral force P is applied at height h above the base, determine whether or not the block remains in static equilibrium. If the block moves, indicate whether it slides or tips. If the block remains static, determine the factors of safety against sliding and overturning.

SOLUTION

Sliding

By summing forces in the y direction, the normal reaction at the base equals the weight of the block:

$$\Sigma F_y = 0$$

$$N = W$$

By summing forces in the x direction, the friction force at the base equals the applied lateral force P:

$$\Sigma F_x = 0$$

$$F = P$$

The maximum available friction force for resisting the lateral force is the normal reaction times the coefficient of static friction:

$$\text{Max. } F = \mu N$$

If the lateral force exceeds the maximum available resisting force

$$P > \text{max. } F$$

sliding will occur.
   On the other hand, if

$$P < \text{max. } F$$

the object does not slide, and the *factor of safety against sliding* can be computed as

$$\boxed{\text{FS}_\text{sliding} = \frac{\text{max.} F}{P} = \frac{\mu N}{P} = \frac{\mu W}{P}}$$

Moment Stability

If the block tips, the normal force N shifts to the pivot point at O. By summing moments about point O, N drops out, and the maximum overturning moment OM is the lateral force times the height above the base:

$$\Sigma M_O$$

$$\text{OM} = Ph$$

Also by summing moments about point O, the resisting moment RM is the weight of the block times the lever arm to the line of action of this force:

$$\text{RM} = W\left(\frac{b}{2}\right)$$

If the maximum overturning moment exceeds the available resisting moment

$$\text{OM} > \text{RM}$$

tipping will occur.
   On the other hand, if

$$\text{OM} < \text{RM}$$

the object remains in static equilibrium, and the *factor of safety against overturning* can be computed as

$$\boxed{\text{FS}_\text{overturning} = \frac{\text{RM}}{\text{OM}}}$$

   Although the example of a block sliding or tipping is an elementary problem, it is helpful to begin a review of overturning for a wood-frame shearwall with these basic principles in mind.
   Most wood-frame buildings use a combination of diaphragms and shearwalls to resist lateral forces. In checking overturning for these types of structures, the usual practice is to examine the moment stability of the vertical elements

(i.e., the shearwalls) in the primary LFRS of the structure. If the individual shearwalls are stable, the overall moment stability of the structure is ensured.

It has been noted throughout this book that the key to successful lateral force design is to adequately tie the elements together so that a continuous path is provided for the transfer of forces from the roof level down into the foundation. The simple statics problem is expanded to a shearwall overturning analysis by considering the forces acting on a shearwall from a one-story building. See Example 16.4.

In this example, the stability of a free body diagram (FBD) taken at the base of a shearwall is analyzed. Similar problems were considered in Chap. 10. The approach taken here is purely from a statics point of view, and consideration is not given to whether the lateral force is wind or seismic, and the FS required by the code is not defined. The expressions necessary to determine the forces in the *shearwall chords* and *anchorage connections* for overturning are again reviewed.

EXAMPLE 16.4   Overturning at Base of Shearwall

If the lateral forces to a shearwall are relatively small, the dead load of the wall plus any roof or floor dead load supported by the wall may be sufficient to resist overturning (see Fig. 16.7). On the other hand, if the lateral forces are large or if the dead loads above the foundation are small, it may be necessary to attach the shearwall to the foundation with anchorage hardware so that the weight of the foundation (and other structure sitting on the foundation) can be included in the moment stability analysis.

Figure 16.7   Moment stability of a one-story shearwall. FBD cut at base of wall (top of foundation).

The gross overturning moment at the *base of the shearwall* is the lateral force (the reaction from the roof diaphragm) times the height from the top of the foundation to the roof level:

$$\text{Gross OM} = Rh$$

The resisting moment is the sum of any dead loads supported by the wall (including the weight of the wall and any other dead load framing into the wall) times the lever arm from the pivot point O to the line of action of the dead loads. For a uniformly distributed dead load, the lever arm is simply one-half of the shearwall length:

$$\text{RM} = W_D \left(\frac{b}{2} \right)$$

In order for the shearwall to be stable, the maximum available resisting moment must be greater than or equal to the gross overturning moment:

$$\text{RM} \geq \text{OM}$$

Alternatively, this could be expressed by requiring the factor of safety against overturning to be 1.0 or larger:

$$\text{FS} = \frac{\text{RM}}{\text{OM}} \geq +1.0$$

For specific ASD lateral force combinations the IBC requires the RM to be based on 60 percent of the dead load. Thus the code effectively requires the FS to be $\frac{5}{3} = 1.67$.

Because of the relatively small dead loads involved in wood-frame buildings, it is not uncommon for the OM at the base of a shearwall to exceed the RM. The typical solution in such a case is to attach the shearwall to the supporting foundation with hardware (tie-down) of adequate strength so that the *design overturning moment* is transferred to the foundation. According to statics, the design OM is the difference between the gross OM and the dead load RM based on 60 percent of the dead load:

$$\text{Design OM} = \text{gross OM} - 0.6 \text{ RM}$$

$$= Rh - 0.6 \text{ RM}$$

Obviously if the sign of the design OM is negative, the shearwall is stable without special tie-down attachments to the foundation. In this case, the only force to be transferred to the foundation with hardware is the shear force $V = R$. Generally the friction force between the wall and the foundation is ignored, and *anchor bolts* are designed to transfer the shear from the shearwall into the foundation (see Example 10.10 in Sec. 10.10 for additional information).

If the sign of the design OM is positive, additional anchorage hardware is required to transfer the design OM to the foundation. The design OM is resolved into a couple by dividing the moment by the width of the shearwall (distance between shearwall chords). One force from the resulting couple is a tension force T, and the other is a compression force C. Since the force can originate from either direction, anchorage at either end of the shearwall must be designed for both tension and compression forces. See Sec. 10.5 and Fig. 10.12 (Sec. 10.10) for additional information.

When the magnitude of the dead-load-resisting moment is relatively small, the designer may choose to ignore the resisting dead load. In this case, the design OM equals the gross OM, and the tension and compression chord forces obtained by resolving the OM into a couple are known as *gross chord forces:*

$$\text{Gross } T = \text{gross } C = \frac{\text{gross OM}}{b}$$

On the other hand, when the resisting dead load is considered, the tension chord force is based on the design OM. The *design tension chord force* is

$$\text{Design } T = \frac{\text{design OM}}{b} = \frac{\text{gross OM} - \text{RM}}{b}$$

It is important to realize that for ASD only 60 percent of the dead load of the structure is used to resist overturning. This again reflects the conservative approach used in structural design. There may or may not be gravity *live* loads acting on the shearwall system when lateral forces are applied. There is a greater chance of having floor live loads present during a major windstorm or earthquake than roof life loads. However, the presence of neither type of live load can be ensured, and consequently only gravity dead loads are considered in checking uplift due to overturning.

The specific provisions for overturning under wind and seismic forces will be given in the next sections. To do this, three general classes of practical moment stability problems are identified, and a summary of the important design criteria for each type is given. The intent is to provide the information in a form in which the similarities and differences are clear for each situation. The three classes of overturning problems include one wind and two earthquake problems:

1. Wind overturning—general wind force requirements (Example 16.5)

2. Seismic overturning
   a. Regular structures (Example 16.6)
   b. Irregular structures (Example 16.7)

Each of these cases produces an overturning situation which, according to the code, is unique in some way. For each overturning problem there are certain basic requirements, including the following:

1. Check the moment stability using the appropriate combination of loads and forces at the base of the shearwall (excluding the weight of the foundation)

2. Determine shearwall chord forces.
   a. Net tension chord force (known also as the *uplift* or *anchorage* force)
   b. Maximum compression chord force (including tributary gravity load)

16.6 Overturning—Wind

The general wind force requirements from the IBC were introduced in Secs. 2.10 to 2.12. The specific requirements for checking overturning under wind forces are summarized in this portion of Chap. 16.

It was previously noted that only 60 percent of the dead load D of the structure is used to resist overturning. This is reflected in IBC basic ASD load combinations 16-14 and 16-15:

$$0.6\,D + W \text{ and } 0.6\,D + 0.7\,E$$

When the dead load of a structure is estimated, designers often take an approach that is conservative for vertical load analysis. This may be done by rounding the summation of gravity loads upward and by slightly overestimating the weight of the members so that it will not be necessary to revise design calculations once member sizes have been determined. The practice of overestimating loads is generally considered to be on the side of greater safety.

However, in the moment stability analysis of a structure, an overestimation of dead loads is *unconservative* because the larger loads will produce a calculated resisting moment that is larger than is actually available. For this and other reasons, a factor of safety is provided by the IBC for checking overturning at the base of a structure. This can be stated mathematically in several different forms. The factor of safety for moment stability can be written as

$$\text{FS} = \frac{\text{RM}}{\text{OM}} \geq 1.67$$

The critical combination of gravity loads and wind forces for the compression chord will be another of the basic load combinations given in Sec. 2.17. Therefore, the design force for the compression chord is the gross chord force caused by the wind overturning moment plus the tributary gravity loads.

It was noted in Chap. 2 that the IBC wind design provisions require wind uplift, if any, to be considered *simultaneously* with horizontal wind forces. Whether or not wind uplift acts on a given shearwall depends on the roof framing system and whether the shearwall is a bearing or nonbearing wall for vertical loads. If the wall supports gravity loads in addition to its own weight, it is classified as a bearing wall. The same framing members that cause additional gravity loads to be supported by the wall may also cause wind uplift to be transferred to the shearwall.

EXAMPLE 16.5   Overturning Requirement for Wind—General

General IBC requirements are that horizontal wind forces and wind uplift are to be considered simultaneously. See Fig. 16.8a. Basic ASD overturning requirements are

Implied FS for base overturning: $\text{FS} = \dfrac{1}{0.6} = 1.67$

Combination of loads and forces incorporating FS = 1.67: 0.6D + horizontal wind + uplift wind

Load duration factor for wood members and fasteners, $C_D = 1.6$

To develop an FS of 1.67 for base overturning, the shearwall tension chord force *T* is determined by using 60 percent of the resisting moment. Dead load only is considered in computing the resisting moment for the tension chord. See Figs. 16.8a and 16.8b. The

Figure 16.8a Loads and forces for wind overturning at foundation—soil interface.

Figure 16.8b Loads and forces for wind overturning at shearwall connection to foundation for *tension chord.*

critical shearwall compression chord force C will be governed by a combination of loads and forces that uses the full dead load D (not 0.6D).

The gross overturning moment is

$$\text{Gross OM} = Rh + U\left(\frac{b}{2}\right)$$

The expression developed in the statics analysis of the overturning problem in Example 16.4 are revised as follows to reflect the appropriate factors of safety:

Tension Chord

Load combination: 0.6D + horizontal wind + uplift wind

Design overturning moment (ΣM_O): design OM = gross OM − 0.6RM

Design tension chord force:

$$\text{Design } T = \frac{\text{design OM}}{b}$$

$$= \frac{\text{gross OM} - 0.6\text{RM}}{b}$$

Load duration factor $C_D = 1.6$

Compression Chord

Load combinations: D + W

$$D + 0.75(L + [L_r \text{ or } S] + W)$$

Compression chord force:

$$C = \frac{\text{gross OM} + \text{RM}}{b}$$

Load duration factor $C_D = 1.6$

Note that the FBD in Fig. 16.8b shows the load and force combination for the tension chord analysis. The D in the sketch represents all of the dead load to the shearwall above the foundation.

For designing the compression chord, the FBD would have to be revised to reflect the two required load cases. The calculation of the shearwall chord forces may be carried out using the expressions given above. A similar approach with a slightly different format was described in Chap. 10.

Example 16.5 deals with the *general* wind overturning requirements, which involve the addition of overturning due to horizontal wind and overturning due to uplift wind. If the uplift force U is large, the requirement to consider overturning due to the two wind effects simultaneously can be a significant force increase over previous codes.

16.7   Overturning—Seismic

The seismic force requirements from the code were introduced in Secs. 2.13 to 2.16. The specific requirements for checking overturning under seismic forces are summarized in this section.

The current seismic code contains detailed design provisions for *regular* and *irregular* structures. Because irregular structures are inherently more susceptible to earthquake damage, the code assigns additional design requirements for these structures. As previously noted, these penalties may take the form of higher design force levels, or other requirements may be imposed. For comparison, the overturning requirements for regular structures will be reviewed first, followed by a summary of the design provisions for irregular structures.

Even in a regular structure the forces generated during an earthquake are erratic, and the magnitude of the force level changes rapidly with time. This is due in part to the changing ground accelerations and in part to the dynamic characteristics of the structure. The transient and reversing nature of seismic forces as compared to wind forces, and observed behavior of buildings in past earthquakes, indicate that the accumulated effects at the base (foundation—soil interface) of a structure are less critical than would be obtained by a direct addition of the seismic overturning forces acting on the structure above. This behavior needs to be balanced against the realization that the seismic forces used for design are not the real forces that will be experienced by the structure. Further study is needed to establish a true understanding of requirements for adequate overturning resistance.

It is interesting to note that very early seismic codes included a coefficient that was used to reduce the seismic overturning moment. Although the previous reduction factor has been deleted for some time, for the equivalent static force procedure, ASCE 7 now reflects this behavior by permitting foundations to be designed for 75 percent of the overturning moment at the foundation-to-soil interface. This can be found in ASCE 7 Sec. 12.13.4. The following example will look at the structure-to-foundation interface rather than the foundation-to-soil interface, so the reduction in overturning will not be included.

It is necessary to look at the required load combinations in order to discuss the implied factor of safety against overturning at design level forces. This will be addressed in Example 16.6. Again, because the seismic forces used for design are not those that will be experienced by the structure, the overturning design and the implied factor of safety, together, are merely a design tool and not a predictor of building behavior.

EXAMPLE 16.6   Seismic Overturning Requirements—Regular Structures

Seismic overturning effects are less critical at the foundation-soil interface (Fig. 16.9) because of the transient and reversing nature of the seismic forces on the structure above. As an indication of this less critical situation, ASCE 7 reflects this behavior for the equivalent lateral force design procedure by permitting foundations to be designed for 75 percent of the overturning moment at the foundation—soil interface.

Figure 16.9   Loads and forces for seismic overturning at foundation–soil interface.

In order to evaluate overturning for the shearwall in Fig. 16.9, the code-required load combinations must be used. Applicable ASD load combinations from IBC Sec. 1605.3.1

$$D + H + F + 0.7E$$

$$D + H + F + 0.75\,(0.7E) + 0.75L + 0.75(L_r \text{ or } S \text{ or } R)$$

$$0.6D + 0.7E + H$$

These equations will be simplified for this example by setting H, F, S, and R to zero, giving:

$$D + 0.7E$$

$$D + 0.75(0.7E) + 0.75L + 0.75L_r$$

$$0.6D + 0.7E$$

These equations need to be combined with the equation for E given in ASCE 7 Sec. 12.4:

$$E = E_h \pm E_v = \rho Q_E \pm 0.2S_{DS}D$$

Combining these equations gives the following worst case equations for uplift and downward loads:

$$D + 0.7(\rho Q_E + 0.2\,S_{DS}D)$$

$$D + 0.75\,(0.7(\rho Q_E + 0.2\,S_{DS}D)) + 0.75L + 0.75L_r$$

$$0.6D + 0.7\,(\rho Q_E - 0.2\,S_{DS}D)$$

If S_{DS} and ρ are taken as 1.0, these equations can be simplified to:

$$1.14D + 0.7Q_E$$

$$1.11D + 0.525Q_E + 0.75L + 0.75L_r$$

$$0.46D + 0.7Q_E$$

Tension Chord

Load and force combination: $0.46D + 0.7Q_E$

Resisting moment:

$$RM = (0.46D)(b/2) = 0.23Db$$

Net overturning moment:

$$\text{Net OM} = \text{gross OM} - RM$$

Net tension chord force:

$$\text{Net } T = \frac{\text{net OM}}{b}$$

$$= \frac{\text{gross OM} - RM}{b}$$

$$= \frac{0.7Q_E h - 0.23Db}{b}$$

Load duration factor $C_D = 1.6$

Compression Chord—First Equation

Load and force combinations: $1.11D + 0.525Q_E + 0.75L + 0.75L_r$

Gross overturning moment:

$$\text{Gross OM} = 0.525Q_E h$$

Resisting moment:

$$RM = (1.11D + 0.75L + 0.75L_r)(b/2) = 0.56Db + 0.38Lb + 0.38L_r b$$

Net overturning moment:

$$\text{Net OM} = \text{Gross OM} + RM$$

Compression chord force:

$$\text{Net } C = \frac{\text{net OM}}{b}$$

$$= \frac{\text{gross OM} + RM}{b}$$

$$= \frac{0.525Q_E h + 0.56Db + 0.38Lb + 0.38L_r b}{b}$$

Load duration factor $C_D = 1.6$

Compression Chord—Second Equation
(either one might control)

Load and force combinations:

$$1.14D + 0.7Q_E$$

Gross overturning moment:

$$\text{Gross OM} = 0.7 Q_E h$$

Resisting moment:

$$RM = 1.14D(b/2) = 0.57b$$

Net overturning moment:

$$\text{Net OM} = \text{Gross OM} + RM$$

Compression chord force:

$$\text{Net C} = \frac{\text{net OM}}{b}$$

$$= \frac{\text{gross OM} + RM}{b}$$

$$= \frac{0.7 Q_E h + 0.57b}{b}$$

Load duration factor = 1.6

Note that the FBD in Fig. 16.10 shows the load and force combination for the tension chord analysis. For designing the compression chord, the FBD is shown in Fig. 16.11.

The calculation of the shearwall chord forces may be carried out by using the expressions given above. A similar but slightly different approach was described in Chap. 10.

Figure 16.10 Loads and forces for seismic overturning at shearwall tension chord connection to foundation—regular structure.

Figure 16.11 Loads and forces for seismic overturning at shearwall compression chord—regular structure.

The seismic overturning requirements for a *regular structure* are summarized in Example 16.6

The design penalties that are assigned to *irregular* structures in areas of high seismic risk were outlined in Sec. 16.3. It will be recalled that two particularly critical types of discontinuities are the out-of-plane offset (Fig. 16.4*b*) and the in-plane-discontinuity (Fig. 16.5). These are *horizontal irregularity Type 4* and *vertical irregularity Type 4*, respectively. For buildings in SDCs B and higher with either Type 4 irregularity, the code requires that members supporting the shearwall system be designed by greatly increased earthquake force levels. The required load and force combinations for the design of these elements are given in Example 16.7.

EXAMPLE 16.7   Seismic Overturning Requirements—Irregular Structures

The overturning of a second-story shear panel with a Type 4 vertical structural irregularity was introduced in Example 16.2. A special seismic force that approximates the peak capacity of the system (rather than the lesser design forces) is to be considered. Figure 16.12 illustrates the loads and forces resulting from the IBC special seismic load combinations. Figure 16.12 shows the special seismic loads and forces applied to the shearwall in order to illustrate the overturning and resisting moments. The special seismic load combinations, however, apply only to the structure (beams and columns) supporting the shearwall and not to the shearwall itself.

Figure 16.12 Overturning of second-story shear-wall with Type 4 vertical or horizontal irregularity. Loads and forces shown are for seismic overturning for tension chord. The special seismic load combinations, however, apply only to the structure supporting the shearwall (i.e., first-story columns) and not to the shearwall itself. Note that the shear from the wall is transferred to the second-floor horizontal diaphragm, as shown in Fig. 16.4c.

Load Combinations

In order to evaluate overturning for the shearwall in Fig. 16.12, the code-required load combinations must be used. For this irregular structure, the IBC special seismic load combinations from Sec. 1605.4 will be used:

$$1.2D + f_1L + E_m$$
$$0.9D + E_m$$

The variable f_1 will be taken as 0.5 for this example. The E_m special seismic forces are found in ASCE 7 Sec. 12.4.3.

$$E_m = \Omega_o Q_E + 0.2S_{DS}D$$
$$E_m = \Omega_o Q_E - 0.2S_{DS}D$$

When combined to reflect the two critical overturning conditions, the following results:

$$1.2D + 0.5L + \Omega_o Q_E + 0.2S_{DS}D$$
$$0.9D + \Omega_o Q_E - 0.2S_{DS}D$$

These are special *strength level* load combinations. It is the intent of the IBC that these load combinations be used with LRFD or strength design provisions. With the dual-format ASD/LRFD NDS (Ref. 16.1) available, this is the recommended approach. For those complying with ASCE 7 and not IBC, ASCE 7 Sec. 12.4 does provide alternate ASD load combinations, to be used in combination with a 1.2 multiplier and a load duration factor C_D.

For this example, for purposes of illustration, S_{DS} will be set to 1.0 and Ω_o to 3, permitting the equations to be simplified to:

$$1.2D + 0.5L + 3Q_E + 0.2D = 1.4D + 0.5L + 3Q_E$$
$$0.9D + 3Q_E - 0.2D = 0.7D + 3Q_E$$

Tension Column at First Floor

Gross overturning moment:

$$\text{Gross OM} = 3Q_E(h)$$

Resisting moment:

$$RM = 0.7D(b/2) = 0.35Db$$

Net overturning moment:

$$\text{Net OM} = \text{gross OM} - RM$$

Net tension force:

$$
\begin{aligned}
\text{Net } T &= \frac{\text{net OM}}{b} \\[6pt]
&= \frac{\text{gross OM} - RM}{b} \\[6pt]
&= \frac{3Q_E h - 0.35Db}{b}
\end{aligned}
$$

This strength level force will be compared to LRFD adjusted reference design values, using a time effect factor λ.

NOTE: The FBD in Fig. 16.12 shows the load and force combination for the tension chord analysis. For the compression chord, the FBD would have to be revised to reflect the combination for the compression chord.

It is hoped that the fairly extensive treatment of the moment stability problem will give the reader a better appreciation of the problem and provide a convenient summary of the design criteria.

16.8 Lateral Analysis of Nonrectangular Buildings

Many of the buildings that have been considered thus far have been rectangular in plan view with exterior and possibly interior shearwalls. These types of buildings have been used to illustrate basic concepts about diaphragm design.

In practice, however, buildings are often not rectangular. Obviously, an unlimited number of plan configurations are possible. However, a simple L-shaped building can be used to illustrate how the lateral forces on a nonrectangular building can be resisted. See Example 16.8.

For seismic design, buildings of this nature are said to have a horizontal irregularity known as a *reentrant corner* (ASCE 7 Table 12.3-1, Type 2). This type of discontinuity was described earlier (Fig. 16.2).

For buildings that are not braced entirely by light-frame wood shearwalls, the design of collectors and their connections to shearwalls will be governed by a requirement for use of the *special seismic load combinations* using the *special seismic (overstrength)* E_m *force level*. This requirement is applied to all collectors of other than all-light-frame buildings in SDCs C and higher and is required irrespective of the existence of any structural irregularities. This requirement was discussed in Sec. 9.10.

EXAMPLE 16.8   Lateral Analysis of Nonrectangular Buildings

A building with a nonrectangular plan may use rectangular diaphragms (see Fig. 16.13). Reactions to the shearwalls are calculated on a tributary width basis. The nonuniform seismic force results from the distribution of the dead load (Example 2.15 in Sec. 2.15). A similar analysis is used for lateral forces in the other direction.

Figure 16.13   Wind and seismic forces to diaphragm in nonrectangular building.

In the approach illustrated, the building is divided into rectangles which are treated as separate diaphragms. The shearwall that is common to both diaphragms receives a portion of its force from the collector (drag strut). The collector collects the roof diaphragm unit shear from the unsupported segments of the two diaphragms. A numerical example using this same building illustrates the method of calculating the design forces in both directions. See Example 16.9.

The simple span diaphragm approach illustrated in Example 16.9 has been used successfully by a number of designers to analyze buildings of nonrectangular configuration. However, Ref. 16.6 recommends that the continuity of the diaphragm at the interior shearwall be taken into account. In this case, the chord in narrower diaphragms will have to be extended (see heavy line along line B in Fig. 16.14b and line 2 in Fig. 16.14c). See Ref. 16.6 for additional information.

EXAMPLE 16.9   L-Shaped Building

Determine the design unit roof shears, chord forces, and maximum collector connection forces for the one-story building in Fig. 16.14a. Consider lateral forces acting in both the N-S and E-W directions, and assume simple span diaphragms.

Figure 16.14a   L-shaped building.

Known Information

$$\text{Roof D} = 20 \text{ psf}$$

$$\text{Wall D} = 75 \text{ psf}$$

$$\text{Wind} = 20 \text{ psf}$$

$$\Omega_o = 3.0$$

$$\text{Basic Seismic} = 0.275W \text{ (Sec. 3.4)}$$

$$\text{Special Seismic} = 0.275\Omega_oW = 0.275\,(3.0)\,W = 0.825W \text{ strength level}$$

$$\text{Tributary wall height to roof level} = 12 \text{ ft}$$

Lateral Forces in N-S Direction

Diaphragm forces (Fig.16.14b):

Wind

$$w = 20 \times 12 = 240 \text{ lb/ft}$$

Figure 16.14b   Diaphragm forces in N-S direction.

Basic Seismic

$$w_{u1} = .275([20 \times 72] + 2[75 \times 12])$$

$$= 891 \text{ lb/ft}$$

$$w_{u2} = .275([20 \times 120] + 2[75 \times 12])$$

$$= 1155 \text{ lb/ft seismic governs}$$

Roof shears:

Line 1 and west of line 2:

$$v_{u1} = \frac{V_u}{b} = \frac{w_{u1}L}{2b} = \frac{891(88)}{2(72)} = 545 \text{ lb/ft}$$

Line 3 and east of line 2:

$$v_{u2} = \frac{w_{u2}L}{2b} = \frac{1155(80)}{2(120)} = 385 \text{ lb/ft}$$

Chord forces:

Lines B and C between lines 1 and 2:

$$T_u = C_u = \frac{M_u}{b} = \frac{w_{u1}L^2}{8b} = \frac{.891(88)^2}{8(72)} = 12 \text{ k}$$

Lines A and C between lines 2 and 3:

$$T_u = C_u = \frac{w_{u2}L^2}{8b} = \frac{1.155(80)^2}{8(120)} = 7.70 \text{ k}$$

Collector (drag strut) force:

Connection on line 2 at line B: Collector resists unsupported roof shear from both diaphragms. The special seismic load combinations are to be used. From Example 16.7, the load combinations are:

$$1.2D + 0.5L + \Omega_oQ_E + 0.2S_{DS}D$$

$$0.9D + \Omega_oQ_E - 0.2S_{DS}D$$

In some instances, collectors will also carry significant vertical loads (such as when a gravity beam is used as a collector), and in other cases the collector only has to support its own weight. For this example, it is assumed that the collector supports essentially no vertical load, in which case both equations are simplified to Ω_oQ_E. For this system Ω_o is taken as 3.0.

$$\text{special seismic } T_u = [(v_{u1} + v_{u2})72]\Omega_o$$

$$\text{special seismic } T_u = [(0.545 + 0.385)72]3.0 = 201 \text{ k}$$

This is a strength level force which can be compared to an LRFD adjusted design value.

Lateral Forces in E-W Direction

Figure 16.14c   Diaphragm forces in E-W direction.

Diaphragm forces (Fig. 16.14c):

Wind does not govern.

Basic Seismic

$$w_{u3} = .275([20 \times 80] + 2[75 \times 12])$$

$$= 935 \text{ lb/ft}$$

$$w_{u4} = .275([20 \times 168] + 2[75 \times 12])$$

$$= 1420 \text{ lb/ft seismic governs}$$

Roof shears:

Line A and north of line B:

$$v_{u3} = \frac{V_u}{b} = \frac{w_{u3}L}{2b} = \frac{935(48)}{2(80)} = 281 \text{ lb/ft}$$

Line C and south of line B:

$$v_{u4} = \frac{w_{u4}L}{2b} = \frac{1420(72)}{2(168)} = 304 \text{ lb/ft}$$

Chord forces:

Lines 2 and 3 between lines A and B:

$$T_u = C_u = \frac{M_u}{b} = \frac{w_{u3}L^2}{8b} = \frac{.935(48)^2}{8(80)} = 3.37 \text{ k}$$

Lines 1 and 3 between lines B and C:

$$T_u = C_u = \frac{w_{u4}L^2}{8b} = \frac{1.420(72)^2}{8(168)} = 5.48 \text{ k}$$

Collector (drag strut) force:

Connection on line B at line 2. Collector resists unsupported roof shear from both diaphragms. The *special seismic load combinations* are to be used with $\Omega_o = 3.0$.

$$\text{special seismic } T_u = ([v_{u3} + v_{u4}]80)\Omega_o$$

$$\text{special seismic } T_u = ([0.281 + 0.304]80)3.0 = 140 \text{ k}$$

This is again a strength level force.

16.9   Rigid Diaphragm Analysis

In Sec. 9.11, diaphragm flexibility was discussed in some detail and the classification of diaphragms as *rigid* or *flexible* was introduced. It was noted that the most prevalent design practice is to classify wood frame diaphragms as flexible.

It should be noted that a rigid diaphragm analysis can be used for both wind and seismic forces. As long as the diaphragm is rigid with respect to the shearwalls and the center of the lateral force does not coincide with the center of rigidity, the diaphragm will have the ability to rotate. As a result of rotation the lateral forces will be redistributed between the shearwalls. Example 16.10

is provided to demonstrate the code calculation criteria for classifying a diaphragm as flexible. The example uses seismic forces at a strength level. For wind design ASD level forces would be used.

EXAMPLE 16.10   Diaphragm Classification

To demonstrate the use of the code classification criteria, the building from Example 9.18 (Sec. 9.11) will be used to compare the diaphragm deflection Δ_D with the shearwall story drift Δ_S (see Fig. 16.15a). It will be assumed that both the diaphragm and the shearwalls are sheathed with wood structural panels.

Figure 16.15b provides dimensions for the diaphragm and shearwalls and gives the diaphragm loading, reactions, and unit shears. In accordance with the code procedure

Figure 16.15a   Shearwall and diaphragm deflections.

Figure 16.15b   Seismic forces and reactions in the longitudinal direction used for diaphragm classification.

for calculating diaphragm deflection, it will be assumed that the shearwalls are infinitely stiff, and the diaphragm spans as a simple beam between shearwalls.

The equations for the diaphragm and shearwall deflections come from the IBC Secs. 2305.2.2 and 2305.3.2 (Ref. 16.10). These equations were developed by the American Plywood Association and further information regarding their use can be found in Refs. 16.4 and 16.5. Both Δ_S and Δ_D are calculated using IBC strength level forces.

Diaphragm and Shear Wall Unit Stresses

Before starting the diaphragm deflection calculations, it is useful to calculate the unit shears in the diaphragm and shearwalls.

$$R_{u1} = R_{u2} = w_u(L)/2 = 300(20)/2 = 3000 \text{ lb}$$

For the diaphragm:

$$v_{u1} = v_{u2} = R_{u1}/b = 3000/30 = 100 \text{ lb/ft}$$

For the shearwalls:

$$v_{u1} = R_{u1}/l_1 = 3000/30 = 100 \text{ lb/ft}$$

$$v_{u0} = R_{u2}/\Sigma l_2 = 3000/(5 + 5) = 300 \text{ lb/ft}$$

Diaphragm Deflection

The first step is to calculate the diaphragm deflection, but first the construction of the diaphragm needs to be defined:

Assume: Diaphragm and shearwall sheathing is $^{15}/_{32}$ in. span-rated wood structural panels with exterior glue; span rating 32/16.

The diaphragm and shearwalls are blocked and nailed with 8d common nails at 6-in. o.c. at panel edges and diaphragm boundaries.

The diaphragm chord is two 2 × 4 top plates of No. 2 DF-L

Top plate splices occur at 10 ft o.c. and the slip at each splice is $^1/_{16}$ in.

The diaphragm deflection equation can be expressed as:

$$\Delta_D = \Delta_b + \Delta_v + \Delta_n + \Delta_a$$

These four terms represent the same deflection sources that are discussed in Sec. 9.8. Conceptually they are:

$$\Delta_b = \text{bending deflection of the diaphragm}$$

$$\Delta_v = \text{shear deflection of the diaphragm}$$

$$\Delta_n = \text{deflection due to nail slip}$$

$$\Delta_a = \text{deflection due to chord member slip}$$

The four terms can be evaluated individually and then summed:

Diaphragm Term 1—Bending Deflection

$$\Delta_b = \frac{5vL^2}{8EAb}$$

The variables are as follows:

$v = 100$ lb/ft maximum diaphragm unit shear

$L = 20$ ft (Fig. 16.15*b*)

   = diaphragm span

$E = 1,700,000$ psi (NDS Supplement Table 4A)

   = diaphragm chord modulus of elasticity

$A = 2(3.5)1.5 = 10.5$ in.^{2} for two 2 × 4

   = diaphragm chord area

$b = 30$ ft (Fig. 16.15*b*) = diaphragm width

This results in:

$$\Delta_b = \frac{5(100)(20)^3}{8(1,700,000)(10.5)(30)} = 0.0009 \text{ in.}$$

Diaphragm Term 2—Panel Shear Deformation

$$\Delta_v = \frac{vL}{4Gt}$$

With:

$v = 100$ lb/ft = maximum diaphragm unit shear

$L = 20$ ft = diaphragm span

$Gt = 27,000$ lb/in. (Ref. 16.10 Table 2305.2.2(2))

   = rigidity through the thickness

This results in:

$$\Delta_v = \frac{100(20)}{4(27,000)} = 0.0185 \text{ in.}$$

Diaphragm Term 3—Panel Nail Slip

$$\Delta_n = 0.188Le_n$$

The variable e_n represents the slip of the sheathing with respect to the supporting framing as a function of the load per nail at design forces. In order to calculate e_n, the load per nail in pounds needs to be calculated. This is the unit shear v, divided by the number of nails per foot. This results in 100 lb/ft ÷ 2 nails per foot which equals 50 lb/nail. Values for e_n are found in IBC Table 2305.2.2(1). Table 2305.2.2(1) provides e_n equations for framing green at time of construction and dry in use (green/dry). The availability of dry framing for new construction varies widely, and local availability should be checked. For this example the green/dry condition will be used. In addition, footnote a to IBC Table 2305.2.2(1) notes that an increase of 1.2 is needed where the sheathing is not STRUCTURAL I. Dougals Fir-Larch framing will be assumed with a specific gravity of 0.5, and the sheathing is not STRUCTURAL I, so the multiplier of 1.2 will need to be used. For 8d common nails:

$$e_n = (0.00)(1.2) = 0.00 \text{ in.} = \text{nail slip coefficient}$$

This results in:

$$\Delta_n = 0.188(20)(0.00) = 0.00 \text{ in.}$$

Diaphragm Term 4—Chord Splice Slip

$$\Delta_a = \sum(\Delta_c X)/2b$$

With:

Δ_c = 0.0625 in (assumed slip) diaphragm chord slip at a single chord splice.

X = 5 ft and 5 ft (assumed)

  = distance from diaphragm chord splice to nearest supporting shearwall.

b = 30 ft = diaphragm width

$$\sum(\Delta_c X) = 0.0625(5) + 0.0625(5) = 0.625 \text{ in-ft}$$
$$\Delta_a = 0.625/2(30) = 0.0104 \text{ in.}$$

Total Diaphragm Deflection

The total diaphragm deflection can be calculated as:

$$\Delta_D = \Delta_b + \Delta_v + \Delta_n + \Delta_a$$
$$= 0.0009 + 0.0185 + 0.00 + 0.0104 = 0.0298 \text{ in.}$$

Story Drift Calculation

The code criteria requires that the diaphragm deflection Δ_D be compared to the average story drift Δ_S, based on the tributary lateral load to the shearwalls. In order to determine the average story drift, the drift at each of the two shearwall lines will need to be calculated. One shearwall line has 30 ft of shearwall length and the other has 10 ft. Like the diaphragm deflection equation, the shearwall deflection equation can be divided into four parts. The calculation of shearwall deflections is treated in more detail in Sec. 10.11.

Shearwall Term 1—Chord Flexural Deflection

$$\Delta_b = \frac{8vh^3}{EAb}$$

The variables are as follows:

v = 100 lb/ft and 300 lb/ft = shearwall unit shears

h = 10 ft = shearwall height

E = 1,700,000 psi (NDS Supplement Table 4A)

= shearwall chord modulus of elasticity

A = 3.5(3.5) = 12.25 in.^{2} for 4×4

= shearwall chord area

b = 30 ft and 5 ft (Fig. 16.15b)

= shearwall widths

This results in:

Δ_b = 8(100)(10)^3/1700000 (12.25)(30) = 0.0013 ft = 0.016 in. for the 30 ft wall

Δ_b = 8(300)(10)^3/1700000 (12.25)(5) = 0.0230 ft = 0.28 in. for the two 5 ft walls

Shearwall Term 2—Panel Shear Deformation

$$\Delta_v = \frac{vh}{Gt}$$

With:

v = 100 lb/ft and 300 lb/ft = shearwall unit shears

h = 10 ft = shearwall height

Gt = 27,000 lb/in. (Ref 16.10 Table 2305.2.2(2))

= rigidity through the thickness

This results in:

Δ_v = 100(10)/27,000 = 0.0370 in. for the 30 ft wall

Δ_v = 300(10)/27,000 = 0.1111 in. for the two 5 ft walls

Shearwall Term 3—Panel Nail Slip

$$\Delta_n = 0.75he_n$$

The variable e_n is the same variable used to calculate the third term of the diaphragm deflection. In order to calculate e_n, the load per nail in pounds needs to be calculated. This is the unit shear v, divided by the number of nails per foot. This results in 100 lb/ft ÷ 2 nails per foot which equals 50 lb/nail for the 30-ft-long wall. This value will be used. For the two 5-ft-long walls, 300/2 = 150 lb/ft.

Values for e_n are again found in IBC Table 2305.2.2(1). For this example, the green/dry condition will again be used. In addition, the footnote a multiplier will again be used since the sheathing is not STRUCTURAL I. For 8d common nails:

$$e_n = (0.00)(1.2) = 0.00 \text{ in.} = 30 \text{ ft wall nail slip coefficient}$$

$$e_n = (0.04)(1.2) = 0.048 \text{ in.} = 5 \text{ ft wall nail slip coefficient}$$

With $h = 10$ ft, this results in:

$$\Delta_n = 0.75(10)(0.00) = 0.00 \text{ in. for the 30 ft wall}$$

$$\Delta_n = 0.75(10)(0.048) = 0.36 \text{ in. for the 5 ft walls}$$

Shearwall Term 4—Chord Anchorage Slip

$$\Delta_a = d_ah/b$$

With:

$d_a = 0.125$ in. based on manufacturer's data = chord anchorage slip

$h = 10$ ft = shearwall height

$b = 30$ and 5 ft = shearwall widths

$\Delta_a = 0.125\,(10)/30 = 0.0417$ in. for the 30 ft wall

$\Delta_a = 0.125\,(10)/5 = 0.2500$ in. for the two 5 ft walls

Total Shearwall Deflection and Average Story Drift

The total shearwall deflection can be calculated as:

$$\Delta_S = \Delta_b + \Delta_v + \Delta_n + \Delta_a$$

$$= 0.0013 + 0.0370 + 0.00 + 0.0417 = 0.08 \text{ in. for the 30 ft wall}$$

$$= 0.0230 + 0.1111 + 0.36 + 0.2500 = 0.7441 \text{ in. for two 5 ft walls}$$

The average story drift is the average of these two shearwall defections:

$$\Delta_S = (0.08 + 0.7441) = 0.412 \text{ in.}$$

The diaphragm deflection was calculated to be 0.0298 in. The code criteria defines the diaphragm as flexible if $\Delta_D > 2(\Delta_S)$. In our case, 0.0298 in. < 2 (0.412 in.) so the diaphragm is not classified as flexible. This example uses strength level forces because seismic loading is being considered. The process would be identical for ASD wind forces.

Example 16.10 merely demonstrates the code criteria for determining the applicability of a flexible diaphragm analysis. Example 16.10 concludes that there is justification for use of a semi-rigid or rigid diaphragm analysis. Example 16.11 demonstrates the rigid diaphragm analysis method. A simple rigid diaphragm analysis with a straightforward solution was demonstrated in Example 9.19. Example 16.11 will demonstrate the more generalized rigid diaphragm analysis approach using the building from Examples 16.10. In Example 16.10, it was already demonstrated that the roof diaphragm can be considered a rigid diaphragm. Now the rigid diaphragm force distribution to the shearwalls will be demonstrated.

See discussion in Chapters 2, 3, and 9 regarding common diaphragm modeling assumptions for design. See Refs. 16.7 and 16.11 for research on the topic of flexible and rigid diaphragm force distribution.

EXAMPLE 16.11   General Rigid Diaphragm Analysis

This example will again look at the building used in Examples 9.18 and 16.10. This building is assumed to have a wood structural panel roof diaphragm and wood structural panel shearwalls. The configuration shown in Fig. 16.16a will be assumed. The seismic forces shown in Fig. 16.16b will be assumed to include all seismic forces tributary to the roof

Figure 16.16a   Direct shear applied through the center of rigidity.

diaphragm including the weight contribution of exterior walls in both directions, as well as any interior partitions and nonstructural components. For this one-story building with wood shearwalls, this means:

$$V = F_x = F_{px} = 300 \text{ lb/ft } (20 \text{ ft}) = 6000 \text{ lb}$$

This example will consider seismic forces in the longitudinal direction.

In Example 16.10, for the purposes of evaluating diaphragm flexibility, a flexible diaphragm with equal reactions at Lines 1 and 2 was assumed. This distribution will now be reevaluated using a rigid diaphragm model.

Locating the Center of Mass

The mass of the building in Fig. 16.15*b* is shown as uniformly distributed and consequently the seismic force of 300 lb/ft in the longitudinal direction is uniformly distributed. Therefore, the center of mass is at the center of the building. The assumption of uniform seismic force is a slight simplification because it ignores lower wall weight at openings. When this type of assumption results in slightly conservative requirements, it is usually acceptable for design purposes. Because not every building will be this simple and symmetrical, the designer should be prepared to calculate the location of the center of mass. This can be thought of as the location of the resultant of the seismic forces. In this case, the distance *m* from Line 1 to the center of mass can be calculated as:

$$m = \frac{\Sigma \text{ Moments of lateral force (lb-ft)}}{\text{Total lateral force (lb)}}$$

$$m = w(\text{SPAN})(\text{SPAN/2})/W$$

$$= (300 \text{ lb/ft})(20 \text{ ft})(10 \text{ ft})/6000 \text{ lb}$$

$$= 10 \text{ ft}$$

Figure 16.16*b* Moment applied through the center of mass. The center of mass rotates about the center of rigidity.

Locating the Center of Rigidity and Calculating the Eccentricity

The stiffness of the wood structural panel shearwalls and resulting rigidity is affected by the construction of the walls. In this case, it will be assumed that the construction (including sheathing nail spacing) of all the walls is the same. This will allow an assumption that the wall stiffness is substantially proportional to the wall length. The distance r from line 1 to the center of rigidity can be calculated as:

$$r = \frac{l_1(0) + \Sigma l_2(\text{SPAN})}{l_1 + \Sigma l_2}$$

$$= \frac{30(0) + (5 + 5)20}{30 + (5 + 5)}$$

$$= \frac{200}{40} = 5.0 \text{ ft}$$

The eccentricity e can then be calculated as:

$$e = m - r = 10.0 - 5.0 = 5.0 \text{ ft}$$

This is called the *actual eccentricity*. For rigid diaphragm design, the code requires that, in addition to the *actual eccentricity*, the center of mass be displaced 5 percent of the building dimension perpendicular to the force direction. This results in additional eccentricity, referred to as the *accidental eccentricity*. The *design eccentricity e'* is then the *actual* and *accidental eccentricities*:

$$e' = e \pm 0.05 \ (L) = 5.0 \pm 0.05 \ (20) = 4.0 \text{ ft and } 6.0 \text{ ft}$$

The design eccentricity of 4.0 ft need not be investigated because 6.0 ft will control. In the perpendicular direction, by observation, the center of rigidity is at the center of the building.

Resulting Longitudinal Forces

As was discussed in Example 9.18, the forces generated in a rigid diaphragm analysis are generally considered in a two-step process. The first step is the distribution of the F_x force to the walls in proportion to the wall stiffness. Graphically, this can be thought of as applying the seismic force to the structure at the center of rigidity. This is illustrated in Fig. 16.16a. Because wall stiffness is being assumed equal to the wall length, the unit shear in all the longitudinal walls will be the same:

$$V = F_x/(l_1 + \Sigma l_2) = 6000 \text{ lb}/(30 + 5 + 5 \text{ ft}) = 150 \text{ lb/ft}$$

The wall line reactions from this first of two steps can then be calculated as:

$$R_{u1} = 150 \text{ lb/ft } (30 \text{ ft}) = 4500 \text{ lb}$$

$$R_{u2} = 150 \text{ lb/ft } (5 + 5 \text{ ft}) = 1500 \text{ lb}$$

The second step in distributing seismic forces involves consideration of the moment that results from the force F_x times the design eccentricity e'. It should be noted that this moment is resisted by the shearwalls on the four sides of the building, not just lines 1 and 2. The distribution of the moment $F_x(e')$ requires that the torsional stiffness of the shearwalls on the four sides of the building be calculated. The following table shows a common approach to distributing this moment.

The second column is the wall stiffness which, as in Example 16.10, will be taken as proportional to the wall length. The third column gives the distance from the wall to the center of rigidity. The fourth column calculates $k(r)$, which is just a convenient intermediate value to be used in the last column. In the fifth column $k(r)^2$ is the torsional stiffness of each wall and the sum for the group of walls.

The last column calculates the shear to individual walls due to the moment. This shear is obtained by multiplying the portion of the moment to each wall $k(r)/\Sigma k(r)^2$ times the moment $F_x(e')$ divided by the moment arm r. This can be accomplished by multiplying the moment by the fourth column and dividing it by the sum of the fifth column. The table shows one approach to distributing this moment to the resisting shearwalls.

Rigid Diaphragm Moment Distribution

Line	Stiffness (ft) k	Distance (ft) r	$k(r)$	$k(r^2)$	Force (lb) $\dfrac{F_x(e')k(r)}{\Sigma k(r^2)}$
1	30	5	150	750	450
2	10	15	150	2250	450
A	20	15	300	4500	900
B	20	15	300	4500	900
				12,000	

The shearwall forces resulting from this moment appear in the last column of the table and are depicted in Fig. 16.16b. It is important to keep track of the direction of the forces due to the moment. The center of applied lateral force can be visualized as rotating about the center of rigidity. In this case, a clockwise rotation of the building will result. The force due to the moment will add to the direct shear forces on line 2 and subtract from the direct shear forces on line 1. The sum of forces on each line is shown in Fig. 16.16c.

Figure 16.16c   Result of direct shear and moment.

The design forces used for lines A and B need to be most critical resulting from either the seismic forces in the transverse direction or due to the rotation of the diaphragm resulting from forces in the longitudinal direction. In the example building, the longitudinal seismic force is 750 lb, which is significantly less than the 3000 lb expected for transverse seismic forces.

Both the roof diaphragm and the shearwalls in the Example 16.11 building would be designed for the forces resulting from the rigid diaphragm analysis. A flexible diaphragm analysis is currently a lot more common in practice. It is recommended that the reader analyze several example buildings using both the flexible and rigid diaphragm analyses to develop a feeling for how much the distribution of shear changes between the two approaches.

While the force distribution to shearwalls will be different in each building studied, the reader will be able to develop an understanding of what type of plan layout results in significant redistribution of forces to shearwalls between the two methods. The layouts resulting in significant differences generally have a substantial distance between their center of mass and center of rigidity. Where this occurs, it is recommended that the diaphragms and shearwalls be examined using a rigid diaphragm analysis. Another approach is to perform both flexible and rigid diaphragm analyses and design each element for the larger of the resulting forces.

16.10 Additional Topics in Diaphragm Design

The lateral design forces required by the code in areas of high seismic risk have increased substantially in recent years. In order to provide design values for heavily loaded diaphragms, APA—The Engineered Wood Association conducted a series of tests on diaphragms that were designed to develop high unit shears. These diaphragms are referred to as *high-load* diaphragms, and test results are published in Ref. 16.4. High-load diaphragms are recognized in IBC Table 2306.3.2.

The IBC high-load diaphragms use relatively thick plywood ($\frac{5}{8}$ in. and $\frac{3}{4}$ in.) and closely spaced fasteners. Tests indicated that closely spaced 10d and 16d nails often caused the framing lumber to split. However, very closely spaced (1-in. o.c.) pneumatically driven wire staples were found to not cause splitting. High unit shears were obtained with these diaphragms.

Among other objectives, these diaphragm tests were designed to measure the effects of openings in diaphragms. See Refs. 16.4, 16.6 and 16.8 for information on openings in diaphragms and other advanced topics in diaphragm design.

16.11 References

[16.1] American Forest and Paper Association (AF&PA). 2005. *National Design Specification for Wood Construction* and Supplement, ANSI/AF&PA NDS-2005, AF&PA, Washington DC.

[16.2] American Forest and Paper Association (AF&PA). 2006. *ASD/LRFD Manual for Engineered Wood Construction* and Supplements and Guidelines, 2006 ed., AF&PA, Washington DC.

[16.3] American Society of Civil Engineers (ASCE). 2006. *Minimum Design Loads for Buildings and Other Structures (ASCE 7-05)*, ASCE, Reston, VA.

[16.4] APA—The Engineered Wood Association. Rev. 2000. *Research Report 138—Plywood Diaphragms*, APA, Tacoma, WA.

[16.5] APA—The Engineered Wood Association. 2004. *Design/Construction Guide—Diaphragms and Shear Walls*, APA, Tacoma, WA.

[16.6] Applied Technology Council (ATC). 1981. *Guidelines for the Design of Horizontal Wood Diaphragms (ATC-7)*, ATC, Redwood City, CA.

[16.7] Cobeen, K., Russell, J. E., and Dolan, J. D. 2003. *Recommendations for Earthquake Resistance in the Design and Construction of Woodframe Buildings* (CUREE Publication W-30), Consortium of Universities for Research in Earthquake Engineering, Richmond, CA.

[16.8] Faherty, Keith F., and Williamson, T. G. (eds.). 1999. *Wood Engineering and Construction Handbook*, 3rd ed., McGraw-Hill, New York, NY.

[16.9] FEMA. 2004. *NEHRP Recommended Provisions for Seismic Regulations for New Buildings and Other Structures* and *Commentary*, 2003 ed. (FEMA 450, Parts 1 and 2), FEMA, Washington DC.

[16.10] International Codes Council (ICC). 2006. *International Building Code*, 2006 ed., ICC, Falls Church, VA.

[16.11] Mosalam, K. 2003. *Seismic Evaluation of an Asymmetric Three-Story Woodframe Building* (CUREE Publication W-19), Consortium of Universities for Research in Earthquake Engineering, Richmond, CA.

Appendix

A

Equivalent Uniform Weights of Wood Framing

Weights are for Douglas Fir-Larch lumber (S4S) used at an equilibrium moisture content of 15 percent (the maximum found in most covered structures). The unit weight of Douglas Fir-Larch equals or exceeds the unit weight of most softwood species, and the dead loads given below are conservative for most designs.

	Spacing					
	12 in. o.c.		16 in. o.c.		24 in. o.c.	
Nominal size	Weight, psf	Board feet* per ft^2	Weight, psf	Board feet* per ft^2	Weight, psf	Board feet* per ft^2
2 × 3	0.9	0.50	0.7	0.38	0.4	0.25
2 × 4	1.2	0.67	0.9	0.50	0.6	0.34
2 × 6	1.9	1.00	1.4	0.75	1.0	0.5
2 × 8	2.5	1.33	1.9	1.00	1.3	0.67
2 × 10	3.2	1.67	2.4	1.25	1.6	0.84
2 × 12	3.9	2.00	2.9	1.50	2.0	1.00
3 × 6	3.2	1.50	2.4	1.13	1.6	0.75
3 × 8	4.2	2.00	3.1	1.50	2.1	1.00
3 × 10	5.4	2.50	4.0	1.88	2.7	1.25
3 × 12	6.6	3.00	4.9	2.25	3.3	1.50
3 × 14	7.7	3.50	5.8	2.63	3.9	1.75
4 × 8	5.9	2.67	4.4	2.00	3.0	1.34
4 × 10	7.5	3.33	5.7	2.50	3.8	1.67
4 × 12	9.2	4.00	6.9	3.00	4.6	2.00
4 × 14	11.0	4.67	8.1	3.50	5.4	2.34
4 × 16	12.4	5.33	9.3	4.00	6.2	2.67

*Lumber is ordered and priced by the board foot. A board foot is a volume of lumber corresponding to 1 in. thick by 12 in. wide by 1 ft long. Nominal dimensions are used to calculate board measure.
SOURCE: *Western Woods Use Book*. 4th ed., 2005 update, Western Wood Products Association, Portland, OR.

Weights of Building Materials

Loads given in Appendix B are typical values. Specific products may have weights which differ considerably from those shown, and manufacturer's catalogs should be consulted for actual loads.

Roof dead loads			
Material	Weight, psf		
Lumber sheathing, 1 in. nominal	2.5		
Plywood, per inch of thickness	3.0		
Timber decking (MC = 15 percent):	2 in. nom.	3 in. nom.	4 in. nom.
DF-Larch	4.2	7.0	9.8
DF (South)	3.9	6.5	9.1
Hem-Fir	3.7	6.1	8.5
Spruce-Pine-Fir	3.7	6.1	8.5
Western Woods	3.5	5.8	8.1
Western Cedars	3.5	5.8	8.1

Roof dead loads.		
Material	Weight, psf	
Aluminum (including laps):	Flat	Corrugated (1½ and 2½ in.)
12 American or B&S gage	1.2	...
14	0.9	1.1
16	0.7	0.9
18	0.6	0.7
20	0.5	0.6
22	...	0.4
Galvanized steel (including laps):	Flat	Corrugated (2½ and 3 in.)
12 U.S. std. gage	4.5	4.9
14	3.3	3.6
16	2.7	2.9
18	2.2	2.4
20	1.7	1.8
22	1.4	1.5
24	1.2	1.3
26	0.9	1.0
Other types of decking (per inch of thickness):		
Concrete plank	6.5	
Insulrock	2.7	
Petrical	2.7	
Porex	2.7	
Poured gypsum	6.5	
Tectum	2.0	
Vermiculite concrete	2.6	
Corrugated asbestos (¼ in.)	3.0	
Felt:		
3-ply	1.5	
3-ply with gravel	5.5	
5-ply	2.5	
5-ply with gravel	6.5	
Insulation (per inch of thickness):		
Expanded polystyrene	0.2	
Fiber glass, rigid	1.5	
Loose	0.5	
Roll roofing	1.0	

Roof dead loads

Material	Weight, psf
Shingles:	
Asphalt ($\frac{1}{4}$ in. approx.)	2.0
Book tile (2 in.)	12.0
Book tile (3 in.)	20.0
Cement asbestos ($\frac{3}{8}$ in. approx.)	4.0
Clay tile (for mortar add 10 psf)	9.0 to 14.0
Ludowici	10.0
Roman	12.0
Slate ($\frac{1}{4}$ in.)	10.0
Spanish	19.0

Ceiling dead loads

Material	Weight, psf
Acoustical fiber tile	1.0
Channel-suspended system	1.0
For gypsum wallboard and plaster, see *Wall and partition dead loads*	

Floor dead loads

Material	Weight, psf
Hardwood (1 in. nominal)	4.0
Plywood (per inch of thickness)	3.0
Asphalt mastic (per inch of thickness)	12.0
Cement finish (per inch of thickness)	12.0
Ceramic and quarry tile ($\frac{3}{4}$ in.)	10.0
Concrete (per inch of thickness)	
Lightweight	6.0 to 10.0
Reinforced (normal weight)	12.5
Stone	12.0
Cork tile ($\frac{1}{16}$ in.)	0.5
Flexicore (6-in. slab)	46.0
Linoleum ($\frac{1}{4}$ in.)	1.0
Terrazo finish ($1\frac{1}{2}$ in.)	19.0
Vinyl tile ($\frac{1}{8}$ in.)	1.4

Wall and partition dead loads

Material	Weight, psf
Wood pancling (1 in.)	2.5
Wood studs (2 × 4 @ 15 percent mc DF-Larch):	
12-in. o.c.	1.2
16-in. o.c.	0.9
24-in. o.c.	0.6
Glass block (4 in.)	18.0
Glass ($\frac{1}{4}$-in. plate)	3.3
Glazed tile	18.0
Marble or marble wainscoting	15.0

Wall and partition dead loads	
Material	Weight, psf
Masonry (per 4 in. of thickness):	
Brick	38.0
Concrete block	30.0
Cinder concrete block	20.0
Hollow clay tile, load bearing	23.0
Hollow clay tile, non-load-bearing	18.0
Hollow gypsum block	13.0
Limestone	55.0
Terra-cotta tile	25.0
Stone	55.0
(The average weights of completed reinforced and grouted concrete block and brick walls can be found in Ref. 12.)	
Plaster (1 in.)	8.0
Plaster (1 in.) on wood lath	10.0
Plaster (1 in.) on metal lath	8.5
Gypsum wallboard (1 in.)	5.0
Porcelain-enameled steel	3.0
Stucco ($\frac{7}{8}$ in.)	10.0
Windows (glass, frame, and sash)	8.0

SOURCE: Weights from *Western Woods Use Book*. 4th ed., 2005 update, Western Wood Products Association, Portland, OR.

Appendix

C

Selected Tables and Figures from The International Building Code, 2006 Edition[*]

Tables and figures for use in determining loads and forces	Table or Figure No.
Deflection Limits	Table 1604.3
Occupancy Category of Buildings and Other Structures	Table 1604.5
Minimum Uniformly Distributed Live Loads and Minimum Concentrated Live Loads	Table 1607.1
Live Load Element Factor, K_{LL}	Table 1607.9.1
Ground Snow Loads, p_g, For Alaskan Locations	Table 1608.2
Ground snow Loads, p_g, For the United States	Figure 1608.2
Basic Wind Speed (3-Second Gust)	Figure 1609
Wind-Borne Debris Protection Fastening Schedule For Wood Structural Panels	Table 1609.1.2
Maximum Considered Earthquake Ground Motion Maps[1]	Figures 1613.5(1)-(14)
Site Class Definitions	Table 1613.5.2
Values of Site Coefficient F_a as a Function of Site Class and Mapped Spectral Response Acceleration at Short Periods (S_S)	Table 1613.5.3(1)
Values of Site Coefficient F_V as a Function of Site Class and Mapped Spectral Response Acceleration at 1-Second Period (S_1)	Table 1613.5.3(2)
Site Classification	Table 1613.5.5
Seismic Design Category Based on Short-Period Response Accelerations	Table 1613.5.6(1)

---

[*]Reproduced from the International Building Code, 2006 ed. Copyright 2006, International Code Council, Inc., Country Club Hills, IL. Reproduced with permission. All rights reserved.

[1]A compact disk of the seismic hazard maps is available from ICC, if requested when purchasing the IBC. The compact disk permits search for mapped spectral acceleration values based on latitude and longitude.

Tables and figures for use in determining loads and forces	Table or Figure No.
Seismic Design Category Based on 1-Second Period Response Acceleration	Table 1613.5.6(2)
Fastening Schedule	Table 2304.9.1
e_n Values For Use in Calculating Diaphragm Deflection Due to Nail Slip (Structural I)	Table 2305.2.2(1)
Values of Gt For Use In Calculating Deflection of Wood Structural Panel Shear Walls and Diaphragms	Table 2305.2.2(2)
Maximum Diaphragm Dimension Ratios Horizontal and Sloping Diaphragm	Table 2305.2.3
Maximum Shear Wall Dimension Ratios	Table 2305.3.4
Shear Resistance Adjustment Factor, C_o	Table 2305.3.8.2
Allowable Shear (Pounds Per Foot) For Wood Structural Panel Diaphragms With Framing of Douglas-Fir-Larch, of Southern Pine for Wind or Seismic Loading	Table 2306.3.1
Allowable Shear (Pounds Per Foot) For Wood Structural Panel Shear Walls with Framing of Douglas-Fir-Larch, or Southern Pine, For Wind Or Seismic Loading	Table 2306.4.1
Allowable Shear For Particleboard Shearwall Sheathing	Table 2306.4.3
Allowable Shear for Wind or Seismic Forces For Shearwalls of Lath and Plaster or Gypsum Board Wood Frames Wall Assemblies	Table 2306.4.5

TABLE 1604.3
DEFLECTION LIMITS[a,b,c,h,i]

Construction	L	S or W^{f}	$D + L^{d,g}$

Roof members:[e]
  Supporting plaster ceiling
  Supporting nonplaster ceiling
  Not supporting ceiling
	
	$l/360$
	$l/240$
	$l/180$
		
		$l/360$
		$l/240$
		$l/180$
			
			$l/240$
			$l/180$
			$l/120$

Floor members	$l/360$	—	$l/240$

Exterior walls and interior partitions:
  With brittle finishes
  With flexible finishes
	
	—
	—
		
		$l/240$
		$l/120$
			
			—
			—

Farm buildings	—	—	$l/180$
Greenhouses	—	—	$l/120$

For SI: 1 ft = 304.8 mm.

a. For structural roofing and siding made of formed metal sheets, the total load deflection shall not exceed $l/60$. For secondary roof structural members supporting formed metal roofing, the live load deflection shall not exceed $l/150$. For secondary wall members supporting formed metal siding, the design wind load deflection shall not exceed $l/90$. For roofs, this exception only applies when the metal sheets have no roof covering.

b. Interior partitions not exceeding 6 ft in height and flexible, folding and portable partitions are not governed by the provisions of this section. The deflection criterion for interior partitions is based on the horizontal load defined in Sec. 1607.13.

c. See Sec. 2403 for glass supports.

d. For wood structural members having a moisture content of less than 16 percent at time of installation and used under dry conditions, the deflection resulting from L + 0.5D is permitted to be substituted for the deflection resulting from L + D.

e. The above deflections do not ensure against ponding. Roofs that do not have sufficient slope or camber to assure adequate drainage shall be investigated for ponding. See Sec. 1611 for rain and ponding requirements and Sec. 1503.4 for roof drainage requirements.

f. The wind load is permitted to be taken as 0.7 times the "component and cladding" loads for the purpose of determining deflection limits herein.

g. For steel structural members, the dead load shall be taken as zero.

h. For aluminum structural members or aluminum panels used in skylights and sloped glazing framing, roofs or walls of sunroom additions or patio covers, not supporting edge of glass or aluminum sandwich panels, the total load deflection shall not exceed $l/60$. For aluminum sandwich panels used in roofs or walls of sunroom additions or patio covers, the total load deflection shall not exceed $l/120$.

i. For cantilever members, l shall be taken as twice the length of the cantilever.

TABLE 1604.5
OCCUPANCY CATEGORY OF BUILDINGS AND OTHER STRUCTURES

OCCUPANCY CATEGORY	NATURE OF OCCUPANCY
I	Buildings and other structures that represent a low hazard to human life in the event of failure, including but not limited to: • Agricultural facilities. • Certain temporary facilities. • Minor storage facilities.
II	Buildings and other structures except those listed in Occupancy Categories I, III, and IV
III	Buildings and other structures that represent a substantial hazard to human life in the event of failure, including but not limited to: • Covered structures whose primary occupancy is public assembly with an occupant load greater than 300. • Buildings and other structures with elementary school, secondary school or day care facilities with an occupant load greater than 250. • Buildings and other structures with an occupant load greater than 500 for colleges or adult education facilities. • Health care facilities with an occupant load of 50 or more resident patients, but not having surgery or emergency treatment facilities. • Jails and detention facilities. • Any other occupancy with an occupant load greater than 5000. • Power-generating stations, water treatment for potable water, waste water treatment facilities, and other public utility facilities not included in Occupancy Category IV. • Buildings and other structures not included in Occupancy Category IV containing sufficient quantities of toxic or explosive substances to be dangerous to the public if released.
IV	Buildings and other structures designated as essential facilities, including but not limited to: • Hospitals and other health care facilities having surgery or emergency treatment facilities. • Fire, rescue and police stations, and emergency vehicle garages. • Designated earthquake, hurricane, or other emergency shelters. • Designated emergency preparedness, communication, and operation centers, and other facilities required for emergency response. • Power-generating stations and other public utility facilities required as emergency backup facilities for Occupancy Category IV structures. • Structures containing highly toxic materials as defined by Sec. 307 where the quantity of the material exceeds the maximum allowable quantities of Table 307.1.(2). • Aviation control towers, air traffic control centers and emergency aircraft hangars. • Buildings and other structures having critical national defense functions. • Water treatment facilities required to maintain water pressure for fire suppression.

TABLE 1607.1
MINIMUM UNIFORMLY DISTRIBUTED LIVE LOADS AND MINIMUM CONCENTRATED LIVE LOADS[g]

OCCUPANCY OR USE	UNIFORM (psf)	CONCENTRATED (lbs.)
1. Apartments (see residential)	—	—

2. Access floor systems
   Office use
   Computer use
	
	50
	100
		
		2,000
		2,000

3. Armories and drill rooms	150	—

4. Assembly areas and theaters
   Fixed seats (fastened to floor)
   Follow spot, projections and
     control rooms
   Lobbies
   Movable seats
   Stages and platforms
	
	60
	
	50
	100
	100
	125
		
		—

5. Balconies
   On one- and two-family residences only,
     and not exceeding 100 sq ft
	100
	60
		
		—

6. Bowling alleys	75	—
7. Catwalks	40	300
8. Dance halls and ballrooms	100	—
9. Decks	Same as occupancy served[h]	—
10. Dining rooms and restaurants	100	—
11. Dwellings (see residential)	—	—
12. Cornices	60	
13. Corridors, except as otherwise indicated	100	—
14. Elevator machine room grating (on area of 4 in^2)	—	300
15. Finish light floor plate construction (on area of 1 in^2)	—	200
16. Fire escapes On single-family dwellings only	100 40	—
17. Garages (passenger vehicles only) Trucks and buses	40	Note a
17. Garages (passenger vehicles only) Trucks and buses	See Section 1607.6	
18. Grandstands (see stadium and arena bleachers)	—	—
19. Gymnasiums, main floors and balconies	100	—
20. Handrails, guards and grab bars	See Section 1607.7	

21. Hospitals
   Corridors above first floor
   Operating rooms, laboratories
   Patient rooms
	
	80
	60
	40
		
		1,000
		1,000
		1,000

22. Hotels (see residential)	—	—

(continued)

TABLE 1607.1 (Continued)
MINIMUM UNIFORMLY DISTRIBUTED LIVE LOADS AND MINIMUM CONCENTRATED
LIVE LOADS[g]

Occupancy or use	Uniform (psf)	Concentrated (lbs.)
23. Libraries		
Corridors above first floor	80	1,000
Reading rooms	60	1,000
Stack rooms	150[b]	1,000
24. Manufacturing		
Heavy	250	3,000
Light	125	2,000
25. Marquees	75	—
26. Office buildings		
Corridors above first floor	80	2,000
File and computer rooms shall be designed for heavier loads based on anticipated occupancy	—	—
Lobbies and first-floor corridors	100	2,000
Offices	50	2,000
27. Penal institutions		
Cell blocks	40	—
Corridors	100	
28. Residential		
One- and two-family dwellings		
Uninhabitable attics without storage[i]	10	
Uninhabitable attics with limited storage[i, j, k]	20	
Habitable attics and sleeping areas	30	—
All other areas except balconies and decks	40	
Hotels and multiple-family dwellings		
Private rooms and corridors serving them	40	
Public rooms and corridors serving them	100	
29. Reviewing stands, grandstands and bleachers	Note c	
30. Roofs		
All roof surfaces subject to maintenance workers		300
Awnings and canopies		
Fabric construction supported by a lightweight rigid skeleton structure	5 nonreduceable	
All other construction	20	
Ordinary flat, pitched, and curved roofs	20	
Primary roof members, exposed to a work floor		
Single panel point of lower chord of roof trusses or any point along primary structural members supporting roofs:		
Over manufacturing, storage warehouses, and repair garages		2,000

(continued)

TABLE 1607.1 *(Continued)*
MINIMUM UNIFORMLY DISTRIBUTED LIVE LOADS AND MINIMUM CONCENTRATED LIVE LOADS[g]

Occupancy or use	Uniform (psf)	Concentrated (lbs.)
All other occupancies		300
Roofs used for other special purposes	Note 1	Note 1
Roofs used for promenade purposes	60	
Roofs used for roof gardens or assembly purposes	100	
31. Schools		
Classrooms	40	1,000
Corridors above first floor	80	1,000
First-floor corridors	100	1,000
32. Scuttles, skylight ribs and accessible ceilings	—	200
33. Sidewalks, vehicular driveways and yards, subject to trucking	250[d]	8,000[e]
34. Skating rinks	100	—
35. Stadiums and arenas		
Bleachers	100[c]	—
Fixed seats (fastened to floor)	60[c]	
36. Stairs and exits		Note f
One- and two-family dwellings	40	
All other	100	
37. Storage warehouses (shall be designed for heavier loads if required for anticipated storage)		
Heavy	250	
Light	125	
38. Stores		
Retail		
First floor	100	1,000
Upper floors	75	1,000
Wholesale, all floors	125	1,000
39. Vehicle barriers	See Section 1607.7.3	
40. Walkways and elevated platforms (other than exitways)	60	—
41. Yards and terraces, pedestrians	100	—

For SI: 1 in. = 25.4 mm, 1 square inch = 645.16 mm^2,
       1 ft^2 = 0.0929 m^2,
       1 pound per square foot = 0.0479 kN/m^2, 1 pound = 0.004448 kN,
       1 pound per cubic foot = 16 kg/m^3

a. Floors in garages or portions of buildings used for the storage of motor vehicles shall be designed for the uniformly distributed live loads of Table 1607.1 or the following concentrated loads: (1) for garages restricted to vehicles accommodating not more than nine passengers, 3000 pounds acting on an area of 4.5 in. by 4.5 in.; (2) for mechanical parking structures without slab or deck which are used for storing passenger vehicles only, 2250 pounds per wheel.

b. The loading applies to stack room floors that support nonmobile, double-faced library bookstacks, subject to the following limitations:

(continued)

   1. The nominal bookstack unit height shall not exceed 90 in.;
   2. The nominal shelf depth shall not exceed 12 in. for each face; and
   3. Parallel rows of double-faced bookstacks shall be separated by aisles not less than 36 in. wide.
   c. Design in accordance with the ICC *Standard on Bleachers, Folding and Telescopic Seating and Grandstands.*
   d. Other uniform loads in accordance with an approved method which contains provisions for truck loadings shall also be considered where appropriate.
   e. The concentrated wheel load shall be applied on an area of 20 in.^{2}
   f. Minimum concentrated load on stair treads (on area of 4 in.^{2}) is 300 pounds.
   g. Where snow loads occur that are in excess of the design conditions, the structure shall be designed to support the loads due to the increased loads caused by drift buildup or a greater snow design determined by the building official (see Sec. 1608). For special-purpose roofs, see Sec. 1607.11.2.2.
   h. See Section 1604.8.3 for decks attached to exterior walls.
   i. Attics without storage are those where the maximum clear height between the joist and rafter is less than 42 in., or where there are not two or more adjacent trusses with the same web configuration capable of containing a rectangle 42 in. high by 2 ft wide, or greater, located within the plane of the truss. For attics without storage, this live load need not be assumed to act concurrently with any other live load requirements.
   j. For attics with limited storage and constructed with trusses, this live load need only be applied to those portions of the bottom chord where there are two or more adjacent trusses with the same web configuration capable of containing a rectangle 42 in. high by 2 ft wide or greater, located within the plane of the truss. The rectangle shall fit between the top of the bottom chord and the bottom of any other truss member, provided that each of the following criteria is met:
        i. The attic area is accessible by a pull-down stairway or framed opening in accordance with Section 1209.2, and
        ii. The truss shall have a bottom chord pitch less than 2:12.
        iii. Bottom chords of trusses shall be designed for the greater of actual imposed dead load or 10 psf, uniformly distributed over the entire span.
   k. Attic spaces served by a fixed stair shall be designed to support the minimum live load specified for habitable attics and sleeping rooms.
   l. Roofs used for other special purposes shall be designed for appropriate loads as approved by the building official.

TABLE 1607.9.1
LIVE LOAD ELEMENT FACTOR, K_{LL}

Element	K_{LL}
Interior columns	4
Exterior columns without cantilever slabs	4
Edge columns with cantilever slabs	3
Corner columns with cantilever slabs	2
Edge beams without cantilever slabs	2
Interior beams	2

All other members not identified above including:
    Edge beams with cantilever slabs
    Cantilever beams
    Two-way slabs
    Members without provisions for continuous shear
        transfer normal to their span
	1

TABLE 1608.2
GROUND SNOW LOADS, p_g, FOR ALASKAN LOCATIONS

Location	Pounds per square foot	Location	Pounds per square foot	Location	Pounds per square foot
Adak	30	Galena	60	Petersburg	150
Anchorage	50	Gulkana	70	St. Paul Islands	40
Angoori	70	Homer	40	Seward	50
Barrow	25	Juneau	60	Shemya	25
Barter Island	35	Kenai	70	Sitka	50
Bethel	40	Kodiak	30	Talkeetna	120
Big Delta	50	Kotzebue	60	Unalakleet	50
Cold Bay	25	McGrath	70	Valdez	160
Cordova	100	Nenana	80	Whittier	300
Fairbanks	60	Nome	70	Wrangell	60
Fort Yukon	60	Palmer	50	Yakutat	150

For SI: 1 pound per square foot = 0.0479 kN/m^2.

FIGURE 1608.2
GROUND SNOW LOADS, p_g, FOR THE UNITED STATES (psf)

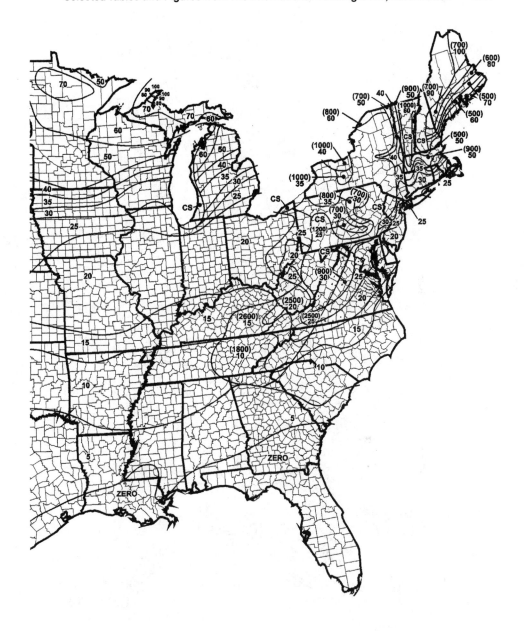

FIGURE 1608.02–continued
GROUND SNOW LOADS, p_g, FOR THE UNITED STATES (psf)

FIGURE 1609
BASIC WIND SPEED (3-SECOND GUST)

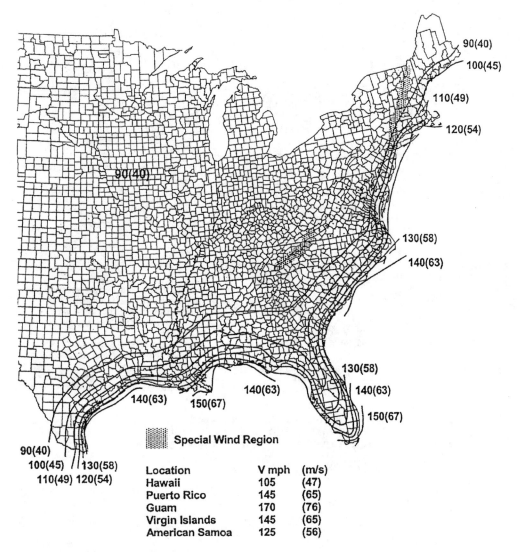

Location	V mph	(m/s)
Hawaii	105	(47)
Puerto Rico	145	(65)
Guam	170	(76)
Virgin Islands	145	(65)
American Samoa	125	(56)

Notes:
1. Values are nominal design 3-second gust wind speeds in miles per hour (m/s) at 33 ft (10 m) above ground for Exposure C category.
2. Linear interpolation between wind contours is permitted.
3. Islands and coastal areas outside the last contour shall use the last wind speed contour of the coastal area.
4. Mountainous terrain, gorges, ocean promontories, and special wind regions shall be examined for unusual wind conditions.

FIGURE 1609—continued
BASIC WIND SPEED (3-SECOND GUST)

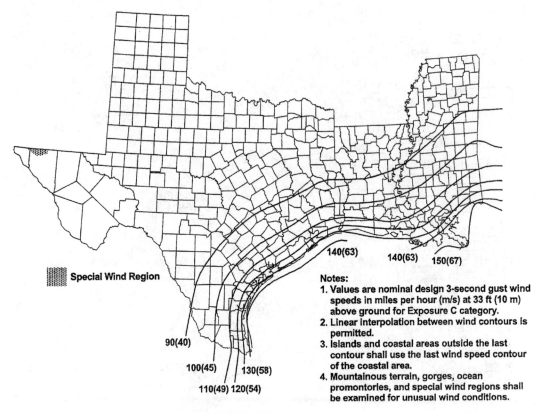

Special Wind Region

140(63)   140(63)   150(67)

90(40)

100(45)   130(58)

110(49) 120(54)

Notes:
1. Values are nominal design 3-second gust wind speeds in miles per hour (m/s) at 33 ft (10 m) above ground for Exposure C category.
2. Linear interpolation between wind contours is permitted.
3. Islands and coastal areas outside the last contour shall use the last wind speed contour of the coastal area.
4. Mountainous terrain, gorges, ocean promontories, and special wind regions shall be examined for unusual wind conditions.

**FIGURE 1609—continued
BASIC WIND SPEED (3-SECOND ONDGUST)
WESTERN GULF OF MEXICO HURRICANE COASTLINE**

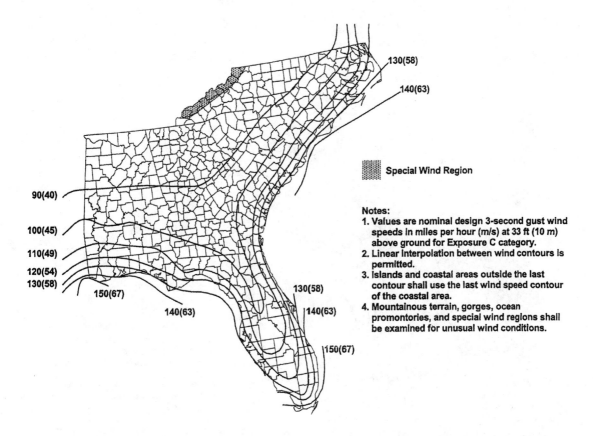

90(40)

100(45)

110(49)

120(54)
130(58)

150(67)

140(63)

130(58)

140(63)

150(67)

130(58)

140(63)

130(58)

140(63)

150(67)

Special Wind Region

Notes:
1. Values are nominal design 3-second gust wind speeds in miles per hour (m/s) at 33 ft (10 m) above ground for Exposure C category.
2. Linear interpolation between wind contours is permitted.
3. Islands and coastal areas outside the last contour shall use the last wind speed contour of the coastal area.
4. Mountainous terrain, gorges, ocean promontories, and special wind regions shall be examined for unusual wind conditions.

FIGURE 1609—continued
BASIC WIND SPEED (3-SECOND GUST)
EASTERN GULF OF MEXICO SOUTHEASTERN U.S. HURRICANE COASTLINE

90(40)

100(45)

110(49)

120(54)

Special Wind Region

Notes:
1. Values are nominal design 3-second gust wind speeds in miles per hour (m/s) at 33 ft (10 m) above ground for Exposure C category.
2. Linear interpolation between wind contours is permitted.
3. Islands and coastal areas outside the last contour shall use the last wind speed contour of the coastal area.
4. Mountainous terrain, gorges, ocean promontories, and special wind regions shall be examined for unusual wind conditions.

FIGURE 1609—continued
BASIC WIND SPEED (3-SECOND GUST)
MID AND NORTHERN ATLANTIC HURRICANE COASTLINE

TABLE 1609.1.2
WIND-BORNE DEBRIS PROTECTION FASTENING SCHEDULE FOR WOOD STRUCTURAL PANELS[a,b,c,d]

Fastener type	Fastener spacing (inches)		
	Panel span ≤ 4 feet	4 feet < panel span ≤ 6 feet	6 feet < panel span ≤ 8 feet
No. 6 screws	16	12	9
No. 8 screws	16	16	12

For SI: 1 in. = 25.4 mm, 1 ft = 304.8 mm, 1 pound = 4.4 N,
          1 mil per hour = 0.44 m/s.

a. This table is based on a maximum wind speed (3-sec gust) of 130 mph and mean roof height of 33 ft or less.
b. Fasteners shall be installed at opposing ends of the wood structural panel. Fasteners shall be located a minimum of 1 inch from the edge of the panel.
c. Fasteners shall be long enough to penetrate through the exterior wall covering a minimum of 1.75 in. into wood wall framing; a minimum of 1.25 in. into concrete block or concrete; or into steel framing by at least three threads. Fasteners shall be located a minimum of 2.5 in. from the edge of concrete block or concrete.
d. Where screws are attached to masonry or masonry/stucco, they shall be attached utilizing vibration-resistant anchors having a minimum withdrawal capacity of 490 pounds.

DISCUSSION

The acceleration values contoured on this map are for the random horizontal component of acceleration. For design purposes, the reference site condition for the map is to be taken as Site Class B.

Selected contours have been deleted for clarity. Regional maps should be used when additional detail is required.

Leyendecker, Frankel, and Rukstales (2001, 2004) have prepared a CD-ROM that contains software to allow determination of Site Class B map values by latitude-longitude. The software on the CD contains site coefficients that allow the user to adjust map values for different Site Classes. Additional maps at different scales are also included on the CD. The CD was prepared using the same data as that used to prepare the Maximum Considered Earthquake Ground Motion maps.

The National Seismic Hazard Mapping Project Web Site, http://eqhazmaps.usgs.gov/, contains electronic versions of this map and others. Documentation, gridded values, and Arc/INFO coverages used to make the maps are also available.

The California portion of the map was produced jointly with the California Geological Survey.

Map prepared by U.S. Geological Survey.

REFERENCES

Building Seismic Safety Council 2004, NEHRP Recommended Provisions for Seismic Regulations for New Buildings and other Structures, Part 1 - Provisions, FEMA 450.
Building Seismic Safety Council 2004, NEHRP Recommended Provisions for Seismic Regulations for New Buildings and other Structures, Part 2 - Commentary, FEMA 450.
Frankel, A., Petersen, M, Mueller, C., Haller, K., Wheeler, R., Leyendecker, E, Wesson, R., Harmsen, S., Cramer, C., Perkins, D., and Rukstales, K., 2002, Documentation for the 2002 Update of the National Seismic Hazard Maps, U.S. Geological Survey Open-File Report 02-420.
Frankel, A., Petersen, M, Mueller, C., Haller, K., Wheeler, R., Leyendecker, E, Wesson, R., Harmsen, S., Cramer, C., Perkins, D., and Rukstales, K., 2004, Seismic-Hazard Maps for the Conterminous United States, Sheet 4 - Horizontal Spectral Response Acceleration for 0.2 Second Period with 2% Probability of Exceedance in 50 Years, U.S. Geological Survey Geologic Investigation Series, scale 1:7,000,000. (in progress)
Leyendecker, E., Frankel, A., and Rukstales, K., 2001, Seismic Design Parameters, U.S. Geological Survey Open-File Report 01-437.
Leyendecker, E., Frankel, A., and Rukstales, K., 2004, Seismic Design Parameters, U.S. Geological Survey Open-File Report (in progress).
National Seismic Hazard Mapping Project Web Site, http://eqhazmaps.usgs.gov, U. S. Geological Survey.

Index of detailed regional map(s) at larger scale(s)

- Region 1 is shown enlarged in figure 1613.5(3)
- Region 2 is shown enlarged in figure 1613.5(5)
- Region 3 is shown enlarged in figure 1613.5(7)
- Region 4 is shown enlarged in figure 1613.5(9)

FIGURE 1613.5(1)
MAXIMUM CONSIDERED EARTHQUAKE GROUND MOTION FOR THE CONTERMINOUS UNITED STATES OF 0.2 SEC SPECTRAL RESPONSE ACCELERATION (5 PERCENT OF CRITICAL DAMPING), SITE CLASS B

FIGURE 1613.5(1)—continued
MAXIMUM CONSIDERED EARTHQUAKE GROUND MOTION FOR THE CONTERMINOUS UNITED STATES OF 0.2 SEC SPECTRAL RESPONSE ACCELERATION (5 PERCENT OF CRITICAL DAMPING), SITE CLASS B

FIGURE 1613.5(2)
MAXIMUM CONSIDERED EARTHQUAKE GROUND MOTION FOR THE CONTERMINOUS UNITED STATES OF 1.0 SEC SPECTRAL RESPONSE ACCELERATION (5 PERCENT OF CRITICAL DAMPING), SITE CLASS B

FIGURE 1613.5(2)—continued
**MAXIMUM CONSIDERED EARTHQUAKE GROUND MOTION FOR THE CONTERMINOUS UNITED STATES
OF 1.0 SEC SPECTRAL RESPONSE ACCELERATION (5 PERCENT OF CRITICAL DAMPING), SITE CLASS B**

FIGURE 1613.5(3)
MAXIMUM CONSIDERED EARTHQUAKE GROUND MOTION FOR REGION 1 OF 0.2 SEC SPECTRAL RESPONSE ACCELERATION (5 PERCENT OF CRITICAL DAMPING), SITE CLASS B

FIGURE 1613.5(3)—continued

MAXIMUM CONSIDERED EARTHQUAKE GROUND MOTION FOR REGION 1 OF 0.2 SEC SPECTRAL RESPONSE ACCELERATION (5 PERCENT OF CRITICAL DAMPING), SITE CLASS B

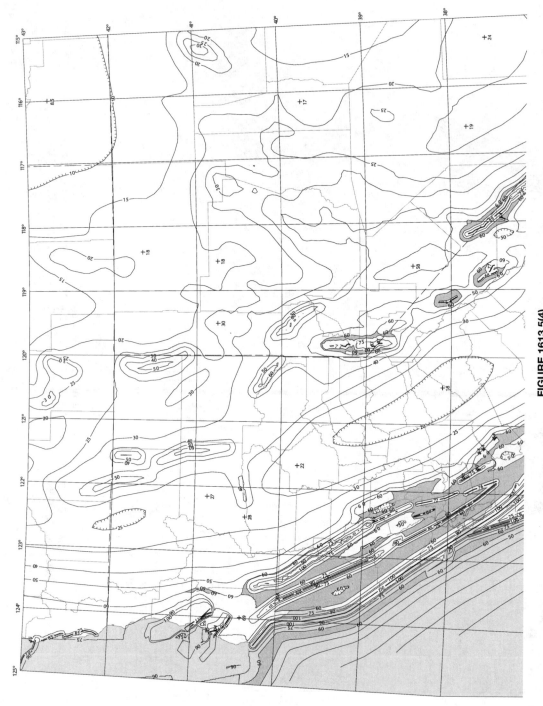

FIGURE 1613.5(4)
MAXIMUM CONSIDERED EARTHQUAKE GROUND MOTION FOR REGION 1 OF 1.0 SEC SPECTRAL RESPONSE ACCELERATION (5 PERCENT OF CRITICAL DAMPING), SITE CLASS B

FIGURE 1613.5(4)—continued

MAXIMUM CONSIDERED EARTHQUAKE GROUND MOTION FOR REGION 1 OF 1.0 SEC SPECTRAL RESPONSE ACCELERATION (5 PERCENT OF CRITICAL DAMPING), SITE CLASS B

FIGURE 1613.5(5)

MAXIMUM CONSIDERED EARTHQUAKE GROUND MOTION FOR REGION 2 OF 0.2 SEC SPECTRAL RESPONSE ACCELERATION (5 PERCENT OF CRITICAL DAMPING), SITE CLASS B

FIGURE 1613.5(6)

MAXIMUM CONSIDERED EARTHQUAKE GROUND MOTION FOR REGION 2 OF 1.0 SEC SPECTRAL RESPONSE ACCELERATION (5 PERCENT OF CRITICAL DAMPING), SITE CLASS B

Index map showing location of study area

Explanation

Contour intervals, % g

— 200
— 175
— 150
— 125
— 100
— 90
— 80
— 70
— 60
— 50
— 40
— 35
— 30
— 25
— 20
— 15
— 10
— 5

Note: contours are irregularly spaced

Areas with a constant spectral response acceleration of 150% g

+ 6.2 Point value of spectral response acceleration expressed as a percent of gravity

—10— Contours of spectral response acceleration expressed as a percent of gravity. Hachures point in direction of decreasing values.

DISCUSSION

Refer to the map of Maximum Considered Earthquake Ground Motion for the Conterminous United States of 0.2 sec Spectral Response Acceleration (Figure 1613.5(1)) for discussion and references.

100 MILES

100 KILOMETERS

FIGURE 1613.5(7)
MAXIMUM CONSIDERED EARTHQUAKE GROUND MOTION FOR REGION 3 OF 0.2 SEC SPECTRAL RESPONSE ACCELERATION (5 PERCENT OF CRITICAL DAMPING), SITE CLASS B

C.28

Explanation

Contour intervals, % g

———150
———125
———100
———90
———75
———60
———50
———40
———30
———25
———20
———15
———10
———8
———6
———4
———2
———0

Note: contours are irregularly spaced

[shaded box] Areas with a constant spectral response acceleration of 60% g

+ Point value of spectral response
6.2   acceleration expressed as a percent of gravity

——10—— Contour of spectral response acceleration expressed as a percent of gravity. Hachures point in direction of decreasing values.

——10——

DISCUSSION

Refer to the map of Maximum Considered Earthquake Ground Motion for the Conterminous United States of 1.0 sec Spectral Response Acceleration (Figure 1613.5(2)) for discussion and references.

Index map showing location of study area

FIGURE 1613.5(8)
MAXIMUM CONSIDERED EARTHQUAKE GROUND MOTION FOR REGION 3 OF 1.0 SEC SPECTRAL RESPONSE ACCELERATION (5
PERCENT OF CRITICAL DAMPING), SITE CLASS B

FIGURE 1613.5(9)
MAXIMUM CONSIDERED EARTHQUAKE GROUND MOTION FOR REGION 4 OF 0.2 AND 1.0 SEC SPECTRAL RESPONSE ACCELERATION (5 PERCENT OF CRITICAL DAMPING), SITE CLASS B

FIGURE 1613.5(10)
MAXIMUM CONSIDERED EARTHQUAKE GROUND MOTION FOR HAWAII OF 0.2 AND 1.0 SEC SPECTRAL RESPONSE ACCELERATION (5 PERCENT OF CRITICAL DAMPING), SITE CLASS B

FIGURE 1613.5(11)

MAXIMUM CONSIDERED EARTHQUAKE GROUND MOTION FOR ALASKA OF 0.2 SEC SPECTRAL RESPONSE ACCELERATION (5 PERCENT OF CRITICAL DAMPING), SITE CLASS B

FIGURE 1613.5(12)

MAXIMUM CONSIDERED EARTHQUAKE GROUND MOTION FOR ALASKA OF 1.0 SEC SPECTRAL RESPONSE ACCELERATION (5 PERCENT OF CRITICAL DAMPING), SITE CLASS B

0.2 SEC SPECTRAL RESPONSE ACCELERATION (5% OF CRITICAL DAMPING)

1.0 SEC SPECTRAL RESPONSE ACCELERATION (5% OF CRITICAL DAMPING)

Explanation

+ 6.2
Point value of spectral response acceleration expressed as a percent of gravity

—— 10 ——
Contours of spectral response acceleration expressed as a percent of gravity. Hachures point in direction of decreasing values.

DISCUSSION

The acceleration values contoured on this map are for the random horizontal component of acceleration. For design purposes, the reference site condition for the map is to be taken as Site Class B.

Leyendecker, Frankel, and Rukstales (2001, 2004) have prepared a CD-ROM that contains software to allow determination of Site Class B map values by latitude-longitude. The software on the CD contains site coefficients that allow the user to adjust map values for different Site Classes. Additional maps at different scales are also included on the CD. The CD was prepared using the same data as that used to prepare the Maximum Considered Earthquake Ground Motion maps.

The National Seismic Hazard Mapping Project Web Site, http://eqhazmaps.usgs.gov, contains electronic versions of this map and others. Documentation, gridded values, and Arc/INFO coverages used to make the maps are also available.

Map prepared by U.S. Geological Survey.

REFERENCES

Building Seismic Safety Council 2004, NEHRP Recommended Provisions for Seismic Regulations for New Buildings and other Structures, Part 1 - Provisions, FEMA 450.
Building Seismic Safety Council 2004, NEHRP Recommended Provisions for Seismic Regulations for New Buildings and other Structures, Part 2 - Commentary, FEMA 450.
Leyendecker, E., Frankel, A., and Rukstales, K., 2001, Seismic Design Parameters, U.S. Geological Survey Open-File Report 01-437.
Leyendecker, E., Frankel, A., and Rukstales, K., 2004, Seismic Design Parameters, U.S. Geological Survey Open-File Report (in progress).
Mueller, C., Frankel, A., Petersen, M., and Leyendecker, E., 2003, Documentation for 2003 USGS Seismic Hazard Maps for Puerto Rico and the U.S. Virgin Islands, U.S. Geological Survey Open-File Report 03-379.
Mueller, C., Frankel, A., Petersen, M., and Leyendecker, E., 2004, Seismic-Hazard Maps for Puerto Rico and the U.S. Virgin Island, Sheet 2 - 2% Probability of Exceedance in 50 Years for Peak Horizontal Acceleration and Horizontal Spectral Response Acceleration for 0.2, 0.3, and 1.0 Second Periods U.S. Geological Survey Geologic Investigation Series (in progress).
National Seismic Hazard Mapping Project Web Site, http://eqhazmaps.usgs.gov, U.S. Geological Survey.

FIGURE 1613.5(13)
MAXIMUM CONSIDERED EARTHQUAKE GROUND MOTION FOR PUERTO RICO, VIEQUES, ST. THOMAS, ST. JOHN AND ST. CROIX OF 0.2 AND 1.0 SEC SPECTRAL RESPONSE ACCELERATION (5 PERCENT OF CRITICAL DAMPING), SITE CLASS B

DISCUSSION

Leyendecker, Frankel, and Rukstales (2001, 2004) have prepared a CD-ROM that contains software to allow determination of Site Class B map values by either latitude-longitude or zip code. The software on the CD contains site coefficients that allow the user to adjust map values for different Site Classes.

Map prepared by U.S. Geological Survey.

0.2 SEC SPECTRAL RESPONSE ACCELERATION (5% OF CRITICAL DAMPING)

REFERENCES

Building Seismic Safety Council 2004, NEHRP Recommended Provisions for Seismic Regulations for New Buildings and other Structures, Part 1 - Provisions, FEMA 450.
Building Seismic Safety Council 2004, NEHRP Recommended Provisions for Seismic Regulations for New Buildings and other Structures, Part 2 - Commentary, FEMA 450.
Leyendecker, E., Frankel, A., and Rukstales, K., 2001, Seismic Design Parameters, U.S. Geological Survey Open-File Report 01-437.
Leyendecker, E., Frankel, A., and Rukstales, K., 2004, Seismic Design Parameters, U.S. Geological Survey Open-File Report (in progress).
National Seismic Hazard Mapping Project Web Site, http://eqhazmaps.usgs.gov, U.S. Geological Survey.

1.0 SEC SPECTRAL RESPONSE ACCELERATION (5% OF CRITICAL DAMPING)

FIGURE 1613.5(14)
MAXIMUM CONSIDERED EARTHQUAKE GROUND MOTION FOR GUAM AND TUTUILLA OF 0.2 AND 1.0 SEC SPECTRAL RESPONSE ACCELERATION (5 PERCENT OF CRITICAL DAMPING), SITE CLASS B

TABLE 1613.5.2
SITE CLASS DEFINITIONS

Site class	Soil profile name	Average properties in top 100 feet, as per section 1615.1.5		
		Soil shear wave velocity, $\bar{v}_s$, (ft/s)	Standard penetration resistance, $\bar{N}$	Soil undrained shear strength, $\bar{s}_y$, (psf)
A	Hard rock	$\bar{v}_s > 5,000$	N/A	N/A
B	Rock	$2,500 < \bar{v}_s \leq 5,000$	N/A	N/A
C	Very dense soil and soft rock	$1,200 < \bar{v}_s \leq 2,500$	$\bar{N} > 50$	$\bar{s}_a \geq 2,000$
D	Stiff soil profile	$600 \leq \bar{v}_s \leq 1,200$	$15 \leq \bar{N} \leq 50$	$1,000 \leq \bar{s}_y \leq 2,000$
E	Soft soil profile	$\bar{v}_s < 600$	$\bar{N} < 50$	$\bar{s}_y < 1,000$
F	—	Any profile with more than 10 feet of soil having the following characteristics: 1. Plasticity index $PI > 20$, 2. Moisture content $w \geq 40\%$, and 3. Undrained shear strength $\bar{s}_a < 500$ psf		
G	—	Any profile containing soils having one or more of the following characteristics: 1. Soils vulnerable to potential failure or collapse under seismic loading such as liquefiable soils, quick and highly sensitive clays, collapsible weakly cemented soils. 2. Peats and/or highly organic clays ($H > 10$ feet of peat and/or highly organic clay where H = thickness of soil) 3. Very high plasticity clays ($H > 25$ feet with plasticity index $PI > 75$) 4. Very thick soft/medium stiff clays ($H > 120$ feet)		

For SI: 1 foot = 304.8 mm, 1 square foot = 0.0929 m^2, 1 pound per square foot = 0.0479 kPa. N/A = Not applicable

TABLE 1613.5.3(1)
VALUES OF SITE COEFFICIENT F_a[a]

Site class	Mapped spectral response acceleration at short period				
	$S_s \leq 0.25$	$S_s = 0.50$	$S_s = 0.75$	$S_s = 1.00$	$S_s \geq 1.25$
A	0.8	0.8	0.8	0.8	0.8
B	1.0	1.0	1.0	1.0	1.0
C	1.2	1.2	1.1	1.0	1.0
D	1.6	1.4	1.2	1.1	1.0
E	2.5	1.7	1.2	0.9	0.9
F	Note b	Note b	Note b	Note b	Note b

a. Use straight-line interpolation for intermediate values of mapped spectral response acceleration at short period, S_s.
b. Values shall be determined in accordance with Sec. 11.4.7 of ASCE 7.

TABLE 1613.5.3(2)
VALUES OF SITE COEFFICIENT F_v^a

Site class	Mapped spectral response acceleration at 1-second period				
	$S_1 \leq 0.1$	$S_1 = 0.2$	$S_1 = 0.3$	$S_1 = 0.4$	$S_1 \geq 0.5$
A	0.8	0.8	0.8	0.8	0.8
B	1.0	1.0	1.0	1.0	1.0
C	1.7	1.6	1.5	1.4	1.3
D	2.4	2.0	1.8	1.6	1.5
E	3.5	3.2	2.8	2.4	2.4
F	Note b	Note b	Note b	Note b	Note b

a. Use straight-line interpolation for intermediate values of mapped spectral response acceleration at 1-second period, S_1.
b. Values shall be determined in accordance with Sec. 11.4.7 of ASCE 7.

TABLE 1613.5.5
SITE CLASSIFICATION^a

Site class	$\overline{v}_s$	$\overline{N}$ or $\overline{N}_{ch}$	$\overline{s}_v$
E	< 600 ft/s	< 15	< 1,000 psf
D	600 to 1,200 ft/s	15 to 50	1,000 to 2,000 psf
C	1,200 to 2,500 ft/s	> 50	> 2,000

For SI: 1 foot per second = 304.8 mm per second, 1 pound per square foot = 0.0479 kN/m^2.
a. If the $\overline{s}_v$ method is used and the $\overline{N}_{ch}$ and $\overline{s}_v$ criteria differ, select the category with the softer soils (for example, use Site Class E instead of D).

TABLE 1613.5.6(1)
SEISMIC DESIGN CATEGORY BASED ON
SHORT-PERIOD RESPONSE ACCELERATIONS

Value of S_{DS}	Occupancy category		
	I or II	III	IV
$S_{DS} < 0.167$g	A	A	A
0.167g $\leq S_{DS} < 0.33$g	B	B	C
0.33g $\leq S_{DS} < 0.50$g	C	C	D
0.50g $\leq S_{DS}$	D	D	D

TABLE 1613.5.6(2)
SEISMIC DESIGN CATEGORY BASED ON
1-SECOND PERIOD RESPONSE ACCELERATION

Value of S_{D1}	Occupancy category		
	I or II	III	IV
$S_{D1} < 0.067$g	A	A	A
0.067g $\leq S_{D1} < 0.133$g	B	B	C
0.133g $\leq S_{D1} < 0.20$g	C	C	D
0.20g $\leq S_{D1}$	D	D	D

TABLE 2304.9.1
FASTENING SCHEDULE

Connection	Fastening[a,m]	Location
1. Joist to still or girder	3 - 8d common ($2^1/_2$ in. × 0.131 in.) 3 - 3 in. × 0.131 in. nails 3 - 3 in. 14 gage staples	toenail
2. Bridging to joist	2 - 8d common ($2^1/_2$ in. × 0.131 in.) 2 - 3 in. × 0.131 in. nails 2 - 3 in. 14 gage staples	toenail each end
3. 1 in. × 6 in. subfloor or less to each joist	2 - 8d common ($2^1/_2$ in. × 0.131 in.)	face nail
4. Wider than 1 in. × 6 in. subfloor to each joist	3 - 8d common ($2^1/_2$ in. × 0.131 in.)	face nail
5. 2 in. subfloor to joist or girder	2 - 16d common ($3^1/_2$ in. × 0.162 in.)	blind and face nail
6. Sole plate to joist or blocking	16d ($3^1/_2$ in. × 0.135 in.) at 16 in. o.c. 3 in. × 0.131 in. nails at 8 in. o.c. 3 in. 14 gage staples at 12 in. o.c.	typical face nail
Sole plate to joist or blocking at braced wall panel	3 in. - 16d ($3^1/_2$ in. × 0.135 in.) at 16 in. 4 - 3 in. × 0.131 in. nails at 16 in. 4 - 3 in. 14 gage staples per 16 in.	braced wall panels
7. Top plate to stud	2 - 16d common ($3^1/_2$ in. × 0.162 in.) 3 - 3 in. × 0.131 in. nails 3 - 3 in. 14 gage staples	end nail
8. Stud to sole plate	4 - 8d common ($2^1/_2$ in. × 0.131 in.) 4 -3 in. × 0.131 in. nails 3 - 3 in. 14 gage staples	toenail
	2 - 16d common ($2^1/_2$ in. × 0.162 in.) 3 - 3 in. × 0.131 in. nails 3 - 3 in. 14 gage staples	end nail
9. Double studs	16d ($3^1/_2$ in. × 0.135 in.) at 24 in. o.c. 3 in. × 0.131 in. nail at 8 in. o.c. 3 in. 14 gage staple at 8 in. o.c.	face nail
10. Double top plates	16d ($3^1/_2$ in. × 0.135 in.) at 16 in. o.c. 3 in. × 0.131 in. nail at 12 in. o.c. 3 in. 14 gage staple at 12 in. o.c.	typical face nail
Double top plates	8-16d common ($3^1/_2$ in. × 0.162 in.) 12-3 in. × 0.131 in. nails 12-3 in. 14 gage staples	lap splice
11. Blocking between joists or rafters to top plate	3 - 8d common ($2^1/_2$ in. × 0.131 in.) 3 - 3 in. × 0.131 in. nails 3 - 3 in. 14 gage staples	toenail
12. Rim joist to top plate	8d ($2^1/_2$ in. × 0.131 in.) at 6 in. o.c. 3 in. × 0.131 in. nail at 6 in. o.c. 3 in. 14 gage staple at 6 in. o.c.	toenail
13. Top plates, laps and intersections	2 - 16d common ($3^1/_2$ in. × 0.162 in.) 3 - 3 in. × 0.131 in. nails 3 - 3 in. 14 gage staples	face nail
14. Continuous header, two pieces	16d common ($3^1/_2$ in. × 0.162 in.)	16 in. o.c. along edge
15. Ceiling joists to plate	3 - 8d common ($2^1/_2$ in. × 0.131 in.) 5 - 3 in. × 0.131 in. nails 5 - 3 in. 14 gage staples	toenail

(continued)

TABLE 2304.9.1 (*Continued*)
FASTENING SCHEDULE

Connection	Fastening[a,m]	Location
16. Continuous header to stud	4 - 8d common ($2^1/_2$ in. × 0.131 in.)	toenail
17. Ceiling joists, laps over partitions (see Section 2308.10.4.1, Table 2308.10.4.1)	3 - 16d common ($3^1/_2$ in. × 0.162 in.) minimum, Table 2308.10.4.1 4 - 3 in. × 0.131 in. nails 4 - 3 in. 14 gage staples	face nail
18. Ceiling joists to parallel rafters (see Section 2308.10.4.1, Table 2308.10.4.1)	3 - 16d common ($3^1/_2$ in. × 0.162 in.) minimum, Table 2308.10.4.1 4 - 3 in. × 0.131 in. nails 4 - 3 in. 14 gage staples	face nail
19. Rafter to plate (see Section 2308.10.1, Table 2308.10.1)	3 - 8d common ($2^1/_2$ in. × 0.131 in.) 3 - 3 in. × 0.131 in. nails 3 - 3 in. 14 gage staples	toenail
20. 1 in. diagonal brace to each stud and plate	2 - 8d common ($2^1/_2$ in. × 0.131 in.) 2 - 3 in. × 0.131 in. nails 3 - 3 in. 14 gage staples	face nail
21. 1 in. × 8 in. sheathing to each bearing	3 - 8d common ($2^1/_2$ in. × 0.131 in.)	face nail
22. Wider than 1 in. × 8 in. sheathing to each bearing	3 - 8d common ($2^1/_2$ in. × 0.131 in.)	face nail
23. Built-up corner studs	16d common ($3^1/_2$ in. × 0.162 in.) 3 in. × 0.131 in. nails 3 in. 14 gage staples	24 in. o.c. 16 in. o.c. 16 in. o.c.
24. Built-up girder and beams	20d common (4 in. × 0.192 in.) 32 in. o.c. 3 in. × 0.131 in. nail at 24 in. o.c. 3 in. 14 gage staple at 24 in. o.c. 2 - 20d common (4 in. × 0.192 in.) 3 - 3 in. × 0.131 in. nails 3 - 3 in. 14 gage staples	face nail at top and bottom staggered on opposite sides face nail at ends and at each splice
25. 2 in. planks	16d common ($3^1/_2$ in. × 0.162 in.)	at each bearing
26. Collar tie to rafter	3 - 10d common (3 in. × 0.148 in.) 4 - 3 in. × 0.131 in. nails 4 - 3 in. 14 gage staples	face nail
27. Jack rafter to hip	3 - 10d common (3 in. × 0.148 in.) 4 - 3 in. × 0.131 in. nails 4 - 3 in. 14 gage staples 2 - 16d common ($3^1/_2$ in. × 0.162 in.) 3 - 3 in. × 0.131 in. nails 3 - 3 in. 14 gage staples	toenail face nail
28. Roof rafter to 2-by ridge beam	2 - 16d common ($3^1/_2$ in. × 0.162 in.) 3 - 3 in. × 0.131 in. nails 3 - 3 in. 14 gage staples 2-16d common ($3^1/_2$ in. × 0.162 in.) 3 - 3 in. × 0.131 in. nails 3 - 3 in. 14 gage staples	toenail face nail
29. Joist to band joist	3 - 16d common ($3^1/_2$ in. × 0.162 in.) 4 - 3 in. × 0.131 in. nails 4 - 3 in. 14 gage staples	face nail

TABLE 2304.9.1 (Continued)
FASTENING SCHEDULE

Connection	Fastening[a,m]		Location
30. Ledger strip	3 - 16d common ($3^1/_2$ in. × 0.162 in.) 4 - 3 in. × 0.131 in. nails 4 - 3 in. 14 gage staples		face nail
31. Wood structural panels and particleboard[b] Subfloor, roof and wall sheathing (to framing)	$^1/_2$ in. and less	6d[e, l] $2^3/_8$ in. × 0.113 in. nail[n] $1^3/_4$ in. 16 gage[o]	
	$^{19}/_{32}$ in. to $^3/_4$ in.	8d[d] or 6d[e] $2^3/_8$ in. × 0.113 in. nail[p] 2 in. 16 gage[p]	
	$^7/_8$ in. to 1 in.	8d[c]	
Single Floor (combination subfloor-underlayment to framing)	$1^1/_8$ in. to $1^1/_4$ in. $^3/_4$ in. and less $^7/_8$ in. to 1 in. $1^1/_8$ in. to $1^1/_4$ in.	10d[d] or 8d[d] 6d[e] 8d[e] 10d[d] or 8d[e]	
32. Panel siding (to framing)	$^1/_2$ in. or less $^5/_8$ in	6d[f] 8d[f]	
33. Fiberboard sheathing[g]	$^1/_2$ in.	No. 11 gage roofing nail[h] 6d common nail (2 in. × 0.113 in.) No. 16 gage staple[i]	
	$^{25}/_{32}$ in.	No. 11 gage roofing nail[h] 8d common nail ($2^1/_2$ in. × 0.131 in.) No. 16 gage staple[i]	
34. Interior paneling	$^1/_4$ in. $^3/_8$ in.	4d[j] 6d[k]	

For SI: 1 in. = 25.4 mm.

a. Common or box nails are permitted to be used except where otherwise stated.

b. Nails spaced at 6 in. on center at edges, 12 in. at intermediate supports except 6 in. at supports where spans are 48 in. or more. For nailing of wood structural panel and particleboard diaphragms and shear walls, refer to Sec. 2305. Nails for wall sheathing are permitted to be common, box or casing.

c. Common or deformed shank (6d - 2 in. × 0.113 in.; 8d - $2^1/_2$ in × 0.131 in.; 10d - 3 in. × 0.148 in.).

d. Common (6d - 2 in. × 0.113 in.; 8d - $2^1/_2$ in. × 0.131 in.; 10d - 3 in. × 0.148 in.).

e. Deformed shank (6d - 2 in. × 0.113 in.; 8d - $2^1/_2$ in. × 0.131 in.; 10d - 3 in. × 0.148 in.).

f. Corrosion-resistant siding (6d - $1^7/_8$ in. × 0.106 in.; 8d - $2^3/_8$ in. × 0.128 in.) or casing (6d - 2 in. × 0.099 in.; 8d - $2^1/_2$ in. × 0.113 in.) nail.

g. Fasteners spaced 3 inches on center at exterior edges and 6 inches on center at intermediate supports, when used as structural sheathing. Spacing shall be 6 in. on center on the edges and 12 in. on center at intermediate supports for nonstructural applications.

h. Corrosion-resistant roofing nails with $^7/_{16}$-in. diameter head and $1^1/_2$-in. length for $^1/_2$-in. sheathing and $1^3/_4$-in. length for $^{25}/_{32}$-in. sheathing.

i. Corrosion-resistant staples with nominal $^7/_{16}$-in. crown and $1^1/_8$-in. length for $^1/_2$-in. sheathing and $1^1/_2$-in. length for $^{25}/_{32}$-in. sheathing. Panel supports at 16 in. (20 in. if strength axis in the long direction of the panel, unless otherwise marked).

j. Casing ($1^1/_2$ in. × 0.080 in.) or finish ($1^1/_2$ in. × 0.072 in.) nails spaced 6 in. on panel edges, 12 in. at intermediate supports.

k. Panel supports at 24 in. Casing or finish nails spaced 6 in. on panel edges, 12 in. at intermediate supports.

l. For roof sheathing applications, 8d nails ($2^1/_2$ in. × 0.113 in.) are the minimum required for wood structural panels.

m. Staples shall have a minimum crown width of $^7/_{16}$ in.

n. For roof sheathing applications, fasteners spaced 4 in. on center at edges, 8 in. at intermediate supports.

o. Fasteners spaced 4 in. on center at edges, 8 in. at intermediate supports for subfloor and wall sheathing and 3 in. on center at edges, 6 in. at intermediate supports for roof sheathing.

p. Fasteners spaced 4 in. on center at edges, 8 in. at intermediate supports.

TABLE 2305.2.2(1)
**e_n VALUES (inches) FOR USE IN CALCULATING DIAPHRAGM
DEFLECTION DUE TO FASTENER SLIP (Structural I)[a,d]**

LOAD PER FASTENER[c] (pounds)	FASTENER DESIGNATIONS[b]			
	6d	8d	10d	14-Ga staple x 2 inches long
60	0.01	0.00	0.00	0.011
80	0.02	0.01	0.01	0.018
100	0.03	0.01	0.01	0.028
120	0.04	0.02	0.01	0.04
140	0.06	0.03	0.02	0.053
160	0.10	0.04	0.02	0.068
180	—	0.05	0.03	—
200	—	0.07	0.47	—
220	—	0.09	0.06	—
240	—	—	0.07	—

For SI: 1 inch = 25.4 mm, 1 foot = 304.8 mm, 1 pound = 4.448 N.

a. Increase e_n values 20 percent for plywood grades other than Structural I.

b. Nail values apply to common wire nails or staples identified.

c. Load per fastener = maximum shear per foot divided by the number of fasteners per foot at interior panel edges.

d. Decrease e_n values 50 percent for seasoned lumber (moisture content < 19 percent).

TABLE 2305.2.3
**MAXIMUM DIAPHRAGM DIMENSION RATIOS
HORIZONTAL AND SLOPED DIAPHRAGM**

TYPE	MAXIMUM LENGTH - WIDTH RATIO
Wood structural panel, nailed all edges	4:1
Wood structural panel, blocking omitted at intermediate joints	3:1
Diagonal sheathing, single	3:1
Diagonal sheathing, double	4:1

TABLE 2305.2.2(2)
VALUES OF *Gt* FOR USE IN CALCULATING DEFLECTION OF WOOD STRUCTURAL PANEL SHEAR WALLS AND DIAPHRAGMS

PANEL TYPE	SPAN RATING	VALUES OF *Gt* (lb/in. panel depth or width)							
		OTHER				STRUCTURAL I			
		3-ply Plywood	4-ply Plywood	5-ply Plywood[a]	OSB	3-ply Plywood	4-ply Plywood	5-ply Plywood[a]	OSB
Sheathing	24/0	25,000	32,500	37,500	77,500	32,500	42,500	41,500	77,500
	24/16	27,000	35,000	40,500	83,500	35,000	45,500	44,500	83,500
	32/16	27,000	35,000	40,500	83,500	35,000	45,500	44,500	83,500
	40/20	28,500	37,000	43,000	88,500	37,000	48,000	47,500	88,500
	48/24	31,000	40,500	46,500	96,000	40,500	52,500	51,000	96,000
Single Floor	16 o.c.	27,000	35,000	40,500	83,500	35,000	45,500	44,500	83,500
	20 o.c.	28,000	36,500	42,000	87,000	36,500	47,500	46,000	87,000
	24 o.c.	30,000	39,000	45,000	93,000	39,000	50,500	49,500	93,000
	32 o.c.	36,000	47,000	54,000	110,000	47,000	61,000	59,500	110,000
	48 o.c.	50,500	65,500	76,000	155,000	65,500	85,000	83,500	155,000

	Thickness (In.)	OTHER			STRUCTURAL I		
		A-A, A-C	Marine	All Other Grades	A-A, A-C	Marine	All Other Grades
Sanded Plywood	$1/4$	24,000	31,000	24,000	31,000	31,000	31,000
	$11/32$	25,500	33,000	25,500	33,000	33,000	33,000
	$3/8$	26,000	34,000	26,000	34,000	34,000	34,000
	$15/32$	38,000	49,500	38,000	49,500	49,500	49,500
	$1/2$	38,500	50,000	38,500	50,000	50,000	50,000
	$19/32$	49,000	63,500	49,000	63,500	63,500	63,500
	$5/8$	49,500	64,500	49,500	64,500	64,500	64,500
	$23/32$	50,500	65,500	50,500	65,500	65,500	65,500
	$3/4$	51,000	66,500	51,000	66,500	66,500	66,500
	$7/8$	52,500	68,500	52,500	68,500	68,500	68,500
	1	73,500	95,500	73,500	95,500	95,500	95,500
	$1 1/8$	75,000	97,500	75,000	97,500	97,500	97,500

For SI: 1 inch = 25.4 mm, 1 pound/inch = 0.1751 N/mm.
a. Applies to plywood with five or more layers; for five-ply/three-layer plywood, use values for four ply.

TABLE 2305.3.4
MAXIMUM SHEAR WALL DIMENSION RATIOS

TYPE	MAXIMUM HEIGHT-WIDTH RATIO
Wood structural panels or particleboard, nailed edges	For other than seismic: $3^1/_2$:1 For seismic: 2:1[a]
Diagonal sheathing, single	2:1
Fiberboard	$1^1/_2$:1
Gypsum board, gypsum lath, cement plaster	$1^1/_2$:1[b]

a. For design to resist seismic forces, shear wall height-width ratios greater than 2:1, but not exceeding $3^1/_2$:1, are permitted provided the allowable shear values in Table 2306.4.1 are multiplied by $2w/h$.

b. Ratio shown is for unblocked construction. Height-to-width ratio is permitted to be 2:1 where the wall is installed as blocked construction in accordance with Section 2306.4.5.1.2.

TABLE 2305.3.8.2
SHEAR RESISTANCE ADJUSTMENT FACTOR, C_o

WALL HEIGHT, H	MAXIMUM OPENING HEIGHT[a]				
	H/3	H/2	2H/3	5H/6	H
8' wall	2'-8"	4'-0"	5'-4"	6'-8"	8'-0"
10' wall	3'-4"	5'-0"	6'-8"	8'-4"	10'-0"
Percentage of full-height sheathing[b]	Shear resistance adjustment factor				
10%	1.00	0.69	0.53	0.43	0.36
20%	1.00	0.71	0.56	0.45	0.38
30%	1.00	0.74	0.59	0.49	0.42
40%	1.00	0.77	0.63	0.53	0.45
50%	1.00	0.80	0.67	0.57	0.50
60%	1.00	0.83	0.71	0.63	0.56
70%	1.00	0.87	0.77	0.69	0.63
80%	1.00	0.91	0.83	0.77	0.71
90%	1.00	0.95	0.91	0.87	0.83
100%	1.00	1.00	1.00	1.00	1.00

For SI: 1 inch = 25.4 mm, 1 foot = 304.8 mm.
a. See Section 2305.3.8.2.2, Item 2.
b. See Section 2305.3.8.2.2, Item 1.

TABLE 2306.3.1
ALLOWABLE SHEAR (POUNDS PER FOOT) FOR WOOD STRUCTURAL PANEL DIAPHRAGMS WITH FRAMING OF DOUGLAS FIR-LARCH, OR SOUTHERN PINE[a] FOR WIND OR SEISMIC LOADING[h]

Panel grade	Common nail size or staple[f] length and gage	Minimum fastener penetration in framing (inches)	Minimum nominal panel thickness (inch)	Minimum nominal width of framing members at adjoining panel edges and boundaries[g] (inches)	Blocked diaphragms — Fastener spacing (inches) at diaphragm boundaries (all cases) at continuous panel edges parallel to load (Cases 3, 4), and at all panel edges (Cases 5, 6)[b]; at other panel edges (Cases 1, 2, 3 and 4)[b]				Unblocked diaphragms — Fasteners spaced 6 in. max. at supported edges[b]	
					boundary 6 / other 6	4 / 6	2½[c] / 4	2[c] / 3	Case 1 (No unblocked edges or continuous joints parallel to load)	All other configurations Cases 2, 3, (4, 5 and 6)
Structural I Grades	6d[e] (2 in. × 0.113 in.)	1¼	5/16	2	185	250	375	420	165	125
				3	210	280	420	475	185	140
	1½ 16 Gage	1		2	155	205	310	350	135	105
				3	175	230	345	390	155	115
	8d[d] (2½ in. × 0.131)	1⅜	3/8	2	270	360	530	600	240	180
				3	300	400	600	675	265	200
	1½ 16 Gage	1		2	175	235	350	400	155	115
				3	200	265	395	450	175	130
	10d[d] (3 in. × 0.148 in.)	1½	15/32	2	320	425	640	730	285	215
				3	360	480	720	820	320	240
	1½ 16 Gage	1		2	175	235	350	400	155	120
				3	200	265	395	450	175	130
Sheathing, single floor and other grades	6d[e] (2 in. × 0.113)	1¼	5/16	2	170	225	335	380	150	110
				3	190	250	380	430	170	125
	1½ 16 Gage	1		2	140	185	275	315	125	90
				3	155	205	310	350	140	105

(continued)

TABLE 2306.3.1 (Continued)

ALLOWABLE SHEAR (POUNDS PER FOOT) FOR WOOD STRUCTURAL PANEL DIAPHRAGMS WITH FRAMING OF DOUGLAS FIR-LARCH, OR SOUTHERN PINE[a] FOR WIND OR SEISMIC LOADING[h]

Panel grade	Common nail size or staple[f] length and gage	Minimum fastener penetration in framing (inches)	Minimum nominal panel thickness (inch)	Minimum nominal width of framing members at adjoining panel edges and boundaries[g] (inches)	Blocked diaphragms — Fastener spacing (inches) at diaphragm boundaries (all cases) at continuous panel edges parallel to load (Cases 3, 4), and at all panel edges (Cases 5, 6)[b]				Unblocked diaphragms	
					6	4	2½[c]	2[c]	Case 1 (No unblocked edges or continuous joints parallel to load)	All other configurations Cases 2, 3, (4, 5 and 6)
					Fastener spacing (inches) at other panel edges (Cases 1, 2, 3 and 4)[b]				Fasteners spaced 6 in. max. at supported edges[b]	
					6	6	4	3		
covered in DOC PS 1 and PS 2	6d[e] (2 in. × 0.113)	1¼	3/8	2	185	250	375	420	165	125
				3	210	280	420	475	185	140
	8d (2½ in. × 0.131)	1⅜	3/8	2	240	320	480	545	215	160
				3	270	360	540	610	240	180
	1½ 16 Gage	1		2	160	210	315	360	140	105
				3	180	235	355	400	160	120
	8d (2½ in. × 0.131 in.)	1⅜	7/16	2	255	340	505	575	230	170
				3	285	380	570	645	255	190
	1½ 16 Gage	1		2	165	225	335	380	150	110
				3	190	250	375	425	165	125
	8d (2½ in. × 0.131 in.)	1⅜	15/32	2	270	360	530	600	240	180
				3	300	400	600	675	265	200
	10d[d] (3 in. × 0.148 in.)	1½		2	290	385	575	655	255	190
				3	325	430	650	735	290	215
	1½ 16 Gage	1		2	160	210	315	360	140	105
				3	180	235	355	405	160	120
	10d[d] (3 in. × 0.148 in.)	1½	19/32	2	320	425	640	730	285	215
				3	360	480	720	820	320	240
	1¾ 16 Gage	1		2	175	235	350	400	155	115
				3	200	265	395	450	175	130

(continued)

TABLE 2306.3.1 *(Continued)*
ALLOWABLE SHEAR (POUNDS PER FOOT) FOR WOOD STRUCTURAL PANEL DIAPHRAGMS WITH FRAMING OF DOUGLAS FIR-LARCH, OR SOUTHERN PINE[a] FOR WIND OR SEISMIC LOADING[h]

For SI: 1 in. = 25.4 mm, 1 pound per foot = 14.5939 N/m.

a. For framing of other species: (1) Find specific gravity for species of lumber in AF&PA NDS. (2) For staples find shear value from table above for Structural I panels (regardless of actual grade) and multiply value by 0.82 for species with specific gravity of 0.42 or greater, or 0.65 for all other species. (3) For nails find shear value from table above for nail size for actual grade and multiply value by the following adjustment factor: Specific Gravity Adjustment Factor = (1-[0.5-SG]), where SG = Specific Gravity of the framing lumber. This adjustment factor shall not be greater than 1.

b. Space fasteners maximum 12 in. o.c. along intermediate framing members (6 inches o.c. where supports are spaced 48 in. o.c.).

c. Framing at adjoining panel edges shall be 3 in. nominal or wider, and nails shall be staggered where nails are spaced 2 or $2^{1}/_{2}$ in. o.c.

d. Framing at adjoining panel edges shall be 3 in. nominal or wider, and nails shall be staggered where both of the following conditions are met: (1) 10d nails having penetration into framing of more than $1^{1}/_{2}$ inches and (2) nails are spaced 3 in. o.c. or less.

e. 8d is recommended minimum for roofs due to negative pressures of high winds.

f. Staples shall have a minimum crown width of $^{7}/_{16}$ in. and shall be installed with their crowns parallel to the long dimension of the framing members.

g. The minimum nominal width of framing members not located at boundaries or adjoining panel edges shall be 2 in.

h. For shear loads of normal or permanent load duration as defined by the AF&PA NDS, the values in the table above shall be multiplied by 0.63 or 0.56, respectively.

TABLE 2306.4.1

ALLOWABLE SHEAR (POUNDS PER FOOT) FOR WOOD STRUCTURAL PANEL SHEAR WALLS WITH FRAMING OF DOUGLAS FIR-LARCH OR SOUTHERN PINE[a] FOR WIND OR SEISMIC LOADING[b,h,i,j,l]

Panel grade	Minimum nominal panel thickness (inch)	Minimum fastener penetration in framing (inches)	Panels applied direct to framing — Nail (common or galvanized box or staple size[k])	Direct 6	4	3	2[e]	Panels applied over 1/2 in. or 5/4 in. gypsum sheathing — NAIL (common or galvanized box or staple size[k])	Over 6	4	3	2[e]
	5/16	1 1/4	6d (2 × 0.113 in. common, 2 in. × 0.099 in. galvanized box)	200	300	390	510	8d (2 1/2 in. × 0.131 in. common, 2 1/2 in.× 0.113 in. galvanized box)	200	300	390	510
		1	1 1/2 16 Gage	165	245	325	415	2 16 Gage	125	185	245	315
	3/8	1 3/8	8d (2 1/2 in. × 0.131 in. common, 2 1/2 in. × 0.113 in. galvanized box)	230[d]	360[d]	460[d]	610[d]	10d (3 in. × 0.148 in. common, 3 in. × 0.128 in. galvanized box)	280	430	550[f]	730
		1	1 1/2 16 Gage	155	235	315	400	2 16 Gage	155	235	310	400
Structural 1 Sheathing	7/16	1 3/8	8d (2 1/2 in. × 0.131 in. common, 2 1/2 in. × 0.113 in. galvanized box)	255[d]	395[d]	505[f]	670[d]	10d (3 in. × 0.148 in. common, 3 in. × 0.128 in. galvanized box)	280	430	550[f]	730
		1	1 1/2 16 Gage	170	260	345	440	2 16 Gage	155	235	310	400
	15/32	1 3/8	8d (2 1/2 in. × 0.131 in. common, 2 1/2 in. × 0.113 in. galvanized box)	280	430	550	730	10d (3 in. × 0.148 in. common, 3 in. × 0.128 in. galvanized box)	280	430	550[f]	730
		1	1 1/2 16 Gage	185	280	375	475	2 16 Gage	155	235	300	400
		1 1/2	10d (3 in. × 0.148 in. common, 3 in. × 0.128 in. galvanized box)	340	510	665[f]	870	10d (3 in. × 0.148 in. common, 3 in. × 0.128 in. galvanized box)	—	—	—	—
Sheathing plywood siding[g] except Group 5 Species	5/16 or 1/4[c]	1 1/4	6d (2 in. × 0.113 in. common, 2 in. × 0.099 in. galvanized box)	180	270	350	450	8d (2 1/2 in. × 0.131 in. common, 2 1/2 × 0.113 in. galvanized box)	180	270	350	450
		1	1 1/2 16 Gage	145	220	295	375	2 16 Gage	110	165	220	285
	3/8	1 1/4	6d (2 in. × 0.113 in. common, 2 in. × 0.099 in. galvanized box)	200	300	390	510	8d (2 1/2 in. × 0.131 in. common,(2 1/2 × 0.113 in. galvanized box)	200	300	390	510

(continued)

TABLE 2306.4.1 (Continued)

ALLOWABLE SHEAR (POUNDS PER FOOT) FOR WOOD STRUCTURAL PANEL SHEAR WALLS WITH FRAMING OF DOUGLAS FIR-LARCH OR SOUTHERN PINE[a] FOR WIND OR SEISMIC LOADING[b,h,i,j,l]

Panel grade	Minimum nominal panel thickness (inch)	Minimum fastener penetration in framing (inches)	Panels applied direct to framing — NAIL (common or galvanized box or staple size[k])	6	4	3	2[e]	Panels applied over 1/2 in. or 5/4 in. gypsum sheathing — NAIL (common or galvanized box or staple size[k])	6	4	3	2[e]
				Fastener spacing at panel edges (inches)					Fastener spacing at panel edges (inches)			
Sheathing, plywood siding[g] except Group 5 Species	3/8	1 3/8	8d (2½ in. × 0.131 in. common, 2½ in. × 0.113 in. galvanized box)	220[d]	320[d]	410[d]	530[d]	10d (3 in. × 0.148 in. common, 3 in. × 0.128 in. galvanized box)	260	380	490[f]	640
		1	1½ 16 Gage	140	210	280	360	2 16 Gage	140	210	280	360
	7/16	1 3/8	8d (2½ in. × 0.131 in. common, 2½ in. × 0.113 in. galvanized box)	240[d]	350[d]	450[d]	585[d]	10d (3 in. × 0.148 in. common, 3 in. × 0.128 in. galvanized box)	260	380	490[f]	640
		1	1½ 16 Gage	155	230	310	395	2 16 Gage	140	210	280	360
	15/32	1 3/8	8d (2½ in. × 0.131 in. common, 2½ in. × 0.113 in. galvanized box)	260	380	490	640	10d (3 in. × 0.148 in. common, 3 in. × 0.128 in. galvanized box)	260	380	490[f]	640
		1	10d (3 in. × 0.148 in. common, 3 in. × 0.128 in. galvanized box)	310	460	600[f]	770	—	—	—	—	—
	19/32	1½	10d (3 in. × 0.148 in. common, 3 in. × 0.128 in. galvanized box)	340	510	665[f]	870	2 16 Gage	140	210	280	360
		1	1¾ 16 Gage	185	280	375	475	—	—	—	—	—
	5/16[c]	1¼	Nail Size (galvanized casing) 6d (2 in. × 0.099 in.)	140	210	275	360	Nail Size (galvanized casing) 8d (2½ in. × 0.113 in.)	140	210	275	360
	3/8	1 3/8	8d (2½ in. × 0.113 in.)	160	240	310	410	10d (3 in. × 0.128 in.)	160	240	310[f]	410

Notes to Table 2306.4.1

For SI: 1 in. = 25.4 mm, 1 pound per foot = 14.5939 N/m.

a. For framing of other species: (1) Find specific gravity for species of lumber in AF&PA NDS. (2) For staples find shear value from table above for Structural I panels (regardless of actual grade) and multiply value by 0.82 for species with specific gravity of 0.42 or greater, or 0.65 for all other species. (3) For nails find shear value from table above for nail size for actual grade and multiply value by the following adjustment factor: Specific Gravity Adjustment Factor = [1-(0.5 - SG)], where SG = Specific Gravity of the framing lumber. This adjustment factor shall not be greater than 1.

b. Panel edges backed with 2-in. nominal or wider framing. Install panels either horizontally or vertically. Space fasteners maximum 6 in. on center along intermediate framing members for $^3/_8$ and $^7/_{16}$-in. panels installed on studs spaced 24 in. on center. For other conditions and panel thickness, space fasteners maximum 12 in. on center on intermediate supports.

c. $^3/_8$-in. panel thickness or siding with a span rating of 16 in. on center is the minimum recommended where applied direct to framing as exterior siding.

d. Allowable shear values are permitted to be increased to values shown for $^{15}/_{32}$-in. sheathing with same nailing provided (a) studs are spaced a maximum of 16 in. on center, or (b) panels are applied with long dimension across studs.

e. Framing at adjoining panel edges shall be 3 in. nominal or wider, and nails shall be staggered where nails are spaced 2 in. on center.

f. Framing at adjoining panel edges shall be 3 in. nominal or wider, and nails shall be staggered where both of the following conditions are met: (1) 10d (3 in. × 0.148 in.) nails having penetration into framing of more than $1^1/_2$ in. and (2) nails are spaced 3 in. on center.

g. Values apply to all-veneer plywood. Thickness at point of fastening on panel edges governs shear values.

h. Where panels applied on both faces of a wall and nail spacing is less than 6 in. o.c. on either side, panel joints shall be offset to fall on different framing members, or framing shall be 3 in. nominal or thicker at adjoining panel edges and nails on each side shall be staggered.

i. In Seismic Design Category D, E or F, where shear design values exceed 350 pounds per linear foot, all framing members receiving edge nailing from abutting panels shall not be less than a single 3-in. nominal member, or two 2-in. nominal members fastened together in accordance with Sec. 2306.1 to transfer the design shear value between framing members. Wood structural panel joint and sill plate nailing shall be staggered in all cases. See Sec. 2305.3.11 for sill plate size and anchorage requirements.

j. Galvanized nails shall be hot dipped or tumbled.

k. Staples shall have a minimum crown width of $^7/_{16}$ in. and shall be installed with their crowns parallel to the long dimension of the framing members.

l. For shear loads of normal or permanent load duration as defined by the AF&PA NDS, the values in the table above shall be multiplied by 0.63 or 0.56, respectively.

TABLE 2306.4.3
ALLOWABLE SHEAR FOR PARTICLEBOARD SHEAR WALL SHEATHING[b]

Panel grade	Minimum nominal panel thickness (inch)	Minimum nail penetration in framing (inches)	Panels applied direct to framing				
			Nail size (common or galvanized box)	Allowable shear (pounds per foot) nail spacing at panel edges (inches)[a]			
				6	4	3	2
M-S "Exterior Glue" and M-2 "Exterior Glue"	$^3/_8$	$1^1/_2$	6d	120	180	230	300
	$^3/_8$	$1^1/_2$	8d	130	190	240	315
	$^1/_2$			140	210	270	350
	$^1/_2$	$1^5/_8$	10d	185	275	360	460
	$^5/_8$			200	305	395	520

For SI: 1 in. = 25.4 mm, 1 pound per foot = 14.5939 N/m.

a. Values are not permitted in Seismic Design Category D, E or F.
b. Galvanized nails shall be hot-dipped or tumbled.

TABLE 2306.4.5
ALLOWABLE SHEAR FOR WIND OR SEISMIC FORCES FOR SHEAR WALLS OF LATH AND PLASTER OR GYPSUM BOARD WOOD FRAMED WALL ASSEMBLIES

Type of material	Thickness of material	Wall construction	Fastener spacing[b] maximum (inches)	Shear value[a,e] (plf)	Minimum fastener size[c,d,j,k]
1. Expanded metal or woven wire lath and Portland cement plaster	$7/8$ in.	Unblocked	6	180	No. 11 gage $1\frac{1}{2}$ in. long, $7/16$ in. head 16 Ga. Galv. Staple, $7/8$ in. legs
2. Gypsum lath, plain or perforated	$3/8$ in. lath and $1/2$ in. plaster	Unblocked	5	100	No. 13 page, $1\frac{1}{8}$ in. long, $19/64$ in. head, plasterboard nail 16 Ga. Galv. Staple, $1\frac{1}{8}$ in. long 0.120 in. Nail, min. $3/8$ in. head, $1\frac{1}{4}$ in. long
3. Gypsum sheathing	$1/2$ in. $\times$ 2 ft. $\times$ 8 ft.	Unblocked	4	75	No. 11 gage, $1\frac{3}{4}$ in. long, $7/16$ in. head, diamond-point, galvanized
	$1/2$ in. $\times$ 4 ft.	Blocked[f]	4	175	16 Ga. Galv. Staple, $1\frac{3}{4}$ in. long
		Unblocked	7	100	
	$5/8$ in. $\times$ 4	Blocked	4 in. edge/ 7 in. field	200	6d galvanized 0.120 in. Nail, min. $3/8$ in. head, $1\frac{3}{4}$ in. long
4. Gypsum board, gypsum veneer base or water resistant gypsum backing board	$1/2$ in.	Unblocked[f]	7	75	5d cooler ($1\frac{5}{8}$ in. $\times$ .086 in.) or wallboard 0.120 in. nail, min. $3/8$ in. head, $1\frac{1}{2}$ in. long 16 Gage Staple, $1\frac{1}{2}$ long
		Unblocked[f]	4	110	
		Unblocked	7	100	
		Unblocked	4	125	
		Blocked[g]	7	125	
		Blocked[g]	4	150	
		Unblocked	8/12[h]	60	No. 6 $1\frac{1}{4}$ screws[i]
		Blocked[g]	4/16[h]	160	
		Blocked[g]	4/12[h]	155	
		Blocked[f,g]	8/12[h]	70	
		Blocked[g]	6/12[h]	90	

(continued)

TABLE 2306.4.5 (Continued)

ALLOWABLE SHEAR FOR WIND OR SEISMIC FORCES FOR SHEAR WALLS OF LATH AND PLASTER OR GYPSUM BOARD WOOD FRAMED WALL ASSEMBLIES (Continued)

Type of material	Thickness of material	Wall construction	Fastener spacing[b] maximum (inches)	Shear value[a,e] (plf)	Minimum fastener size[c,d,j,k]
	$5/8$ in.	Unblocked[f]	7	115	6d cooler ($1^7/_8$ in. × 0.092 in.) or wallboard 0.120 in. Nail, min. $^3/_8$ in. head, $1^3/_4$ in. long 16 Gage staple, $1^1/_2$ in. legs, $1^5/_8$ in. long
		Unblocked[f]	4	145	
		Blocked[g]	7	145	
		Blocked[g]	4	175	
		Blocked[g] Two-ply	Base ply: 9 Face ply: 7	250	Base ply-6d cooler ($1^7/_8$ in. × 0.092 in.) or wallboard $1^3/_4$ in. × 0.120 in. Nail, min. $^3/_8$ in. head $1^5/_8$ in. 16 Ga. Galv. Staple $1^5/_8$ in. 16 Gage Galv. Staple Face ply-8d cooler ($2^3/_8$ in. × 0.113 in.) or wallboard 0.120 in. Nail, min. $^3/_8$ in. head, $2^3/_8$ in. long 15 Ga. Galv. Staple, $2^1/_4$ in. long
		Unblocked	8/12[h]	70	No. 6-$1^1/_4$ in. screws[i]
		Blocked[g]	8/12[h]	90	

For SI: 1 in. = 25.4 mm, 1 foot = 304.8 mm, 1 pound per foot = 14.5939 N/m.

a. These shear walls shall not be used to resist loads imposed by masonry or concrete construction (see Sec. 2305.1.5). Values shown are for short-term loading due to wind or seismic loading. Walls resisting seismic loads shall be subject to the limitations in Sec. 12.2.1 of ASCE 7. Values shown shall be reduced 25 percent for normal loading.

b. Applies to fastening at studs, top and bottom plates and blocking.

c. Alternate fasteners are permitted to be used if their dimensions are not less than the specified dimensions. Drywall screws are permitted to substitute for the 5d ($1^5/_8$ in. × 0.086 in.), and 6d ($1^7/_8$ in. × 0.092 in.) (cooler) nails listed above, and No. 6 $1^1/_4$ in. Type S or W screws for 6d ($1^7/_8$ in. × 0.092 in.) (cooler) nails.

d. For properties of cooler nails, see ASTM C 514.

e. Except as noted, shear values are based on a maximum framing spacing of 16 inches on center.

f. Maximum framing spacing of 24 in. on center.

g. All edges are blocked, and edge fastening is provided at all supports and all panel edges.

h. First number denotes fastener spacing at the edges; second number denotes fastener spacing at intermediate framing members.

i. Screws are Type W or S.

j. Staples shall have a minimum crown width of $^7/_{16}$ in., measured outside the legs, and shall be installed with their crowns parallel to the long dimension of the framing members.

k. Staples for the attachment of gypsum lath and woven-wire lath shall have a minimum crown width of $^3/_4$ in., measured outside the legs.

SI Units

Introduction

In 1960 the Eleventh General Conference of Weights and Measures adopted the name International System of Units (accepted abbreviation SI, from Système International d'Unites) for a practical and consistent set of units of measure. Rules for common usage, notation, and abbreviation were adopted. Most industrial nations in the world have adopted and converted to SI.

A large number of code and industry design tables are required to carry out a timber structural design. Many of these tables now have dual units which includes both U.S. Customary units and SI (metric) units. However, the primary units are still U.S. Customary units, and the SI equivalent dimensions are generally provided for information purposes ("soft" conversion). Most wood products are still produced in traditional U.S. Customary unit sizes, and a "hard" conversion to SI or metric sizes has not taken place. Consequently most practical structural design in the U.S. is still done in U.S. Customary units.

However, the future trend will be toward the use of SI units. The following brief introduction to SI is included to aid the structural engineer in conversion between U.S. Customary (USC) units and SI units. Additional information can be found in ASTM E380 "Metric Practice Excerpts."

Notation

SI is made up of seven base units, two supplementary units, and many consistent derived units. The following units are pertinent to structural design.

	Quantity	Unit	SI symbol
Base unit	Length	meter	m
	Mass	kilogram	kg
	Time	second	s
Supplementary unit	Plane angle	radian	rad
Derived unit	Density (weight)		N/m^3
	Density (mass)		kg/m^3
	Energy	joule	$J\ (N \cdot m)$
	Force	newton	$N\ (kg \cdot m/s^2)$
	Frequency	hertz	$Hz\ (1/s)$
	Moment of force		$N \cdot m\ (kg \cdot m^2/s^2)$
	Moment of inertia		m^4
	Stress, pressure	pascal	$Pa\ (N/m^2)$

Prefixes

The following are SI-approved prefixes and their abbreviations used to denote very small or very large quantities.

Multiplication factor	Name	Symbol
$0.000\ 000\ 001 = 10^{-9}$	nano	n
$0.000\ 001 = 10^{-6}$	micro	μ
$0.001 = 10^{-3}$	milli	m
$1000 = 10^3$	kilo	k
$1\ 000\ 000 = 10^6$	mega	M
$1000\ 000\ 000 = 10^9$	giga	G

Conversion Factors

The following is a list of structurally pertinent conversions between U.S. Customary units and SI units.

Area

$$1\ \text{in.}^2 = 645.2\ \text{mm}^2$$

$$1\ \text{ft}^2 = 92.90 \times 10^{-3}\ \text{m}^2$$

Bending Moment or Torque

$$1 \text{ in.-lb} = 0.1130 \text{ N} \cdot \text{m}$$
$$1 \text{ ft-lb} = 1.356 \text{ N} \cdot \text{m}$$
$$1 \text{ ft-kb} = 1.356 \text{ kN} \cdot \text{m}$$

Lengths or Displacements

$$1 \text{ in.} = 25.40 \text{ mm}$$
$$1 \text{ ft} = 0.3048 \text{ m}$$

Loads

$$1 \text{ lb} = 4.448 \text{ N}$$
$$1 \text{ k} = 4.448 \text{ kN}$$
$$1 \text{ lb/ft} = 14.59 \text{ N/m}$$
$$1 \text{ k/ft} = 14.59 \text{ kN/m}$$

Moment of Inertia

$$1 \text{ in.}^4 = 0.4162 \times 10^6 \text{ mm}^4$$
$$1 \text{ ft}^4 = 8.631 \times 10^{-3} \text{ m}^4$$

Section Modulus or Volume

$$1 \text{ in.}^3 = 16.39 \times 10^3 \text{ mm}^3$$
$$1 \text{ ft}^3 = 28.32 \times 10^{-3} \text{ m}^3$$

Stress and Modulus of Elasticity

$$1 \text{ psi} = 6.895 \text{ kPa}$$
$$1 \text{ ksi} = 6.895 \text{ MPa}$$
$$1 \text{ psf} = 47.88 \text{ Pa}$$

Unit Weight, Density

$$1 \text{ lb/ft}^3 = 0.157 \text{ kN/m}^3$$

Index

ABOUT THE AUTHORS

DONALD E. BREYER, P.E., is Professor Emeritus of Engineering Technology and Civil Engineering, California State Polytechnic University.

KENNETH J. FRIDLEY, Ph.D., is Professor and Head, Department of Civil, Construction, and Environmental Engineering, University of Alabama. He is an active member of the Wood Design Committee of the American Forest & Paper Association (AF&PA), which oversees the development of all design guides and specification, including the NDS, related to wood construction. He is also a member of the American Society of Civil Engineers' LRFD Engineered Wood Standards Committee.

KELLY E. COBEEN, S.E., is Principal, Cobeen & Associates Structural Engineering, Lafayette, California. She is an active participant in the development of codes, standards, and guideline documents.

DAVID G. POLLOCK, Ph.D., P.E., is Associate Professor, Department of Civil and Environmental Engineering, Washington State University. Prior to joining the faculty at WSU, he was Director of Engineering for the AF&PA American Wood Council.